Oxidative Damage to Plants

Antioxidant Networks and Signaling

Edited by

Parvaiz Ahmad

S.P. College, Srinagara
Jammu and Kashmir, India

ELSEVIER

AMSTERDAM • BOSTON • HEIDELBERG • LONDON
NEW YORK • OXFORD • PARIS • SAN DIEGO
SAN FRANCISCO • SINGAPORE • SYDNEY • TOKYO
Academic Press is an imprint of Elsevier

Academic Press is an imprint of Elsevier
525 B Street, Suite 1800, San Diego, CA 92101-4495, USA
32 Jamestown Road, London NW1 7BY, UK
225 Wyman Street, Waltham, MA 02451, USA

British Library Cataloguing-in-Publication Data
A catalogue record for this book is available from the British Library

Library of Congress Cataloging-in-Publication Data
A catalog record for this book is available from the Library of Congress

ISBN: 978-0-12-799963-0

For information on all Academic Press publications
visit our website at elsevierdirect.com

Typeset by MPS Limited, Chennai, India
www.adi-mps.com

Printed and bound in the United States of America

14 15 16 17 18 10 9 8 7 6 5 4 3 2 1

Dedication

This book is dedicated to

Hakim Abdul Hameed
(1908–1999)

Founder of Jamia Hamdard (Hamdard University)
New Delhi, India

Contents

Foreword

Nearly 2.7 billion years ago, the introduction of molecular oxygen (O_2) into our atmosphere, resulted in the reactive oxygen species (ROS) as unwelcome companions in the ecosystem. Although they control many different processes in plants, their toxic nature is also capable of injuring cells. How plants solve this problem is still unknown. However, we do know that the steady-state level of ROS in cells needs to be tightly regulated. Recent studies have suggested some key players in this connection, but many questions regarding its mode of regulation, protective role and modulation of signaling networks controlling growth, development and stress responses still need an answer. The studies suggest a dual role for ROS in plant biology as both toxic byproducts and key regulators.

Plants live in continuously changing environment. As a result they have been driven to evolve a very flexible growth and development cycle within the environment surrounding them. This is led by the cell regulation process involving an important indicator such as the redox state. The antioxidants present within the plant cells give a response to the changing environment continuously as they are affiliated to their metabolic state. As plants do not show mobility they develop phtorespiratory, enzymatic and non-enzymatic pathways at different stages in order to face different stresses like drought, salinity, low temperature, and UV-B, thereby producing more ROS. The responses also include gene regulation as well as anatomical aspects. At present, responses to both abiotic as well as biotic stresses have been investigated under controlled conditions following the challenges singly, not the way it happens in the natural world of plants. The metabolic activities in a plant cell are perfectly regulated to follow different biosynthetic pathways with photodynamic or reductive activation of molecular oxygen to produce ROS. Later changes are programmed genetically, in many cases following developmental and environmental changes which in turn have an impact on primary, as well as secondary, metabolism. The transmission of information in plants following the environmental changes is also expressed in the form of production of bursts of superoxide at the plasma membrane. The levels of ROS determine the type of response, and enhanced ROS production has been related to oxidative stress – a negative term depicting a harmful process however, in several cases it helps in sending signals for proper adjustment of gene expression. In fact, the antioxidants and ROS interact with each other, leading towards different functions in higher plants, giving them a high flexibility.

The steps in the biosynthetic pathways leading to antioxidant accumulation in plants have basically been characterized both at a physiological and molecular level. However, all participants involved are still not yet identified. A better understanding of antioxidant degradation and transport in plants is needed, together with an elucidation of the molecular events involved. The future will determine the involvement of ascorbate, glutathione, and tocopherol, as well as how they initiate and control redox signal transduction and trigger the gene expression for optimization of different survival strategies. Other questions are related to the coordination of antioxidants in plant growth and development in the changing environment, linkage of redox signaling with hormonal regulation, nutrient status and redox potential in plants, as metabolism in plants must be highly regulated for an effective integration of biosynthetic pathways.

This first chapter of this book covers reactive oxygen species and photosynthesis, presenting an overview of the general characteristics of various stress factors and their capability to increase the generation of ROS in the cells, as its generation in photosynthesizing tissues is significantly exacerbated under environmental stress conditions. It discusses the complex set of mechanisms for avoidance of ROS production or scavenging the ROS. The main concept is presented as the signaling role of ROS as well as electron transport components as redox molecules, which may precisely regulate the expression of nucleus-encoded genes through the redox status of plastids. Accordingly, chloroplasts have been regarded as sensors for environmental signals that link the effect of stress factors to the plant metabolism in general and carbon reactions in particular.

The first chapter of the book, on reactive oxygen species and plant hormones, discusses the evidence related to ROS and plant hormone signaling pathways regulating plant growth and development, in relation to the responses to environmental factors. Plant hormones and ROS are intrinsically interlinked in plant biology and development, as well as in stress responses. The discussion includes interactions between ROS and plant hormones during the physiological events of seed germination, under environmental stress and during plant development. It is mentioned that the identification of new plant hormones and their functions, as well as identification of ROS receptors, will help in clarifying the roles and signal interactions between ROS and hormones in plant physiology.

In Chapter 3, *Superoxide Dismutase and Abiotic Stress Tolerance in Plants: An Overview*, an attempt has been made to highlight the facts about the influence of abiotic stress factors on plants, because their direct or indirect presence may affect the development, growth, and basic metabolism in plants. The factors considered are heavy metals, salt stress, chilling, drought, and UV-B radiation. These may induce the formation or the overproduction of ROS. The restoration of redox homestasis through cells' evolution of large scale enzymatic and non-enzymatic antioxidants to scavenge ROS in plants is discussed. An attempt has been made to summarize the characteristics of SOD enzymes and show the differences and similarities in the effects of different abiotic stress factors on SOD activity.

Chapter 4 presents a discussion on catalase, an antioxidant enzyme present in all aerobic organisms, catalysing H_2O_2 in an energy-efficient manner in the cells exposed to environmental stress. Its deficiency in plants develops anomalies. Multiple molecular forms of catalase isozymes indicate its versatile role within plants. The expression of *cat* gene shows time, species and stress specificity. The chapter compiles the up-to-date information on catalase structure, localization, biochemistry, genes and function in plants.

In Chapter 5, the role of glutathione (GSH) in abiotic stress tolerance is discussed at length. GSH is a non-protein low molecular weight tripeptide found in most plant tissues, that plays a role in biosynthetic pathways, detoxification of xenobiotics, antioxidant chemistry, whilst at the same time protects plants against oxidative stress, acting as a storage and transport form for reduced sulfur.

Chapter 6 on glutathione metabolism in plants under environmental stress looks at the essential role of glutathione as a part of defense system in plants and human beings exposed to various environmental stresses, mechanisms involving metal sequestration and scavenging of ROS. The chapter contains information regarding the mechanism of a glutathione mediated protection system, metabolism, transport and biosynthesis of glutathione in plants.

Non-enzymatic Antioxidants in Plants, Chapter 7, relays information about the oxidative stress caused by abiotic and biotic factors that include salinity, pathogen colonization, UV stress, herbicide activity and oxygen deficiency, which affect biochemical, physiological, developmental and structural processes within individual plant and plant communities. The defence mechanisms for protection from the spectrum of harmful effects are given.

Chapter 8 is a discussion of ascorbic acid, a reduced form of vitamin C synthesized from hexose sugar. The text provides insight on various aspects of ascorbic acid such as biosynthesis, diverse functions with a particular reference to environmental stresses, an update on new roles and manipulation of ascorbic acid levels in the stresses of plants through transgenic approaches.

A discussion of carotenoids involved in chloroplasts is presented in Chapter 9. The properties of carotenoids as effective non-enzymatic plant antioxidants have been described. The chemical structure of carotenoids in relation to their antioxidant properties is explained, followed by photoprotective role of two all-*trans*β-carotene molecules existing in PSII reaction center. The role of ascorbate as an antioxidant and as a reductant required to carry out de-epoxidation has been considered. In the last part of the chapter, the significance of xanthophylls in photoprotection is discussed. Special attention has been paid on the role of carotenoids involved in several types of the xanthophyll cycle.

In Chapter 10, lipophilic molecules occurring in plant cell with strong antioxidant properties have been given, with the prenyllipids being the most efficient. In the first section, characterization of the mechanisms leading to production of ROS at different sites of chloroplasts during exposure to abiotic stress factors are given together with their negative effects. The second section of the chapter outlines the current knowledge about antioxidant properties of prenyllipids, followed by a discussion on the function of ROS and prenyllipids (vitamin E) as signaling molecules.

In Chapter 11, the discussion revolves around the effects of abiotic factors on the metabolic activities of plants, in order to understand how water deficit can significantly enhance the production of crop plants and quality of the environment. An attempt has been made to describe the mechanism of drought resistance in plants on the basis of antioxidants, physiology, enzymatics, and proteomics.

Chapter 12 considers thermotolerance mechanisms in plants to minimize damage and ensure protection of cellular homeostasis. ROS detoxification mediated by antioxidant compounds and antioxidant enzyme systems is one of the mechanisms for high temperature stress (HTS) tolerance. Very little information is available on the cellular factors essential for this response. Previous efforts to improve HTS tolerance through breeding and genetic engineering has had limited success. Comparative genomics, high-throughput sequencing, gene expression analysis, and gene function validation may provide necessary insights in the mechanisms of stress tolerance and critical genes for enhancement of genetic improvement in crops.

Chapter 13, entitled *Reactive Oxygen Species and Antioxidants in Response to Pathogens and Wounding*, discusses the development of mechanisms in plants to help cope with biotic and abiotic stresses. In the case of an injury, wound healing cascades activate that help secure the wound and keep the pathogens at bay, to speed up the healing process. The accumulation of ROS is highest around the wounds. Due to ROS's high toxicity, pathogens trying to access the wounds are killed. Attempts have been made to discuss in detail the role of different types of ROS in evading stress and injury, the independent roles of each antioxidant in wound healing, the genes involved in their synthesis and the pathways of wound healing.

The role of ascorbate peroxidase on postharvest treatments in horticultural crops is covered in Chapter 14. Proper postharvest handling is of utmost importance for increasing food availability. Plants have evolved an efficient antioxidant defense system which prevents the accumulation of ROS. Ascorbate peroxidase (APX) is involved in ascorbate-glutathione cycle and utilizes ascorbate as the electron donor and plays the most essential role in scavenging ROS. APX plays an important role in controlling the concentration of ROS that participates in the signal transduction in many physiological processes, and the management of ROS during stress in horticultural crops. There are some connections between APX induction and regulation of metabolism. Postharvest treatments can therefore be very important for improving shelf life and quality retention during postharvest handling of horticultural crops.

Chapter 15, entitled *Mycorrhizal Association and ROS in Plants* reviews beneficial microorganisms distributed in the soil ecosystem, such as the arbuscular mycorrhizal fungi. In view of the fact that relations between the mycorrhizal associations and ROS are widely concerned with plants, mycorrhizal symbiosis is generally restricting the oxidative burst under environmental stresses. Therefore, arbuscule is associated with accumulation of H_2O_2, whose overaccumulation in arbuscules may predicate the collapse or degradation of arbuscules during mycorrhizal development. The mycorrhizal associations can enhance antioxidant enzyme activities and increase antioxidant contents of the host plant, partly alleviating oxidative stress. The mycorrhizal symbiosis, ROS occurrence under mycorrhizal symbiosis, and role of mycorrhizal association in antioxidant enzymes and antioxidants of the host plant have been discussed, together with some future prospects.

Chapter 16 sees authors presenting their views on proline protecting against abiotic oxidative stress. They have summarized and discussed the current understanding of ROS formation, proline biosynthesis and its accumulation in plants under various abiotic stresses. New insights gained about the molecular mechanisms of proline mediated oxidative stress tolerance have also been discussed.

Trace elements tolerance modulated by the plant antioxidant system is detailed in Chapter 17. The authors have analyzed the information available regarding trace element induction of oxidative stress and trace element tolerance modulation in seeds and seedlings through antioxidant activities.

Chapter 18 deals with plant signaling under environmental stress. The role of small RNAs is of significance in plants under stress. The review highlights the fact that the use of the proteomic technique can be useful for evaluating plant responses under stress. Plants, with the higher ability to manage their signaling pathways under stress, are more likely to survive. Recognition of the signaling pathways under stress can be important for the production of tolerant plants.

In Chapter 19, entitled *Hydrogen Peroxide Generation, Scavenging, and Signaling in Plants,* authors present the biochemistry of H_2O_2 and their production sites, H_2O_2 scavenging antioxidant defense machinery, the mechanisms of H_2O_2 perception and signal transduction, the factors that act downstream and H_2O_2 interference with other signaling pathways such as calcium and protein phosphorylation networks. Reviewing this biochemistry will provide new insights into how plants balance H_2O_2.

Finally, Chapter 20 looks at the role of ROS as signaling molecules in plants. An attempt has been made to discuss two of the main roles of ROS: signaling molecules regulating many processes in plants under normal and adverse conditions, and the production of toxic byproducts from aerobic metabolisms. The authors have also tried to bring some insights of ROS generation, and its role in plant development together with the signaling roles of ROS and their interactions with other molecules.

I hope that the chapters presented in this book will be useful for scientists and graduate students in the area towards encouraging further discussion, research and development on Oxidative Damage to Plants: Antioxidant Networks and Signaling.

Dr Münir Öztürk (MSc, PhD, DSc, FIAS)
Professor (Emeritus) of Botany, Ege University, Izmir, Turkey
Consultant Fellow, Faculty of Forestry, Universiti Putra Malaysia (UPM), Selangor-Malaysia
Distinguished Visiting Scientist, ICCBS, Karachi University, Pakistan

Preface

Industrialization and urbanization in the name of growth and development is rapidly proceeding on a global level. This growth and development has deteriorated our environment to a great extent. Pollution of soil, water, air is creating havoc in our living environment. Environmental stresses, such as high levels of salinity, drought, floods, temperatures, and UV damage is increasing at an alarming rate, resulting conversion of agricultural land to non-agricultural. Increases in population and decreases in agricultural and horticultural production are leading towards starvation and malnutrition for nearly one billion stomachs around the world. It will definitely prove more dangerous for our future generations if steps are not taken.

Environmental stress has led to oxidative stress, caused by the generation of reactive oxygen species (ROS) in plants and animals. ROSs react with cell biomolecules, leading to organelle dysfunction. However, Nature has equipped plants with defense mechanisms, including enzymatic and non-enzymatic antioxidants. Upregulation and downregulation of these antioxidants helps plants tolerate stress. ROSs also have a significant role as signaling molecules under stress.

Oxidative Damage to Plants: Antioxidant Networks and Signaling is an attempt to gain more insight of the devastating role of oxidative stress in plants and the defense mechanisms required. Chapter 1 describes the protection of photosynthetic plants against ROS and role of ROS in signaling. Chapter 2 discusses reactive oxygen species and plant hormones, highlighting the role of ROS in plant signaling and seed germination. Chapters 3 and 4 discuss superoxide dismutase and catalase in detail and their role under stress. Chapters 5 and 6 deal with glutathione metabolism and its role under stress. Chapters 7–10 deal with the role of different non-enzymatic antioxidants, such as ascorbic acid, carotenoids, and lipophylic molecules under different stresses.

Chapters 11–13 deal with drought, temperature stress and pathogens and wounding. The generation of ROS, ROS sites, antioxidants, proteomics and transcriptional regulation are also discussed in these chapters. In Chapter 14 the role of ascorbate peroxidase under environmental stress and postharvest produce is reviewed. Mycorrhizal association and ROS in plants is explained in Chapter 15. Chapters 16 and 17 discuss the role of proline and trace elements in helping plants to withstand stress. Chapters 18–20 review the different roles of ROS and signaling under stress.

This book is the compilation of different chapters on oxidative stress and signaling in plants. Each topic has been discussed to its full potential to ensure that readers receive the maximum information for each topic. There may be some errors in the book even after my best efforts. I need your feedback and suggestions for this.

I would like to thank all the contributors of this volume for their timely submission. Their collaboration and patience during the preparation of this book is unforgettable. I would also like to thank Kristi S. Gomez (Acquisition Editor, Elsevier), Patricia Gonzalez (Editorial Project Manager, Elsevier) and other team members of Elsevier for their help, suggestions and timely publication of this volume.

Parvaiz Ahmad
Jammu and Kashmir, India

Acknowledgments

I acknowledge all the contributors of this volume for their valuable contribution. I also acknowledge the Higher Education Department, Government of Jammu and Kashmir, India for their support. I am also thankful to Prof. Dr. Munir Ozturk for writing foreword for this volume.

List of Contributors

Elsayed Fathi Abd-Allah Plant Production Department, College of Food and Agricultural Sciences, King Saud University, Riyadh, Saudi Arabia

Fakiha Afzal Atta-ur-Rahman School of Applied Biosciences, National University of Sciences and Technology, H-12, Islamabad, Pakistan

Parvaiz Ahmad Department of Botany, S.P. College, Srinagar, Jammu & Kashmir, India

Muhammad Ashraf Atta-ur-Rahman School of Applied Biosciences, National University of Sciences and Technology, H-12, Islamabad, Pakistan

MM Azooz Department of Biological Sciences, Faculty of Science, King Faisal University, Saudi Arabia

Monika Beszterda Department of Chemistry, Poznań University of Life Sciences, Poznań, Poland

David J. Burritt Department of Botany, University of Otago, Dunedin, New Zealand

Vishwanathan Chinnusamy Division of Plant Physiology, Indian Agricultural Research Institute, New Delhi, India

Masayuki Fujita Laboratory of Plant Stress Responses, Department of Applied Biological Science, Faculty of Agriculture, Kagawa University, Takamatsu, Kagawa, Japan

Olimpia Gładysz Department of Inorganic Chemistry, Wroclaw Medical University, Wrocław, Poland

Piotr Goliński Department of Chemistry, Poznań University of Life Sciences, Poznań, Poland

Alvina Gul Kazi Atta-ur-Rahman School of Applied Biosciences, National University of Sciences and Technology, H-12, Islamabad, Pakistan

Ghader Habibi Department of Biology, Payame Noor University, Iran

Roghieh Hajiboland Plant Science Department, University of Tabriz, Iran

Asiya Hameed Department of Botany, Hamdard University, New Delhi, India

Mirza Hasanuzzaman Department of Agronomy, Faculty of Agriculture, Sher-e-Bangla Agricultural University, Bangladesh

Md. Anamul Hoque Department of Soil Science, Bangladesh Agricultural University, Bangladesh

Mohammad Anwar Hossain Department of Genetics and Plant Breeding, Bangladesh Agricultural University, Bangladesh

Yoshihiro Imahori Laboratory of Postharvest Physiology, Graduate School of Life and Environmental Sciences, Osaka Prefecture University, Japan

Sumiya Jamsheed Department of Botany, Hamdard University, New Delhi, India

Rohit Joshi School of Plant, Environmental, and Soil Sciences, Louisiana State University Agricultural Center, Baton Rouge, Louisiana, USA

Philippe Juneau Department of Biological Sciences, TOXEN, Ecotoxicology of Aquatic Microorganisms Laboratory and Institut des Sciences de l'environnement, Université du Québec à Montréal, Montréal, Québec, Canada

Ramanpreet Kaur Department of Botany, Panjab University, Chandigarh, India

Rabia Khurshid Atta-ur-Rahman School of Applied Biosciences, National University of Sciences and Technology, H-12, Islamabad, Pakistan

Ashwani Kumar Department of Botany, School of Biological Sciences, Dr. H.S. Gour Central University, Sagar, India

Dariusz Latowski Department of Plant Physiology and Biochemistry, Faculty of Biochemistry, Biophysics and Biotechnology, Jagiellonian University, Krakow, Poland

Helal Ahmad Lone Department of Botany, A.S. College, Jammu and Kashmir, India

Marília Mércia Lima de Carvalho Carneiro Departamento de Biologia, Universidade Federal de Lavras (UFLA), Lavras, Brazil

Mohammad Miransari Department of Book & Article, AbtinBerkeh Limited Co., Isfahan, Iran

Harsh Nayyar Department of Botany, Panjab University, Chandigarh, India

Beatrycze Nowicka Department of Plant Physiology and Biochemistry, Faculty of Biochemistry, Biophysics and Biotechnology, Jagiellonian University, Krakow, Poland

Hassan Pakniyat Crop Production and Plant Breeding Department, College of Agriculture, Shiraz University, Shiraz, Iran

Marcelo Pedrosa Gomes Department of Biological Sciences, TOXEN and Institut des Sciences de l'environnement, Université du Québec à Montréal, Montréal, Québec, Canada

Hadi Pirasteh-Anosheh Crop Production and Plant Breeding Department, College of Agriculture, Shiraz University, Shiraz, Iran

Saima Rasool Department of Botany, Hamdard University, New Delhi, India

Armin Saed-Moucheshi Crop Production and Plant Breeding Department, College of Agriculture, Shiraz University, Shiraz, Iran

Iti Sharma Department of Botany, Dr. Mohanlal Piramal (PG) College, Rajasthan, India

Élise Smedbol Department of Biological Sciences, TOXEN Université du Québec à Montréal, Montréal, Québec, Canada

Queila Souza Garcia Departamento de Botânica, Instituto de Ciências Biológicas, Universidade Federal de Minas Gerais, Belo Horizonte, Brazil

Kazimierz Strzalka Department of Plant Physiology and Biochemistry, Faculty of Biochemistry, Biophysics and Biotechnology, Jagiellonian University, Krakow, Poland

Kinga Szentner Department of Chemistry, Poznań University of Life Sciences, Poznań, Poland

Réka Szőllősi Department of Plant Biology, University of Szeged, Hungary

Renata Szymanska Department of Plant Physiology and Biochemistry, Faculty of Biochemistry, Biophysics and Biotechnology, Jagiellonian University, Krakow, Poland

Agnieszka Waśkiewicz Department of Chemistry, Poznań University of Life Sciences, Poznań, Poland

Qiang-Sheng Wu College of Horticulture and Gardening, Yangtze University, People's Republic of China

Ying-Ning Zou College of Horticulture and Gardening, Yangtze University, People's Republic of China

Chapter 1

Reactive Oxygen Species and Photosynthesis

Roghieh Hajiboland

1.1 INTRODUCTION

Photosynthesis is comprised of a series of redox reactions, in which light produces NADPH, which acts as the reducing molecule for CO_2 fixation via the Calvin cycle. During the electron transport reactions, ATP is produced by a proton gradient and is consumed in carbon reductions.

Drought, salinity, chilling, nutrient deficiencies and other environmental stresses reduce the activity of the Calvin cycle directly or indirectly, i.e. by feedback regulation following impaired sink utilization of photoassimilates. Both these regulatory mechanisms result in a decline of $NADP^+$ regeneration, thus, overreduction of the electron transport chain. Under such conditions, reactive oxygen species (ROS) are generated via transfer of excess electrons to oxygen. Similar events can occur under the changing light environment, which directly affects the photosynthetic light reactions. High light conditions, particularly in combination with low temperature, leads to an overexcitation of the photosynthetic apparatus, production of excess excitation energy (EEE) and photoinhibition.

In this chapter, first the main steps of light and dark reactions are introduced with an emphasis on the components that are influenced and/or regulated under environmental stresses. In the following sections, the two main components influencing photosynthesis under environmental stresses, i.e. EEE and ROS, regarding their production and scavenging are discussed. Finally, the role of redox status and ROS in the regulation of photosynthesis through their contribution in the signaling pathways is presented. The evolution of plant adaptation mechanisms to counteract the damaging effects of EEE and ROS is also discussed briefly in the final section of the chapter.

P. Ahmad (Ed): Oxidative Damage to Plants. DOI: http://dx.doi.org/10.1016/B978-0-12-799963-0.00001-0

1.2 PHOTOSYNTHESIS: LIGHT REACTIONS

An important part of the light energy trapped by chlorophylls (Chl) and carotenoids is converted to chemical energy. This process depends on collaboration between pigment molecules and some proteins acting as components of electron transfer reactions.

1.2.1 Photosystems

Two supramolecular complexes, called photosystem I (PS I) and II (PS II) carry out the early energy transfer processes during light reactions. These two photosystems are physically and chemically distinct, and are linked together by an electron transport chain (Fig. 1.1). The PS II reaction center, along with its antenna Chl and associated electron transport proteins, is located predominantly in the grana lamellae. The PS I reaction center and its associated antenna pigments and electron transfer proteins together with ATP synthase are localized in the stroma lamellae. The cytochrome b_6f (Cyt b_6f) complex of the electron transport chain that connects the two photosystems is evenly distributed between stroma and grana. Due to a spatial separation of PS I and II, intermediate electron carrier molecules, i.e. plastoquinone (PQ) and plastocyanin (PC), are required for delivery of electrons to PS I (Allen and Forsberg, 2001).

1.2.2 Light-Absorbing Antenna Systems

Antenna systems, comprising 200 to 300 Chl molecules, deliver energy efficiently to the associated reaction centers. Approximately 95–99% of the energy

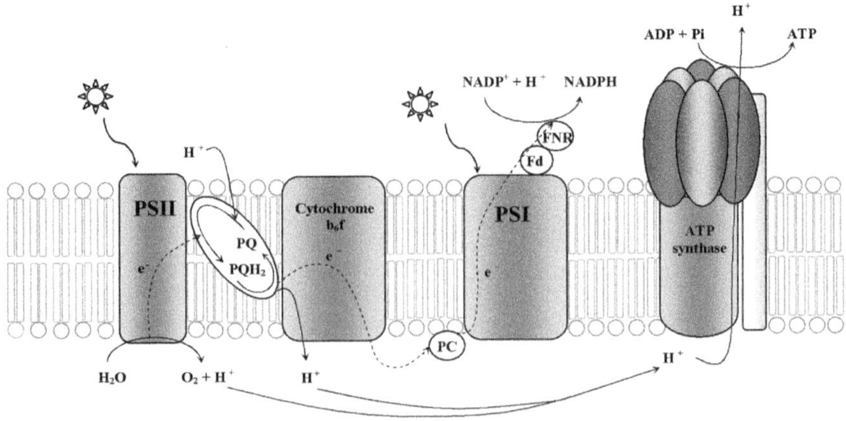

FIGURE 1.1 Photosynthetic electron transport chain in the thylakoid membranes with the contribution of four major protein complexes: PSII, the Cyt b_6f complex, PSI and the ATP synthase. Abbreviations: Fd: ferredoxin; FNR: Fd–NADP reductase; PSI and PSII: photosystems I, II; PQ: plastoquinone; PC: plastocyanin. The dashed and solid lines indicate electron and proton transports, respectively.

of photons absorbed by the antenna pigments is transferred to the reaction center and is used in photochemistry. The maximum absorption of the antenna pigments toward the reaction center is continuously shifted toward longer wavelengths, and thus the difference in excitation energy between two neighboring pigments is lost as heat (Lawlor, 2001; Taiz and Zeiger, 2010).

1.2.2.1 Structure of Antenna Systems

In photosynthetic cells of eukaryotic organisms containing both Chla and Chlb, the antenna proteins are called light-harvesting complex (LHC) proteins and are associated either with PS II (LHC II) or PS I (LHC I). The structure and sequence of the LHC I are similar to those of the LHCII proteins. Light absorbed by carotenoids or Chlb in the LHC proteins is rapidly transferred to Chla and then to other antenna pigments that are closely associated with the reaction center (Lawlor, 2001; Taiz and Zeiger, 2010).

Photons excite a specialized Chl in the reaction center (P_{680} for PS II, and P_{700} for PS I), either directly by absorption or via energy transfer from an antenna pigment. Then an electron is ejected from P_{680} and P_{700}, is transferred through electron carriers and reduces P_{700} and $NADP^+$, respectively (Lawlor, 2001; Taiz and Zeiger, 2010).

1.2.3 Photosystem II Structure and Function

PS II is a multisubunit protein-pigment complex. In higher plants, it contains two reaction centers and some antenna complexes (Fig. 1.2). The core of the PS II reaction center is composed of two membrane proteins, i.e. D1 and D2 proteins and some other polypeptides. The primary electron donor Chl (P_{680}), other Chl molecules, carotenoids, pheophytins and PQs are associated with the D1 and D2 proteins. Other proteins function as antenna molecules or components of water splitting complex. Some other molecules, e.g. Cyt b_{559}, may have a role in cyclic electron flow around PS II (see Section 1.5.1.5) (Lawlor, 2001; Taiz and Zeiger, 2010).

1.2.3.1 Water Oxidation and O_2 Evolution

Water is oxidized in a chemical reaction in which four electrons are removed from two water molecules, generating an oxygen molecule and four hydrogen ions:

$$2H_2O \rightarrow O_2 + 4H^+$$

Water is a very stable molecule and the photosynthetic oxygen-evolving complex is the only known biochemical system that carries out this reaction. The protons produced by water oxidation are released into the lumen of the thylakoid, because of localization of the oxygen-evolving complex on the internal surface of the thylakoid membrane. Four Mn ions are associated

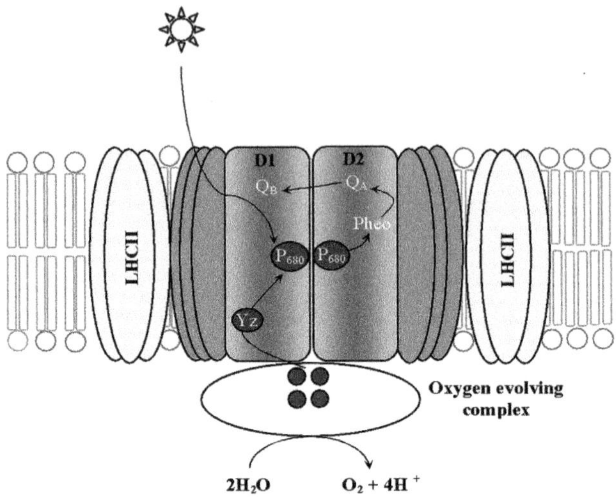

FIGURE 1.2 Structural model of the PSII reaction center consisting of two complete reaction centers and some antenna complexes. The core of the reaction center consists of two membrane proteins known as D1 and D2, as well as other proteins. Abbreviations: LHCII: light harvesting complexes of PSII; P680: the reaction center chlorophyll; Pheo: pheophytin; QA and QB: primary and secondary quinones, respectively; Yz: a tyrosine radical. Small dark circles in the oxygen evolving complex indicate Mn atoms.

with each oxygen-evolving complex, and Cl^- and Ca^{2+} ions are also essential for O_2 evolution (Barber et al., 1999).

1.2.4 Electron Flow through the PS II and Cyt b_6f Complex

One electron carrier with a high tendency to retain its electrons, identified as Y_z, operates between the oxygen-evolving complex and P_{680}. Y_z is indeed a radical generated from a tyrosine residue of the D1 protein. Pheophytin is an early electron acceptor in PS II, followed by a complex of two PQs (Q_A and Q_B) that are associated with the reaction center and accept electrons from pheophytin. Transfer of the two electrons to Q_B and formation of $Q_B{}^{2-}$ is followed by taking two protons from the stroma and formation of a fully reduced molecule named plastohydroquinone (QH_2). The QH_2 then transfers its electrons to the Cyt b_6f complex. The Cyt b_6f complex is a large multi-subunit protein with several prosthetic groups containing two b-type hemes and one c-type heme (Cyt f). The complex also contains a Rieske iron–sulfur protein (FeS_R), in which two Fe atoms are bridged by two S atoms (Berry et al., 2000).

1.2.4.1 Q Cycle

The pathway of electrons and protons flowing through the Cyt b_6f complex has been described by a mechanism known as the Q cycle. In this mechanism, QH_2 is oxidized, and one of the two electrons is transferred via a linear electron transport chain toward PS I, while the other electron flows through a cyclic pathway. The number of protons pumped per electron is higher in the cyclic electron pathway through the Cyt b and PQ compared with a linear flow (Lawlor, 2001; Taiz and Zeiger, 2010).

1.2.5 PS I Reaction Center and Reduction of $NADP^+$

The PS I reaction center is a large multi-subunit complex. The core antenna and P_{700} are associated with two proteins, PsaA and PsaB (Fig. 1.3). One of the early electron acceptors (A_0) is a Chl molecule, and another (A_1) is a quinone, e.g. phylloquinone or vitamin K_1. Further electron acceptors include three membrane-associated Fe−S proteins, i.e. bound ferredoxins, including FeS_X, FeS_A, and FeS_B. Electrons are transferred through centers A and B to ferredoxin (Fd), a small, water-soluble Fe−S protein. The membrane-associated flavoprotein Fd−NADP reductase (FNR) is responsible for reduction of $NADP^+$ to NADPH. The last reaction completes the linear (noncyclic) electron flow (the Z scheme) that begins with the oxidation of water (Chitnis, 2001).

1.2.5.1 ATP Synthesis

Another fraction of the captured light energy is applied for ATP synthesis, e.g. photophosphorylation. During proton translocation, the stroma becomes alkaline while the lumen becomes acidic. This proton gradient across the thylakoid membrane creates a power, i.e. proton motive force, for ATP synthesis; the

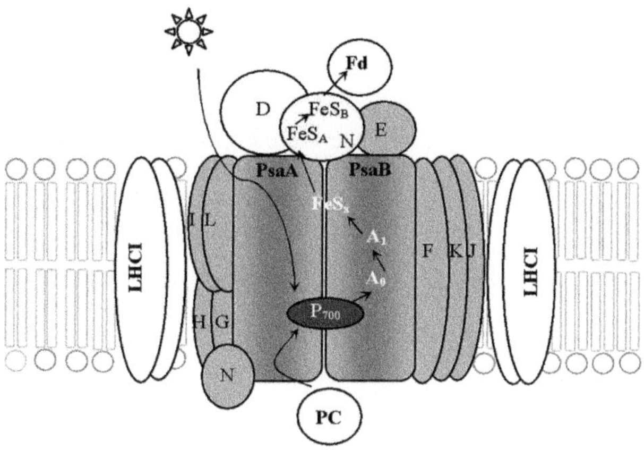

FIGURE 1.3 Components of the PSI reaction center. These molecules are organized around two major proteins, PsaA and PsaB; other Psa proteins are labeled C to N. Abbreviations: A_0 and A_1: chlorophyll a and phyloquinone molecules, respectively; FeS_x, FeS_A and FeS_B: iron-sulfur proteins; P700: the reaction center chlorophyll.

stoichiometry of protons translocated per ATP synthesized is four H^+ ions per ATP. ATP synthesis is catalyzed by a large enzyme complex, e.g. ATP synthase consisting of (i) a hydrophobic membrane-bound portion (CF_0) that forms a channel to allow protons to be passed through the membrane and (ii) a portion that protrudes into the stroma (CF_1) and is made up of several peptides (McCarty et al., 2000).

1.3 PHOTOSYNTHESIS: CARBON REACTIONS AND PHOTORESPIRATION

The photochemical events in the chloroplast thylakoid membrane result in the generation of NADPH and ATP. Reduction of CO_2 to carbohydrate is undertaken by enzymes localized in the stroma and needs NADPH and ATP. These stroma reactions not only depend on the NADPH and ATP as products of the photochemical reactions, but are also directly regulated by light.

1.3.1 The Calvin Cycle

Plants reduce CO_2 to carbohydrate via the photosynthetic carbon reduction cycle, i.e. Calvin cycle (reductive pentose phosphate cycle). This cycle was originally described for C_3 species, but other metabolic pathways associated with the photosynthetic fixation of CO_2, such as the C_4 photosynthetic carbon assimilation cycle and the photorespiratory carbon oxidation cycle, are also dependent on the basic Calvin cycle (Lawlor, 2001; Taiz and Zeiger, 2010).

1.3.1.1 Light-regulated Enzymes in Calvin Cycle

Five enzymes in the Calvin cycle are regulated by light including ribulose 1,5 bisphosphate (RuBP) carboxylase/oxygenase (Rubisco), NADP:glyceraldehyde-3-phosphate dehydrogenase, fructose-1,6-bisphosphatase, Sedoheptulose-1,7-bisphosphatase and ribulose-5-phosphate kinase. Activity of the last four enzymes is controlled by light via the ferredoxin–thioredoxin system. These enzymes contain one or more disulfide groups that exist in the oxidized state (–S–S–) in dark; thus the enzyme is inactive or subactive (Fig. 1.4). In the light, the –S–S– group is subjected to a redox change and reduced to the sulfhydryl state (–SH HS–) resulting in activation of the enzyme. Regulation of Rubisco by this system is indirect, i.e. via a thioredoxin accessory enzyme, Rubisco activase (Berg et al., 2002)

1.3.2 Photorespiration

Rubisco is able to catalyze both the carboxylation and the oxygenation of RuBP. Oxygenation is the first reaction of photorespiration. The 2-phosphoglycolate formed in the chloroplast by oxygenation of RuBP is hydrolyzed to glycolate by a phosphatase. Subsequent metabolism of the glycolate

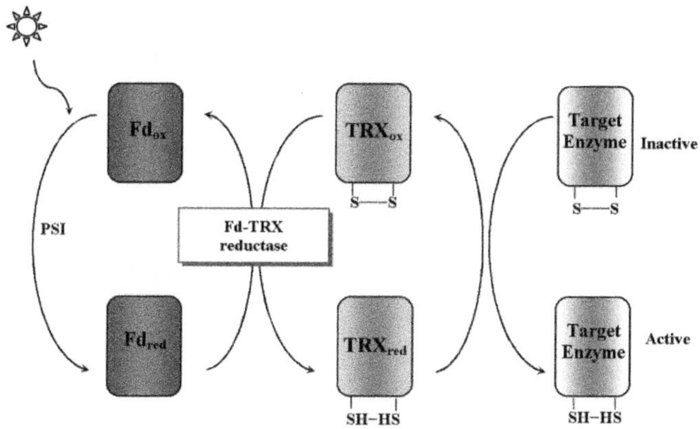

FIGURE 1.4 The Fd–TRX system for regulation of photosynthetic enzymes. The activation process starts in the light by a reduction of Fd by PSI. Formed Fd_{red} plus two protons are used to reduce a catalytically active disulfide (–S–S–) group of the Fe-S enzyme Fd-TRX reductase, which in turn reduces the disulfide bond of the TRX. The reduced form (–SH HS–) of TRX then reduces the critical disulfide bond of a target enzyme and activates that enzyme. Abbreviations: TRX: thioredoxin.

occurs in peroxisomes and mitochondria (Fig. 1.5). In photorespiration, 75% of the carbon lost by the oxygenation of RuBP is recovered and returned to the Calvin cycle (Lawlor, 2001; Taiz and Zeiger, 2010).

The relative rate of two carboxylation and oxidation reactions is influenced by CO_2 and O_2 partial pressure in the environment and temperature. The concentration ratio of CO_2 to O_2 decreases as the temperature rises; thus, photorespiration (oxygenation) increases relative to photosynthesis (carboxylation) under higher temperatures. Regarding kinetic properties of Rubisco, a relative increase in oxygenation occurs at higher temperatures (Lawlor, 2001; Taiz and Zeiger, 2010).

Recycling of phosphoglycollate to phosphoglycerate (in order to reenter the Calvin cycle) results in a considerable loss of assimilated carbon. In addition, large amounts of H_2O_2 are produced during the oxidation of the glycollate in the peroxisomes. Nevertheless, photorespiration has an adaptive role under excess light conditions as well as at lower intercellular CO_2 concentration. Under conditions of stomatal limitation imposed, for example, by drought stress, photorespiration is involved in dissipation of surplus ATP and reducing power produced during the light reactions, thus protecting photosynthetic apparatus from photoinhibition and damage (see Section 1.5.1.7). Mutants unable to photorespire grow normally under higher (2%) CO_2 conditions, while ceasing to grow and dying rapidly after transferring to ambient air (0.35% CO_2). Both C_4 and Crassulacean acid metabolism (CAM) plants are able to concentrate CO_2 around Rubisco to avoid strong photorespiration (Taiz and Zeiger, 2010).

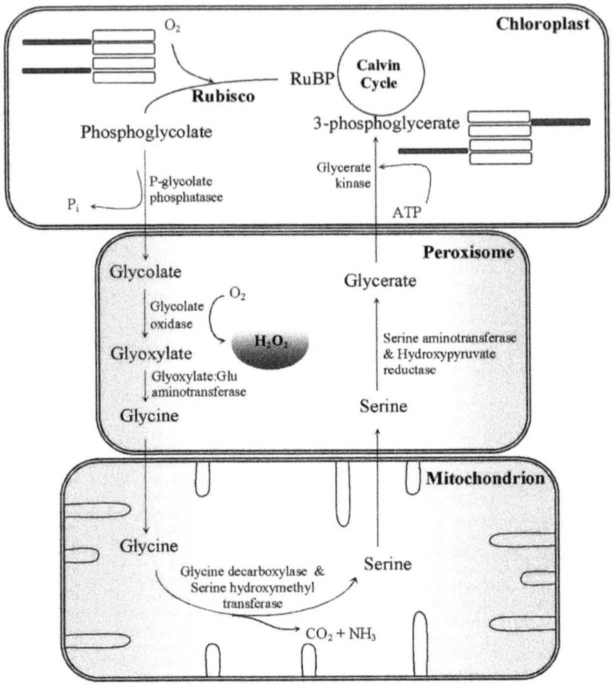

FIGURE 1.5 Photorespiration or C_2 oxidative photosynthetic cycle and production of H_2O_2. The cycle involves three organelles: chloroplasts, mitochondria, and peroxisomes. Abbreviations: RuBP: ribulose 1,5 bisphosphate, Rubisco: ribulose 1,5-bisphosphate carboxylase/oxygenase.

1.4 REACTIVE OXYGEN SPECIES (ROS)

Oxygen in the atmosphere and water is required for sustaining aerobic life on Earth. However, aerobic organisms must cope with the adverse effects of oxygen. As atmospheric concentrations of O_2 rise, it may inhibit or inactivate certain enzymes and it also competes with photosynthetic CO_2 fixation by Rubisco, increasing the energy cost of photosynthesis. However, the toxic effect of oxygen is mainly exerted by its reactive derivatives, whereas ground state O_2 is rather unreactive and does not damage organic molecules. This characteristic is explained by the parallel spins of two unpaired electrons of O_2, imposing an energetic barrier on its reaction with nonradical compounds (the spin restriction). Activation of O_2 is an inevitable result of the photosynthetic electron transfer under aerobic conditions (Apel and Hirt, 2004).

1.4.1 Chemical Forms of ROS

In order to become reactive, O_2 must be physically or chemically activated (Fig. 1.6). Physical activation of oxygen takes place by direct transfer of excitation

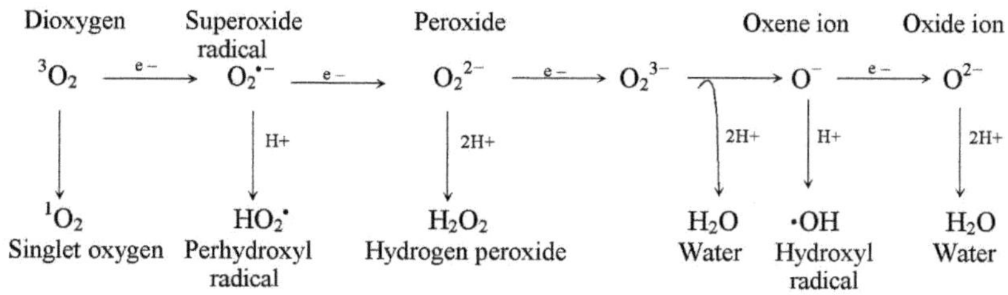

FIGURE 1.6 Generation of different types of reactive oxygen species from ground-state dioxygen.

energy from a photo-activated pigment, i.e. an excited Chl molecule to O_2. The latter absorbs sufficient energy and as a result the spin of one electron is inverted. The singlet state of oxygen (designated as $^1O_2$) is a prevalent reactive species, highly diffusible through the membranes and able to react with many organic molecules (whose electrons are usually paired), thus damaging the photosynthetic apparatus (Arora et al., 2002).

Chemical activation occurs by univalent reduction of O_2, i.e. addition of electrons one by one. Four electrons (and four protons) are needed for the complete reduction of O_2 to water. All three intermediates of univalent reduction including superoxide ($O_2^{\bullet -}$), hydrogen peroxide (H_2O_2) and the hydroxyl radical ($^{\bullet}OH$) are chemically reactive and biologically toxic. This toxicity is reflected by their short half-lives before reacting with cellular compounds, as compared to that of O_2 (>100 s). Reactive oxygen species may extract an electron from an organic molecule, rendering it a radical, e.g. the peroxyl ($RCOO^{\bullet}$) and alkoxyl ($RO^{\bullet}$) radicals capable of propagating a chain reaction (Møller et al., 2007).

Superoxide is the first reduction product of ground state oxygen and could be subjected to both oxidation and reduction reactions. Superoxide is able to react with some molecules to produce other reactive species or may be dismutated to H_2O_2 spontaneously or enzymatically (Arora et al., 2002).

Hydrogen peroxide could not be considered a free radical, but operates as oxidant or reductant in many cellular reactions. Unlike $O_2^{\bullet -}$, H_2O_2 is very diffusible through membranes and it can directly inactivate some sensitive enzymes even at very low concentration. Hydrogen peroxide is rather stable and therefore less toxic than other ROS. The main threat imposed by both $O_2^{\bullet -}$ and H_2O_2 lies in their ability to generate highly reactive $^{\bullet}OH$ (Møller et al., 2007).

The hydroxyl radical is the most effective ROS species in the cells for oxidation of various molecules. It is capable to react non-specifically with any biological molecules and this limits its diffusion within the cell to a distance of two molecular diameters from its site of production. No specific scavengers of $^{\bullet}OH$ are known, although several metabolites, such as urea and glucose, were proposed as hydroxyl scavengers in animal systems (Møller et al., 2007).

1.4.2 Varied Sources of ROS in Plant Cells

Reactive oxygen species are not only produced by nonenzymatic pathways, but also enzymes such as NADPH-oxidase, xanthine oxidase, peroxidases and amine oxidase have been reported as ROS sources in plants under particular conditions such as pathogen attacks. In green plant parts in the light, however, photochemical events as well as photorespiration are the main ROS sources. Accordingly, although in nongreen plant parts or in darkness the mitochondria appear to be the main ROS producers, in the light and green tissues the chloroplasts and peroxisomes are the most important ROS-producing cellular compartments (Apel and Hirt, 2004; Møller et al., 2007).

1.4.2.1 Chloroplasts

The photosynthetic electron transport system is the major source of ROS in plant tissues, having potential to generate both $^1O_2$ and $O_2^{\bullet -}$.

1.4.2.1.1 Singlet Oxygen

The excited state of Chl is the primary source of $^1O_2$ in photosynthetic cells. Singlet oxygen may also be generated by lipoxygenase as a by-product of its reaction (Krieger-Liszkay, 2005).

The triplet state of oxygen ($^3O_2$) is relatively stable and is considered ground state oxygen. Although the specific electron configuration of $^3O_2$ prevents it from reacting with many other molecules, the very reactive $^1O_2$ can be formed by providing extra energy. In this case, the energy input increases considerably the oxidizing power of the oxygen (Krieger-Liszkay, 2005).

Chlorophyll is very efficient in absorbing light and its excited state (^{1}Chl*) has a long half-life that allows the excitation energy to be converted into an electrochemical potential. However, every disturbance in the balance between light harvesting and utilization of energy extends the half-life of ^{1}Chl*, thereby inducing the possibility of generating the triplet state Chl (^{3}Chl*) by intersystem crossing. The latter state has an even longer lifetime (a few microseconds) and can react with ground state oxygen to give up the very highly destructive ROS, $^1O_2$:

$$^1\text{Chl}^* \rightarrow {}^3\text{Chl}^*$$

$$^3\text{Chl}^* + {}^3O_2 \rightarrow \text{Chl} + {}^1O_2$$

Formation of $^1O_2$ is favored under certain conditions such as exposure to high light intensities or low CO_2 availability following closure of stomata under different environmental stresses such as salinity and drought. Under such conditions the PQ pool exists in a highly reduced state and the forward electron transfer is very limited.

Singlet oxygen has a short half-time in cells and reacts with many molecules, including proteins and lipids, and is the primary cause for light-induced loss of PS II activity, degradation of the D1 protein and pigment bleaching (Apel and Hirt, 2004; Krieger-Liszkay, 2005; Møller et al., 2007).

For preventing $^1O_2$ damages there are mechanisms for minimizing $^3Chl^*$ generation and/or $^3Chl^*$ and $^1O_2$ quenching in thylakoid membranes. On the other hand, reduction of $^1Chl^*$ half-time via an efficient electron transport in the reaction centers as well as thermal dissipation of EEE quenches $^1Chl^*$ to its ground state (Arora et al., 2002).

The triplet state of Chl can be quenched directly by carotenoids in close proximity (see Section 1.5.2.2). The distance between the two molecules must be less than the van der Waals distance (3.6 Å); this possibility is given in the antenna system, but not in the reaction center, although two β-carotene molecules are present in the PS II reaction center. In the reaction center, the distance between the carotenes and the ^3Chl is too large to allow a direct triplet quenching. Hence the primary function of these β-carotenes is probably the quenching of $^1O_2$ (Arora et al., 2002; Møller et al., 2007).

Another important antioxidant located in the thylakoid membrane is α-tocopherol (see Section 1.5.2.2). Tocopherol is an efficient scavenger and is oxidized in the reaction with $^1O_2$. It has been shown that inhibition of tocopherol biosynthesis in *Chlamydomonas* causes an enhancement in the loss of PS II activity and degradation of D1 protein under higher light intensities (Trebst et al., 2002).

In the absence of an efficient $^1O_2$ scavenging by carotenoids and tocopherol, this ROS reacts with the D1 protein as a target molecule. Damaged D1 protein is degraded and PS II is repaired efficiently by the assembly of newly synthesized D1, i.e. D1 protein damage–repair cycle (Aro et al., 1993). Interestingly, the rapid turnover of the D1 protein was observed even under low illuminations (Keren et al., 1995) suggesting the existence of a background $^1O_2$ generation even under low light intensities. Such "controlled" degradation of the D1 protein is considered to be a safety valve that operates for detoxification of $^1O_2$ directly at the site of its generation. It has been regarded a physiological defense system to prevent "uncontrolled" damage of PS II (Trebst, 2003).

1.4.2.1.2 Production of $^1O_2$ in PS I

When the acceptor side of PS I is fully reduced, recombination between the radical pairs $P_{700}{}^+/A_0{}^-$ or $P_{700}/A_1{}^-$ can generate the triplet state of P_{700}. Chl triplets can react with the molecular oxygen to create $^1O_2$ (Yordanov and Velikova, 2000). However, $^1O_2$ is believed not to be produced in PS I (Hideg and Vass, 1995). Some experimental evidence suggests also that PS I is not a major source of $^1O_2$. Exciting PS I with far-red light did not induce the formation of the β-carotene endoperoxide, whereas preferential excitation of PS II with blue-green light was associated with a substantial increase in this compound after 8 h. These findings are in line with the idea that PS I is not a major source of $^1O_2$ in leaves (Ramel et al., 2012).

The importance of $^1O_2$ generation in plants' response to lighting conditions was clearly demonstrated by analyzing *Arabidopsis flu* mutants (Meskauskiene et al., 2001). In contrast to Chl, which is associated with proteins and transfers energy to the downstream components, protochlorophyllide (Pchlide) occurs mostly in the free form and under light conditions produces $^1O_2$. Accordingly, plants require an efficient mechanism for regulation of their concentration. This regulatory mechanism acts at the step of δ-aminolevulinic acid (ALA).

Arabidopsis flu mutant with damaged regulation of Pchlide synthesis shows over-accumulation of excess Pchlide in etiolated seedlings. Upon light exposure, Pchlide acts as a photosensitizer and triggers the release of $^1O_2$ that results in the rapid bleaching and death of *flu* seedlings. However, under continuous light Pchlide is immediately reduced via NADPH:protochlorophyllide oxidoreductase (POR) to chlorophyllide (Chlide) and does not reach critical levels for production of $^1O_2$ (Meskauskiene et al., 2001).

Accordingly, *flu* mutant can survive only either under continuous darkness or continuous light. Under the latter conditions, the plant is able to convert all produced Pchlide into Chl and prevents overaccumulation of Pchlide despite lacking regulation of its formation. In barley *Tigrina* mutant (mutated on the same gene) illumination causes cell death in the majority of the leaf tissue that has developed in the darkness, while areas of the leaf developed during the day survive (Lee et al., 2003). The suppressor mutants (*ex1* and *ex2*) of the *flu* mutant have lost the capability for perceiving the presence of $^1O_2$ in chloroplasts and are not able to activate $^1O_2$-mediated signaling pathway (see Section 1.6.5) (Lee et al., 2007).

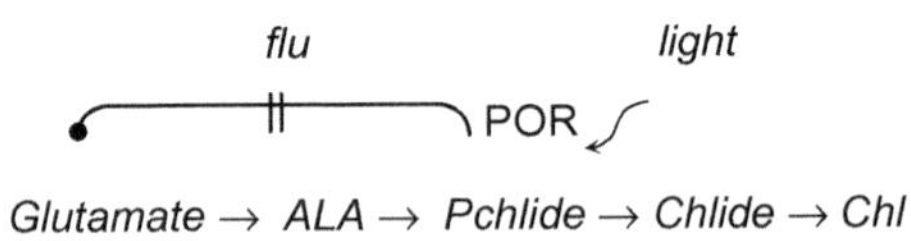

The FLU is encoded by the nuclear genome and contains some protein–protein interaction sites and is located in the thylakoid membrane. It affects the Mg-branch of tetrapyrrole biosynthesis by direct interaction with glutamyl-tRNA reductase. Inactivation of FLU in *flu* mutant impedes negative feedback control of protochlorophyllide on glutamyl-tRNA reductase (Meskauskiene and Apel, 2002).

1.4.2.1.3 Superoxide Radical

Ferredoxin and the electron carriers on the reducing side of PS I have sufficiently negative electrochemical potentials and can deliver electrons to oxygen resulting in the generation of $O_2^{\bullet-}$. Activity of Calvin cycle and the rate of electron flow have important roles in the redox state of the ferredoxin.

The reduced ferredoxin (Fd_{red}) is the main molecule *in vivo* which reduces molecular oxygen to the $O_2^{\bullet -}$ (photoreduction of O_2) (Asada, 1999):

$$O_2 + Fd_{red} \rightarrow O_2^{\bullet -} + Fd_{ox} \quad \text{(Mehler reaction)}$$

The formation of $O_2^{\bullet -}$ by O_2 reduction in PS I by Fd is a rate-limiting step. Once formed, $O_2^{\bullet -}$ may be protonated to HO_2 (perhydroxyl radical):

$$H^+ + O_2^{\bullet -} \rightleftharpoons HO_2 \quad (pK = 4.8)$$

or accept one electron and two protons and dismutated nonenzymatically or by SOD to H_2O_2:

$$O_2^{\bullet -} + 2H^+ \rightarrow H_2O_2 + O_2$$

However, $O_2^{\bullet -}$ mainly has a long lifetime because protons are not available within the interior spaces of thylakoid membranes and the rate of spontaneous or enzymatic disproportionation of $O_2^{\bullet -}$ is low. At a neutral pH the HO_2 concentration is very low, but increases at a lower pH. The diffusion rate of both protonated, neutral species (HO_2) and H_2O_2 through the membranes is higher than that of the anionic ($O_2^{\bullet -}$) species (Apel and Hirt, 2004).

The photoreduction of O_2 to $O_2^{\bullet -}$ has also been observed in Fd-free thylakoids. The thylakoid-bound [4Fe-4S] clusters X on psaA and psaB or A/B on psaC donate electrons to O_2 (Asada, 1999). In close proximity to the Fe-S centers, where Fe^{2+} is available for Fenton reaction, H_2O_2 is transformed into the $^{\bullet}OH$ (see below).

Photoreduction of $O_2^{\bullet -}$ has also been observed in PS II when the membranes are not intact. Since the generation of $O_2^{\bullet -}$ in intact thylakoids is inhibited by 3-(3,4-dichlorophenyl)-1,1-dimethylurea (DCMU) or dibromothymoquinone (DBMIB), the role of PS II and PQ in the $O_2^{\bullet -}$ production could be ruled out. In PS I-free mutants of *Oenothera* and *Scenedesmus* photoreduction of O_2 could not be detected supporting further the participation of only PS I in $O_2^{\bullet -}$ formation (Halliwell and Gutteridge, 2007).

1.4.2.1.4 Hydrogen Peroxide

Hydrogen peroxide is the product of $O_2^{\bullet -}$ dismutation reaction. Catalase, an H_2O_2 scavenging enzyme is localized solely in peroxisomes and transport of H_2O_2 from chloroplasts to peroxisome is improbable. Thus, H_2O_2 produced in chloroplasts is scavenged by a peroxidase reaction, the electron donor is ascorbate (Asc) (Arora et al., 2002; Apel and Hirt, 2004; Cheng and Song, 2006).

Hydrogen peroxide is produced under non-stress conditions, however, stress factors such as drought, low temperatures, higher light intensities and UV radiation increase its generation. In addition of plastidial and mitochondrial electron transport chains, β-oxidation pathway and photorespiration are major sources of H_2O_2 in plant cells. NADPH oxidase and xanthine oxidase also have a role in H_2O_2 production in plants (Cheng and Song, 2006; Sharma et al., 2012).

Hydrogen peroxide has moderate reactivity and is relatively stable with a half-life of about 1 ms and can readily move across membranes either directly or through aquaporins. Because of these properties, H_2O_2 has been considered as a common signaling molecule involved in the regulation of many biological processes and triggering plants' responses to various environmental stresses. Nevertheless, at higher concentrations, i.e. 10 μM and higher, H_2O_2 can oxidize the cysteine and methionine residues in the enzymes and inactivate them. Some enzymes in the Calvin cycle, Cu/Zn-SOD and Fe-SOD are targets for such reactions (Sharma et al., 2012).

1.4.2.1.5 Hydroxyl Radicals

Hydrogen peroxide and $O_2^{\bullet-}$ can produce $^{\bullet}OH$, a highly damaging ROS. The oxidizing potential of H_2O_2 with ferrous salts is the basis of Fenton reaction that produces $^{\bullet}OH$ as the oxidizing species:

$$Fe^{2+} + H_2O_2 \rightarrow Fe^{3+} + OH^- + {}^{\bullet}OH \quad \text{(Fenton reaction)}$$

Metal ions such as Cu^+, Cu^{2+} can replace Fe^{2+}, Fe^{3+} in these reactions. The rate of reaction in the absence of metal catalysis is very low (Halliwell and Gutteridge, 2007).

Although the availability of Fe^{2+} may limit the reaction, this ion can be reproduced by reducing agents such as $O_2^{\bullet-}$:

$$Fe^{3+} + O_2^{\bullet-} \rightarrow Fe^{2+} + O_2$$

Therefore, the net reaction is:

$$O_2^{\bullet-} + H_2O_2 \rightarrow OH^- + O_2 + {}^{\bullet}OH \quad \text{(Haber-Weiss reaction)}$$

Hydroxyl radical is the most reactive ROS with a single unpaired electron and is capable of reacting with almost all biological molecules and damages cellular components. There is no enzymatic mechanism to scavenge $^{\bullet}OH$, accordingly, its excess generation usually leads to cell death (Møller et al., 2007). Generation of $^{\bullet}OH$ by the Fenton reaction at the active site of the Rubisco under light conditions leads to the fragmentation of rbcL (large subunit of Rubisco). The free $^{\bullet}OH$ has a short lifetime and a highly positive redox potential. Organic oxygen radicals such as alkoxy and peroxy are products of reaction of free $^{\bullet}OH$ with organic molecules (Arora et al., 2002).

1.4.2.2 Other Organelles

Mitochondria use oxygen during respiratory electron transport and there are different sites of $O_2^{\bullet-}$ and H_2O_2 generation in the mitochondrial respiratory chain (Blokhina and Fagerstedt, 2010). Direct reduction of oxygen to $O_2^{\bullet-}$ in the flavoprotein region of NADH dehydrogenase leads to the production of $O_2^{\bullet-}$. The responsible component is likely the flavoprotein (of either internal or external dehydrogenase) or perhaps an Fe–S center. Oxygen reduction to $O_2^{\bullet-}$ in the respiratory chain between ubiquinone and cytochrome is another $O_2^{\bullet-}$ generating point in the pathway (Arora et al., 2002).

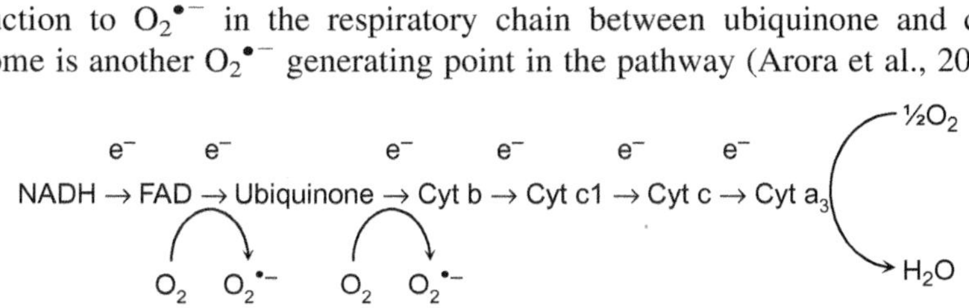

It has been proposed that, in contrast to animal cells having mitochondria as the main source of ROS, the role of this organelle in ROS production in green tissues is very low. The alternative oxidase (AOX) that catalyzes the tetravalent reduction of O_2 by ubiquinone is the reason for low ROS production in plant mitochondria. The AOX competes with the cytochrome bc1 complex for electrons and thus reduces ROS production in mitochondria. This assumption is confirmed by evidence showing that overproduction of AOX in transgenic cell lines reduces ROS generation, while antisense cell lines with lowered AOX activity accumulate higher ROS compared with control cells (Apel and Hirt, 2004).

1.4.2.2.1 Endoplasmic Reticulum, Peroxisomes and Glyoxysomes

The smooth ER and the microsomes derived from it harbor various oxidative processes. Mixed-function oxygenases, such as cytochrome P450, perform important hydroxylation reactions in the mevalonic acid pathway, adding oxygen atoms to substrate molecules. NAD(P)H is the electron donor and $O_2^{\bullet-}$ may be released by such reactions (Elstner, 1991).

Peroxisomes and glyoxysomes are single membrane organelles that compartmentalize enzymes involved in the β-oxidation of fatty acids, and the C_2 photorespiratory cycle. Xanthine oxidase, urate oxidase and NADH oxidase generate $O_2^{\bullet-}$ (Elstner, 1991). Peroxisomes are probably the major sites of intracellular H_2O_2 production. The oxidation of glycolate by glycolate oxidase during photorespiration is the major source of H_2O_2 in peroxisomes (see Section 1.3.2). Hydrogen peroxide is generated through four metabolic pathways in microsomes including glycolate oxidase reaction, fatty acid β-oxidation, enzymatic reaction of flavin oxidases and disproportionation of $O_2^{\bullet-}$ radicals. Similar with mitochondria and chloroplasts, $O_2^{\bullet-}$ is also generated during normal metabolism in peroxisomes (del Río et al., 2006).

1.4.2.2.2 Plasmamembrane and the Apoplast Compartment

NAD(P)H oxidase are ubiquitous components of plasma membrane and may produce $O_2^{\bullet-}$ and H_2O_2. Oxygen activation occurs also in the apoplast and some enzymes associated with the cell wall are responsible for apoplastic ROS production. NADH peroxidase, amine oxidases and oxalate oxidase are three major cell wall-associated enzymes responsible for H_2O_2 generation in the apoplast. The latter enzyme is involved in apoplastic H_2O_2 accumulation during interaction of fungal pathogens with cereal species (Elstner, 1991; Denness et al., 2011; Heyno et al., 2011; Sharma et al., 2012).

1.4.3 Effect of Environmental Factors on ROS Production during Photosynthesis

Generation and accumulation of ROS in plants is low under nonstress conditions. Under various environmental stresses, however, ROS production is increased and causes a drastic disturbance in the cellular balance of $O_2^{\bullet -}$, $^{\bullet}OH$, and H_2O_2 levels. The effects of various environmental stresses such as drought, salinity, chilling, metal toxicity, UV-B radiation, and pathogen attack on ROS are well documented (Apel and Hirt, 2004).

On the other hand, most environmental conditions that impose constraints on plant growth and development promote an increase in EEE, thus influencing the efficiency of light energy fixation (Bechtold et al., 2005; Zhou et al., 2007). Under these stressful conditions, imbalance between light energy absorbed by photosystems and the ultimate consumption of the photosynthetic electrons through metabolic pathways such as the Calvin cycle, photorespiration and nutrient assimilation occurs. This leads to an increased formation of ROS and to photooxidative stress. Stomatal limitation imposed by drought, salinity or low temperature inhibits CO_2 assimilation and $NADP^+$ regeneration by the Calvin cycle, while at the same time the light-driven photosynthetic electron transfer proceeds at high rates. This leads to an overreduction of the electron transport chain. Consequently, the formation of ROS is initiated by the transfer of electrons to alternative acceptors, predominantly molecular oxygen. Similarly, excitation pressure may be induced by the lack of essential nutrients because of limitations in the availability of electron acceptors such as NO_3^- or SO_4^{2-} (Wilson et al., 2006).

1.5 PROTECTION OF PHOTOSYNTHETIC PLANTS AGAINST ROS

The different reactive species described here will cause to various extents inhibition of sensitive enzymes, Chl degradation or bleaching and lipid peroxidation. Free radicals, H_2O_2 and $^1O_2$ readily attack unsaturated fatty acids, yielding lipid hydroperoxides, and in the presence of metal catalysts, alkoxyl and peroxyl radicals that propagate chain reactions in the membranes, changing and disrupting lipid structure and membrane organization and integrity. Indiscriminate attack by hydroxyl radicals of organic molecules, including DNA, damages this molecule. A variety of oxidatively altered DNA species can be identified following $^{\bullet}OH$ attack, including base alterations and strand breaks that may be difficult to repair. Proteins exposed to $^{\bullet}OH$ undergo typical modifications, including specific amino acid alterations, polypeptide fragmentation, aggregation, denaturation and susceptibility to proteolysis (Asada, 2006; Mittler, 2006).

Generation and scavenging of ROS are both regulated processes in plant cells and the balance between the oxidative and antioxidative capacities determines the plant's response. Under nonstress conditions the antioxidant defense system provides adequate protection against ROS, while under stress conditions production of ROS exceeds the scavenging capacity of plants and oxidative stress occurs (Apel and Hirt, 2004).

As mentioned above, light energy in excess can lead to the production of ROS and damage can occur if the EEE is not dissipated safely (Horton et al., 1996). Photosynthetic organisms contain complex mechanisms that regulate energy flow in the antenna system in order to prevent induction of EEE in the reaction centers, i.e. photoprotection (Niyogi, 1999). Although very effective, these mechanisms are not efficient enough and ROS are produced. Thus, additional processes are required not only for ROS scavenging but also for repairing damaged photosynthetic components (Arora et al., 2002).

Arrival of EEE at the reaction centers leads to their inactivation and damage, i.e. photoinhibition. Photoinhibition is composed of various molecular processes generally described as the inhibition of photosynthesis by excess light. Photoinhibition is reversible in early stages; however, prolonged inhibition results in damage to the photosystems and other electron transfer components (Long et al., 1994; Niyogi, 1999). The main target of this injury is the D1 protein as one part of the PS II reaction center complex. All three ROS – $O_2^{\bullet -}$, $^{\bullet}OH$, and H_2O_2 – are involved in the degradation of oxidized D1 protein (Chen et al., 2012). Phosphorylation of D1 protein is required for efficient migration of damaged PS II complexes from grana to stroma lamellae (Tikkanen et al., 2008). The damaged D1 protein is then dephosphorylated and degraded by a specific protease. Without phosphorylation of D1 protein and its migration, D1 degradation is impaired and accumulation of photo-damaged molecule strengthens the photo-oxidative damages in the thylakoid membranes. After a successful removal from the membrane, D1 could be replaced with a newly synthesized molecule (Zhang and Aro, 2002).

Because photoinhibition involves photochemical inactivation mainly of PS II, all photosynthesizing organisms are potentially susceptible to damage under some radiation incidence. However, as mentioned above, the degree of susceptibility is influenced by several types of factors including environmental factors, e.g. light, temperature, water, CO_2, O_2, and nutritional status; genotypical factors, e.g. sun or shade plants; and physiological factors, e.g. the rate of carbon metabolism. The imposition of additional stress factors during exposure to high radiance exacerbates these adverse effects. Low and high temperature, water stress and low CO_2 availability (stomatal closing) in combination with high light conditions may cause a higher degree of photoinhibition in plants (Niyogi, 1999).

1.5.1 Protection from Excess Excitation Energy: Photoprotection

The photoprotection mechanisms are considered to be safety valves, venting EEE before it can damage the photosynthetic cells and/or before ROS production (Niyogi, 2000). Photoprotection mechanisms function either for antenna quenching or for reaction center quenching. It has been proposed that the extent of these quenching mechanisms in the photoprotection *in vivo* depends on the environmental conditions. These quenching strategies are involved in the photoprotection of PS II or PS I.

1.5.1.1 Nonphotochemical Quenching

When the energy is stored in Chl molecules in the excited state, it is rapidly dissipated by excitation transfer or photochemistry, and thus the excited state is quenched. Dissipation of EEE is accomplished by both photochemical quenching and nonphotochemical quenching (NPQ) processes. Photochemical quenching of EEE refers not only to the application of electrons in the dark reactions in the Calvin cycle but also to the events that raise energy usage through some supplementary metabolic sinks such as photorespiration and chlororespiration (see Section 1.5.1.7) as well as increased N and S assimilatory metabolism (Müller et al., 2001).

Among mechanisms for dissipation of excess energy absorbed by Chl, thermal energy dissipation within the photosynthetic apparatus is the most flexible and fast mechanism for photoprotection. The consequence of operation of such a mechanism is a decline of the Chl fluorescence yield, e.g. NPQ (Niyogi, 1999).

Nonphotochemical quenching is the main procedure that regulates transfer of excitation energy for adjustment of the flow of excitations to the PS II reaction center based on the light intensity and usage of excitation energy (Fig. 1.7). As the consequence of NPQ, the main portion of the excitation energy in the antenna system produced under intensive illumination is quenched by conversion of energy into heat. During NPQ processes, the ^{3}Chl excitation energy is transferred to the carotenoids. Subsequently, during return to a nonexcited ground state, carotenoids dissipate the excess energy as heat, i.e. thermal energy dissipation or heat dissipation (Niyogi, 1999). Carotenoids apply their photoprotective role by rapidly quenching the excited state of Chl, thus preventing $^1O_2$ generation. Since carotenoids at an excited state do not have enough energy for $^1O_2$ generation, they decay back to the ground state and lose their energy as heat. Mutants lacking carotenoids do not survive under a combination of aerobic conditions and illumination (Müller et al., 2001).

Three xanthophyll types of carotenoids are involved in nonphotochemical quenching: violaxanthin (V), antheraxanthin (A) and zeaxanthin (Z). The xanthophylls cycle is catalyzed by two enzymes in the thylakoid membrane. Violaxanthin de-epoxidase (VDE), located in the thylakoid lumen, catalyzes de-epoxidation half of the cycle under low pH and in the presence of Asc.

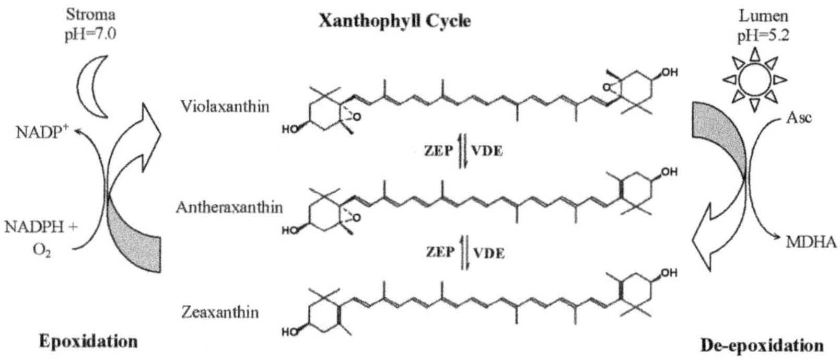

FIGURE 1.7 The xanthophylls cycle. Abbreviations: VDE: Violaxanthin de-epoxidase; ZEP: Zeaxanthin epoxidase.

Zeaxanthin epoxidase (ZEP), located on the stromal side of the thylakoid membrane, catalyzes the reverse reaction for regenerating V. This epoxidase reaction is carried out in the dark or under low illumination with a pH optimum of 7.5. The light intensity-dependent inter-conversion of V and Z are known the V xanthophyll cycle (or V-cycle) (Eskling et al., 1997).

Binding of protons and Z to the light-harvesting antenna proteins induces some conformational changes resulting in energy quenching and heat dissipation. By means of a photophysical mechanism, Z stabilizes and amplifies NPQ at different quenching sites (Jahns and Holzwarth, 2012), regulating it over longer time periods than the proton gradient, ranging from minutes to hours. Nonphotochemical quenching is related to a peripheral antenna protein of PS II, PsbS. The nuclear-encoded PsbS protein does not affect photosynthetic efficiency of plants but is definitely necessary for NPQ (Li et al., 2004). Mutants of *PsbS* lack the main component of NPQ (Ivanov et al., 2008). Results on the possible binding of PsbS protein to the pigments are controversial, but the role of this protein in the sensing lumen pH has received much support (Li et al., 2004).

Nonphotochemical quenching is regulated by three distinctly acting components. These include: (i) the build-up of a proton gradient (ΔpH) across thylakoid membranes (Horton et al., 1996), which is generated by photosynthetic proton pumping; (ii) the activity of the xanthophylls cycle with the corresponding formation of Z (Demmig-Adams et al., 1990); and (iii) the protein PsbS (Li et al., 2000). Both PsbS and Z act independently and have distinct regulatory functions (Bonente et al., 2008; Jahns and Holzwarth, 2012). The magnitude of ΔpH is usually proportional to the excitation pressure, acting as a signal that activates NPQ, while PsbS and certain LHCs act as sensors of proton concentration (García-Plazaola et al., 2012).

Depending on the relaxation kinetics in darkness after a period of light and response to different inhibitors, three different components in the NPQ process have been described (Nilkens et al., 2010): (i) a fast and ΔpH dependent component (qE or $NPQ_{\Delta pH}$), which can be turned on and off in seconds, tracking changes in light conditions (Nichol et al., 2012); (ii) the second component is photoinhibitory quenching (qI), which is irreversible in the short-term (hours or days), resulting in the sustained downregulation of photochemical efficiency; and (iii) a third component of NPQ, which is ΔpH-independent and entirely dependent on the Z content (qZ) (Nilkens et al., 2010). The role of the ΔpH- and Z-dependent shifts in the oligomerization state of LHCII (Horton et al., 1996) and activation state of Z (Ruban et al., 2002) in development of the rapidly relaxing and energy dependent component (qE) of NPQ has been well described and is considered a reflection of an indirect, allosteric mechanism for antenna quenching (Ivanov et al., 2008). Another component of NPQ that relaxes within minutes is named qT and related to the state transition (see Section 1.5.1.3). State transition quenching is crucial for algae, but is of minor importance in higher plants under excess illumination and does not seem to be important for photoprotection under saturating light conditions (Niyogi, 1999; Nilkens et al., 2010).

Based on the main chemical form of the carotenoids involved in thermal dissipation reactions in higher plants, two different xanthophyll cycles have been described, the violaxanthin cycle (V-cycle) and the lutein-epoxide cycle (Lx-cycle) (see Section 1.5.1.1.2) (García-Plazaola et al., 2007).

1.5.1.1.1 Flexible versus Sustained Energy Dissipation

Nonphotochemical quenching may occur at different temporal domains and it can be regulated by diverse mechanisms. Under severe environmental stresses such as low temperature and desiccation that are associated with high irradiance, rapidly reversible $NPQ_{\Delta pH}$ mechanisms are not enough to compensate for the excess light energy. Under these conditions, a set of ΔpH-independent NPQ mechanisms exists (García-Plazaola et al., 2012).

Excess energy may appear during diurnal changes in sun elevation and cloudiness. Under such conditions, the photosystems use a reversible NPQ mechanism, i.e. flexible NPQ to adjust the fraction of excitation energy that is being thermally dissipated, thus maintaining the energy balance between light and carbon reactions.

Evergreen species growing at high altitude or latitude, however, face an additional excess energy during the cold months. Low temperatures increase the levels of excess energy and the demand for photoprotection because of inactivating the enzymatic carbon reactions as well as impairing enzymatic ROS scavenging systems (Demmig-Adams and Adams, 2006). Under these conditions generation of the ΔpH or enzymatic de-epoxidation reaction necessary for the optimal functioning of the $NPQ_{\Delta pH}$ are very slow (Eskling et al., 2001). Thus a simple upregulation of $NPQ_{\Delta pH}$ PQ would not suffice during winter.

Overwintering evergreens require mechanisms with sustained thermal energy dissipation capacities in order to cope with the particularly high levels of excess energy during the cold season (García-Plazaola et al., 2012). Sustained NPQ mechanisms are not simply a slow version of $NPQ_{\Delta pH}$ but require the regulation of gene expression and the subsequent structural and biochemical reorganization of the photosystems (Demmig-Adams and Adams, 2006; Zarter et al., 2006).

1.5.1.1.2 Lutein-related Thermal Dissipation

Most forests with dense vegetation possess a closed canopy, in which only a small fraction of the incident solar radiation (0.5−5 %) reaches the understory. A significant proportion of this energy penetrates the canopy during brief periods of direct sunlight, the "sunflecks" that are one of the key factors for plant survival in the understory (García-Plazaola et al., 2012).

Understory species must be able to cope with the random variation of light and being simultaneously efficient for light harvesting and carbon reactions. A rapid engagement of Z and A in energy dissipation during sunfleck activity and rapid disengagement upon return to low light play a central role in plant response to this dynamic light environment (Fig. 1.8). Another mechanism is different pigment composition. Lutein (L), a carotenoid that is found mostly in plants growing in shady environments, is involved in the lutein epoxide (Lx)-cycle (Lx xanthophylls cycle) in these species (García-Plazaola et al., 2007).

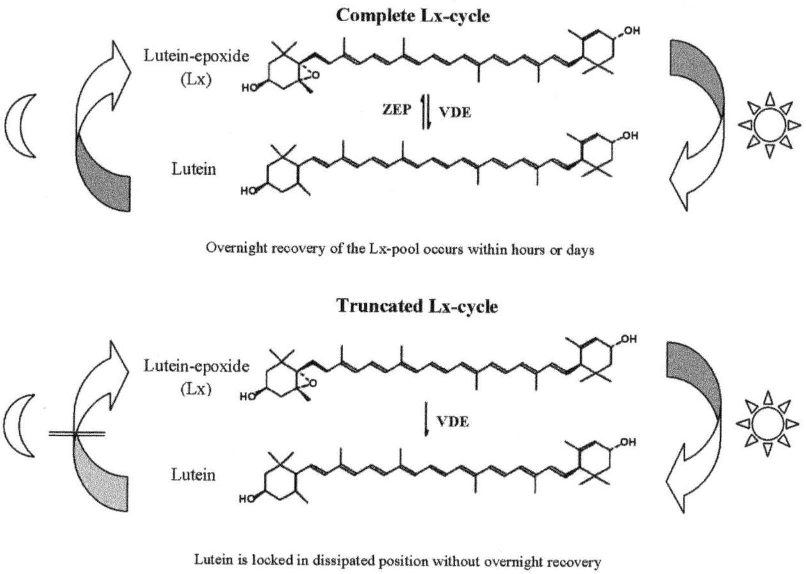

FIGURE 1.8 Lutein-epoxide (Lx) cycle. In the case of overnight recovery of Lx-pool, the cycle is completed, while in the absence of Lx recovery a truncated cycle takes place.

Unlike the V-cycle, which is widespread in higher plants, the Lx-cycle is confined to taxonomically distinct groups (Esteban et al., 2009). This cycle operates through the light-driven de-epoxidation of Lx into L, a process that occurs in parallel with the de-epoxidation of V in the V-cycle, both cycles being catalyzed apparently by the same enzymes (García-Plazaola et al., 2007). The presence of Lx in prolonged low-light conditions may increase light-harvesting efficiency by facilitating excitation energy transfer to Chl in light limiting conditions. The Lx-cycle is completed when the reversal of the epoxidation of L to Lx occurs. However, this step is frequently missing or occurs at a very slow rate. Two types of Lx-cycles exist: (i) a "complete" cycle that is the epoxidation of L to Lx with full recovery of the Lx-pool in the dark and (ii) a "truncated" cycle, with no overnight recovery of the initial Lx-pool. In the case of plants in which the inner canopy leaves initially grow under strong light, the leaves might acclimate to progressive shading by the accumulation of Lx that enhances light-harvesting capacity. However, when a gap forms in the forest, a "truncated" cycle may represent an emergency mechanism of sustained energy dissipation after abrupt changes in the light environment. The operation of this cycle would then facilitate the shift from highly efficient LHCs to excitation dissipating centers, stabilizing the capacity to rapidly engage these functions (García-Plazaola et al., 2007) (Fig. 1.8).

The existence of both cycles (V and Lx) in some plant species suggests an ecological or physiological role for the Lx-cycle that is different or complementary to the V-cycle. It has been shown (Matsubara et al., 2011) that the L formed from Lx was able to enhance $NPQ_{\Delta pH}$ and, therefore, the Lx-cycle contributes together with the V-cycle to the regulation of NPQ.

1.5.1.1.3 Thermal Dissipation in Desiccation-tolerant Plants

In contrast to the majority of plant species, desiccation-tolerant (DT) plants can lose more than 90% of their water content and recover normal metabolic functions upon rewatering (Vertucci and Farrant, 1995). Most of the DT plants, i.e. in the homoiochlorophyllous DT species, preserve an intact photosynthetic apparatus in the dry state (Tuba, 2008).

One of the main strategies in these DT species is their capability to regulate photosynthetic activity during dehydration. At very low leaf water content, photosynthetic electron transport is totally inhibited and PS II is switched off in DT plants, while it remains active in non-DT plants. One of the mechanisms that preserves the functionality of DT-photosynthetic organisms is the desiccation-induced NPQ (NPQ_{DT}), which is different from that induced by light (Nabe et al., 2007).

The NPQ_{DT} follows a regular pattern during dehydration-rehydration events in most DT plants (Fig. 1.9). NPQ_{DT} increases during desiccation until it reaches a maximum value, which is usually maintained until the next rehydration. At the end of the stress, rehydration is also a critical step

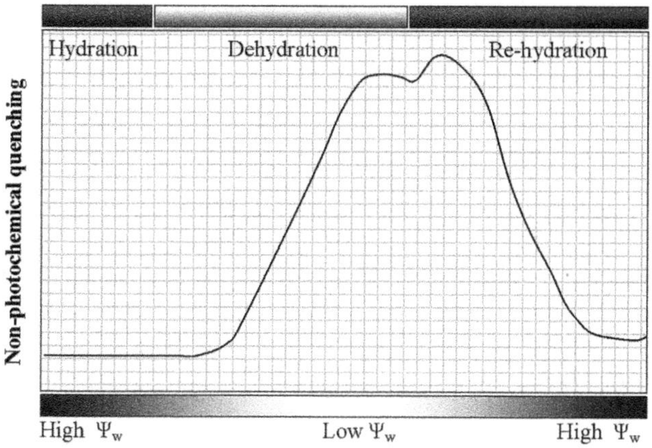

FIGURE 1.9 Change in the extent of nonphotochemical quenching in drought-tolerant plants during dehydration–rehydration cycles. Abbreviation: Ψ_w: water potential. Redrawn according to García-Plazaola et al. (2012).

because the sudden influx of water and resumption of metabolic activity results in oxidative stress. Thus, on rewatering the photosynthetic apparatus needs to be protected and NPQ_{DT} undergoes a transient increase before decreasing to the initial basal values during rehydration (Csintalan et al., 1999; Marschall and Proctor, 1999; Peeva and Cornic, 2009; García-Plazaola et al., 2012).

Experimental evidence shows that dissipation of excess light energy is carried out and regulated by separate mechanisms in DT compared with other plant species during dehydration (Nabe et al., 2007). The NPQ_{DT} is not triggered either by light or a protonation reaction, but is triggered by the dehydration process and is rapidly reversed upon rehydration. The molecular basis of NPQ_{DT} is still obscure.

1.5.1.2 Quenching by Inactive PS II Reaction Centers

In addition to ΔpH and Z-dependent NPQ, quenching of Chl fluorescence may be the consequence of the conversion of PS II α-centers (dimers) to PS II β-centers (monomers). The monomerization of PS II centers triggered under high illuminations can efficiently reduce the absorption cross-section of PS II. The balance between the active PS II centers and inactive centers is dependent both on the intra-thylakoid ΔpH and the relative amount of closed reaction centers (Krause, 1988; Ivanov et al., 2008). On the other hand, photo-inactivated PS II complexes are also efficient in thermal dissipation, being more effective under severe photoinactivation (Lee et al., 2001). Several reports have confirmed that the conversion of photochemically active into photochemically inactive PS II reaction centers prevents further damage to the photo-inactivated reaction centers as well as neighboring active PS II reaction centers (Lee et al., 2001; Matsubara and Chow, 2004).

1.5.1.3 State I/State II Transition

State transitions involve reversible phosphorylation/dephosphorylation of the major LHC II by a thylakoid-bound kinase that phosphorylates a particular threonine residue in the LHC II. This kinase is regulated by the redox state of the PQ pool. It is activated when reduced PQ accumulates, i.e. when PS II is activated more repeatedly than PS I. Phosphorylated LHCII (pLHC II) move from the stacked areas of the thylakoid membranes into the unstacked areas. Lateral movement of pLHC II from PS II towards PS I results in a redistribution of the energy in favor of PS I and shifts the energy balance toward PS I, i.e. state II. Under excess excitation of PS I and when PQ is highly oxidized, the kinase is deactivated and a membrane-bound phosphatase reduces the phosphorylation level of LHC II. As the consequence, LHC II migrates back to the stacked areas, i.e. state I. State transition provides a possibility for an exact regulation of the energy allocation to the photosystems, and consequently, an effective usage of the incident light energy (Allen, 1995).

State transition is considered to be a strategy for optimizing the allocation of excitation energy to the two photosystems, and thus likely plays an important role in PS II protection from overexcitation (Lunde et al., 2000; Haldrup et al., 2001). Cyt b_6f has been identified as the redox sensor of the PQ pool and has been involved in controlling the phosphorylation of LHC II. The small PsaH subunit of the PS I complex has a critical role in state transition in *Arabidopsis thaliana* (Lunde et al., 2000).

1.5.1.4 Aggregation of LHCII

The spectroscopic data indicate that the xanthophyll cycle pigments, Z and V, exist outside the PS II antenna system and induce aggregation of proteins that is stabilized by hydrogen bonds among molecules (Gruszecki et al., 2006). Aggregation of LHC II has been considered the main reason for protection of LHC II complex against overexcitation, because quenching of EEE is associated with the protein oligomerization and crystallization (Horton et al., 2005; Pascal et al., 2005).

The LHCII aggregation is associated with a decline in the extinction coefficient of pigments and causes light scattering and reduction of absorption cross-section (Gruszecki et al., 2006). The new excitonic band energy level facilitates also the energy equilibrium between photosystems through state I—state II transition (Allen, 2003). Accordingly, the LHCII aggregation induced by the xanthophyll pigments could be regarded as a mechanism for switching LHCII between light-harvesting and energy dissipation systems (Gruszecki et al., 2006).

1.5.1.5 Cyclic Electron Flow around PS II and PS I

1.5.1.5.1 Cyclic Electron Flow around PS II

Electron delivery to P_{680}^+ by Cyt b_{559} via a cyclic electron flow is considered to be one dissipating mechanism functioning within the PS II reaction center (Allakhverdiev et al., 1997; Ivanov et al., 2008):

$$\text{Cyt } b_{559} \rightarrow \text{dChlz} \rightarrow \beta\text{-Car} \rightarrow P_{680}^+$$

The conversion of Cyt b_{559} from its high potential form (HP) to the low-potential form (LP) may play a key role as a molecular switch, allowing it to act as an electron donor to P_{680}^+ (HP form) or an electron acceptor (LP form). Additionally, β-carotene can be photo-oxidized and it has been suggested to facilitate the electron flow from Cyt b_{559} and dChlz to P_{680}^+. Cyclic electron flow around PS II in intact chloroplasts has been proposed to be practically comparable to the water–water cycle (see Section 1.5.1.6), and thus can participate efficiently in dissipation of EEE and PS II photoprotection (Allakhverdiev et al., 1997; Miyake and Yokota, 2001; Ivanov et al., 2008).

1.5.1.5.2 Cyclic Electron Flow around PS I

Cyclic electron flow around PS I is much more important in photoprotection compared with that around PS II. In this cyclic flow, the electrons produced by light at PS I return to the photosynthetic electron transport system through

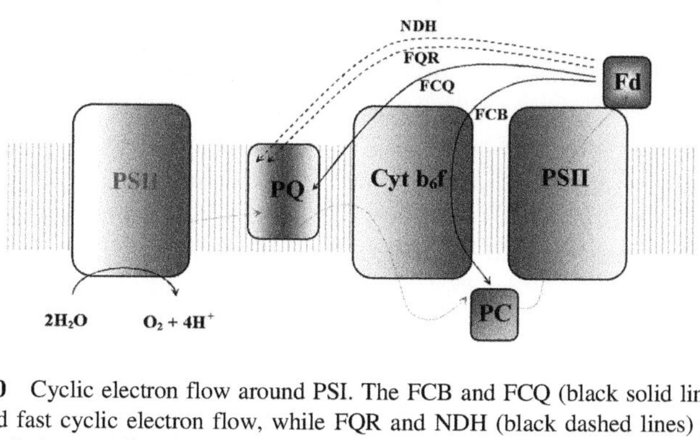

FIGURE 1.10 Cyclic electron flow around PSI. The FCB and FCQ (black solid lines) pathways are considered fast cyclic electron flow, while FQR and NDH (black dashed lines) pathways are components of slow cyclic electron flow. Abbreviations: FCB: Fd-heme c-hem b pathway; FCQ: Fd-heme c-PQ pathway; FQR: Fd-quinone oxidoreductase pathway; NDH: NAD(P)H dehydrogenase pathway. The gray dotted lines indicate liner electron flow.

Fd or NADPH, and then are delivered to the PQ or Cyt b_6f (Joliot and Joliot, 2002, 2005, 2006).

Four mechanisms of cyclic electron flow in PS I have been proposed (Miyake, 2010) (Fig. 1.10). Considering the extent of electron flow, cyclic electron flow is categorized into two types: (i) fast cyclic electron flow, and (ii) slow cyclic electron flow. Two subtypes are defined within the fast cyclic electron, FCQ (Fd-heme c-PQ) and FCB (Fd-heme c-hem b). In the FCQ pathway, electrons produced by light at PS I flow from Fd to the PQ pool through heme c in the Cyt b_6f complex. In the FCB pathway, electrons flow from Fd to the heme b in the Cyt b_6f complex through heme c without development of ΔpH (Laisk et al., 2010). Slow cyclic electron flow is similarly categorized into two types, i.e. Fd-quinone oxidoreductase (FQR) and NAD(P)H dehydrogenase (NDH) pathways. In the FQR pathway, electrons derived from PS I flow from Fd to the PQ pool through FQR. The FQR pathway probably involves two related proteins, PGR5 (Protin Gradient Regulation 5) and PGRL1 (PGR Like 1) (Dalcorso et al., 2008). In the NDH pathway, electrons originated from PS I flow from NAD(P)H to the PQ pool through FNR, Fd and NDH (see also Section 1.5.1.7) (Endo and Asada, 2002).

Some physiological functions have been attributed to the cyclic electron flow in thylakoid membranes (Joliot and Joliot, 2006; Livingston et al., 2010; Miyake, 2010): (i) provision of ATP needed for RuBP recovery for sustaining the Calvin cycle (this function is much more important under stress conditions because of increased ATP demand); (ii) induction of NPQ when the rate of $NADP^+$ recovery acts as a limiting factor for linear electron flow; and (iii) suppression of $O_2^{\bullet -}$ production in the Mehler reaction (see Section 1.5.1.6).

1.5.1.6 Water–Water Cycle

Excess energy excitation that cannot be converted into chemical energy can also be dissipated by electron transport to O_2 in the Mehler-peroxidase (water–water cycle) pathway. In contrast to the isolated thylakoids, intact chloroplasts under light conditions and in the presence of $^{18}O_2$ evolve $^{16}O_2$ from water and reduce $^{18}O_2$ without $H_2{}^{18}O_2$ accumulation though lacking catalase activity. These evidences demonstrated that chloroplasts use a mechanism for incorporating $^{18}O_2$ to $H_2{}^{18}O$ by employment of the electrons derived from water (Asada, 1999; Miyake, 2010). Experimental data suggested that the H_2O_2 produced after dismutation of PS I-derived $O_2^{\bullet -}$ is further reduced to water via a peroxidase reaction. The presence of an Asc specific peroxidase (APX) in chloroplasts as well as other enzymes required for Asc regeneration revealed the existence and function of an O_2-dependent sequential electron flow in the chloroplasts. This cyclic pathway of electrons is termed the Mehler–Asc peroxidase pathway (Neubauer and Schreiber, 1989) or water–water cycle (Asada, 1999; Miyake, 2010). In this cycle, the electrons derived from water in PS II reduce atmospheric O_2 to water in PS I without a net change of O_2 (Asada, 1999):

$$2H_2O \rightarrow 4e^- + 4H^+ + O_2 \quad \text{(Photooxidation of water in PS II)}$$
$$2O_2 \rightarrow 2e^- + 2O_2^{\bullet -} \quad \text{(Photoreduction of } O_2 \text{ in PS I, Mehler reaction)}$$
$$2O_2^{\bullet -} + 2H^+ \rightarrow H_2O_2 + O_2 \quad \text{(SOD} - \text{catalyzed disproportionation of } O_2^{\bullet -}\text{)}$$
$$H_2O_2 + 2Asc \rightarrow 2H_2O + 2MDA \quad \text{(APX} - \text{catalyzed reduction of } H_2O_2 \text{ by Asc)}$$
$$2MDA(\text{or DHA}) + 2e^- + 2H^+ \rightarrow 2\,Asc\ (\text{or } 1\,Asc) \quad \text{(Regeneration of Asc)}$$
$$\sum 2H_2O + O_2^* \rightarrow O_2 + 2H_2O^*$$

The functional significance of the water–water cycle is mainly related to an immediate scavenging of $O_2^{\bullet -}$ and H_2O_2 before generation of the hydroxyl radical ($^{\bullet}OH$) (see Section 1.4.2.1). Every disturbance in the function of various enzymatic components of the water–water cycle, i.e. in aged leaves, may result in $^{\bullet}OH$ generation. Thus, this cycle can contribute significantly in dissipation of EEE and protection of photosynthetic cells from photoinhibition (Miyake, 2010).

1.5.1.7 Chlororespiration and Photorespiration

Chlororespiration is also an O_2 uptake reaction in chloroplasts. In this process, electrons transfer from NAD(P)H to a terminal oxidase via PQ, as intermediate carrier between photosynthetic and respiratory electron transfer systems (Fig. 1.11). Chloroplasts of algae and higher plants are derived from cyanobacteria through an endosymbiosis event. Since cyanobacteria was initially competent for both photosynthesis and respiration, it is well expected that the competence for oxidizing NAD(P)H at the expense of O_2 is partially preserved in plant plastids (Rumeau et al., 2007).

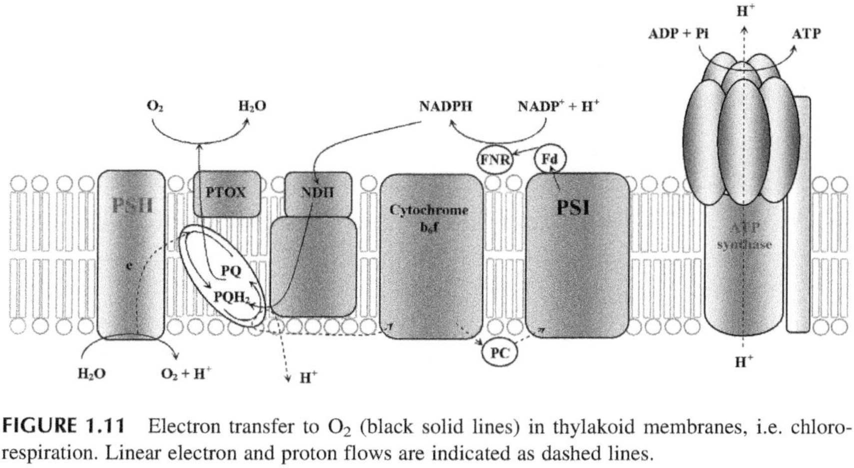

FIGURE 1.11 Electron transfer to O_2 (black solid lines) in thylakoid membranes, i.e. chlororespiration. Linear electron and proton flows are indicated as dashed lines.

The occurrence of a respiratory pathway in the plastids received supports after the discovery of some new proteins in the thylakoid membranes including a plastidial NAD(P)H-dehydrogenase complex (NDH), an alternative plastid terminal oxidase (PTOX) and PGR5 and PGRL1 proteins (see Section 1.5.1.5). Plastidial genes encoding some subunits of NDH are homologous to mitochondrial complex I; PTOX also shows sequence similarity with the mitochondrial alternative oxidase (AOX) (Carol et al., 1999).

Physiological and molecular studies demonstrated that some of the previously mentioned components are involved in protection and adaptation of plants to environmental stress such as high light and heat stress and water deficit (Suorsa et al., 2012). PTOX operates as an electron sink, but its role in cyclic electron flow (see Section 1.5.1.5) is rather indirect. By regulating the redox state of PQ, PTOX plays a role in an efficient operation of cyclic electron flow (Joet et al., 2002). Operation of PTOX is considered to be a safety valve that prevents overreduction of not only PS I acceptors (Rumeau et al., 2007) but also PS II acceptors (Streb et al., 2005). The role of PTOX in photoprotection is studied in *Arabidopsis* lines overexpressing *AtPTOX* (Joet et al., 2002). It has been observed that in these lines overreduction of Q_A and intermediate electron carriers is significantly lower than wild type lines, although this effect was only detected during transition from dark to light (Joet et al., 2002). On the other hand, over-expression of *AtPTOX* in transgenic *Arabidopsis* lines lacking catalase and ascorbate peroxidase indicated that PTOX may replace functionally H_2O_2 scavenging activity (Rizhsky et al., 2002). Although the latter effect has not been tested *in vivo*, a parallel increase in the amount of NDH complex, another component of the chlororespiratory pathway and thylakoid ascorbate peroxidase observed in response to excess light energy, may support indirectly the role of chlororespiration in H_2O_2 detoxification (Casano et al., 2000).

Photorespiration may also contribute to dissipation of EEE and acts as an alternative sink for excess electrons due to consumption of NADPH and ATP. It has been demonstrated that the photorespiratory pathway may play an indicative role in protection of plants against photoinhibition (Rumeau et al., 2007). Function of various EEE quenching mechanisms is summarized in Fig. 1.12.

1.5.1.8 Reaction Center Quenching

The interest of researchers in photoprotection has generally been focused on the role of antenna quenching in the NPQ process. However, there is evidence for an alternative mechanism as nonradiative energy dissipation within the reaction center of PS II (Krause, 1988; Matsubara and Chow, 2004; Zulfugarov et al., 2007; Ivanov et al., 2008). Reaction center quenching operates only under conditions when reaction centers are closed. Thus, overreduction of Q_A is an important prerequisite for efficient dissipation of EEE within the reaction center of PS II. The Z-independent dissipation of EEE

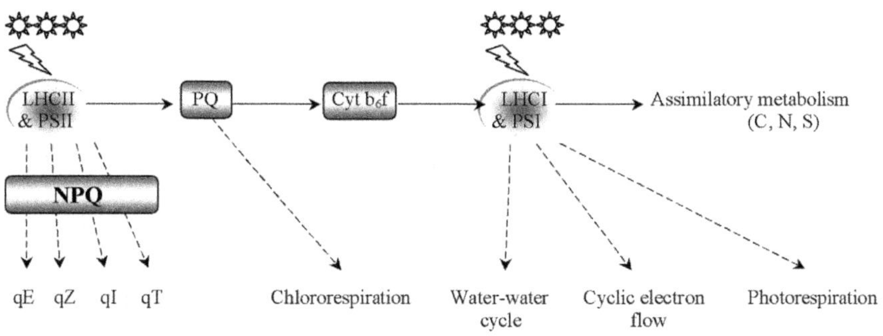

FIGURE 1.12 An overview of the various mechanisms for quenching excess excitation energy in photosynthesizing cells. Abbreviations: NPQ: nonphotochemical quenching; qE: energy dependent NPQ; qI: photoinhibitory NPQ; qT: transition-state-dependent NPQ; qZ: zeaxanthin-dependent NPQ.

within the reaction center of PS II involves some putative pathways for non-radiative Q_A^- quenching (Öquist and Huner, 2003).

Acclimation to low temperatures is strongly correlated with an increased proportion of reduced Q_A at the given growth temperature. Accordingly, the increased population of Q_A^- due to the altered redox potentials of Q_A and Q_B during acclimation to chilling enhances dissipation of EEE within the reaction center of PS II via nonradiative $P_{689}{}^+Q_A^-$ recombination (Öquist and Huner, 2003). Nonradiative dissipation of excess energy (reaction center quenching) increases not only in cold acclimated plants, but also in plants acclimated to high growth irradiance. Since chilling and high illumination both induce formation of Q_A^-, it could be suggested that reaction center quenching is related to the excess of excitation energy. Thus, environmental conditions that induce formation of excess Q_A^- will enhance the contribution of reaction center quenching to photoprotection (Huner et al., 2006).

1.5.2 Protection by Scavenging Systems: Antioxidative Defense

Under nonstress conditions, there is a stable equilibrium between generation rate and scavenging ROS. Under various stress conditions, however, this equilibrium is perturbed, leading to a considerable rise in cellular ROS levels. Plants employ some defense mechanisms relying on the function of several metabolites and enzymes to quench ROS. These ROS scavenging systems are found in different cell organelles such as chloroplasts, mitochondria, and peroxisomes (Apel and Hirt, 2004; Møller et al., 2007).

1.5.2.1 Enzymes

Several antioxidant enzymes including superoxide dismutase (SOD), catalase (CAT), guaiacol peroxidase (POD), ascorbate peroxidase (APX), monodehydroascorbate reductase (MDHAR), dehydroascorbate reductase (DHAR),

TABLE 1.1 The Scavenging of Reactive Oxygen Species by Various Antioxidant Enzymes.

Enzymes	Reactions
SOD	$O_2^{\bullet -} + 2H^+ \rightarrow H_2O_2$
CAT	$H_2O_2 \rightarrow H_2O + ½O_2$
POD	$H_2O_2 \rightarrow H_2O + ½O_2$
APX	$H_2O_2 + Asc \rightarrow H_2O + MDA$
MDAR	$MDA + NAD(P)H \rightarrow Asc + NAD(P)^+$
DHAR	$DHA + GSH \rightarrow Asc + GSSG$
GR	$GSSG + NAD(P)H \rightarrow GSH + NAD(P)^+$
GPX	$H_2O_2 + GSH \rightarrow H_2O + GSSG$

Abbreviations: SOD: superoxide dismutase; CAT: catalase; POD: peroxidase; APX: ascorbate peroxidase, MDAR: monodehydroascorbate reductase; DHAR: dehydroascorbate reductase; GR: glutathione reductase; GPX: glutathione peroxidase; GSH and GSSG: reduced and oxidized glutathione, respectively; Asc: ascorbate.

glutathione reductase (GR) and glutathione peroxidase (GPX) are responsible for scavenging different types of ROS (Apel and Hirt, 2004; Møller et al., 2007). These enzymes and their related isozymes are localized in different cell compartments and activated to various extents upon exposure to stress (Table 1.1).

1.5.2.1.1 Superoxide Dismutase (EC 1.15.1.1)

Superoxide dismutase (SOD) belongs to the family of metalloenzymes and catalyzes disproportionation of $O_2^{\bullet -}$ into H_2O_2 and O_2. SODs are classified into three types based on their metal cofactor: (i) Fe-SOD (localized to chloroplasts); (ii) Mn-SOD (localized to mitochondria) and (iii) Cu/Zn-SOD (localized to chloroplasts, peroxisomes, and cytosol). These isozymes are encoded by the nucleus and have differential susceptibility to H_2O_2 and potassium cyanide (KCN) and are activated by various abiotic stresses such as water deficiency, chilling, heat, hypoxia and toxic concentrations of heavy metals (Mittler, 2002; Apel and Hirt, 2004; Asada, 2006; Halliwell and Gutteridge, 2007; Karuppanapandian et al., 2011).

1.5.2.1.2 Catalase (EC 1.11.1.6)

Catalase (CAT) is a heme-containing enzyme that catalyses the disproportionation of H_2O_2 into H_2O and O_2. The main function of this enzyme in aerobic organisms including plants is scavenging H_2O_2 that is produced in peroxisomes during β-oxidation, photorespiration and purine catabolism. In addition, H_2O_2 produced in excess under stress conditions may diffuse from the cytosol and be metabolized by CAT. Various isoforms of CAT have been found in plants and their related genes are found to be differentially expressed and independently regulated. Catalase is very sensitive to light, likely because of absorption of light by the heme moiety of the enzyme (Mittler, 2002; Blokhina et al., 2003; Karuppanapandian et al., 2011).

1.5.2.1.3 Peroxidase (EC 1.11.1.7)

Peroxidase (POD) is also a heme-containing enzyme, oxidizes various substrates using H_2O_2 and prevents excess accumulation of H_2O_2 generated by normal metabolism or under stress conditions. Various isozymes of POD, particularly those in the extracellular spaces, have a role in the lignin synthesis, and thus in the plant's defense against pathogens. Various POD isozymes accept preferentially some aromatic electron donors such as guaiacol and syringaldazine and oxidize Asc at very low rate. Peroxidase is activated rapidly by various stress factors and, because of a wide range of subcellular localization of its isoforms, is highly efficient in the metabolism of H_2O_2. Peroxidase is considered a "stress enzyme" and the level of its activity is used as an index for evaluation of the intensity of stress (Blokhina et al., 2003; Karuppanapandian et al., 2011).

1.5.2.1.4 Ascorbate Peroxidase (EC 1.11.1.11)

Disproportionation of H_2O_2 in peroxisomes is accomplished by CAT while ascorbate peroxidase (APX) catalyzes this reaction in chloroplast and cytosol using Asc as a hydrogen donor and produces monodehydroascorbate (MDHA) (Asada, 2000). APX is a component of the Asc-GSH cycle (Foyer-Halliwell-Asada cycle) (Fig. 1.13). The mRNA of cytosolic APX shows upregulation during stress (Naya et al., 2007) and overexpression of cytosolic APXs in tomato enhances chilling and salinity tolerance of plants (Wang et al., 2005).

In contrast to cytosolic APX, chloroplastic APX has higher substrate (Asc) specificity and exists as two isoforms, thylakoid-bound (tAPX) and soluble stromal (sAPX) enzymes. tAPX is an important component of the

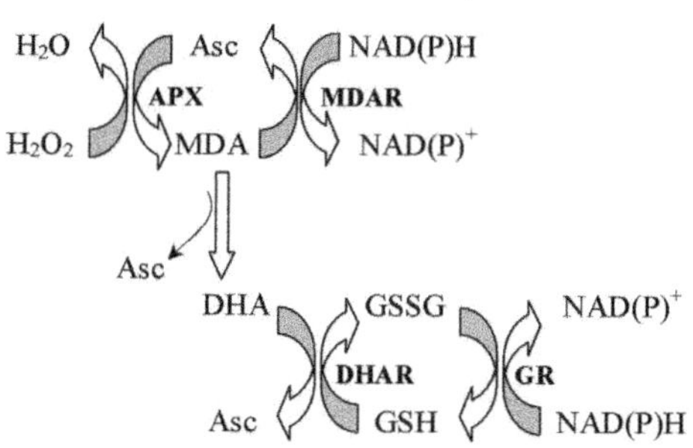

FIGURE 1.13 The Asc-GSH (Foyer-Halliwell-Asada) cycle. For abbreviations see legend of Table 1.1.

water–water cycle (see Section 1.5.1.6). *Arabidopsis* plants deficient in sAPX and tAPX show higher sensitivity to excess light and rapidly develop necrosis spots under low concentrations of cellular Asc (Giacomelli et al., 2007). Simultaneous overexpression of Cu/Zn-SOD and APX genes in chloroplasts of transgenic tall fescue plants results in higher tolerance to abiotic stresses (Lee et al., 2007). $O_2^{\bullet -}$ produced at the surface of the thylakoid membrane in PS I can be metabolized immediately to H_2O_2 by membrane-bound CuZn-SOD or Fe-SOD, which is further scavenged by tAPX (see the water–water cycle, Section 1.5.1.6) (Asada 2000, 2006; Ishikawa and Shigeoka, 2008).

1.5.2.1.5 Monodehydroascorbate Reductase (EC 1.6.5.4)

Together with dehydroascorbate reductase (DHAR), monodehydroascorbate reductase (MDHAR) is involved in the Asc regeneration reaction (Noctor and Foyer, 1998; Asada, 2000). In addition to chloroplasts, MDHAR is also located in mitochondria and peroxisomes (Mittler, 2002). The activity of MDHAR (and other enzymes involved in regeneration of Asc) is higher in drought-stressed plants, the increase in MDHAR activity contributes towards chilling tolerance in tomato and overexpression of MDHAR in transgenic tobacco increased tolerance against salt and osmotic stresses (Karuppanapandian et al., 2011).

1.5.2.1.6 Dehydroascorbate Reductase (EC 1.8.5.1)

DHAR is an important component of Asc recycling reactions. The oxidation of Asc at the first step produces MDHA that, in turn, is reduced to dehydroascorbate (DHA) enzymatically or spontaneously. Dehydroascorbate is then reduced to Asc by DHAR using GSH (Asada, 2000; 2006; Ishikawa and Shigeoka, 2008). DHAR regulates Asc pool size in both symplasm and apoplasm and its overexpression increases plant tolerance to stresses (Karuppanapandian et al., 2011).

1.5.2.1.7 Glutathione Reductase (EC 1.6.4.2)

Glutathione reductase (GR) catalyzes reduction of glutathione disulphide (GSSG) to its sulfhydryl form (GSH) using NADPH. The role of GR in H_2O_2 scavenging has been demonstrated in the Foyer-Halliwell-Asada pathway (Noctor and Foyer, 1998; Asada, 2000) (Fig. 1.13). This enzyme is important for maintaining the cellular pool of GSH and an increase in GR activity under stress conditions enhances plant performance under unfavorable conditions (Karuppanapandian et al., 2011).

1.5.2.1.8 Glutathione Peroxidase (EC 1.11.1.9)

Glutathione peroxidase (GPX) catalyzes H_2O_2 scavenging using GSH. Its role in ROS homeostasis in plants has been doubted by some researchers (Jung et al., 2002). However, changes in the expression of plant *GPX* genes in response to abiotic stresses and a role in limiting oxidative burst and programmed cell death (Chen et al., 2004) as well as in protection of plants during acclimation to photooxidative stress (Chang et al., 2009) have been reported. In *Arabidopsis*, *GPX*s are encoded by a gene family of eight members (*AtGPX1* to *AtGPX8*), which are assigned to the cytosol, chloroplast, mitochondria and endoplasmic reticulum (Milla et al., 2003). Lipid hydroperoxides as products of H_2O_2 and $^1O_2$ attack on unsaturated fatty acids are toxic cellular metabolites and the major candidate for their scavenging is the phospholipid hydroperoxide glutathione peroxidase (PHGPx). PHGPx is a member of the GPx family and is involved in direct reduction of phospholipid hydroperoxides and complex hydroperoxy lipids (Faltin et al., 2010). In plants, the thioredoxin-regenerating system is more effective than the glutathione system, and therefore the plant PHGPx is actually a thioredoxin peroxidase, i.e. peroxiredoxin, PRx (Jung et al., 2002; Tanaka et al., 2005). In plants, the PHGPx family plays a role in H_2O_2 scavenging, signaling events and photoprotection (Miao et al., 2006; Chang et al., 2009).

1.5.2.2 Metabolites

In addition to an array of enzymes, there are metabolites that act as ROS scavengers either in conjunction with the antioxidative enzymes or independently. Nonenzymatic components of the antioxidative defense system include the major cellular redox buffers Asc and GSH as well as tocopherol, carotenoids and phenolic compounds. They are involved in many cellular processes and not only have critical roles in plant tolerance and act as enzyme cofactors, but also affect plant growth and development from earlier growth stages to senescence. Mutants with reduced levels of these compounds have higher susceptibility to stresses (Sharma et al., 2012).

1.5.2.2.1 Ascorbic Acid

Asc is one of the most important antioxidants and exists in various cell organelles and in the apoplast (Horemans et al., 2000; Smirnoff, 2000). Under normal physiological conditions, Asc exists mainly in its reduced form in chloroplasts. The ability of Asc to donate electrons in a wide range of enzymatic and nonenzymatic reactions makes Asc the main ROS-detoxifying compound. Asc is able to scavenge $O_2^{\bullet -}$, $^{\bullet}OH$, and $^1O_2$ directly and is also capable of reduction of H_2O_2 to H_2O via the APX reaction. In chloroplasts, Asc acts as a cofactor of violaxanthin de-epoxidase (VDE), and thus is involved in thermal dissipation of EEE (Smirnoff, 2000). Asc is also responsible for regeneration of tocopherol (TOC) from the tocoperoxyl radical ($TOC^{\bullet}$) (Horemans et al., 2000; Smirnoff, 2000; Foyer and Noctor, 2011). Accordingly, apart from its crucial role in regulating various metabolic processes, a rise in the cellular level of Asc is mainly a prerequisite for attenuation of oxidative stress in plants (Smirnoff, 2000). The regeneration system of Asc within the chloroplast is considered a protective mechanism against damaging effects of excess light energy, i.e. the water–water cycle (see Section 1.5.1.6).

1.5.2.2.2 Glutathione

In addition to its role as the main storage form of reduced nonprotein sulfur in plants, GSH (γ-glutamylcysteinglycine) has critical roles in protection of plants from oxidative damages. This multifunctional tripeptide is abundant in plant tissues, exists in all cellular components including chloroplasts, mitochondria, ER, vacuoles, and cytosol (Noctor and Foyer, 1998). In conjunction with its oxidized form (GSSG), GSH is involved in maintaining redox homeostasis that permits regulation of the cellular metabolism under both nonstress and stressful conditions. This function of GSH is the basis of its role in stress signaling.

GSH is synthesized through two subsequent reactions using ATP. These reactions are catalyzed by γ-glutamyl cysteine synthetase (γ-ECS) and glutathione synthetase. These enzymes have cytosolic and chloroplastic isoforms. A cysteine residue with a nucleophilic nature is the reason for a high reducing power of GSH and its ability for scavenging H_2O_2 and reacting with other ROS, such as $^1O_2$, $O_2^{\bullet-}$ and $^{\bullet}OH$ (Noctor and Foyer, 1998). Its major function in the antioxidative defense is its contribution to the Asc regeneration through the Asc-GSH (Foyer-Halliwell-Asada) cycle (Noctor and Foyer, 1998). The concentration of GSH changes considerably under abiotic stress conditions (Horemans et al., 2000; Smirnoff, 2000; Foyer and Noctor, 2011).

1.5.2.2.3 Tocopherols

Tocopherols (TOCs) as lipophilic antioxidants are also constituents of biomembranes (Kiffin et al., 2006). In the chloroplast of higher plants, TOCs in the membranes are involved in protection of lipids and other membrane molecules against $^1O_2$ via quenching. α-TOC reacts with and repairs oxidizing lipid radicals and thus prevents the chain reactions of lipid autooxidation. α-TOC reacts with lipid radicals in the membrane-water interface and α-TOC donates a hydrogen atom to lipid radicals. Consequently, $TOH^{\bullet}$ is produced that in turn is converted to α-TOC using Asc or GSH. α-TOC also acts as chemical scavenger of $^1O_2$ by charge transfer mechanism (Blokhina et al., 2003). Like other antioxidants, the amount of α-TOC is enhanced in response to environmental stresses and a rise in the expression of genes related to α-TOC synthesis has been observed (Munné-Bosch, 2005; 2007; Karuppanapandian et al., 2011).

1.5.2.2.4 Carotenoids

Carotenoids (CARs) are lipophilic compounds that exist in the plastids. Besides the function of CARs as antenna molecules, in capturing light in the blue light region (450–570 nm) and transferring photons energy to the Chl, the main crucial role of CARs is its ROS detoxifying function in chloroplasts (Cazzonelli, 2011). Regarding antioxidant properties, CARs can protect photosystems (i) via interacting with lipid radicals and breaking chain reactions, (ii) by scavenging $^1O_2$, (iii) by reacting with $^1Chl^*$ or $^3Chl^*$ to impair generation of $^1O_2$, or (iv) by dissipating EEE through the xanthophyll cycle (Blokhina et al., 2003; Fanciullino et al., 2013). The major role of β-carotene in green tissues is quenching of $^3Chl^*$, thus providing inhibition of $^1O_2$ production and damage. As a quenching mechanism, energy is transferred from $^3Chl^*$ to CAR, that in turn is dissipated as nonradiative, thermal energy (heat) (Collins, 2001). The conversion of $^3Chl^*$ to $^1Chl^*$ by zeaxanthin, another important CAR, is more effective than that by β-carotene (Young, 1991; Cunningham and Gantt, 1998; Della Penna, 1999; Cazzonelli and Pogson, 2010).

1.5.2.2.5 Phenolic Compounds

Polyphenols possess a proper structural chemistry for free radical scavenging, and are more efficient in antioxidative activity than TOCs and Asc *in vitro*. The antioxidative nature of phenolics is due to their high reactivity as electron donors and their ability to chelate transition metal ions and terminate the Fenton reaction (Blokhina et al., 2003). Another mechanism underlying the antioxidative properties of phenolics is the ability of the polyphenolic compounds, flavonoids, to alter peroxidation kinetics by decreasing the fluidity of the membranes. These changes could sterically hinder the diffusion of ROS and restrict peroxidative reactions (Arora et al., 2000; Karuppanapandian et al., 2011).

1.5.2.2.6 Differential Intercellular Partitioning of Antioxidants in C_4 Plans

The antioxidant compounds are not equally distributed between all photosynthetic cells in some C_4 plants (Fig. 1.14). In maize, two types of photosynthetic cells exist with very different function. In the leaves of this species, GR and DHAR were almost exclusively localized in the mesophyll cells whereas the majority of the APX and SOD activities were localized in the bundle sheath tissue (Pastori et al., 2000). Catalase and MDHR were observed to be evenly allocated to the two cell types. Hydrogen peroxide was detected only in the mesophyll cells under nonstress conditions. These observations are interesting because the enzymes of the Calvin cycle, which are very sensitive to inhibition by H_2O_2, are found only in bundle sheath chloroplasts (Doulis et al., 1997).

The localization of GR and DHAR in the mesophyll tissues results from the requirement of these enzymes for reducing power. Bundle sheath cells are deficient in PS II and may not produce adequate NADPH required for

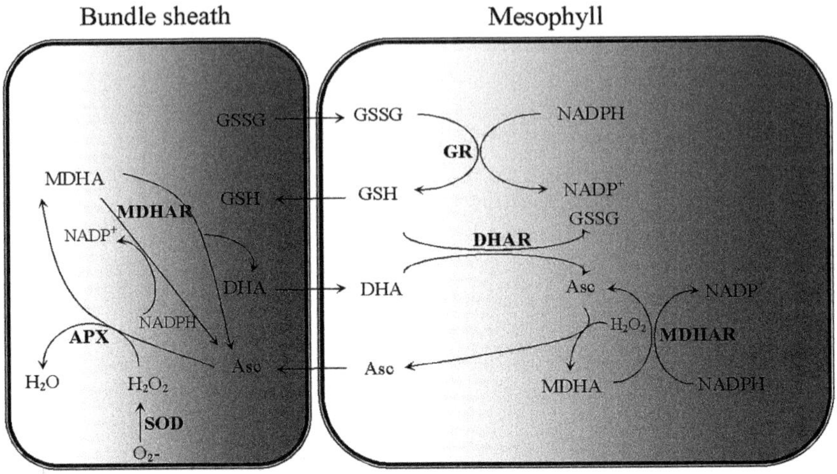

FIGURE 1.14 Differential localization of various antioxidant enzymes in the bundle sheath and mesophyll cells in maize as a C_4 species. For abbreviations see legend of Table 1.1.

reduction of GSSG and DHA. GSSG and DHA generated in the bundle sheath must be moved to the mesophyll for reduction. Because of this requirement for cycling of reduced and oxidized forms of Asc and GSH, the bundle sheath cells may be less protected against oxidative damage than the mesophyll cells. This assumption has been confirmed by the results of an experiment with methyl viologen. In maize leaves treated with methyl viologen, oxidative damage was localized almost exclusively in the bundle sheath as the consequence of an insufficient antioxidant protection during stress in these cells (Doulis et al., 1997; Pastori et al., 2000).

1.6 REDOX AND ROS SIGNALING

In the photosynthetic light reactions, changes in environmental factors result in a change of the redox potential of electron transport components or of the pool size of related redox molecules. Excess excitation energy generates a status of high energy in the chloroplast by switching redox couples such as PQH_2/PQ, Fd_{red}/Fd_{ox} and $NADPH/NADP^+$ to more negative potentials. Thus, the photosynthetic apparatus acts as a global redox sensor that detects and processes the incoming environmental signals. It has been widely accepted that numerous plastid processes generate these signals and that "exchanging information" between plastid and nucleus has a crucial role in plant stress responses. Messages originating from chloroplasts influence the expression of defense and regulatory genes, thus modulating either the acclimatory process or the execution of the Calvin cycle (Fernández and Strand, 2008; Foyer and Noctor, 2012; Suzuki et al., 2012; Bykova and Rampitsch, 2013; Konert et al., 2013).

Various compounds are involved in the exchange of information between chloroplasts and nucleus and cytosol. One of the major ways by which plants transmit information about the changing environmental factors is the ROS sensing, producing and scavenging system. Environmental stresses are perceived not only as a source of oxidative stress, but also as a mechanism controlling the main aspects of plant adaptation to various growth conditions. The interaction between ROS production and detoxification regulates the steady-state level of ROS in cells; in addition, the ROS localization and signature, i.e. the duration and amplitude of ROS signals, determine plant ultimate responses (Foyer and Noctor, 2003; 2005; 2009).

On the other hand, abiotic and biotic stressors affect the cellular balance between different redox buffers and oxidants, called redox homeostasis. It is now widely accepted that redox signals are key regulators of plant metabolism, morphology, development, growth, and eventual death. The redox state of the electron transport chain in chloroplasts is known to be involved in some posttranslational modifications in order to adjust light harvesting capacity with the metabolism of sink (Foyer and Allen, 2003; Kornas et al., 2010).

The redox potential (E) of an electron- or hydrogen-transferring molecule is defined usually by the Nernst equation (Pfannschmidt, 2003). The redox potential is determined by the midpoint potential (E_m) which is a component-specific value, the number of transferred electrons (n) and the concentration ratio of oxidized to reduced forms of the molecule:

$$\mathrm{E} = \mathrm{Em} - \frac{RT}{nF} \times \mathrm{Ln}\frac{[Red]}{[Ox]}$$

Environmental factors that have an impact on or are involved in the electrochemistry of a molecule will cause a change in its redox potential. Since most redox molecules are functionally active within a narrow range of their redox potential, plants possess several strategies in order to maintain the potential of the important redox molecules in their steady-state levels (Pfannschmidt, 2003).

Retrograde signaling regulates the expression of nuclear organelle genes in response to the metabolic and developmental state of the organelle. Besides the cross-talk between chloroplasts/mitochondria and the nucleus, chloroplast–mitochondrion redox communication has been established during plant evolution to coordinate the activities of these two bioenergetic organelles to enable an optimized acclimation response (Fernández and Strand, 2008; Pfannschmidt et al., 2009; Schwarzländer and Finkemeier, 2013). *Arabidopsis* mutants with disruption in the communication between chloroplast and nucleus were identified and referred to as the genome-uncoupled (*gun*) mutants. These mutants provided valuable information on the components and function of redox and ROS signaling in plants. Five *gun* mutants were identified (*gun1–5*) that express nuclear-encoded photosynthetic genes in the absence of proper chloroplast development (Susek et al., 1993).

1.6.1 Redox Signaling through PQ

The redox state of the PQ pool has been proposed to connect the electron transport events to the gene expression in chloroplasts. The redox state of the PQ delivers the information on the electron flux in the photosynthetic electron transport chain to response regulatory proteins that switch the photosystem genes on and off (Puthiyaveetil and Allen, 2009). In diatoms, the redox state of the PQ pool plays the central role in the high light acclimation confirming the existence of a plastid-to-nucleus retrograde signaling mechanism in algae (Lepetit et al., 2013). Detailed analyses in higher plants, however, suggest that the redox state of PQ could not regulate the nuclear-encoded photosynthetic genes; consequently, the redox state of the PQ pool is not involved in the chloroplast-to-nucleus signaling pathway (Fey et al., 2005). Recent works suggest that the generation of downstream metabolites or signaling molecules, as well as their redox state, are most probable intermediates for the information exchange between chloroplasts and nucleus. In this model rather than the redox state of the PQ itself, the redox state of the downstream components of the electron transport chain is closely related to the energy balance of the cell and can operate as sensor of environmental stresses (Fernández and Strand, 2008).

1.6.2 Redox Signaling and NPQ

There is a complex interaction between redox parameters and the extent of NPQ. As described before (see Section 1.5.1.1), low pH in the lumen activates VDE that catalyzes conversion of V to A and Z using Asc as cosubstrate. Accordingly, function of the xanthophyll cycle is related to the Asc pool as well as the electron transport activity in general and the PQ pool in particular, which are involved in acidification of the lumen. On the other hand, operation of cyclic electron flow around PS I increases proton flow into the lumen and enhances related proton motive force (Avenson et al., 2004). Induction of cyclic electron transport is mediated by Fd as the main distributor of high energy electrons at PS I. It has been observed that efficient dissipation of EEE by NPQ is linked to metabolic redox cues. However, the relative contribution of each component, i.e. Asc availability, Fd and PQ redox state, pH and stromal redox state, has not been fully understood (Oelze et al., 2008).

Induction of enzymes responsible for NPQ processes may also start signaling pathways. Abscisic acid (ABA) is synthesized from Z in the chloroplasts. Transcript regulation of 9-cisepoxycarotenoid dioxygenase (NCED) catalyzing one of the important steps in ABA biosynthesis in leaves depends on light (Thompson et al., 2000). This implies the likely existence of a putative regulatory link between the ABA biosynthesis pathway and photosynthesis and/or probably redox signals. Availability of Asc likely acts as an intermediate that links the ABA biosynthesis and chloroplast redox state. An Asc biosynthetic mutant, *vtc1*, is characterized by increased leaf ABA levels (Pastori et al., 2003).

1.6.3 Chlorophyll Synthesis and Redox Signaling

The tetrapyrrole biosynthetic pathway exists in plastids and produces Chl, heme and the chromophore of phytochrome. Regulation of this pathway is necessary in order not only to produce adequate ligands for proteins and enzymes but also to avoid accumulation of phototoxic intermediates of the pathway such as Pchlide (see Section 1.4.2.1). Mutations in the genes of tetrapyrrole biosynthesis revealed that the intermediates of this pathway are involved in retrograde signaling from chloroplast to nucleus (Strand et al., 2003). Tetrapyrroles are produced in the chloroplast and to act as plastid signals they must leave the chloroplast. Kropat et al. (2000) hypothesized that Chl precursors are exported actively from the chloroplast; this transport is facilitated under stress conditions.

The Chl precursor, Mg-protoporphyrinIX (Mg-ProtoIX) is accumulated under stress conditions such as low temperature (Wilson et al., 2003) and could be considered as an indicator of disturbance in the chloroplast metabolism. Further support for a role of Mg-ProtoIX in plastid-nucleus signaling under stress conditions has been obtained by molecular genetics approaches and analyzing the *gun* mutants. It has been demonstrated that some *gun* genotypes have defects in specific steps of tetrapyrrole biosynthesis. It is likely that accumulation of Mg-ProtoIX provides a signal transferring information from the chloroplast to the nucleus (Strand et al., 2003; Fernández and Strand, 2008).

Mg-ProtoIX as a signaling metabolite not only regulates nuclear-encoded photosynthetic genes, but also influences the expression of the plastid-encoded photosynthetic genes *psbA*, *psbD*, *psaA*, *psaC*, and *rbcL* by controlling the expression of the sigma factors necessary for the function of the plastid-encoded RNA-polymerase (Ankele et al., 2007; Fernández and Strand, 2008). Transgenic tobacco plants with either overexpression or underexpression of CHLM sequence encoding Mg-ProtoIX methyl transferase have either lower or increased levels of Mg-ProtoIX and in parallel show either elevated or reduced expression of nuclear-encoded photosynthetic genes. An *Arabidopsis* mutant that accumulates Mg-ProtoIX due to a TDNA insertion in CHLM showed repression of the nuclear-encoded light harvesting Chl a/b binding protein (LHCB protein, PS II CAB polypeptide) gene (Pontier et al., 2007). In *Chlamydomonas reinhardtii*, Mg-ProtoIX acts as plastid signals increasing the nuclear-encoded heat shock protein, HSP70A (von Gromoff et al., 2008).

The CHLH, one of the three subunits of Mg-chelatase (D, H, and I subunits), has been reported as GUN5 (Mochizuki et al., 2001), and is an ABA-specific binding protein that mediates ABA-signaling pathways including gene expression and stomatal closure (Wu et al., 2009). The key component of the circadian clock, TIMING OF CAB (CAB is the LHCB polypeptide of PS II) 1 (TOC1) inhibits the expression of *CHLH* via interacting with its promoter. Plants overexpressing *TOC1*, similarly with RNAi-mediated knockdown of *CHLH*, were ABA insensitive regarding stomatal response (Legnaioli et al., 2009).

1.6.4 Redox Signaling and Dithiol/Disulphide Exchanges

Thioredoxins (TRXs) are small disulfide proteins responsible for oxidoreduction of disulfide bonds in different target proteins (Vlamis-Gardikas and Holmgren, 2002; Jacquot et al., 2013). Various members of the TRX family exist in all cellular compartments; in chloroplasts TRXs have important roles in the regulation of photosynthesis. Under light conditions, Fd is reduced by electrons derived from the photosynthetic electron transport chain, and further transfers electrons to the downstream acceptors, including TRXs through Fd-TRX reductase (FTR) (Michelet et al., 2005). In *Arabidopsis* all 11 enzymes of the Calvin cycle can be regulated by TRXs (Meyer et al., 2005; 2008).

Because H_2O_2 can oxidize thiol groups, H_2O_2 generated under stress could be detected via modification of disulfide bonds in target proteins. The Asc-GSH cycle serves as the main pathway coupling of the ROS scavenging reactions with the redox signaling in plants (Foyer and Noctor, 2005; Pitzschke et al., 2006). A high GSH concentration is required not only for protection against ROS, but also for preventing oxidation of thiol groups in the target proteins (Oelze et al., 2008). Under stress conditions high rates of ROS generation shift the chloroplast redox state to a more oxidized state. Under these conditions, accumulation of GSSG and reduction of GSH:GSSG ratio acts as a signal that triggers response reactions in the cell. It has been demonstrated that oxidative stress drives oxidation of GSH and induces vacuolar sequestration of GSSG. This process plays an important role in redox homeostasis and signaling, key factors in determining the outcome of plant responses to stress (Noctor et al., 2013).

Reaction of ROS with cysteine residues of proteins make them sensitive to be irreversibly converted to sulfinic or sulfonic acids. Glutathionylation of protein thiols can protect thiols from oxidation (Oelze et al., 2008). Besides this protective function, glutathionylation plays a prominent role in the regulation of activity and stabilization of proteins and is considered a redox signaling mechanism (Zaffagnini et al., 2012). The glutathionylation of four enzymes of the Calvin cycle – phosphoribulokinase, glyceraldehyde-3-phosphate dehydrogenase, ribose-5-phosphate isomerase and phosphoglycerate kinase – has been considered a major mechanism of regulation of the Calvin cycle under oxidative stress conditions (Zaffagnini et al., 2012).

The chloroplastidic f-type thioredoxin (TRX-f) is also glutathionylated and it has been observed that glutathionylated TRX-f is a poor substrate for FTR (Michelet et al., 2005). Exposure of cells to ROS results in glutathionylation of TRX-f; thus, glutathionylation is a mechanism for reduction of TRX-f activity under conditions of increased ROS generation. As the result of reduction in the activity of TRX-f, activity of its target proteins including Calvin-cycle enzymes, will also diminish (Michelet et al., 2005). It could be concluded that glutathionylation is a redox-dependent mechanism of regulation of dark reactions under stress conditions (Oelze et al., 2008).

As the result of reaction with ROS, thylakoid membranes release oxygenated fatty acid derivatives, i.e. oxylipins (Orozco-Cardenas et al., 2001). These compounds in *Arabidopsis* induce expression of *GLUTATHIONE-S-TRANSFERASE1* (*GST1*). The expression of this enzyme increases in response to various stress conditions, including overreduction of the photosynthetic electron transport chain (Vollenweider et al., 2000).

Three proteins in tetrapyrrole biosynthesis are regulated by thioredoxin-mediated mechanism, including glutamate-1-semialdehyde-2,1-aminomutase, uroporphyrinogen decarboxylase and Mg-chelatase. A key step in Chl biosynthesis is the insertion of Mg^{2+} into protoporphyrin IX by Mg-chelatase, whose activity is stimulated upon reduction (Ikegami et al., 2007).

The chloroplastic O-acetylserine(thiol)lyase isoform has an S-sulfocysteine synthase activity and is located in the thylakoid lumen. Its S-sulfocysteine activity is essential for the performance of the chloroplast under long-day growth conditions. S-sulfocysteine synthase acts as a sensor for detecting thiosulfate. Under inadequate detoxification of ROS, i.e. excess light conditions, thiosulfate is accumulated; thus, the production of S-sulfocysteine molecule by this enzyme triggers protection mechanisms of the photosynthetic apparatus (Gotor and Romero, 2013).

1.6.5 ROS Signaling

An important part of stress signaling mechanisms is mediated by ROS, which are unavoidable by-products of photosynthesis. Generation of H_2O_2 and $^1O_2$ at PS I and PS II respectively, as well as photorespiratory H_2O_2, is associated with the control of nuclear gene expression (Op den Camp et al., (2003); Vandenabeele et al., 2004; Pitzschke et al., 2006; Pfannschmidt et al., 2009; Sunil et al., 2013; Fischer et al., 2013; Foyer and Noctor, 2013).

Although cellular damages evoked by various types of ROS are similar, the different ROS start definite signaling pathways (Laloi et al., 2007); the compartment of ROS origin is also decisive. A characteristic role for $^1O_2$ in retrograde signaling was discovered by *Arabidopsis flu* mutant (see Section 1.4.2.1) (Meskauskiene et al., 2001). The release of $^1O_2$ induces mainly genes encoding some proteins responsible for cell death. Only 15% of the $^1O_2$-responsive genes encode plastid proteins (Wagner et al., 2004). Accordingly, the $^1O_2$-derived signaling events target mainly regular responses of plants to stresses rather than modulation of gene expression patterns (Meskauskiene and Apel, 2002; Wagner et al., 2004; Fernández and Strand, 2008).

Nevertheless, analysis of gene expression in the *flu* mutant and in wild plants treated with the herbicide paraquat showed that $^1O_2$ induces expression of a distinct group of genes that is different from that activated by $O_2^{\bullet-}$ and/or H_2O_2. In addition, it was demonstrated that H_2O_2 antagonizes the $^1O_2$-derived stress responses in the *flu* mutant. This crosstalk between H_2O_2- and $^1O_2$-dependent signaling pathways may provide the possibility for a highly precise regulation of the response to environmental stresses (Laloi et al., 2007).

It was found that topoisomerase VI (Topo VI) is an important component of $^1O_2$ retrograde signaling (Šimková et al., 2012). In *Arabidopsis*, *CAA39* encodes the A-subunit of Topo VI. Expression analysis of *flu* and *flu/caa39* mutants indicated that Topo VI is required for activation of a $^1O_2$-responsive gene, *AAA-ATPase*, and some of $^1O_2$-responsive transcripts in response to $^1O_2$. AAA (ATPase Associated Activities) or AAA-ATPase proteins are involved in diverse cellular processes, including DNA replication, protein degradation, signal transduction and the regulation of gene expression. Topo VI directly regulates the expression of *AAA-ATPase* via direct binding to the promoter of this and other $^1O_2$-responsive genes. Under excess light conditions that induce generation of $^1O_2$ and H_2O_2, Topo VI regulates $^1O_2$- and H_2O_2-responsive genes with clearly different patterns. These results indicate that Topo VI is involved in the integration of diverse signaling pathways started by ROS (Šimková et al., 2012).

Because of a very short half-life (200 ns), $^1O_2$ must produce some signals that could leave the chloroplast. Two chloroplast proteins, EXECUTER1 (EX1) and EXECUTER2 (EX2) were identified through a screen for *flu* suppressor mutants. The EX1 and EX2 proteins are putative sensors and/or mediators of $^1O_2$ in the chloroplast and are associated with the thylakoid membrane, i.e. in close proximity to the generation place of $^1O_2$. The *ex1/flu* double mutant, and particularly the *ex1/ex2/flu* triple mutant, overaccumulates $^1O_2$ but completely suppresses the $^1O_2$-induced genes (Lee et al., 2007). The blue-light–absorbing protein cryptochrome, cry1, is also associated with the $^1O_2$-derived stress signaling. The *flu/cry1* double mutant lost cell death response (Danon et al., 2006). In addition, some *gun* mutants were identified as *cry1* alleles. Under some circumstances, cry1 operates for downregulation of LHCB expression via converting the HY5 (transcription factor) from a positive to a negative regulator (Ruckle et al., 2007). This evidence indicates that plastid signals have interactions with the light-signaling networks.

In contrast to $^1O_2$, H_2O_2 has a long half-time and moves readily across membranes. H_2O_2 plays a distinct role as a second messenger in multiple signal transduction pathways. In order to act as a second messenger and to diffuse from the chloroplast to the cytosol, H_2O_2 needs to be present at high concentrations. Such high concentration levels may only be realized under highly imbalanced redox conditions (Neill et al., 2002; Stone and Yang, 2006).

Under conditions of EEE, $^1O_2$ and H_2O_2 are produced via electron transfer to oxygen and transcription of cytosolic APX (cAPX) is increased (Yabuta et al., 2004). Treatment with DCMU (3-(3,4-dichlorophenyl)-1,1-dimethylurea) inhibited *cAPX* expression under excess light, while DBMIB (dibromothymoquinone) stimulate transcription under both low and high light conditions (Chang et al., 2004). These results suggest that stimulation of *cAPX* expression under excess illumination is related to the redox state of the PQ pool as well as H_2O_2 concentration (Fryer et al., 2003).

Although the constituents of ROS signal transduction pathways are not fully identified, the role of kinases, mitogen-activated protein (MAP) kinase cascades and receptor-like kinases has been hypothesized (see Section 1.6.6). Increases in leaf H_2O_2 concentrations is involved in the stimulation of *APX2* expression in *Arabidopsis* under excess light, but the cellular site for H_2O_2 production was unknown (Yabuta et al., 2004). Data on the activity of Rubisco and CO_2 assimilation rate under different temperature regimes suggested that the cellular H_2O_2 level is an important signal for the GSH-dependent regulation of redox-sensitive enzymes of CO_2 assimilation (Li et al., 2013).

A chloroplast membrane protein LTO1/*At*VKOR-DsbA is involved in ROS homeostasis and in redox regulation of cysteine-containing proteins in chloroplast (Lu et al., 2013). In *lto1-2* plants the activity of APX, CAT and DHAR is reduced simultaneous with ROS accumulation. The soluble DsbA-like domain of LTO1 has reduction, oxidation and isomerization activities, and its potential target is involved in chlorophyll degradation and photooxidative stress response (Lu et al., 2013).

Intracellular Ca^{2+} fluxes and signaling molecules, such as phosphoinositides and jasmonic acid (JA), respond to oxidative stress and could serve as regulatory cross-talk between pathways (Knight and Knight, 2001). The activation of Ca^{2+}-dependent protein kinase provides likely a mechanism for integration of a chloroplast-originated ROS signal into a common regulatory pathway. The induction of a MAP kinase pathway in the protoplasts of *Arabidopsis* upon H_2O_2 treatment resulted in the expression of antioxidative genes while inhibiting genes for normal plant growth (Kovtun et al., 2000). These examples indicate the likely action mode of these networks (Mullineaux and Karpinski, 2002).

The photorespiratory pathway is induced significantly in response to EEE and, consequently, H_2O_2 production is increased via the oxidation of glycolate (Willekens et al., 1997). Thus, in addition to the role of H_2O_2 in eliciting cellular antioxidant defenses, photorespiratory metabolites are also candidates for an EEE-related signal derived from a compartment other than chloroplast (Mullineaux and Karpinski, 2002).

1.6.6 Involvement of Kinases in the Perception Mechanisms of EEE and Redox Signaling

Plants growing under dense canopies or other changing light environments are faced with a strong light quality gradient that causes imbalanced excitation of the two photosystems and consequently reduction of efficiency of light reactions. In order to maintain optimal photosynthetic activity plants use a process of structural rearrangement of photosystems, called state transitions (see Section 1.5.1.3). The action of THYLAKOID-ASSOCIATED KINASE1 (TAK1) is crucial for the state transition process. The phosphorylation of LHCB protein is closely associated with the redox changes of the constituents of photosynthetic electron transport chain. A direct interaction occurs between the Cyt b_6f complex and TAK1; TAK1 then is released from the complex and performs the phosphorylation of thylakoid proteins including LHCB protein (Snyders and Kohorn, 2001). This mode of action may be similar to signaling events that target modulations in the expression of nuclear genes in response to EEE.

Another kinase involved in the EEE signaling is NPH1-like1 (NPL1) that is related with the NONPHOTOTROPHIC HYPOCOTYL1 (NPH1), the blue-light receptor involved in phototropism. NPL1 is responsible for the chloroplast movement under excess light conditions in *Arabidopsis* and encodes an extra-plastidial serine/threonine protein kinase. It has been suggested that sensitivity to the redox state and activation of both NPH1 and NPL1 molecules are related to their FMN moiety (Huala et al., 1997; Kagawa et al., 2001).

Recent evidence suggests the involvement of receptor-like kinases (RLKs) in the redox and ROS signaling in plants (Munné-Bosch et al., 2013). The expression of RLKs is influenced by various oxidants (H_2O_2 and O_3) and antioxidants (Asc) (Munné-Bosch et al., 2013). RLKs are Ser/Thr protein kinases that are involved in normal development via cell-to-cell signaling, act as hormone receptors and regulate plants responses to abiotic and biotic stresses (Shiu and Bleecker, 2003; Gish and Clark, 2011).

1.6.7 The Transmission of Signals across the Chloroplast Envelope

Not only different types of ROS but also the main hormonal signals related to stresses, such as jasmonic acid (JA) and abscisic acid (ABA), are synthesized within the chloroplast (Müller, 1997; Qin and Zeevaart, 1999). In the processes such as stomatal closure under drought stress the function of ABA is mediated by H_2O_2 (Pei et al., 2000), presenting an example for interaction of hormonal signaling with the products of EEE.

The mechanisms through which these molecules leave the chloroplast are mainly not known. It is likely that H_2O_2 moves freely across the membranes (Willekens et al., 1997); thus, H_2O_2 produced in the chloroplasts could directly interact with other signaling molecules in the cytosol. The putative role of H_2O_2 as an intracellular signaling molecule is evidenced by its role in the systemic responses of plants to higher light intensities, pathogens and physical damage (Karpinski et al., 1999; Mullineaux and Karpinski, 2002).

Using spectroscopic methods it has been demonstrated that some components in the chloroplast envelope, e.g. iron-sulphur proteins, semiquinones, flavins and α-tocopherol, are likely responsible for transferring electrons (Jäger-Vottero et al., 1997). These molecules could be involved in the passage of signals out of the chloroplasts. A model has been proposed in which an electron transport chain initiates with NADPH in the stroma, passes through the chloroplast envelope and terminates with O_2 as the final electron acceptor on the chloroplast outer surface (Jäger-Vottero et al., 1997). The stoichiometry of such an electron transport chain implicates that ROS could be produced outside the chloroplast using electrons derived from the inside. Under certain conditions such as high illumination, electron transport rates increase, which could induce the activity of an electron transport chain located at the envelope (Mullineaux and Karpinski, 2002).

1.6.8 Redox and ROS Signaling: The Molecular Approach

The expression of both nuclear and chloroplast genes encoding constituents of photosynthesis and antioxidant defense system is linked to the redox changes in light-driven electron transport chains (Mullineaux and Karpinski, 2002). The expression of nuclear genes for *Cab* (encoding a Chl a/b binding protein), *Lhc* (encoding the LHCB protein), *RbcS* (encoding the small unit of Rubisco), *APX1* and *APX2* is associated with the redox state (Karpinski et al., 1997; Oswald et al., 2000). Examples for plastid genes that are related to the redox state of chloroplasts are *psbA* (encoding the D1 protein of the PS II reaction center) and *psaAB* (encoding the PS I reaction center protein) (Pfannschmidt et al., 1999). However, the actual mechanism for "translation" of the redox state and energy balance to the gene expression in the chloroplast and the nucleus is still obscure. Various mutants of *Arabidopsis* indicated some links between photosynthesis, redox homeostasis and EEE.

A light and redox-mediated protein phosphorylation system exists in plant thylakoid membranes. The thylakoid protein kinase, STATE TRANSITION 7 (STN7), is necessary for state transitions (see Section 1.5.1.3) and photosynthetic acclimation (Bellafiore et al., 2005). The *stn7* mutant exhibits an altered expression pattern for nuclear-encoded photosynthetic genes (Bonardi et al., 2005). STN7 is presumably involved in utilization of the chloroplast-derived redox signal for expression of genes in the nucleus.

In the redox imbalanced mutants (*rimb*), the expression of the nuclear-encoded genes for the antioxidant enzyme 2-cys-peroxiredoxin (2-CPA) is uncoupled from the redox state of the PS I acceptor side (Heiber et al., 2007). The lesion stimulating disease 1 (*lsd1*) mutant is hyper-responsive to cell death initiators and exhibits a "runaway cell death" phenotype paralleled by ROS production in a long-day photoperiod. The Zn-finger motif transcription factor LSD1 was defined as a negative regulator of plant cell death and basal disease resistance (Aviv et al., 2002; Wituszynska et al., 2013). Interestingly, *lsd1* fail to acclimate to both excess light conditions and to low temperature. LSD1 function is necessary for optimum catalase activity and thus determines the efficiency of photorespiration in protection of plants from overreduction of electron transport chain (see Section 1.5.1.7) (Mateo et al., 2004). An *Arabidopsis* mutant with hypersensitive cell death and constitutively activated defense responses has been found lacking the function of the *FZL* gene (Landoni et al., 2013). This gene encodes a membrane-remodeling GTPase with an essential role in the determination of thylakoid and chloroplast morphology. Since the chloroplasts are a major source of ROS, the characterization of this mutant suggests that ROS accumulation, triggered by damage to the chloroplast membranes, is a signal for starting the defense signaling cascade (Landoni et al., 2013).

The molecular responses to oxidative stress are regulated by redox-sensitive transcription factors, as was observed for NPR1 (Non-expressor of Pathogenesis-Related 1) protein. An increase in GSH content and simultaneous shift in the cellular redox status toward reducing conditions (Mou et al., 2003) results in reduction of disulphide bonds in NPR1 protein. Such reduction switches this molecule from an inactive, oligomeric complex with cytosolic localization to an active monomeric state moving to the nucleus. In the nucleus it interacts with transcription factors of the TGA-type bZIP transcription factor (Mou et al., 2003).

One of the *GUN* genes, *GUN1*, encodes a chloroplast-targeted pentatricopeptide repeat (PPR) protein (Koussevitzky et al., 2007). In contrast to the other *gun* mutants, *gun1* shows multiple alterations in responses mediated by redox status, Mg-ProtoIX and the organellar gene expression. The phenotype of *gun1* mutant implied that different plastid signals are integrated within this organelle, and then use GUN1 to produce and/or transfer a universal signal to the nucleus (Koussevitzky et al., 2007). Treatment with norflurazon that causes accumulation of Mg-ProtoIX influences also the transcription of plastid-encoded polymerase (PEP)-dependent plastid encoded genes (Ankele et al., 2007). It could be concluded that Mg-ProtoIX may also regulate the organellar gene expression as an alternate effect.

In the absence of protein synthesis in plastids following application of inhibitors, the *abi4* mutant displays a *gun* phenotype. ABA-INSENSITIVE (ABI)4 is a transcription factor implicated in response to ABA. In response to the GUN1-derived signal, ABI4 binds the promoter of the LHCB in close proximity to a light-regulatory element required for retrograde signaling (Strand et al., 2003; Koussevitzky et al., 2007). This prevents binding of related transcription factors and leads to inhibition of the expression of nuclear-encoded photosynthetic genes in response to light (Koussevitzky et al., 2007).

Studies on the leaf transcriptome profiles of *Arabidopsis* mutants (for a review see Munné-Bosch et al., 2013) that are deficient in the scavenging photorespiratory (*cat2*) or chloroplastic (*tapx*) H_2O_2 as well as in Asc (*vitamin c 1,2* [*vtc1* and *vtc2*]) and GSH (*rootmeristemless1* [*rml1*]) biosynthesis revealed that changes in cellular levels of these oxidants and antioxidants influence the transcription pattern of various RLKs, including LRR (Leucine rich repeat) receptor kinases, proline rich extensin like receptor kinases, receptor like cytoplasmatic kinases, legume-lectin receptor kinase, WAK (Wall-Associated-Kinases) receptor kinase and DUF26 (Domain of Unknown Function) receptor kinases. These receptor kinases are involved in hormone signaling, redox signaling and stress responses (Munné-Bosch et al., 2013). Fig. 1.15 provides an overview of the redox and ROS signaling pathways between chloroplasts and the nucleus.

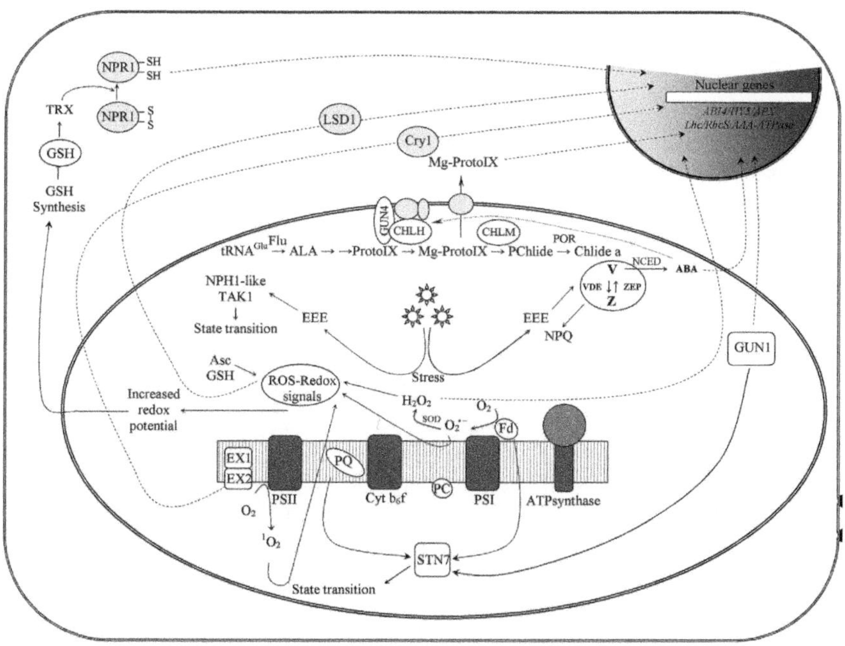

FIGURE 1.15 A model for plastid-nucleus signaling pathway during stress responses in photosynthesizing cells. Solid lines indicate interaction of various plastidial components of the pathway and dashed lines indicate their effects on the expression of nuclear genes. Abbreviations: *AAA-ATPase:* genes encoding AAA (ATPase Associated Activities) proteins; *ABI4: Abscisic Acid Insensitive 4;* ALA: 5-Aminolevulinic acid; *APX: Ascorbate Peroxidase;* CHLH, CHLM: enzymes involved in chlorophyll synthesis; Cry1: Cryptochrome 1; EEE: excess excitation energy; EX1 & EX2: Executer 1,2; GUN: Genome uncoupled; *HY5:* LONG *Hypocotyl 1; LSD1*: *Lesion Stimulating Disease* 1; Lhc and RbcS: genes encoding the small subunits of the Rubisco and the light-harvesting chlorophyll *a/b*-binding proteins (CAB), respectively; NCED: 9-cis-epoxycarotenoid dioxygenase; NPH1: Nonphototropic, NPR1: Nonexpressor of Pathogenesis-Related Protein 1; ProtoIX: protoporphyrinIX; PChlide: proto*chlorophyllide*; POR: *NADPH*: *protochlorophyllide oxidoreductase;* STN7: State Transition 7; TAK1: Thylakoid-Associated Kinase 1; TRX: thioredoxins.

1.7 ROS METABOLISM IN PHOTOSYNTHESIZING ORGANISMS: FROM AN EVOLUTIONARY POINT OF VIEW

The emergence of oxygenic photosynthesis was a huge revolution in the young Earth environment, because a new, relatively active chemical compound, i.e. molecular oxygen, appeared. Since photosynthesis released oxygen to the environment, the generation of ROS would have occurred more commonly. Therefore, in aerobic organisms an enzymatic antioxidant system against ROS has evolved in order to sufficiently detoxify ROS (Halliwell, 2006). According to the commonly accepted view, aerobic metabolism was possible only after oxygen was released into the Earth's atmosphere by oxygenic photosynthesis performed by cyanobacteria, and antioxidant cellular machinery has evolved at the same time as aerobic metabolism and oxygenic photosynthesis. This statement is based on the fact that ROS is a toxic by-product of both respiratory and photosynthetic electron transport chains (Halliwell, 2006; Shaw, 2008; Kornas et al., 2010).

Regarding the co-evolution of the antioxidant system and oxygenic photosynthesis, two possible hypotheses concerning the sequence of these two events were formulated. According to the first hypothesis, the oxygenic photosynthesis evolved first, and because of a steep gradient existing in the anaerobic environment, oxygen diffused out of the cells before inducing ROS production. Accordingly, no antioxidant system was required until the oxygen concentration reached its higher levels. The second hypothesis states that an antioxidant system would have evolved first in response to ROS generated following some abiotic factors (Thomas et al., 2008; Kornas et al., 2010).

Local environments on the young Earth, especially shallow oceans, could be enriched in oxygen and ROS induced by UV and cosmic rays. H_2O_2 formation on pyrite surfaces was shown in the absence of oxygen (Borda et al., 2001). Additionally, the widespread occurrence of basic antioxidant enzymes, such as SOD, superoxide reductases (SOR), CAT and POD in contemporary species from Bacteria, Archaea and Eucarya domains, and even in organisms belonging to obligate anaerobes (Brioukhanov and Netrusov, 2004) might indicate that the Last Universal Common Ancestor (LUCA) was not an obligate anaerobe. LUCA was rather a facultative anaerobe able to remove ROS if it was necessary for its own metabolism. The existence of an antioxidant system would have protected ancient cells carrying out nonoxygenic photosynthesis, and provided a preadaptation possibility for the subsequent evolution of oxygenic photosynthesis (Thomas et al., 2008). For this reason, most probably ancient cyanobacterial cells were already equipped with some crucial antioxidant enzymes, which they had inherited from moderately anaerobic ancestors (Kornas et al., 2010).

One of the most important antioxidant enzymes of the first line of defense against ROS is SOD. Three main classes of SOD have been identified: Fe-SOD, Mn-SOD and Cu/Zn-SOD (Alscher et al., 2002). The Mn-SOD and Fe-SOD are phylogenetically related to each other and they are very similar in their primary and tertiary structures, whereas Cu/Zn-SOD shows different structural features (Fink and Scandalios, 2002). Fe-SOD has been postulated as an "archaic enzyme" (Schäfer and Kardinahl, 2003). The increasing content of oxygen in the atmosphere and the occurrence of transition metals on the young Earth might indicate that an iron form is the most "ancient" SOD. In this scenario, when the early Earth atmosphere was anoxic, Fe was abundant in the reduced soluble form Fe(II). For this reason it would seem that Fe(II) was the first transition metal present at the active site of the first SOD. Later, during biological evolution and an increasing level of O_2 in the Earth's atmosphere, Fe ions were replaced by Mn, and a new SOD using Cu/Zn as metal cofactors appeared (Konras et al., 2010).

The mechanism that couples electron transfer with gene expression is two-component redox signal transduction, i.e. the redox-controlled kinase that phosphorylates proteins in the LHCII and thus regulates distribution of excitation energy between PS I and PS II (Allen, 2003). A two-component signal transduction system, comprising sensor kinases and response regulators, was originated from prokaryotes, introduced through early bacterial ancestors, endosymbiotic organelles as well as lateral gene transfer to the eukaryotes, and evolved further in this domain of life (Puthiyaveetil and Allen, 2009). Foyer and Allen (2003) suggest that redox signaling was the first type of sensory regulation that evolved in nature. Sequence similarities in higher plants to cyanobacterial redox signaling components indicate homology and suggest conserved sensory and signaling functions (Forsberg et al., 2001). Photosynthetic organisms have perfected the art of redox control. Plastids are the result of cyanobacterial symbiosis which occurred over 1.2 billion years ago, the present phylogenomic data pointing to filamentous, heterocyst-forming (nitrogen-fixing) cyanobacteria as plastid ancestors (Deusch et al., 2008). There is evidence that chloroplast genome encodes proteins whose function and biogenesis are particularly regulated by electron transfer. It has been suggested that the main cause for the retention of key proteins of photosynthetic (and respiratory) electron transport in the respective organelles is that the chloroplast (and mitochondrial) genetic system facilitates a fast and direct response to the changing redox environment. Such rapid response is required for attenuation of the destructive effects of ROS (Allen and Allen, 2008). Regarding other group of genes, on the other hand, Allen and Raven (1996) suggest that it had been a selective pressure favoring movement of some genes from organelles to the nucleus. Using this strategy, plants were able to avoid harmful consequences of a high rate of mutations in organellar genes as the main ROS generating cell compartments.

1.8 CONCLUSIONS

Environmental stress conditions cause reduction in the activity of assimilate sinks that produces EEE in the photosynthesizing tissues. In the presence of oxygen, EEE leads to the generation of ROS. Plants possess various mechanisms, which could be divided into quenching EEE or scavenging ROS.

Quenching of EEE and scavenging of ROS could be considered strategies for attenuating stress and thus for survival of plants under unfavorable conditions. Other aspects of these events is the starting of the signaling pathways by ROS and other redox molecules. The production of $O_2^{\bullet -}$ and H_2O_2 at PS I and the formation of $^1O_2$ at PS II contribute in the development of a signaling network that in turn provides information on the redox state of chloroplasts. Like ROS, the concentrations of low molecular weight antioxidants such as Asc, GSH, tocopherols and carotenoids, and the activity of antioxidant enzymes are modified by environmental cues.

Redox and ROS signaling trigger modulation and regulation of various metabolic pathways and are involved in backward signaling from chloroplasts to nucleus. Accordingly, chloroplasts have been regarded as sensors for environmental stress factors that link the redox status of the cells to gene expression.

Redox molecules and ROS do not behave as isolated signals in linear pathways. They act rather as members of the stress signaling network that integrates information from other pathways, e.g. hormone and sugar signaling, in order to regulate whole plant growth and stress responses. Redox metabolites and associated signaling events are involved also in the cross-talk pathways of biotic and abiotic stresses and have crucial roles in tolerance of plants to both of these stresses.

Nevertheless, each redox molecule within a given cell compartment is associated with an individual signaling event. Evidence suggests that redox gradient across the plasma membrane is a sensor and regulator of gene expression. Accordingly, existing models of redox and ROS signaling are indeed oversimplified pathways and do not provide any information on the balance between oxidants and antioxidants in different cellular compartments.

The major topics for future research may include not only the identification of molecular components of the pathways of backward redox signaling from the chloroplast to the nucleus, but must also focus on the mechanisms for transmission of the information across the chloroplast envelope and the combination of different signaling events into one pathway in order to induce an acclimatory response.

REFERENCES

Allakhverdiev, S.I., Klimov, V.V., Carpentier, R., 1997. Evidence for the involvement of cyclic electron transport in the protection of photosystem II against photoinhibition: influence of a new phenolic compound. Biochemistry 36, 4149–4154.

Allen, C.A., Allen, J.F., 2008. Redox effects on chloroplast protein synthesis and phosphorylation. In: Allen, J.F., Gantt, E., Golbeck, J.H., Osmond, B. (Eds.), Photosynthesis: Energy from the Sun. Springer, Heidelberg, Germany, pp. 903–907. Proceedings of the 14th International Congress on Photosynthesis July 2007, Glasgow, UK.

Allen, J.F., 1995. Thylakoid protein phosphorylation, state 1-state 2 transitions, and photosystem stoichiometry adjustment: redox control at multiple levels of gene expression. Physiol. Plant. 93, 196–205.

Allen, J.F., 2003. State transitions – A question of balance. Science 299, 1530–1532.

Allen, J.F., Forsberg, J., 2001. Molecular recognition in thylakoid structure and function. Trends Plant Sci. 6, 317–326.

Allen, J.F., Raven, J.A., 1996. Free-radical-induced mutation vs. redox regulation: costs and benefits of genes in organelles. J. Mol. Evol. 42, 482–492.

Alscher, R.G., Erturk, N., Heath, L.S., 2002. Role of superoxide dismutases (SODs) in controlling oxidative stress in plants. J. Exp. Bot. 53, 1331–1341.

Ankele, E., Kindgren, P., Pesquet, E., Strand, A., 2007. *In vivo* visualization of Mg-protoporphyrin IX, a coordinator of photosynthetic gene expression in the nucleus and the chloroplast. Plant Cell 19, 1964–1979.

Apel, K., Hirt, H., 2004. Reactive oxygen species: metabolism, oxidative stress, and signal transduction. Annu. Rev. Plant Biol. 55, 373–399.

Aro, E-M., Virgin, I., Andersson, B., 1993. Photoinhibition of photosystem II. Inactivation, protein damage and turnover. Biochim. Biophys. Acta 1143, 113–134.

Arora, A., Byrem, T.M., Nair, M.G., Strasburg, G.M., 2000. Modulation of liposomal membrane fluidity by flavonoids and isoflavonoids. Arch. Biochem. Biophys. 373, 102–109.

Arora, A., Sairam, R.K., Srivastava, G.C., 2002. Oxidative stress and antioxidative system in plants. Curr. Sci. 82, 1227–1238.

Asada, K., 1999. The water–water cycle in chloroplasts: scavenging of active oxygens and dissipation of excess photons. Annu. Rev. Plant. Physiol. Plant Mol. Biol. 50, 601–639.

Asada, K., 2000. The water–water cycle as alternative photon and electron sinks. Phil. Trans R. Soc. Lond. B. 355, 1419–1431.

Asada, K., 2006. Production and scavenging of reactive oxygen species in chloroplasts and their functions. Plant Physiol. 141, 391–396.

Avenson, T.J., Cruz, J.A., Kramer, D.M., 2004. Modulation of energy-dependent quenching of excitons in antennae of higher plants. Proc. Natl. Acad. Sci. (USA) 101, 5530–5535.

Aviv, D.H., Rustérucci, C., Holt 3rd, B.F., Dietrich, R.A., Parker, J.E., Dangl, J.L., 2002. Runaway cell death, but not basal disease resistance, in lsd1 is SA- and NIM1/NPR1-dependent. Plant J. 29, 381–391.

Barber, J., Nield, N., Morris, E.P., Hankamer, B., 1999. Subunit positioning in photosystem II revisited. Trends Biochem. Sci. 24, 43–45.

Bechtold, U., Karpinski, S., Mullineaux, P.M., 2005. The influence of the light environment and photosynthesis on oxidative signaling responses in plant-biotrophic pathogen interactions. Plant Cell Environ. 28, 1046–1055.

Bellafiore, S., Barneche, F., Peltier, G., Rochaix, J.D., 2005. State transitions and light adaptation require chloroplast thylakoid protein kinase STN7. Nature 433, 892–895.

Berg, J.M., Tymoczko, J.L., Stryer, L., 2002. Biochemistry, Fifth edn. Freeman, New York, USA.

Berry, E.A., Guergova-Kuras, M., Huang, L.S., Crofts, A.R., 2000. Structure and function of cytochrome bc complexes. Annu. Rev. Biochem. 69, 1005–1075.

Blokhina, O., Fagerstedt, K.V., 2010. Reactive oxygen species and nitric oxide in plant mitochondria: origin and redundant regulatory systems. Physiol. Plant. 138, 447–462.

Blokhina, O., Virolainen, E., Fagerstedt, K.V., 2003. Antioxidants, oxidative damage and oxygen deprivation stress: a review. Ann. Bot. 91, 179–194.

Bonardi, V., Pesaresi, P., Becker, T., Schleiff, E., Wagner, R., Pfannschmidt, T., et al., 2005. Photosystem II core phosphorylation and photosynthetic acclimation require two different protein kinases. Nature 437, 1179–1182.

Bonente, G., Howes, B.D., Caffari, S., Smulevich, G., Bassi, R., 2008. Interactions between the photosystem II subunit PsbS and xanthophylls studied *in vivo* and *in vitro*. J. Biol. Chem. 283, 8434–8445.

Borda, M.J., Elsetinow, A.R., Schooen, M.A., Strongin, D.R., 2001. Pyrite-induced hydrogen peroxide formation as a driving force in the evolution of photosynthetic organisms on an early Earth. Astrobiology 1, 283–288.

Brioukhanov, A.L., Netrusov, A.I., 2004. Catalase and superoxide dismutase: distribution, properties, and physiological role in cells of strict anaerobes. Biochemistry (Moscow) 69, 949–962.

Bykova, N.V., Rampitsch, C., 2013. Modulating protein function through reversible oxidation: redox-mediated processes in plants revealed through proteomics. Proteomics 13, 579–596.

Carol, P., Stevenson, D., Bisanz, C., Breitenbach, J., Sandmann, G., Mache, R., et al., 1999. Mutations in the *Arabidopsis* gene *immutans* cause a variegated phenotype by inactivating a chloroplast terminal oxidase associated with phytoene desaturation. Plant Cell 11, 57–68.

Casano, L.M., Zapata, J.M., Martin, M., Sabater, B., 2000. Chlororespiration and poising of cyclic electron transport – plastoquinone as electron transporter between thylakoid NADH dehydrogenase and peroxidase. J. Biol. Chem. 275, 942–948.

Cazzonelli, C.I., 2011. Carotenoids in nature: insights from plants and beyond. Funct Plant Biol. 38, 833–847.

Cazzonelli, C.I., Pogson, B.J., 2010. Source to sink: regulation of carotenoid biosynthesis in plants. Trends Plant Sci. 15, 266–274.

Chang, C.C., Ball, L., Fryer, M.J., Baker, N.R., Karpinski, S., Mullineaux, P.M., 2004. Induction of *ASCORBATE PEROXIDASE 2* expression in wounded *Arabidopsis* leaves does not involve known wound-signalling pathways but is associated with changes in photosynthesis. Plant J. 38, 499–511.

Chang, C.C.C., Ślesak, I, Jordá, L, Sotnikov, A, Melzer, M., Miszalski, Z., 2009. Arabidopsis chloroplastic glutathione peroxidases play a role in cross talk between photooxidative stress and immune responses. Plant Physiol. 150, 670–683.

Chen, L., Jia, H., Tian, Q., Du, L., Gao, Y., Miao, X., et al., 2012. Protecting effect of phosphorylation on oxidative damage of D1 protein by down-regulating the production of superoxide anion in photosystem II membranes under high light. Photosynth. Res. 112, 141–148.

Chen, S., Vaghchhipawala, Z., Li, W., Asard, H., Dickman, M.B., 2004. Tomato phospholipid hydroperoxide glutathione peroxidase inhibits cell death induced by bax and oxidative stresses in yeast and plants. Plant Physiol. 135, 1630–1641.

Cheng, Y., Song, C., 2006. Hydrogen peroxide homeostasis and signaling in plant cells. Science in China, Series C- Life Sciences 49, 1–11.

Chitnis, P.R., 2001. Photosystem I: function and physiology. Annu. Rev. Plant Physiol. Plant Mol. Biol. 52, 593–626.

Collins, A.R., 2001. Carotenoids and genomic stability. Mutant Res. 18, 21–28.

Csintalan, Z., Proctor, M.C.F., Tuba, Z., 1999. Chlorophyll fluorescence during drying and rehydration in the mosses *Rhytidiadelphus loreus* (Hedw.) Warnst., *Anomondon viticulosus* (Hedw.) Hook. & Tayl. and *Grimmia pulvinata* (Hedw.) Sm. Ann. Bot. 84, 235–244.

Cunningham Jr., F.X., Gantt, E., 1998. Genes and enzymes of carotenoid biosynthesis in plants. Annu. Rev. Plant Physiol. Plant Mol. Biol. 49, 557–583.

DalCorso, G., Pesaresi, P., Masiero, S., Aseeva, E., Nemann, D.S., Finazzi, G., et al., 2008. A complex containing PGRL1 and PGR5 is involved in the switch between linear and cyclic electron flow in *Arabidopsis*. Cell 132, 273–285.

Danon, A., Coll, N.S., Apel, K., 2006. Cryptochrome-1-dependent execution of programmed cell death induced by singlet oxygen in *Arabidopsis thaliana*. Proc. Natl. Acad. Sci. (USA) 103, 17036–17041.

Del Río, L.A., Sandalio, L.M., Corpas, F.J., Palma, J.M., Barroso, J.B., 2006. Reactive oxygen species and reactive nitrogen species in peroxisomes. Production, scavenging, and role in cell signaling. Plant Physiol. 141, 330–335.

Della Penna, D., 1999. Carotenoid synthesis and function in plants: insights from mutant studies in *Arabidopsis*. Pure Appl. Chem. 71, 2205–2212.

Demmig-Adams, B., 1990. Carotenoids and photoprotection in plants: a role for the xanthophylls zeaxanthin. Biochim. Biophys. Acta 1020, 1–24.

Demmig-Adams, B., Adams, W.W., 2006. Photoprotection in an ecological context: the remarkable complexity of thermal energy dissipation. New Phytol. 172, 11–21.

Denness, L., McKenna, J.F., Segonzac, C., Wormit, A., Madhou, P., Bennett, M., et al., 2011. Cell wall damage-induced lignin biosynthesis is regulated by a reactive oxygen species- and jasmonic acid-dependent process in *Arabidopsis*. Plant Physiol. 156, 1364–1374.

Deusch, O., Landan, G., Roettger, M., Gruenheit, N., Kowallik, K.V., Allen, J.F., et al., 2008. Genes of cyanobacterial origin in plant nuclear genomes point to a heterocyst-forming plastid ancestor. Mol. Biol. Evol. 25, 748–761.

Doulis, A.C., Debian, N., Kingston-Smith, A.H., Foyer, C.H., 1997. Differential localization of antioxidants in maize leaves. Plant Physiol. 114, 1031–1037.

Elstner, E.F., 1991. Mechanisms of oxygen activation in different compartments of plant cells. In: Pell, E.J., Steffen, K.L. (Eds.), Active Oxygen/oxidative Stress and Plant Metabolism. American Society of Plant Physiologists, Rockville, Md, USA, pp. 13–25.

Endo, T., Asada, K., 2002. Photosystem I and photoprotection: Cyclic electron flow and water–water cycle. In: Demming-Adams, B., Adams III, W.W., Mattoo, A.K. (Eds.), Advances in Photosynthesis and Respiration: Photoprotection, Photoinhibition, Gene Regulation and Environment, Vol. 21. Springer, Dordrecht, pp. 205–221.

Eskling, M., Arvidsson, P-O., Åkerlund, H.E., 1997. The xanthophyll cycle, its regulation and components. Physiol. Plant 100, 806–816.

Eskling, M., Emanuelsson, A., Åkerlund, H.E., 2001. Enzymes and mechanisms for violaxanthin-zeaxanthin conversion. In: Aro, E.M., Andersson, B. (Eds.), Regulation of Photosynthesis. Kluwer Academic Publishers, Dordrecht, pp. 433–452.

Esteban, R., Olano, J.M., Castresana, J., Fernández-Marín, B., Hernández, A., Becerril, J.M., et al., 2009. Distribution and evolutionary trends of photoprotective isoprenoids (xanthophylls and tocopherols) within the plant kingdom. Physiol. Plant 135, 379–389.

Faltin, Z., Holland, D., Velcheva, M., Tsapovetsky, M., Roeckel-Drevet, P., Handa, A.K., et al., 2010. Glutathione peroxidase regulation of reactive oxygen species level is crucial for in vitro plant differentiation. Plant Cell Physiol. 51, 1151–1162.

Fanciullino, A.L., Bidel, L.P.R., Urban, L. (2013). Carotenoid responses to environmental stimuli: integrating redox and carbon controls into a fruit model. Plant Cell Environ. doi:10.1111/pce.12153.

Fernández, A.P., Strand, Å., 2008. Retrograde signaling and plant stress: plastid signals initiate cellular stress responses. Curr. Opin. Plant Biol. 11, 509–513.

Fey, V., Wagner, R., Brautigam, K., Wirtz, M., Hell, R., Dietzmann, A., et al., 2005. Retrograde plastid redox signals in the expression of nuclear genes for chloroplast proteins of *Arabidopsis thaliana*. J. Biol. Chem. 280, 5318–5328.

Fink, R.C., Scandalios, J.G., 2002. Molecular evolution and structure–function relationships of the superoxide dismutase gene families in angiosperms and their relationship to other eukaryotic and prokaryotic superoxide dismutases. Arch. Biochem. Biophys 399, 19–36.

Fischer, B.B., Hideg, E., Krieger-Liszkay, A., 2013. Production, detection, and signaling of singlet oxygen in photosynthetic organisms. Antiox. Redox Signal. 18, 2145–2162.

Forsberg, J., Rosenquist, M., Fraysse, L., Allen, J.F., 2001. Redox signaling in chloroplasts and mitochondria: genomic and biochemical evidence for two-component regulatory systems in bioenergetic organelles. Biochem. Soc. Trans. 29, 403–407.

Foyer, C.H., Allen, J.F., 2003. Lessons from redox signaling in plants. Antiox. Redox Signal. 5, 3–5.

Foyer, C.H., Noctor, G., 2003. Redox sensing and signalling associated with reactive oxygen in chloroplasts, peroxisomes and mitochondria. Physiol. Plant 119, 355–364.

Foyer, C.H., Noctor, G., 2005. Oxidant and antioxidant signalling in plants: a re-evaluation of the concept of oxidative stress in a physiological context. Plant Cell Environ. 28, 1056–1071.

Foyer, C.H., Noctor, G., 2005. Redox homeostasis and antioxidant signaling: a metabolic interface between stress perception and physiological responses. The Plant Cell 17, 1866–1875.

Foyer, C.H., Noctor, G., 2009. Redox regulation in photosynthetic organisms: signaling, acclimation, and practical implications. Antiox. Redox Signal. 11, 862–905.

Foyer, C.H., Noctor, G., 2011. Ascorbate and glutathione: the heart of the redox hub. Plant Physiol. 155, 2–18.

Foyer, C.H., Noctor, G., 2012. Managing the cellular redox hub in photosynthetic organisms. Plant Cell Environ. 35, 199–201.

Foyer, C.H., Noctor, G., 2013. Redox signaling in plants. Antiox. Redox Signal 18, 2087–2090.

Fryer, M.J., Ball, L., Oxborough, K., Karpinski, S., Mullineaux, P.M., Baker, N.R., 2003. Control of ascorbate peroxidase 2 expression by hydrogen peroxide and leaf water status during excess light stress reveals a functional organisation of Arabidopsis leaves. Plant J. 33, 691–705.

García-Plazaola, J.I., Esteban, R., Fernández-Marín, B., Kranner, I., Porcar-Castell, A., 2012. Thermal energy dissipation and xanthophyll cycles beyond the *Arabidopsis* model. Photosynth. Res. 113, 89–103.

García-Plazaola, J.I., Matsubara, S., Osmond, B., 2007. The lutein epoxide cycle in higher plants: its relationship to other xanthophylls cycles and possible functions. Funct. Plant Biol. 34, 759–773.

Giacomelli, L., Masi, A., Ripoll, D.R., Lee, M.J., van Wijk, K.J., 2007. *Arabidopsis thaliana* deficient in two chloroplast ascorbate peroxidases shows accelerated light-induced necrosis when levels of cellular ascorbate are low. Plant Mol. Biol. 65, 627–644.

Gish, L.A., Clark, S.E., 2011. The RLK/Pelle family of kinases. Plant J. 66, 117–127.

Gotor, C., Romero, L.C., 2013. S-sulfocysteine synthase function in sensing chloroplast redox status. Plant Signal Behav. 8, 23313.

Gruszecki, W.I., Grudzinski, W., Gospodarek, M., Patyra, M., Maksymiec, W., 2006. Xanthophyll-induced aggregation of LHCII as a switch between light-harvesting and energy dissipation systems. Biochim. Biophys. Acta 1757, 1504–1511.

Haldrup, A., Jensen, P.E., Lunde, C., Scheller, H.V., 2001. Balance of power: a view of the mechanism of photosynthetic state transitions. Trends Plant Sci. 6, 301–305.

Halliwell, B., 2006. Reactive species and antioxidants. Redox biology is a fundamental theme of aerobic life. Plant Physiol. 141, 312–322.

Halliwell, B., Gutteridge, J.M.C., 2007. Free Radicals in Biology and Medicine, Fourth edition. Oxford University Press, Oxford, UK.

Heiber, I., Stroher, E., Raatz, B., Busse, I., Kahmann, U., Bevan, M.W., et al., 2007. The redox imbalanced mutants of *Arabidopsis* differentiate signaling pathways for redox regulation of chloroplast antioxidant enzymes. Plant Physiol. 143, 1774–1788.

Heyno, E., Mary, V., Schopfer, P., Krieger-Liszkay, A., 2011. Oxygen activation at the plasma membrane: relation between superoxide and hydroxyl radical production by isolated membranes. Planta 234, 35–45.

Hideg, E., Vass, I., 1995. Singlet oxygen is not produced in photosystem I under photoinhibitory conditions. Photochem. Photobiol. 62, 949–952.

Horemans, N, Foyer, CH, Potters, G, Asard, H, 2000. Ascorbate function and associated transport systems in plants. Plant Physiol. Biochem. 38, 531–540.

Horton, P., Ruban, A.V., Walters, R.G., 1996. Regulation of light harvesting in green plants. Annu. Rev. Plant Physiol. Plant Mol. Biol. 47, 665–684.

Horton, P., Wentworth, M., Ruban, A., 2005. Control of the light harvesting function of chloroplast membranes: the LHCII-aggregation model for non-photochemical quenching. FEBS Lett. 579, 4201–4206.

Huala, E., Oeller, P.W., Liscum, E., Han, I-S., Larsen, E., Briggs, W.R., 1997. *Arabidopsis* NPH1: a protein kinase with a putative redox-sensing domain. Science 278, 2120–2123.

Huner, N.P.A., Ivanov, A.G., Sane, P.V., Pocock, T., Krol, M., Balseris, A., et al., 2006. Photoprotection of photosystem II: reaction centre quenching versus antenna quenching. In: Demming-Adams, B., Adams III, W.W., Mattoo, A.K. (Eds.), Advances in Photosynthesis and Respiration: Photoprotection, Photoinhibition, Gene Regulation and Environment, Vol. 21. Springer, Dordrecht, pp. 155–173.

Ikegami, A., Yoshimura, N., Motohashi, K., Takahashi, S., Romano, P.G., Hisabori, T., et al., 2007. The CHLI1 subunit of *Arabidopsis thaliana* magnesium chelatase is a target protein of the chloroplast thioredoxin. J. Biol. Chem. 282, 19282–19291.

Ishikawa, T., Shigeoka, S., 2008. Recent advances in ascorbate biosynthesis and the physiological significance of ascorbate peroxidase in photosynthesizing organisms. Biosci. Biotechnol. Biochem. 72, 1143–1154.

Ivanov, A.G., Hurry, V., Prafullachandra, V., Öquist, G., Huner, N.P.A., 2008. Reaction centre quenching of excess light energy and photoprotection of photosystem II. J. Plant Biol. 51, 85–96.

Jacquot, J.P., Dietz, K.J., Rouhier, N., Meux, E., Lallement, P.A., Selles, B., et al., 2013. Redox regulation in plants: Glutathione and "redoxin" related families. In: Jakob, U., Reichmann, D. (Eds.), Oxidative Stress and Redox Regulation. Springer Science and Business Media, pp. 213–231.

Jäger-Vottero, P., Dorne, A-J., Jordanov, J., Douce, R., Joyard, J., 1997. Redox chains in chloroplast envelope membranes: spectroscopic evidence for the presence of electron carriers, including iron sulfur centers. Proc. Natl. Acad. Sci. (USA) 94, 1597–1602.

Jahns, P., Holzwarth, A.R., 2012. The role of xanthophyll cycle and of lutein in photoprotection of photosystem II. Biochim. Biophys. Acta 1817, 182–193.

Joet, T., Genty, B., Josse, E.M., Kuntz, M., Cournac, L., Peltier, G., 2002. Involvement of a plastid terminal oxidase in plastoquinone oxidation as evidenced by expression of the *Arabidopsis thaliana* enzyme in tobacco. J. Biol. Chem. 277, 31623–31630.

Joliot, P., Joliot, A., 2002. Cyclic electron transport in plant leaf. Proc. Natl. Acad. Sci. (USA) 99, 10209–10214.

Joliot, P., Joliot, A., 2005. Quantification of cyclic and linear flows in plants. Proc. Natl. Acad. Sci. (USA) 102, 4913–4918.

Joliot, P., Joliot, A., 2006. Cyclic electron flow in C_3 plants. Biochim. Biophys. Acta 1757, 362–368.

Jung, B.G., Lee, K.O., Lee, S.S., Chi, Y.H., Jang, H.H., Kang, S.S., et al., 2002. A Chinese cabbage cDNA with high sequence identity to phospholipid hydroperoxide glutathione peroxidases encodes a novel isoform of thioredoxin-dependent peroxidase. J. Biol. Chem. 277, 12572–12578.

Kagawa, T., Sakai, T., Suetsugo, N., Oikawa, K., Ishiguro, S., Kato, T., et al., 2001. *Arabidopsis* NPL1: a phototropin homolog controlling the chloroplast high-light avoidance response. Science 291, 2138–2141.

Karpinski, S., Escobar, C., Karpinska, B., Creissen, G., Mullineaux, P.M., 1997. Photosynthetic electron transport regulates the expression of cytosolic ascorbate peroxidase genes in *Arabidopsis* during excess light stress. Plant Cell 9, 627–640.

Karpinski, S., Reynolds, H., Karpinska, B., Wingsle, G., Creissen, G., Mullineaux, P.M., 1999. Systemic signalling and acclimation in response to excess excitation energy in *Arabidopsis*. Science 284, 654–657.

Karuppanapandian, T., Moon, J-C., Kim, C., Manoharan, K., Kim, W., 2011. Reactive oxygen species in plants: their generation, signal transduction, and scavenging mechanisms. AJCS 5, 709–725.

Keren, N., Gong, H., Ohad, I., 1995. Oscillations of reaction center II-D1 protein degradation *in vivo* induced by repetitive light flashes. Correlation between the level of RCII-QB- and protein degradation in low light. J. Biol. Chem. 270, 806–814.

Kiffin, R., Bandyopadhyay, U., Cuervo, A.M., 2006. Oxidative stress and autophagy. Antiox. Redox Signal 8, 152–162.

Knight, H., Knight, M.R., 2001. Abiotic stress signaling pathways: specificity and cross-talk. Trends Plant Sci. 6, 262–267.

Konert, G., Rahikainen, M., Andrea Trotta, A., Kangasjärvi, S., 2013. Systemic signaling in light acclimation of leaves. In: Baluška, F. (Ed.), Long-Distance Systemic Signaling and Communication in Plants: Signaling and Communication in Plants, Vol. 19. Springer Science and Business Media, pp. 231–250.

Kornas, A., Kuźniak, E., Ślesak, I., Miszalski, Z, 2010. The key role of the redox status in regulation of metabolism in photosynthesizing organisms. Acta Biochim. polonica 57, 143–151.

Koussevitzky, S., Nott, A., Mockler, T.C., Hong, F., Sachetto-Martins, G., Surpin, M., et al., 2007. Signals from chloroplasts converge to regulate nuclear gene expression. Science 316, 715–719.

Kovtun, Y., Chiu, W-L., Tena, G., Sheen, J., 2000. Functional analysis of oxidative stress-activated mitogen-activated protein kinase cascade in plants. Proc. Natl. Acad. Sci. (USA) 97, 2940–2945.

Krause, G.H., 1988. Photoinhibition of photosynthesis. an evaluation of damaging and protective mechanisms. Physiol. Plant 74, 566–574.

Krieger-Liszkay, 2005. Singlet oxygen production in photosynthesis. J. Exp. Bot. 56, 337–346.

Kropat, J., Oster, U., Rudiger, W., Beck, C.F., 2000. Chloroplast signalling in the light induction of nuclear *HSP70* genes requires the accumulation of chlorophyll precursors and their accessibility to cytoplasm/nucleus. Plant J. 24, 523–531.

Laisk, A., Talts, E., Oja, V., Eichelmann, H., Peterson, R.B., 2010. Fast cyclic electron transport around photosystem I in leaves under far-red light: a proton-uncoupled pathway? Photosynth. Res. 103, 75–95.

Laloi, C., Stachowiak, M., Pers-Kamczyc, E., Warzych, E., Murgia, I., Apel, K., 2007. Cross-talk between singlet oxygen- and hydrogen peroxide-dependent signaling of stress responses in *Arabidopsis thaliana*. Proc. Natl. Acad. Sci. (USA) 104, 672–677.

Landoni, M., De Francesco, A., Bellatti, S., Delledonne, M., Ferrarini, A., Venturini, L., et al. (2013). A mutation in the *FZL* gene of *Arabidopsis* causing alteration in chloroplast morphology results in a lesion mimic phenotype. J. Exp. Bot. 64, 4313–4328.

Lawlor, D.W., 2001. Photosynthesis, Third edition. BioS Scientific Publishers Limited, UK.

Lee, H.Y., Hong, Y.N., Chow, W.S., 2001. Photoinactivation of photosystem II complexes and photoprotection by non-functional neighbours in *Capsicum annuum* L. leaves. Planta 212, 332–342.

Lee, K.P., Kim, C., Landgraf, F., Apel, K., 2007. EXECUTER1- and EXECUTER2-dependent transfer of stress-related signals from the plastid to the nucleus of *Arabidopsis thaliana*. Proc. Natl. Acad. Sci. (USA) 104, 10270–10275.

Lee, K.P., Kim, C., Lee, D.W., Apel, K., 2003. TIGRINA d, required for regulating the biosynthesis of tetrapyrroles in barley, is an ortholog of the FLU gene of *Arabidopsis thaliana*. FEBS Lett. 553, 119–124.

Lee, S.H., Ahsan, N., Lee, K.W., Kim, D.H., Lee, D.G., Kwak, S.S., et al., 2007. Simultaneous overexpression of both CuZn superoxide dismutase and ascorbate peroxidase in transgenic tall fescue plants confers increased tolerance to a wide range of abiotic stresses. J. Plant Physiol. 164, 1626–1638.

Legnaioli, T., Cuevas, J., Mas, P., 2009. TOC1 functions as a molecular switch connecting the circadian clock with plant responses to drought. EMBO J. 28, 3745–3757.

Lepetit, B., Sturm, S., Rogato, A., Gruber, A., Sachse, M., Falciatore, A., et al., 2013. High light acclimation in the secondary plastids containing diatom *Phaeodactylum tricornutum* is triggered by the redox state of the plastoquinone pool. J. Plant Physiol. 161, 853–865.

Li, H., Wang, X.M., Chen, L., Ahammed, G.J., Xia, X.J., Shi, K., et al., 2013. Growth temperature-induced changes in biomass accumulation, photosynthesis and glutathione redox homeostasis as influenced by hydrogen peroxide in cucumber. Plant Physiol. Biochem. 71, 1–10.

Li, X.P., Björkman, O., Shih, C., Grossman, A.R., Rosenquist, M., Jansson, S., et al., 2000. A pigment-binding protein essential for regulation of photosynthetic light harvesting. Nature 403, 391–395.

Li, X.P., Gilmore, A.M., Caffarri, S., Bassi, R., Golan, T., Kramer, D., et al., 2004. Regulation of photosynthetic light harvesting involves intrathylakoid lumen pH sensing by the PsbS protein. J. Biol. Chem. 279, 22866–22874.

Livingston, A.K., Kanazawa, A., Cruz, J.A., Kramer, D.M., 2010. Regulation of cyclic electron flow in C_3 plants: differential effects of limiting photosynthesis at ribulose-1,5-bisphosphate carboxylase/oxygenase and glyceraldehyde-3-phosphate dehydrogenase. Plant Cell Environ. 33, 1779–1788.

Long, S.P., Humphries, S., Falkowski, P.G., 1994. Photoinhibition of photosynthesis in nature. Annu. Rev. Plant Physiol. Plant Mol. Biol. 45, 633–662.

Lu, Y., Wang, H.R., Li, H., Cui, H.R., Feng, Y.G., Wang, X.Y., 2013. A chloroplast membrane protein LTO1/AtVKOR involving in redox regulation and ROS homeostasis. Plant Cell Rep. 32, 1427–1440.

Lunde, C., Jensen, P.E., Haldrup, A., Knoetzel, J., Scheller, H.V., 2000. The PSI-H subunit of photosystem I is essential for state transitions in plant photosynthesis. Nature 408, 613–615.

Marschall, M., Proctor, M.C.F., 1999. Desiccation tolerance and recovery of the leafy liverwort *Porella platyphylla* (L.) Pfeiff.: chlorophyll-fluorescence measurements. J. Bryol. 21, 257–262.

Mateo, A., Mühlenbock, P., Rustérucci, C., Chang, C.C., Miszalski, Z., Karpinska, B., et al., 2004. LESION SIMULATING DISEASE 1 is required for acclimation to conditions that promote excess excitation energy. Plant Physiol. 136, 2818–2830.

Matsubara, S., Chen, Y.C., Caliandro, R., Govindjee, C.R.M., 2011. Photosystem II fluorescence lifetime imaging in avocado leaves: contributions of the lutein epoxide and violaxanthin cycles to fluorescence quenching. J. Photochem. Photobiol. B. 104, 271–284.

Matsubara, S., Chow, W.S., 2004. Populations of photoinactivated photosystem II reaction centers characterized by chlorophyll a fluorescence lifetime *in vivo*. Proc. Natl. Acad. Sci. (USA) 101, 18234–18239.

McCarty, R.E., Evron, Y., Johnson, E.A., 2000. The chloroplast ATP synthase: a rotary enzyme? Annu. Rev. Plant Physiol. Plant Mol. Biol. 51, 83–109.

Meskauskiene, R., Apel, K., 2002. Interaction of FLU, a negative regulator of tetrapyrrole biosynthesis, with the glutamyl-tRNA reductase requires the tetratricopeptide repeat domain of FLU. FEBS Lett. 532, 27–30.

Meskauskiene, R., Nater, M., Goslings, D., Kessler, F., op den Camp, R, Apel, K, 2001. FLU: a negative regulator of chlorophyll biosynthesis in *Arabidopsis thaliana*. Proc. Natl. Acad. Sci. (USA) 98, 12826–12831.

Meyer, Y., Reichheld, J.P., Vignols, F., 2005. Thioredoxins in *Arabidopsis* and other plants. Photosynth. Res. 86, 419–433.

Meyer, Y., Siala, W., Bashandy, T., Riondet, C., Vignols, F., Reichheld, J.P., 2008. Glutaredoxins and thioredoxins in plants. Biochim. Biophys. Acta 1783, 589–600.

Miao, Y., Lv, D., Wang, P., Wang, X.C., Chen, J., Miao, C., et al., 2006. An *Arabidopsis* glutathione peroxidase functions as both a redox transducer and a scavenger in abscisic acid and drought stress responses. Plant Cell 18, 2749–2766.

Michelet, L., Zaffagnini, M., Marchand, C., Collin, V., Decottignies, P., Tsan, P., et al., 2005. Glutathionylation of chloroplast thioredoxin f is a redox signaling mechanism in plants. Proc. Natl. Acad. Sci. (USA) 102, 16478–16483.

Milla, M.A.R., Maurer, A., Huete, A.R., Gustafson, J.P., 2003. Glutathione peroxidase genes in *Arabidopsis* are ubiquitous and regulated by abiotic stresses through diverse signaling pathways. Plant J. 36, 602–615.

Mittler, R., 2006. Abiotic stress, the field environment and stress combination. Trends Plant Sci. 11, 15–19.

Miyake, C., 2010. Alternative electron flows (water–water cycle and cyclic electron flow around PSI) in photosynthesis: molecular mechanisms and physiological functions. Plant Cell Physiol. 51, 1951–1963.

Miyake, C., Yokota, A., 2001. Cyclic flow of electrons within PSII in thylakoid membranes. Plant Cell Physiol. 42, 508–515.

Mochizuki, N., Brusslan, J.A., Larkin, R., Nagatani, A., Chory, J., 2001. *Arabidopsis* genomes uncoupled 5 (GUN5) mutant reveals the involvement of Mg-chelatase H subunit in plastid-to-nucleus signal transduction. Proc. Natl. Acad. Sci. (USA) 98, 2053–2058.

Møller, I.M., Jensen, P.E., Hansson, A., 2007. Oxidative modifications to cellular components in plants. Annu. Rev. Plant Biol. 58, 459–481.

Mou, Z., Fan, W., Dong, X., 2003. Inducers of plant systemic acquired resistance regulate NPR1 function through redox changes. Cell 113, 935–944.

Müller, M.J., 1997. Enzymes involved in jasmonic acid biosynthesis. Physiol. Plant 100, 653–663.

Müller, P., Li, X-P., Niyogi, K.K., 2001. Non-photochemical quenching. A response to excess light energy. Plant Physiol. 125, 1558–1566.

Mullineaux, P., Karpinski, S., 2002. Signal transduction in response to excess light: getting out of the chloroplast. Curr. Opin. Plant Biol. 5, 43–48.

Munné-Bosch, S., 2005. The role of alpha-tocopherol in plant stress tolerance. J. Plant Physiol. 162, 743–748.

Munné-Bosch, S., 2007. Alpha-tocopherol: a multifaceted molecule in plants. Vitam. Horm. 76, 375–392.

Munné-Bosch, S., Queval, G., Foyer, C.H., 2013. The impact of global change factors on redox signaling underpinning stress tolerance. Plant Physiol. 161, 5–19.

Nabe, H., Funabiki, R., Kashino, Y., Koike, H., Satoh, K., 2007. Responses to desiccation stress in bryophytes and important role of dithiothreitol-insensitive non-photochemical quenching against photoinhibition in dehydrated states. Plant Cell Physiol. 48, 1548–1557.

Naya, L., Ladrera, R., Ramos, J., González, E.M., Arrese-Igor, C., Minchin, F.R., et al., 2007. The response of carbon metabolism and antioxidant defenses of alfalfa nodules to drought stress and to the subsequent recovery of plants. Plant Physiol. 144, 1104–1114.

Neill, S., Desikan, R., Hancock, J., 2002. Hydrogen peroxide signaling. Curr. Opin. Plant Biol. 5, 388–395.

Neubauer, C., Schreiber, U., 1989. Photochemical and non-photochemical quenching of chlorophyll fluorescence induced by hydrogen peroxide. Z. Naturforsch. C. 44, 262–270.

Nichol, C., Pieruschka, T.K., Förster, B., Kolber, Z., Rascher, U., Grace, J., et al., 2012. Canopy conundrums: building on the Biosphere 2 experience to scale measurements of inner and outer canopy photoprotection from the leaf to the landscape. Funct. Plant Biol. 39, 1–24.

Nilkens, M., Kress, E., Lambrev, P., Miloslavina, Y., Müller, M., Holzwarth, A.R., et al., 2010. Identification of a slowly inducible zeaxanthin-dependent component of non-photochemical quenching of chlorophyll fluorescence generated under steady-state conditions in *Arabidopsis*. Biochim. Biophys. Acta 1797, 466–475.

Niyogi, K.K., 1999. Photoprotection revisited: genetic and molecular approaches. Annu. Rev. Plant Physiol. Plant Mol. Biol. 50, 333–359.

Niyogi, K.K., 2000. Safety valves for photosynthesis. Curr. Opin. Plant Biol. 3, 455–460.

Noctor, G., Foyer, C.H., 1998. Ascorbate and glutathione: keeping active oxygen under control. Annu. Rev. Plant Physiol. Plant Mol. Biol. 49, 249–279.

Noctor, G., Mhamdi, A., Queval, G., Foyer, C.H., 2013. Regulating the redox gatekeeper: vacuolar sequestration puts glutathione disulfide in its place. Plant Physiol. 163, 665–671.

Oelze, M-L., Kandlbinder, A., Dietz, K-J., 2008. Redox regulation and overreduction control in the photosynthesizing cell: complexity in redox regulatory networks. Biochim. Biophys. Acta 1780, 1261–1272.

Op den Camp, R.G.L., Przybyla, D., Ochsenbein, C., Laloi, C., Kim, C., Danon, A., et al., 2003. Rapid induction of distinct stress responses after the release of singlet oxygen in *Arabidopsis*. Plant Cell 15, 2320–2332.

Öquist, G., Huner, N.P.A., 2003. Photosynthesis of overwintering evergreen plants. Annu. Rev. Plant Biol. 54, 329–355.

Orozco-Cardenas, M.L., Narvaez-Vasquez, J., Ryan, C.A., 2001. Hydrogen peroxide acts as a secondary messenger for the induction of defense genes in tomato plants in response to wounding, systemin and methyl jasmonate. Plant Cell 13, 179–191.

Oswald, O., Martin, T., Dominy, P.J., Graham, I.A., 2000. Plastid redox state and sugars: interactive regulators of nuclear-encoded photosynthetic gene expression. Proc. Natl. Acad. Sci. (USA) 98, 2047–2052.

Pascal, A.A., Liu, Z., Broess, K., van Oort, B., van Amerongen, H., Wang, C., et al., 2005. Molecular basis of photoprotection and control of photosynthetic light-harvesting. Nature 436, 134–137.

Pastori, G., Foyer, C.H., Mullineaux, P., 2000. Low temperature-induced changes in the distribution of H_2O_2 and antioxidants between the bundle sheath and mesophyll cells of maize leaves. J. Exp. Bot. 51, 107–113.

Pastori, G.M., Kiddle, G., Antoniw, J., Bernard, S., Veljovic-Jovanovic, S., Verrier, P.J., et al., 2003. Leaf vitamin C contents modulate plant defense transcripts and regulate genes that control development through hormone signaling. Plant Cell 15, 939–951.

Peeva, V., Cornic, G., 2009. Leaf photosynthesis of *Haberlea rhodopensis* before and during drought. Environ. Exp. Bot. 65, 310–318.

Pei, Z-M., Murata, Y., Benning, G., Thomine, S., Klusener, B., Allen, G.J., et al., 2000. Calcium channels activated by hydrogen peroxide mediate abscisic acid signalling in guard cells. Nature 406, 731–734.

Pfannschmidt, T., 2003. Chloroplast redox signals: how photosynthesis controls its own genes. Trends Plant Sci. 8, 33–41.

Pfannschmidt, T., Bräutigam, K., Wagner, R., Dietzel, L., Schröter, Y., Steiner, S., et al., 2009. Potential regulation of gene expression in photosynthetic cells by redox and energy state: approaches towards better understanding. Ann. Bot. 103, 599–607.

Pfannschmidt, T., Nilsson, A., Allen, J.F., 1999. Photosynthetic control of chloroplast gene expression. Nature 397, 625–628.

Pitzschke, A., Forzani, C., Hirt, H., 2006. Reactive oxygen species signaling in plants. Antiox. Redox Signal 8, 1757–1764.

Pontier, D., Albrieux, C., Joyard, J., Lagrange, T., Block, M.A., 2007. Knock-out of the magnesium protoporphyrin IX methyltransferase gene in *Arabidopsis*. Effects on chloroplast development and on chloroplast-to-nucleus signaling. J. Biol. Chem. 282, 2297–2304.

Puthiyaveetil, S., Allen, J.F., 2009. Chloroplast two-component systems: evolution of the link between photosynthesis and gene expression. Proc. R. Soc. B. 276, 2133–2145.

Qin, X.Q., Zeevaart, J.A.D., 1999. The 9-cis-epoxycarotenoid cleavage reaction is the key regulatory step of abscisic acid biosynthesis in water-stressed bean. Proc. Natl. Acad. Sci. (USA) 96, 15354–15361.

Ramel, F., Birtic, S., Cuiné, S., Triantaphylidés, C., Ravanat, J-L., Havaux, M., 2012. Chemical quenching of singlet oxygen by carotenoids in plants. Plant Physiol. 158, 1267–1278.

Rizhsky, L., Hallak-Herr, E., Van Breusegem, F., Rachmilevitch, S., Barr, J.E., Rodermel, S., et al., 2002. Double antisense plants lacking ascorbate peroxidase and catalase are less sensitive to oxidative stress than single antisense plants lacking ascorbate peroxidase or catalase. The Plant J. 32, 329–342.

Ruban, A.V., Pascal, A.A., Robert, B., Horton, P., 2002. Activation of zeaxanthin is an obligatory event in the regulation of photosynthetic light harvesting. J Biol. Chem. 277, 7785–7789.

Ruckle, M.E., DeMarco, S.M., Larkin, R.M., 2007. Plastid signals remodel light signaling networks and are essential for efficient chloroplast biogenesis in *Arabidopsis*. Plant Cell 19, 3944–3960.

Rumeau, D., Peltier, G., Cournac, L., 2007. Chlororespiration and cyclic electron flow around PSI during photosynthesis and plant stress response. Plant Cell Environ. 30, 1041–1051.

Schäfer, G., Kardinahl, S., 2003. Iron superoxide dismutase: structure and function of an archaic enzyme. Biochem. Soc. Trans. 31, 1330–1334.

Schwarzländer, M., Finkemeier, I., 2013. Mitochondrial energy and redox signaling in plants. Antiox. Redox Signal. 18, 2122–2144.

Sharma, P., Bhushan, J.A., Dubey, R.S., Pessarakli, M., 2012. Reactive oxygen species, oxidative damage, and antioxidative defense mechanism in plants under stressful conditions. J. Bot. Available from: http://dx.doi.org/10.1155/2012/217037.

Shaw, G.H., 2008. Earth's atmosphere – Hadean to early Proterozoic. Chem. Erde Geochem. 68, 235–264.

Shiu, S.H., Bleecker, A.B., 2003. Expansion of the receptor-like kinase/Pelle gene family and receptor-like proteins in *Arabidopsis*. Plant Physiol. 132, 530–543.

Simková, K., Moreau, F., Pawlak, P., Vriet, C., Baruah, A., Alexandre, C., et al., 2012. Integration of stress-related and reactive oxygen species mediated signals by Topoisomerase VI in *Arabidopsis thaliana*. Proc. Natl. Acad. Sci. (USA) 109, 16360–16365.

Smirnoff, N., 2000. Ascorbic acid: metabolism and functions of a multi-faceted molecule. Curr. Opin. Plant Biol. 3, 229–235.

Snyders, S., Kohorn, B.D., 2001. Disruption of thylakoid-associated kinase 1 leads to alteration of light harvesting in *Arabidopsis*. J. Biol. Chem. 276, 32169–32176.

Stone, J.R., Yang, S., 2006. Hydrogen peroxide: a signaling messenger. Antiox. Redox Signal. 8, 243–270.

Strand, A., Asami, T., Alonso, J., Ecker, J.R., Chory, J., 2003. Chloroplast to nucleus communication triggered by accumulation of Mg-protoporphyrinIX. Nature 421, 79–83.

Streb, P., Josse, E-M., Gallouet, E., Baptist, F., Kuntz, M., Cornic, G, 2005. Evidence for alternative electron sinks to photosynthetic carbon assimilation in the high mountain plant species *Ranunculus glacialis*. Plant Cell Environ. 28, 1123–1135.

Sunil, B., Talla, S.K., Aswani, V., Raghavendra, A.S., 2013. Optimization of photosynthesis by multiple metabolic pathways involving interorganelle interactions: resource sharing and ROS maintenance as the bases. Photosynth. Res. 117, 61–71.

Suorsa, M., Grieco, M., Järvi, S., Gollan, P.J., Kangasjärvi, S., Tikkanen, M., et al., 2012. PGR5 ensures photosynthetic control to safeguard photosystem I under fluctuating light conditions. Plant Signal. Behav. 8, pii: e22741.

Susek, R.E., Ausubel, F.M., Chory, J., 1993. Signal transduction mutants of *Arabidopsis* uncouple nuclear CAB and RBCS gene expression from chloroplast development. Cell 74, 787–799.

Suzuki, N., Koussevitzky, S., Mittler, R., Miller, G., 2012. ROS and redox signalling in the response of plants to abiotic stress. Plant Cell Environ. 35, 259–270.

Taiz, L., Zeiger, E., 2010. Plant Physiology, 5th Edition. Sinauer Associates, MA, USA.

Tanaka, T., Izawa, S., Inoue, Y., 2005. GPX2, encoding a phospholipid hydroperoxide glutathione peroxidase homologue, codes for an atypical 2-Cys peroxiredoxin in *Saccharomyces cerevisiae*. J. Biol. Chem. 280, 42078–42087.

Thomas, D.J., Boling, J., Crowell, C.M., Eubanks, L.M., McCarthy, N., McSpadden, T., et al., 2008. A test of the oxygen paradox using antioxidant-deficient cyanobacteria. Gravitation Space Biol. 21, 27–28.

Thompson, A.J., Jackson, A.C., Parker, R.A., Morpeth, D.R, Burbidge, A., Taylor, I.B, 2000. Abscisic acid biosynthesis in tomato: regulation of zeaxanthin epoxidase and 9- cis-epoxycarotenoid dioxygenase mRNAs by light/dark cycles, water stress and abscisic acid. Plant Mol. Biol. 42, 833–845.

Tikkanen, M., Nurmi, M., Kangasjärvi, S., Aro, E.M., 2008. Core protein phosphorylation facilitates the repair of photodamaged photosystem II at high light. Biochim. Biophys. Acta 1777, 1432–1437.

Trebst, A., 2003. Function of b-carotene and tocopherol in photosystem II. Z. Naturforsch. C. 58, 609–620.

Trebst, A., Depka, B., Holländer-Czytko, H., 2002. A specific role for tocopherol and of chemical singlet oxygen quenchers in the maintenance of photosystem II structure and function in *Chlamydomonas reinhardtii*. FEBS Lett. 516, 156–160.

Tuba, Z., 2008. Notes on the poikilochlorophyllous desiccation tolerant plants. Acta Biol. Szeged 52, 111–113.

Vandenabeele, S., Vanderauwera, S., Vuylsteke, M., Rombauts, S., Langebartels, C., Seidlitz, H. K., et al., 2004. Catalase deficiency drastically affects gene expression induced by high light in *Arabidopsis thaliana*. Plant J. 39, 45–58.

Vertucci, C.W., Farrant, J.M., 1995. Acquisition and loss of desiccation tolerance. In: Galili, K.L. (Ed.), Seed Development and Germination. Marcel Dekker Press, New York, pp. 237–271.

Vlamis-Gardikas, A., Holmgren, A., 2002. Thioredoxin and glutaredoxin isoforms. Methods Enzymol. 347, 286–296.

Vollenweider, S., Weber, H., Stolz, S., Chetelet, A., Farmer, E.E., 2000. Fatty acid ketodiences and fatty acid ketotrienes: michael addition acceptors that accumulate in wounded and diseased *Arabidopsis* leaves. Plant J. 24, 467–476.

von Gromoff, E.D., Alawady, A., Meinecke, L., Grimm, B., Beck, C.F., 2008. Heme, a plastid-derived regulator of nuclear gene expression in *Chlamydomonas*. Plant Cell 20, 552–567.

Wagner, D., Przybyla, D., Op den Camp, R., Kim, C., Landgraf, F., Lee, K.P., et al., 2004. The genetic basis of singlet oxygen-induced stress responses of *Arabidopsis thaliana*. Science 306, 1183–1185.

Willekens, H., Chamnongpol, S., Davey, M., Schraudner, M., Langebartels, C., Van Montagu, M., et al., 1997. Catalase is a sink for H_2O_2 and is indispensable for stress defence in C_3 plants. Embo J. 16, 4806–4816.

Wilson, K.E., Ivanov, A.G., Öquist, G., Grodzinski, B., Sarhan, F., Huner, N.P.A., 2006. Energy balance, organellar redox status, and acclimation to environmental stress. Can. J. Bot. 84, 1355–1370.

Wilson, K.E., Sieger, S.M., Huner, N.P.A., 2003. The temperature-dependent accumulation of Mg-protoporphyrin IX and reactive oxygen species in *Chlorella vulgaris*. Physiol. Plant 119, 126–136.

Wituszynska, W., Slesak, I., Vanderauwera, S., Szechynska-Hebda, M., Kornas, A., Van Der Kelen, K., et al., 2013. LSD1, EDS1 and PAD4 conditionally regulate cellular signaling homeostasis, photosynthesis, water use efficiency and seed yield in *Arabidopsis*. Plant Physiol. 161, 1795–1805.

Wu, F.Q., Xin, Q., Cao, Z., Liu, Z.Q., Du, S.Y., Mei, C., et al., 2009. The magnesium-chelatase H subunit binds abscisic acid and functions in abscisic acid signaling: new evidence in *Arabidopsis*. Plant Physiol. 150, 1940–1954.

Yabuta, Y., Maruta, T., Yoshimura, K., Ishikawa, T., Shigeoka, S., 2004. Two distinct redox signaling pathways for cytosolic APX induction under photooxidative stress. Plant Cell Physiol. 161, 1795–1805.

Yordanov, I., Velikova, V., 2000. Photoinhibition of photosystem I. Bulg. J. Plant Physiol. 26, 70–92.

Young, J., 1991. The photoprotective role of carotenoids in higher plants. Physiol. Plant 83, 702–708.

Zaffagnini, M., Bedhomme, M., Groni, H., Marchand, C.H., Puppo, C., Gontero, B., et al., 2012. Glutathionylation in the photosynthetic model organism *Chlamydomonas reinhardtii: a proteomic survey*. Mol. Cell Proteomics 11, M111.014142.

Zarter, C.R., Adams III, W.W., Ebbert, V., Adamska, I., Jansson, S., Demmig-Adams, B., 2006. Winter acclimation of PsbS and related proteins in the evergreen *Arctostaphyllos uva-ursi* as influenced by altitude and light environment. Plant Cell Environ. 29, 869–878.

Zhang, L., Aro, E.M., 2002. Synthesis, membrane insertion and assembly of the chloroplast-encoded D1 protein into photosystem II. FEBS Lett. 512, 13–18.

Zhou, Y.H., Lam, H.M., Zhang, J.H., 2007. Inhibition of photosynthesis and energy dissipation induced by water and high light stresses in rice. J. Exp. Bot. 58, 1207–1217.

Zulfugarov, I.S., Ham, O-K., Misra, S.R., Kim, J-Y., Nath, K., Koo, H-Y., et al., 2007. Dependence of reaction center-type energy-dependent quenching on photosystem II antenna size. Biochim. Biophys. Acta 1767, 773–780.

Chapter 2

Reactive Oxygen Species and Plant Hormones

Marcelo Pedrosa Gomes, Élise Smedbol, Marília Mércia Lima Carvalho Carneiro, Queila Souza Garcia and Philippe Juneau

2.1 INTRODUCTION

Our understanding of the roles of reactive oxygen species (ROS) in biological processes has significantly evolved in the last few decades. Although these compounds were initially viewed as hazardous to plants, recent studies have recognized that derivates of oxygen reduction (superoxide [$O_2^{\bullet-}$], hydrogen peroxide [H_2O_2], hydroxyl radicals [$^{\bullet}OH$], and singlet oxygen [$^1O_2$]) are important cellular molecules (Gomes and Garcia, 2013). ROS have been implicated in several plant biological processes, especially as second messengers in plant hormone responses. Their functions in programmed cell death (PCD) in the aleurone layer during seed germination, their role in seed germination and dormancy release, as well as in processes related to hydrogen peroxide (H_2O_2) interactions with gibberellic acid (GA) and abscisic acid (ABA) have been well documented (Fath et al., 2002, 2001; Oracz et al., 2009; Barba-Espín et al., 2011; Leymarie et al., 2012). ROS have also been implicated in auxin signaling and gravitropism (Joo et al., 2001), and ethylene and ROS are required in root nodule initiation and functioning (D'Haeze et al., 2003) and in stomatal opening and closing (Desikan et al., 2005; Tanaka et al., 2005). ROS are also known to induce gene expression and the activities of ROS-scavenging enzymes via ROS-dependent ABA signaling pathways (Hu et al., 2005, 2006, 2007). The roles of ROS in biological processes such as those cited here show them to be important biological compounds, often acting as signaling molecules under stress conditions.

2.2 ROS IN PLANT SIGNALING

ROS production is a common event during normal cell metabolism: 1) mitochondrial complexes I and II are sources of $O_2^{\bullet-}$ that can be further reduced

P. Ahmad (Ed): Oxidative Damage to Plants. DOI: http://dx.doi.org/10.1016/B978-0-12-799963-0.00002-2

to H_2O_2 by superoxide dismutase (SOD); 2) the oxygen generated during photosynthesis can be reduced by additional electrons from the photosystems, forming $O_2^{\bullet-}$, $^1O_2$ and H_2O_2 (Gill and Tuteja, 2010); 3) peroxisomes are known to produce $O_2^{\bullet-}$ (by xanthine oxidase in the organelle matrix, and using the peroxisome electron transport chain) as well as H_2O_2 (by lipid oxidation, enzymatic reactions of flavin oxidases, and the deprotonation of $O_2^{\bullet-}$) (Corpas et al., 2013); 4) the plasma membrane is an additional site of ROS production by NADP oxidases and cytochrome P450 activity (Grant and Loake, 2000). Thus, ROS are continuously produced by plant cells during aerobic metabolism. To cope with ROS production and avoid its accumulation and associated oxidative bursts, the cell contains numerous antioxidant systems. The functions of these systems are essential to regulate ROS – as oxidative damagers or oxidative messengers.

H_2O_2 appears to be the principal compound responsible for ROS signaling in plant cells (Neill et al., 2002; Petrov and Van Breusegem, 2012), as the other ROS have properties that make them improbable cellular messengers. Superoxides, for example, have negative charges at neutral pHs (Halliwell and Gutteridge, 2007) and their transport through the cell membrane is mediated by voltage-dependent anion channels (porins) found only in the outer mitochondrial membrane (Han et al., 2003). Inner mitochondrial membrane, chloroplast, and thylakoid membranes do not have these kinds of porins (Han et al., 2003) and superoxides cannot leave these compartments for signaling functions. Furthermore, the high reactivities of singlet oxygen and hydroxyl radicals make them inappropriate for signal transmission over long distances (μm) (Møller and Sweetlove, 2010). The relative stability of H_2O_2 and its ability to cross membranes using aquaporines (Bienert et al., 2007) therefore pose it as the best candidate for a plant messenger.

Ca-signaling cascades, protein phosphorylation, and MAP kinases are well-established ROS-signaling mechanisms (Hardin et al., 2009; Horbinski and Chu, 2005; Nakagami et al., 2005; Wrzaczek et al., 2013), and the perception of increased levels of H_2O_2 has been attributed to redox-sensitive transcription factors in downstream signaling cascades (Miller et al., 2008). Meta-analyses have shown, however, that stress situations are sensed in specific cellular compartments and that transcriptional responses are restricted to gene clusters specific to the source of that stress (Gadjev et al., 2006), and selective secondary ROS messengers (i.e. oxidized peptides derived from oxidatively damaged proteins) have therefore been proposed as mediating ROS-signaling in plant cells (Møller and Sweetlove, 2010).

H_2O_2 can directly modulate plant hormone biosynthesis through interactions with their precursors. In the biosynthesis of salicylic acid (SA), for example, the conversion of the immediate precursor (benzoic acid) to SA through the activation of benzoic acid 2-hydroxylase is H_2O_2-mediated (Leon et al., 1995). Plant hormones are intrinsically involved in important interactions in plant development processes and stress responses that will be presented in the following sections. ROS and hormone signaling in seed germination are discussed here first, followed by ROS and hormone interactions during plant development and in response to stress conditions.

2.3 ROS AND HORMONE SIGNALING IN SEED GERMINATION

The roles of ROS as well as their interactions with plant hormones have been well documented in seeds (Diaz-Vivancos et al., 2013). ROS and hormones actively participate in cell elongation, reserve mobilization, endosperm weakening, as well as in seed germination and dormancy. ROS can also act as messengers or transmitters of environmental signals during seed germination, assuring germination only under adequate conditions as well as protecting the seeds from infections. These processes will be discussed in the following sections.

2.3.1 Cell Elongation

After imbibition, the active embryo initiates a stage of high metabolic activity that generates high energetic demands and leads to increased cell elongation and cell divisions, with cell wall loosening continuously occurring in this context – a process in which auxins (particularly indole-3-acetic acid [IAA]) are involved.

Auxin-induced cell elongation is a consequence of cell wall loosening mediated by hydroxyl radicals ($^{\bullet}OH$). In a reaction catalyzed by cell wall-associated peroxidase (POX), auxin reacts with dioxygen (O_2) producing $O_2^{\bullet-}$, which is, in turn, readily converted to H_2O_2 (mainly through superoxide dismutase – SOD activity) (Kawano, 2003). $^{\bullet}OH$ can then easily be produced from H_2O_2 by Fenton reactions, which are catalyzed by metal ions. $^{\bullet}OH$ can also be produced by NADPH oxidase (NOX) in the plasma membrane (Mojović et al., 2004), and will contribute to cell wall loosening. As a consequence of their high reactivity, $^{\bullet}OH$ will react with organic compounds and cleave cell wall polysaccharides (Schopfer et al., 2002) (i.e. xyloglucan polymers), allowing the walls to stretch (Fry, 1998). The cell wall will be mechanically weakened and wall tensions relaxed, leading to irreversible cell extension driven by turgor pressure (Schopfer et al., 2002).

Kawano (2003) proposed a hypothetical model of the auxin signaling pathways resulting in cell elongation (Fig. 2.1). According to this author, in addition to cell wall loosening induced by auxin in the apoplast, there is a cytosolic signaling pathway in which auxin binds to the auxin receptor ubiquitine ligase E3 SCF^{TR} (the F-box protein). Downstream control of IAA/AUX transcription factors (IAA/AUX TF) through ubiquitination then occurs, leading to the regulation of auxin-inducible genes. At the same time, the H^+-pump associated with the plasma membrane is activated, resulting in

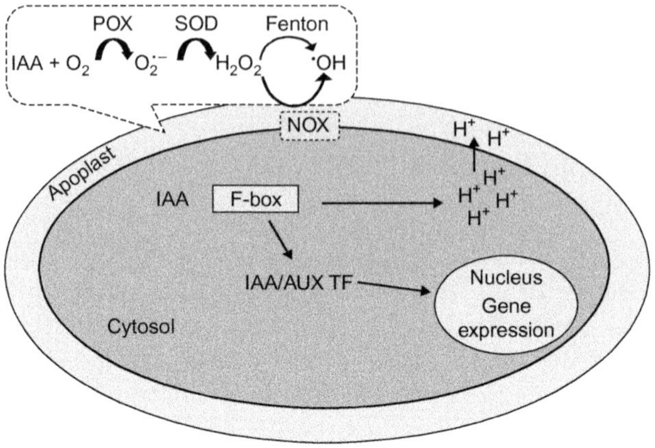

FIGURE 2.1 A model of the auxin signaling pathways resulting in plant cell elongation growth (according to Kawano, 2003). Apoplastic cell elongation induced by auxin is a consequence of cell wall loosening mediated by hydroxyl radicals (•OH). In the cytosolic path, auxin binds to the auxin receptor ubiquitine ligase E3 SCF^{TR} (F-box protein) in downstream control of the IAA/AUX transcription factors (IAA/AUX TF), leading to the regulation of auxin-inducible genes. There is simultaneously an activation of a plasma membrane associated H^+-pump, resulting in cell wall acidification.

cell wall acidification. This process may be influenced by intracellular auxin, although its exact mechanisms have not yet been identified. One hypothesis concerning this regulation is associated with ROS/Ca^{2+} signal cassette activation through the modification of the intracellular signaling pathway, using Ca^{2+} signalization (Demidchik et al., 2007). Together, the intra- and extracellular pathways of ROS/auxin signaling result in cell elongation growth.

2.3.2 Reserve Mobilization and Endosperm Weakening

Due to their high reactivity, ROS can directly react with cell components and break down polysaccharides (Schweikert et al., 2002), DNA, RNA, and fatty acids (Buetler, 2004) and facilitate their mobilization during seed germination. ROS also participate in protein carbonylation, an irreversible oxidative process that leads to losses of protein function but increases protein susceptibility to proteolytic cleavage, thus increasing their mobilization (Job et al., 2005). The interactions of ROS with plant hormones are also responsible for reserve mobilization. As discussed above, ROS and auxin are involved in cell wall loosening, which aids in polysaccharide mobilization.

ROS and plant hormones are also involved in the programmed cell death (PCD) of aleurone cells. The aleurone is the outermost layer of the endosperm and an important tissue for reserve storage. The PCD of aleurone cells provides the embryo with an energy source and nutrients. GA and ABA have antagonistic effects in this process, with GA initiating PCD and ABA

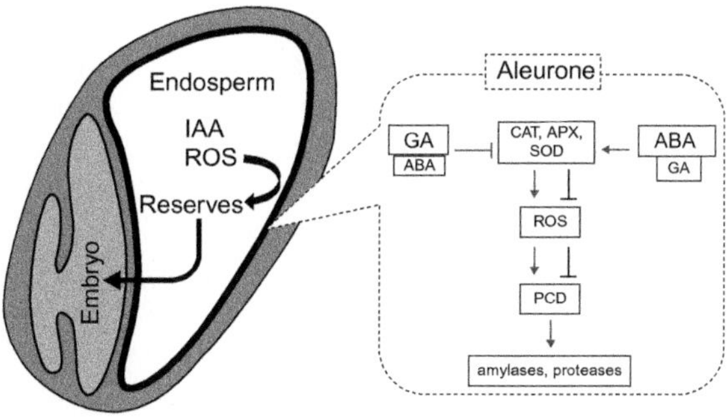

FIGURE 2.2 Mobilization of endosperm reserves and the programmed cell death (PCD) of aleurone cells in germinating seeds. ROS can directly react with cellular components and mobilize reserves (polysaccharides, DNA, RNA, fatty acids, and proteins). ROS can also interact with auxin (IAA) in cell wall loosening. ROS levels increase and PCD is initiated with high GA/ABA rations, while low GA/ABA ratios are related to low ROS levels and the blockage of PCD.

inhibiting it (Fath et al., 2001, 2002). While the activities of ROS scavenging enzymes: catalase (CAT), ascorbate peroxidase (APX) and superoxide dismutase (SOD) are downregulated by GA, they are maintained by ABA. GA therefore induces the accumulation of ROS (by inhibiting antioxidant systems), which leads to an oxidative burst and then cell death. ABA, on the other hand, maintains genic expression of antioxidant enzymes, keeping ROS levels under control (Fig. 2.2). Fath et al. (2002) found reductions in the numbers of lipid-storing oleosomes as well as increased activities of the enzymes involved in the glyoxylate cycle during PCD, suggesting that mitochondria and glyoxysomes are major sources of ROS in aleurone cells. In response to PCD initiation by GA, amylase and protease excretions are stimulated, thus mobilizing nutrients for the growing embryo (Fath et al., 2001).

The endosperm can limit germination in some seeds by acting as a mechanical barrier to radicle protrusion, so that a lessening of the mechanical resistance of the endosperm covering the radicle tip (the micropylar endosperm) is needed before radicle emergence can occur. In addition to weakening the endosperm, ROS can act in cell wall loosening, breaking down cell wall polysaccharides, regulating cellular calcium channels and calcium signalizing pathways (Carol and Dolan, 2006), and in mediating auxin functions in cell growth (Schopfer et al., 2002), as discussed earlier.

2.3.3 Seed Germination and Dormancy

H_2O_2 production has been observed during the early imbibition phase of seeds (El-Maarouf-Bouteau and Bailly, 2008; Ishibashi et al., 2013), and increased ROS concentrations have been reported during seed germination (El-Maarouf-Bouteau and Bailly, 2008) that contributed to the transition from a quiescent to a metabolically active state (Gomes and Garcia, 2013). The resumption of respiration in imbibed seeds leads to increased ROS production, making mitochondrial activity the major source of their ROS production (Pergo and Ishii-Iwamoto, 2011). ROS accumulation has been linked with the ability of a seed to germinate, as nondormant imbibed seeds showed higher H_2O_2 content than dormant imbibed seeds (Bailly et al., 2008).

Specific proteins are targeted for protein carbonylation by ROS during seed germination (Job et al., 2005; Tanou et al., 2009; Barba-Espín et al., 2011). In addition to the carbonylation of seed storage proteins (which facilitates their mobilization), other important proteins are likewise targeted for oxidation. Barba-Espín et al. (2011) reported the carbonylation of enzymes involved in glycolysis, blocking the cell's reducing power, and the stimulation of the oxidative phase of the pentose phosphate pathway (oxPPP), and the generation of another source of reducing power in the form of NADPH. The inhibition of the cytochrome respiratory pathway followed by switching to the oxPPP are both involved in breaking seed dormancy, suggesting that they are prerequisites for dormancy relief (Finch-Savage et al., 2007). Besides maintaining the redox potential necessary for protection against oxidative stress (Kruger and von Schaewen, 2003), oxPPP activation provides NADPH for the thioredoxin system involved in seed germination (Lozano et al., 1996). Specific carbonylation can also cause structural changes in proteins, altering their activities and converting them to new conformations that interact with different compounds and initiating events that stimulate seed germination (Hancock et al., 2005; Barba-Espín et al., 2011).

Germinating seeds of *Arabidopsis thaliana* showed twice the H_2O_2 concentrations of nongerminating seeds, and seed germination has been associated with the accumulation of superoxides in the radicle (Leymarie et al., 2012). While superoxide accumulation in dormant seeds is irregular, this compound specifically accumulates in the embryo radicle in nondormant seeds, indicating the presence of an efficient cell-to-cell ROS-propagation system in germinating seeds (Leymarie et al., 2012). The increases observed in the activities of ROS sources (such as NADPH-oxidase) in germinating seeds reinforce their role in seed germination (Oracz et al., 2009). The modulation of ROS-scavenging activities can also be a source of ROS during germination, and the inhibitions of CAT and SOD activities, leading to ROS accumulation, have been reported in germinating seeds (Oracz et al., 2009).

Mitogen-activated protein kinase cascade activation (MAPKs) is involved in ROS transduction pathways and has been reported to modulate seed germination (Xu et al., 2010; Barba-Espín et al., 2011; Meng and Zhang, 2013). Barba-Espín et al. (2011) noted a significant correlation between internal H_2O_2 contents and MAPK gene expression. As MAPK cascades have important roles in regulating plant growth and development, their activation could induce ROS (H_2O_2) functions during seed germination (Barba-Espín et al., 2011) (Fig. 2.3). MAPK cascades are also involved in hormone-mediated

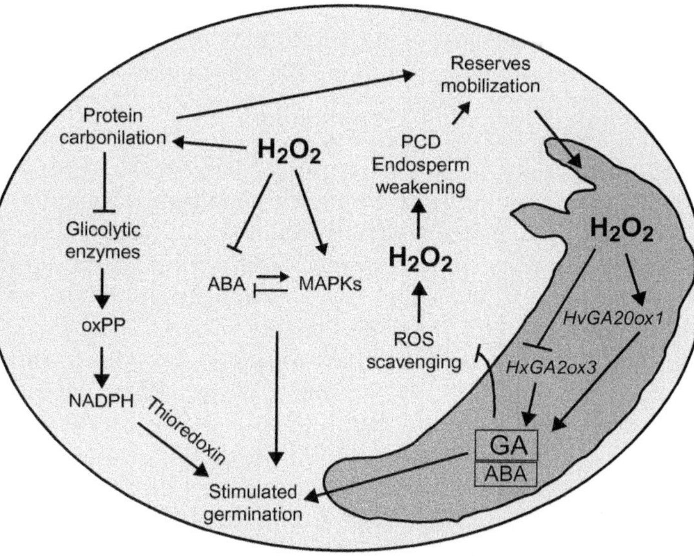

FIGURE 2.3 The role of H_2O_2 and seed hormone interactions in seed germination (modified from Braba-Espín et al., 2011). H_2O_2 decreases ABA content by way of a MAPK-dependent pathway. The carbonylation of the enzymes involved in glycolysis stimulates the oxidative phase of the pentose phosphate pathway (oxPPP), which is a source of reducing power (NADPH) for the thioredoxin system critical to seed germination (Lozano et al., 1996). H_2O_2 inhibited the expression of *HxGA2ox3* (involved in GA catabolism) in embryos and enhanced the expression of *HvGA20ox1* (implicated in GA synthesis) (Bazin et al., 2011). ROS levels increased under high GA/ABA rations, leading to PCD, endosperm weakening, and reserve mobilization.

reactions during seed germination (Xu et al., 2010). Treating dormant seeds with H_2O_2 resulted in decreased ABA levels and increased germination (Wang et al., 1998), and the interactions between H_2O_2 and ABA have been studied by Bazin et al. (2011). Although ROS accumulations have not been reported in barley embryos during after-ripening, their interaction with seed hormones have been observed (Bazin et al., 2011). The exogenous application of H_2O_2 leads to small increases in embryonic ABA content (and seed dormancy alleviation) that is related to transcription induction associated with primary dormancy (*HvNCED*). H_2O_2 did not affect ABA sensitivity, but upregulated the expression of *HvExpA11*, inhibited the expression of *HxGA2ox3,* and enhanced the expression of *HvGA20ox1* (a GA-induced gene involved in GA catabolism) as well as a gene implicated in GA synthesis. Thus, H_2O_2 was implicated in dormancy alleviation by activating GA signaling and synthesis pathways (Bazin et al., 2011).

In soybean, H_2O_2 produced in the embryonic axis after imbibitions induced the production of endogenous ethylene, promoting cell elongation in the root tip and seed germination (Ishibashi et al., 2013). Ethylene and ROS interactions were also verified in seeds of *Vigna radiata* (Chaudhuri et al., 2013). According to Chaudhuri et al. (2013), ROS produced during germination act downstream of ethylene action, allowing its positive role on seed germination.

ROS are known to take on the role of messengers or transmitters of environmental conditions during seed germination (Bailly et al., 2008). ROS production can prevent seed germination during unfavorable conditions by (for example) preventing radicle emergence (Bailly et al., 2008), thus assuring that germination occurs only under optimal conditions. Some of the functions of ROS as environmental transmitters are due to their interactions with plant hormones to alleviate seed dormancy, initiate seed germination (as previously discussed), or increase seed stress tolerance. Brassinosteroids (BRs), for example, were shown to alleviate the inhibitory effects of salinity on seed germination (Anuradha and Rao, 2001), and ROS were identified as critical components in BR-induced stress tolerance (Xia et al., 2011). BRs stimulate ROS production – which is important, in turn, to the subsequent establishment of stress tolerance (Xia et al., 2011). ROS are also used as "weapons" (toxic compounds) in protecting seeds against pathogens, and may trigger hypersensitive reactions and PCD in infected sites (Grant and Loake, 2000).

2.4 ROS, HORMONES, AND ABIOTIC STRESSES

The biosynthesis of antioxidants following the generation of oxidants in plant cells reflects the ability of plants to react to changing environments. This redox balance can be compromised by abiotic stresses such as salinity, drought, or pesticide exposure, leading to increased production of ROS causing cell damage and death (Gill and Tuteja, 2010). ROS are also known to have roles in inducing stress tolerance in plants, especially through H_2O_2, acting as secondary messengers in signaling pathways (Bhattacharjee, 2005). In addition to hormones, ROS have been co-implicated in multiple pathways among the different plant components and plant cells (Munné-Bosch et al., 2013). In the following sections, we will explore the roles of ROS and hormones in plant development under abiotic stress.

2.4.1 ABA and ROS in Guard Cells – Their Roles under Abiotic Stresses

Abscisic acid (ABA) is a key hormone for plant development and for responding to abiotic stress. In addition to its role in seed germination, ABA regulates stomatal closure in response to water stress caused by droughts, salinity, or cold temperatures, and leaf expansion (Zhang et al., 2006). In order to regulate the transpiration rate and plant-atmosphere gaseous exchanges, stomatal movements are induced through turgor pressure variations. In the case of water stress, plant roots detect drought conditions through histidine kinase osmoreceptors (ATHK1) that activate ABA biosynthesis (Harrison, 2012); this hormone is then transported to the leaves through the xylem (Osakabe et al., 2013; Zhang et al., 2006) and binds to an ATP-binding cassette transporter (Harrison, 2012). The ABA is then transferred to the stomatal guard cells where it activates ion channels on the guard cell membranes, transcription factors, and gene expression (Umezawa et al., 2010; Harrison, 2012). This activation involves several processes such as protein phosphorylation, ABA-induced ROS production (Marten et al., 2007), and hormone cross-talking; these processes are known together as the ABA signalosome or PYR/RCAR-PP2C-SnRK2 pathway (Hubbard et al., 2010; Umezawa et al., 2010; Harrison, 2012).

As the first step in this pathway, ABA binds directly to soluble receptors in the cytoplasm of guard cells – the PYR/RCAR receptors or Pyrabactin (4-bromo-N-[pyridin-2-yl methyl] naphthalene-1-sulfonamide) Resistance (PYR)/Regulatory Component of ABA Receptor (RCAR) (Hubbard et al., 2010; Harrison, 2012). These components have been investigated by Ma et al. (2009) and Park et al. (2009), and have furnished new insights into the ABA signaling pathway. In the second step, this complex binds to type 2C protein phosphatases (PP2Cs) ABI1 and ABI2, which are negative regulators of ABA signalosome (Ma et al., 2009). In the absence of ABA, ABI1 and ABI2 will inactivate the phosphorylation activity of SNF1-related kinase 2 (SnRK2). While PP2Cs are inactivated by the binding of ABA to PYR/RCAR, SnRK2 activity can occur, resulting in the activation of ion channels in the membranes of the guard cells (Harrison, 2012; Hubbard et al., 2010; Umezawa et al., 2010). This process also activates transcription factors such as AREB/ABF bZIP proteins that are involved in drought stress tolerance (Bartoli et al., 2013; Yoshida et al., 2010). The opening of anionic channels allows Cl^- and $malate^{2-}$ ions to leak out of the guard cells. The depolarization of the guard cell membranes then leads to the opening of K^+ channels, resulting in a net K^+ flux from the cytosol. Simultaneously, Ca^{2+} flows into the cytosol through vacuole and endoplasmic reticulum Ca^{2+} channels via the Ca^{2+}-dependent ABA signal transduction pathway (Campalans et al., 1999), which involves the second messenger inositol 1, 4, 5-triphosphate (IP_3) (Campalans et al., 1999; Harrison, 2012). The Ca^{2+} ions will also activate Ca^{2+}-dependent protein kinases (CDPKs), leading to the activation of anionic channels on the guard cell membranes (Harrison, 2012). The overall process reduces the negative osmotic potential of the guard cell cytosol and the turgor pressure necessary to maintain them open, resulting in stomatal closure (Raven et al., 2007).

H_2O_2 and nitric oxide (NO) are also required for the regulation of stomatal movements induced by ABA (Bright et al., 2006; Hu et al., 2006; Harrison, 2012). Indeed, H_2O_2 is implicated in the opening of Ca^{2+} channels in a manner that mimics activation by ABA, and it can also stimulate stomatal closure (Mori et al., 2001; Kohler et al., 2003). H_2O_2 will inhibit PP2Cs ABI1 and ABI2 (Kar, 2011) (the negative regulators of ABA signaling, as seen above) and also plays a role in mitogen-activated protein kinase (MAPK) cascades, leading to gene expression implicated in drought tolerance (Campalans et al., 1999; Kar, 2011). Kar (2011) proposed that H_2O_2 and NO are both capable of direct interactions with proteins via thiol modifications and may have the same targets and involvement in ABA signaling, as both induce stomatal closure. This ABA-dependent process can be compromised by removing the H_2O_2 and NO sources, as demonstrated by (Bright et al., 2006). While all of the details of H_2O_2 and NO ABA-induced interactions in the ABA signalosome are not fully known, they all appear to play major roles in drought tolerance in plants.

2.4.2 Salicylic Acid and Stomatal Closure

Salicylic acid (SA) is an important hormone mediating the induction of systemic acquired resistance (SAR – see more in Section 2.4.5) in pathogen attacks. Stomatal closure likewise appears to be important in preventing pathogen invasion through the aperture, and SA appears to have a role in that process (Mori et al., 2001). It has been proposed that SA induces the production of superoxide anions catalyzed by extracellular peroxidases in tobacco cell suspension culture following increases in cytosolic free calcium concentrations triggered by $O_2^{\bullet-}$ (Kawano and Muto, 2000).

SA induces ROS production through the following reactions: firstly, extracellular secreted guaiacol peroxidase (GPX) decomposes H_2O_2 into water and an intermediary compound I, which can be reduced by SA to compound II, with the formation of an SA radical ($SA^{\bullet}$). $SA^{\bullet}$ then reacts with O_2 dissolved in water, leading to the formation of $O_2^{\bullet-}$. By inhibiting CAT activity, SA induces H_2O_2 accumulation, therefore acting as an $O_2^{\bullet-}$ and H_2O_2 generator (Mori et al., 2001). $O_2^{\bullet-}$ generation by SA takes place in extracellular spaces and in epidermal and guard cells (which also show GPX activity) (Mori et al., 2001). H_2O_2 involved in stomatal closure can also be supplied by diffusion through the cell walls of adjacent mesophyll and epidermal cells. SA activation of Ca^{2+} channels in the plasma membrane therefore leads to increasing cytosolic concentrations of Ca^{2+} (Mori et al., 2001) and stomatal closure, as discussed in Section 2.4.1.

Due to their action in stomatal closure, SA as ABA have been proposed to mediate drought tolerance and induced ROS accumulation has been proposed as an integrator between SA and ABA signaling on the regulation of stomatal closure (Miura et al., 2012). Either SA or ABA causes a reduction of transpiration by inducing stomatal closure, thus allowing the storage of water in leaves for survival under drought conditions. Moreover, SA can stimulate ABA accumulation, and so both hormones may have synergistic effects on stomatal closure and drought tolerance acquisition (Miura et al., 2012).

2.4.3 Ethylene in PCD, Pathogen Susceptibility and Stomatal Closure

Ethylene (ET – C_2H_4) is a gaseous plant hormone that has roles in many important physiological processes, including germination, seed dormancy, ripening, senescence, stomatal closure, in defenses against biotic and abiotic stress (Bleecker and Kende, 2000; Lin et al., 2009; Barna et al., 2012a), plant survival under adverse conditions, in the control of plant growth, and in stress-induced PCD (Achard et al., 2003).

While ET induces ROS generation, hydrogen peroxide stimulates the expression of ET-responsive proteins and ET biosynthesis (by stimulating the expression and activity of enzymes involved in ET synthesis) (Vandenabeele et al., 2003). A role for ET in PCD has been reported in leaf abscission (Sakamoto et al., 2008). During this process, the cells of the abscission layer must be degraded through PCD in order for the leaves to be shed. The H_2O_2 generated through NADPH oxidase during this process is triggered by ET (Sakamoto et al., 2008). ET acts upstream in ROS-dependent signaling of PCD (Chae and Lee, 2001), and the application of antioxidants, ET inhibitors, or NADPH oxidases will block leaf abscission (Sakamoto et al., 2008). ET has been implicated in modulating plant sensitivity to stress induced by UV-B radiation (A.-H.-Mackerness et al., 1999), metals (Bueso et al., 2007), and ozone (Overmyer et al., 2003). ET-insensitive mutants are less sensitive to lithium, which triggers H_2O_2 accumulation in plants (Bueso et al., 2007). Ozone treatments induce the formation of lesions and ET accumulation by increasing the expression of ethylene-synthesizing enzymes, and the application of an ethylene precursor (1-aminocyclopropane-1-carboxylic acid – ACC) to ET insensitive mutants did not appear to enhance the ozone damage seen in wide-type plants (Overmyer et al., 2003).

Ethylene and ROS have also been observed to act on pathogen susceptibility of plants (Wi et al., 2012). After shoot inoculation with the hemibiotrophic species *Phytophthora parasitica*, tobacco plants showed a biphasic production of ethylene and ROS (48 and 72 h after pathogen inoculation) mediated by NADPH oxidase. Transgenic plants with inhibited biphasic production of ROS were more tolerant against *P. parasitica* infection. This tolerance was attributed to the inhibition of ROS and ethylene production, which further resulted in lower activation of ROS-detoxifying enzymes (Wi et al., 2012). Accordingly, Barna et al. (2012b) verified a larger amount of ethylene on leaves of tobacco mosaic virus (TMV) infected plants, which was induced by ROS. ET or ABA can activate NADPH oxidase, the source of H_2O_2 in guard cells (Desikan et al., 2005; Grefen et al., 2008). The NADPH oxidase isoforms AtrbohD and AtrbohF have been shown to be largely responsible for H_2O_2 generation in the ET-induced stomatal closure process. Mutants deficient in these isoforms show decreased ET-induced stomatal closure. Similarly, plants with mutated ET receptors *etr1-1-* and *etr1-3* are ET-insensitive and showed less H_2O_2-dependent stomatal closure (Desikan et al., 2005; Grefen et al., 2008). Although ABA-induced H_2O_2 accumulation was observed in *etr1-1* mutants, the H_2O_2 formed did not lead to stomatal closure, indicating that this mutant is H_2O_2-insensitive (Desikan et al., 2005). ET or H_2O_2 do not lead to stomatal closure in the ET signaling mutants *ein2-1* and *arr2*, although the mutants are able to generate H_2O_2 following ET stimulation. The activation of AtrbohD and AtrbohF by EIN2 and ARR2 may therefore involve upstream components of ET signaling (Bright et al., 2006). ABA-induced synthesis of NO in guard cells requires the prior activation of AtrbohD and AtrbohF as well as NADPH oxidase-dependent H_2O_2 synthesis, demonstrating another interaction between the ABA and ET signaling pathways (Bright et al., 2006).

2.4.4 Jasmonic Acid in Plant Defenses and Stomatal Closure

Jasmonic acid (JA) belongs to a group of compounds formed by the oxygenation of fatty acids (Bartoli et al., 2013) in response to wounding or attacks by necrotrophic pathogens (Devoto and Turner, 2005), as JA (like ABA) mediates stomatal closure via the ROS-dependent signaling pathway (Suhita et al., 2004) (Fig. 2.4).

JA and ET have synergistic effects, with JA and ET-dependent defense pathways being activated by chewing insects and necrotrophic pathogens, while SA-defense pathways are activated by biotrophic pathogens (Kerchev et al., 2012). JA restricts the spread of PCD induced by SA in an antagonistic relationship through the suppression of SA biosynthesis and signaling, and also by attenuating ET-sensitivity (Overmyer et al., 2003). Overmyer et al. (2003) likewise reported that JA can participate in the containment of ROS-dependent lesion propagation, as JA mutants demonstrated increased ozone sensitivity that may be linked to a reduction of ET-dependent ROS generation, leading to ET-dependent lesion propagation. JA can also control growth through its effects on DELLA proteins, which physically interact with JASMONATE ZIM-domain (JAZ) proteins that activate JA signaling (for further details, see Section 2.4.7) (Robert-Seilaniantz et al., 2011).

Tissue JA content and the expression of JA-associated defensive proteins can increase in response to environmental stimuli (Rao et al., 2000). Koo et al. (2009) have shown that JA signaling pathways can be systemic and that JA accumulation and the expression of JA-responsive genes in a single leaf after wounding can be sensed by other leaves located far from the wounded site. JA-response genes encode for antioxidant molecules (including ascorbate and glutathione) and associated defense proteins (Wolucka et al., 2005; Xiang and Oliver, 1998). JA has also been implicated in the regulation of PCD triggered in response to abiotic stress. Sasaki-Sekimoto et al.

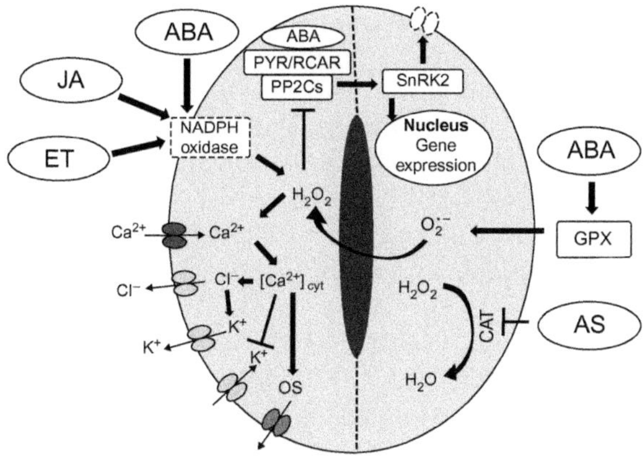

FIGURE 2.4 A schematic model of hormonal triggering of stomatal closure through ROS production. Stomatal closing is induced by net ion release and turgor pressure reductions. Abscisic acid (ABA), jasmonic acid (JA), and ethylene (ET) induce H_2O_2 production via plasma membrane NADPH oxidase. When ABA enters the cell, it forms a complex with PYR/RCAR proteins, inhibiting type 2C phosphotase (PP2Cs) proteins, as does H_2O_2. The inhibition of PP2Cs liberates SnRK2 activity in ionic channels and gene expression. In parallel, salicylic acid (SA) induces $O_2^{\bullet-}$ production via the activation of cell wall guaiacol peroxidase (GPX) and H_2O_2 accumulation, due to the inhibition of catalase (CAT) activity. H_2O_2 activates plasma membrane calcium channels, resulting in enhanced concentrations of cytosolic Ca^{2+} ($[Ca^{2+}]_{cyt}$). As a consequence, outward Cl^- channels are activated and inward K^+ channels are inactivated – resulting in membrane depolarization. Membrane polarization activates outward K^+ channels and facilitates the efflux of organic solutes (OS).

(2005) showed that JA-deficient mutants did not show the same induction of antioxidant genes observed in the wild type of *Arabidopsis*.

Like ABA, methyl jasmonate (MeJA, a biologically active derivative of JA) can induce stomatal closure mediated by ROS-dependent signaling pathways (Suhita et al., 2004). MeJA has been implicated in ROS generation in cell guards through NADPH oxidase activation (Kwak et al., 2003), although MeJA and ABA signaling pathways are transduced separately upstream of NADPH oxidase activation. MeJA did not induce stomatal closure in the JA-insensitive *Arabidopsis* mutants *jar1-1* (Suhita et al., 2004) or *coi1* (Munemasa et al., 2007), and elevations of ROS contents were not seen in these plants; ABA did, however, induce stomatal closure in these plants. Similarly, the guard cells of the ABA-insensitive mutant *ost1-2* showed both stomatal closure and ROS accumulation in response to MeJA, but were insensitive to ABA (Suhita et al., 2004).

2.4.5 Salicylic Acid and Systemic Acquired Resistance

Salicylic acid (SA) is a phenolic compound (monohydroxybenzoic acid) required in signal transduction cascades involved in plant defense mechanisms in response to stress (Bartoli et al., 2013). SA and its derivate methyl salicylate (MeSA) induce systemic acquired resistance (SAR) and hypersensitive response cell death in response to pathogen attacks (Loake and Grant, 2007). SA acts in the cellular life cycle by regulating gene expression in a process involving plasma membrane-bound NADPH-oxidases. Under stress conditions, NADPH-oxidases and cell wall peroxidases are responsible for ROS accumulations (oxidative bursts) in the extracellular spaces and the accompanying cytosolic Ca^{2+} release (Kawano and Muto, 2000).

The SA-binding protein (SABP) SABP2 of tobacco plants shows high affinity for SA (Du and Klessig, 1997). Once SA is bound to SABP2, the SABP2's MeSA esterase is inhibited, thus facilitating MeSA accumulation in the infected tissues for subsequent transport to uninfected regions (Forouhar et al., 2005; Park et al., 2007). Catalase (CAT) has also been suggested as an SA-binding protein (Dempsey et al., 1999) and, according to these authors, CAT activity is specifically inhibited by SA or by biologically active SA analogs. SA also inhibited cytosolic ascorbate peroxidase (APX) activity, leading to increased levels of H_2O_2 in SA-treated tobacco leaves. These ROS function directly as "weapons" to kill invading pathogens and/or activate cell wall cross-linking and lignification that aid in finding the invaders at the infected site (Dempsey et al., 1999). H_2O_2 was also shown to induce the expression of pathogenesis-related (PR) proteins (Chen et al., 1993). SA was thought to facilitate H_2O_2 accumulation during oxidative bursts induced by pathogen infections, although H_2O_2 accumulation was not found in tobacco during SAR (Neuencchwander et al., 1995), and H_2O_2-induced PR proteins were shown to be SA-dependent, suggesting that SA acts downstream of H_2O_2 generation (Neuencchwander et al., 1995). The relationship between SA and ROS is obviously complicated, sometimes leading to apparently contradictory observations. Indeed, several researchers have proposed that SA and H_2O_2 form a self-amplifying feedback loop, which would reconcile the conflicting views of whether SA is upstream of ROS or vice versa (Vlot et al., 2009). SA, cell death, and H_2O_2 may be regulated via an oxidative cell death loop (Overmyer et al., 2003) in which: 1) upon pathogen infection, initial H_2O_2 increases and SA synthesis is activated; then, 2) SA and ROS work together to potentiate cell death and defense gene expression; and 3) SA also potentiates H_2O_2 production, which in turn activates the synthesis of more SA and cell death in a self-amplifying loop.

2.4.6 ABA and Abiotic Stress Associated with ROS Generation and Scavenging

Abscisic acid (ABA) is a mediator of plant tolerance to a wide range of abiotic stress situations. ABA acts as a regulator of leaf senescence in response to stress provoked by water deficits (caused by factors such as drought, salt, or extreme temperatures) by regulating stomatal closure, inducing the expression of stress responsive genes, and the accumulation of osmo-compatible solutes (Sairam and Srivastava, 1999).

ROS production induced by ABA in extracellular spaces is important in inducing plant defense responses (Torres et al., 2002; Torres and Dangl, 2005). ABA induces H_2O_2 production by the plasma membrane NADPH oxidases RbohD and RbohF, followed by the activation of calcium-permeable channels and increased cytosolic Ca^{2+} concentrations – mechanisms involved in stomatal closure (Kwak et al., 2006). OST1 protein kinase plays an important role in ABA-induced ROS generation, as *ost1* mutant plants demonstrate inhibition of ABA-induced ROS production in guard cells (Kwak et al., 2006). SNF1-related protein kinase 2 (SnRK2) is an important regulator of ABA signaling in the control of plant development and responses to water stress. Phosphorylated SnRK2 is required for the activation of ABA-induced genes and often forms complexes with 2C protein phosphatase (PP2Cs) (especially in ABI1 and ABI2). PP2Cs are inactivated by H_2O_2 in ABA signaling, and PP2Cs are therefore targets for ROS signaling in the coordinated regulation of ABA-mediated responses (Kwak et al., 2006; Bailly et al., 2008). Within this context, PYR/PYL/proteins, which are important ABA receptors, function upstream of the PP2Cs-SnRK2 complexes, inhibiting the phosphatase activities of PP2Cs in the following manner: 1) ABA binds to PYR/PYL/proteins, forming a PP2Cs-sequester complex; 2) The phosphatase activities of PP2Cs are inhibited; and 3) Phosphorylated SnRK2 protein kinases activate downstream targets, including transcription factors and anion channels (Bartoli et al., 2013).

Similar to its functions in PCD, the ROS-dependent ABA signal pathway induces antioxidant enzyme activity. ABA treatment increases the activity of antioxidant enzymes (CAT, APX and SOD) and has beneficial effects on the content of the redox signaling molecules (ascorbate, glutathione, α-tocopherol and carotenoids) of maize seedlings (Jiang and Zhang, 2002). ABA also induces gene expression and SOD activities, chloroplast and cytosolic APXs, and chloroplast and cytosolic glutathione reductase (GR) (Hu et al., 2005, 2006, 2007). The upregulation of antioxidant systems could have an important role in ABA signaling.

2.4.7 GA Functions in Response to Environmental Changes

GA is also a translator of extrinsic environmental signals, such as changes in light conditions, temperature, and other environmental stress situations (Yamaguchi, 2008). Ascorbate (AsA) is a co-factor in GA synthesis (Arrigoni and De Tullio, 2000), and as AsA contents are often altered under stress conditions, GA biosynthesis and functions could likewise be affected.

GA has an important relationship with nuclear growth-repressing DELLA proteins, which are transcriptional regulators that restrain plant growth (Harberd et al., 2009). GA modulates growth in response to environmental signals by alleviating the limitations on gene expression imposed by DELLA (Harberd et al., 2009). DELLA proteins regulate plant growth and tolerance to abiotic and biotic stress conditions. DELLA proteins accumulate in plants under adverse conditions to restrain growth and enhance plant survival, and GA promotes growth by stimulating their destruction (Harberd et al., 2009). Increased DELLA contents and lower ROS accumulations due to the enhanced expression of genes encoding for antioxidant systems were demonstrated in plants under salt stress (Archard et al., 2008). *Rht-B1d* and *Rht-D1b* mutants, which show reduced sensitivity to endogenous GAs, were observed to increase their antioxidant systems and maintain high chlorophyll contents even when subjected to potassium deficiency (Moriconi et al., 2012).

Since DELLA protein accumulations were observed under stress conditions that also enhanced ABA/GA ratios (Finkelstein et al., 2008), DELLA protein effects may be associated with ABA content (as an antagonist of GA). ABA decreases ROS production, which leads to a repression of ascorbate and GA accumulation (Ye et al., 2012) and allows DELLA functions. DELLA proteins also inhibit seed germination by stimulating ABA synthesis (Kanai et al., 2010). The ratio of ABA to GA is therefore a fundamental determinant of growth or quiescence (Ross et al., 2011).

The GA/DELLA pathways can alter the balance between SA and JA/ethylene signaling, so that DELLA proteins can also contribute to pathogen resistance (Archard et al., 2008). *Arabidopsis* mutants deficient in GA show higher levels of DELLA proteins, increased resistance to necrotrophic pathogens, and much less susceptibility to ROS-dependent cell death (Achard et al., 2008). Increased SA biosynthesis and signaling were observed in DELLA protein mutants or GA (GA3) treated plants (Robert-Seilaniantz et al., 2011). A role for GA in SA-dependent responses has also been reported in plants subjected to salt, oxidative, and heat stresses (Alonso-Ramírez et al., 2009). GA promotes JA biosynthesis in a DELLA-dependent process in stamens (Robert-Seilaniantz et al., 2011).

2.4.8 Brassinosteroids in Plant Stress Responses

Brassinosteroids (BRs) are important plant hormones that regulate several processes of plant growth and development (Choudhary et al., 2012). BRs can also improve plant tolerance to stress (Xia et al., 2011) by stimulating ROS production. BRs stimulate NADPH oxidase activity, resulting in the production of ROS, and also alter the cellular redox status – leading to gene expression and stress responses (Xia et al., 2011). During development of systemic acquired resistance (SAR), ROS accumulation activates the NPR1 pathway to upregulate the *PR1* gene (Mou et al., 2003). BR-deficient plants showed downregulation of the *PR1* gene while overexpression of a BR biosynthetic gene increased *PR1* gene expression, thus demonstrating the role of BRs in regulating *PR1* expression through the alteration of the cell redox state (Szekeres et al., 1996). BRs were recently classified as hormones, but little information is currently available concerning how their signaling is linked to ROS.

2.5 CONCLUSIONS

Accumulated evidence suggests that ROS and plant hormone signaling pathways form an integrated network that regulates plant growth, development, and their responses to environmental factors. This network involves a number of crosstalk pathways, including the similar roles of ABA, ethylene, and JA in stimulating NADPH oxidase as well as ROS formation and signaling during stomatal closure. Additionally, two NADPH oxidase isoforms (AtrbohD and AtrbohF) are under the direct control of two different hormones, ABA and MeJA. Other examples of cross-interactions are: the similar degradation of AUX and SA by ROS produced by cell peroxidases; and the induction of ROS accumulation by SA, ethylene, and GA during PCD. From seed germination to stress tolerance acquisition, ROS and plant hormones are intrinsically interwoven, and the identification of new compounds as plant hormones (such as brassinosteroids), together with their functions, and the identification of ROS receptors have helped to clarify the roles and the signal interactions between ROS and hormones in plant physiology. Research in this area will be essential, as environmental stressors related to climate change and droughts, salinity and temperature variations will almost certainly grow in importance in the near future.

ACKNOWLEDGMENTS

The authors are grateful to Dr. Fabio Vieira for providing and improving the figures.

REFERENCES

A.-H.-Mackerness, S., Surplus, S.L., Blake, P., John, C.F., Buchanan-Wollaston, V., Jordan, B.R., et al., 1999. Ultraviolet-B-induced stress and changes in gene expression in *Arabidopsis thaliana*: role of signalling pathways controlled by jasmonic acid, ethylene and reactive oxygen species. Plant, Cell and Environ. 22, 1413–1423.

Achard, P., Vriezen, W.H., Straeten, D., Van Der, Harberd, N.P, 2003. Ethylene regulates Arabidopsis development via the modulation of DELLA protein growth repressor function. The Plant Cell 15, 2816–2825.

Achard, P., Renou, J.-P., Berthomé, R., Harberd, N.P., Genschik, P., 2008. Plant DELLAs restrain growth and promote survival of adversity by reducing the levels of reactive oxygen species. Curr. Biol. 18, 656–660.

Alonso-Ramírez, A., Rodríguez, D., Reyes, D., Jiménez, J.A., Nicolás, G., López-Climent, M., et al., 2009. Cross-talk between gibberellins and salicylic acid in early stress responses in *Arabidopsis thaliana* seeds. Plant Signal. Behav. 4, 750–751.

Anuradha, S., Rao, S.S.R., 2001. Effect of brassinosteroids on salinity stress induced inhibition of seed germination and seedling growth of rice (*Oryza sativa* L.). Plant Growth Regul. 151–153.

Arrigoni, O., De Tullio, M.C., 2000. The role of ascorbic acid in cell metabolism: between gene-directed functions and unpredictable chemical reactions. J. Plant Physiol. 157, 481–488.

Bailly, C., El-Maarouf-Bouteau, H., Corbineau, F., 2008. From intracellular signaling networks to cell death: the dual role of reactive oxygen species in seed physiology. C. R. Biol. 331, 806–814.

Barba-Espín, G., Diaz-Vivancos, P., Job, D., Belghazi, M., Job, C., Hernández, J.A., 2011. Understanding the role of H_2O_2 during pea seed germination: a combined proteomic and hormone profiling approach. Plant, Cell & Environ. 34, 1907–1919.

Barna, B., Fodor, J., Harrach, B.D., Pogány, M., Király, Z., 2012a. The Janus face of reactive oxygen species in resistance and susceptibility of plants to necrotrophic and biotrophic pathogens. Plant Physiol. Biochem. 59, 37–43.

Barna, B., Pogány, M., Koehl, J., Heiser, I., Elstner, E.F., 2012b. Induction of ethylene synthesis and lipid peroxidation in damaged or TMV-infected tobacco leaf tissues by light. Acta Physiologiae Plantarum 34, 1905–1914.

Bartoli, C.G., Casalongué, C.A., Simontacchi, M., Marquez-Garcia, B., Foyer, C.H., 2013. Interactions between hormone and redox signalling pathways in the control of growth and cross tolerance to stress. Environ. Exp. Bot. 94, 73–88.

Bazin, J., Langlade, N., Vincourt, P., Arribat, S., Balzergue, S., El-Maarouf-Bouteau, H., et al., 2011. Targeted mRNA oxidation regulates sunflower seed dormancy alleviation during dry after-ripening. The Plant Cell 23, 2196–2208.

Bhattacharjee, S., 2005. Reactive oxygen species and oxidative burst: roles in stress, senescence and signal transduction in plants. Curr. Sci. 89, 1113–1121.

Bienert, G.P., Møller, A.L.B., Kristiansen, K.A., Schulz, A., Møller, I.M., Schjoerring, J.K., et al., 2007. Specific aquaporins facilitate the diffusion of hydrogen peroxide across membranes. J. Biol. Chem. 282, 1183–1192.

Bleecker, A.B., Kende, H, 2000. Ethylene: a gaseous signal molecule in plants. Annu. Rev. Cell Dev. Biol. 16, 1–18.

Bright, J., Desikan, R., Hancock, J.T., Weir, I.S., Neill, S.J., 2006. ABA-induced NO generation and stomatal closure in *Arabidopsis* are dependent on H_2O_2 synthesis. The Plant J. 45, 113–122.

Bueso, E., Alejandro, S., Carbonell, P., Perez-Amador, M.A., Fayos, J., Bellés, J.M., et al., 2007. The lithium tolerance of the *Arabidopsis* cat2 mutant reveals a cross-talk between oxidative stress and ethylene. The Plant J. 52, 1052–1065.

Buetler, T.M., 2004. Role of superoxide as a signaling molecule. News Physiol. Sci. 19, 120–123.

Campalans, A., Messeguer, R., Goday, A., Pagès, M., 1999. Plant responses to drought, from ABA signal transduction events to the action of the induced proteins. Plant Physiol. Biochem. 37, 327–340.

Carol, R.J., Dolan, L., 2006. The role of reactive oxygen species in cell growth: lessons from root hairs. J. Exp. Bot. 57, 1829–1834.

Chae, H.., Lee, W., 2001. Ethylene- and enzyme-mediated superoxide production and cell death in carrot cells grown under carbon starvation. Plant Cell Rep. 20, 256–261.

Chaudhuri, A., Singh, K.L., Kar, R.K., S., 2013. Interaction of hormones with reactive oxygen species in regulating seed germination of *Vigna radiata* (L.) Wilczek. Plant Biochem. Physiol. 1, 1–5.

Chen, Z., Silva, H., Klessig, D.F., 1993. Active oxygen species in the induction of plant systemic acquired resistance by salicylic acid. Science 262, 1883–1886.

Choudhary, S.P., Yu, J.-Q., Yamaguchi-Shinozaki, K., Shinozaki, K., Tran, L.-S.P., 2012. Benefits of brassinosteroid crosstalk. Trends in Plant Science 17, 594–605.

Corpas, F.J., Barroso, J.B., Palma, M., Luis, A.R., 2013. Peroxisomes and their key role in cellular signaling and metabolism. In: del Río, L.A. (Ed.), Peroxisomes and their Key Role in Cellular Signaling and Metabolism. Springer, Netherlands, Dordrecht, pp. 283–298.

Demidchik, V., Shabala, S.N., Davies, J.M., 2007. Spatial variation in H_2O_2 response of *Arabidopsis thaliana* root epidermal Ca^{2+} flux and plasma membrane Ca^{2+} channels. The Plant J. 49, 377–386.

Dempsey, D.A., Shah, J., Klessig, D.F., 1999. Salicylic acid and disease resistance in plants. Crit. Rev. Plant Sci. 18, 547–575.

Desikan, R., Hancock, J.T., Bright, J., Harrison, J., Weir, I., Hooley, R., et al., 2005. A role for ETR1 in hydrogen peroxide signaling in stomatal guard cells 1. Plant Physiol. 137, 831–834.

Devoto, A., Turner, J.G., 2005. Jasmonate-regulated *Arabidopsis* stress signalling network. Physiologia Plantarum 123, 161–172.

Diaz-Vivancos, P., Barba-Espín, G., Hernández, J.A., 2013. Elucidating hormonal/ROS networks during seed germination: insights and perspectives. Plant Cell Rep. 32, 1491–1502.

Du, H., Klessig, D.F., 1997. Identification of a soluble, high-affinity salicylic acid-binding protein in tobacco. Plant Physiol. 113, 1319–1327.

D'Haeze, W., De Rycke, R., Mathis, R., Goormachtig, S., Pagnotta, S., Verplancke, C., et al., 2003. Reactive oxygen species and ethylene play a positive role in lateral root base nodulation of a semiaquatic legume. Proc. Nat. Acad. Sci. U.S.A. 100, 11789–11794.

El-Maarouf-Bouteau, H., Bailly, C., 2008. Oxidative signaling in seed germination and dormancy. Plant Signal. Behav, 175–182.

Fath, A., Bethke, P.C., Jones, R.L., 2001. Down-regulated prior to gibberellic acid-induced programmed cell death in barley aleurone 1. Plant Physiol. 126, 156–166.

Fath, A., Bethke, P., Beligni, V., Jones, R., 2002. Active oxygen and cell death in cereal aleurone cells. J. Exp. Bot. 53, 1273–1282.

Finch-Savage, W.E., Cadman, C.S.C., Toorop, P.E., Lynn, J.R., Hilhorst, H.W.M., 2007. Seed dormancy release in *Arabidopsis* Cvi by dry after-ripening, low temperature, nitrate and light shows common quantitative patterns of gene expression directed by environmentally specific sensing. The Plant J. 51, 60–78.

Finkelstein, R., Reeves, W., Ariizumi, T., Steber, C., 2008. Molecular aspects of seed dormancy. Annu. Rev. Plant. Biol. 59, 387–415.

Forouhar, F., Yang, Y., Kumar, D., Chen, Y., Fridman, E., Park, S.W., et al., 2005. Structural and biochemical studies identify tobacco SABP2 as a methyl salicylate esterase and implicate it in plant innate immunity. Proc. Nat. Acad. Sci. U.S.A. 102, 1773–1778.

Fry, S.C., 1998. Oxidative scission of plant cell wall polysaccharides by ascorbate-induced hydroxyl radicals. Biochem. J. 515, 507–515.

Gadjev, I., Vanderauwera, S., Gechev, T.S., Laloi, C., Minkov, I.N., Mittler, R., et al., 2006. Transcriptomic footprints disclose specificity of reactive oxygen species signaling in Arabidopsis. Plant Physiol. 141, 436–445.

Gill, S.S., Tuteja, N., 2010. Reactive oxygen species and antioxidant machinery in abiotic stress tolerance in crop plants. Plant Physiol. Biochem. 48, 909–930.

Gomes, M.P., Garcia, Q.S., 2013. Reactive oxygen species and seed germination. Biologia 68, 351–357.

Grant, J.J., Loake, G.J., 2000. Role of reactive oxygen intermediates and cognate redox signaling in disease resistance. Plant Physiol. 124, 21–30.

Grefen, C., Städele, K., Růzicka, K., Obrdlik, P., Harter, K., Horák, J., 2008. Subcellular localization and in vivo interactions of the *Arabidopsis thaliana* ethylene receptor family members. Mol. Plant 1, 308–320.

Halliwell, B., Gutteridge, J., 2007. Free Radicals in Biology and Medicine, fourth ed. Oxford University Press, Oxford.

Han, D., Antunes, F., Canali, R., Rettori, D., Cadenas, E., 2003. Voltage-dependent anion channels control the release of the superoxide anion from mitochondria to cytosol. J. Biol. Chem. 278, 5557–5563.

Hancock, J.T., Henson, D., Nyirenda, M., Desikan, R., Harrison, J., Lewis, M., et al., 2005. Proteomic identification of glyceraldehyde 3-phosphate dehydrogenase as an inhibitory target of hydrogen peroxide in *Arabidopsis*. Plant Physiol. Biochem. 43, 828–835.

Harberd, N.P., Belfield, E., Yasumura, Y., 2009. The angiosperm gibberellin-GID1-DELLA growth regulatory mechanism: how an "inhibitor of an inhibitor" enables flexible response to fluctuating environments. The Plant Cell 21, 1328–1339.

Hardin, S.C., Larue, C.T., Oh, M.-H., Jain, V., Huber, S.C., 2009. Coupling oxidative signals to protein phosphorylation via methionine oxidation in *Arabidopsis*. Biochem. J. 422, 305–312.

Harrison, M.A., 2012. Cross-talk between phytohormone signaling pathways under both optimal and stressful environmental conditions. In: Khan, N.A., Nazar, R., Iqbal, N., Anjum, N.A. (Eds.), Phytohormones and Abiotic Stress Tolerance in Plants SE - 2. Springer, Berlin Heidelberg, pp. 49–76.

Horbinski, C., Chu, C.T., 2005. Kinase signaling cascades in the mitochondrion: a matter of life or death. Free Radical Biol. Med. 38, 2–11.

Hu, X., Jiang, M., Zhang, A., Lu, J., 2005. Abscisic acid-induced apoplastic H_2O_2 accumulation up-regulates the activities of chloroplastic and cytosolic antioxidant enzymes in maize leaves. Planta 223, 57–68.

Hu, X., Zhang, A., Zhang, J., Jiang, M., 2006. Abscisic acid is a key inducer of hydrogen peroxide production in leaves of maize plants exposed to water stress. Plant Cell Physiol. 47, 1484–1495.

Hu, X., Jiang, M., Zhang, J., Zhang, A., Lin, F., Tan, M., 2007. Calcium-calmodulin is required for abscisic acid-induced antioxidant defense and functions both upstream and downstream of H_2O_2 production in leaves of maize (*Zea mays*) plants. New Phytol. 173, 27–38.

Hubbard, K.E., Nishimura, N., Hitomi, K., Getzoff, E.D., Schroeder, J.I., 2010. Early abscisic acid signal transduction mechanisms: newly discovered components and newly emerging questions. Genes Dev. 24, 1695–1708.

Ishibashi, Y., Koda, Y., Zheng, S.-H., Yuasa, T., Iwaya-Inoue, M., 2013. Regulation of soybean seed germination through ethylene production in response to reactive oxygen species. Ann. Bot. 111, 95–102.

Jiang, M., Zhang, J., 2002. Role of abscisic acid in water stress-induced antioxidant defense in leaves of maize seedlings. Free Radical Res. 36, 1001–1015.

Job, C., Rajjou, L., Lovigny, Y., Belghazi, M., Job, D., 2005. Patterns of protein oxidation in *Arabidopsis* seeds and during germination. Plant Physiol. 138, 790–802.

Joo, J.H., Bae, Y.S., Lee, J.S., 2001. Role of auxin-induced reactive oxygen species in root gravitropism. Plant Physiol. 126, 1055–1060.

Kanai, M., Nishimura, M., Hayashi, M., 2010. A peroxisomal ABC transporter promotes seed germination by inducing pectin degradation under the control of ABI5. The Plant J. 62, 936–947.

Kar, R.K., 2011. Plant responses to water stress: role of reactive oxygen species. Plant Signal. Behav. 6, 1741–1745.

Kawano, T., 2003. Roles of the reactive oxygen species-generating peroxidase reactions in plant defense and growth induction. Plant Cell Rep. 21, 829–837.

Kawano, T., Muto, S., 2000. Mechanism of peroxidase actions for salicylic acid-induced generation of active oxygen species and an increase in cytosolic calcium in tobacco cell suspension culture. J. Exp. Bot. 51, 685–693.

Kerchev, P.I., Fenton, B., Foyer, C.H., Hancock, R.D., 2012. Plant responses to insect herbivory: interactions between photosynthesis, reactive oxygen species and hormonal signalling pathways. Plant, Cell Env. 35, 441–453.

Kohler, B., Hills, A., Blatt, M.R., 2003. Control of guard cell ion channels by hydrogen peroxide and abscisic acid indicates their action through alternate signaling pathways 1. Plant Physiol. 131, 385–388.

Koo, A.J.K., Gao, X., Jones, A.D., Howe, G.A., 2009. A rapid wound signal activates the systemic synthesis of bioactive jasmonates in *Arabidopsis*. The Plant J. 59, 974–986.

Kruger, N.J., von Schaewen, A., 2003. The oxidative pentose phosphate pathway: structure and organisation. Curr. Opin. Plant Biol. 6, 236–246.

Kwak, J.M., Mori, I.C., Pei, Z.-M., Leonhardt, N., Torres, M.A., Dangl, J.L., et al., 2003. NADPH oxidase AtrbohD and AtrbohF genes function in ROS-dependent ABA signaling in *Arabidopsis*. The EMBO J. 22, 2623–2633.

Kwak, J.M., Nguyen, V., Schroeder, J.I., 2006. The role of reactive oxygen species in hormonal responses 1. Plant Physiol. 141, 323–329.

Leon, J., Lawton, M.A., Raskin, I., 1995. Hydrogen peroxide stimulates salicylic acid biosynthesis in tobacco. Plant Physiol. 108, 1673–1678.

Leymarie, J., Vitkauskaité, G., Hoang, H.H., Gendreau, E., Chazoule, V., Meimoun, P., et al., 2012. Role of reactive oxygen species in the regulation of *Arabidopsis* seed dormancy. Plant Cell Physiol. 53, 96–106.

Lin, Z., Zhong, S., Grierson, D., 2009. Recent advances in ethylene research. J. Exp. Bot. 60, 3311–3336.

Loake, G., Grant, M., 2007. Salicylic acid in plant defence – the players and protagonists. Curr. Opin. Plant Biol. 10, 466–472.

Lozano, R., Wong, J., Yee, B., Peters, A., Kobrehel, K., Buchanan, B., 1996. New evidence for a role for thioredoxin h in germination and seedling development. Planta 200, 100–106.

Ma, Y., Szostkiewicz, I., Korte, A., Moes, D., Yang, Y., Christmann, A., et al., 2009. Regulators of PP2C phosphatase activity function as abscisic acid sensors. Science 324, 1064–1068.

Marten, H., Konrad, K.R., Dietrich, P., Roelfsema, M.R.G., Hedrich, R., 2007. Ca^{2+}-dependent and -independent abscisic acid activation of plasma membrane anion channels in guard cells of *Nicotiana tabacum*. Plant Physio. 143, 28–37.

Meng, X., Zhang, S., 2013. MAPK cascades in plant disease resistance signaling. Ann. Rev. Phytopathol. 51, 245–266.

Miller, G., Shulaev, V., Mittler, R., 2008. Reactive oxygen signaling and abiotic stress. Physiologia Plantarum 133, 481–489.

Miura, K., Okamoto, H., Okuma, E., Shiba, H., Kamada, H., Hasegawa, P.M., et al., 2012. SIZ1 deficiency causes reduced stomatal aperture and enhanced drought tolerance via controlling salicylic acid-induced accumulation of reactive oxygen species in *Arabidopsis*. The Plant J. 91–104.

Mojović, M., Vuletić, M., Bacić, G.G., Vucinić, Z., 2004. Oxygen radicals produced by plant plasma membranes: an EPR spin-trap study. J. Exp. Bot. 55, 2523–2531.

Mori, I.C., Pinontoan, R., Kawano, T., Muto, S., 2001. Involvement of superoxide generation in salicylic acid-induced stomatal closure in *Vicia faba*. Plant Cell Physiol. 42, 1383–1388.

Moriconi, J.I., Buet, A., Simontacchi, M., Santa-María, G.E., 2012. Near-isogenic wheat lines carrying altered function alleles of the Rht-1 genes exhibit differential responses to potassium deprivation. Plant Sci. 185-186, 199–207.

Mou, Z., Fan, W., Dong, X., 2003. Inducers of plant systemic acquired resistance regulate NPR1 function through redox changes. Cell 113, 935–944.

Munemasa, S., Oda, K., Watanabe-Sugimoto, M., Nakamura, Y., Shimoishi, Y., Murata, Y., 2007. The coronatine-insensitive 1 mutation reveals the hormonal signaling interaction between abscisic acid and methyl jasmonate in *Arabidopsis* guard cells. Specific impairment of ion channel activation and second messenger production. Plant Physiol. 143, 1398–1407.

Munné-Bosch, S., Queval, G., Foyer, C.H., 2013. The impact of global change factors on redox signaling underpinning stress tolerance. Plant Physiol. 161, 5–19.

Møller, I.M., Sweetlove, L.J., 2010. ROS signalling – specificity is required. Trends Plant Sci. 15, 370–374.

Nakagami, H., Pitzschke, A., Hirt, H., 2005. Emerging MAP kinase pathways in plant stress signalling. Trends in Plant Sci. 10, 339–346.

Neill, S.J., Desikan, R., Clarke, A., Hurst, R.D., Hancock, J.T., 2002. Hydrogen peroxide and nitric oxide as signalling molecules in plants. J. Exp. Bot. 53, 1237–1247.

Neuencchwander, U., Vernooij, B., Friedrich, L., Uknes, S., Kessmann, H., Ryals, J., 1995. Is hydrogen peroxide a second messenger of salicylic acid in systemic acquired resistance? The Plant J. 8, 227–233.

Oracz, K., El-Maarouf-Bouteau, H., Kranner, I., Bogatek, R., Corbineau, F., Bailly, C., 2009. The mechanisms involved in seed dormancy alleviation by hydrogen cyanide unravel the role of reactive oxygen species as key factors of cellular signaling during germination. Plant Physiol. 150, 494–505.

Osakabe, Y., Yamaguchi-Shinozaki, K., Shinozaki, K., Tran, L.-S.P., 2013. Sensing the environment: key roles of membrane-localized kinases in plant perception and response to abiotic stress. J. Exp. Bot. 64, 445–458.

Overmyer, K., Brosché, M., Kangasjärvi, J., 2003. Reactive oxygen species and hormonal control of cell death. Trends Plant Sci. 8, 335–342.

Park, S.-W., Kaimoyo, E., Kumar, D., Mosher, S., Klessig, D.F., 2007. Methyl salicylate is a critical mobile signal for plant systemic acquired resistance. Science 318, 113–116.

Park, S.-Y., Fung, P., Nishimura, N., Jensen, D.R., Fujii, H., Zhao, Y., et al., 2009. Abscisic acid inhibits type 2C protein phosphatases via the PYR/PYL family of START proteins. Science 324, 1068–1071.

Pergo, E.M., Ishii-Iwamoto, E.L., 2011. Changes in energy metabolism and antioxidant defense systems during seed germination of the weed species *Ipomoea triloba L.* and the responses to allelochemicals. J. Chem. Ecol. 37, 500–513.

Petrov, V.D., Van Breusegem, F., 2012. Hydrogen peroxide – a central hub for information flow in plant cells. AoB Plants 2012, pls014.

Rao, M.V, Lee, H., Creelman, R.A., Mullet, J.E., Davis, K.R., 2000. Jasmonic acid signaling modulates ozone-induced hypersensitive cell death. The Plant Cell 12, 1633–1646.

Robert-Seilaniantz, A., Grant, M., Jones, J.D.G., 2011. Hormone crosstalk in plant disease and defense: More than just jasmonate-salicylate antagonism. Ann. Rev. Phytopathol. 49, 317–343.

Ross, J.J., Weston, D.E., Davidson, S.E., Reid, J.B., 2011. Plant hormone interactions: how complex are they? Physiol. Plant. 141, 299–309.

Sairam, V.C.R.K., Srivastava, G.C., 1999. Physiological and biochemical responses of hexaploid and tetraploid wheat to drought stress. J. Agron. Crop Sci. 108–116.

Sakamoto, M., Munemura, I., Tomita, R., Kobayashi, K., 2008. Involvement of hydrogen peroxide in leaf abscission signaling, revealed by analysis with an *in vitro* abscission system in *Capsicum* plants. The Plant J. 56, 13–27.

Sasaki-Sekimoto, Y., Taki, N., Obayashi, T., Aono, M., Matsumoto, F., Sakurai, N., et al., 2005. Coordinated activation of metabolic pathways for antioxidants and defence compounds by jasmonates and their roles in stress tolerance in *Arabidopsis*. The Plant J. 44, 653–668.

Schopfer, P., Liszkay, A., Bechtold, M., Frahry, G., Wagner, A., 2002. Evidence that hydroxyl radicals mediate auxin-induced extension growth. Planta 214, 821–828.

Schweikert, C., Liszkay, A., Schopfer, P., 2002. Polysaccharide degradation by Fenton reaction – or peroxidase-generated hydroxyl radicals in isolated plant cell walls. Phytochem. 61, 31–35.

Suhita, D., Raghavendra, A.S., Kwak, J.M., Vavasseur, A., Cadarache-dsv-devm, C.E.A., Membranaires, E., 2004. Cytoplasmic alkalization precedes reactive oxygen species production during methyl jasmonate- and abscisic acid-induced stomatal closure 1. Plant Physiol. 134, 1536–1545.

Szekeres, M., Németh, K., Koncz-Kálmán, Z., Mathur, J., Kauschmann, A., Altmann, T., et al., 1996. Brassinosteroids rescue the deficiency of CYP90, a cytochrome P450, controlling cell elongation and de-etiolation in *Arabidopsis*. Cell 85, 171–182.

Tanaka, Y., Sano, T., Tamaoki, M., Nakajima, N., Kondo, N., 2005. Ethylene inhibits abscisic acid-induced stomatal closure in *Arabidopsis* 1. Plant Physiol. 138, 2337–2343.

Tanou, G., Job, C., Rajjou, L., Arc, E., Belghazi, M., Diamantidis, G., et al., 2009. Proteomics reveals the overlapping roles of hydrogen peroxide and nitric oxide in the acclimation of citrus plants to salinity. The Plant J.: For Cell Mol. Biol. 60, 795–804.

Torres, M.A., Dangl, J.L., 2005. Functions of the respiratory burst oxidase in biotic interactions, abiotic stress and development. Curr. Opin. Plant Biol. 8, 397–403.

Torres, M.A., Dangl, J.L., Jones, J.D.G., 2002. *Arabidopsis* gp91 phox homologues AtrbohD and AtrbohF are required for accumulation of reactive oxygen intermediates in the plant defense response. Proc. Nat. Acad. Sci. U.S.A. 99.

Umezawa, T., Nakashima, K., Miyakawa, T., Kuromori, T., Tanokura, M., Shinozaki, K., et al., 2010. Molecular basis of the core regulatory network in ABA responses: sensing, signaling and transport. Plant Cell Physiol. 51, 1821–1839.

Vandenabeele, S., Van Der Kelen, K., Dat, J., Gadjev, I., Boonefaes, T., Morsa, S., et al., 2003. A comprehensive analysis of hydrogen peroxide-induced gene expression in tobacco. Proc. Nat. Acad. Sci. U.S.A. 100, 16113–16118.

Vlot, A.C., Dempsey, D.A., Klessig, D.F., 2009. Salicylic acid, a multifaceted hormone to combat disease. Ann. Rev. Phytopathol. 47, 177–206.

Wang, M., van der Meulen, R.M., Visser, K., Van Schaik, H.-P., Van Duijn, B., de Boer, A.H, 1998. Effects of dormancy-breaking chemicals on ABA levels in barley grain embryos. Seed Sci. Res. 8, 129–137.

Wi, S.J., Ji, N.R., Park, K.Y., 2012. Synergistic biosynthesis of biphasic ethylene and reactive oxygen species in response to hemibiotrophic *Phytophthora parasitica* in tobacco plants. Plant Physiol. 159, 251–265.

Wisniewski, J.P., Cornille, P., Agnel, J.P., Montillet, J.L., 1999. The extensin multigene family responds differentially to superoxide or hydrogen peroxide in tomato cell cultures. FEBS Lett. 447, 264–268.

Wolucka, B.A., Goossens, A., Inzé, D., 2005. Methyl jasmonate stimulates the *de novo* biosynthesis of vitamin C in plant cell suspensions. J. Exp. Bot. 56, 2527–2538.

Wrzaczek, M., Brosché, M., Kangasjärvi, J., 2013. ROS signaling loops – production, perception, regulation. Curr. Opin. Plant Biol. 1–8.

Xia, X.-J., Zhou, Y.-H., Ding, J., Shi, K., Asami, T., Chen, Z., et al., 2011. Induction of systemic stress tolerance by brassinosteroid in *Cucumis sativus*. The New Phytol. 191, 706–720.

Xiang, C., Oliver, D.J., 1998. Glutathione metabolic genes coordinately respond to heavy metals and jasmonic acid in *Arabidopsis*. The Plant Cell 10, 1539–1550.

Xu, H., Li, K., Yang, F., Shi, Q., Wang, X., 2010. Overexpression of CsNMAPK in tobacco enhanced seed germination under salt and osmotic stresses. Mol. Bio. Rep. 37, 3157–3163.

Yamaguchi, S., 2008. Gibberellin metabolism and its regulation. Ann. Rev. Plant Bio. 59, 225–251.

Ye, N., Zhu, G., Liu, Y., Zhang, A., Li, Y., Liu, R., et al., 2012. Ascorbic acid and reactive oxygen species are involved in the inhibition of seed germination by abscisic acid in rice seeds. J. Exp. Bot. 63, 1809–1822.

Yoshida, T., Fujita, Y., Sayama, H., Kidokoro, S., Maruyama, K., Mizoi, J., et al., 2010. AREB1, AREB2, and ABF3 are master transcription factors that cooperatively regulate ABRE-dependent ABA signaling involved in drought stress tolerance and require ABA for full activation. The Plant J. 61, 672–685.

Zhang, J., Jia, W., Yang, J., Ismail, A.M., 2006. Role of ABA in integrating plant responses to drought and salt stresses. Field Crops Res. 97, 111–119.

Chapter 3

Superoxide Dismutase (SOD) and Abiotic Stress Tolerance in Plants: An Overview

Réka Szőllősi

3.1 INTRODUCTION

At the beginning of the 21st century mankind has to answer many questions in connection with decreasing fossil energy sources, increasing population, and changeable environmental circumstances, such as extreme temperatures or decreasing water supply. Plants are always exposed to unpredictable environmental factors, and as they cannot escape from the unfavorable conditions, they have to tolerate them or they will die. Since plants are the primary producers of the food chain due to photosynthesis, it is useful for the consumers that plants have evolved several strategies to survive the threat caused by biotic (e.g. pathogens) or abiotic stress factors such as water stress, salt stress or heavy metals (Maksymiec, 2007). Moreover, stress factors often overlap and may enhance each other's effects, potentially provoking remarkable damage to crops (Mittler, 2006). Naturally, this acclimation comprises biochemical and/or morphological changes in plants, but researchers have also developed numerous transgenic varieties to improve stress tolerance (Potters et al., 2007; reviewed by Gill and Tuteja, 2010). In addition, there are encouraging results concerning, for example, the application of salicylic acid (SA) pretreatment in salt-stressed tomato plants (Szepesi et al., 2008). Furthermore, nitric oxide (NO), an endogenous signaling factor, has been much investigated due to its key role in plant growth and development and abiotic or biotic stress responses (Kolbert et al., 2011). So, although we have a lot of data, it is not enough; most of the basic studies still focus on the consequences of abiotic stresses, namely physiological alterations and/or the accumulation and the regulation of reactive oxygen species (ROS).

P. Ahmad (Ed): Oxidative Damage to Plants. DOI: http://dx.doi.org/10.1016/B978-0-12-799963-0.00003-4

3.2 ABIOTIC STRESS FACTORS AND OXIDATIVE STRESS

Abiotic stress factors like salinity, heavy metals or UV-B radiation are similar in terms of resulting in some kind of change in the plant body. Depending on the strength of the stress, positive (e.g. acclimation) or negative effects (e.g. decreased activity of photosynthesis, growth inhibition, accelerated senescence or damage of the plant organs) can be observed (Mittler, 2006; Maksymiec, 2007). These influences are usually the consequences of the redox imbalance and the overproduction of the ROS in the cells. ROS that can be nonradical types like hydrogen peroxide (H_2O_2) and free radicals such as superoxide ($O_2^{\bullet -}$) or hydroxyl radical ($^{\bullet}OH$) are basically present in the different compartments but due to the stress and consequently the absence of the redox homeostasis, they might accumulate at higher concentrations, resulting in damage to numerous molecules like DNA, proteins, lipids, photosynthetic pigments, etc. (Dat et al., 2000; Arora et al., 2002; Jaleel et al., 2009).

As mentioned previously, ROS are produced not only in stressful conditions but at low levels, and they are used in signaling pathways including cell wall lignification. The level of ROS is strictly controlled within the plant cells since some of them can be highly reactive and toxic to biomolecules (Arora et al., 2002; Mittler, 2002; Maksymiec, 2007; Gill and Tuteja, 2010). The most familiar ROS derived from molecular oxygen by its reduction are superoxide ($O_2^{\bullet -}$), hydrogen peroxide (H_2O_2) and hydroxyl radical ($^{\bullet}OH$).

Superoxide ($O_2^{\bullet -}$) is the first species to be generated after the reduction of molecular O_2 and is considered to have strong reactivity and oxidizing ability. It arises in several cell compartments but the main superoxide producers are the photosynthetic electron transport chain in the chloroplast and the mitochondrial electron transport (respiration). Moreover, cell-wall-bound peroxidases and NADPH oxidase found in the plasma membrane can also generate $O_2^{\bullet -}$ radicals (Arora et al., 2002; Mittler, 2002). Although it has a short half-life time (2–4 μs) and is unable to diffuse, it can provoke damage to the cell membranes due to lipid peroxidation (LP) or when reacting with nitric oxide (NO) it can form peroxynitrite. Moreover, with protonation it may result in the generation of perhydroxyl radicals ($^{\bullet}HO_2$) (Dat et al., 2000; Mittler, 2002; Gill and Tuteja, 2010).

Hydrogen peroxide (H_2O_2) is not a radical, it has longer half-time (1 ms) and can originate from superoxide by dismutation mainly in the peroxisome and the apoplast. This reaction may take place spontaneously at low pH or is catalyzed enzymatically (superoxide dismutase, SOD; Arora et al., 2002). Due to its diffusibility it can reach numerous biomolecules and affect the activity of proteins oxidizing the thiol groups. Together with the superoxide radicals H_2O_2 can also be the participant of the Fenton- and Haber-Weiss-reactions producing the highly reactive hydroxyl radical ($^{\bullet}OH$) (Fig. 3.1). At the same time, H_2O_2 may act as a signaling factor, e.g. in early senescence

Fenton-reaction:

$$H_2O_2 + Cu^+/Fe^{2+} \rightarrow \bullet OH + OH^- + Cu^{2+}/Fe^{3+}$$

$$Cu^{2+}/Fe^{3+} + \bullet O_2^- \rightarrow Cu^+/Fe^{2+} + O_2$$

Haber-Weiss-reaction:

$$H_2O_2 + \bullet O_2^- \rightarrow \bullet OH + OH^- + O_2$$

FIGURE 3.1 The pathways of hydroxyl radicals ($^\bullet OH$) generation.

or in the cell wall thickening process (Inzé and Van Montagu, 1995; Dat et al., 2000; Arora et al., 2002; Mittler, 2002; Maksymiec, 2007).

Hydroxyl radical ($^\bullet OH$) is considered to be the most reactive oxidant with a very short half-life time (less than 1 μs) in the plant cells and can be easily formed due to the above-mentioned Fenton- and Haber-Weiss-reactions using H_2O_2 and $O_2^{\bullet -}$ radicals (Fig. 3.1). Its destructive ability includes the initiation of LP and the impairment of biomolecules like DNA or proteins (Dat et al., 2000; Arora et al., 2002; Gill and Tuteja, 2010).

3.3 ANTIOXIDATIVE RESPONSES, STRESS TOLERANCE IN PLANTS

Plants, like other living organisms, have evolved a large scale of chemical components and strategies to take control over the physiological ROS production or the increased level of ROS under stress conditions like drought, salinity, UV-B radiation or heavy metal contamination (Yousuf et al., 2012). These molecules, known as antioxidants, can be found in almost all cell organelles and work in a well-organized system (Gill and Tuteja, 2010). Antioxidants, which are functionally divided into two groups, namely nonenzymatic and enzymatic components, are the participants in the ROS-scavenging pathways, such as the water–water cycle in the chloroplast, the Halliwell-Asada (or ascorbate–glutathione) cycle in the chloroplast, mitochondria, peroxisomes, apoplast and the cytosol (Fig. 3.2; Inzé and Van Montagu, 1995; Arora et al., 2002; Mittler, 2002; Zagorchev et al., 2013), and the peroxisomal glutathione peroxidase cycle, as well (Mittler, 2002).

3.3.1 Nonenzymatic Antioxidants

Nonenzymatic antioxidants usually have low molecular weights and augment the enzymatic part of the defense system. For example, the lipid-soluble vitamin E (α-tocopherol) takes part in the elimination of lipid peroxides and the neutralization of $^\bullet OH$ and $O_2^{\bullet -}$ radicals or singlet oxygen ($^1O_2$), which is also highly destructive to biomolecules. However, the most important reductants that play a key role in the Halliwell-Asada cycle are water-soluble

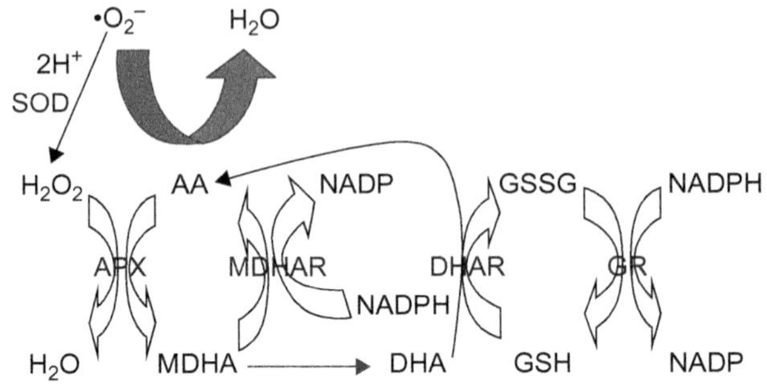

FIGURE 3.2 The ascorbate-glutathione cycle (Halliwell–Asada cycle). Blue arrows signify the nonenzymatic reactions. Abbreviations: AA- ascorbic acid, MDHA- monodehydro ascorbate, DHA- dehydroascorbate, GSH- reduced glutathione, GSSG- oxidized glutathione, APX- ascorbate peroxidase, MDHAR- monodehydroascorbate reductase, DHAR- dehydro-ascorbate reductase, GR- glutathione reductase, SOD- superoxide dismutase.

ascorbate (vitamin C) and reduced glutathione (GSH). In addition, we must mention the phenolics (flavonoids) and carotenoids (Arora et al., 2002; Blokhina et al., 2003; Jaleel et al., 2009).

3.3.1.1 Vitamin E (Tocopherols)

Tocopherols, like tocotrienols, are lipid-soluble antioxidants and membrane-associated, so therefore are good scavengers of not only $^1O_2$, $^\bullet OH$ and $O_2^{\bullet -}$ but lipid peroxides as well (Arora et al., 2002; Blokhina et al., 2003). Tocopherols have four isomers (α-, β-, γ- and δ-) but α-tocopherol has the highest antioxidant capacity. Tocopherols are synthesized by plants and are localized in the chloroplast, acting as a membrane stabilizer. Their protective role has been investigated and proved under drought, salinity or heavy metal stress (Jaleel et al., 2009; Gill and Tuteja, 2010).

3.3.1.2 Vitamin C (Ascorbate, Ascorbic Acid)

Ascorbate (vitamin C) is abundant and occurs in many cellular compartments such as chloroplasts, cytoplasm, vacuoles and apoplasts. Higher amounts of ascorbate can be found in mature leaves, mainly in mesophyll cells. It is also known that about 40% of the ascorbate is in the chloroplast and 90% of the ascorbate pool is in reduced form under physiological conditions (Arora et al., 2002; Blokhina et al., 2003; Gill and Tuteja, 2010). Vitamin C has numerous functions including cell growth regulation, division, and differentiation. Due to its ability to be an electron donor, it can detoxify singlet oxygen ($^1O_2$), $^\bullet OH$ and $O_2^{\bullet -}$ radicals, similarly to vitamin E. As the cofactor of ascorbate peroxidase (APX, see Fig. 3.2) it reduces H_2O_2 to water in the ascorbate-glutathione cycle and, furthermore, regenerates α-tocopherol from tocopheroxyl radicals (Blokhina et al., 2003; Jaleel et al., 2009; Gill and Tuteja, 2010).

3.3.1.3 Reduced Glutathione

Glutathione is a tripeptide (γ-glutamyl cysteinyl glycine) that occurs predominantly in the reduced form (GSH) while only a small amount is oxidized (GSSG) under physiological circumstances (Noctor et al., 1998). GSH is abundant in the chloroplast (1–4 mM) but is also found in the cytosol, vacuole, mitochondria, peroxisomes, endoplasmic reticulum and the apoplast. This nonprotein thiol has multiple functions such as scavenging $^1O_2$, $O_2^{\bullet -}$ and $^{\bullet}OH$ radicals and H_2O_2, hence it is the glutathione pool that determines the redox state of the plant cell and the GSH/GSSG ratio is suggested as a good indicator of the redox balance (Blokhina et al., 2003; Jaleel et al., 2009; Yadav, 2010). In addition, GSH is a remarkable attendant of the regeneration of ascorbate as the cofactor of dehydroascorbate reductase (DHAR, see ascrobate-glutathione cycle, Fig. 3.2); it then becomes reduced by glutathione reductase (GR; Noctor et al., 1998; Arora et al., 2002). Moreover, it can detoxify herbicides by conjugation, can inactivate xenobiotics as the substrate of glutathione-S-transferase and acts as a precursor of phytochelatins (PC), which form complexes with toxic heavy metals like cadmium. Last but not least, it participates in the regulation of the cell cycle, cell division, differentiation and senescence (Blokhina et al., 2003; Jaleel et al., 2009; Gill and Tuteja, 2010; Zagorchev et al., 2013).

3.3.1.4 Phenols

Phenols or polyphenols include secondary metabolites with different functions, e.g. flavonoids, which are abundant either in leaves or reproductive organs and are localized mainly in the vacuole and the apoplast; they can be more efficient antioxidants *in vitro* than ascorbate or tocopherols due to their phenolic–OH groups (Sakihama et al., 2002). Therefore, it is not surprising that plants with high flavonoid (e.g. anthocyanin) content possess higher antioxidant capacity (Moyer et al., 2002; Papp et al., 2010; Bolda et al., 2011). They are capable of being either proton or electron donors, hence they not only can participate in the quenching procedure of hydrogen peroxide but also can act as metal chelators (Hässig et al., 1999; Blokhina et al., 2003; Pourcel et al., 2007; Gill and Tuteja, 2010). As a result of this scavenging activity phenolics become oxidized to phenoxyl radicals. The regeneration of the phenoxyl radicals can be carried out via a nonenzymatic reaction with ascorbate or is catalyzed by monodehydro ascorbate reductase (MDHAR), an enzyme of the Halliwell-Asada cycle. In this case, MDHAR is primarily responsible for the reduction of monodehydro-ascorbate radicals (Sakihama et al., 2002). It has also been reported that abiotic stress factors (drought, UV-B radiation, cold or heavy metal excess) increase flavonoid synthesis in plants, which supports the protective role of these phenolic compounds (Chalker-Scott, 1999; Sakihama et al., 2002; Michalak, 2006).

3.3.1.5 *Carotenoids*

Carotenoids are lipophilic isoprenoids with the most familiar β-carotene, lycopene or zeaxanthin found mainly in leaves, flowers and fruits (Stahl and Sies, 2003). They have an accessory light-absorbing role in photosynthesis and are important for photoprotection of chlorophyll (DellaPenna and Pogson, 2006; Gill and Tuteja, 2010). Similar to tocopherols, ascorbate and GSH, carotenoids are known to have antioxidative properties by scavenging singlet oxygen in the chloroplast or peroxyl radicals produced by lipid peroxidation (Inzé and Van Montagu, 1995; Arora et al., 2002; Stahl and Sies, 2003). Since both β-carotene and α-tocopherol are lipid soluble molecules, they can work in cooperation in the antioxidative defense system, namely carotene can reduce tocopheroxyl radicals followed by the regeneration of carotenoid radicals by vitamin C (Stahl and Sies, 2003; Sharma et al., 2012).

3.3.2 Enzymatic Antioxidants

In order to control the amount of ROS, plants have evolved a large scale of enzymes, too. Some of the antioxidant enzymes like SOD or APX can be found almost everywhere in the plant cell, i.e. in the chloroplasts, cytosol, mitochondria, peroxisomes and the apoplast, while others are localized in a few compartments, e.g. catalase is definitely peroxisomal (Mittler, 2002; Blokhina et al., 2003; Jaleel et al., 2009; Sharma et al., 2012).

3.3.2.1 *Superoxide Dismutase (SOD, EC 1.15.1.1.)*

This enzyme is needed to catalyze the transformation of $O_2^{\bullet -}$ to H_2O_2 because hydrogen peroxide is less reactive and harmful, as was first published in 1969 (Giannopolitis and Ries, 1977), although it can react with superoxide radicals in the presence of transition metals, resulting in more reactive $^{\bullet}OH$ radicals (Fenton and Haber-Weiss reaction; see Fig. 3.1). SODs are metalloproteins working with Cu, Zn, Mn or Fe as cofactors and occur in the chloroplasts, mitochondria, cytosol, peroxisomes and the apoplast (Mittler, 2002). Accordingly, three isoenzymes can be separated in plants having different structures and functions: Cu/Zn-SOD, Mn-SOD and Fe-SOD (Table 3.1; Alscher et al., 2002; Gill and Tuteja, 2010; Mahanty et al., 2012). Mn-SOD is present in the mitochondria and peroxisomes, while Cu/Zn-SOD is mainly cytosolic, mitochondrial and plastidic. Similarly, Fe-SOD seems to be frequent as its localization has been described in the chloroplasts, cytosol, mitochondria and peroxisomes (Giannopolitis and Ries, 1977; Inzé and Van Montagu, 1995; Arora et al., 2002). Mn- and Fe-SODs are thought to be the most ancient types for their occurrence in both prokaryotes and eukaryotes. However, Cu/Zn-SOD has been detected predominantly in eukaryotes. The sensitivity of the isoenzymes is different, too. Fe-SOD is resistant to KCN but can be inhibited by H_2O_2. Structurally it has

TABLE 3.1 Types of SOD, Subcellular Location, Sensitivity

SOD Isoenzymes	Isolated from (Species)	Structure	Subcellular Location	Sensitivity to
Fe-SOD	**Prokaryotes** *Escherichia coli* *Mycobacterium tuberculosis* *Thiobacillus denitrificans* **Eukaryotes** *Euglena gracilis* *Nuphar luteum* *Arabidopsis thaliana* *Glycine max* *Oryza sativa* *Zea mays* *Brassica campestris* *Ginkgo biloba*	homodimeric and tetrameric	Chloroplast Cytosol Mitochondria Peroxisome	H_2O_2 but not to KCN
Mn-SOD	**Prokaryotes** *Cyanobacteria* **Eukaryotes** *Pisum sativum* *Spinacia oleracea* *Vigna mungo* *Zea mays* *Nicotiana plumbaginifolia*	homodimeric and homotetrameric	Mitochondria Peroxisome	not to KCN and H_2O_2
Cu/Zn-SOD	**Prokaryotes** *Photobacterium leiognathi* *Caulobacter crescentus* **Eukaryotes** *Avena sativa* *Citrullus lanatus* *Pisum sativum* *Spinacia oleracea* *Zea mays*	homodimeric and homotetrameric	Cytosol Chloroplast Mitochondria Peroxisome	to KCN and H_2O_2

homodimer and tetramer forms. Mn-SOD can be homodimeric and homotetrameric and is resistant to both KCN and H_2O_2, while Cu/Zn-SOD can be activated by both inhibitors (Alscher et al., 2002; Gill and Tuteja, 2010; Sharma et al., 2012). It has been reported that, like other antioxidant enzymes such as GR or APX, the activity of SOD also can be regulated by abscisic acid via gene expression (Gepstein and Glick, 2013). Many studies have reported that abiotic stress factors provoke the increased activation of SOD but we also can find contradictory data. These contradictions will be discussed later.

3.3.2.2 Catalase (CAT, EC 1.11.1.6.)

The elimination of less reactive but lipid-soluble H_2O_2 is important to avoid the reaction with superoxide radicals or transition metals. The enzymes that participate in this process are the peroxidases found in the cytosol or are bound to the cell wall and the peroxisomal or glyoxisomal catalase (Mittler, 2002; Blokhina et al., 2003; Jaleel et al., 2009). They have a tetrameric structure and many isoforms have been isolated from higher plant species like *Nicotiana plumbaginifolia, Arabidopsis thaliana* or *Zea mays*. For example, in maize three isoforms have been identified that differ with respect to the sensitivity to different environmental stresses (Dat et al., 2000; Gill and Tuteja, 2010).

3.3.2.3 Ascorbate Peroxidase (APX, EC 1.11.1.11.)

APX plays a key role in the removal of H_2O_2; therefore it is present in almost all cell organelles, similar to its substrate, ascorbic acid. Thus, it is involved both in the water–water cycle in the chloroplast and the Halliwell–Asada cycle and shows a higher affinity to hydrogen peroxide than catalase (Arora et al., 2002; Mittler, 2002; Jaleel et al., 2009; Gill and Tuteja, 2010).

3.3.2.4 Glutathione Peroxidase (GPX, EC 1.11.1.9.)

GPXs are versatile enzymatic antioxidants because they reduce not only H_2O_2 but neutralize lipid hydroperoxides, as well. Although earlier studies identified it only in the cytosol, it seems that it also exists in mitochondria, chloroplasts and endoplasmic reticulum. GPX is one of the GSH consumers, using reduced glutathione to scavenge H_2O_2, and then the regeneration of GSSG is executed by glutathione reductase (GR; Mittler, 2002; Gill and Tuteja, 2010).

3.3.2.5 Monodehydro-ascorbate Peroxidase (MDHAR, EC 1.6.5.4.) and Dehydro-ascorbate Peroxidase (DHAR, EC 1.8.5.1.)

MDHAR and DHAR are essential participants of the ascorbate-glutathione cycle as they both help ascorbate to get reduced by either monodehydro ascorbate (MDHA) or dehydroascorbate (DHA; Inzé and Van Montagu, 1995; Asada, 2006). MDHAR is found in the chloroplasts, cytosol, mitochondria and it has peroxisomal localization, too, and works with only NADH or NADPH (Becana et al., 1998; Corpas et al., 2001; Gill and Tuteja, 2010). Due to its reducing ability it is also able to detoxify phenoxyl radicals, as well (Asada, 2006). Similar to MDHAR, DHAR occurs in the previously mentioned cellular compartments but uses GSH to reduce ascorbate; consequently it also contributes to the regulation of redox balance (Mittler, 2002, Jaleel et al., 2009; Gill and Tuteja, 2010).

3.3.2.6 Glutathione Reductase (GR, EC 1.6.4.2.)

GR, located in chloroplasts, mitochondria and the cytosol, is responsible for the transformation of oxidized glutathione (GSSG) into its reduced form (GSH) and therefore it is regarded as playing an important role in maintaining the GSH pool of cells (Arora et al., 2002; Gill and Tuteja, 2010). This flavo-protein enzyme is involved in either the ascorbate–glutathione cycle or the glutathione peroxidase cycle, so it is not surprising that increase in GR activity can be observed in some types of abiotic stresses (Noctor et al., 1998; Mittler, 2002; Zagorchev et al., 2013).

3.4 SOD AND ABIOTIC STRESS FACTORS

3.4.1 Salinity

Salt stress is one of the most serious abiotic stress factors, sometimes combined with drought in arid and semi-arid fields, which may influence the quality and the amount of crops (Mittler, 2006). Several plants (halophytes) can tolerate and have been acclimated to high salt concentration of the soil or the water but most of the plants (including crops) are nontolerant and show, among other things, decreased photosynthetic activity or morphological disorders, as has been well documentated up to now (Dat et al., 2000; Meloni et al., 2003; Sheokand et al., 2008; Szepesi et al., 2008; Zsigmond et al., 2012; Duque et al., 2013). All the studies confirm that salinity, like other abiotic stressors, evokes the overproduction of ROS and changes the activities of the ROS-scavenging enzymes. Some of the results confirmed that salt stress triggers an increase in the total SOD activity while other data showed no change or the opposite tendency with respect to either total SOD activity or that of the isozymes (Table 3.2). For instance, SOD activity showed an increase in the leaves of *Cicer arietinum*, *Beta vulgaris*, and *Brassica juncea* but was diminished in those of *Vigna unguiculata* (Hernandez et al., 2006; Bor et al., 2003; Sheokand et al., 2008; Kumar et al., 2013; Rasool et al., 2013). Moreover, even in the seedlings of *Arabidopsis* and *Nicotiana* elevated SOD activity was detected (Zsigmond et al., 2012; Lee et al., 2013). These contradictions suggest that SOD activity is affected by several conditions, depending on the plant species (tolerant or not), the duration and the concentration applied and, furthermore, the plant organ used for the measurements.

3.4.2 Drought

Water deficit, in addition to salinity, may have an effect on the productivity of crops. These days it is not a marginal question, as the weather has become more changeable and unpredictable. Thus, acclimation to drought is a great challenge for nontolerant plants. It is well known that drought stress

TABLE 3.2 Changes of SOD Activity under Salt Stress, Drought, Chilling and Heat Stress

Type of Abiotic Stress Factor	Plant Name	Concentration or Degree of the Stress	Time of Exposure	Type of SOD Activated	Plant Organ Investigated	Change in SOD Activity*	Reference
Salinity (NaCl)	*Arabidopsis thaliana* L.	150 mM	0, 6 and 24 h	Total SOD	Seedling	↑	Zsigmond et al., 2012
	Beta vulgaris L. cv. Ansa (moderately tolerant)	150 and 500 mM	6–12 days	Total SOD	Leaves	↑	Bor et al., 2003
	Beta maritima TR 51196 (tolerant)	151 and 500 mM	13 days	Total SOD	Leaves	↑	Bor et al., 2003
	Brassica juncea L.	200 mM	45 days	Total SOD	Leaves	↑ ns	Kumar et al., 2013
	Cicer arietinum L. cv. HC-3	100 mM	48 h	Total SOD	Leaves	↑	Sheokand et al., 2008
	Cicer arietinum L. cv. SKUA-06 and -07 (tolerant genotypes)	25–100 mM	1–5 days	Total SOD	Leaves	↑	Rasool et al., 2013
	Gossypium hirsutum L. cv. Pora	50–200 mM	21 days	Total SOD	Leaves	↑	Meloni et al., 2003
	Hydrilla verticillata L.	1–3%	24 h	Total SOD	Entire plant	No change	Panda and Khan, 2004
	Jatropha curcas L.	100 mM	10 days	Total SOD	Leaves	↑	Silva et al., 2013
	Lycopersicon esculentum Mill. cvar. Rio Fuego	100 mM	7 days	Total SOD	Roots	↑	Szepesi et al., 2008
	Nicotiana tabacum cv. Xanthi	100 mM	55 days	Cu/Zn-SOD	Seedling	↑	Lee et al., 2013
	Phaseolus vulgaris L. genotype BAT 477	50 mM	No data	Total SOD	Roots	↓ ns	Jebara et al., 2005
				Mn-SOD		No change	
				Cu/Zn-SOD		↓	
				Fe-SOD		↑	

	Pennisetum glaucum L.	250 mM	6–24 h	Total SOD	Leaves	↑	Mahanty et al., 2012
	Pistacia vera L.	100–350 mM	21 days	Total SOD	Leaves	↑	Abbaspour, 2012
	Vigna unguiculata (L.) Walp.	35 and 100 mM	15 days	Total SOD	Leaves	↓ ns	Hernandez et al., 2006
				Mn-SOD		↓	
				Cu/Zn-SOD		↓ and no change	
				Mn-SOD	Leaf protoplast	↓	
				Cu/Zn-SOD		↓	
Drought (water deficit)	*Allium cepa* L.	suspended watering	7 days	Total SOD	Shoot	↓ ns and ↑ ns	Csiszár et al., 2007
					Roots	↓, ↑ and ↑ ns	
	Allium ascalonicum auct. hort.	suspended watering	7 days	Total SOD	Shoot	↑ and ↑ ns	
					Roots	↓ and ↑ ns	
	Allium sativum L.	suspended watering	7 days	Total SOD	Shoot	↓ ns	
					Roots	↑ and ↑ ns	
	Beta vulgaris L. cv. Felicita	10–80 g L^{-1} PEG-6000	7–28 days	Total SOD	Shoot tip tissue culture	↑	Sen and Alikamanoglu, 2013
	Brassica juncea L.	200 mM mannitol	45 days	Total SOD	Leaves	↑ ns	Kumar et al., 2013
	Cicer arietinum L. cv. ICC 4958	0.6 % PEG-6000	3 days	Cu/Zn-SOD	Roots	↑ and ↓	Ceylan et al., 2013
				Mn-SOD		↓ ns	
				Fe-SOD		↓ ns	

(*Continued*)

TABLE 3.2 (Continued)

Type of Abiotic Stress Factor	Plant Name	Concentration or Degree of the Stress	Time of Exposure	Type of SOD Activated	Plant Organ Investigated	Change in SOD Activity*	Reference
	Festuca arundinacea L.	withholding irrigation	30 days	Total SOD	Leaves	↑, later ↓	Jiang and Huang, 2001
	Gossypium herbaceum genotype RAHS 187	suspended watering	No data	Total SOD	Leaves	↑, later ↓	Deeba et al., 2012
	Hordeum vulgare L. cv. Jubilant	20% PEG-6000	3–24 h	Total SOD	Root tip	↑	Tamás et al., 2010
				Mn-SOD	Root tip	↑	
	Hydrilla verticillata L.	10–30% PEG-6000	24 h	Total SOD	Entire plant	↓ ns	Panda and Khan, 2004
	Lycopersicon esculentum Mill. cv. Lucullus (sensitive)	No data	3 weeks	Total SOD	Leaves	↑	Ünyayar et al., 2005
	Lycopersicon peruvianum (L.) Mill. (tolerant)			Total SOD	Leaves	↑	
	Nicotiana tabacum cv. Xanthi	4% PEG-8000	55 days	Cu/Zn-SOD	Seedling	↑	Lee et al., 2013
	Olea europaea L. cv. Koroneiki	withholding irrigation	37 days	Total SOD	Leaves	↑	Doupis et al., 2013
	Oryza sativa L. cv. Daeya	30% PEG-6000	2 days	Total SOD	Shoot	↓ ns	Boo and Jung, 1999
				Mn-SOD	Shoot	↓ ns	
				Cu/Zn-SOD	Shoot	↓ ns	
	Oryza sativa L. cv. Malviya-36 and Pant-12	17% and 41.2% PEG-6000	24 h	Total SOD	Roots	↑	Sharma and Dubey, 2005

				Shoot	↑	
			Mn-SOD	Shoot	↑ and ↑ ns	
			Cu/Zn-SOD	Shoot	↑ and ↑ ns	
			Fe-SOD	Shoot	↑ and ↑ ns	
Phaseolus vulgaris nanus L.	suspended watering	7 days	Total SOD	Primary leaves	↑ ns and ↓ ns	Tari et al., 2008
Pennisetum glaucum L.	withholding irrigation	8 days	Total SOD	Leaves	↑	Mahanty et al., 2012
Pinus halepensis Mill.	suspended watering	11 days	KCN-resistant SOD (Fe- + Mn-SOD)	Younger leaves	No change	Alonso et al., 2001
				Older leaves	↓	
			Cu/Zn-SOD	Younger leaves	↑	
				Older leaves	↓ ns	
Pisum sativum L. cv. Citrina	10% PEG-6000	7 days	Total SOD	Leaves	↓	Alexieva et al., 2001
Pisum sativum L. cv. Frilene	no watering	10 days	Mn-SOD	Leaves	No change	Moran et al., 1994
		Cu/Zn-SOD	Leaves	↓ ns		
Pisum sativum L. cv. Lincoln	no watering	7–9 days	Total SOD	Leaves	↑, later ↓	Iturbe-Ormaetxe et al., 1998
			Mn-SOD		↑	
			Cu/Zn-SOD		↑, later ↓	
Poa pratensis L.	withholding irrigation	30 days	Total SOD	Leaves	↑, later ↓	Jiang and Huang, 2001

(*Continued*)

TABLE 3.2 (Continued)

Type of Abiotic Stress Factor	Plant Name	Concentration or Degree of the Stress	Time of Exposure	Type of SOD Activated	Plant Organ Investigated	Change in SOD Activity*	Reference
	Triticum aestivum L. cv. Centauro	10% PEG-6000	7 days	Total SOD	Leaves	↓	Alexieva et al., 2001
	Triticum aestivum L. cv. GK Öthalom (medium drought tolerant)	100–400 mOsm (PEG- 6000)	14 days	Total SOD	Roots	↓	Gallé et al., 2013
	and cv. Kobomugi (drought tolerant)					↓	
Low temperature (chilling)	*Cucumis sativus* L. cv. Pyunggangnaebyungsamchuk	4°C	6 and 12 h	Total SOD	Leaves	↑	Lee and Lee, 2000
	Hordeum vulgare L. cv. Akhisar (sensitive)	5–7°C	3 days	Total SOD	Leaves	↓	Mutlu et al., 2013
	cv. Tokak (tolerant)			Total SOD	Leaves	↓	
	Musa acuminata coll. cv. Williams 8818 (sensitive)	5°C	3 days	Total SOD	Leaves	↓	Kang et al., 2003
	Nicotiana tabacum cv. MSk326 (sensitive)	5°C	2 and 4 days	Total SOD	Shoot	↑	Xu et al., 2010
					Root	↓	
	cv. HHDJY (tolerant)			Total SOD	Shoot	No change	
					Root	No change	
	Nicotiana tabacum cv. Petit Havana SR1	5°C	1–4 days	Total SOD	No data	No change	Gechev et al., 2003

	Nicotiana tabacum cv. Xanthi	15°C	55 days	Cu/Zn-SOD	Seedling	↑	Lee et al., 2013
	Oryza sativa L. cv. Xiangnuo-1 (tolerant) cv. IR-50 (sensitive)	8°C	1–5 days	Total SOD	Shoot	No change ↓	Huang and Guo, 2005
	Pennisetum glaucum L.	4°C	6–24 h	Total SOD	Leaves	↑	Mahanty et al., 2012
	Zea mays L. cv. Pioneer G50	4°C	7 days	Total SOD	Seedling	No change	Prasad et al., 1994
High temperature (heat)	*Cicer arietinum* L. cv. ICC 4958	37–44°C	3 h	Cu/Zn-SOD	Roots	↑ and ↓	Ceylan et al., 2013
				Mn-SOD		↑ and ↑ ns	
				Fe-SOD		↑ ns	
	Festuca arundinacea L.	30–35°C	30 days	Total SOD	Leaves	No change	Jiang and Huang, 2001
	Hydrilla verticillata L.	35–55°C	24 h	Total SOD	Entire plant	↑, later ↓	Panda and Khan, 2004
	Jatropha curcas L.	43°C	6 h	Total SOD	Leaves	↑	Silva et al., 2013
	Morus alba L. cultivars	40°C	2–6 h	Total SOD	Leaves	↑	Chaitanya et al., 2002
	Pennisetum glaucum L.	48°C	6–24 h	Total SOD	Leaves	↑	Mahanty et al., 2012
	Poa pratensis L.	30–35°C	30 days	Total SOD	Leaves	No change	Jiang and Huang, 2001

*↑ indicates significant and ↑ ns indicates nonsignificant increase, while ↓ refers to significant decrease and ↓ ns signals nonsignificant reduction in enzyme activity

generates numerous physiological alterations in plants such as a decline in CO_2 fixation due to abscisic acid-mediated stomatal closure, which leads to decreased photosynthetic activity, and/or morphological disorders (mainly reduced growth of organs) can be observed (Alexieva et al., 2001). Moreover, elevated ROS production arises, accompanied by lipid peroxidation and changes in both enzymatic and nonenzymatic antioxidants (Dat et al., 2000; Arora et al., 2002; Csiszar et al., 2007; Tari et al., 2008; Gill and Tuteja, 2010; Dinakar et al., 2012; Sen and Alikamanoglu, 2013). In plants exposed to water stress the activity of total SOD and its isoforms usually seemed to increase in the leaves of *Olea europaea* or in the shoots of *Oryza sativa* cultivars in both tolerant and nontolerant tomato species (Table 3.2; Sharma and Dubey, 2005; Ünyayar et al., 2005; Doupis et al., 2013). But in the leaves of *Gossypium* or two grasses (*Festuca arundinacea* and *Poa pratensis*) elevated activity of total SOD was followed by diminution (Jiang and Huang, 2001; Deeba et al., 2012). At the same time, there are available data on a remarkable decrease in SOD activity in pea and wheat (Alexieva et al., 2001). These results suggest that, beyond the total SOD activity, it would be more informative to investigate activity of the isozymes, too.

3.4.3 Chilling

Extreme temperatures, both high and low, might cause decreased crop yields, though some plants have acclimated. For example, pine trees can survive chilling due to their waxy cuticle on the needles. However, most of the commercial species can barely tolerate extremely low temperatures. Several morphological injuries and physiological symptoms of damage caused by low temperature have been described. Chilling can provoke functional disorders in the chloroplasts or mitochondria in sensitive plants, which unequivocally results in ROS-overproduction, i.e. oxidative stress (Prasad et al., 1994; Rivero et al., 2001). Earlier experiments indicated that remarkable H_2O_2 formation was found to occur in stressed plants during chilling and higher antioxidant enzyme activities seemed to correlate with higher tolerance (Dat et al., 2000; Lee and Lee, 2000; Huang and Guo, 2005).

In tolerant cultivars of tobacco and rice there were no changes in the activity of total SOD, while in the shoots or leaves of the sensitive types of banana, tobacco and rice it significantly decreased, similar to that of both sensitive and tolerant cultivars of barley (Table 3.2; Kang et al., 2003; Huang and Guo, 2005; Xu et al., 2010; Mutlu et al., 2013), which also supports the positive correlation between higher antioxidant enzyme activity (including SOD) and chilling tolerance.

3.4.4 Heat Stress

Extremely high temperatures may cause significant physiological and morphological destructions in plants combined with water deficit or salt stress (Mittler, 2006). Though there is a complex system for thermotolerance in plants comprising heat shock proteins regulated by heat stress transcription factors, elevated temperatures usually trigger oxidative stress as well (Dat et al., 2000; Kotak et al., 2007; Ceylan et al., 2013). It has been reported that accumulation of H_2O_2 occurs during heat stress and total phenol content increases while the activity of peroxidase diminishes in tomato exposed to higher temperatures (Dat et al., 2000; Rivero et al., 2001). We have relatively few data on the changes of SOD activity in heat-stressed plants. In two grasses (*Festuca arundinacea* and *Poa pratensis*) total SOD activity showed no alterations, while in the leaves of mulberry and *Jatropha curcas* it increased significantly, which resulted in the accumulation of H_2O_2 as it was published earlier (Table 3.2; Chaitanya et al., 2002; Jiang and Huang, 2001; Silva et al., 2013).

3.4.5 Heavy Metals

In recent years the amount of heavy metals has been elevated due to mining, industrial, and agricultural activities and the output of municipal wastes, which may result in the overproduction of ROS. Although aluminum (Al) is not a heavy metal as regards its density (2.70 g cm^{-3}), due to its toxicity it is often mentioned among the classical heavy metals such as Fe or Pb. Heavy metal stress is generally followed by physiological and/or morphological alterations in plants. Typical symptoms are chlorosis, reduced germination rate, retarded growth of vegetative organs and at toxic level necrosis of plant tissues (Dat et al., 2000; Arora et al., 2002). At the cellular level, not only toxic metals such as Pb, Hg, and Cd can cause oxidative stress in the plant cells but also essential Fe, Cu or Zn at higher concentrations than physiological level might trigger the increase of ROS generation.

Copper (Cu) is an important biometal and its mean concentration in the soil is 2–40 mg kg^{-1} DM (dry matter), but it can be accumulated in agricultural soils at 10–15-fold higher levels as a result of applying sewage sludge and fertilizers, and may trigger remarkable changes in photosynthetic activity, accompanied by reduced growth and leaf chlorosis (Hirt and Shinozaki, 2004; Janas et al., 2010; Yadav, 2010). Cu ions can cause lipid peroxidation; because of their redox activity they catalyze $^{\bullet}OH$ radical formation through the Fenton reaction and alter the activity of antioxidant enzymes (Arora et al., 2002; reviewed by Schützendübel and Polle, 2002; Szőllősi et al., 2011). Literature data show that the activity of either total SOD or Cu/Zn-SOD in leaves or roots is generally induced by excess copper (Table 3.3). For instance, in the leaves of *Arabidopsis*, *Brassica juncea* and sunflower and the roots of *Brassica juncea* and rice, Cu treatment resulted in the increase of SOD activity (Gallego et al., 1996; Chen et al., 2000; Drążkiewicz et al., 2004; Wang et al., 2004; Devi and Prasad, 2005).

Zinc (Zn) is also an essential microelement for plants; its mean concentration in soil is 10–80 mg kg^{-1} DM but in contaminated soils the range can be 150–300 mg kg^{-1} DM (Hirt and Shinozaki, 2004; Yadav, 2010). Zn is involved in N-metabolism, cell growth regulation, auxin biosynthesis and acts as a cofactor of several enzymes such as oxidoreductases, tranferases, and ligases (Rout and Das, 2003; Broadley et al., 2007). At the same time, it may cause oxidative stress, and consequently cell wall and membrane damage, lipid peroxidation or disorders in various metabolic processes (Broadley et al., 2007; Janas et al., 2010; Yadav, 2010). Plants exposed to relatively high zinc contents (1.12 mM) show chlorosis and growth inhibition (Wang et al., 2009). Similar to Cu, zinc is also a bivalent cation but redox inactive; thus, it evokes ROS overproduction indirectly, not via the Fenton reaction causing lipid peroxidation as well (Dat et al., 2000; Cuypers et al., 2002). According to data published in the literature, SOD acivity seems to be dose- and time-dependent. It was significantly induced in the shoots of *Cajanus cajan* and *Brassica juncea* but its activity was reduced in the leaves of *Brassica napus* and *Eruca sativa*, which are close relatives of Indian Mustard (Table 3.3; Prasad et al., 1999; Rao and Sresty, 2000; Wang et al., 2009; Ozdener and Aydin, 2010).

Iron (Fe) is an essential heavy metal for plants, involved in photosynthesis, N-fixation, hormone biosynthesis, and respiration and is the cofactor of antioxidant enzymes such as Fe-SOD, catalase, ascorbate peroxidase or guaiacol peroxidase (Becana et al., 1998; Shainberg et al., 2000; Arora et al., 2002). Excess Fe may result in bronzing and brownish pigmentation of leaves as visual symptoms. As a transitional metal, it induces ROS generation through the Fenton reaction and lipid peroxidation (Dat et al., 2000; Fang et al., 2001). Results which have been reported up to now show that total SOD activity may be reduced or constant in leaves but Fe-SOD can become induced after excess Fe, such as in those of bean (Table 3.3; Kampfenkel et al., 1995; Gallego et al., 1996; Shainberg et al., 2000; Sinha and Saxena, 2006).

There are relatively few data available on manganese (Mn) toxicity in plants, although this transitional metal is the cofactor of Mn-SOD and plays a key role in respiration, tricarboxylic acid cycle and N-metabolism (Alscher et al., 2002; Bueno and Piqueras, 2002; Xue et al., 2004). Relatively high (3 and 6 mM) Mn concentrations induced the activity of total SOD and SOD isoforms in the shoot of rice while 60 mM Mn partially inhibited SOD (Table 3.3; Mou et al., 2011; Srivastava and Dubey, 2011).

Nickel (Ni) is also a transitional metal and an essential microelement for plants (Gajewska et al., 2006). In natural soils its concentration range is 10–1000 mg kg^{-1} DM but it can reach 20–30-fold levels due to mining, output of sewage sludge or usage of fertilizers (Yadav, 2010). At the same

TABLE 3.3 Changes of SOD Activity under Heavy Metal Stress

Type of Abiotic Stress Factor	Plant Name	Concentration or Degree of the Stress	Time of Exposure	Type of SOD Activated	Plant Organ Investigated	Change in SOD Activity*	Reference
Heavy metals							
Cu	*Arabidopsis thaliana* L.	5–300 mM	7 days	Total SOD	Leaves	↑	Drążkiewicz et al., 2004
	Brassica juncea L.	2–16 mM	4 days	Total SOD	Roots	↑	Wang et al., 2004
		8 mM	2–8 days			↑, later ↓ ns	
	Brassica juncea L.	50–200 mM	2 days	Total SOD	Leaves	↑ ns	Devi and Prasad, 2005
					Roots	↑ ns	
	Brassica juncea L.	0.78–3.15 mM	12–96 h	Total SOD	Seedling	↑	Szőllősi et al., 2012
	Elsholtzia haichowensis Sun	0.32–500 mM	6 days	Total SOD	Roots	↑	Zhang et al., 2008
				Mn-SOD	Roots	No change	
				Cu/Zn-SOD	Roots	↑	
	Helianthus annuus L. cv. Mycosol 2	0.5 mM	12 h	Total SOD	Leaf disc	↑	Gallego et al., 1996
	Helianthus annuus L.	50 mM	5 days	Total SOD	Roots	↓	Jouili and El Ferjani, 2003
	Hydrilla verticillata L.	10–1000 mM	24 h	Total SOD	Entire plant	↑	Panda and Khan, 2004
	Oryza sativa L. cv. Taichung Native 1	30 mM	5 days	Total SOD	Roots	↑	Chen et al., 2000
	Pisum sativum L.	20 mM	4 days	Total SOD	Leaves	↑ ns	Chaoui and El Ferjani, 2005
		100 mM	4 days		Leaves	No change	

(*Continued*)

TABLE 3.3 (Continued)

Type of Abiotic Stress Factor	Plant Name	Concentration or Degree of the Stress	Time of Exposure	Type of SOD Activated	Plant Organ Investigated	Change in SOD Activity*	Reference
Zn	*Brassica napus* L. cv. Suyou No.1	0.07–1.12 mM	7 days	Total SOD	Leaves	↓	Wang et al., 2009
					Roots	↓	
	Brassica juncea L.	0.007–10 mM	10 days	Total SOD	Shoot	↑	Prasad et al., 1999
	Cajanus cajan (L.) Millspaugh	0.5–1.5 mM	6 days	Total SOD	Shoot	↑	Rao and Sresty, 2000
					Roots	↑	
	Eruca sativa L. (Rocket)	250–2000 mg kg^{-1} soil	20 days	Total SOD	Leaves	↓	Ozdener and Aydin, 2010
	Hydrilla verticillata L.	10–1000 mM	24 h	Total SOD	Entire plant	↓ ns	Panda and Khan, 2004
Fe	*Bacopa monnieri* L.	10–160 μM	24–72 h	Total SOD	Leaves	↓	Sinha and Saxena, 2006
				Total SOD	Roots	↑, later ↓ at 80 and 160 μM	
	Helianthus annuus L. cv. Mycosol 2	0.5 mM	12 h	Total SOD	Leaf disc	↓	Gallego et al., 1996
	Hydrilla verticillata (l.f.) Royle	0.5–5 μg mL^{-1}	2–7 days	Total SOD	Entire plant	↑ ns	Sinha et al., 1997
	Nicotiana plumbaginifolia L.	30 μM	12 h	Total SOD	Leaves	No change	Kampfenkel et al., 1995
	Oryza sativa L. cv. Taichung Native 1	0.01–10 mM	1–3 days	Total SOD	Leaves	↓, later ↑	Fang et al., 2001

	Phaseolus vulgaris L. cv. Pinto	900 μM	24 h	Fe-SOD	Leaves	↑	Shainberg et al., 2000
				Mn-SOD	Leaves	No change	
				Cu/Zn-SOD	Leaves	No change	
Mn	*Oryza sativa* L. cv. Pant-12	3 and 6 mM	5–20 days	Total SOD	Shoot	↑	Srivastava and Dubey, 2011
				Mn-SOD	Shoot	↑	
				Cu/Zn-SOD	Shoot	↑	
				Fe-SOD	Shoot	↑	
	Pavlova viridis (microalga)	20–200 μM	1–16 days	Total SOD	Entire plant	No change	Li et al., 2007
	Nicotiana tabacum L. cv. Bright Yellow 2	10 and 100 mM	7 days	Total SOD	Cell suspension	↓ns at 10 mM, but ↑ at 100 mM	Bueno and Piqueras, 2002
				Cu/Zn-SOD		↓ns at 10 mM, but ↑ at 100 mM	
				Fe-SOD		↓	
	Vitis vinifera L.	15–60 mM	105 days	Total SOD	Leaves	↓ at 60 mM	Mou et al., 2011
Ni	*Alyssum argenteum* L.	0.001–1.0 mM	14 days	Total SOD	Shoot	↓	Schickler and Caspi, 2002
	Alyssum maritimum L. ssp. Benthami					↑	
	Cajanus cajan (L.) Millspaugh	0.5–1.5 mM	6 days	Total SOD	Shoot	↑	Rao and Sresty, 2000
					Roots	↑	
		10–200 mM	1–9 days	Total SOD	Leaves	↓ and ↓ ns	

(*Continued*)

TABLE 3.3 (Continued)

Type of Abiotic Stress Factor	Plant Name	Concentration or Degree of the Stress	Time of Exposure	Type of SOD Activated	Plant Organ Investigated	Change in SOD Activity*	Reference
	Pisum sativum L. cv. Cud Kelvedonu						Gajewska and Skłodowska, 2005
					Roots	↓ and ↓ ns	
				Cu/Zn-SOD	Leaves	↓ and ↓ ns	
					Roots	↓ and ↓ ns	
	Triticum aestivum L. cv. Zyta	10–200 μM	6 days	Total SOD	Shoot	↓ and ↓ ns	Gajewska and Skłodowska, 2008
					Roots	↓	
	Zea mays L. cv. LG 23/01	250 μM	5 days	Total SOD	Shoot	↑, later ↓	Baccouch et al., (1998)
	Zea mays L. cv. 777	100 μM	1–14 days	Total SOD	Leaves	↑	Kumar et al., 2007
				Mn-SOD	Leaves	↑	
				Cu/Zn-SOD	Leaves	↑	
				Fe-SOD	Leaves	↑	
Co	*Brassica campestris* L. cv. T-59	100–500 μM	6 days	Total SOD	Leaves	↑	Sinha et al., 2012
					Roots	↑	
	Brassica juncea L. cv. PBR 91	0.5–2 μM	30 and 60 days	Total SOD	Leaves	↑, later ↓	Arora et al., 2012
	Pavlova viridis (microalga)	20–200 μM	1–16 days	Total SOD	Entire plant	↑, later ↓ with conc.	Li et al., 2007

	Phaseolus aureus Roxb.	50–400 μM	10 days	Total SOD	Leaves	↑ ns	Tewari et al., 2002
Cd	*Arabidopsis thaliana* L.	300 and 500 mM	21 days	Total SOD	Leaves	↓	Cho and Seo, 2005
	Brassica juncea L. cv. Varuna	25–200 mM	10 days	Total SOD	Seedling	↑ but at 100–200 mM ↓	Shekhawat et al., 2010
					Callus	↑ but at 200 mM ↓	
	Brassica juncea L.	20 mM	45 days	Total SOD	Leaves	↑ ns	Kumar et al., 2013
	Cucumis sativus L.	100–1000 mM	10 days	Total SOD	Seedling	↑	Gonçalves et al., 2007[j]
	Helianthus annuus L. cv. Mycosol 2	0.5 mM	12 h	Total SOD	Leaf disc	↓	Gallego et al., 1996
	Helianthus annuus L. cv. Mycosol 2	50–200 mM	4 days	Total SOD	Cotyledon	No change	Gallego et al., 1999
	Hydrilla verticillata L.	10–1000 mM	24 h	Total SOD	Entire plant	↑	Panda and Khan, 2004
	Hordeum vulgare L.	5 mM	1–25 days	Total SOD	Leaves	↑, later ↓	Chen et al., 2010
					Roots	↓, later ↑	
				Mn-SOD	Leaves	↓	
					Roots	↑	
				Cu/Zn-SOD	Leaves	↑	
					Roots	↑	
	Lepidium sativum L.	25–100 mg kg^{-1} soil	30 days	Total SOD	Leaves	↑	Gill et al., 2012
	Oryza sativa L. cv. Taichung Native 1	5 mM	24 h	Total SOD	Leaves	↑	Hsu and Kao, 2004
	Pisum sativum L. cv. Azad	4 and 40 mM	1, 3, 5 and 7 days	Total SOD	Leaves	↑	Dixit et al., 2001
					Roots	↓, later ↑	
	Pisum sativum L.	20 mM	4 days	Total SOD	Leaves	↓ ns	Chaoui and El Ferjani, 2005

(*Continued*)

TABLE 3.3 (Continued)

Type of Abiotic Stress Factor	Plant Name	Concentration or Degree of the Stress	Time of Exposure	Type of SOD Activated	Plant Organ Investigated	Change in SOD Activity*	Reference
		100 mM	4 days	Total SOD	Leaves	↓	
	Pisum sativum L. cv. Lincoln	50 mM	14 days	Total SOD	Roots	↑	Rodríguez-Serrano et al., 2006
				Mn-SOD	Roots	↑	
				Cu/Zn-SOD	Roots	↓	
				Fe-SOD	Roots	↑	
	Phragmites australis L.	50 mM	21 days	Total SOD	Leaves	↑	Iannelli et al., 2002
					Roots	↑	
					Stolon	↑	
	Triticum durum Desf. cv. Adamello (Cd-sensitive)	2–4 mM	10 days	Total SOD	Leaves	↓	Milone et al., 2003
					Roots	↓	
	Vigna radiata L.	25–100 mg kg^{-1} soil	30 days	Total SOD	Leaves	↑	Anjum et al., 2011
Hg	*Atriplex codonocarpa* Paul G.Wilson	0.05–1 mg L^{-1}	1–4 weeks	Total SOD	Shoot	↑	Lomonte et al., 2010
					Roots	↑, later ↓	
	Lycopersicon esculentum Mill. cv. Seokwang	10 and 50 mM	10 days	Total SOD	Leaves	↑	Cho and Park, 2000

					Roots	↑	
			20 days		Leaves	↑	
					Roots	↑	
	Sesbania drummondii	10–100 mg L^{-1}	10 days	Total SOD	Seedling	↑, later ↓	Israr et al., 2006
Cr	*Brassica juncea* L. cv. Pusa Jaikisan	0.2–20 μM	1–15 days	Total SOD	Leaves	↑ ns	Pandey et al., 2005
					Roots	↑	
	Brassica juncea L. cv. Pusa Jaikisan	50–200 mM	1–7 days	Total SOD	Shoot	↑	Diwan et al., 2010
	Hydrilla verticillata L.	10–1000 mM	24 h	Total SOD	Entire plant	↓ ns	Panda and Khan, 2004
	Ocimum tenuiflorum L.	10–100 mM	24–72 h	Total SOD	Leaves	↑, later ↓	Rai et al., 2004
	Oryza sativa L. cv. Disang	50–100 mM	24–48 h	Total SOD	Roots	↑	Panda, 2007
	Triticum aestivum L. cv. HD 2329	0.10–0.25 mM	20 days	Total SOD	Shoot	↑ ns	Subrahmanyam, 2008
	Vigna radiata L. Cv. Pusa Ratna	50–200 mM	1–7 days	Total SOD	Shoot	↑, later no change	Diwan et al., 2010
	Zea mays var. Sartaj and Deccan	50–300 μM	7 days	Total SOD	Leaves	↑	Maiti et al., 2012
Pb	*Cassia angustifolia* Vahl.	0–500 mM	120 days	Total SOD	Leaves	↑, later ↓	Qureshi et al., 2007
	Hordeum vulgare L. cv. Jubilant	0.5 mM	3–24 h	Total SOD	Root tip	↑	Tamás et al., 2010
				Mn-SOD	Root tip	↑	
	Lupinus luteus L. cv. Ventus	50–350 mg L^{-1}	1–48 h	Total SOD	Root tip	↑	Rucinska et al., 1999
	Oryza sativa L. cvs. Ratna and Jaya	500 and 1000 mM	5–20 days	Total SOD	Shoot	↑	Verma and Dubey, 2003

(*Continued*)

TABLE 3.3 (Continued)

Type of Abiotic Stress Factor	Plant Name	Concentration or Degree of the Stress	Time of Exposure	Type of SOD Activated	Plant Organ Investigated	Change in SOD Activity*	Reference
					Roots	↑	
	Sesbania drummondii	500 mg L^{-1}	4 weeks	Total SOD	Seedling	↑	Ruley et al., 2004
	Triticum aestivum L. cv. Sonalika	200–2000 μM	6 days	Total SOD	Shoot	↑	Dey et al., 2007
					Roots	↑	
Al	*Camellia sinensis* L. cv. Yabukita	400 mM	12 weeks	Total SOD	Roots	↑ ns	Ghanati et al., 2005
	Coffea arabica L.	300 mM	3 days	Total SOD	Cell suspension	↑	Bottcher et al., 2012
	Oryza sativa L. cv. Taichung Native 1	5 mM	3 days	Total SOD	Leaves	↓	Kuo and Kao, 2003
	Pisum sativum L. cv. Alaska	10 and 50 mM	24–48 h	Total SOD	Shoot	↓ and ↓ ns	Panda and Matsumoto, 2010
					Roots	↓ and ↓ ns	
				Mn-SOD	Shoot	↑, later ↓	
					Roots	↑, later ↓	
				Cu/Zn-SOD	Shoot	↑, later ↓	
					Roots	↑	
				Fe-SOD	Shoot	↑	
					Roots	↑	
	Zea mays L. S1587-17 (sensitive)	6–60 mM	48 h	Total SOD	Root tip	↑	Boscolo et al., 2003

*↑ indicates significant and ↑ ns indicates nonsignificant increase, while ↓ refers to significant decrease and ↓ ns signs nonsignificant reduction in enzyme activity

time, high levels of Ni might evoke growth inhibition, chlorosis, necrosis and lipid peroxidation or increased ROS production in the plant cells (Gajewska et al., 2006; Kumar et al., 2007). Micromolar concentrations of Ni seemed to induce SOD in maize shoots and leaves, while millimolar levels triggered a decrease in vegetative organs of *Alyssum argenteum* and pea (Table 3.3; Schickler and Caspi, 2002; Gajewska and Skłodowska, 2005; Kumar et al., 2007).

Cobalt (Co), similar to Ni, is a transitional element and a constituent of several enzymes. Cobalt toxicity has been less studied even though it may influence growth, drought tolerance, ethylene biosynthesis, as well as photosynthetic activity or the uptake and translocation of other macro- and microelements (Li et al., 2007; Yadav, 2010; Arora et al., 2012). In the vegetative parts of Co-stressed plants elevated SOD activity was exhibited, which suggests that Co-toxicity generally enhances H_2O_2 formation (Table 3.3; Tewari et al., 2002; Sinha et al., 2012).

Cadmium (Cd), known as a nonessential toxic heavy metal, generates various morphological and physiological changes and disorders in plants. Reduction of plant growth and photosynthesis, chlorosis in leaves or inhibition of seed germination are typical signs of Cd toxicity (Chaoui et al., 1997; Arora et al., 2002; Benavides et al., 2005; Yadav, 2010; Kranner and Colville, 2011). These symptoms are the probable consequences of oxidative stress, i.e. formation of ROS, increased lipid peroxidation (LP), and decreased or increased activity of enzymatic and nonenzymatic antioxidants (Gallego et al., 1999; Hasan et al., 2009; Szőllősi et al., 2009). In natural soils Cd content is estimated to be about 0.06–0.50 mg kg^{-1} DM but due to industrial, agricultural and municipal activities, cadmium emission to agricultural environment has increased (Qadir et al., 2000). Cd is not redox active, so it causes indirectly elevated ROS generation and has great affinity to bind to –SH groups of proteins, which leads to the inhibition of the enzymes (Arora et al., 2002). Schützendübel and Polle (2002) have reviewed the changes of antioxidant enzyme activities (including SOD) induced by Cd exposure. Total SOD activity occurrence is somewhat contradictory as, for example, it increases in the leaves of *Brassica juncea*, *Lepidium sativum*, pea, rice, *Phragmites australis* and *Vigna radiata* but lessens in those of *Arabidopsis* or durum wheat (Table 3.3; Dixit et al., 2001; Ianelli et al., 2002; Milone et al., 2003; Hsu and Kao, 2004; Cho and Seo, 2005; Anjum et al., 2011; Kumar et al., 2013). SOD activity is strongly influenced by Cd concentration and time of exposure, as well.

Mercury (Hg) is one of the most hazardous heavy metals for almost all living organisms, although Hg-accumulator plant species have already been described (Yadav, 2010). Nevertheless, as mercury does not have any physiological functions, it is basically phytotoxic: namely, binding to water channel proteins of root cells it induces stomatal closure in the leaves and destroys water flow. Moreover, it can disturb the uptake and translocation of nutrients, such as Co or Cd. Being a toxic metal, it also generates oxidative stress via the Fenton reaction and might cause the malfunction of enzymes by its affinity to their thiol groups or the loss of plasma membrane integrity due to lipid peroxidation (Israr et al., 2006; Lomonte et al., 2010). Not surprisingly, either at lower or higher concentrations Hg resulted in a significant increase of SOD activity in all plant organs investigated (Table 3.3; Cho and Park, 2000; Lomonte et al., 2010).

Chromium (Cr), a redox active nonessential metal, is present in soils and groundwater, hence all living organisms are threatened. Many phytotoxic effects of Cr are known including inhibited seed germination, pigment degradation, nutritional imbalance, and leaf chlorosis. Even at lower concentrations Cr may induce ROS-overgeneration via the Fenton reaction and peroxidative damage to membranes (Panda and Choudhury, 2005; Yadav, 2010). After Cr-stress SOD activity was shown to be induced in plants, which supports its toxicity for plants (Table 3.3; Rai et al., 2004; Pandey et al., 2005; Panda, 2007, Diwan et al., 2010; Maiti et al., 2012).

The concentration of lead (Pb) has risen to toxic levels in soils as a result of mining, industrial and municipal activities. Some species are tolerant of Pb and can accumulate it, but most of the plants show symptoms of toxicity such as inhibition of root growth or chlorosis. These symptoms are the consequences of water imbalance, disturbed mineral nutrition or malfunction of enzymes (Seregin and Ivanov, 2001; Qureshi et al., 2007; Yadav, 2010). Also, its presence leads to the accumulation of ROS in plant cells, so consequently Pb induces ROS-eliminating antioxidant enzymes like SOD. All reported data show that total SOD activity grows after Pb exposure (Table 3.3; Rucińska et al., 1999; Verma and Dubey, 2003; Ruley et al., 2004; Dey et al., 2007; Tamás et al., 2010).

Aluminum (Al) is abundant in the soil but it is not essential for plants. At acidic pH it forms a trivalent cation (Al^{3+}), and hence becomes toxic as well. Several studies have supported that Al toxicity results in root inhibition and severe damage to root epidermal cells, which is due to peroxidation of lipids and elevated production of $O_2^{\bullet -}$ radicals catalyzed by NADPH-dependent oxidase (Dat et al., 2000; Ezaki et al., 2008; Matsumoto and Motoda, 2012). Plants exposed to Al-stress exhibited variable changes of SOD activity (Table 3.3). At lower Al stress in pea shoots and roots, all SOD isoforms had increased activities, while the activity of total SOD decreased (Panda and Matsumoto, 2010). At the same time, much higher concentrations of Al induced SOD activity in tea and coffee plants (Ghanati et al., 2005; Bottcher et al., 2012).

3.4.6 UV B-radiation

Depletion of stratospheric ozone (O_3) has led to an increase in UV-B radiation (280−320) and its harmful effects on living organisms (Mackerness, 2000). It is well-known that UV-B influences plant growth and other metabolic processes; it may alter the pigment composition and therefore the efficiency of photosynthetic activity (Balakumar et al., 1997; Mackerness, 2000; Costa et al., 2002). Moreover, it can cause damage to proteins and DNA since it induces ROS formation and affects the activity of antioxidant enzymes such as GR or GPX (Dat et al., 2000). In accordance with this, plants irradiated by UV-B generally have significantly elevated SOD (total and isoforms) activity in the leaves, except for *Nicotiana plumbaginifolia* (Table 3.4; Willekens et al., 1994; Rao et al., 1996; Dai et al., 1997; Alexieva et al., 2001).

3.4.7 Air Pollutants

It was proven at the end of the 20th century that air pollutants may be very harmful not only to humans but to plants as well. Up to now this problem has not been solved, and as a result of industrial activities, usage of machines working with fossil fuel, etc., the emission of air pollutants has not been reduced. Ozone (O_3) and sulphur dioxide (SO_2) are considered to be the main atmopheric pollutants. Both can enter plants through the stomata and can easily cause physiological and morphological injuries in the tissues and finally in the crop yield (Dat et al., 2000; Alonso et al., 2001).

Ozone toxicitiy is related to the formation and accumulation of ROS such as $^{\bullet}O_2$ radicals and H_2O_2 (Rao et al., 1996). Enzymatic antioxidants are involved in the avoidance of oxidative burst caused by ozone. SOD is increased in the leaves of *Arabidopsis*, rice and the younger needles of *Pinus halepensis*, which suggests that the defense system is probably affected by the age of the plant (Table 3.4; Rao et al. 1996; Alonso et al., 2001; Ismail and Suroto, 2012).

SO_2 entering the leaf becomes hydrated and forms sulphur trioxide (HSO_3-), $^{\bullet}O2-$ and $^{\bullet}OH$ radicals (Mehler reaction; Dat et al., 2000; Bernardi et al., 2001). Thus, sulphur dioxide can be regarded as a hazardous pollutant. In order to scavange these reactive radicals, antioxidant molecules are involved. For instance, the induction of total SOD and Fe-SOD activities has been recorded in the leaves of bean and tolerant pea (Table 3.4; Madamanchi and Alscher, 1991; Bernardi et al., 2001).

3.5 CONCLUSIONS

Of recent times, air pollution, heavy metal contamination of water and soil, increasing drought in arid and semiarid regions, and UV-B radiation are still problems to be solved for mankind. Plants are sessile and always exposed to abiotic stress factors, so it is a challenge for them to be tolerant of and acclimate to the unfavorable conditions, since these factors often overlap and result in more serious damage of crops (Mittler, 2006). Many studies have already proved that ROS generation and redox imbalance caused by different

TABLE 3.4 Changes of SOD Activity under UV B-radiation, Ozone and Sulphur Dioxide Stress

Type of Abiotic Stress Factor	Plant Name	Concentration or Degree of the Stress	Time of Exposure	Type of SOD Activated	Plant Organ Investigated	Change in SOD Activity*	Reference
UV-B	*Arabidopsis thaliana* L.	18 kJ m^{-2} d^{-1}	5 days	Total SOD Mn-SOD Cu/Zn-SOD	Leaves	↑ ↑ ns ↑	Rao et al., 1996
	Cassia auriculata L.	7.5 and 15.0 kJ m^{-2} d^{-1}	50 and 100 min.	Total SOD	Seedling	↑	Agarwal, 2007
	Helianthus annuus L. Cv. Paraiso 5	15 and 30 kJ m^{-2} d^{-1}	50 and 100 min.	Total SOD	Cotyledon	↓	Costa et al., 2002
	Lycopersicon esculentum Mill.	6.3 kJ m^{-2} d^{-1}	15 days	Total SOD	Leaves	↑	Balakumar et al., 1997
	Nicotiana plumbaginifolia L.	8.5 kJ m^{-2} d^{-1}	3–33 h	Fe-SOD	Leaves	↓ ns	Willekens et al., 1994
	Oryza sativa L. cv. IR 74	6 and 13 kJ m^{-2} d^{-1}	28 days	Total SOD	Leaves	↑, later ↓	Dai et al., 1997
	Pisum sativum L. cv. Citrina	49 kJ m^{-2} d^{-1}	7 days	Total SOD	Leaves	↑	Alexieva et al., 2001
	Triticum aestivum L. cv. Centauro	49 kJ m^{-2} d^{-1}	7 days	Total SOD	Leaves	↑	Alexieva et al., 2001
Atmospheric pollutants							
Ozone (O_3)	*Arabidopsis thaliana* L.	220 ppb	8 days	Total SOD Mn-SOD Cu/Zn-SOD	Leaves	↑ ↑ ns ↑	Rao et al., 1996

	Nicotiana plumbaginifolia L.	120 nL L^{-1}	3–33 h and 2–6 days	Fe-SOD	Leaves	↓	Willekens et al., 1994
	Oryza sativa cv. MR 263 and MR 84	120 ppb	3 days (8h/day)	Total SOD	Leaves	↑	Ismail and Suroto, 2012
	Pinus halepensis Mill.	40 nL L^{-1}	20 months	KCN-resistant SOD (Fe- + Mn-SOD)	Younger leaves	↑	Alonso et al., 2001
					Older leaves	↑ ns	
				Cu/Zn-SOD	Younger leaves	↑	
					Older leaves	↓	
Sulphur dioxide (SO2)	*Nicotiana plumbaginifolia* L.	120 nL L^{-1}	3–33 h and 2–5 days	Fe-SOD	Leaves	No change, later ↓	Willekens et al., 1994
	Phaseolus vulgaris L. cv. Groffy	30, 60 and 90 nL L^{-1}	2–7 days	Fe-SOD	Leaves	↑	Bernardi et al., 2001
	Pisum sativum L. cv. Progress (tolerant)	0.8 mL L^{-1}	210 min	Total SOD	Leaves	↑	Madamanchi and Alscher, 1991
	cv. Nugget (sensitive)				No change		

*↑ indicates significant and ↑ ns indicates nonsignificant increase, while ↓ refers to significant decrease and ↓ ns signals nonsignificant reduction in enzyme activity

abiotic stress factors are the precedents of enzymatic disorders, reduced photosynthesis, and damage of proteins, lipids and DNA, which finally lead to retarded growth of plants and low crop yield.

The complexity of the antioxidant defense system requires basic research to continue, but it is also important to summarize the results from time to time. Change of SOD activity under abiotic stresses seems to be contradictory. This phenomenon shows that enzyme activity is dependent on numerous parameters such as plant species, plant organs investigated, concentration applied or time of exposure.

Although there are always popular topics in plant science and stress physiology is one of them, it would still be useful to refresh the data on, for example, oxidative stress triggered by air pollutants or UV-B radiation. Naturally, it is indispensable to work on transgenic varieties to understand and to improve stress tolerance (Potters et al., 2007; Gill and Tuteja, 2010). Furthermore, due to the application of salicylic acid pretreatment in salt-stressed plants or the investigation of transgenic species, we can get more information about the regulation of oxidative and nitrosative stress, which might help us to grow crops that are more tolerant of either abiotic or biotic stressors (Szepesi et al., 2008; Kolbert et al., 2011; Duque et al., 2013; Ziogas et al., 2013).

ACKNOWLEDGMENTS

I thank Dr. Gábor Laskay for correcting the English of the manuscript and Dr. Ilona Sz. Varga for scientific reviewing.

REFERENCES

Abbaspour, H., 2012. Effect of salt stress on lipid peroxidation, antioxidative enzymes, and proline accumulation in pistachio plants. J. Med. Plant. Res. 6, 526–529.

Agarwal, S., 2007. Increased antioxidant activity in *Cassia* seedlings under UV-B radiation. Biol. Plant. 51, 157–160.

Alexieva, V., Sergiev, I., Mapelli, S., Karanov, E., 2001. The effect of drought and ultraviolet radiation on growth and stress markers in pea and wheat. Plant Cell Environ. 24, 1337–1344.

Alonso, R., Elvira, S., Castillo, F.J., Gimeno, B.S., 2001. Interactive effects of ozone and drought stress on pigments and activities of antioxidative enzymes in *Pinus halepensis*. Plant Cell Environ. 24, 905–916.

Alscher, R.G., Erturk, N., Heath, L.S., 2002. Role of superoxide dismutases (SODs) in controlling oxidative stress in plants. J. Exp. Bot. 53, 1331–1341.

Anjum, N.A., Umar, S., Iqbal, M., Khan, N.A., 2011. Cadmium causes oxidative stress in mung bean by affecting the antioxidant enzyme system and ascorbate-glutathione cycle metabolism. Russ. J. Plant Physiol. 58, 92–99.

Arora, A., Sairam, R.K., Srivastava, G.C., 2002. Oxidative stress and antioxidative system in plants. Curr. Sci. 82, 1227–1238.

Arora, P., Bhardwaj, R., Kanwar, M.K., 2012. Effect of 24-epibrassinolide on growth, protein content and antioxidative defense system of *Brassica juncea* L. subjected to cobalt ion toxicity. Acta Physiol. Plant. 34, 2007–2017.

Asada, K., 2006. Production and scavenging of reactive oxygen species in chloroplasts and their functions. Plant Physiol. 141, 391–396.

Baccouch, S., Chaoui, A., El Ferjani, E., 1998. Nickel-induced oxidative damage and antioxidant responses in *Zea mays* shoots. Plant Physiol. Biochem. 36, 689–694.

Balakumar, T., Gayathri, B., Anbudurai, P.R., 1997. Oxidative stress injury in tomato plants induced by supplemental UV-B radiation. Biol. Plant. 39, 215–221.

Becana, M., Moran, J.F., Iturbe-Ormaetxe, I., 1998. Iron-dependent oxygen free radical generation in plants subjected to environmental stress: toxicity and antioxidant protection. Plant Soil 201, 137–147.

Benavides, M.P., Gallego, S.M., Tomaro, M.L., 2005. Cadmium toxicity in plants. Braz. J. Plant Physiol. 17, 21–34.

Bernardi, R., Nali, C., Gargiulo, R., Pugliesi, C., Lorenzini, G., Durante, M., 2001. Protein pattern and Fe-superoxide dismutase activity of bean plants under sulphur dioxide stress. J. Phytopathol. 149, 477–480.

Blokhina, O., Virolainen, E., Fagerstedt, K.V., 2003. Antioxidants, oxidative damage and oxygen deprivation stress: a review. Ann. Bot. 91, 179–194.

Bolda, V.V., Botau, D., Szőllősi, R., Pető, A., Gallé, Á., Tari, I., 2011. Studies on elemental composition and antioxidant capacity in callus cultures and native plants of *Vaccinium myrtillus* L. local populations. Acta Biol. Szeged. 55, 255–259.

Boo, C.Y., Jung, J., 1999. Water deficit-induced oxidative stress and antioxidative defenses in rice plants. *J. Plant Physiol.* 155, 255–261.

Bor, M., Özdemir, F., Türkan, I., 2003. The effect of salt stress in lipid peroxidation and antioxidants in leaves of sugar beet *Beta vulgaris* L. and wild beet *Beta maritima* L. Plant Sci. 164, 77–84.

Boscolo, P.R., Menossi, M., Jorge, R.A., 2003. Aluminum-induced oxidative stress in maize. Phytochemistry 62, 181–190.

Bottcher, A., Nobile, P.M., Martins, P.F., Conte, F.F., Azevedo, R.A., Mazzafera, P., 2012. Antioxidative responses of cell suspension cultures of two *Coffea arabica* varieties to low aluminum levels at pH 5.8. Hoehnea 39, 01–10.

Broadley, M.R., White, P.J., Hammond, J.P., Zelko, I., Lux, A., 2007. Zinc in plants. New Phytol. 173, 677–702.

Bueno, P., Piqueras, A., 2002. Effect of transition metals on stress, lipid peroxidation and antioxidant enzyme activities in tobacco cell cultures. Plant Growth Regul. 36, 161–167.

Ceylan, H.A., Türkan, I., Sekmen, A.H., 2013. Effect of coronatine on antioxidant enzyme response of chickpea roots to combination of PEG-induced osmotic stress and heat stress. J. Plant Growth Regul. 32, 72–82.

Chaitanya, K.V., Sundar, D., Masilamani, S., Ramachandra Reddy, A., 2002. Variation in heat stress-induced antioxidant enzyme activities among three mulberry cultivars. Plant Growth Regul. 36, 75–180.

Chalker-Scott, L., 1999. Environmental significance of anthocyanins in plant stress responses. Photochem. Photobiol. 70, 1–9.

Chaoui, A., El Ferjani, E., 2005. Effects of cadmium and copper on antioxidant capacities, lignification and auxin degradation in leaves of pea (*Pisum sativum* L.) seedlings. C. R. Biol. 328, 23–31.

Chaoui, A., Mazhoudi, S., Ghorbal, M.H., El Ferjani, E., 1997. Cadmium and zinc induction of lipid peroxidation and effects on antioxidant enzyme activities in bean (*Phaseolus vulgaris* L.). Plant Sci. 127, 139–147.

Chen, F., Wang, F., Sun, H., Cai, Y., Mao, W., Zhang, G., et al., 2010. Genotype-dependent effect of exogenous nitric oxide on Cd-induced changes in antioxidative metabolism, ultrastructure, and photosynthetic performance in barley seedlings (*Hordeum vulgare*). J. Plant Growth Regul. 29, 394–408.

Chen, L.M., Lin, C.C., Kao, C.H., 2000. Copper toxicity in rice seedlings: changes in antioxidative enzyme activities, H_2O_2 level, and cell wall peroxidase activity in roots. Bot. Bull. Acad. Sin. 41, 99–103.

Cho, U.H., Park, J.O., 2000. Mercury-induced oxidative stress in tomato seedlings. Plant Sci. 156, 1–9.

Cho, U.H., Seo, N.H., 2005. Oxidative stress in *Arabidopsis thaliana* exposed to cadmium is due to hydrogen peroxide accumulation. Plant Sci. 168, 113–120.

Corpas, F.J., Barroso, J.B., del Rio, L.A., 2001. Peroxisomes as a source of reactive oxygen species and nitric oxide signal molecules in plant cells. Trends Plant Sci. 6, 145–150.

Costa, H., Gallego, S.M., Tomaro, M.L., 2002. Effect of UV-B radiation on antioxidant defense system in sunflower cotyledons. Plant Sci. 162, 939–945.

Cuypers, A., Vangronsveld, J., Clijsters, H., 2002. Peroxidases on roots and primary leaves of *Phaseolus vulgaris* copper and zinc phytotoxicity: a comparison. J. Plant Physiol. 159, 869–876.

Csiszár, J., Lantos, E., Tari, I., Madosa, E., Wodala, B., Vashegyi, Á., et al., 2007. Antioxidant enzyme activities in *Allium* species and their cultivars under water stress. Plant Soil Environ. 53, 517–523.

Dai, Q., Yan, B., Huang, S., Liu, X., Peng, S., Miranda, M.L.L., et al., 1997. Response of oxidative stress defense systems in rice (*Oryza sativa*) leaves with supplemental UV-B radiation. Physiol. Plant. 101, 301–308.

Dat, J., Vandenabeele, S., Vranova, E., Van Montagu, M., Inzé, D., Van Breusegem, F., 2000. Dual action of the active oxygen species during plant stress responses. Cell. Mol. Life Sci. 57, 779–795.

Deeba, F., Pandey, A.K., Ranjan, S., Mishra, A., Singh, R., Sharma, Y.K., et al., 2012. Physiological and proteomic responses of cotton (*Gossypium herbaceum* L.) to drought stress. Plant Physiol. Biochem. 53, 6–18.

DellaPenna, D., Pogson, B.J., 2006. Vitamin synthesis in plants: tocopherols and carotenoids. Ann. Rev. Plant Biol. 57, 711–738.

Devi, S.R., Prasad, M.N.V., 2005. Antioxidant capacity of *Brassica juncea* plants exposed to elevated levels of copper. Russ. J. Plant Physiol. 52, 205–208.

Dey, S.K., Dey, J., Patra, S., Pothal, D., 2007. Changes in the antioxidative enzyme activities and lipid peroxidation in wheat seedlings exposed to cadmium and lead stress. Braz. J. Plant Physiol. 19, 53–60.

Dinakar, C., Djilianov, D., Bartels, D., 2012. Photosynthesis in desiccation tolerant plants: energy metabolism and antioxidative stress defense. Plant Sci. 182, 29–41.

Diwan, H., Khan, I., Ahmad, A., Iqbal, M., 2010. Induction of phytochelatins and antioxidant defence system in *Brassica juncea* and *Vigna radiata* in response to chromium treatments. Plant Growth Regul. 61, 97–107.

Dixit, V., Pandey, V., Shyam, R., 2001. Differential antioxidative responses to cadmium in roots and leaves of pea (*Pisum sativum* L. cv. Azad). J. Exp. Bot. 52, 1101–1109.

Doupis, G., Bertaki, M., Psarras, G., Kasapakis, I., Chartzoulakis, K., 2013. Water relations, physiological behavior and antioxidant defence mechanism of olive plants subjected to different irrigation regimes. Sci. Hortic. 153, 150–156.

Drążkiewicz, M., Skórzyńska-Polit, E., Krupa, Z., 2004. Copper-induced oxidative stress and antioxidant defence in *Arabidopsis thaliana*. BioMetals 17, 379–387.

Duque, A.S., de Almeida, A.M., da Silva, A.B., da Silva, J.M., Farinha, A.P., Santos, D., et al., 2013. Abiotic stress responses in plants: unraveling the complexity of genes and networks to survive. In: Vahdati, K., Lesliepp, C. (Eds.), Abiotic Stress – Plant Responses and Applications in Agriculture. InTech, pp. 49–101.

Ezaki, B., Nagao, E., Yamamoto, Y., Nakashima, S., Enomoto, T., 2008. Wild plants, *Andropogon virginicus* L. and *Miscanthus sinensis* Anders, are tolerant to multiple stresses including aluminum, heavy metals and oxidative stresses. Plant Cell Rep. 27, 951–961.

Fang, W.C., Wang, J.W., Lin, C.C., Kao, C.H., 2001. Iron induction of lipid peroxidation and effects on antioxidative enzyme activities in rice leaves. Plant Growth Regul. 35, 75–80.

Gajewska, E., Skłodowska, M., 2005. Antioxidative responses and proline level in leaves and roots of pea plants subjected to nickel stress. Acta Physiol. Plant. 27, 329–340.

Gajewska, E., Skłodowska, M., 2008. Differential biochemical responses of wheat shoots and roots to nickel stress: antioxidative reactions and proline accumulation. Plant Growth Regul. 54, 179–188.

Gajewska, E., Skłodowska, M., Słaba, M., Mazur, J., 2006. Effect of nickel on antioxidative enzyme activities, proline and chlorophyll contents in wheat shoots. Biol. Plant. 50, 653–659.

Gallé, Á., Csiszár, J., Benyó, D., Leviczky, T., Erdei, L., Tari, I., 2013. Isohydric and anisohydric strategies of wheat genotypes under osmotic stress: biosynthesis and function of ABA in stress responses. *J. Plant Physiol* 170, 1389–1399.

Gallego, S.M., Benavídes, M.P., Tomaro, M.L., 1996. Effect of heavy metal ion excess on sunflower leaves: evidence for involvement of oxidative stress. Plant Sci. 121, 151–159.

Gallego, S.M., Benavides, M.P., Tomaro, M.L., 1999. Effect of cadmium ions on antioxidant defense system in sunflower cotyledons. Biol. Plant. 42, 49–55.

Gechev, T., Willekens, H., Van Montagu, M., Inzé, D., Van Camp, W., Toneva, V., et al., 2003. Different responses of tobacco antioxidant enzymes to light and chilling stress. J. Plant Physiol. 160, 509–515.

Gepstein, S.H., Glick, B.R., 2013. Strategies to ameliorate abiotic stress-induced plant senescence. Plant Mol. Biol. 82, 623–633.

Ghanati, F., Moritam, A., Yokota, H., 2005. Effects of aluminum on the growth of tea plant and activation of antioxidant system. Plant Soil 276, 133–141.

Giannopolitis, C.N., Ries, S.K., 1977. Superoxide dismutases I. Occurrence in higher plants. Plant Physiol. 59, 309–314.

Gill, S.S., Tuteja, N., 2010. Reactive oxygen species and antioxidant machinery in abiotic stress tolerance in crop plants. Plant Physiol. Biochem. 48, 909–930.

Gill, S.S., Khan, N.A., Tuteja, N., 2012. Cadmium at high dose perturbs growth, photosynthesis and nitrogen metabolism while at low dose it up regulates sulfur assimilation and antioxidant machinery in garden cress (*Lepidium sativum* L.). Plant Sci. 182, 112–120.

Gonçalves, J.F., Becker, A.G., Cargnelutti, D., Tabaldi, L.A., Pereira, L.B., Battisti, V., et al., 2007. Cadmium toxicity causes oxidative stress and induces response of the antioxidant system in cucumber seedlings. Braz. J. Plant Physiol. 19, 223–232.

Hasan, S.A., Fariduddin, Q., Ali, B., Hayat, S., Ahmad, A., 2009. Cadmium: toxicity and tolerance in plants. J. Environ. Biol. 30, 164–174.

Hässig, A., Linag, W.X., Schwabl, H., Stampfli, K., 1999. Flavonoids and tannins: plant-based antioxidants with vitamin character. Med. Hypotheses 52, 479–481.

Hernandez, J.A., del Río, L.A., Sevilla, F., 2006. Salt stress-induced changes in superoxide dismutase isozymes in leaves and mesophyll protoplasts from *Vigna unguiculata* (L.) Walp. New Phytol. 126, 37–44.

Hirt, H., Shinozaki, K., 2004. Plant responses to abiotic stress. In: Hirt, H., Shinozaki, K. (Eds.), Topics in Current Genetics, vol. 4. Springer-Verlag, Berlin-Heidelberg, pp. 188–189.

Hsu, Y.T., Kao, C.H., 2004. Cadmium toxicity is reduced by nitric oxide in rice leaves. Plant Growth Regul. 42, 227–238.

Huang, M., Guo, Z., 2005. Responses of antioxidative system to chilling in two rice cultivars differing in sensitivity. Biol. Plant. 49, 81–84.

Iannelli, M.A., Pietrini, F., Fiore, L., Petrilli, L., Massacci, A., 2002. Antioxidant response to cadmium in *Phragmites australis* plants. Plant Physiol. Biochem. 40, 977–982.

Inzé, D., Van Montagu, M., 1995. Oxidative stress in plants. Curr. Opin. Biotechnol. 6, 153–158.

Ismail, M., Suroto, A., 2012. Response of biochemical activities on selected rice plants towards ozone stress. International Conference on Nanotechnology and Chemical Engineering, December, 2002, Thailand, pp. 21–22.

Israr, M., Sahi, S., Datta, R., Sarkar, D., 2006. Bioaccumulation and physiological effects of mercury in *Sesbania drummondii*. Chemosphere 65, 591–598.

Iturbe-Ormaetxe, I., Escuredo, P.R., Arrese-Igor, C., Becana, M., 1998. Oxidative damage in pea plants exposed to water deficit or paraquat. Plant Physiol. 116, 173–181.

Jaleel, C.A., Riadh, K., Gopi, R., Manivannan, P., Inès, J., Al-Juburi, H.J., et al., 2009. Antioxidant defense responses: physiological plasticity in higher plants under abiotic constraints. Acta Physiol. Plant. 31, 427–436.

Janas, K.M., Zielińska-Tomaszewska, J., Rybaczek, D., Maszewski, J., Posmyk, M.M., Amarowicz, R., et al., 2010. The impact of copper ions on growth, lipid peroxidation, and phenolic compound accumulation and localization in lentil (*Lens culinaris* Medic.) seedlings. J. Plant Physiol. 167, 270–276.

Jiang, Y., Huang, B., 2001. Drought and heat stress injury to two cool-season turfgrasses in relation to antioxidant metabolism and lipid peroxidation. Crop Sci. 41, 436–444.

Jebara, S., Jebara, M., Limam, F., Aouani, M.E., 2005. Changes in ascorbate peroxidase, catalase, guaiacol peroxidase and superoxide dismutase activities in common bean (*Phaseolus vulgaris*) nodules under salt stress. J. Plant Physiol. 162, 929–936.

Jouili, H., El Ferjani, E., 2003. Changes in antioxidant and lignifying enzyme activities in sunflower roots (*Helianthus annuus* L.) stressed with copper excess. C. R. Biol. 326, 639–644.

Kampfenkel, K., Van Montagu, M., Inzé, D., 1995. Effects of iron excess on *Nicotiana plumbaginifolia* plants (implications to oxidative stress). Plant Physiol. 107, 725–735.

Kang, G., Wang, C., Sun, G., Wang, Z., 2003. Salicylic acid changes activities of H_2O_2-metabolizing enzymes and increases the chilling tolerance of banana seedlings. Environ. Exp. Bot. 50, 9–15.

Kolbert, Z.S., Pető, A., Szőllősi, R., Erdei, L., Tari, I., 2011. Nitric oxide (NO) generation during vegetative/generative transition of the apical meristem in wheat. Acta Biol. Szeged. 55, 95–97.

Kotak, S., Larkindale, J., Lee, U., von Koskull-Döring, P., Vierling, E., Scharf, K.D., 2007. Complexity of the heat stress response in plants. Curr. Opin. Plant Biol. 10, 310–316.

Kranner, I., Colville, L., 2011. Metals and seeds: biochemical and molecular implications and their significance for seed germination. Environ. Exp. Bot. 72, 93–105.

Kumar, D., Yusuf, M.A., Singh, P., Sardar, M., Sarin, N.B., 2013. Modulation of antioxidant machinery in α-tocopherol-enriched transgenic *Brassica juncea* plants tolerant to abiotic stress conditions. Protoplasma 250, 1079–1089.

Kumar, P., Kumar, R.T., Sharma, P.N., 2007. Excess nickel-induced changes in antioxidative processes in maize leaves. J. Plant Nutr. Soil Sci. 170, 796–802.

Kuo, M.C., Kao, C.H., 2003. Aluminum effects on lipid peroxidation and antioxidative enzyme activities in rice leaves. Biol. Plant. 46, 149–152.

Lee, D.H., Lee, C.B., 2000. Chilling stress-induced changes of antioxidant enzymes in the leaves of cucumber: in gel enzyme activity assays. Plant Sci. 159, 75–85.

Lee, Y.P., Ahmad, R., Lee, H.S., Kwak, S.S., Shafqat, M.N., Kwon, S.Y., 2013. Improved tolerance of Cu/Zn superoxide dismutase and ascorbate peroxidase expressing transgenic tobacco seeds and seedlings against multiple abiotic stresses. Int. J. Agri. Biol. 15, 725–730.

Li, M., Zhu, Q., Hu, C.W., Chen, L., Liu, Z.L., Kong, Z.M., 2007. Cobalt and manganese stress in the microalga *Pavlova viridis* (Prymnesiophyceae): effects on lipid peroxidation and antioxidant enzymes. J. Environ. Sci. 19, 1330–1335.

Lomonte, C., Sgherri, C., Baker, A.J., Kolev, S.D., Navari-Izzo, F., 2010. Antioxidative response of *Atriplex codonocarpa* to mercury. Environ. Exp. Bot. 69, 9–16.

Mackerness, S.H.A., 2000. Plant responses to ultraviolet-B (UV-B: 280–320 nm) stress: what are the key regulators? Plant Growth Regul. 32, 27–39.

Madamanchi, N.R., Alscher, R.G., 1991. Metabolic bases for differences in sensitivity of two pea cultivars to sulfur dioxide. Plant Physiol. 97, 88–93.

Mahanty, S., Kaul, T., Pandey, P., Reddy, R.A., Mallikarjuna, G., Reddy, C.S., et al., 2012. Biochemical and molecular analyses of copper–zinc superoxide dismutase from a C_4 *Pennisetum glaucum* reveals an adaptive role in response to oxidative stress. Gene 505, 309–317.

Maiti, S., Ghosh, N., Mandal, C., Das, K., Dey, N., Adak, M.K., 2012. Responses of the maize plant to chromium stress with reference to antioxidation activity. Braz. J. Plant Physiol. 24, 203–212.

Maksymiec, W., 2007. Signaling responses in plants to heavy metal stress. Acta Physiol. Plant. 29, 177–187.

Matsumoto, H., Motoda, H., 2012. Aluminum toxicity recovery processes in root apices. Possible association with oxidative stress. Plant Sci. 185, 1–8.

Meloni, D.A., Oliva, M.A., Martinez, C.A., Cambraia, J., 2003. Photosynthesis and activity of superoxide dismutase, peroxidase and glutathione reductase in cotton under salt stress. Environ. Exp. Bot. 49, 69–76.

Michalak, A., 2006. Phenolic compounds and their antioxidant activity in plants growing under heavy metal stress. Pol. J. Environ. Stud. 15, 523–530.

Milone, M.T., Sgherri, C., Clijsters, H., Navari-Izzo, F., 2003. Antioxidative responses of wheat treated with realistic concentration of cadmium. Environ. Exp. Bot. 50, 265–276.

Mittler, R., 2002. Oxidative stress, antioxidants and stress tolerance. Trends Plant Sci. 7, 405–410.

Mittler, R., 2006. Abiotic stress, the field environment and stress combination. Trends Plant Sci. 11, 15–19.

Moran, J.F., Becana, M., Iturbe-Ormaetxe, I., Frechilla, S., Klucas, R.V., Aparicio-Tejo, P., 1994. Drought induces oxidative stress in pea plants. Planta 194, 346–352.

Mou, D., Yao, Y., Yang, Y., Zhang, Y., Tian, C., Achal, V., 2011. Plant high tolerance to excess manganese related with root growth, manganese distribution and antioxidative enzyme activity in three grape cultivars. Ecotoxicol. Environ. Saf. 74, 776–786.

Moyer, R.A., Hummer, K.E., Finn, C.E., Frei, B., Wrolstad, R.E., 2002. Anthocyanins, phenolics, and antioxidant capacity in diverse small fruits: *Vaccinium, Rubus*, and *Ribes*. J. Agri. Food Chem. 50, 519–525.

Mutlu, S., Karadağoğlu, Ö., Atici, Ö., Nalbantoğlu, B., 2013. Protective role of salicylic acid applied before cold stress on antioxidative system and protein patterns in barley apoplast. Biol. Plant. 57, 507–513.

Noctor, G., Arisi, A.C.M., Jouanin, L., Kunert, K.J., Rennenberg, H., Foyer, C.H., 1998. Glutathione: biosynthesis, metabolism and relationship to stress tolerance explored in transformed plants. J. Exp. Bot. 49, 623–647.

Ozdener, Y., Aydin, B.K., 2010. The effect of zinc on the growth and physiological and biochemical parameters in seedlings of *Eruca sativa* (L.) (Rocket). Acta Physiol. Plant. 32, 469–476.

Panda, S.K., 2007. Chromium-mediated oxidative stress and ultrastructural changes in root cells of developing rice seedlings. J. Plant Physiol. 164, 1419–1428.

Panda, S.K., Choudhury, S., 2005. Chromium stress in plants. Braz. J. Plant Physiol. 17, 95–102.

Panda, S.K., Khan, M.H., 2004. Changes in growth and superoxide dismutase activity in *Hydrilla verticillata* L. under abiotic stress. Braz. J. Plant Physiol. 16, 115–118.

Panda, S.K., Matsumoto, H., 2010. Changes in antioxidant gene expression and induction of oxidative stress in pea (*Pisum sativum* L.) under Al stress. BioMetals 23, 753–762.

Pandey, V., Dixit, V., Shyam, R., 2005. Antioxidative responses in relation to growth of mustard (*Brassica juncea* cv. Pusa Jaikisan) plants exposed to hexavalent chromium. Chemosphere 61, 40–47.

Papp, N., Szilvássy, B., Abrankó, L., Szabó, T., Pfeiffer, P., Szabó, Z., et al., 2010. Main quality attributes and antioxidants in Hungarian sour cherries: identification of genotypes with enhanced functional properties. Int. J. Food Sci. Technol. 45, 395–402.

Potters, G., Pasternak, T.P., Guisez, Y., Palme, K.J., Jansen, M.A.K., 2007. Stress-induced morphogenic responses: growing out of trouble? Trends Plant Sci. 12, 98–105.

Pourcel, L., Routaboul, J.M., Cheynier, V., Lepiniec, L., Debeaujon, I., 2007. Flavonoid oxidation in plants: from biochemical properties to physiological functions. Trends Plant Sci. 12, 29–36.

Prasad, K.V.S.K., Paradha, S.P., Sharmila, P., 1999. Concerted action of antioxidant enzymes and curtailed growth under zinc toxicity in *Brassica juncea*. Environ. Exp. Bot. 42, 1–10.

Prasad, T.K., Anderson, M.D., Martin, B.A., Stewart, C.R., 1994. Evidence for chilling-induced oxidative stress in maize seedlings and a regulatory role for hydrogen peroxide. Plant Cell 6, 65–74.

Qadir, A.M., Ghafoor, G., Murtaza, G., 2000. Cadmium concentration in vegetables grown on urban soils irrigated with untreated municipial sewage. Environ. Dev. Sustainability 2, 11–19.

Qureshi, M.I., Abdin, M.Z., Qadir, S., Iqbal, M., 2007. Lead-induced oxidative stress and metabolic alterations in *Cassia angustifolia* Vahl. Biol. Plant. 51, 121–128.

Rai, V., Vajpayee, P., Singh, S.N., Mehrotra, S., 2004. Effect of chromium accumulation on photosynthetic pigments, oxidative stress defense system, nitrate reduction, proline level and eugenol content of *Ocimum tenuiflorum* L. Plant Sci. 167, 1159–1169.

Rao, K.M., Sresty, T.V.S., 2000. Antioxidative parameters in the seedlings of pigeonpea (*Cajanus cajan* (L.) Millspaugh) in response to Zn and Ni stresses. Plant Sci. 157, 113–128.

Rao, M.V., Paliyath, G., Ormrod, D.P., 1996. Ultraviolet-B-and ozone-induced biochemical changes in antioxidant enzymes of *Arabidopsis thaliana*. Plant Physiol. 110, 125–136.

Rasool, S., Ahmad, A., Siddiqi, T.O., Ahmad, P., 2013. Changes in growth, lipid peroxidation and some key antioxidant enzymes in chickpea genotypes under salt stress. Acta Physiol. Plant. 35, 1039–1050.

Rivero, R.M., Ruiz, J.M., Garcia, P.C., Lopez-Lefebre, L.R., Sánchez, E., Romero, L., 2001. Resistance to cold and heat stress: accumulation of phenolic compounds in tomato and watermelon plants. Plant Sci. 160, 315–321.

Rodríguez-Serrano, M., Romero-Puertas, M.C., Zabalza, A.N.A., Corpas, F.J., Gómez, M., Del Río, L.A., et al., 2006. Cadmium effect on oxidative metabolism of pea (*Pisum sativum* L.) roots. Imaging of reactive oxygen species and nitric oxide accumulation *in vivo*. Plant Cell Environ. 29, 1532–1544.

Rout, G.R., Das, P., 2003. Effect of metal toxicity on plant growth and metabolism: I. Zinc. Agronomie 23, 3–11.

Rucinska, R., Waplak, S., Gwozdz, E.A., 1999. Free radical formation and activity of antioxidant enzymes in lupin roots exposed to lead. Plant Physiol. Biochem. 37, 187–194.

Ruley, T.A., Sharma, N.C., Sahi, S.V., 2004. Antioxidant defense in a lead accumulating plant, *Sesbania drummondii*. Plant Physiol. Biochem. 42, 899–906.

Sakihama, Y., Cohen, M.F., Grace, S.C., Yamasaki, H., 2002. Plant phenolic antioxidant and prooxidant activities: phenolics-induced oxidative damage mediated by metals in plants. Toxicology 177, 67–80.

Schickler, H., Caspi, H., 2002. Response of antioxidative enzymes to nickel and cadmium stress in hyperaccumulator plants of the genus *Alyssum*. Physiol. Plant. 105, 39–44.

Schützendübel, A., Polle, A., 2002. Plant responses to abiotic stresses: heavy metal-induced oxidative stress and protection by mycorrhization. J. Exp. Bot. 53, 1351–1365.

Sen, A., Alikamanoglu, S., 2013. Antioxidant enzyme activities, malondialdehyde, and total phenolic content of PEG-induced hyperhydric leaves in sugar beet tissue culture. In Vitro Cell. Dev. Biol.-Plant 49, 396–404.

Seregin, I.V., Ivanov, V.B., 2001. Physiological aspects of cadmium and lead toxic effects on higher plants. Russ. J. Plant Physiol. 48, 523–544.

Shainberg, O., Rubin, B., Rabinowitch, H.D., Libal, Y., Tel-Or, E., 2000. Acclimation of beans to oxidative stress by treatment with sublethal iron levels. J. Plant Physiol. 157, 93–99.

Sharma, P., Dubey, R.S., 2005. Drought induces oxidative stress and enhances the activities of antioxidant enzymes in growing rice seedlings. Plant Growth Regul. 46, 209–221.

Sharma, P., Jha, A.B., Dubey, R.S., Pessarakli, M., 2012. Reactive oxygen species, oxidative damage, and antioxidative defense mechanism in plants under stressful conditions. J. Bot. 2012, 1–26.

Shekhawat, G.S., Verma, K., Jana, S., Singh, K., Teotia, P., Prasad, A., 2010. In vitro biochemical evaluation of cadmium tolerance mechanism in callus and seedlings of *Brassica juncea*. Protoplasma 239, 31–38.

Sheokand, S., Kumari, A., Sawhney, V., 2008. Effect of nitric oxide and putrescine on antioxidative responses under NaCl stress in chickpea plants. Physiol. Mol. Biol. Plants 14, 355–362.

Silva, E.N., Vieira, S.A., Ribeiro, R.V., Ponte, L.F., Ferreira-Silva, S.L., Silveira, J.A., 2013. Contrasting physiological responses of *Jatropha curcas* plants to single and combined stresses of salinity and heat. J. Plant Growth Regul. 32, 159–169.

Sinha, S., Saxena, R., 2006. Effect of iron on lipid peroxidation, and enzymatic and non-enzymatic antioxidants and bacoside-A content in medicinal plant *Bacopa monnieri* L. Chemosphere 62, 1340–1350.

Sinha, S., Gupta, M., Chandra, P., 1997. Oxidative stress induced by iron in *Hydrilla verticillata* (lf) Royle: response of antioxidants. Ecotoxicol. Environ. Saf. 38, 286–291.

Sinha, P., Khurana, N., Nautiyal, N., 2012. Induction of oxidative stress and antioxidant enzymes by excess cobalt in mustard. J. Plant Nutr. 35, 952–960.

Srivastava, S., Dubey, R.S., 2011. Manganese-excess induces oxidative stress, lowers the pool of antioxidants and elevates activities of key antioxidative enzymes in rice seedlings. Plant Growth Regul. 64, 1–16.

Stahl, W., Sies, H., 2003. Antioxidant capacity of carotenoids. Mol. Aspects Med. 24, 345–351.

Subrahmanyam, D., 2008. Effects of chromium toxicity on leaf photosynthetic characteristics and oxidative changes in wheat (*Triticum aestivum* L.). Photosynthetica 46, 339–345.

Szepesi, Á., Poór, P., Gémes, K., Horváth, E., Tari, I., 2008. Influence of exogenous salicylic acid on antioxidant enzyme activities in the roots of salt stressed tomato plants. Acta Biol. Szeged. 52, 199–200.

Szőllősi, R., Sz. Varga, I., Erdei, L., Mihalik, E., 2009. Cadmium-induced oxidative stress and antioxidative mechanisms in germinating Indian mustard (*Brassica juncea* L.) seeds. Ecotoxicol. Environ. Saf. 72, 1337–1342.

Szőllősi, R., Kálmán, E., Medvegy, A., Pető, A., Sz. Varga, I., 2011. Studies on oxidative stress caused by Cu and Zn excess in germinating seeds of Indian mustard (*Brassica juncea* L.). Acta Biol. Szeged. 55, 175–178.

Szőllősi, R., Kálmán, E., Medvegy, A., Sz Varga, I., 2012. Survival to early toxic copper excess: biochemical and anatomical changes during germination of Indian Mustard. Eur. Chem. Bull. 1, 466–469.

Tamás, L., Mistrík, I., Huttová, J., Halušková, L.U., Valentovičová, K., Zelinová, V., 2010. Role of reactive oxygen species-generating enzymes and hydrogen peroxide during cadmium, mercury and osmotic stresses in barley root tip. Planta 231, 221–231.

Tari, I., Camen, D., Coradini, G., Csiszár, J., Fediuc, E., Gémes, K., et al., 2008. Changes in chlorophyll fluorescence parameters and oxidative stress responses of bush bean genotypes for selecting contrasting acclimation strategies under water stress. Acta Biol. Hung. 59, 335–345.

Tewari, R.K., Kumar, P., Sharma, P.N., Bisht, S.S., 2002. Modulation of oxidative stress responsive enzymes by excess cobalt. Plant Sci. 162, 381–388.

Ünyayar, S., Keleş, Y., Çekiç, F.Ö., 2005. The antioxidative response of two tomato species with different drought tolerances as a result of drought and cadmium stress combinations. Plant Soil Environ. 51, 57–64.

Verma, S., Dubey, R.S., 2003. Lead toxicity induces lipid peroxidation and alters the activities of antioxidant enzymes in growing rice plants. Plant Sci. 164, 645–655.

Wang, C., Zhang, S.H., Wang, P.F., Hou, J., Zhang, W.J., Li, W., et al., 2009. The effect of excess Zn on mineral nutrition and antioxidative response in rapeseed seedlings. Chemosphere 75, 1468–1476.

Wang, S.H., Yang, Z.M., Yang, H., Lu, B., Li, S.Q., Lu, Y.P., 2004. Copper-induced stress and antioxidative responses in roots of *Brassica juncea* L. Bot. Bull. Acad. Sin. 45, 203–212.

Willekens, H., Van Camp, W., Van Montagu, M., Inzé, D., Langebartels, C., Sandermann Jr., H., 1994. Ozone, sulfur dioxide, and ultraviolet B have similar effects on mRNA accumulation of antioxidant genes in *Nicotiana plumbaginifolia* L. Plant Physiol. 106, 1007–1014.

Xu, S.C., Li, Y.P., Hu, J., Guan, Y.J., Ma, W.G., Zheng, Y.Y., et al., 2010. Responses of antioxidant enzymes to chilling stress in tobacco seedlings. Agri. Sci. China 9, 1594–1601.

Xue, S.G., Chen, Y.X., Reeves, R.D., Baker, A.J., Lin, Q., Fernando, D.R., 2004. Manganese uptake and accumulation by the hyperaccumulator plant *Phytolacca acinosa* Roxb. (Phytolaccaceae). Environ. Poll. 131, 393–399.

Yadav, S.K., 2010. Heavy metals toxicity in plants: an overview on the role of glutathione and phytochelatins in heavy metal stress tolerance of plants. S. Afr. J. Bot. 76, 167–179.

Yousuf, P.Y., Hakeem, K.U.R., Chandna, R., Ahmad, P., 2012. Role of glutathione reductase in plant abiotic stress. In: Ahmad, P., Prasad, M.N.V. (Eds.), Abiotic Stress Responses in Plants: Metabolism, Productivity and Sustainability. Springer, New York, pp. 149–158.

Zagorchev, L., Seal, C.E., Kranner, I., Odjakova, M., 2013. A central role for thiols in plant tolerance to abiotic stress. Int. J. Mol. Sci. 14, 7405–7432.

Zhang, H., Xia, Y., Wang, G., Shen, Z., 2008. Excess copper induces accumulation of hydrogen peroxide and increases lipid peroxidation and total activity of copper–zinc superoxide dismutase in roots of *Elsholtzia haichowensis*. Planta 227, 465–475.

Ziogas, V., Tanou, G., Filippou, P., Diamantidis, G., Vasilakakis, M., Fotopoulos, V., et al., 2013. Nitrosative responses in citrus plants exposed to six abiotic stress conditions. Plant Physiol. Biochem. 68, 118–126.

Zsigmond, L., Szepesi, Á., Tari, I., Rigó, G., Király, A., Szabados, L., 2012. Overexpression of the mitochondrial PPR40 gene improves salt tolerance in *Arabidopsis*. Plant Sci. 182, 87–93.

Chapter 4

Catalase

A Versatile Antioxidant in Plants

Iti Sharma and Parvaiz Ahmad

4.1 INTRODUCTION

Catalase (CAT, 1.11.1.6) is present in all living beings, ranging from unicellular prokaryotes to multicellular eukaryotes. The evolutionary design of catalytic enzymes started about 3.5 billion years ago when the ancestral planctonic bacteria began aerobic respiration (Lenton, 2003). Thereafter, the catalase evolution proceeded with development of the aerobic biosphere on Earth (Drews, 2011). All aerobic organisms performed both photosynthesis and respiration within cells with generation of harmful reactive oxygen species (ROS), namely superoxide anion ($O_2^{\bullet -}$), hydroxyl radical ($^{\bullet}OH$), hydrogen peroxide (H_2O_2) and singlet oxygen ($^1O_2$). The enhanced production of ROS can pose oxidative stress to cells by causing peroxidation of lipids, oxidation of proteins, and damage to nucleic acids, enzyme inhibition, activation of the programmed cell death (PCD) pathway and ultimately leading to cell death (Ahmad et al., 2010a,b, 2011; Ahmad and Umar, 2011; Koyro et al., 2012; Sharma, 2012). Further, the enzymatic and nonenzymatic antioxidant defense system had developed for detoxification of ROS in aerobic organisms. As an enzymatic antioxidant, catalase has a key role to prevent cellular oxidative damage by degrading hydrogen peroxide (H_2O_2) into water and oxygen with high efficiency (Ahmad et al., 2011). It is a predominant peroxisomal enzyme, but it also exists in the mitochondria and cytoplasm of cells. In plants, catalase scavenges H_2O_2 generated during mitochondrial electron transport, oxidation of the fatty acids, and most importantly, photorespiratory oxidation during normal and stress conditions. Catalase always draws the attention of researchers due to its efficient catalytic and regulatory properties among all antioxidant enzymes of the plant system. Thus, it has been purified and extensively characterized at the genetic, biochemical, and molecular level of plants. Dounce (1983) proposed the monofunctional and

P. Ahmad (Ed): Oxidative Damage to Plants. DOI: http://dx.doi.org/10.1016/B978-0-12-799963-0.00004-6

bifunctional mechanism for the catalytic action of catalase. Recently it has been established that catalase is present as multiple isoforms (CAT1, CAT2, CAT3) encoded by multiple genes (*Cat1*, *Cat2*, *Cat3*) expressed in organelle, temporal and stress specific manners. Specific evolutionary events including horizontal gene transfer, paralog formation and gene fusion have been revealed by phylogenetic study of the catalase isozymes (Zamocky et al., 2012). Although much effort has been made to understand catalase among prokaryotes and animals, very little information is available on the versatility of catalase in plants. The chapter presents a review of the overall aspects of catalase in plants.

4.2 OXIDATIVE STRESS

Atomic oxygen is the most abundant element in aerobic organisms. The oxygen molecule has the potential to be partially reduced and form reactive oxygen species through physical or chemical activation (Halliwell and Gutteridge, 1992). Physical activation occurs mainly by transfer of excitation energy from photo-activated pigments, such as the excited chlorophyll molecule to dioxygen, which absorbs sufficient energy and, as a result, the spin of one e^- is inverted and the singlet state of oxygen ($^1O_2$) is formed (Perl et al., 1992). Chemical activation is the other mechanism to circumvent spin restriction. It occurs when molecular oxygen absorbs activation energy of 22 Kcal/mol to rise from its ground state and by a chain of reduction reactions various stable intermediates are formed in a stepwise manner. Four e^- (and 4 protons) are required for the full reduction of dioxygen to water; all the intermediates of univalent reduction of oxygen, namely superoxide ($O_2^{\bullet -}$), hydroxyl radical ($^\bullet OH$) and hydrogen peroxide (H_2O_2), are known as reactive oxygen species (ROS) (Yu et al., 1992). In a biological system reactive oxygen species are mainly generated in chloroplast, mitochondria and peroxisomes during respiration, photosynthesis and metabolic reactions (Ahmad et al., 2011).

4.3 H_2O_2 AN EFFECTIVE ROS PRODUCED IN PLANTS: PRODUCTION AND TOXICITY

Though all reactive oxygen species are extremely reactive and cytotoxic, H_2O_2 is unique among them. H_2O_2 is formed during reactions like dismutation of $O_2^{\bullet -}$ by superoxide dismutase (SOD) activity, photorespiration, β-oxidation of fatty acids and proton-induced decomposition of $O_2^{\bullet -}$ and defense against pathogens (Scandalios, 1994; Sagi and Fluhr, 2006; Andre et al., 2013). It is produced in various subcellular organelles and can diffuse through aquaporins of intracellular and intercellular membranes of cells (Bienert et al., 2007).

In peroxisomes H_2O_2 can be formed by glycolate oxidase during the photorespiratory glyoxylate cycle (Foyer et al., 2009) and by xanthine oxidase with the superoxide dismutase enzyme system (Corpas et al., 2008). Moreover, chloroplast also generates ample amounts of H_2O_2 by electron leakage during electron flow through the photosystem (I), which results in production of the superoxide radical. Further, $O_2^{\bullet -}$ reduces to form H_2O_2 by the action of superoxide dismutase (SOD). Superoxide radical can also react with NAD(P)H molecules to produce H_2O_2. H_2O_2 may also be generated in chloroplast during synthesis of glycolate from 3-Phosphoglycolate and conversion of glycolate back to 3-Phosphoglycolate in peroxisome.

In mitochondria the rate of electron transfer is directly proportional to H_2O_2 production during respiration. The electron-rich form of ubiquinon, ubisemiquinon, can reduce molecular oxygen to form superoxide radical. The radical, being charged, accumulates in the mitochondria and SOD degrades it to H_2O_2 and O_2 (Fridovich, 1975).

Another source of H_2O_2 production is the conversion of ribonucleotide diphospates to deoxyribonucleotide diphosphates through riboncleotide reductase. An enzyme, namely NAD (P)H:flavin oxidoreductase, produces tyrosyl free radical for the action of ribonucleotide reductase. During production of tyrosyl free radical, the enzyme also liberates superoxide ion which readily changes to H_2O_2 via superoxide dismutase.

$$2O_2^{-\bullet} + 2H^+ \rightarrow H_2O_2 + O_2$$

Chance et al. (1979) also reported the production of H_2O_2 in cytosol, endoplasmic reticulum (ER) and nucleus during different metabolic processes.

H_2O_2 is a potentially dangerous by-product of oxygen metabolism, a process that occurs in most living things, from aerobic bacteria to plants and animals. H_2O_2 is quite reactive with molecules containing Fe^{2+} or other transition metals through the Fenton reaction, which results in homolysis of H_2O_2 to two harmful $^{\bullet}OH$ radicals (Becana et al., 1998; Sharma, 2013a). Moreover, direct reaction of H_2O_2 with the $-SH$ groups inactivates enzymes, e.g. fructose bisphosphatase (Charles and Halliwell, 1981), sedoheptulose bisphosphatase, and glyceraldehyde 3-phosphate dehydrogenase. H_2O_2 can react with metalloproteins, leading to dissociation of their metal ion and hindering their biological activity.

In addition, H_2O_2 has an important role in processes from signal transduction, cell wall lignification, plant growth, development of root hair, and xylem differentiation, to root shoot coordination and stomatal control (Cheesman, 2007).

4.4 ANTIOXIDANT SYSTEM AGAINST H_2O_2

A well-maintained balance between H_2O_2 production and its removal is essential for the survival of plants. To minimize the damaging effects of H_2O_2, plant cells evolved both enzymatic and nonenzymatic defenses.

Nonenzymatic consist of glutathione, ascorbate and flavonoids. The enzymatic defense system includes the ascorbate glutathione cycle (i.e. ascorbate peroxidase (APX), ascorbate reductase, glutathione reductase) and catalase (Ahmad et al., 2010a,b, 2011; Ahmad and Umar, 2011; Singh et al., 2012; Koyro et al., 2012). Among these, catalase is the most active catalyst produced in all aerobic organisms. It decomposes H_2O_2 at an enormously fast rate with an energy-efficient approach.

4.5 CATALASE: A POTENT ANTIOXIDANT IN PLANTS

Catalase plays a central role in maintaining balance of cellular hydrogen peroxide in plants. It was the first antioxidant enzyme documented and it appears in all prokaryotes and eukaryotes. Degradation of hydrogen peroxide (H_2O_2) through catalase was first noticed by Loew (1901). Then Warburg (1923) proposed that catalase is inhibited by cyanide so it may be an iron-containing enzyme. Next, Zeile and Hellstrom (1930) discovered that catalase has a hematin prosthetic group. Catalase was initially isolated and crystallized from beef extract by Sumner and Dounce (1937). Thereafter, a great deal of information on catalase has been generated and, recently, Ray et al. (2012) have used *E. coli* as a heterologous expression system and have isolated a soluble bioactive recombinant catalase-A in rice plant. The new protein has molecular and biochemical features of a typical monofunctional plant catalase. Today, isozymes of catalase and phylogenic relationships among various species of catalase have been identified by researchers.

The significance of catalases in the antioxidant defense system of plants has been proven by various studies (Shim et al., 1999; Sharma et al., 2007; Beulah and Ramana, 2013). Willekens et al. (1997) observed that CAT-deficient mutants are more prone to oxidative stress as compared to wild plants treated with salt, ozone, paraquat and H_2O_2. Increased CAT activity is crucial for survival of plants under moderate metal stress, while severe metal stress leads to irreversible damage of the catalase enzyme (Youssef and Azoo, 2013; Bocova et al., 2012; Sharma and Travlos, 2012).

4.5.1 Location in Cellular Environment

Location and import of catalase in peroxisomes has been confirmed by studies of subcellular fractionation and in-situ activity staining in different plant species (Mullen et al., 1997). Catalase activity was also detected in isolated chloroplasts. However, no catalase has been found in the stroma of chloroplast, but in lumen a PSII membrane associated heme catalase was reported by Sheptovitsky and Brudig (1996). Catalase activity in the cytosol and mitochondria has also been reported in photosynthetic organisms (Kato et al., 1997; Spanou et al., 2012; Gu et al., 2013).

4.5.2 Structure of Catalase

Catalase is a heme-containing redox enzyme present in all organisms exposed to oxygen. The enzyme has a dumbbell-shaped tetrameric structure with four identical monomer subunits of 220,000 to 350,000 kD. The heme group is the key component for enzymatic activity. The heme prosthetic group of each monomer unit consists of a central iron atom attached with a protoporphyrin ring (Fig. 4.1A,B). The protoporphyrin ring has four pyrrole rings linked by methene bridges. Side chains made up of four methyl, two vinyl, and two propionate are also attached. The central iron atom of the prosthetic group may present in ferrous (Fe^{+2}) or the ferric (Fe^{+3}) valance state. A specific nonpolypeptide unit is also bound at the catalytic center of some plant catalase. In certain species catalase monomer also contains one tightly bound NADP (nicotinamide adenine dinucleotide phosphate) for each subunit. The NADP acts as the main electron carrier in reductive biosynthesis and shields the enzyme oxidation by its H_2O_2 substrate (Regelsberger et al., 2002).

Catalase has the highest turnover rates among all antioxidant enzymes. It has been reported that one molecule of catalase can convert 6 million molecules of H_2O_2 into H_2O and O_2 per minute.

On the basis of subunit sizes, quaternary structures, prosthetic groups, and sequences, catalase can be divided into three main groups (I - Monofunctional catalases, II - Catalase peroxidases and III - Nonheme catalases) (Loewen et al., 2002). Most of the plant species have I - Monofunctional catalase, which executes peroxidase activity on small organic substrates.

4.5.3 Molecular Isoforms of Catalase

In recent years, the role of catalases in plant metabolism, antioxidant defense system, and signaling has been reported by various studies. Multiple molecular forms of catalase have been reported in different plant species, e.g. *Nicotiana tobacco* (Havir and McHale, 1987), cotton (Ni et al., 1990), *Nicotiana plumbaginlfolia* (Willekens et al., 1994b), *Arabidopsis thaliana* (Zhong et al., 1994), *Pinus taeda* (Mullen and Gifford, 1993), sunflower (Eising et al., 1989), pumpkin (Yamaguchi et al., 1986) and tomato (Gianinetti et al., 1993). Catalase nomenclature is also based on its isoforms in different plant species. According to a classification suggested by Willekens et al. (1995), Class I, Class II and Class III catalases are specifically expressed in photosynthetic tissues, vascular tissues and reproductive tissues, respectively.

The presence of multiple catalase isozymes suggests structural and functional versatility of catalases in a variety of plant species. The cDNA of various catalases has been isolated and characterized from different plant species to understand genes and their regulatory components (Scandalios, 1992). The

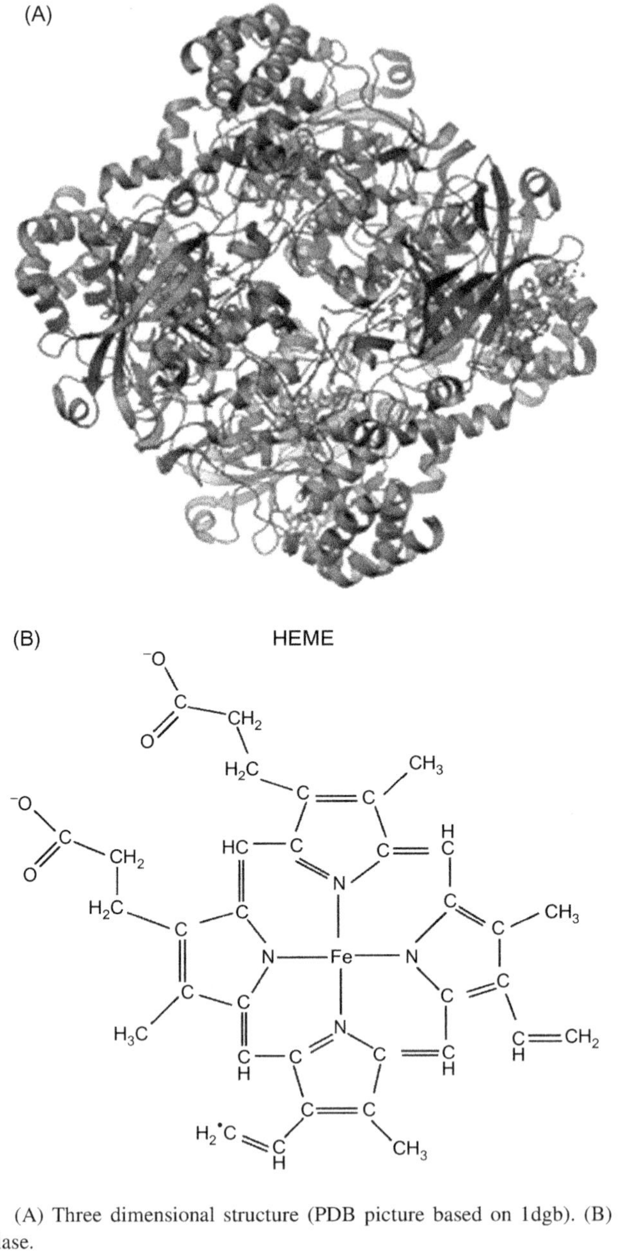

FIGURE 4.1 (A) Three dimensional structure (PDB picture based on 1dgb). (B) Single heme group of catalase.

isozymes of catalase exhibit developmental stage and organ specificity in plants (Table 4.1).

On the basis of unrooted parsimony and codon analysis of plant catalase sequences, Guan et al. (1996) have established the level of phylogenetic and

TABLE 4.1 Site and Tissue Specific Gene Expressions of *Cat* Gene

S. No.	Developmental Stage of Plant	Plant Organ	Catalase Isozyme Expressed	Reference
1.	Anthesis	Pollen	CAT-1	Scandalios et al., 1984
2.	Ovule development and early postpollination kernel development	Pericarp	CAT-3 and CAT-1	
3.	Postpollination	Scutellum	CAT-1	
4.	Early kernel development	Mature pollen Milky endosperm Aleurone Scutellum	CAT- 1	Scandalios, 1983; Wadsworth and Scandalios 1989; Acevedo and Scandalios, 1990
5.	Late kernel development	Aleurone	CAT-1 and CAT 2	Scandalios, 1974
6.	Post imbibitions	Seeds	CAT-2	
7.	Dark grown tissues	Leaf	CAT-1 and CAT-3	
8.	Light grown tissues	Green leaf	CAT-2	
9.	Early sporophytic development	Scutellum Cleoptile	CAT-2 CAT-1 and CAT-3	
10.	Mature green leaves	Mesophyll cells Bundle sheath cells	CAT- 1 and CAT-3 CAT-2	Tsaftaris et al., 1983
11.	Root development Shoot development	Young roots Mature stem	CAT-1 and CAT-3 CAT-3	

evolutionary relations among the various plant catalases. These workers divided plant catalase into three groups. Group I includes the monocot maize CAT-1 and most of the dicot catalases, while group II includes *N. plumbaginifolia* CAT-3 and tomato catalase, and group III contains the monocot specific catalases maize CAT-3, rice CAT-A, and barley CAT-2. Similarly, the three catalase genes of monocot plants have specific nucleotide preferences at the third codon position (Redinbaugh et al., 1990). Frugoli et al. (2012) reported that CAT1, CAT2 and CAT3 map to chromosome 1, 3 and 1, respectively. Phylogenetic studies reveal that the majority of the dicot catalases and several monocot catalases are closely linked to each other.

4.5.4 Biochemistry of Catalase (Antioxidant Activity of Catalase)

Catalase is an excellent H_2O_2-degrading antioxidant enzyme that can reduce H_2O_2 at a rapid rate and without consuming cellular energy. The enzymatic efficiency (frequency of enzyme substrate collision) of catalase (k_{cat}/K_M = $4.0 \times 10^8\ M^{-1}\ s^{-1}$) has a very high value. The enzymatic efficiency of catalase is at its diffusion limit, and therefore catalase shows catalytic excellence. A detailed mechanism for catalatic action has been proposed by Boon et al. (2007) who showed interaction of H_2O_2 with amino acids on heme active sites leads to a proton (hydrogen ion) transfer between the oxygen atoms. The free oxygen atom coordinates release the newly formed water molecule and E-Fe^{+4} = O (Compound I). Newly formed E-Fe^{+4} = O reacts with a second H_2O_2 molecule to reform E-Fe^{+3-}E and produce water and oxygen. In the fifth iron ligand a phenolate ligand of Tyr357 has been observed, which helps in conversion of Fe^{+3} to Fe^{+4} and improves the reaction efficiency of iron center. Spectrophotometric and kinetic evidence suggests that, depending on the concentration of H_2O_2, catalase can function in the peroxidatic or catalytic mode of activities (Deisseroth and Dounce, 1970).

$$2H_2O_2 \xrightarrow{\text{Catalase}} 2H_2O + O_2$$

Catalase can either catalyze the direct conversion of H_2O_2 into H_2O and O_2 (catalatic mode) or oxidize substrates (such as methanol, ethanol, formaldehyde, formate, or nitrite) by using hydrogen peroxide (peroxidatic mode). The ratio of enzymatic reactions is calculated as Rpic = (mU peroxidatic/U catalatic) × 10 (Havir and McHale, 1987).

Initially, the interaction of substrate H_2O_2 and the catalase iron center leads to production of oxygen-rich iron peroxide.

$$\text{E-Fe-OH} + H_2O_2 \rightarrow \underset{\text{(intermediate)}}{\text{E-Fe-OOH}} + H_2O$$

(I) Peroxidatic mode of action: (At low substrate concentrations)

At low concentration of H_2O_2 (<10−6 mol L^{-1}), catalase reacts in the peroxidatic mode and various hydrogen donors (e.g. methanol, ethanol, phenols, formaldehyde, formate, nitrite and ascorbic acid) can be oxidized in the following mode.

$$\text{RH2} + H_2O_2 \rightarrow \text{R} + 2H_2O$$

$$\text{CAT-Fe-OOH} + C_2H_5OH \rightarrow \text{CAT-Fe-OH} + H_2O + \text{CH3CHO (Peroxidatic Rx)}$$

At low substrate concentrations E-Fe-OOH can be reduced by hydrogen donors (e.g. ethanol including phenols, formic acid, formaldehyde and alcohols) peroxidatically. The enzyme produces useful products and water from toxins and hydrogen peroxide, respectively.

(II) Catalatic mode of action: (At high substrate concentrations)

At high concentrations of the substrate, H_2O_2 acts as both acceptor and donor of hydrogen molecules for an extremely rapid catalatic reaction.

$$2H_2O_2 \rightarrow 2H_2O + O_2$$

$$\text{CAT-Fe-OOH} + H_2O_2 \rightarrow \text{CAT-Fe-OH} + H_2O + O_2 \text{ (Catalatic Rx)}$$

Intermediate compound CAT-Fe-OOH reacts with H_2O_2 to produce water and molecular oxygen (Scandalios, 1994). Catalase has a very high Michaelis constant, so the enzyme activity increases linearly with H_2O_2 concentrations and maintains a balanced intracellular H_2O_2 concentration. Finally, catalase certainly consumes hydrogen peroxide during execution of catalytic and peroxidative modes of reaction in a biological system.

On the basis of reaction pattern, catalase isozymes fall into two main classes:

(i) HPI Catalase: This isoenzyme is bifunctional and has both catalytic and peroxidative activities.
(ii) HPII Catalase: Catalase catalyzes only disproportionation of H_2O_2 by peroxidative reactions.

Many reports are available that show a prominent role of catalase as an antioxidant in plants exposed to different environmental stresses (Isah and Mujib, 2012). Xu et al. (2013a) have reported that coexpression of MeCu-ZnSOD and MeCAT1 in cassava could dramatically improve ROS scavenging ability, leading to reduced H_2O_2 accumulation, improved abiotic stress resistance and delayed post-harvest physiological deterioration in cassava storage roots. Table 4.2 illustrates the biochemical and physical factors affecting the catalytic activity of catalase. Moreover, Xu et al. (2013a,b) has also reported that implementation of Cu/Zn superoxide dismutase (MeCu/ZnSOD) and catalase (MeCAT1) leads to extending the life of plant storage-roots by the stabilization of reactive oxygen species (ROS) homeostasis subsequent to harvest.

4.5.5 Factors Affecting Catalase Activities

The expression of the catalase gene is subject to various exogenous environmental factors. These physical and chemical factors are also known to cause

TABLE 4.2 Factors Affecting the Catalytic Activity of Catalase

S. No.	Factors	Details
1	Concentration of substrate	Amount of H_2O_2 defines the type of reaction as catalytic and peroxidative activities.
2	Temperature	Should be below 40–50°C
3	pH	6.8 and 7.5
4	Presence of inhibitor	Inhibitors strongly binding to the heme of catalase and stopping the enzyme's action. Noncompetitive inhibitor: e.g. copper Cations in copper (II) sulfate. Competitive inhibitor: e.g. cyanide
5	Osmolytes	Sucrose induces a stimulatory characteristic to CAT

oxidative stress and induction of the antioxidant defense system in different plant species. Table 4.2 shows physical factors affecting the catalytic activity of catalase.

According to a report by Scandalios et al. (1997) catalase genes demonstrate alterations in response to fungal toxin (Cercosporin), exogenously applied H_2O_2 and temperature (−4°C to 40°C).

Similarly, plant hormones like salicylic acid (acts as an endogenous signal molecule in the induction of systemic acquired resistance (SAR) in plants) can bind to catalase and inhibit its activities in dicot plants (Chen et al., 1993). However, the catalytic activities increased in monocot plants exposed to salicylic acid (Scandalios et al., 1997). Shim et al. (2003) have also reported that a fall in catalase activity under oxidative stress is related to accumulation of salicylic acid in treated plants. Other phytohormones, abscisic acid (ABA) and IAA, are responsible for increased catalase activities (Scandalios et al., 1997).

The multiple catalase genes respond differentially to light intensity, color and spectrum. It has been reported that a constant dark period is involved in inhibition of translocation of *Cat2* mRNA, whereas exposure of that leaf to white light or UV light (290–400 nm) leads to induction of CAT-2 isozymes (Fig. 4.2).

The transcription of *Cat3* gene in maize (encoding Class II catalase), *Cat2* (encoding class I catalase) and *Cat 3* (encoding Class II catalase) in *Arabidopsis*, and *Cat1* gene of *Nicotiana plumbaginifolia* (encoding class III catalase) have been regulated by circadian rhythm of corresponding plants (Willekens et al., 1994a). In C3 plants (*Arabidopsis* and *N. plumbaginlfolia*) the phase of accumulation of catalase transcripts early in the light period coincides with that of photosynthetic genes, which indicates a role of catalase in

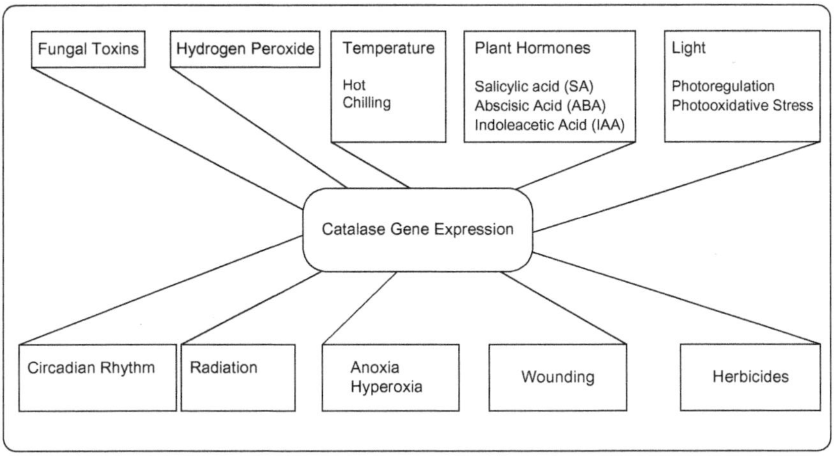

FIGURE 4.2 Environmental factors affecting catalase gene expression.

decomposing H_2O_2 generated during the photorespiratory oxidation of glycolate in the peroxisome (Ogren, 1984). However, phasing of *Cat3* gene with maximum expression in maize during photorespiratory conditions indicates that this gene is not associated with photorespiration in C4 plants.

The inhibitory effect of radiation on CAT activity was also reported by several researchers (Liang et al., 2000; Al-Rumaih and Al-Rumaih, 2008). Moreover, osmolytes such as sorbitol, mannitol and sucrose facilitate maintenance of high activity during salt stress (Shau et al., 2010).

4.5.6 Catalase Deficiency in Plants

A complete deficiency of catalase leads to autosomal recessive peroxisomal disorder acatalasemia (also known as Takahara's disease) in humans (Zamocky et al., 2008; Patnaik et al., 2013). Similarly, in plants catalase-mediated tolerance has been reported in *Pennisetum typhoides* exposed to arsenic-induced oxidative stress (Sharma, 2013b). Prominent catalase deficiency symptoms have been characterized in catalase-deficient mutants (Woodson and Kerdnaimongkol, 1999). Kendall et al. (1983) reported that catalase-deficient barley mutant was unable to survive under photorespiratory conditions, but stayed alive in a high CO_2 atmosphere. High catalase activities in mutants of *Nicotiana tabacum* balance the photorespiratory CO_2 loss more than wild type plants. High catalase activity inhibits the chemical decarboxylation of *a*-keto acids by peroxisomal H_2O_2, which may cause the decrease in photorespiration. So, mutants with cloned catalase genes into C3 plants can handle more photorespiratory carbon loss than wild plants. Plants deficient in catalase activity are able to grow to maturity but sustain chlorosis and significant head sterility. The important role of catalase in C3 photosynthetic metabolism has been well-established by analysis through antisense technology, as spontaneous lesion formation and oxidative stress are observed in catalase deficient mutant (Willekens et al., 1997; Chamnongpol et al., 1998). An undetectable increase in H_2O_2 during *Cat2* deficiency generates significant effects on gene expression and cell redox state (Chaouch et al., 2010; Mhamdi et al., 2010). Michelet et al. (2013) proposed that under high light intensity, the redox state of the photosynthetic electron transport chain is sensed and transmitted to the cytosol to regulate the catalase activity. This allows a temporary accumulation of H_2O_2, inducing a signalling event that is transmitted to the nucleus to modulate the expression of chloroplast-directed protection enzymes. Catalase-deficient barley plants exhibit disturbed glutathione status due to the enhanced burden on catalase-independent pathways to metabolize intracellular H_2O_2 (Smith et al., 1984). Similar alteration in glutathione level has also been observed in catalase-deficient tobacco and *Arabidopsis* plants (Rizhsky et al., 2002; Queval et al., 2007). In *Nicotiana tabacum* and barley, Cat1 deficiency caused white necrotic lesions on the leaves when plants were exposed to higher light intensities (300−1000 mmol/m^2/s to increased H_2O_2 production rather than to a reduced PPFR) (Kendall et al., 1983; Willekens et al., 1997). The reduced ability of Cat1-deficient plants to remove exogenous H_2O_2 manipulates the other components of the defense system. At least three central players, namely APX, CAT, and GPX, are involved in the removal of H_2O_2 in plants. Mittler (2002) has reported that CAT deficiency results in the induction of APX mitAOX, MDA reductase and GPX, suggesting that these enzymes may be induced to compensate for CAT suppression. Reduced peroxisomal catalase activity leads to increased sensitivity towards ozone and photorespiratory H_2O_2 induced cell death in transgenic *Arabidopsis thaliana*. It has been observed that deleterious effects of salinity may be reduced by H_2O_2 pretreatment and elevated activity of catalase in the maize plant (Gondim et al., 2012). Reports suggest that alteration of thiol-disulphide status appears to balance catalase deficiency and establish a more oxidized cellular redox state in mutant plants (Queval et al., 2009). Catalase-deficient tobacco being shifted to high light induces photorespiration and accumulation of GPX proteins along with APX (Willekens et al., 1997). The induction of GPX transcript has also been observed in *Cat2* deficient mutants that act as alternative paths for H_2O_2 decomposition in *Arabidopsis* (Queval et al., 2007; Mhamdi et al., 2010). Thus, catalase deficiency is a noteworthy loss to plants, but plants have a well-defined alternative pathway to manage the situation.

4.6 CONCLUSIONS

Plants often experience different stress conditions and a basic theme in most of their physiological challenges is the generation of harmful reactive oxygen species. To combat the deleterious effects of ROS, plants are equipped with an antioxidant defense system. Catalase, which is an important antioxidant enzyme, has excellent enzyme kinetics and plays a key role in removal of H_2O_2 from cells. It occurs in peroxisomes, mitochondria, cytosol and chloroplast. Catalase isozymes exhibit developmental stages and tissue specificity in plants. Moreover, differential expression of *cat* genes against different chemical and physical stimulations has been reported in various plant species, indicating the role of catalase as a versatile antioxidant enzyme present in aerobic organisms. On the basis of *cat* gene isolation and characterization, phylogenic relationships may be established among different plant species. Catalase is extensively studied in animal and microbial cells but information is restricted to a narrow range among plant systems. It would be desirable to explore various aspects of enzyme kinetics, enzyme structure and mechanism in plant systems.

ACKNOWLEDGMENTS

This work was financially supported by the Department of Science and Technology (DST), New Delhi, Govt. of India in the form of Women Scientist-A Scheme (SR/WOS-A/LS-192/2007).

REFERENCES

Acevedo, A., Scandalios, J.G., 1990. Expression of the catalase and superoxide dismutase genes in mature pollen in maize. Theor. Appl. Genet. 80, 705–711.

Ahmad, P., Jaleel, C.A., Salem, M.A., Nabi, G., Sharma, S., 2010a. Roles of enzymatic and non-enzymatic antioxidants in plants during abiotic stress. Crit. Rev. Biotechnol. 30, 161–175.

Ahmad, P., Nabi, G., Jeleel, C.A., Umar, S., 2011. Free radical production, oxidative damage and antioxidant defense mechanisms in plants under abiotic stress. In: Ahmad, P., Umar, S. (Eds.), Oxidative Stress: Role of Antioxidants in Plants. Studium Press Pvt. Ltd., New Delhi, India, pp. 19–53.

Ahmad, P., Umar, S., 2011. Antioxidants: Oxidative Stress Management in Plants. Studium Press Pvt. Ltd., New Delhi, India.

Ahmad, P., Umar, S., Sharma, S., 2010b. Mechanism of free radical scavenging and role of phytohormones during abiotic stress in plants. In: Ashraf, M., Ozturk, M., Ahmad, M.S.A. (Eds.), Plant Adaptation and Phytoremediation. Springer, Dordrecht Heidelberg London, NY, pp. 99–108.

Al-Rumaih, M.M., Al-Rumaih, M.M., 2008. Influence of ionizing radiation on antioxidant enzymes in three species of *Trigonella*. American Journal of Environmental Sciences 4, 151–156.

Andre, C., Kim, S.W., Yu, X.-H. Shanklin, J. (2013). Fusing catalase to an alkane-producing enzyme maintains enzymatic activity by converting the inhibitory byproduct H_2O_2 to the cosubstrate O_2. Proc. Natl Acad. Sci. USA 110, 3191–3196.

Becana, M., Moran, J.F., Iturbe-Ormaetxe, I., 1998. Iron-dependent oxygen free radical generation in plants subjected to environmental stress: toxicity and antioxidant protection. Plant Soil 201, 137–147.

Beulah, K., Ramana, T., 2013. Purification, properties and kinetic studies of catalase from leaves of *phyllanthus reticulatu*. Int. J. Pharm. Chem. Bio. Sci. 3, 940–948.

Bienert, G.P., Moller, A.L.B., Kristiansen, K.A., Schulz, A., Moller, I.M., Schjoerring, J.K., et al., 2007. Specific aquaporins facilitate the diffusion of hydrogen peroxide across membrane. J. Biol. Chem. 282, 1183–1192.

Bočová, B., Huttová, J., Liptáková, L'., Mistrík, I., Ollé, M., Tamás, L., 2012. Impact of short-term cadmium treatment on catalase and ascorbate peroxidase activities in barley root tips. Biol. Plant. 56, 724–728.

Boon, E.M., Downs, A., Marcey, D., 2007. Proposed Mechanism of Catalase *Catalase:* H_2O_2: H_2O_2 Oxidoreductase: Catalase Structural Tutorial. Available from: http://biology.kenyon.edu/BMB/Chime/catalase/frames/cattx.htm

Chamnongpol, S., Willekens, H., Moeder, W., Langebartels, C., Sandermann, J.r.H., Van Montagu, M., et al., 1998. Defense activation and enhanced pathogen tolerance induced by H_2O_2 in transgenic plants. Proc. Natl. Acad. Sci. USA 95, 5818–5823.

Chance, B., Sies, H., Boveris, A., 1979. Hydroperoxide metabolism in mammalian organs. Physiol. Rev. 59, 527–605.

Chaouch, S., Queval, G., Vanderauwera, S., Mhamdi, A., Vandorpe, M., Langlois–Meurinne, M., et al., 2010. Genetic reversion of cell death in the *Arabidopsis* cat2 knockout mutant shows that peroxisomal H_2O_2 is coupled to biotic defense responses by isochorismate synthase 1 in a daylength-related manner. Plant Physiol. 153, 1692–1705.

Charles, S.A., Halliwell, B., 1981. Light activation of fructose bisphosphate in isolated spinach chloroplasts and deactivation by H_2O_2. Planta 151, 242–246.

Cheeseman, J.M., 2007. Hydrogen peroxide and plant stress: A challenging relationship. Plant Stress 1, 4–15.

Chen, Z., Silva, H., Klessig, D.F., 1993. Active oxygen species in the indication of plant systemic acquired resistance by salicylic acid. Science 262, 1883–1886.

Corpas, F.J., Palma Sandalio, L.M., Valderrama, R., Barroso, J.B., Del Río, L.A., 2008. Peroxisomal xanthine oxidoreductase: characterization of the enzyme from pea (*Pisum sativum L.*) leaves. J. Plant Physiol. 165, 1319–1330.

Deisseroth, A., Dounce, A., 1970. Catalase: Physical and chemical properties, mechanism of catalysis and physiological role. Physiol. Rev. 50, 319–375.

Dounce, A.L., 1983. A proposed mechanism for the catalatic action of catalase. J. Theor. Biol. 105, 553–567.

Drews, G., 2011. The evolution of cyanobacteria and photosynthesis. In: Peschek, G.A., Obinger, C., Renger, G (Eds.), Bioenergetic Processes of Cyanobacteria. Springer, Dordrecht, pp. 265–284.

Eising, R., Trelease, R., Ni, W., 1989. Biogenesis of catalase in glyoxysomes and leaf-type peroxisomes of sunflower cotyledons. Arch. Biochem. Biophys. 278, 258–264.

Foyer, C.H., Noctor, G., 2009. Redox regulation in photosynthetic organisms: signaling, acclimation, and practical implications. Antioxidants Redox Signal 11, 861–905.

Fridovich, I., 1975. Superoxide dismutases. Annu. Rev. Biochem. 44, 146–159.

Frugoli, J.A., Zhong, H.H., Nuccio, M.L., McCourt, P., McPeek, M.A., Thomas, T.L., et al., 2012. Catalase is encoded by a multigene family in *Arabidopsis thaliana* (L.) Heynh. Plant Physiol. 112, 327–336.

Gianinetti, A., Cantoni, M., Lorenzoni, C., Salamini, F., Marocco, A., 1993. Altered levels of antioxidant enzymes associated with two mutations in tomato. Physiol. Plant. 89, 157–164.

Gondim, F.A., Gomes-Filho, E., Costa, J.H., Mendes, Alencar, N.L., Prisco, J.T., 2012. Catalase plays a key role in salt stress acclimation induced by hydrogen peroxide pretreatment in maize. Plant Physiol. Biochem. 56, 62–71.

Gu, Y., Sattayasamitsathit, S., Kaufmann, K., Vazquez-Duhalt, R., Gao, W., Wangb, C., et al., 2013. Self-propelled chemically-powered plant-tissue biomotors. Chemical Communication 49, 7307–7309.

Guan, L., Polidoros, A.N., Scandalios, J.G., 1996. Isolation, characterization and expression of the maize *Cat2* catalase gene. Plant Mol. Biol. 30, 913–924.

Halliwell, B., Gutteridge, J.M.C., 1992. Free radicals, antioxidants and human diseases: where are we now? J. Lab. Clin. Med. 119, 598–620.

Havir, A.E., McHale, N.A., 1987. Biochemical and developmental characterization of multiple forms of catalase in tobacco leaves. Plant Physiol. 84, 450–455.

Isah, T., Mujib, A., 2012. Studies on antioxidant enzymes activity during in vitro morphogenesis of *Caladium bicolor Linn.* Int. J. Modern Cell. Mol. Biol. 1, 1–9.

Kato, J., Yamahara, T., Tanaka, K., Takio, S., Satoh, T., 1997. Characterization of catalase from green algae *Chlamydomonas reinhardtii*. J. Plant Physiol. 151, 262–268.

Kendall, A.C., Keys, A.J., Turner, J.C., Lea, P.J., Miflin, B.J., 1983. The isolation and characterization of a catalase-deficient mutant of barley. Planta 159, 505–511.

Koyro, H.W., Ahmad, P., Geissler, N., 2012. Abiotic stress responses in plants: an overview. In: Ahmad, P., Prasad, M.N.V. (Eds.), Environmental Adaptations and Stress Tolerance of Plants in the Era of Climate Change. Springer, New York, USA, pp. 1–28.

Lenton, T.M., 2003. The coupled evolution of life and atmospheric oxygen. In: Rothschild, L., Lister, A. (Eds.), Evolution of Planet Earth. Elsevier, London, UK, pp. 35–53.

Liang, Y.E., Hui, Y.G., Qi, Z., 2000. Response of the antioxidant systems and xanthophylls cycle in *Phaseolus vulgaris* to the combined stress of high irradiation and high temperature. Photosynthetica 38, 205–210.

Loew, O., 1901. Catalase, new enzyme of general occurrence, with special reference to the tobacco plant. US Dep Agric Rep. 68, 47.

Loewen, P.C., Klotz, M.G., Hassett, D.J., 2002. Catalase – an "old" enzyme that continues to surprise us. ASM News 66, 76–82.

Mhamdi, A., Hager, J., Chaouch, S., Queval, G., Han, Y., Taconnat, L., et al., 2010. *Arabidopsis* Glutathione Reductase 1 plays a crucial role in leaf responses to intracellular H_2O_2 and in ensuring appropriate gene expression through both salicylic acid and jasmonic acid signaling pathways. Plant Physiol. 153, 1144–1160.

Michelet, L., Roach, T., Fischer, B.B., Bedhomme, M., Lemaire, S.D., Krieger-Liszkay, A., 2013. Down-regulation of catalase activity allows transient accumulation of a hydrogen peroxide signal in *Chlamydomonas reinhardtii*. Plant Cell Environ. 36, 1204–1213.

Mittler, R., 2002. Oxidative stress, antioxidants, and stress tolerance. Trends in Plant Sci. 7, 405–410.

Mullen, R.T., Gifford, D.J., 1993. Purification and characterization of catalase from Loblolly pine *(Pinus taeda* L.) mega gametophytes. Plant Physiol. 103, 477–483.

Mullen, R.T., Lee, M.S., Trelease, R.N., 1997. Identification of the peroxisomal targeting signal for cottonseed catalase. Plant J. 12, 313–322.

Ni, W., Turley, R., Trelease, R., 1990. Characterization of a cDNA encoding cottonseed catalase. Biochim. Biophys. Acta. 1049, 219–222.

Ogren, W.L., 1984. Photorespiration: Pathways, regulation and modification. Annu. Rev. Plant Physiol. 35, 415–442.

Patnaik, S.C., Sahoo, D.K., Chainy, G.B., 2013. A comparative study of catalase activities in different vertebrates. Webmed Central Zoology 4, WMC004270.

Perl, A., Perl-Treves, R., Galili, S., Aviv, D., Shalgi, E., Malkin, S., et al., 1992. Enhanced oxidative-stress defense in transgenic potato expressing tomato Cu, Zn superoxide dismutases. Theor. Appl. Genet. 85, 568–576.

Queval, G., Issakidis-Bourguet, E., Hoeberichts, F.A., Vandorpe, M., Gakiere, B., Vanacker, H., et al., 2007. Conditional oxidative stress responses in the *Arabidopsis* photorespiratory mutant cat2 demonstrate that redox state is a key modulator of daylength-dependent gene expression, and define photoperiod as a crucial factor in the regulation of H_2O_2-induced cell death. Plant J. 52, 640–657.

Queval, G., Thominet, D., Vanacker, H., Miginiac-Maslow, M., Gakiére, B., Noctor, G., 2009. H_2O_2-activated up-regulation of glutathione in *Arabidopsis* involves induction of genes encoding enzymes involved in cysteine synthesis in the chloroplast. Mol. Plant 2, 344–356.

Ray, M., Mishra, P., Das, P., Sabat, S.C., 2012. Expression and purification of soluble bio-active rice plant catalase-A from recombinant *Escherichia coli*. J. Biotechnol. 157, 12–19.

Redinbaugh, M.G., Sabre, M., Scandalios, J.G., 1990. Expression of the maize Cat3 catalase gene is under the influence of a circadian rhythm. Proc. Natl Acad. Sci. USA 87, 6853–6857.

Regelsberger, G., Jakopitsch, C., Plasser, L., Schwaiger, H., Fürtmuller, P.G., Peschek, G.A., et al., 2002. Occurrence and biochemistry of hydroperoxidase in oxygenic phototrophic prokaryotes (cyanobacteria). Plant Physiol. Biochem. 40, 479–490.

Rizhsky, L., Hallak–Herr, E., Van Breusegem, F., Rachmilevitch, S., Barr, J., Rodermel, S., et al., 2002. Double antisense plants lacking ascorbate peroxidase and catalase are less sensitive to oxidative stress than single antisense plants lacking ascorbate peroxidase or catalase. Plant J. 32, 329–342.

Sagi, M., Fluhr, R., 2006. Production of reactive oxygen species by plant NADPH oxidases. Plant Physiol. 141, 336–340.

Sahu, S., Das, P., Ray, M., Sabat, S.C., 2010. Osmolyte modulated enhanced rice leaf catalase activity under salt-stress. Adv. Biosci. Biotechnol. 1, 39–46.

Scandalios, J.G., 1974. Subcellular localization of catalase variants coded by two genetic loci during maize development. J. Hered. 65, 28–32.

Scandalios, J.G., 1983. Multiple varieties of isozymes and their role in studies of gene regulation and expression during eukaryote development. Curr. Top. Biol. Med. Res. 9, 1–31.

Scandalios, J.G. (Ed.), 1992. Current Communications in Cell And Molecular Biology: Molecular Biology of Free Radical Scavenging Systems (5). Cold Spring Harbor Laboratory Press, Cold Spring Harbor, New York.

Scandalios, J.G., 1994. Regulation and properties of plant catalases. In: Foyer, C.H., Mullineaux, P.M. (Eds.), Causes of Photooxidative Stress and Amelioration of Defense Systems in Plants. CRC Press, Boca Raton, Florida, pp. 275–315.

Scandalios, J.G., Guan, L., Polidoros, A.N., 1997. Catalases in plants: gene structure, properties, regulation, and expression. Oxidative Stress and the Molecular Biology of Antioxidant Defenses. Cold Spring Harbor Laboratory Press, pp. 343–406.

Scandalios, J.G., Tsaftaris, A.S., Chandlee, J.M., Skadsen, R.W., 1984. Expression of the developmentally regulated catalase *(Cat)* genes in maize. Dev. Genet. 4, 2–293.

Sharma, I., 2012. Arsenic induced oxidative stress in plants. Biologia. (Bratisl). 67, 447–453.

Sharma, I., 2013a. Arsenic stress in plants: An inside story. In: Khalid, R.H., Ahmad, P., Munir, O. (Eds.), Crop Improvement New Approaches and Modern Techniques. Springer, New York Dordrecht Heidelberg London, pp. 379–400.

Sharma, I., 2013b. Arsenic induced oxidative stress and antioxidant defense system of *Pisum sativum* and *Pennisetum typhoides*: A comparative study. Res. J. Biotechnol. 8, 48–56.

Sharma, I., Singh, R., Tripathi, B.N., 2007. Biochemistry of arsenic toxicity and tolerance in plants. Biochem. Cell. Arch. 7, 165–170.

Sharma, I., Travlos, I.S., 2012. Phosphate supply as a promoter of tolerance to arsenic in pearl millet. Int. J. Plant Production 6, 443–456.

Sheptovitsky, Y.G., Brudig, G.W., 1996. Isolation and characterization of spinach photosystem II membrane-associated catalase and polyphenol oxidase. Biochemistry 35, 16255–16263.

Shim, I.S., Momose, Y., Yamamoto, A., Kim, D.W., Usui, K., 2003. Inhibition of catalase activity by oxidative stress and its relationship to salicylic acid accumulation in plants. Plant Growth Regul. 39, 285–292.

Shim, I.S., Naruse, Y., Kim, Y.H., Kobayashi, K., Usui, K., 1999. Scavenging activity of NaCl-induced activated oxygen in two rice (*Oryza sativa* L.) cultivars differing in salt tolerance. Japanese J. Tropical Agricul. 43, 32–41.

Singh, S., Braus-Stromeyer, S.A., Timpner, C., Valerius, O., Tiedemann, A., Karlovsky, P., et al., 2012. The plant host *Brassica napus* induces in the pathogen *Verticillium longisporum* the expression of functional catalase peroxidase which is required for the late phase of disease. Mol. Plant Microbe Interact. 25 (4), 569–581.

Smith, I.K., Kendall, A.C., Keys, A.J., Turner, J.C., Lea, P.J., 1984. Increased levels of glutathione in a catalase-deficient mutant of barley (*Hordeum vulgare* L.). Plant Sci. Letters 37, 29–33.

Spanou, C.I., Veskoukis, A.S., Stagos, D., Liadaki, K., Aligiannis, N., Angelis, A., et al., 2012. Effects of Greek legume plant extracts on xanthine oxidase, catalase and superoxide dismutase activities. J. Physiol. Biochem. 68, 37–45.

Sumner, J.B., Dounce, A.L., 1937. Crystalline catalase. J. Biol. Chem. 121, 417–424.

Tsaftaris, A.S., Bosabalidis, A.M., Scandalios, J.G., 1983. Cell-type-specific gene expression and acatalasemic peroxisomes in a null *Cat2* catalase mutant of maize. Proc. Natl. Acad. Sci. 80, 4455–4459.

Wadsworth, G.J., Scandalios, J.G., 1989. Differential expression of the maize catalase genes during kernel development: The role of steady-state mRNA levels. Dev. Genet. 10, 304–310.

Warburg, O., 1923. Ueber die antikatalytische Wirkung der Blaus ure. Biochem. Z. 136, 266–272.

Willekens, H., Langebartels, C., Tire, C., Van Montagu, M., Inze, D., Van Camp, W., 1994a. Differential expression of catalase genes in *Nicotiana plumbaginifolia*. Proc. Natl. Acad. Sci. USA. 91, 10450–10454.

Willekens, H., Villarroel, R., Inze, D., Van Montagu, M., Van Camp, W., 1994b. Molecular identification of catalases from *Nicotiana plumbaginifolia*. FEBS Lett. 352, 79–83.

Willekens, H., Inze, D., Van Montagu, M., Van Camp, W., 1995. Catalases in plants. Mol. Breeding 1, 207–228.

Willekens, H., Chamnongpol, S., Davey, M., Schraudner, M., Langebartels, C., Van Montagu, M., et al., 1997. Catalase is a sink for H_2O_2 and is indispensable for stress defense in C3 plants. EMBO J. 16, 4806–4816.

Woodson, W.R., Kerdnaimongkol, K., 1999. Inhibition of catalase by antisense RNA increases susceptibility to oxidative stress and chilling injury in transgenic tomato plants. J. Amer. Soc. Hort. Sci. 124, 330–336.

Xu, J., Duan, X., Yang, J., Beeching, J.R., Zhang, P., 2013a. Enhanced ROS scavenging by over-production of superoxide dismutase and catalase delays post-harvest physiological deterioration of cassava storage roots. Plant Physiol. 161, 1517–1528.

Xu, J., Duan, X., Yang, J., Beeching, J.R., Zhang, P., 2013b. Coupled expression of Cu/Zn-superoxide dismutase and catalase in cassava improves tolerance against cold and drought stresses. Plant Signaling Behav. 8 (6), 24525.

Yamaguchi, J., Nishimura, M., Akazawa, T., 1986. Purification and characterization of heme-containing low-activity form of catalase from greening pumpkin cotyledons. Fur. J. Biochem. 159, 315–322.

Youssef, M.M., Azooz, M.M., 2013. Biochemical studies on the effects of zinc and lead on oxidative stress, antioxidant enzymes and lipid peroxidation in Okra (*Hibiscus esculentus* cv. *Hassawi*). Sci. Int. 1, 12–16.

Yu, L., Lamb, C.J., Dixon, R.A., 1992. Purification and biochemical characterization of proteins which bind to the H-box *cis*-element implicated in the transcriptional activation of plant defense genes. Plant J. 3, 805–816.

Zamocky, M., Furtmüller, P.G., Obinger, C., 2008. Evolution of catalases from bacteria to humans. Antioxidants Redox Signal. 10, 1527–1547.

Zamocky, M., Gasselhuber, B., Furtmüller, P.G., Obinger, C., 2012. Molecular evolution of hydrogen peroxide degrading enzymes. Arch. Biochem. Biophys. 525, 131–144.

Zeile, K., Hellstrom, H., 1930. Ueber die aktive Gruppe der Leberkatalase. Z. Physiol. Chem. 192, 171–174.

Zhong, H.H., Young, J.C., Pease, E.A., Hangarter, R.P., McClung, R.C., 1994. Interactions between light and the circadian clock in the regulation of the *CAT2* expression in *Arabidopsis*. Plant Physiol. 104, 889–898.

Chapter 5

Role of Glutathione in Abiotic Stress Tolerance

Agnieszka Waśkiewicz, Olimpia Gładysz, Kinga Szentner and Piotr Goliński

5.1 INTRODUCTION

Plants are constantly exposed to environmental changes, and when these changes are fast and extreme, they cause stress. Physically, stress is defined as the mechanical force per unit area on a subject (Mahajan and Tuteja, 2005).

In general, stress can be divided into biotic and abiotic. Biotic stresses include pathogens (fungi, bacteria and viruses) and herbivorous insects. In turn, abiotic stress factors include high temperatures, cold (cooling and freezing), drought, salinity, UV radiation, heavy metals and herbicides.

Abiotic stress, such as salinity, drought, extreme temperature, herbicides and heavy metals, is a major factor affecting plant growth, development and yield (Meloni et al., 2003; Salama and Al-Mutawa, 2009). Salinity is the main environmental stress, which can cause both hypertonic and hyperosmotic stress and may indirectly affect growth through reduced photosynthesis (Meloni et al., 2003). The effect of other stresses such as heat or drought is associated with reduced water uptake leading to cell dehydration. Thus, changes in cellular metabolism connected with osmotic adjustment can be expected (Ruelland et al., 2009). Cold as a stress factor is associated with significant alterations in energy metabolism, namely a decrease in the rate of enzyme-catalyzed reactions resulting in metabolic imbalances associated with an oxidative stress (Apel and Hirt, 2004). Heavy metal stress may be caused by different toxic metals that generate the overproduction of reactive oxygen species in plants (Lewis et al., 2001; Gallego, 2005; Xu et al., 2010; Yadav, 2010). Exposure of plants to high levels of metals causes inhibition of photosynthesis as well as problems in distribution of water and nutrients (Wójcik and Tukiendorf, 2004; Mohanpuria et al., 2007).

P. Ahmad (Ed): Oxidative Damage to Plants. DOI: http://dx.doi.org/10.1016/B978-0-12-799963-0.00005-8

In most environmental stresses reactive oxygen species (ROS) are generated and play a dual role: as plant toxic compounds and as regulators of biological processes (Meloni et al., 2003; Foyer and Noctor, 2005; Parida and Das, 2005; Mittler, 2006; Fujita et al., 2006). ROS levels are controlled both by enzymatic (superoxide dismutase, peroxidase, catalase, and glutathione reductase) and nonenzymatic antioxidants such as glutathione, tocopherols, ascorbic acid, vitamin, phenolic and carotenoids (Jaleel et al., 2009). Glutathione plays a key role as an antioxidant in environmental stresses, which constitutes the basis for further discussion.

5.2 PLANTS UNDER VARIOUS ABIOTIC STRESSES

Plants are continuously exposed to numerous abiotic stresses that adversely affect plant growth, development and yield, causing crop loss worldwide (Gao et al., 2007; Edmeades, 2009). In addition, more than one abiotic stress can occur at one time (Tester and Bacic, 2005; Mittler, 2006). Plants respond to stresses as individual cells and synergistically as whole organisms and in relation to changes in morphological, physiological and biochemical processes (Bhatnagar-Mathur et al., 2009; Aroca et al., 2012). The stress signal is first perceived by the receptors present on the membrane of the plant cells. Afterwards the signal information is transduced downstream, resulting in the activation of various stress responsive genes (Seki, 2003; Avni Óktem et al., 2008). Advances in physiology, genetics, and molecular biology have greatly improved our understanding of plant responses to stresses (Amudha and Balasubramani, 2011).

5.3 TOLERANCE TO ABIOTIC STRESSES

Tolerance and susceptibility to abiotic stresses are very complex and may affect multiple stages of plant development (Chinnusamy et al., 2004; Cramer et al., 2011; Zagorchev et al., 2013). Plant traits that are associated with resistance mechanisms are multigenic and difficult to predict and control (Sreenivasulu et al., 2007; Roy et al., 2011). Plant adaptation to environmental stresses is dependent upon the activation of cascades of molecular networks involved in stress perception, signal transduction, and the expression of specific stress-related genes and metabolites. Resistance or sensitivity of plants to stress depends on species, genotypes and development age (Singh et al., 2012). Antioxidants play an important role in preventing the stress-induced accumulation of toxic concentrations of reactive oxygen species (ROS) such as superoxide anion ($O_2^{\bullet -}$), hydroxyl radicals ($^{\bullet}OH$), singlet oxygen (O_2), hydrogen peroxide (H_2O_2) or oxylipins from unsaturated fatty acids (Bhattacharjee, 2005; Halliwell and Gutteridge, 2007; Kellos et al., 2008; Eltayeb et al., 2010; Gill and Tuteja, 2010; Suzuki et al., 2012), which levels need to be controlled and various antioxidants have evolved for this purpose

(Il-Sup et al., 2012; Szarka et al., 2012). Plants may protect themselves against oxidative stress by enhancing activities of enzymatic (superoxide dismutase, peroxidase, catalase and glutathione peroxidase) and nonenzymatic antioxidants (ascorbic acid, glutathione, α-tocopherol, β-carotene, flavonoids and hydroquinone) (Moller et al., 2007; Wang et al., 2012). Glutathione (GSH), as a nonenzymatic antioxidant, is involved in both the direct and the indirect control of ROS concentrations (Mittler, 2006; Szalai et al., 2009).

5.4 GLUTATHIONE

Glutathione as a common component of plant and animal cells plays a key role as a cellular antioxidant (Pompella et al., 2003; Elias et al., 2008) and protects organisms against toxic reactive oxygen species (Anjum et al., 2012; García-Giménez et al., 2013; Han et al., 2013a). Moreover, glutathione is involved in numerous aspects of the plant's life (Akter et al., 2013; Bossio et al., 2013). Multiple actions of glutathione in plant metabolism concern signaling of sulfur status, resistance to xenobiotics, heavy metal tolerance and pathogen response (Tausz et al., 2004; Han et al., 2013b). It also maintains redox homeostasis in cells (Szalai et al., 2009).

5.4.1 Chemical Structure

The chemical structure of this tripeptide (Fig. 5.1) is based on the peptide bond between glutamic acid, cysteine and glycine (γ-L-glutamyl-L-cysteinyl-glycine). The sequence and bonds of aminoacids define chemical properties of glutathione. There is a rare linkage between the amine group of cysteine and the γ-carboxyl group of glutamic acid. This isopeptide bond determines a high resistance to enzymatic activity, while a thiol group influences reactivity of glutathione. The enzyme γ-glutamyltransferase (GGT) is capable of degrading glutathione molecules (Huseby et al., 2003). Initially γ-glu-cys bonds in GSH are hydrolyzed by γ-glutamyltransferase (GGT) and then the remaining cysteinylglycine is hydrolyzed by other peptidases.

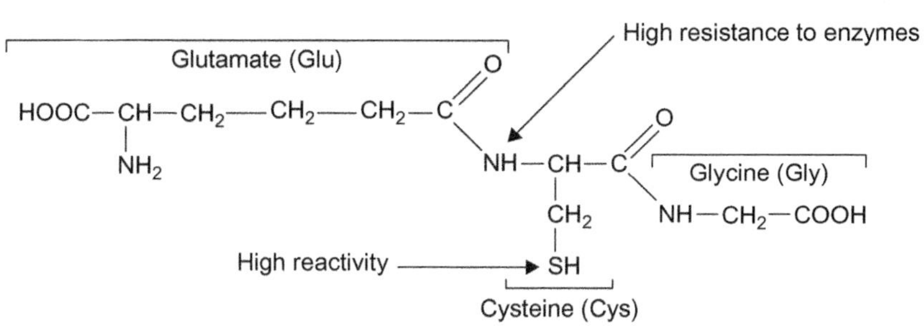

FIGURE 5.1 The chemical structure of glutathione.

Molecular dynamics (MD) simulations and NMR spectroscopy reveal that glutathione is very flexible in aqueous solutions and can form hydrogen bonds. There are bridge hydrogen bonds between water molecules and GSH that make conformations more stable. Such a hydrogen bonding network may influence glutathione activity under physiological conditions (Zhang and Wu, 2011).

5.4.2 Biosynthesis

Glutathione biosynthesis relies on two consecutive reactions of proper amino acids (Martin 2003; Mullineaux and Rausch, 2005; Noctor et al., 2012):

$$Glu + Cys + ATP \xrightarrow{\gamma - ECS} \gamma - GluCys + Gly + ATP \xrightarrow{GS} GSH \tag{5.1}$$

The first step of glutathione biosynthesis is γ-glutamylcysteine formation. The reaction between L-glutamate and L-cysteine is catalyzed by γ-glutamylcysteine synthetase (γ-ECS, GSH1) and energy is consumed as adenosine triphosphate (ATP). There is a peptide bond formed between the carboxyl group of glutamate and the amino group of cysteine. During the next step glycine is added to γ-glutamylcysteine. There is a peptide bond between the N-terminal end of glycine and the C-terminal end of γ-glutamylcysteine. Glutathione formation is catalyzed by glutathione synthetase (GS, GSH-S, GSH2). The reaction is adenosine triphosphate (ATP) dependent as well. The GSH1 enzyme substantially controls glutathione biosynthesis and it is concentrated in plastids, while the GSH2 enzyme is located in plastids and mainly in the cytosol. Glutathione (GSH) is found in the cytosol, chloroplasts, vacuoles, mitochondria and the endoplasmic reticulum (Yadav, 2010).

5.4.3 Properties

Glutathione is freely soluble in water and accumulates to millimolar concentrations, relatively high in comparison to other cellular thiols (Szarka et al., 2012).

The thiol group (−SH) in glutathione molecules acts as a reducing and nucleophilic agent and may undergo oxidation, resulting in glutathione disulfide (GSSG) formation:

$$2GSH \rightarrow GSSG + 2H^{+} + 2e \tag{5.2}$$

Besides nonenzymatic oxidation, an enzymatic process also occurs. It is catalyzed by glutathione peroxidase (GPx) in the presence of hydrogen peroxide H_2O_2 and organic peroxides:

$$2GSH \xrightarrow{H_2O_2, GPx} GSSG \tag{5.3}$$

It is claimed that the physiological function of glutathione peroxidase (GPx) is to remove hydrogen peroxide and other oxyradicals in animals, but the existence of this enzyme in plant cells was also reported (Eshdat et al., 1997). This phospholipid hydroperoxide glutathione peroxidase (PHGPX) contains cysteine instead of selenocysteine; however, there are reports on selenium-dependent GPx homologues in plants (Fu et al., 2002). Contrary to animal metabolism of hydrogen peroxide (H_2O_2), which is associated with selenium-dependent glutathione peroxidase (GPx), plant metabolism of hydrogen peroxide is linked to ascorbate (Noctor et al., 2012) and ascorbate peroxidase (APX) (Pinto et al., 2006). Phospholipid hydroperoxide glutathione peroxidase (PHGPx) is a member of the glutathione peroxidase (GPx) family and uses glutathione to its regeneration. It needs to be added that thioredoxins (Trx) are much more efficient for PHGPx regeneration in comparison to glutathione (Faltin et al., 2010). In fact, plant glutathione peroxidase (GPx) was initially characterized as glutathione dependent, but it does not consume glutathione to its regeneration. Thioredoxin (Trx) is a substrate for plant glutathione peroxidase (GPx) instead of glutathione (GSH) (Navrot et al., 2006).

The oxidized form of glutathione is glutathione disulfide (GSSG). Glutathione oxidation is a reversible process. Glutathione disulfide (GSSG) reduction is catalyzed by glutathione reductase (GR) and nicotinamide adenine dinucleotide phosphate (NADPH) and a solvent proton is consumed:

$$GSSG + NADPH + H^+ \xrightarrow{GR} 2GSH + NADP^+ \tag{5.4}$$

The thiol groups (−SH) of glutathione reduced state (GSH) react easily with free radicals such as reactive oxygen species (ROS) and protein radicals:

$$OH^\bullet + GSH \rightarrow H_2O + GS^\bullet \tag{5.5}$$

$$protein^\bullet + GSH \rightarrow protein + GS^\bullet \tag{5.6}$$

$$2GS^\bullet \rightarrow GSSG \tag{5.7}$$

One of the main purposes of glutathione is to maintain protein stability. Oxidized protein states (protein disulphide) are nonenzymatically and enzymatically reduced. An enzyme glutathione transhydrogenase induces protein disulphide reduction to reduced thiol groups of protein. As a result, the physiological protein function may be preserved.

Moreover, thiol groups (−SH) of glutathione are reactive towards electrophilic compounds and maintain cellular detoxication. Covalent bond formation between the sulfur atom of glutathione and an electrophilic agent is catalyzed by glutathione S-transferase (GST):

$$GSH + X \xrightarrow{GST} GS - X \tag{5.8}$$

The products of their S-coniugation are removed from cells. Glutathione S-transferase (GST) induces this process in relation to both endogenous and exogenous substances (particularly toxic xenobiotics X) (Dixon et al., 2002;

Frova, 2003; Edwards and Dixon, 2004; Moons, 2005; Basantani and Srivastava, 2007; Oztetik, 2008; Dixon et al., 2010; Cummins et al., 2011; Mohsenzadeh et al., 2011).

One of the detoxification mechanisms removes toxic methylglyoxal and formaldehyde (Noctor et al., 2012). Methylglyoxal detoxification is based on the glyoxalase system and it involves glutathione to form a hemithioacetal. Glyoxalase I (GLY I) converts this conjugation into S-lactoyl-glutathione and it is further metabolized to lactate by glyoxalase II (GLY II):

$$GSH + CHOCOCH_3 \rightarrow GSCHOHCOCH_3 \tag{5.9}$$

$$GSCHOHCOCH_3 \xrightarrow{GLY_I} GSCOCHOHCH_3 \xrightarrow{GLY_{II}} GSH + CH_3CHOHCOO^- \tag{5.10}$$

Similarly, S-conjugation is formed during formaldehyde detoxification and two enzymes: formaldehyde dehydrogenase (FDH) and S-formylglutathione hydrolase (FGH), catalyze this process:

$$GSH + HCHO \xrightarrow{FDH} GS - COH \xrightarrow{FGH} GSH + HCOO^- \tag{5.11}$$

Moreover, glutathione is able to form chelates with metals (Dimitrova et al., 2010; Yadav, 2010) such as cadmium (Delalande et al., 2010), copper (Shtyrlin et al., 2005) and chromium (Levina et al., 2003). Accumulation of metal ions M^+ in plants is caused by exposure to heavy metal stress. Cellular detoxication of heavy metals is based on the conjugation between GSH and heavy metals. This process is catalyzed by glutathione S-transferase (GST):

$$2GSH + M^+ \xrightarrow{GST} GS - M - SG \tag{5.12}$$

Secondly, glutathione takes part in phytochelatin synthesis at the excess of heavy metals. Phytochelatins (PCs) are glutathione oligochetatins ($[\gamma$-Glu-Cys$]_n$Gly) able to bind heavy metals (Cd, Hg, Zn, Pb, Ni) (Yadav, 2010). The process is catalyzed by phytochelatin synthase (PCS). Complexes of metal and glutathione or phytochelatins are formed in the cytosol and they are transported to vacuoles.

Additionally, reactive oxygen species (ROS) are produced under heavy metal stress conditions and they are utilized through the vital ascorbate-glutathione cycle. The antioxidant cooperation between glutathione and ascorbate has been known since 1967 as the Halliwell and Foyer cycle (Foyer and Noctor, 2011). Glutathione reduces the oxidized ascorbate state (dehydroascorbate DHAsc), then glutathione disulphide (GSSG) is reduced to glutathione (GSH) using nicotinamide adenine dinucleotide phosphate (NADPH). While ascorbate (Asc) is chemically able to reduce H_2O_2, ROOH or organic peroxides, an enzymatic link is provided by ascorbate peroxidase (APX). Fig. 5.2 presents the ascorbate-glutathione pathway for peroxide metabolism, but glutathione can be involved in ascorbate independent cycles

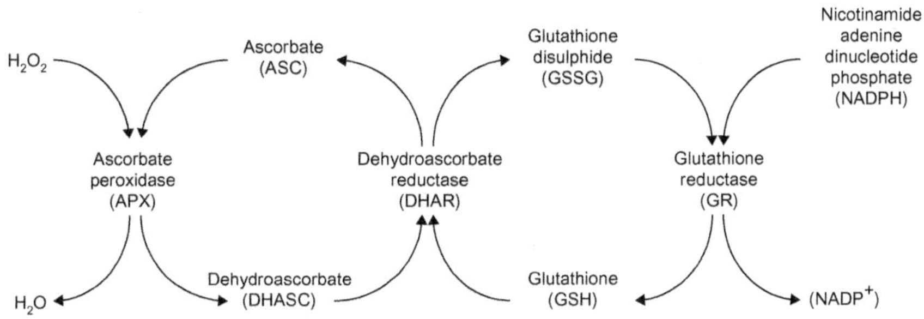

FIGURE 5.2 Glutathione–ascorbate cycle, aka Halliwell and Foyer cycle.

of H_2O_2 or peroxide reduction. It requires additional proteins to relate GSH oxidation and peroxide reduction (Noctor et al., 2012). Particularly glutathione S-transferase (GST) seems to play a direct role of glutathione peroxidase (GPX) when glutathione (GSH) acts as a reductant.

Furthermore, glutathione, ascorbate and α-tocopherol constitute the triad of nonenzymatic low molecular weight antioxidants in living organisms (Szarka et al., 2012). A decrease of glutathione concentration leads to imbalance in the interdependent ROS scavenging network of the triad. The loss of ascorbate-glutathione-α-tocopherol cooperation results in higher oxidative stress in plant cells.

Numerous redox reactions, where glutathione acts as a reducing agent and antioxidant, are possible thanks to the value of approximate redox potential $E_{GSSH/GSH} = -0.24$ V (Foyer and Noctor, 2005). According to the Nernst equation, glutathione redox potential is a function of GSSG/GSH concentration ratio:

$$E = E^o + \frac{RT}{nF}\ln\frac{[GSSG]}{[GSH]^2} \tag{5.13}$$

Values of redox potential $E_{GSSH/GSH}$ can be linked to the status of a cell's life (Schafer and Buettner, 2001), because cells proliferate when $E_{GSSH/GSH} = -0.24$ V, differentiate at -0.20 V and apoptosis occurs at -0.17 V. This parameter can be a universal marker to estimate plant viability during plant stress (Kranner and Birtic, 2006) and the value above -0.16 V seems to initiate programmed cell death (PCD). Glutathione can be regarded as a main cellular redox buffer and buffering capacity of the GSH/GSSG couple maintains cellular homeostasis. The glutatione redox buffer is a part of signal transmission during environmental, biotic and abiotic stress in cells (Meyer, 2008).

Generally glutathione accumulates to millimolar concentrations, but it varies depending on specific cell types and subcellular compartments. For example, total foliar glutathione concentrations in higher plants are in the range 0.1–1.5 mM, mainly as reduced GSH, but it depends on cell types

(Mullineaux and Rausch, 2005). There is 0.2–0.3 mM GSH in poplar mesophyll and epidermal cells. In *Arabidopsis* epidermal, basement and trichome cells of leaves GSH concentrations are estimated as 0.14, 0.08, 0.24 mM, respectively, in contrast to 2.7 mM in root hair cells and 5.5 mM in nonroot hair cells (Mullineaux and Rausch, 2005). Regardless of GSH concentration there is a higher concentration of the reduced glutathione state (GSH) than the oxidized state (GSSG) in healthy cell. In the case of nonstress conditions, the GSH/GSSG ratio is about 20:1 (Mhamdi et al., 2010; Noctor et al., 2012; Szarka et al., 2012) or even 100:1 (Maughan and Foyer, 2006). The change of this relation is an essential signal that oxidative stress takes place. When there is excess glutathione disulfide (GSSG), it reacts with thiol groups of proteins and as a result they become inactivated:

$$protein - SH + GSSG \rightarrow protein - SSG + GSH \tag{5.14}$$

$$HS - protein - SH + GSSG \rightarrow protein\ disulphide + 2GSH \tag{5.15}$$

Glutathione reductase (GR) prevents such an undesirable process (Peerzada et al., 2012) (see reaction 4). If glutathione reductase (GR) fails, then multidrug resistance proteins (MRP) drop out GSSG of cells (Klein et al., 2006).

All key features of glutathione, as well as high water solubility and stability confirm that glutathione is an important molecule in cellular metabolism, as well as redox signaling. When plants are exposed to a variety of abiotic and biotic stresses then reactive oxygen species (ROS) are formed and oxidative stress occurs. Potentially reactive oxygen species (ROS) can cause damage to plant cells, but they are also able to initiate and transmit signals in cellular compartments (Galvez-Valdivieso and Mullineaux, 2010). Signals are generated in response to environmental changes and they include either programmed cell death (PCD) or adaptation to different conditions. Changes in glutathione content and redox state are the source of signaling. Low levels of reduced glutathione (GSH) are caused by a deficiency in γ-glutamylcysteine synthetase (γ-ECS) and this phenomenon can be associated with the exposure to abiotic plant stress. GSH concentration is responsive to environmental changes, e.g. high light exposure, and changes in GSH content can carry a signal out of the chloroplast. This signal induces a redox change in the cytosol.

Glutathione is involved in both ROS detoxification and sensing and transmission of the redox signal. Consequently, ROS concentration influences the $[GSSG]/[GSH]^2$ ratio. The model of redox signaling by cellular glutathione redox buffer is based on the transmission of ROS-dependent redox signals to redox-sensitive target proteins involved in transcription, metabolism and development (Meyer, 2008). Glutaredoxins (GRXs) are mediators for reversible electron transfer between glutathione and target proteins. There is full reversibility of electron transfer, thus glutathione redox buffer and native target proteins are in equilibrium.

Glutathione participation in cell signaling is based on at least two mechanisms: protein S-glutathionylation and cysteine S-nitrosylation (Zhang and Forman, 2012). Protein S-glutathionylation unfolds when GSH reacts with cysteine residues in proteins and protein disulphides (PSSG) are formed. The second mechanism relies on a reaction of thiolate ($-S^-$) with nitrosoglutathione (GSNO).

5.4.4 Analytical Methods for Determination of Glutathione

The total glutathione concentration is represented by free glutathione forms (glutathione GSH and glutathione disulphide GSSG) as well as glutathione fractions bound to proteins. In order to determine free glutathione concentration, proteins should be precipitated using such strong acids as, e.g., sulfosalicylic, perchloric or trichloroacetic (Pastore et al., 2003). After centrifugation the solution consists of free glutathione forms without proteins and it is prepared for further measurements by fluorimetry, spectrophotometry or high performance liquid chromatography (HPLC). In general glutathione determination requires the reduction of disulphide bonds between glutathione and other thiols or proteins.

Almost all methods are based on products of glutathione derivatization. Such analytical methods use colorimetric reagents, e.g. Ellman's reagent (5,5′-dithiobis-2-nitrobenzoic acid, DNTB) (Koehler, 2003; Chen et al., 2008) or Sanger's reagent (1-fluoro-2,4-dinitrobenzene, DNFB) (Kranner and Grill, 1996; Wang et al., 2012) and then high-performance liquid chromatography (HPLC) is applied.

Glutathione may also be determined as an *o*-phthalaldehyde derivative (Bohmer et al., 2011), because *o*-phthalaldehyde (OPA) reacts with sulphydryl and amino groups of glutathione. GSH-OPA species are highly stable and fluorescent, facilitating sensitive and specific quantitative determination of GSH in biological systems by HPLC with fluorescence (FL) detection (Kandar et al., 2007; Michaelsen et al., 2009).

Monochlorobimane (MCB) and glutathione (GSH) create a highly specific and fluorescent product, which is another derivative used in fluorimetric analysis (Sakhi et al., 2007; Hepel and Stobiecka, 2011).

Other glutathione derivatives are formed with the benzofurazan reagent, SBD-F (7-fluoro-2,1,3-benzoxadiazole-4-sulfonate) and ABD-F (4-aminosulfonyl-7-fluoro-2,1,3 benzoxadiazole). They were analyzed by high-performance liquid chromatography with fluorescence detection (Santa et al., 2006) and fluorescence microscopy (Li et al., 2012).

To sum up, there are numerous papers concerning glutathione determination in biological samples published since the 1960s (Pastore et al., 2003). They are based on spectrophotometry (Hormozi-Nezhad et al., 2012), fluorometry (Rousar et al., 2012), chemiluminescence (Wang et al., 2003), capillary electrophoresis (Mendoza et al., 2004; Mendoza et al., 2008), mass spectrometry with liquid chromatography (Gucek et al., 2002; Zhu et al., 2008; Harwood et al., 2009), as well as Raman spectroscopy (Huang et al., 2009). High performance liquid chromatography (HPLC) can be used together with fluorometric detection (Nolin et al., 2007; Janes et al., 2010), chemiluminescence (McDermott et al., 2011) dansylation (Ma et al., 2010) and electrochemical detection (Potesil et al., 2005).

Direct, selective and sensitive methods are based on electrochemical properties of glutathione (Harfield et al., 2012). Voltammetric methods such as cyclic voltammetry (CV), differential pulse voltammetry (DPV) and the glassy carbon electrode technique can be used (Wang et al., 2009). The glutathione oxidation peak is concentration dependent and can be registered on a modified glassy carbon electrode (Hou and Ai, 2011). Glutathione oxidation can be studied on a platinum electrode (Budnikov et al., 2004), a modified carbon paste electrode (Raoof et al., 2009), a mesoporous carbon electrode (Ndamanisha et al., 2009), a nanoscale copper hydroxide carbon ionic liquid composite electrode (Safavi et al., 2009) or a powder microelectrode (Chen et al., 2006).

Also a hanging mercury drop electrode (HMDE) and the Brdicka procedure can be applied (Vacek et al., 2004), which is a polarographic glutathione determination method in ammonia buffered Co (III) solution. Glutathione electroreduction is rare as an electrochemical determination method, but it was registered on a glassy carbon electrode that was modified by mercury film (Sezginturk and Dinckaya, 2011).

5.5 CHANGES OF GLUTATHIONE CONTENT IN VARIOUS PLANTS UNDER ABIOTIC STRESS

Glutathione is a key factor for cellular redox homeostasis and tolerance against abiotic and biotic stresses (Chen et al., 2003; Liedschult et al., 2010; Ramirez et al., 2013). The accumulation of glutathione (and other antioxidants) is commonly observed in plants under stress, as a protective response that may determine whether stress causes only slight and temporary deviations from the normal state, or more severe and permanent damage (Freeman et al., 2004; Tausz et al., 2004; Dempsey et al., 2011; Gill et al., 2013). A major function of glutathione in protection against oxidative stress is the reduction of H_2O_2 (Millar et al., 2003; Foyer and Noctor, 2005; Mittler, 2006; Shao et al., 2008; Szalai et al., 2009; Yadav, 2010). The change in the ratio of its reduced (GSH) to oxidized (GSSG) form during the degradation of H_2O_2 is important in certain redox signaling pathways (Pastori and Foyer, 2002; Kocsy et al., 2004a; Kocsy et al., 2004b; Rouhier et al., 2008; Yadav, 2010).

5.5.1 Glutathione Content under Heavy Metal Stress

Trace elements, also called heavy metals, represent a density higher than $5\ g\ cm^{-3}$. Among all heavy metals, 17 can be available and significant for living cells (Yadav, 2010). Exposure to these heavy metals induces responses from the antioxidative system, dependent upon plant species, tissue, metal/metals and intensity of stress (Schútzendúbel and Polle, 2002).

Glutathione plays a fundamental role in many cellular heavy-metal detoxification processes through activation and conjugation with these toxic molecules by glutathione *S*-transferase (Edwards and Dixon, 2005; Rouhier, 2008; Szalai et al., 2009; Garg and Aggarwal, 2011; Anjum et al., 2012; Saenen et al., 2013). GSH is also a precursor of phytochelatins, which are synthesized by phytochelatin synthase. Formation of phytochelatins (PCs) is induced by a range of heavy metals (Mendoza-Cozatl and Moreno-Sanchez, 2006; Seth et al., 2012; Zeng et al., 2012). Many toxic metals such as nickel, cadmium, aluminum, mercury and arsenate are successfully detoxified (chelated) via GSH and PCs (Cobbett and Goldsbrough, 2002; Rouhier et al., 2008).

Among heavy metals, cadmium (Cd) contamination is a serious problem for growth plants (Mendoza-Cozatl and Moreno-Sanchez, 2006; Anjum et al., 2008; Chen et al., 2010; Ju et al., 2011; Wang et al., 2011; López-Climent et al., 2013; Nakamura et al., 2013).

Anjum et al. (2008) studied the application of sulfur (S) on the reduction of toxicity at different levels of cadmium (0, 25, 50 and 100 mg Cd kg^{-1} soil) in mustard (*Brassica campestris* L.). The highest reduction in growth, chlorophyll content and ascorbate and glutathione levels was observed at 100 mg Cd kg^{-1} soil. Addition of sulfur caused decreasing of Cd toxicity. The highest effect was noticed in the case of 25 and 50 mg Cd kg^{-1} soil.

Application of exogenous glutathione (20 mg L^{-1} GSH in 5 μM Cd culture medium) under cadmium stress in two barley genotypes (Cd-sensitive and Cd-tolerant) was evaluated (Chen et al., 2010; Wang et al., 2011). Addition of GSH significantly alleviated Cd-induced growth inhibition and decreased cadmium levels in roots and leaves, especially in sensitive genotypes. Moreover, GSH efficiently reduced Cd-induced ROS and played a protective role under cadmium stress.

High aluminum (Al) level in agricultural regions is one of the important phytotoxicity problems limiting crop yield. Ruiz et al. (2006) examined effectiveness of the boron (B) in stimulation of glutathione metabolism in sunflower plants under high aluminum level. Glutathione content was analyzed in leaves and roots after 15 days of treatment with two different levels of boron (50 and 100 μM H_3BO_3). Similar concentrations of GSH were found in leaves at both doses of boron. It is worthy to note that the effect of aluminum phytotoxicity is primarily in roots and the highest glutathione level (at 100 μM H_3BO_3) in this part of plants was observed.

Copper (Cu) is essential for optimal plant growth and participates in several physiological processes (Yruela 2005). Phytotoxicity of this metal is associated with visible symptoms such as reduction in root and shoot growth and chlorosis (Tausz et al., 2004; Gratao et al., 2005). To explain whether determination of GSH and GSSG levels could be used as an early indicator of copper stress in tomato plants, a short period (acute treatment – 0.4, 9, 36 μM Cu for 48 h) and a long period (chronic treatment – 0.4, 3, 12 μM Cu for 28 days) exposure were compared (Garrido et al., 2010). Toxicity symptoms were observed in plants with chronic treatment, such as chlorosis and drastic reduction of root biomass. These effects were not found in the group of plants with acute treatment. The GSH and GSSG contents were dependent on copper concentration in both treatments. Variability of GSH level in roots was the most visible.

Metal toxicity depends on concentrations in the plant; zinc (Zn) is an essential microelement in many physiological processes and at high concentrations is very toxic for plants (Nan et al., 2002). In order to explain the role of GSH in the defense of poplar clone against high zinc toxicity (1, 5 and 10 mM of Zn) the total (GSH + GSSG) and GSSG contents in the young leaves were investigated (Baccio et al., 2005). Results demonstrated that treatment with Zn decreased the total GSH content in the young leaves with a simultaneous increase of GSSG, which indicates a protective role of GSH metabolism under zinc toxic levels.

Toxicity of chromium (Cr) is associated with inhibition of seed germination and plant growth, disruption of nutrient, inactivation of mitochondrial electron transport, decrease of the activity of antioxidant enzymes, and damage of the chloroplast and cell membrane (Panda and Choudhury, 2005; Shanker et al., 2005; Panda, 2007). Zeng et al. (2012) investigated effects of exogenous glutathione on alleviating Cr-stress in rice plants. The studies used six different experimental blocks (I – control (without GSH and Cr); II – 50 μM GSH; III – 100 μM GSH; IV – 100 μM Cr, V – 100 μM Cr + 50 μM GSH and VI – 100 μM Cr + 100 μM GSH). The highest reduction in plant growth and chlorophyll content was shown in plants exposed to 100 μM Cr, whereas the application of exogenous GSH decreased reduction of these parameters.

Plants can be exposed to several metals simultaneously. This can lead to different interactions between them, such as synergistic, antagonistic, additive or independent (Sun et al., 2005; Son et al., 2013). The effects of two metal (cadmium – Cd and lead (plumbum) – Pb) interactions on the production of phytochelatins and glutathione were monitored (over 14 days) in wheat. After exposure to Cd, Pb and two metals simultaneously (Cd and Pb), the levels of PCs and GSH in different parts of the plant (roots, stems and leaves) were measured. The studies showed that cadmium was more effective than lead in increasing glutathione and phytochelatins production, whereas the combination of Cd and Pb was antagonistic for GSH formation in the growth period. The results indicate that metal interactions should be considered in the application of PCs and GSH as potential biomarkers for the evaluation of metal toxicity (Sun et al., 2005).

In other studies conducted by Huang et al. (2010), the effect of multiple heavy metal stress (cadmium, lead and mercury) on glutathione content in two mangrove plant seedlings (*Kandelia candel* and *Bruguiera gymnorrhiza*) was investigated. This experiment was carried out in five groups with different concentrations of metals (I group – 1.0 mg L^{-1} Pb^{2+}, 0.1 mg L^{-1} Cd^{2+} and 0.1 mg L^{-1} Hg^{2+}; II, III and IV groups contained metal levels that were $\times 5$, $\times 10$ and $\times 15$ higher than I group, respectively, and V group was control without metals). After one month the GSH and AsA contents in leaves, stems and roots were analyzed. During ROS detoxification both studied species used GSH under low heavy metal stress. In the case of higher levels of metals, the antioxidant activity was decreased (Huang et al., 2010).

5.5.2 Glutathione Content of Plants Treated with Herbicides

Xenobiotics, such as synthetic herbicides, are factors causing abiotic stress in plants and they are part of the agricultural ecosystem. The reduced form of glutathione (GSH) accumulates in plant cells to high concentrations under stress conditions and one of its functions is connected with xenobiotic detoxification in plants. According to Maughan and Foyer (2006), a typical GSH/GSSG concentration ratio is maintained even at the level of about 100:1. When exposure to abiotic stress takes place, glutathione is oxidized to glutathione disulphide (GSSG) and a low GSH/GSSG ratio can cause negative effects such as programmed cell death (PCD). Although glutathione response to stress conditions is not fully established, it is known that glutathione reacts with toxic xenobiotics such as herbicides and their conjugation is catalyzed by glutathione S-transferases (GST) (Hu et al., 2009). Glutathione S-transferase (GST) catalyses the conjugation between glutathione (GSH) and xenobiotics (see 5.2) and GST is regarded as a herbicide-detoxifying enzyme (Del Buono et al., 2011).

Paraquat (PQ) is one of the herbicides that inhibit photosynthesis and influence reactive oxygen species (ROS) formation, leading to oxidative stress. Higher resistance to paraquat in pea varieties (pretreated with sulfur dioxide) is connected with the amount of glutathione and glutathione reductase (GR) activity (Madamanchi et al., 1994). GR plays an important role in glutathione disulphide (GSSG) reduction to glutathione (GSH) (Fig. 5.2). In case of pea varieties, exposure to paraquat leads to an increase in total glutathione concentration (GSH and GSSG) and resistance to leaf injury was observed. Similarly, when *Cucumis sativus* was treated with paraquat the activity of antioxidant enzymes, such as glutathione reductase (GR), was altered (Gao et al., 2011). GR activity depends on plant species treated with herbicides.

When atrazine and fluorodifen were used as herbicides, glutathione reductase (GR) turned out to be more active in wheat (*Triticum aestivum*) than Italian ryegrass (*Lolium multiflorum*). Not surprisingly, oxidative damage was observed only in ryegrass. Moreover, glutathione S-transferase (GST) activity as well as accumulation of these two herbicides (atrazine and fluorodifen) resulted in higher detoxification rates in wheat than in ryegrass

(Del Buono et al., 2011). An excess of glutathione reductase (GR) protects against oxidative stress and both glutathione accumulation and the GSH/GSSG ratio are significant as well. However, glutathione reductase (GR) seems to determine herbicide resistance more strongly than glutathione content (Noctor et al., 1998).

There is a promising prospect of some forest trees, especially poplars, with a strong capability to synthesize GSH being used in xenobiotics detoxification by glutathione dependent conjugations (Noctor et al., 1998). Nevertheless, resistance to oxidative stress, mediated by a herbicide, e.g. paraquat, in terms of GR overexpression is not clear. When glutathione reductase (GR) activity was analyzed it turned out that tobacco plants were more resistant to paraquat in comparison to poplar plants (Will et al., 2001). The analysis of hybrid poplar (*Populus tremula* × *Populus alba*) showed that young poplar leaves are more resistant than mature due to the lower glutathione concentration in mature leaves, even though glutathione synthease (GS) activity was stronger for older leaves. Total glutathione (GSH and GSSG) was about 330 nmol g^{-1} f.wt. in young wild-type poplars and about 256 nmol g^{-1} f.wt. in mature ones. In the case of the transgenic poplar line it was about 583 nmol g^{-1} f.wt. The two-fold elevated GSH content was caused by γ-glutamylcysteine synthease (γ-ECS) overexpression. Surprisingly, transgenic poplars did not reveal paraquat resistance. Neither γ-ECS overexpression nor glutathione synthease (GS) overexpression change the resistance to paraquat exposure. Paraquat tolerance was observed in the case of wild poplars and GSH concentration in young leaves was about 30% higher than in mature leaves. In order to gain paraquat resistance, the glutathione level should increase about five-fold in poplar leaves as a result of feeding plants with glutathione. Such glutathione concentration is difficult to achieve using molecular biology techniques (Will et al., 2001).

When the influence of the nonselective herbicide glyphosate on pea plants was studied (Miteva et al., 2010) two days after herbicide treatment, total glutathione content (GSH and GSSG) in leaves was 800–1000 nmol g^{-1} f.wt. and in roots below 100 nmol g^{-1} f.wt.. The content was below 800 nmol g^{-1} f.wt. in untreated leaves and about 100 nmol g^{-1} f.wt. in untreated roots. After two days of leaf exposure to glyphosate, glutathione disulphide (GSSG) content in leaves increased two-fold up to about 150 nmol g^{-1} f.wt. and after the second day a decreasing tendency was observed. In case of root application of the herbicide a strong GSSG accumulation was detected. Enzyme activity also occurred: glutathione reductase (GR) and glutathione S transferase (GST) showed an increasing tendency during the experimental period.

Metribuzin and pretilachlor are other herbicides that produce reactive oxygen species and generate oxidative stress. These herbicides were applied to maize leaves (*Zea mays*) and GSH contents were measured (Nemat et al., 2008). GSH content was the highest in the case of pretilachlor (below 30 mg g^{-1} f.wt. after 4 days of herbicide exposure). GSSG content was the highest in the case of metribuzin (about 175 mg g^{-1} f.wt. after 4 days of herbicide exposure). The GSH/GSSG ratio was the highest when pretilachlor was used and after 8 days it was above 300 mg g^{-1} f.wt. Also activities of glutathione synthetase GS and γ-glutamylcysteine synthase (γ-ECS) increased due to herbicide exposure. Pretilachlor seems to be less toxic to maize in comparison to metribuzin. Different herbicide resistance can be connected with GSH induction and enzymes associated with GSH. The increase in GSH content depends on GSSG reduction as well as GSH biosynthesis.

Some populations of black grass (*Alopecurus myosuroides*) are known as grass weed resistant to graminicides. Herbicide tolerance might be linked to herbicide metabolism and glutathione S-transferase (GST) activity. A change in temperature from 10°C to 25°C showed an increasing tendency of GST activity in case of resistant and sensitive grass. In turn, GST activity was lower for the sensitive biotype (Milner et al., 2007). The average GSH content in the herbicide sensitive black grass was about 249 nmol g^{-1} f.wt (at 10°C) and 138 nmol g^{-1} f.wt (at 25°C). GSH content in herbicide resistant biotype was about 242 (at 10°C) or 164 nmol g^{-1} (at 25°C). Temperature may influence the development and metabolism of resistant black grass. Increasing GST activity and GSH content facilitate adaptation to environmental conditions such as different temperature. The lower GST activity in sensitive grass can suggest less efficient adaptation to herbicide exposure.

In order to enhance herbicide tolerance in cereal crops safeners, such as fenclorim (Brazier-Hicks et al., 2008), can be used. Safeners act by increasing the activity of herbicide detoxifying enzymes, e.g. cytochrome P450, glutathione S-transferase (GST) and xenobiotic transporters (ATP-binding cassette proteins) (Cummins et al., 2009). When *Arabidopsis thaliana* was treated with fenclorim, GSH content changed from 12.4 to 7.4 nmol g^{-1} f.wt. (Skipsey et al., 2011). It showed a reduction of GSH content by about 40%, but in the case of 1-chloro-2,4-dinitrobenzene (CDNB) the change was 93%. Although CDNB increases GST activity in plants, it does not influence herbicide tolerance. Herbicide-resistant plants developed defensive mechanisms based on detoxifying enzymes and glutathione, but they are not fully recognized and understood yet.

5.5.3 Glutathione Content in Drought

Drought results in a water deficit in plant tissue and is one of the main environmental factors that negatively affect plant growth and productivity (Ramachandra et al., 2004; Wang et al., 2010; Ma et al., 2011; Wang et al., 2011; Cia et al., 2012). There are significant differences in the tolerance of plants to drought stress depending upon intensity and duration of stress, plant species and the plant's stage of development (Rampino et al., 2006; Li et al., 2009; Uzilday et al., 2012). Plants have developed appropriate mechanisms to cope with temporary water limitations to ensure their survival and reproduction

(Bian and Jiang, 2009). Increasing crop tolerance to drought stress would be the most economical approach to improve agricultural productivity and to reduce agricultural use of freshwater resources. Understanding the mechanisms of drought tolerance has been a major goal of plant biologists and crop breeders (Xiong et al., 2006; Islam et al., 2011; Baloglu et al., 2012). Exposure to such an unfavorable environment is often associated with increased levels of ROS such as superoxide radicals, singlet oxygen, hydrogen peroxide and hydroxyl radicals, which are toxic for plant cells (Chaves et al., 2003; Cruz de Carvalho et al., 2010; Shan et al., 2012; Singh et al., 2012). An increased antioxidant capacity is a well-known adaptive mechanism for responding to drought conditions (Khanna-Chopra and Selote, 2007; Farfan-Vignolo and Asard, 2012; Ratnayaka et al., 2012; Uzilday et al., 2012).

It is also difficult to compare directly the results of studies concerning glutathione levels, since various experimental conditions are applied by researchers and respective experiments were carried out in different parts of plants and at various growth stages (Chen et al., 2003; Sofo et al., 2005; Liu et al., 2009; Cruz de Carvalho et al., 2010; Shan et al., 2011; Wang et al., 2011; Sánchez-Rodríguez, 2012; Wang et al., 2012). Most studies focus on cereals such as wheat, rice or maize (Chen et al., 2004; Wang et al., 2010; Shan et al., 2011), while others concern apple leaves and rootstocks (Ma et al., 2011; Wang et al., 2012), tomato (Sánchez-Rodríguez et al., 2012), cucumber (Liu et al., 2009) and grasses (Shan et al., 2012).

Glutathione levels in plants are strongly dependent on the duration of stress, as well as the effects of various exogenous factors used in experiments. Chen et al. (2004) revealed that although levels of total and reduced GSH in leaves of spring wheat were decreased, the ratio of GSH/GSSG was markedly increased by drought. Moreover, levels of GSH biosynthetic precursors, i.e. cysteine and c-glutamylcysteine, and the activities of their biosynthetic enzymes, c-glutamylcysteine synthetase and glutathione synthetase, were also significantly increased in stressed plants. Simultaneously, the authors showed that glutathione reductase activity, which is responsible for the conversion of GSSG to GSH, was also increased under this field stress. The results of a study by Wang et al. (2010) on the effects of water stress on the ROS, GSH, ascorbic acid production and antioxidant enzyme activities in three rice hybrids with differing drought resistance showed that a decrease of GSH content in more resistant cultivars was less obvious. These changes of glutathione content correlated significantly to drought resistance of rice hybrids. In another study the effects of exogenous hydrogen sulfide (H_2S) on glutathione metabolism were investigated in wheat seedling leaves under water stress (Shan et al., 2011). So far it has been documented that H_2S is involved in the antioxidant response against stress conditions (Zhang et al., 2010a; b). Pretreatment with sodium hydrosulfide (NaHS), a hydrogen sulfide donor, significantly increases the contents of GSH and total glutathione under water stress from 4 h to 24 h of treatment. These results suggest that exogenous hydrogen sulfide alleviated oxidative damage by regulating glutathione metabolism. Wang et al. (2012) and Ma et al. (2011) conducted research on apple leaves and rootstocks. The first research team compared two apple rootstocks (drought tolerant *Malus prunifolia* and drought sensitive *Malus hupehensis*) in their tolerance to drought stress. The level of glutathione was the highest after 6 days of water deficit in both species, but changes in GSH concentration were more pronounced in drought tolerant *M. prunifolia* (Wang et al., 2012). Ma et al. (2011) used the leaves of 2-year-old potted plants to investigate changes in each component of the ascorbate-glutathione cycle under drought stress. The contents of total glutathione and glutathione were obviously upregulated in apple leaves when the soil water content was 40–45%. Moreover, a stress-induced significant increase in the activities of enzymes was used to regenerate glutathione and ascorbic acid. In similar studies the regulation of ascorbate and glutathione metabolism by nitric oxide (NO) in *Agropyron cristatum* leaves was investigated (Shan et al., 2012). NO is a signal molecule in plants and plays an important role in the responses to biotic and abiotic stresses (Neill et al., 2008; Song et al., 2008). The obtained results showed that NO increased GR activity and GSH content under water deficit.

To verify the effectiveness of the grafting technique the response to water stress was studied in various combinations of rootstocks ungrafted, grafted and self-grafted differing in water stress tolerance tomato plants (Sánchez-Rodríguez et al., 2012). It has been shown that leaf extract grafted plants achieved greater tolerance to water stress apparently by developing a better antioxidant system.

Liu et al. (2009) examined whether exogenous paraquat (PQ) can modify antioxidant levels in drought-stressed cucumber cultivars. In that experiment drought stress was induced by polyethylene glycol (PEG). Results showed that PQ and PEG significantly increase the content of GSH and AsA in cucumber leaves, which can increase the ability of this plant to eliminate ROS.

5.5.4 Glutathione Content and Heat Stress

Recent reports concerning global warming have aroused interest in the effects of high temperatures on crop production and crop yields. Various studies suggest that increasing temperatures will pose a major limitation to crop production in the future (Lobell and Asner, 2003; Battisti and Naylor, 2009). In light of the anticipated temperature increases, cultivation of thermotolerant plants is seen as a key strategic target in maintaining crop yields (Wahid et al., 2007). As is the case with other stresses, heat stress induces the overproduction of ROS and in order to alleviate the impact of ROS plants have evolved enzymatic and nonenzymatic antioxidants (Wahid et al., 2007; Xie et al., 2008).

To compare the effect of heat stress on the synthesis of glutathione and hydroxymethylglutathione (hmGSH) wheat genotypes with different freezing tolerance (freezing-tolerant, moderately tolerant, freezing-sensitive) were used in experiments (Kocsy et al., 2004). Previous studies indicate that a lower GSH content may be found at higher temperatures in freezing-tolerant plant species, while at extremely high temperatures the GSH level will be higher in freezing-sensitive genotypes than in tolerant species (Kurganova et al., 1999; Kocsy et al., 2002). Results recorded by Koscy et al. (2004) confirm previous ones. In studies conducted by Kumar et al. (2012) two genotypes of each maize and rice were compared for their response to different degrees of temperature stress (35/30, 40/35 and 45/40°C) with controls (30/25°C). The glutathione contents increase with the application of higher temperatures in both plant types; in maize genotypes they were by 64–74% higher than in the control in comparison with 27–33% higher than in the control in rice genotypes. These observations on an increase in GSH level at 40°C were similar to those in the high temperature (37°C) treatment of *Lilium longiflorum* plants (Yin et al., 2008). On the other hand, a decrease in GSH levels was found in wheat (Sairam et al., 2000) and apple (Ma et al., 2008). The contents of GSH were reported to be higher in the heat-tolerant genotypes of wheat (Sairam et al., 2000; Dash and Mohanty, 2002) that suggest a positive association with the endogenous content of GSH with heat tolerance.

To verify effectiveness of cinnamic acid (CA) pretreatment (50 μM) seedlings of cucumber cultivar were investigated under normal (25/18°C) and higher (42/38°C) temperatures for 3 days (Dai et al., 2012). In comparison to the control, an increase in GSH content and the ratio of GSH/GSSG was found in leaves of the CA pretreatment. Both heat stress and CA pretreatment increased the contents of GSH and the ratio of GSH/GSSG in cucumber leaves, which enhances heat tolerance in cucumber seedlings.

5.5.5 Glutathione Content under Chilling

Each plant has its own temperature requirements, which are optimum for its proper growth and development. Some temperature conditions that are most favorable for one plant may be stressogenic for another (Mahajan and Tuteja, 2005; Lukatkin et al., 2012). Cold stress is a major factor limiting the geographical distribution and productivity of chilling-sensitive plant species, especially vegetables such as tomato or cucumber (Li et al., 2011; Shu et al., 2011). Chilling induces oxidative stress, during which ROS are accumulated at levels higher than is necessary for optimal metabolism (Kocsy et al., 2001).

In an experiment carried out by Shu et al. (2011) the 7-week old wild-type and transgenic plants were exposed to low temperature (4°C) for 24 h and all measurements were performed on the youngest leaves. Chilling treatment induced a greater decrease in GSH contents and the GSH/GSSG ratio and an increase in GSSG levels in transgenic plants than the wild-type cultivar. In other studies the tolerance of *Citrus* and *Fortunella* species was tested under natural chilling stress (Santini et al., 2013). Glutathione contents measured in leaves of 8-year-old trees were varied, depending on the species. For some tested genotypes an increase of the GSH/GSSG ratio could suggest de novo glutathione synthesis (Queval et al., 2007). Ao et al. (2013) investigated chilling tolerance induced by chill hardening of *Jatropha curcas* seedlings. Low temperature for this plant with a great potential for biodiesel production is an important limiting factor for its distribution and development. Results showed that chill hardening at 12°C for 2 days could increase GSH content in plants and then in hardened seedlings under chilling stress (1°C for 1–7 days) the glutathione level was significantly higher compared to the control (nonhardening plants).

To verify the effectiveness of the protective role of exogenous cinnamic acid (CA) pretreatment was investigated in cucumber leaves under chilling stress at two temperatures (15/8 and 25/18°C) (Li et al., 2011). Such an exogenously applied molecule could enhance the efficiency of the antioxidative system and alleviate chilling stress of plants (Hayat et al., 2010). Similar studies concerning the role of exogenous CA were also performed on cucumber under heat stress (Dai et al., 2012). In analyzed leaves the contents of GSH and the GSH/GSSG ratio increased not only in the CA pretreatment group, but also in the chilling treatment group, which indicates enhanced tolerance of cucumber leaves to chilling stress.

5.5.6 Glutathione Content under Salinity

Salinity, like other stresses, also leads to generation of reactive oxygen species, which are detoxified either directly by nonenzymatic antioxidants (GSH, AsA, tocopherols, and carotenoids) or by antioxidative enzymes (Meloni et al., 2003; Parida and Das, 2005; Liang et al., 2009; Chen et al., 2012; Ferreira-Silva et al., 2012).

To explain the role of GSH biosynthesis and GSH enzymes under salinity (successively increasing the NaCl concentration by 25 mM per day over a period of 4 days, to 100 mM) two tomato species (*Lycopersicon esculentum* – Lem – and *Lycopersicon pennellii* – Lpa) with different salt stress tolerances were carried out (Mittova et al., 2003). GSH content and the GSH/GSSG ratio were higher in leaves and roots of salt-tolerant Lpa than in Lem. In addition, increasing NaCl concentration induced glutathione peroxidase and glutathione S-transferase but not glutathione reductase activities in Lpa.

The changes of antioxidant levels in rice roots under different salt concentrations (0, 0.5, 1.0, 2.0 3.0% of NaCl for 4 h) were studied by Khan et al., (2002). Both GSH and AsA levels were increased in salinity, which indicates their role in detoxification of ROS and protection of rice roots under salinity.

In other studies, the role of cysteine and glutathione biosynthesis were investigated in leaves of wild-type and salt-tolerant transgenic canola (*Brassica napus* L.) under salinity (75 and 150 mM NaCl for 15 days) (Ruiz and Blumwald, 2002). Under salinity, increase in γ-ECS activity correlates with a 230% increase in leaf GSH content in wild-type canola, while cysteine and glutathione levels did not show significant changes in the transgenic plants. Taking into consideration the results, it can be concluded that salinity induces a sulfur demand, necessary for cysteine synthesis, which finally is required for GSH biosynthesis.

To better understand long-term salt-induced oxidative stress some physiological and enzymatic and nonenzymatic parameters in marigold (*Calendula officinalis*) under salinity (50 and 100 mM NaCl for 3 weeks) were investigated (Chaparzadeh et al., 2004). The main protective function of GSH against oxidative stress is its involvement in the ascorbate-glutathione cycle and participation in the regulation of protein thiol-disulphide redox status (Ma et al., 2011). The total glutathione content and GSH/GSSG ratio decreased significantly both in leaves and roots exposed to NaCl, which is connected with enhanced glutathione reductase activity in leaves.

5.5.7 Glutathione Content under High Light Stress

Light is essential for plant growth and development, but when the amount of absorbed light exceeds the amount required for photosynthesis, the excess light can be harmful because it causes the accumulation of reactive oxygen species, which in turn leads to damage of membrane nucleic acids, lipids and proteins (Cadet et al., 2005; Asada et al., 2006; Xu et al., 2010; Chen et al., 2011). In the last decade studies were conducted concerning the influence of light variation on plants such as lettuce (Zhou et al., 2009) and algae (Shiu and Lee, 2005; Tian and Yu, 2009).

In studies conducted by Zhou et al. (2009) lettuce plants were grown at different light conditions (low, middle and high irradiance). Results showed that in lettuce exposed to high light the nutritional value was higher as indicated by the high content of nonenzymatic antioxidants such as glutathione, ascorbate, carotenoids and α-tocopherol.

To explain the impact of high-light stress on glutathione forms content, marine algae *Ulva fasciata* were treated with ultraviolet-B (UV-B, 280–320 nm). Plants were exposed to low, medium and high UV-B irradiance. The contents of total glutathione, glutathione and glutathione disulphide were increased with an increasing UV-B intensity. The GSH/GSSG ratio as an indicator of the glutathione regeneration rate decreased with an increasing UV-B dose because the magnitude of glutathione disulphide increments was higher than that of glutathione increments (Shiu and Lee, 2005). Similar research was carried out on algae (*Dunaliella salina*), which was treated with high doses of UV-B radiation. In *D. salina* GSH and GSSG contents increased with enhanced UV-B radiation, while the GSH/GSSG ratio decreased, showing that oxidative glutathione – GSSG – increased evidently in comparison with reduced glutathione – GSH (Tian and Yu, 2009).

5.6 CONCLUSIONS

Drought, low and high temperatures, as well as salinity are major abiotic stresses that reduce the yield of food crops worldwide. In the last decade significant progress has been made in understanding the mechanisms regulating abiotic stress tolerance in crop plants. However, this situation is complicated when plants have to simultaneously cope with various biotic and abiotic stresses. Among the different methods, techniques and technologies to induce abiotic stress tolerance in high-yielding cultivars we should mention exogenous application of molecules (e.g. caffeic acid, salicylic acid, cinnamic acid) or nitric oxide, hydrogen sulfide and paraquat as a pretreatment method; chill hardening; or development of transgenic plants. Applications of genomics, proteomics and trascriptomics approaches are necessary to provide better understanding of the molecular basis of plant response to stresses. Molecular knowledge of the response and tolerance mechanisms will pave the way for the production of crops that can produce high, but economic, yields under stress conditions.

Increasing human population causes reduction of agriculture areas and, on the other hand, demand for food is growing. Future farming requires more stress-tolerant crops, in particular resistance to temperature stress, heavy metals, salt stress and water deficit. Glutathione, as a nonenzymatic antioxidant, plays a key role in plant stress tolerance. Despite the progress in understanding glutathione synthesis and metabolism, knowledge of its biochemical interactions is still not complete. More detailed studies require clarification of: (i) changes in redox state GSH/GSSG couple, (ii) interactions with other signalling molecules and (iii) function of GSH in improvement of the nutritional quality of food.

REFERENCES

Akter, N., Okuma, E., Sobahan, M.A., Uraji, M., Munemasa, S., Nakamura, Y., et al., 2013. Negative regulation of methyl jasmonate-induced stomatal closure by glutathione in *Arabidopsis*. J. Plant Growth Regul. 32, 208–215.

Amudha, J., Balasubramani, G., 2011. Recent molecular advances to combat abiotic stress tolerance in crop plants. Biotechnol. Mol. Biol. Rev. 6, 31–58.

Anjum, N.A., Ahmad, I., Mohmood, I., Pacheco, M., Duarte, A.C., Pereira, E., et al., 2012. Modulation of glutathione and its related enzymes in plants' responses to toxic metals and metalloids – a review. Environ. Exp. Bot. 75, 307–324.

Anjum, N.A., Umar, C.S., Ahmad, C.A., Iqbal, C.M., Nafees, C., Khan, A., 2008. Sulphur protects mustard (*Brassica campestris* L.) from cadmium toxicity by improving leaf ascorbate and glutatione. Plant Growth Regul. 54, 271–279.

Ao, P.-X., Li, Z.-G., Fan, D.-M., Gong, M., 2013. Involvement of antioxidant defense system in chill hardening-induced chilling tolerance in *Jatropha curcas* seedlings. Acta Physiol. Plant. 35, 153–160.

Apel, K., Hirt, H., 2004. Reactive oxygen species: metabolism, oxidative stress, and signal transduction. Annu. Rev. Plant Biol. 55, 373–399.

Aroca, R., Porcel, R., Ruiz-Lozano, J., 2012. Regulation of root water uptake under abiotic stress conditions. J. Exp. Bot. 63, 43–57.

Asada, K., 2006. Production and scavenging of reactive oxygen species in chloroplasts and their functions. Plant Physiol. 141, 391–396.

Avni Öktem, H., Eyidoğan, F., Selçuk, F., Tufan Öz, M., da Silva, J.A.T., Yücel, M., 2008. Revealing response of plants to biotic and abiotic stresses with microarray technology. Genes, Genomes and Genomics, 14–48.

Baloglu, M.C., Kavas, M., Aydin, D., Oktem, H.A., Yucel, A.M., 2012. Antioxidative and physiological responses of two sunflower (*Helianthus annuus*) cultivars under PEG-mediated drought stress. Turk. J. Bot. 36, 707–714.

Basantani, M., Srivastava, A., 2007. Plant glutathione transferases – a decade falls short. Can. J. Bot. Rev. Can. Bot. 85, 443–456.

Battisti, D.S., Naylor, R.L., 2009. Historical warnings of future food insecurity with unprecedented seasonal heat. Sci. 323, 240–244.

Bhatnagar-Mathur, P., Vadez, V., Sharma, K.K., 2009. Transgenic approaches for abiotic stress tolerance in plants: retrospect and prospects. Plant Cell Rep. 27, 411–424.

Bhattacharjee, S., 2005. Reactive oxygen species and oxidative burst: roles in stress, senescence and signal transduction in plants. Curr. Sci. 89, 1113–1121.

Bian, S., Jiang, Y., 2009. Reactive oxygen species, antioxidant enzyme activities and gene expression patterns in leaves and roots of Kentucky bluegrass in response to drought stress and recovery. Sci. Hortic. 120, 264–270.

Bohmer, A., Jordan, J., Tsikas, D., 2011. High-performance liquid chromatography ultraviolet assay for human erythrocytic catalase activity by measuring glutathione as o-phthalaldehyde derivative. Anal. Biochem. 410, 296–303.

Bossio, E., Diaz Paleo, A., del Vas, M., Baroli, I., Acevedo, A., Ríos, R.D., 2013. Silencing of the glutathione biosynthetic pathway inhibits somatic embryogenesis in wheat. Plant Cell Tissue Organ. Cult. 112, 239–248.

Brazier-Hicks, M., Evans, K.M., Cunningham, O.D., 2008. Catabolism of glutathione conjugates in *Arabidopsis thaliana* – Role in metabolic reactivation of the herbicide safener fenclorim. J. Biol. Chem. 283, 21102–21112.

Budnikov, G.K., Ziyatdinova, G.K., Valitova, Y.R., 2004. Electrochemical determination of glutathione. J. Anal. 59, 573–576.

Cadet, J., Sage, E., Douki, T., 2005. Ultraviolet radiation-mediated damage to cellular DNA. Mutation Res. 571, 3–17.

Chaparzadeh, N., D'Amico, M.L., Khavari-Nejad, R.-A., Izzo, R., Navari-Izzo, F., 2004. Antioxidative responses of *Calendula officinalis* under salinity conditions. Plant Physiol. Biochem. 42, 695–701.

Chaves, M.M., Maroco, J.P., Pereira, J.S., 2003. Understanding plant responses to drought – from genes to the whole plant. Funct. Plant Biol. 30, 239–264.

Chen, F., Wang, F., Wu, F., Mao, W., Zhang, G., Zhou, M., 2010. Modulation of exogenous glutathione in antioxidant defense system against Cd stress in the two barley genotypes differing in Cd tolerance. Plant Physiol. Biochem. 48, 663–672.

Chen, J., He, Z.Y., Liu, H., 2006. Electrochemical determination of reduced glutathione (GSH) by applying the powder microelectrode technique. J. Electroanal. Chem. 588, 324–330.

Chen, J.-H., Jiang, H.-W., Hsieh, E.-J., Chen, H.-Y., Chien, C.-T., Hsieh, H.-L., et al., 2012. Drought and salt stress tolerance of an *Arabidopsis* glutathione S-transferase U17 knockout mutant are attributed to the combined effect of glutathione and abscisic acid. Plant Physiol. 158, 340–351.

Chen, K.-M., Gong, H.-J., Chen, G.-C., Wang, S.-M., Zhang, C.-L., 2003. Up-regulation of glutathione metabolism and changes in redox status involved in adaptation of reed (*Phragmites communis*) ecotypes to drought-prone and saline habitats. J. Plant Physiol. 160, 293–301.

Chen, K.-M., Gong, H.-J., Chen, G.-C., Wang, S.-M., Zhang, C.-L., 2004. Gradual drought under field conditions influences the glutathione metabolism, redox balance and energy supply in spring wheat. J. Plant Growth Regul. 23, 20–28.

Chen, W., Zhao, Y., Seefeldt, T., 2008. Determination of thiols and disulfides via HPLC quantification of 5-thio-2-nitrobenzoic acid. J. Pharm. Biomed. Anal. 48, 1375–1380.

Chen, X., Li, W., Lua, Q., Wena, X., Li, H., Kuanga, T., et al., 2011. The xanthophyll cycle and antioxidative defense system are enhanced in the wheat hybrid subjected to high light stress. J. Plant Physiol. 168, 1828–1836.

Chinnusamy, V., Schumaker, K., Zhu, J.K., 2004. Molecular genetic perspectives on cross-talk and specificity in abiotic stress signalling in plants. J. Exp. Bot. 55, 225–236.

Cia, M.C., Guimaraes, A.C.R., Medici, L.O., Chabregas, S.M., Azevedo, R.A., 2012. Antioxidant responses to water deficit by drought-tolerant and -sensitive sugarcane varieties. Ann. Appl. Biol. 161, 313–324.

Cobbett, C., Goldsbrough, P., 2002. Phytochelatins and metallothioneins: roles in heavy metal detoxification and homeostasis. Annu. Rev. Plant Biol. 53, 159–182.

Cramer, G.R., Urano, K., Delrot, S., Pezzotti, M., Shinozaki, K., 2011. Effects of abiotic stress on plants: a systems biology perspective. BMC Plant Biol. 11, 163–176.

Cruz de Carvalho, M.H., Brunet, J., Bazin, J., Kranner, I., d'Arcy-Lameta, A., Zuily-Fodil, Y., et al., 2010. Homoglutathione synthetase and glutathione synthetase in drought-stressed cowpea leaves: expression patterns and accumulation of low-molecular-weight thiols. J. Plant Physiol. 167, 480–487.

Cummins, I., Bryant, D.N., Edwards, R., 2009. Safener responsiveness and multiple herbicide resistance in the weed black-grass (*Alopecurus myosuroides*). Plant Biotechnol. J. 7, 807–820.

Cummins, I., Dixon, D.P., Freitag-Pohl, S., 2011. Multiple roles for plant glutathione transferases in xenobiotic detoxification. Drug Metab. Rev. 43, 266–280.

Dai, A.-H., Nie, Y.-X., Yu, B., Li, Q., Lu, L.-Y., Bai, J.-G., 2012. Cinnamic acid pretreatment enhances heat tolerance of cucumber leaves through modulating antioxidant enzyme activity. Environ. Exp. Bot. 79, 1–10.

Dash, S., Mohanty, N., 2002. Response of seedlings to heat-stress in cultivars of wheat: growth temperature-dependent differential modulation of photosystem 1 and 2 activity and foliar antioxidant defense capacity. J. Plant Physiol 159, 49–59.

Del Buono, D., Ioli, G., Nasini, L., Proietti, P., 2011. Comparative study on the interference of two herbicides in wheat and Italian ryegrass and on their antioxidant activities and detoxification rates. J. Agric. Food Chem. 59, 12109–12115.

Delalande, O., Desvaux, H., Godat, E., 2010. Cadmium – glutathione solution structures provide new insights into heavy metal detoxification. Febs J. 277, 5086–5096.

Dempsey, R.W., Merchant, A., Tausz, M., 2011. Increasing leaf glutathione through stem feeding does not acclimate *Eucalyptus camaldulensis* seedlings towards high-light stress. Acta Physiol. Plant 33, 221–225.

Di Baccio, D., Kopriva, S., Sebastiani, L., Rennenberg, H., 2005. Does glutathione metabolism have a role in the defence of poplar against zinc excess? New Phytologist 167, 73–80.

Dimitrova, M., Turmanova, S., Vassilev, K., 2010. Complexes of glutathione with heavy metals as catalysts for oxidation. React. Kinet. Mech. Catal. 99, 69–78.

Dixon, D.P., Lapthorn, A., Edwards, R., 2002. Plant glutathione transferases. Genome Biol. 3, 1–10.

Dixon, D.P., Skipsey, M., Edwards, R., 2010. Roles for glutathione transferases in plant secondary metabolism. Phytochem 71, 338–350.

Edmeades, G.O., 2009. Drought tolerance in maize: an emerging reality, "Global Status of Commercial Biotech/GM Crops", ISAAA Brief,, 39. ISAAA, Ithaca, NY, 197–217.

Edwards, R., Dixon, D., 2005. Plant glutathione transferases. Methods Enzymol. 401, 169–186.

Edwards, R., Dixon, D.P., 2004. Metabolism of natural and xenobiotic substrates by the plant glutathione S-transferase superfamily. Ecol. Stud.-Anal. Synth. 170, 17–50.

Elias, R.J., Kellerby, S.S., Decker, E.A., 2008. Antioxidant activity of proteins and peptides. Crit. Rev. Food Sci. Nutr. 48, 430–441.

Eltayeb, A.E., Yamamoto, S., Habora, M.E.E., Matsukubo, Y., Aono, M., Tsujimoto, H., et al., 2010. Greater protection against oxidative damages imposed by various environmental stresses in transgenic potato with higher level of reduced glutatione. Breeding Sci. 60, 101–109.

Eshdat, Y., Holland, D., Faltin, Z., 1997. Plant glutathione peroxidases. Physiol. Plant 100, 234–240.

Faltin, Z., Holland, D., Velcheva, M., 2010. Glutathione peroxidase regulation of reactive oxygen species level is crucial for in vitro plant differentiation. Plant Cell Physiol. 51, 1151–1162.

Farfan-Vignolo, E.R., Asard, H., 2012. Effect of elevated CO_2 and temperature on the oxidative stress response to drought in *Lolium perenne* L. and *Medicago sativa* L. Plant Physiol. Biochem. 59, 55–62.

Ferreira-Silva, S.L., Voigt, E.L., Silva, E.N., Maia, J.M., Aragão, T.C.R., Silveira, J.A.G., 2012. Partial oxidative protection by enzymatic and non-enzymatic components in cashew leaves under high salinity. Biol. Plant 56, 172–176.

Foyer, C.H., Noctor, G., 2005. Redox homeostasis and antioxidant signaling. A metabolic interface between stress perception and physiological responses. Plant Cell 17, 1866–1875.

Foyer, C.H., Noctor, G., 2011. Ascorbate and glutathione: the heart of the redox hub. Plant Physiol. 155, 2–18.

Freeman, J.L., Persans, M.W., Nieman, K., Albrecht, C., Peer, W., Pickering, I.J., et al., 2004. Increased glutathione biosynthesis plays a role in nickel tolerance in *Thlaspi* nickel hyperaccumulators. Plant Cell 16, 2176–2191.

Frova, C., 2003. The plant glutathione transferase gene family: genomic structure, functions, expression and evolution. Physiol. Plant 119, 469–479.

Fu, L.H., Wang, X.F., Eyal, Y., 2002. A selenoprotein in the plant kingdom. Mass spectrometry confirms that an opal codon (UGA) encodes selenocysteine in *Chlamydomonas reinhardtii* glutathione peroxidase. J. Biol. Chem. 277, 25983–25991.

Fujita, M., Fujita, Y., Noutoshi, Y., Takahashi, F., Narusaka, Y., Yamaguchi-Shinozaki, K., et al., 2006. Crosstalk between abiotic and biotic stress responses: a current view from the points of convergence in the stress signaling networks. Curr. Opin. Plant Biol. 9, 436–442.

Gallego, S., Kogan, M.J., Azpilicueta, C.E., Pena, C., Tomaro, M.L., 2005. Glutathione mediated antioxidative mechanism in sunflower (*Helianthus annuus* L.) cells in response to cadmium stress. Plant Growth Reg. 46, 267–276.

Galvez-Valdivieso, G., Mullineaux, P.M., 2010. The role of reactive oxygen species in signaling from chloroplasts to the nucleus. Physiol. Plant 138, 430–439.

Gao, J.-P., Chao, D.-Y., Lin, H.-X., 2007. Understanding abiotic stress tolerance mechanisms: recent studies on stress response in rice. J. Integr. Plant Biol. 49, 742–750.

Gao, Y., Guo, Y., Dai, A., 2011. Paraquat pretreatment alters antioxidant enzyme activity and protects chloroplast ultrastructure in heat-stressed cucumber leaves. Biol. Plant 55, 788–792.

García-Giménez, J.L., Markovic, J., Dasí, F., Queval, G., Schnaubelt, D., Foyer, C.H., et al., 2013. Nuclear glutathione. Biochim. Biophys. Acta 1830, 3304–3316.

Garg, N., Aggarwal, N., 2011. Effects of interactions between cadmium and lead on growth, nitrogen fixation, phytochelatin and glutathione production in mycorrhizal *Cajanus cajan* (L.) Millsp. J. Plant Growth Regul. 30, 286–300.

Garrido, T., Mendoza, J., Riveros, R., Saez, L., 2010. Acute and chronic effect of copper on levels of reduced and oxidized glutathione and nutrient uptake of tomato plants. J. Plant Nutr. Soil Sci. 173, 920–926.

Gill, S.S., Anjum, N.A., Hasanuzzaman, M., Gill, R., Trivedi, D.K., Ahmad, I., et al., 2013. Glutathione and glutathione reductase: a boon in disguise for plant abiotic stress defense operations. Plant Physiol. Biochem. 70, 204–212.

Gill, S.S., Tuteja, N., 2010. Reactive oxygen species and antioxidant machinery in abiotic stress tolerance in crop plants. Plant Physiol. Biochem. 48, 909–930.

Gratao, P., Polle, A., Lea, P., Azevedo, R., 2005. Making the life of heavy metal-stressed plants a little easier. Funct. Plant Biol. 32, 481–494.

Gucek, M., Makuc, S., Mlakar, A., 2002. Determination of glutathione in spruce needles by liquid chromatography/tandem mass spectrometry. Rapid Commun. Mass Spectr. 16, 1186–1191.

Halliwell, B., Gutteridge, J.M.C., 2007. Oxygen is a toxic gas – an introduction to oxygen toxicity and reactive species. In: Halliwell, B., Gutteridge, J.M.C. (Eds.), Free Radicals in Biology and Medicine. Oxford University Press, New York, pp. 1–29.

Han, Y., Chaouch, S., Mhamdi, A., Queval, G., Zechmann, B., Noctor, G., 2013a. Functional analysis of *Arabidopsis* mutants points to novel roles for glutathione in coupling H_2O_2 to activation of salicylic acid accumulation and signaling. Antioxidants Redox Signal. 18, 2106–2121.

Han, Y., Mhamdi, A., Chaouch, S., Noctor, G., 2013b. Regulation of basal and oxidative stress-triggered jasmonic acid-related gene expression by glutathione. Plant Cell Environ. 36, 1135–1146.

Harfield, J.C., Batchelor-McAuley, C., Compton, R.G., 2012. Electrochemical determination of glutathione: a review. Analyst 137, 2285–2296.

Harwood, D.T., Kettle, A.J., Brennan, S., 2009. Simultaneous determination of reduced glutathione, glutathione disulphide and glutathione sulphonamide in cells and physiological fluids by isotope dilution liquid chromatography-tandem mass spectrometry. J. Chromatogr. B – Anal. Technol. Biomed. Life Sci. 877, 3393–3399.

Hayat, Q., Hayat, S., Irfan, M., Ahmad, A., 2010. Effect of exogenous salicylic acid under changing environment: a review. Environ. Exp. Bot. 68, 14–25.

Hepel, M., Stobiecka, M., 2011. Comparative kinetic model of fluorescence enhancement in selective binding of monochlorobimane to glutathione. J. Photochem. Photobiol. Chem. 225, 72–80.

Hormozi-Nezhad, M.R., Seyedhosseini, E., Robatjazi, H., 2012. Spectrophotometric determination of glutathione and cysteine based on aggregation of colloidal gold nanoparticles. Sci. Iran. 19, 958–963.

Hou, J., Ai, S., 2011. Electrochemical determination of reduced glutathione at multiwalled carbon nanotubes/poly(bromocresol green) modified glassy carbon electrode. Chem. Res. Chin. Univ. 27, 934–938.

Hu, T., Qv, X., Xiao, G., 2009. Enhanced tolerance to herbicide of rice plants by over-expression of a glutathione S-transferase. Mol. Breed. 24, 409–418.

Huang, G.G., Han, X.X., Hossain, M.K., 2009. Development of a heat-induced surface-enhanced Raman scattering sensing method for rapid detection of glutathione in aqueous solutions. Anal. Chem. 81, 5881–5888.

Huang, G.-Y., Wang, Y.-S., Sun, C.C., Dong, J.D., Sun, Z.-X., 2010. The effect of multiple heavy metals on ascorbate, glutathione and related enzymes in two mangrove plant seedlings (*Kandelia candel* and *Bruguiera gymnorrhiza*). Int. J. Oceanogr. Hydrobiol. 1, 11–25.

Huseby, N.E., Asare, N., Wetting, S., 2003. Role of gamma-glutamyltransferase in the homeostasis of glutathione during oxidative and nitrosative stress. Biofactors. 17, 151–160.

Il-Sup, K., Kim, Y.-S., Yoon, H.-S., 2012. Glutathione reductase from *Oryza sativa* increases acquired tolerance to abiotic stresses in a genetically modified *Saccharomyces cerevisiae* strain. J. Microbiol. Biotechnol. 22, 1557–1567.

Islam, M.R., Xue, X., Mao, S., Ren, C., Eneji, A.E., Hu, Y., 2011. Effect (*Avena sativa* L.) under drought stress. J. Sci. Food Agric. 91, 680–686.

Jaleel, C.A., Manivannan, P., Wahid, A., Farooq, M., Al-Juburi, H.J., Somasundaram, R., et al., 2009. Drought stress in plants: a review on morphological characteristics and pigments composition. Int. J. Agric. Biol. 11, 100–105.

Janes, L., Lisjak, K., Vanzo, A., 2010. Determination of glutathione content in grape juice and wine by high-performance liquid chromatography with fluorescence detection. Anal. Chim. Acta 674, 239–242.

Ju, X.H., Tang, S., Jia, Y., Guo, J., Ding, Y., Song, Z., et al., 2011. Determination and characterization of cysteine, glutathione and phytochelatins (PC2-6) in *Lolium perenne* L. exposed to Cd stress under ambient and elevated carbon dioxide using HPLC with fluorescence detection. J. Chromatogr. B. 1717–1724.

Kandar, R., Zakova, P., Lotkova, H., 2007. Determination of reduced and oxidized glutathione in biological samples using liquid chromatography with fluorimetric detection. J. Pharm. Biomed. Anal. 43, 1382–1387.

Kellos, T., Timar, I., Szilagyi, V., Szalai, G., Galiba, G., Kocsy, G., 2008. Effect of abiotic stress on antioxidants in maize. Acta Biol. Szeged 52, 173–174.

Khan, M.H., Singha, L.B., Panda, S.K., 2002. Changes in antioxidant levels in *Oriza sativa* L. roots subjected to NaCl-salinity stress. Acta Physiol. Plant. 24, 145–148.

Khanna-Chopra, R., Selote, D.S., 2007. Acclimation to drought stress generates oxidative stress tolerance in drought resistant than susceptible wheat cultivar under field conditions. Environ. Exp. Bot. 60, 276–283.

Klein, M., Burla, B., Martinoia, E., 2006. The multidrug resistance-associated protein (MRP/ABCC) subfamily of ATP-binding cassette transporters in plants. FEBS Lett. 580, 1112–1122.

Kocsy, G., Szalai, G., Galiba, G., 2002. Effect of heat stress on glutathione biosynthesis in wheat. Acta Biol. Szeged 46, 71–72.

Kocsy, G., Szalai, G., Galiba, G., 2004a. Effect of osmotic stress on glutathione and hydroxymethylglutathione accumulation in wheat. J. Plant Physiol. 161, 785–794.

Kocsy, G., Szalai, G., Sutka, J., Paldi, E., Galiba, G., 2004b. Heat tolerance together with heat stress-induced changes in glutathione and hydroxymethylglutathione levels is affected by chromosome 5 A of wheat. Plant. Sci. 166, 451–458.

Koehler, P., 2003. Concentrations of low and high molecular weight thiols in wheat dough as affected by different concentrations of ascorbic acid. J. Agric. Food Chem. 51, 4948–4953.

Kranner, I., Birtic, S., Anderson, K.M., 2006. Glutathione half-cell reduction potential: a universal stress marker and modulator of programmed cell death? Free Radic. Biol. Med. 40, 2155–2165.

Kranner, I., Grill, D., 1996. Determination of glutathione and glutathione disulphide in lichens: a comparison of frequently used methods. Phytochem. Anal. 7, 24–28.

Kumar, S., Gupta, D., Nayyar, H., 2012. Comparative response of maize and rice genotypes to heat stress: status of oxidative stress and antioxidants. Acta Physiol. Plant. 34, 75–86.

Kurganova, L.N., Veselov, A.P., Sinitsina, Y.V., Elikova, E.A., Kulaeva, O.N., 1999. Lipid peroxidation products as possible mediators of heat stress response in plants. Russ. J. Plant Physiol. 46, 181–185.

Levina, A., Zhang, L.B., Lay, P.A., 2003. Structure and reactivity of a chromium (V) glutathione complex. Inorg. Chem. 42, 767–784.

Lewis, S., Donkin, M.E., Depledge, M.H., 2001. Hsp70 expression in *Enteromorpha intestinalis* (Chlorophyta) exposed to environmental stressors. Aquat. Toxicol. 51, 277–291.

Li, Q., Yu, B., Gao, Y., Dai, A.-H., Bai, J.-G., 2011. Cinnamic acid pretreatment mitigates chilling stress of cucumber leaves through altering antioxidant enzyme activity. J. Plant Physiol. 168, 927–934.

Li, Y., Sun, C., Huang, Z., Pan, J., Wang, L., Fan, X., 2009. Mechanisms of progressive water deficit tolerance and growth recovery of Chinese maize foundation genotypes Huangzao 4 and Chang 7-2, which are proposed on the basis of comparison of physiological and transcriptomic responses. Plant Cell Physiol. 50, 2092–2111.

Li, Y., Yang, Y., Guan, X., 2012. Benzofurazan sulfides for thiol imaging and quantification in live cells through fluorescence microscopy. Anal. Chem. 84, 6877–6883.

Liang, W., Li, C., Liu, F., Jiang, H., Li, S., Sun, J., et al., 2009. The *Arabidopsis* homologs of CCR4-associated factor 1 show mRNA deadenylation activity and play a role in plant defence responses. Cell Res. 19, 307–316.

Liedschulte, V., Wachter, A., Zhigang, A., Rausch, T., 2010. Exploiting plants for glutathione (GSH) production: uncoupling GSH synthesis from cellular controls results in unprecedented GSH accumulation. Plant Biotechnol. J. 8, 1–14.

Liu, Z.-J., Zhang, X.-L., Bai, J.-G., Suo, B.-X., Xu, P.-L., Wang, L., 2009. Exogenous paraquat changes antioxidant enzyme activities and lipid peroxidation in drought-stressed cucumber leaves. Sci. Hortic. 121, 138–143.

Lobell, D.B., Asner, G.P., 2003. Climate and management contributions to recent trends in U.S. agricultural yields. Sci. 299, 1032.

López-Climent, M.F., Arbona, V., Pérez-Clemente, R.M., Zandalinas, S.I., and Gómez-Cadenas, A., 2013. Effect of cadmium and calcium treatments on phytochelatin and glutathione levels in citrus plants. Plant Biol. doi: 10.1111/plb.12006.

Lukatkin, A.S., Brazaityte, A., Bobinas, C., Duchovski, P., 2012. Chilling injury in chilling-sensitive plants: a review. Agriculture 99, 111–124.

Ma, Y.H., Ma, F.W., Zhang, J.K., Li, M.J., Wang, Y.H., Liang, D., 2008. Effects of high temperature on activities and gene expression of enzymes involved in ascorbate–glutathione cycle in apple leaves. Plant Sci. 175, 761–766.

Ma, Y., Xiang, F., Jin, W., 2010. Determination of total glutathione in yeasts by high-performance liquid chromatography with dansylation. Z. Nat. Sect. C- J. Biosci. 65, 391–394.

Ma, Y.-H., Ma, F.-W., Wang, Y.-H., Zhang, J.-K., 2011. The responses of the enzymes related with ascorbate–glutathione cycle during drought stress in apple leaves. Acta Physiol. Plant. 33, 173–180.

Madamanchi, N.R., Doulis, X.Y.A., Alscher, R.G., Hatzios, K.K., Cramer, C.L., 1994. Acquired resistance to herbicides in pea cultivars through pretreatment with sulfur dioxide. Pest. Biochem. Physiol. 48, 31–40.

Mahajan, S., Tuteja, N., 2005. Cold, salinity and drought stresses: an overview. Arch. Biochem. Biophys. 444, 139–158.

Martin, M.N., 2003. Biosynthesis and metabolism of glutathione in plants. Genet. Eng. 25, 163–188.

Maughan, S., Foyer, C.H., 2006. Engineering and genetic approaches to modulating the glutathione network in plants. Physiol. Plant 126, 382–397.

McDermott, G.P., Francis, P.S., Holt, J.K., 2011. Determination of intracellular glutathione and glutathione disulfide using high performance liquid chromatography with acidic potassium permanganate chemiluminescence detection. Analyst 136, 2578–2585.

Meloni, D.A., Oliva, M.A., Martinez, C.A., Cambraia, J., 2003. Photosynthesis and activity of superoxide dismutase, peroxidase and glutathione reductase in cotton under salt stress. Environ. Exp. Bot. 49, 69–76.

Mendoza, J., Garrido, T., Riveros, R., 2008. Simultaneous determination of glutathione and glutathione disulfide in an acid extract of plant shoot and root by capillary electrophoresis. J. Chil. Chem. Soc. 53, 1626–1630.

Mendoza, J., Soto, P., Ahumada, I., 2004. Determination of oxidized and reduced glutathione, by capillary zone electrophoresis, in *Brassica juncea* plants treated with copper and cadmium. Electrophoresis 25, 890–896.

Mendoza-Cozatl, D.G., Moreno-Sanchez, R., 2006. Control of glutathione and phytochelatin synthesis under cadmium stress. Pathway modeling for plants. J. Theor. Biol. 238, 919–936.

Meyer, A.J., 2008. The integration of glutathione homeostasis and redox signaling. J. Plant Physiol. 165, 1390–1403.

Mhamdi, A., Hager, J., Chaouch, S., Queval, G., Han, Y., Taconnat, L., et al., 2010. *Arabidopsis* glutathione reductase1 plays a crucial role in leaf responses to intracellular hydrogen peroxide and in ensuring appropriate gene expression through both salicylic acid and jasmonic acid signaling pathways. J. Plant Physiol. 153, 1144–1160.

Michaelsen, J.T., Dehnert, S., Giustarini, D., 2009. HPLC analysis of human erythrocytic glutathione forms using OPA and N-acetyl-cysteine ethyl ester: evidence for nitrite-induced GSH oxidation to GSSG. J. Chromatogr. Anal. Technol. Biomed. Life Sci. 877, 3405–3417.

Millar, A.H., Mittova, V., Kiddle, G., 2003. Control of ascorbate synthesis by respiration and its implications for stress responses. Plant Physiol. 133, 443–447.

Milner, L.J., Reade, J.P.H., Cobb, A.H., 2007. The effect of temperature on glutathione S-transferase activity and glutathione content in *Alopecurus myosuroides* (black grass) biotypes susceptible and resistant to herbicides. Weed Res. 47, 106–112.

Miteva, P.-E., Ivanov, S.V., Alexieva, V.S., 2010. Alterations in glutathione pool and some related enzymes in leaves and roots of pea plants treated with the herbicide glyphosate. Russ. J. Plant Physiol. 57, 131–136.

Mittler, R., 2006. Abiotic stress, the field environment and stress combination. Trends Plant Sci. 11, 15–20.

Mohanpuria, P., Rana, N.K., Yadav, S.K., 2007. Cadmium induced oxidative stress influence on glutathione metabolic genes of *Camellia sinensis* (L.) O. Kuntze. Environ. Toxicol. 22, 368–374.

Mohsenzadeh, S., Esmaeili, M., Moosavi, F., 2011. Plant glutathione S-transferase classification, structure and evolution. Afr. J. Biotech. 10, 8160–8165.

Moller, I.M., Jensen, P.E., Hansson, A., 2007. Oxidative modifications to cellular components in plants. Annu. Rev. Plant Biol. 58, 459–481.

Moons, A., 2005. Regulatory and functional interactions of plant growth regulators and plant glutathione S-transferases (GSTS). Vitam. Horm. 72, 155–202.

Mullineaux, P.M., Rausch, T., 2005. Glutathione, photosynthesis and the redox regulation of stress-responsive gene expression. Photosynth. Res. 86, 459–474.

Nakamura, S.-I., Suzui, N., Nagasaka, T., Komatsu, F., Ishioka, N.S., Ito-Tanabata, S., et al., 2013. Application of glutathione to roots selectively inhibits cadmium transport from roots to shoots in oilseed rape. J. Exp. Bot. 64, 1073–1081.

Nan, Z., Li, J., Zhang, J., Cheng, G., 2002. Cadmium and zinc interactions and their transfer in soil-crop system under actual field conditions. Sci Total Environ. 285, 187–195.

Navrot, N., Collin, V., Gualberto, J., 2006. Plant glutathione peroxidases are functional peroxiredoxins distributed in several subcellular compartments and regulated during biotic and abiotic stresses. Plant. Physiol. 142, 1364–1379.

Ndamanisha, J.C., Bai, J., Qj, B., 2009. Application of electrochemical properties of ordered mesoporous carbon to the determination of glutathione and cysteine. Anal. Biochem. 386, 79–84.

Neill, S., Barros, R., Bright, J., Desikan, R., Hancock, J., Harrison, J., 2008. Nitric oxide, stomatal closure, and abiotic stress. J. Exp. Bot. 59, 165–176.

Nemat Alla, M.M., Badawi, A.M., Hassan, N.M., 2008. Herbicide tolerance in maize is related to increased levels of glutathione and glutathione-associated enzymes. Acta Physiol. Plant 30, 371–379.

Noctor, G., Arisi, A.C.M., Jouanin, L., Kunert, K.J., Rennenberg, H., Foyer, C.H., 1998. Glutathione: biosynthesis, metabolism and relationship to stress tolerance explored in transformed plants. J. Exp. Bot. 49, 623–647.

Noctor, G., Mhamdi, A., Chaouch, S., Han, Y., Neukermans, J., Marquez-Garcia, B., et al., 2012. Glutathione in plants: an integrated overview. Plant Cell Environ. 35, 454–484.

Nolin, T.D., McMenamin, M.M., Himmelfarb, J., 2007. Simultaneous determination of total homocysteine, cysteine, cysteinylglycine, and glutathione in human plasma by high-performance liquid chromatography: application to studies of oxidative stress. J. Chromatogr. B Anal. Technol. Biomed. Life Sci. 852, 554–561.

Oztetik, E., 2008. A tale of plant Glutathione S-transferases: since 1970. Bot. Rev. 74, 419–437.

Panda, S.K., 2007. Chromium-mediated oxidative stress and ultrastructural changes in root cells of developing rice seedlings. J. Plant Physiol. 164, 1419–1428.

Panda, S.K., Choudhury, S., 2005. Chromium stress in plants. Braz. J. Plant Physiol. 17, 95–102.

Parida, A.K., Das, A.B., 2005. Salt tolerance and salinity effects on plants: a review. Ecotoxicol. Environ. Saf. 60, 324–349.

Pastore, A., Federici, G., Bertini, E., Piemonte, F., 2003. Analysis of glutathione: implication in redox and detoxification. Clin. Chim. Acta 333, 19–39.

Pastori, G.M., Foyer, C.H., 2002. Common components, networks, and pathways of cross-tolerance to stress. The central role of "redox" and abscisic acid-mediated controls. Plant. Physiol. 129, 460–468.

Peerzada, Y.Y., Khalid, H.U.R., Chandna, R., 2012. Role of glutathione reductase in plant abiotic stress. In: Ahmad, P., Prasad, M.N.V. (Eds.), Abiotic Stress Responses in Plants: Metabolism, Productivity and Sustainability. Springer, New York, pp. 149–158.

Pinto, M.C., Paradiso, A., Leonetti, P., 2006. Hydrogen peroxide, nitric oxide and cytosolic ascorbate peroxidase at the crossroad between defence and cell death. Plant J. 48, 784–795.

Pompella, A., Visvikis, A., Paolicchi, A., De Tata, V., Casini, A.F., 2003. The changing faces of glutathione, a cellular protagonist. Biochem. Pharmacol. 66, 1499–1503.

Potesil, D., Petrlova, J., Adam, V., 2005. Simultaneous femtomole determination of cysteine, reduced and oxidized glutathione, and phytochelatin in maize (*Zea mays* L.) kernels using high-performance liquid chromatography with electrochemical detection. J. Chromatogr. A 1084, 134–144.

Queval, G., Issakidis-Bourguet, E., Hoeberichts, F.A., Vandorpe, M., Gakiere, B., Vanacker, H., et al., 2007. Conditional oxidative stress responses in the *Arabidopsis* photorespiratory mutant cat2 demonstrate that redox state is a key modulator of daylength-dependent gene expression, and define photoperiod as a crucial factor in the regulation of H_2O_2-induced cell death. Plant J. 52, 640–657.

Ramachandra, R.A., Chaitanya, K.V., Vivekanandan, M., 2004. Drought induced responses of photosynthesis and antioxidant metabolism in higher plants. J. Plant Physiol. 161, 1189–1202.

Ramírez, L., Bartoli, C.G., and Lamattina, L., 2013. Glutathione and ascorbic acid protect *Arabidopsis* plants against detrimental effects of iron deficiency. J. Exp. Bot. 64, 3169–78.

Rampino, P., Pataleo, S., Gerardi, C., Mita, G., Perrotta, C., 2006. Drought stress response in wheat: physiological and molecular analysis of resistant and sensitive genotypes. Plant Cell Environ. 29, 2143–2152.

Raoof, J.B., Ojani, R., Baghayeri, M., 2009. Simultaneous electrochemical determination of glutathione and tryptophan on a nano-$TiO_{(2)}$/ferrocene carboxylic acid modified carbon paste electrode. Sens. Actuators B-Chem 143, 261–269.

Ratnayaka, H.H., Molin, W.T., Sterling, T.M., 2012. Comparison of physiological and antioxidant responses of *Anoda cristata* and cotton to progressive drought. Weed Res. 52, 358–366.

Rouhier, N., Lemaire, S.D., Jacquot, J.P., 2008. The role of glutathione in photosynthetic organisms: emerging functions for glutaredoxins and glutathionylation. Annu. Rev. Plant Biol. 59, 143–166.

Rousar, T., Kucera, O., Lotkova, H., 2012. Assessment of reduced glutathione: comparison of an optimized fluorometric assay with enzymatic recycling method. Anal. Biochem. 423, 236–240.

Roy, S.J., Tucker, E.J., Tester, M., 2011. Genetic analysis of abiotic stress tolerance in crops. Curr. Opin. Plant Biol. 14, 232–239.

Ruelland, E., Vaultier, M.-N., Zachowski, A., Hurry, V., 2009. Cold signaling and cold adaptation in plants. Adv. Bot. Res 49, 35–150.

Ruiz, J.M., Blumwald, E., 2002. Salinity-induced glutathione synthesis in *Brassica napus*. Planta 214, 965–969.

Ruiz, J.M., Rivero, R.M., Romero, L., 2006. Boron increases synthesis of glutathione in sunflower plants subjected to aluminum stress. Plant. Soil. 279, 25–30.

Saenen, E., Horemans, N., Vanhoudt, N., Vandenhove, H., Biermans, G., Hees, M.V., et al., 2013. Effects of pH on uranium uptake and oxidative stress responses induced in *Arabidopsis thaliana*. Environ. Toxicol. Chem. 32, 2125–33.

Safavi, A., Maleki, N., Farjami, E., 2009. Simultaneous electrochemical determination of glutathione and glutathione disulfide at a nanoscale copper hydroxide composite carbon ionic liquid electrode. Anal. Chem. 81, 7538–7543.

Sairam, R.K., Srivastava, G.C., Saxena, D.C., 2000. Increased antioxidant activity under elevated temperatures: a mechanism of heat stress tolerance in wheat genotypes. Biol. Plant 43, 245–251.

Sakhi, A.K., Blomhoff, R., Gundersen, T.E., 2007. Simultaneous and trace determination of reduced and oxidized glutathione in minute plasma samples using dual mode fluorescence detection and column switching high performance liquid chromatography. J. Chromatogr. A 1142, 178–184.

Salama, K.H.A., Al-Mutawa, M.M., 2009. Glutathione-triggered mitigation in salt-induced alterations in plasmalemma of onion epidermal cells. Int. J. Agric. Biol. 11, 639–642.

Sánchez-Rodríguez, E., del Mar Rubio-Wilhelmi, M., Blasco, B., Leyva, R., Romero, L., Ruiz, J.M., 2012. Antioxidant response resides in the shoot in reciprocal grafts of drought-tolerant and drought-sensitive cultivars in tomato under water stress. Plant Sci. 188–189, 89–96.

Santa, T., Aoyama, C., Fukushima, T., 2006. Suppression of thiol exchange reaction in the determination of reduced-form thiols by high-performance liquid chromatography with fluorescence detection after derivatization with fluorogenic benzofurazan reagent, 7-fluoro-2,1,3-benzoxadiazole-4-sulfonate and 4-aminosulfonyl-7-fluoro-2,1,3-benzoxadiazole. Biomed. Chromatogr. 20, 656–661.

Santini, J., Giannettini, J., Pailly, O., Herbette, S., Ollitrault, P., Berti, L., et al., 2013. Comparison of photosynthesis and antioxidant performance of several *Citrus* and *Fortunella* species (Rutaceae) under natural chilling stress. Trees 27, 71–83.

Schafer, F.O., Buettner, G.R., 2001. Redox environment of the cell as viewed through the redox state of the glutathione disulphide/glutathione couple. Free Radic. Biol. Med. 1191–1212.

Schützendübel, A., Polle, A., 2002. Plant responses to abiotic stress: heavy metal-induced oxidative stress and protection by mycorrhization. J. Exp. Bot. 53, 1351–1365.

Seki, M., Kamei, A., Yamaguchi-Shinozaki, K., Shinozaki, K., 2003. Molecular responses to drought, salinity and frost: common and different paths for plant protection. Curr. Opin. Biotechnol. 14, 194–199.

Seth, C.S., Remans, T., Keunen, E., Jozefczak, M., Gielen, H., Opdenakker, K., et al., 2012. Phytoextraction of toxic metals: a central role for glutathione. Plant Cell Environ. 35, 334–346.

Sezginturk, M.K., Dinckaya, E., 2011. Glutathione (GSH) determination by a very simple electrochemical method. Inter. J. Pept. Res. Ther. 17, 87–92.

Shan, C., He, F., Xu, G., Han, R., Liang, Z., 2012. Nitric oxide is involved in the regulation of ascorbate and glutathione metabolism in *Agropyron cristatum* leaves under water stress. Biol. Plantarum 56, 187–191.

Shan, C.-J., Zhang, S.-L., Li, D.-F., Zhao, Y.-Z., Tian, X.-L., Zhao, X.-L., et al., 2011. Effects of exogenous hydrogen sulfide on the ascorbate and glutathione metabolism in wheat seedlings leaves under water stress. Acta Physiol. Plant. 33, 2533–2540.

Shanker, A.K., Cervantes, C., Loza-Tavera, H., Avudainayagam, S., 2005. Chromium toxicity in plants. Environ. Int. 31, 739–753.

Shao, H.B., Chu, L.Y., Shao, M.A., Jaleel, C.A., Hong-Mei, M., 2008. Higher plant antioxidants and redox signaling under environmental stresses. Comp. Rend. Biol. 331, 433–441.

Shiu, C.T., Lee, T.M., 2005. Ultraviolet-B-induced oxidative stress and responses of the ascorbate–glutathione cycle in a marine macroalga *Ulva fasciata*. J. Exp. Bot. 56, 2851–2865.

Shtyrlin, V.G., Zyavkina, Y.I., Ilakin, V.S., 2005. Structure, stability, and ligand exchange of copper (II) complexes with oxidized glutathione. J. Inorg. Biochem. 99, 1335–1346.

Shu, D.-F., Wang, L.-Y., Duan, M., Deng, Y.-S., Meng, Q.-W., 2011. Antisense-mediated depletion of tomato chloroplast glutathione reductase enhances susceptibility to chilling stress. Plant Physiol. Biochem. 49, 1228–1237.

Singh, S., Gupta, A.K., Kaur, N., 2012. Differential responses of antioxidative defence system to long-term field drought in wheat (*Triticum aestivum* L.) genotypes differing in drought tolerance. J. Agron. Crop Sci. 198, 185–195.

Skipsey, M., Knight, K.M., Brazier-Hicks, M., 2011. Xenobiotic responsiveness of *Arabidopsis thaliana* to a chemical series derived from a herbicide safener. J. Biol. Chem. 286, 32268–32276.

Son, J.A., Narayanankutty, D.P., Roh, K.S., 2013. Influence of exogenous application of glutathione on rubisco and rubisco activase in heavy metal-stressed tobacco plant grown *in vitro*. Saudi J. Biol. Sci. doi:org/10.1016/j.sjbs.2013.06.002.

Song, L.L., Ding, W., Shen, J., Zhang, Z.G., Bi, Y.R., Zhang, L.X., 2008. Nitric oxide mediates abscisic acid induced thermotolerance in the calluses from two ecotypes of reed under heat stress. Plant Sci. 175, 826–832.

Sreenivasulu, N., Sopory, S.K., Kishor, P.B., 2007. Deciphering the regulatory mechanisms of abiotic stress tolerance in plants by genomic approaches. Gene 388, 1–13.

Sun, Q., Wang, X.-R., Ding, S.-M., Yuan, X.-F., 2005. Effects of interaction between cadmium and plumbum on phytochelatins and glutathione production in wheat (*Triticum aestivum* L.). J. Integr. Plant Biol. 47, 435–442.

Suzuki, N., Koussevitzky, S., Mittler, R., Miller, G., 2012. ROS and redox signalling in the response of plants to abiotic stress. Plant Cell Environ. 35, 259–270.

Szalai, G., Kellos, T., Galiba, G., Kocsy, G., 2009. Glutathione as an antioxidant and regulatory molecule in plants under abiotic stress conditions. J. Plant Growth Regul. 28, 66–80.

Szarka, A., Tomasskovics, B., Bánhegyi, G., 2012. The ascorbate-glutathione-tocopherol triad in abiotic stress response. Inter. J. Mol. Sci 13, 4458–4483.

Tausz, M., Sircelj, H., Grill, D., 2004. The glutathione system as a stress marker in plantecophysiology: is a stress-response concept valid? J. Exp. Bot. Sulphur Metabol. Plant 55, 1955–1962.

Tester, M., Bacic, A., 2005. Abiotic stress tolerance in grasses: from model plants to crop plants. Plant. Physiol. 137, 791–793.

Tian, J., Yu, J., 2009. Changes in ultrastructure and responses of antioxidant systems of algae (*Dunaliella salina*) during acclimation to enhanced ultraviolet-B radiation. J. Photochem. Photobiol. B: Biol. 97, 152–160.

Uzilday, B., Turkan, I., Sekmen, A.H., Ozgur, R., Karakaya, H.C., 2012. Comparison of ROS formation and antioxidant enzymes in *Cleome gynandra* (C4) and *Cleome spinosa* (C3) under drought stress. Plant Sci. 182, 59–70.

Vacek, J., Petrek, J., Kizek, R., 2004. Electrochemical determination of lead and glutathione in a plant cell culture. Bioelectrochem. 63, 347–351.

Wahid, A., Gelani, S., Ashraf, M., Foolad, M.R., 2007. Heat tolerance in plants: an overview. Environ. Exp. Bot. 61, 199–223.

Wang, C.-Q., Xu, H.-J., Liu, T., 2011. Effect of selenium on ascorbate–glutathione metabolism during PEG-induced water deficit in *Trifolium repens* L. J. Plant Growth Regul. 30, 436–444.

Wang, F., Chen, F., Cai, Y., Zhang, G., Wu, F., 2011. Modulation of exogenous glutathione in ultrastructure and photosynthetic performance against Cd stress in the two barley genotypes differing in Cd tolerance. Biol. Trace Elem. Res. 144, 1275–1288.

Wang, H.-Z., Zhang, L.-H., Ma, J., Li, X.-Y., Li, Y., Zhang, R.-P., et al., 2010. Effects of water stress on reactive oxygen species generation and protection system in rice during grain-filling stage. Agr. Sci. China. 9, 633–641.

Wang, J., Lue, H., Sun, Q., 2009. Method of electrochemical determination of glutathione. Acta Chim. Sin. 67, 415–419.

Wang, L., Li, Y.X., Zhao, D.H., 2003. A novel enhancing flow-injection chemiluminescence method for the determination of glutathione using the reaction of luminol with hydrogen peroxide. Microchim. Acta. 141, 41–45.

Wang, S., Liang, D., Li, C., Hao, Y., Maa, F., Shu, H., 2012. Influence of drought stress on the cellular ultrastructure and antioxidant system in leaves of drought-tolerant and drought-sensitive apple rootstocks. Plant Physiol. Biochem. 51, 81–89.

Wang, X., Chi, D., Song, D., Su, G., Li, L., Shao, L., 2012. Quantification of glutathione in plasma samples by HPLC using 4-fluoro-7-nitrobenzofurazan as a fluorescent labeling reagent. J. Chromatogr. Sci. 50, 119–122.

Will, B., Jouanin, L., Rennenberg, H., 2001. Protection from paraquat-mediated photo-oxidative stress by glutathione in poplar (*Populus tremula* × *P. alba*) plants. Plant Biol 3, 272–278.

Wójcik, M., Tukiendorf, A., 2004. Phytochelatin synthesis and cadmium localization in wild type of *Arabidopsis thaliana*. Plant Growth. Regul. 44, 71–80.

Xie, Z., Duan, L., Tian, X., Wang, B., Egrinya-Eneji, A., Li, Z., 2008. Coronatine alleviates salinity stress in cotton by improving the antioxidative defense system sand radical-scavenging activity. J. Plant Physiol. 165, 375–384.

Xiong, L., Wang, R., Mao, G., Koczan, J.M., 2006. Identification of drought tolerance determinants by genetic analysis of root response to drought stress and abscisic acid. Plant Physiol. 142, 1065–1074.

Xu, Y., Sun, X., Jin, J., Zhou, H., 2010. Protective effect of nitric oxide on light-induced oxidative damage in leaves of tall fescue. J. Plant Physiol. 167, 512–518.

Yadav, S.K., 2010. Heavy metals toxicity in plants: an overview on the role of glutathione and phytochelatins in heavy metal stress tolerance of plants. S. Afr. J. Bot. 76, 167–179.

Yin, H., Chen, Q., Yi, M., 2008. Effects of short-term heat stress on oxidative damage and responses of antioxidant system in *Lilium longiflorum*. Plant Growth Regul. 545, 45–54.

Yruela, I. (2005). Copper in plants. Brazilian Journal of Plant Physiology, 17, 145–156.

Zagorchev, L., Seal, C.E., Kranner, I., Odjakova, M., 2013. A central role for thiols in plant tolerance to abiotic stress. Int. J. Mol. Sci. 14, 7405–7432.

Zeng, F., Qiu, B., Wu, X., Niu, S., Wu, F., Zhang, G., 2012. Glutathione-mediated alleviation of chromium toxicity in rice plants. Biol. Trace Elem. Res. 148, 255–263.

Zhang, H., Forman, H.J., 2012. Glutathione synthesis and its role in redox signaling. Semin. Cell Dev. Biol. 23, 722–728.

Zhang, H., Hu, L.Y., Li, P., Hu, K.D., Jiang, C.X., Luo, J.P., 2010a. Hydrogen sulfide alleviated chromium toxicity in wheat. Biol. Plant 54, 743–747.

Zhang, H., Jiao, H., Jiang, C.X., Wang, S.H., Wei, Z.J., Luo, J.P., et al., 2010b. Hydrogen sulfide protects soybean seedlings against drought-induced oxidative stress. Acta Physiol. Plant. 32, 849–857.

Zhang, R., Wu, W., 2011. Studies on the structures and interactions of glutathione in aqueous solution by molecular dynamics simulations and NMR spectroscopy. J. Mol. Liq. 162, 20–25.

Zhou, Y.-H., Zhang, Y.-Y., Zhao, X., Yu, H.-J., Shi, K., Yu, J.-Q., 2009. Impact of light variation on development of photoprotection, antioxidants, and nutritional value in *Lactuca sativa* L. J. Agric. Food Chem. 57, 5494–5500.

Zhu, P., Oe, T., Blair, I., 2008. A determination of cellular redox status by stable isotope dilution liquid chromatography/mass spectrometry analysis of glutathione and glutathione disulfide. Rapid Commun. Mass Spectr. 22, 432–440.

Chapter 6

Glutathione Metabolism in Plants under Environmental Stress

Asiya Hameed, Iti Sharma, Ashwani Kumar, MM Azooz, Helal Ahmad Lone and Parvaiz Ahmad

6.1 INTRODUCTION

Glutathione is an important component of the defense system in plants and animals. The tripeptide glutathione (g-glutamyl-cysteinyl-glycine) is the most copious form of organic sulfur in plants. Glutathione exists in both oxidized form (GSSG) and reduced form (GSH), among which GSH is considered as a low molecular weight thiol tripeptide. GR is an essential enzyme for the reduction of GSSG to GSH by an NADPH-dependent mechanism (Tandogan and Ulusu, 2006). During stress conditions GSH has the potential to accumulate in cells. GSSG and GSH play a key role in the maintenance and regulation of the cell cycle, thereby mitigating the stress (Bashandy et al., 2010). However, GSH accumulation also acts against cancerous cells in mammals, particularly in malnourished patients (Bray and Taylor, 1993). Generally, in humans, glutathione acts as an antioxidant and a potent detoxifier that contains the amino acids produced from foods (Mari et al., 2009). Besides this, it also helps to fight against various diseases such as cancer, heart disease and Alzheimer's disease (Hyman, 2013).

In plants also, a high level of glutathione (i.e. synthesis as well as accumulation) is present in cell organelles, especially in chloroplasts. The amount of glutathione can be modified by means of genetic engineering techniques in plants which help to understand the mechanism of alterations during the synthesis of glutathione, which affects metabolism as well as the defense system (Xiang et al., 2001; Jozefczak et al., 2012). Glutathione plays a vital role as a metal scavenger owing to its high binding affinity with thiol (−SH) group and also as a precursor of phytochelatins (PCs) (Cuypers et al., 2010; Mittler et al., 2004; Smeets et al., 2009; Ahmad et al., 2008, 2010, 2011a; Ahmad and

P. Ahmad (Ed): Oxidative Damage to Plants. DOI: http://dx.doi.org/10.1016/B978-0-12-799963-0.00006-X

Umar, 2011). Thus, glutathione supports plants in coping with stress by chelating or sequestration of metal, thereby restricting the accumulation of free metal concentration in cells. Exogenous application of GSH has been shown to protect the growth and development of mutant embryos, which implies an endogenous GSH is crucial for seed maturation as well as seed germination (Cairns et al., 2006; Lim et al., 2011).

6.2 GLUTATHIONE AND PHOTO-OXIDATIVE STRESS

ROS generated in many biological reactions are toxic to living organisms and are detoxified by defense mechanisms present in plants as they scavenge these reactive radicals and are also involved in photo-oxidative stress tolerance of plants. Glutathione, along with enzymatic and nonenzymatic antioxidants, is one of the vital components for plant photo-protection. This photo-protection mechanism minimizes ROS-mediated cellular damage (Ort, 2001). In addition, the GSSG/GSH redox system signals specific responses under environmental stress and photo-oxidative damage (Foyer and Noctor, 2011; Seth et al., 2012). Genetic manipulation and overproduction of various antioxidant enzymes in chloroplast can enhance tolerance but this has little or no effect on the levels of photoinhibition in stressed plants (Tyystjarvi et al., 1995). Nonetheless, studies suggest that in transgenic cotton the overproduction of chloroplastic glutathione reductase (GR) has less photoinhibition than wild-type cotton on exposure to environmental stress (Korenyeyev et al., 2001). This has been analyzed by Aono et al. (1997) using four types of transgenic tobacco plants; they are those containing:

1. GR of *E. coli* origin in cytosol
2. *E. coli* GR in chloroplast
3. Antisense spinach cDNA for chloroplastic GR, and
4. GR of *E. coli* and rice superoxide dismutase (SOD)

It has been observed that transgenic tobacco with high cytosolic or chloroplastic GR activity showed increased tolerance to photo-oxidative stress caused by abiotic stress, whereas transgenic tobacco with reduced GR activity exhibited enhanced sensitivity to paraquat. Furthermore, transgenic tobacco with simultaneously enhanced activities of GR and SOD showed tremendous tolerance to paraquat. A tolerance activity against these photo-oxidative stresses shows dependency on this cooperation of antioxidant enzymes (Foyer et al., 1995; Aono et al., 1997).

Due to the exchange of the thiol group, GR possesses an important property that regulates the enzymatic activity and maintains the structure of protein as well as helps in the transduction pathway by means of thioredoxin (TRX) (Buchanan and Balmer, 2005; Meyer et al., 2008). Besides this complexity in the antioxidative system, plants have shown other forms of antioxidative peroxides that reduce H_2O_2 to water, which include TRX fusion protein CPSP32 and various types of peroxiredoxin (PRX); chloroplasts and peroxisomes are the important centers for H_2O_2 storage found in most of the photosynthetic cells. In addition, cytosolic APX1 have played a major part in redox homeostasis in *Arabidopsis* implying the cytosolic-ascorbate-glutathione pathway is essential in metabolizing H_2O_2 and has significance in driving specific responses (Foyer and Noctor, 2005a; Rey et al., 2005; Mhamdi et al., 2010). Thus, redox reactions of glutathione have a significant role in mitigating stress in plants.

6.3 GLUTATHIONE AND HEAVY METALS

As nutrients, metals are an essential component for growth and development of the plant body but they become toxic as the concentration exceeds the limit. Almost all the heavy metals are non-redox metals and unable to form Fenton-type reactions but instead give rise to ROS, such as H_2O_2 and superoxide radicals (Romero-Puertas et al., 2004). In this regard, plants safeguard against oxidative damage through regulation of antioxidant enzymes (Davies, 1986; Sharma et al., 2007). Besides metal homeostasis, GSH in plants develops a well-equipped defense mechanism that manages and helps to combat the damage produced by higher concentrations of metals (Hall, 2002; Sharma, 2013a,b).

Glutathione is found abundantly in plants with many enzymatic activities including signal transduction (Gomez et al., 2004; Foyer and Noctor, 2005a; Ahmad et al., 2008, 2010, 2011a; Ahmad and Umar, 2011; Yousuf et al., 2012). In the reduced glutathione, the cys thiol gets oxidized to oxidized glutathione (GSSG) and its reverse reaction, GSSG to GSH, is catalyzed by GR via NADPH (Ranieri et al., 2005). Two genes are annotated to encode GR in plants. The first plastidial protein has been demonstrated in mitochondria of pea chloroplast (Creissen et al., 1995). Dual targeting of this protein also occurs in *Arabidopsis* which represents the plastid/mitochondrial isoform by GR2 (Chew et al., 2003). The second gene GR1 is predicted to encode a cytosolic enzyme (Stevens et al., 2000). Cytosolic GR has been well characterized in pea but only at the biochemical level, whereas its functional role is still under investigation (Ding et al., 2009). Scientists have made major efforts to enhance the levels of GSH in plants that have the capacity to combat any type of environmental stress; these have also been shown to be an intrinsic response of plants against stress (Noctor et al., 1998). Transcriptional and posttranslational mechanisms have been demonstrated in various plant species such as *Brassica juncea*, where stimulation of γ-ECS of transcript accumulation occurs by heavy metals and in *Arabidopsis* γ-ECS is carried out by post-transcriptional process (Schäfer et al., 1997). Tremendous increases in the level of GSH have been reported with an elevated γ-ECS activity in response to heavy metals. This means that a posttranscriptional mechanism provides a fundamental role in the activities of γ-ECS. Molecular and genetic tools associated with biochemical techniques will use a mechanistic approach to understand the regulation of γ-ECS in plants. Moreover, the sequestration of heavy metal in plants comprises cysteine-rich peptides called phytochelatins. They are repeating units of γ-EC that contain glycine, alanine, serine or glutamate (Rauser, 1995; Zenk, 1996; Sharma, 2012a).

Among heavy metals cadmium (Cd) is one of the known inducers of phytochelatins which form complexes in the cytosol and then move to vacuole. Concentration dependent depletion of glutathione has been demonstrated in cultured cells and in roots exposed to cadmium (Rauser et al., 1991; Klapheck et al., 1995). Moreover, reduction of glutathione is always accompanied by elevation of γ-ECS activity and γ-EC contents in maize (Rüegsegger and Brunold, 1992; Noctor et al., 1998).

Buthionine sulphoxine acts as an inhibitor of γ-ECS in PC accumulation, thereby revealing the significance of γ-ECS activity in synthesis of phytochelatins (Grill et al., 1987). Enhanced γ-ECS contents confer Cd resistance in cultured tomato cells (Chen and Goldsborough, 1994). Likewise the inhibition of γ-ECS highlights the reduction of growth against Cd and has been demonstrated in birch (Gussarson et al., 1996). Plants with antioxidants possess hypertolerant capacity to combat the toxicity produced due to oxidative stress, including heavy metals, and thereby protect the cell organelles from any sort of damage (Rascio and Navari-Izzo, 2011; Tunn and Popova, 2013).

6.4 DETOXIFICATION OF TOXIC SUBSTANCES

Accumulation of sulfur forms a basic part of glutathione that provides a means of detoxification against free radicals and toxins, including heavy metals. Specific enzyme catalyzed reactions convert the harmful products into less toxic metabolites by chemical reactions in cells and are considered a chief means of detoxification (Ziglari and Allameh, 2013). GSH, owing to its high sulphydryl (−SH) content, prevents damage to cellular components by neutralizing free radicals. This makes it a sensitive indicator of cell function and its viability (Cancado et al., 2005; Sharma, 2012b).

GST (Glutathione-S-transferase) is mainly applied for intracellular detoxification of xenobiotics and toxic compounds produced within the cell (Mannervik and Danielson, 1988; Edwards et al., 2000). It has been shown to protect plants against different types of stresses, though various GST activities have been expressed in many plants such as maize, tobacco, soybean, sorghum, barley, sugarcane, etc. (Dixon et al., 1998, 2002, 2003; Thom et al., 2002).

GR has been extracted from a number of plant species. A significant increase of GR activity prevents both oxidation of SH compounds, mainly GSH, and peroxidation of lipids, in beans treated with p-nitrodiphenyl ether acifluorfen. Similarly, Kenyon and Duke (1985) have also found a strong decrease of both glutathione and GR activity on exposure of cucumber to acifluorfen in the light. Bean plants shows positive effects, i.e. enhanced production of antioxidants, that play a vital role in preventing peroxidation.

Enhancement of antioxidants is associated with the increased activity of biosynthetic enzymes like GLO, followed by increase of antioxidant recycling enzymes such as GR. A similar response of GR in maize seedlings was found by Foster and Hess (1982) after exposure of seedlings to an $O_2^{\bullet -}$ enriched atmosphere and is connected to GSH-dependent feedback inhibition of the enzyme γ-glutamycysteine synthetase. Depletion of GSH under peroxidative conditions has been found to stimulate the synthase enzyme leading to high production of total glutathione. However, reduction of GSSG due to increased GR activity has been shown to reactivate the feedback inhibition of the synthetase. Finally, inhibition of both the enzyme and catabolic processes leads to a decrease in the amount of total glutathione (Mata et al., 1985). Another enzyme, namely glutathione peroxidase (GPOX), also reduces hydroperoxides (H_2O_2, hydroperoxides of fatty acids and phospholipids), which are generated during oxidative damage of specific alcohols of the plant cell.

6.5 GLUTATHIONE-MEDIATED TOLERANCE IN PLANTS

The change in redox state of GSH by GR is closely related to tolerance mechanisms in plants under stress. Species-dependent variations in GR activities have been reported in different plants. The enzyme has been found prominently accumulated in roots of pea and in leaves of *Crotolaria juncea* on exposure to environmental stress (Pereira et al., 2002). However, higher binding of GR with the enzyme substrate provides a good strategy to protect the roots against oxidative damage owing to its increased recycling of glutathione (Yannarelli et al., 2007). Likewise alterations in GR isozymes in pea leaves have been demonstrated by Edwards et al. (1994). Different intensities of GR bands have been found in maize seedlings under chilling stress (Anderson et al., 1995). GR protects the cell from oxidative damage, and thereby maintains the stability of the GSH/GSSG ratio in the ascorbate-glutathione cycle (Foyer and Noctor, 2005b). The power of ascorbate regeneration due to reduced glutathione is thought to enhance the scavenging capacity in the cell and therefore to prevent photo-oxidative damage. Despite this system, other mechanisms are required to understand the tolerance of the plant against photo-oxidative injury (Aono et al., 1991, 1993, 1997).

The glutathione S-transferases (GSTs) are also known to catalyze the conjugation of electrophilic and hydrophobic toxic compounds with glutathione to form nontoxic peptide derivatives. The principal function of GST is to play a significant role in the detoxification of lipid hydroperoxides generated in membrane peroxidation and hence to protect the plant from undergoing any symptoms such as necrosis. This leads to a slight change in the level of antioxidant enzymes that forms a strong and beneficial interaction between plants and fungi (Alguacil et al., 2003). This promotes a mycorrhizal association and therefore improves resistance to pathogens (Waller et al., 2005).

Colonization of maize plants by *P. indica* has been found to increase with increased antioxidant capacity and as a result the plant develops resistance against abiotic stress. Unlike AMF, *P. indica* can easily be widely grown in axenic culture in absence of a host plant and this can be practiced for the development of agriculture (Varma et al., 1999). Therefore *P. indica* can be utilized not only in crop strategies but also can be used as a replica to analyze the molecular features susceptible to diseases and yield in cereals (Kumar et al., 2009).

GSH has even been occasionally found to increase in response to Cd exposure, as occurred in *S. cerevisiae* (Vido et al., 2001), which is in contrast to the numerous reports of Cd-induced depletion, caused by the GSH withdrawal for PC synthesis (Schützendübel and Polle, 2002). These enzymes provide antioxidant protection and preserve membrane integrity. The removal of H_2O_2 produced in chloroplasts is essential to avoid inhibition of the Calvin cycle enzymes (Tanaka et al., 1982).

Similar results, i.e. an increase in GR activity in plants following exposure to Cd, were noticed in *Phaseolus vulgaris* and Alyssum (Schickler and Caspi, 1999) whereas a decrease has been observed in *Helianthus annuus* and *Cuscuta reflexa* (Srivastava et al., 2004). GR recycles the GSSG back to GSH, which is crucial for the functioning of the ascorbate–glutathione cycle as well as for the synthesis of PCs (Cobbett, 2000). The increased activity of GR could be explained by transcriptional or translational modification to keep an adequate GR level to protect against Cd stress (Romero-Puertas et al., 2002). When the cells were cultured in the presence of mercury, glutathione was the principal thiol-containing compound produced at controlled levels, which might indicate a heavy metal detoxification function for glutathione, as described for other eukaryotic organisms (Prade et al., 1998).

Hg-exposure resulted in an increase in H_2O_2 content in plants. Although the mechanism of Hg-induced H_2O_2 formation is not known, heavy metals are known to produce reactive oxygen species (Luna et al., 1994; Ahmad et al., 2011a; Ahmad and Umar, 2011). The H_2O_2 accumulation after Hg-exposure may be produced in a manner similar to H_2O_2 in cold-stressed plants (Prasad et al., 1994). It is conceivable that a decrease of enzymatic and nonenzymatic free radical scavengers caused by heavy metals (Salt et al., 1995a) may also contribute to the shift in the balance of free radical metabolism towards H_2O_2 accumulation, and H_2O_2 and O_2^- may interact in the presence of certain metal ions or metal chelates to produce the highly reactive hydroxyl radical (•OH). The susceptibility to oxidative stress is a function of overall balance between the factors that increase oxidant generation and those substances that exhibit antioxidant capability (Foyer et al., 1994).

The occurrence of activated oxygen and symptoms of oxidative injury have also been observed in plants exposed to heavy metals, which do not belong to the group of transition metals Cd (Schützendübel et al., 2001; Ahmad et al., 2011b); Zn (Prasad et al., 1999; Rao and Sresty, 2000); and Ni (Rao and Sresty, 2000). Exposure to heavy metals also provoked pronounced responses of antioxidative systems, but the direction of the response was dependent on the plant species, the metal used for the treatment and the intensity of the stress. Exposure to heavy metal (Cd) initially resulted in a severe depletion of GSH (*Rauvolfia serpentina*: Grill et al., 1987; *Arabidopsis*: Xiang and Oliver, 1998; *Pigeonpea*: Rao and Sresty, 2000, *Pine*: Schützendübel et al., 2001). This is a common response to cadmium caused by an increased consumption of glutathione for phytochelatin production (Mehra and Tripathi, 1999). The sequestration of heavy metals for protection is a major significance of phytochelatins. For Cd, the formation of Cd-thiolate (Cd-S) complexes in phytochelatins has been shown (Strasdeit et al., 1991). The chelated metals are transported to the tonoplast, taken up by active transport systems and deposited in the vacuole (Rea, 1999). This mechanism contributes to protection from heavy metal toxicity in several plant species and in some fungi as well (Ishikawa et al., 1997). In general the glutathione pool recovered after prolonged Cd-exposure, frequently to levels above that of controls (Arisi et al., 2000; Schützendübel et al., 2001). The ability to synthesize glutathione appears to be crucial for protection from cadmium by increased tolerance of plants with elevated levels of GSH as well as decreased tolerance in plants with diminished levels of GSH (Zhu et al., 1999a,b). Lipid peroxidation and an oxidized GSSG/GSH ratio have been found to increase in *Silene cucubalis* when treated with Cu. BSO-treatment leads to depletion of GSH and simultaneously shows a considerable increase in the oxidized damage by Cu (De Vos et al., 1992). Similar reports have been observed by Wojcik and Tukendorff (2011) demonstrating GSH adaptation to Cd stress in *Arabidopsis* with or without addition of BSO.

6.6 GLUTATHIONE BIOSYNTHETIC PATHWAY AND ITS REGULATION

Glutathione is formed by the combination of γ-glutamate and cysteine via γ-glutamylcysteine synthetase (GSH1) through peptide bond and at the same time is combined with glycine catalyzed by glutathione synthase catalyzed by (GSH2) (Fig. 6.1) (Noctor et al., 2012). Both GSH1 and GSH2 are coded by a single gene and have been observed in *Arabidopsis spp.* Any alteration in one of the two genes can undergo mutations. GSH biosynthesis by means of compartmentalization and complexation reveals the restriction of GSH1 to plastids and GSH2 to cytosol of plant cells (Meyer and Fricker, 2002). GSH biosynthesis is affected by many factors such as sulfur availability and GSH1 activity. Elevated GSH due to elevation of metal toxicity helps to undergo detoxification and hence enhances survival. The reduction reactions of sulfate form various compounds including cysteine, which is an important

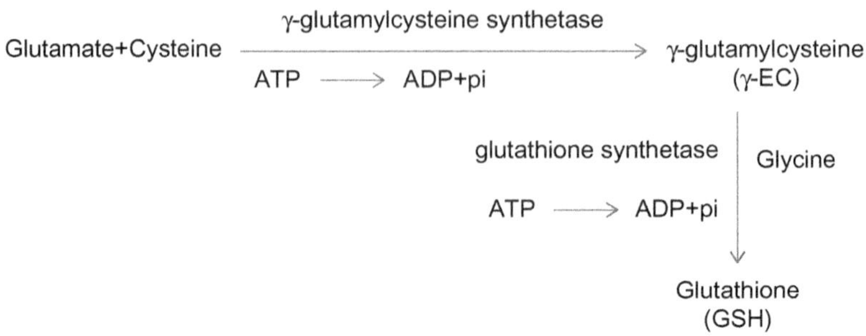

FIGURE 6.1 Biosynthesis of glutathione in plants.

component of glutathione biosynthesis (Davidian and Kopriva, 2010; Queval et al., 2009). GSH also acts as a storage form of reduced sulfur in cells. Increasing metal toxicity enhances the GSH levels and in turn activates uptake of sulfate. Cd reduces the sulfate and simultaneously increases the level of transcription sulfate reductase (ATPs, APSR), which might be related to changes in GSH1 and clear-cut specific isoforms of serine acetyltransferase (SAT) and O-acetylserine (thiol)lyase (OASTL) (Heiss et al., 1999; Nocito et al., 2002, 2006). These studies suggest the concept of the transcriptional regulation of sulfur assimilation along with the GSH synthesis genes being part of the cellular changes to metal stress (Dominguez-Solis et al., 2001; Howarth et al., 2003). GSH is an intracellular thiol as well as a γ-glutamyl compound and cysteine formation of GSH through the −SH linkage plays a major role in various defense pathways. The γ-glutamyl linkage by means of some protease provides the common method of shielding GSH from any oxidative damage, such as:

1. Splitting of GSH by the transfer of γ-glutamate to other dipeptides catalyzed by γ-glutamyl transpeptidase.
2. An enzyme, carboxypeptidase, facilitates the removal of glycine from GSH and the rest of the dipeptides are metabolized by dipeptidase (Foyer et al., 2001; Shaw et al., 2005).

Glutathione is often marked by the group of nonprotein reduced sulfur found in bacteria and higher plants (Kunert and Foyer, 1993; May et al., 1998a). Regulation of the ascorbate-glutathione cycle has been demonstrated in plant cells where oxidation of ascorbate leads to the formation of dehydroascorbate (DHA). This DHA is quickly reduced to ascorbate by GSH in order to prevent a reduction in the total ascorbate level. This reduction is a nonenzymatic reaction and is catalyzed by dehydroascorbate reductase (DHAR) (Foyer and Halliwell, 1976, 1977; Kato et al., 1997). Besides this, the reaction between GSH and DHA is also regulated by other enzymes like glutaredoxins (thioltransferases), protein disulfide isomerases (Wells et al., 1990) and a kunitz-type trypsin inhibitor (Trümper et al., 1994; Morell et al., 1997).

GSH being a water soluble antioxidant seems to reduce reactive oxygen species (ROS) rapidly, unlike hydrogen peroxide which is a slow process. Glutathione peroxidases induced in response to stress detoxifies lipid peroxides whereas the substrate ascorbate is responsible for the reductive detoxification of hydrogen peroxides and seems to be regenerating from its oxidized form (Eshdat et al., 1997; Ziglari and Allameh, 2013). Recycling of glutathione is a significant process in maintaining the levels of the ascorbate-glutathione cycle (Foyer and Halliwell, 1976; Nokano and Asada, 1980). Genes such as GSTM1 and GSTP1 and many more are responsible for the regulation of glutathione metabolism, as these genes produce enzymes that help to maintain homeostasis by glutathione recycling in plants (Hyman, 2010).

Methyl jasmonate expresses a transient increase in transcripts that code for sulfur assimilation enzymes as well as GSH1 and GSH2 mRNA levels and that have been observed in *Arabidopsis* (Harada et al., 2000). Heavy metals and JA increases the transcriptional process; however, its translation mechanism requires oxidative stress, suggesting various control measures at the post-transcriptional level as well as some possible factors like H_2O_2 or modified GSH/GSSG ratio in de-repressing translation of the existing mRNA (Xiang and Oliver, 1998). The 5′ untranslated region (5′ UTR) of the GSH1 gene combines with repressor-binding protein and forms a 5′ UTR-binding complex and is supposed to regulate the translation of γ-ECS mRNA in *Arabidopsis thaliana* (Xiang and Bertrand, 2000).

Posttranslational regulation of γ-ECS has also been observed to control GSH concentrations (May et al., 1998b). In addition this group inhibits γ-ECS activity in plant extracts (Noctor et al., 2002a) as well as in the purified recombinant enzyme (Jez et al., 2004). Such feedback control of γ-ECS activity by GSH concentrations helps to clarify its position in the chloroplast, where limited GSH is accumulated, leading to specific feedback limitations on γ-ECS activity as long as excess GSH is exported to the cytosol (Maughana and Foyer, 2006).

Leguminous plants synthesize analogous tripeptides such as glutamylcysteinyl-β-alanine (homoglutathione) whereas in cereals the production of γ-glutamyl-cysteinyl-serine and γ-glutamyl-cysteinyl-glutamate has been reported by Hell (1997).

6.7 GSH HOMEOSTASIS-TRANSPORT

The γ-EC and GSH transport plays a significant role in glutathione homeostasis and oligopeptide transport proteins play a key role in this transport. Chloroplastic transporters for the γ-EC and GSH export are produced within the chloroplast which is considered a major site of γ-EC productions. Noctor demonstrated that ^{35}S-labelled GSH is transported into wheat chloroplast in a time-dependent manner that shows two independent systems for uptake. Of these, one acts as an active transport system by observing linear uptake of labeled GSH (20–30 μm) up to the saturation point (100–200 μm), followed by an increased uptake rate until the concentration falls to 1 mM. This confers a high capacity for GSH modulation by transport in response to the alteration of GSH, which could be due to changes in the enzymes of cytosol or ascorbate glutathione cycle in the chloroplast (Noctor et al., 2002a; Maughana and Foyer, 2006). GSH transport has also been carried out with heterotrophic tobacco cells, showing the transportation of GSH across the plasma membrane of plants in which both high-affinity as well as low-affinity GSH transport systems have been observed (Schneider et al., 1992).

Yeast cadmium factor (YCF1) protein as a glutathione –S-conjugate transporter has been demonstrated in yeast that transports Cd^{2+} GSH in a vacuolar membrane protein (YCF1), thus conferring Cd^{2+} tolerance (Decottignies and Goffeau, 1997; Morgan et al., 2013). YCF1 belongs to the ABC-type transporter, which has similarities with the multiple drug resistance transporters (MRPs) in humans. One of the common features among MRP proteins is the capacity to transport GS-conjugates and, in some cases, cotransport of GSH and drugs (Rappa et al., 1997). Kolukisaoglu et al. (2002) have also reported MRP homologues in plants. Also there are 14 genes reported in *Arabidopsis* that encode MRP transporters and among these only a few have been characterized, including AtMRP1, AtMRP2, AtMRP3 and ATMRP5. Moreover, AtMRP3 transporters have been proven to detoxify Cd as it is upregulated by Cd treatment, which suggests a key role in transporting GSH-Cd or forming PC-Cd complexes (Bovet et al., 2005).

6.8 CONCLUSIONS

Antioxidants are the key element for the survival of plants, animals and human beings exposed to various stress conditions. Different forms of antioxidants exist, among which glutathione and GR are well-known potent detoxifiers of large numbers of toxic radicals and play a large role in a key adaptive response, thereby maintaining the balance and control mechanisms in various plant species.

A future need is to build up methods for assessing the enhancement of the production of GR to combat stress. Identification of genes provides a mechanism to increase sulfur compounds associated with growth and plant disease response. This could lead not only to improving yield but also to improving the quality of food products. Molecular mechanisms of growth and metabolism of plants are accomplished by the balance of redox reactions appearing during stress. Although genes are identified for detoxification therapy against abiotic stress, combinations of active constituents after isolation as well as identification should be made and further assessment could be done for synergistic effects. In addition, treatment in standard and recommended form may play a vital role in therapeutics for crop plants.

ACKNOWLEDGMENTS

The authors are thankful to the heads of their respective institutes for their support during preparation of this article.

REFERENCES

Ahmad, P., Jaleel, C.A., Salem, M.A., Nabi, G., Sharma, S., 2010. Roles of enzymatic and non-enzymatic antioxidants in plants during abiotic stress. Crit. Rev. Biotechnol. 30 (3), 161–175.

Ahmad, P., Nabi, G., Ashraf, M., 2011b. Cadmium-induced oxidative damage in mustard [*Brassica juncea* (L.) Czern. & Coss.] plants can be alleviated by salicylic acid. S. Afr. J. Bot. 77, 36–44.

Ahmad, P., Nabi, G., Jaleel, C.A., Umar, S., 2011a. Free radical production, oxidative damage and antioxidant defense mechanisms in plants under abiotic stress. In: Ahmad, P., Umar, S. (Eds.), Oxidative Stress: Role of Antioxidants in Plants. Studium Press Pvt. Ltd, New Delhi, India, pp. 19–53.

Ahmad, P., Sarwat, M., Sharma, S., 2008. Reactive oxygen species, antioxidants and signaling in plants. J. Plant Biol. 51 (3), 167–173.

Ahmad, P., Umar, S., 2011. Antioxidants: Oxidative Stress Management in Plants. Studium Press Pvt. Ltd, New Delhi, India.

Alguacil, M.M., Hernandez, J.A., Caravaca, F., Portillo, B., Roldan, A., 2003. Antioxidant enzyme activities in shoots from three mycorrhizal shrub species afforested in a degraded semi-arid soil. Physiol. Plant 118, 562–570.

Anderson, M.D., Prasad, T.K., Stewart, C.R., 1995. Changes in isozyme profiles of catalase, peroxidase, and glutathione reductase during acclimation to chilling in mesocotyls of maize seedlings. Plant Physiol. 109, 1247–1257.

Aono, M., Kubo, A., Saji, H., Natori, T., Tanaka, K., Kondo, N., 1991. Resistance to active oxygen toxicity of transgenic *Nicotiana tabacum* that expresses the gene for glutathione reductase from *Escherichia coli*. Plant Cell Physiol. 32, 691–697.

Aono, M., Kubo, A., Saji, H., Tanaka, K., Kondo, N., 1993. Enhanced tolerance to photooxidative stress of transgenic *Nicotiana tabacum* with high chloroplastic glutathione reductase activity. Plant Cell Physiol. 34, 129–135.

Aono, M., Saji, H., Kondo, N., Tanaka, K., 1997. Paraquat tolerance of transgenic tobacco plants with altered activity of glutathione reductase phyton (Austria). Free Radic. 37, Fasc. 3(13)–(18)1.7.

Arisi, A.C.M., Mocquot, B., Lagriffoul, A., Mench, M., Foyer, C.H., Jouanin, L., 2000. Responses to cadmium in leaves of transformed poplars overexpressing glutamylcysteine synthetase. Physiol. Plant 109, 143–149.

Bashandy, T., Guilleminot, J., Vernoux, T., Caparros-Ruiz, D., Ljung, K., Meyer, Y., et al., 2010. Interplay between the NADP-linked thioredoxin and glutathione systems in *Arabidopsis* auxin signaling. Plant Cell 22, 376–391.

Bovet, L., Feller, U., Martinoia, E., 2005. Possible involvement of plant ABC transporters in cadmium detoxification: a cDNA sub-microarray approach. Environ. Int. 31, 263–267.

Bray, T.M., Taylor, C.G., 1993. Tissue glutathione, nutrition, and oxidative stress. Can. J. Physiol. Pharmacol. 71, 746–751.

Buchanan, B.B., Balmer, Y., 2005. Redox regulation: a broadening horizon. Annu. Rev. Plant Biol. 56, 187–220.

Cairns, N.G., Pasternak, M., Wachter, A., Cobbett, C.S., Meyer, A.J., 2006. Maturation of *Arabidopsis* seeds is dependent on glutathione biosynthesis within the embryo. Plant Physiol. 141, 446–455.

Cancado, G.M.A., De Rosa Jr, V.E., Fernandez, J.H., Maron, L.G., Jorge, R.A., Menossi, M., 2005. Glutathione S-transferase and aluminum toxicity in maize. Funct. Plant Biol. 32, 1045–1055.

Chen, J., Goldsborough, P.B., 1994. Increased activity of γ-glutamylcysteine synthetase in tomato cells selected for cadmium tolerance. Plant Physiol. 106, 233–239.

Chew, O., Whelan, J., Millar, A.H., 2003. Molecular definition of the ascorbate-glutathione cycle in *Arabidopsis* mitochondria reveals dual targeting of antioxidant defenses in plants. J. Biol. Chem. 278, 46869–46877.

Cobbett, C.S., 2000. Phytochelatins and their roles in heavy metal detoxification. Plant Physiol. 123, 825–832.

Creissen, G., Reynolds, H., Xue, Y.B., Mullineaux, P., 1995. Simultaneous targeting of pea glutathione reductase and of a bacterial fusion protein to chloroplasts and mitochondria. Plant J. 8, 167–175.

Cuypers, A., Plusquin, M., Remans, T., Jozefczak, M., Keunen, E., Gielen, H., et al., 2010. Cadmium stress: an oxidative challenge. Biometals 23, 927–940.

Davidian, J.C., Kopriva, S., 2010. Regulation of sulfate uptake and assimilation – The same or not the same? Mol. Plant 3, 314–325.

Davies, K.J., 1986. Intracellular proteolytic systems may function as secondary antioxidant defenses: a hypothesis. Free Radic. Biol. Med. 2, 155–173.

De Vos, C.H.R., Vonk, M.J., Vooijs, R., Schat, H., 1992. Glutathione depletion due to copper-induced phytochelatin synthesis causes oxidative stress in *Silene cucubalus*. Plant Physiol. 98, 853–858.

Decottignies, A., Goffeau, A., 1997. Complete inventory of the yeast ABC proteins. Nat. Genet. 15, 137–145.

Ding, S.H., Lu, Q.T., Zhang, Y., Yang, Z.P., Wen, X.G., Zhang, L.X., et al., 2009. Enhanced sensitivity to oxidative stress in transgenic tobacco plants with decreased glutathione reductase activity leads to a decrease in ascorbate pool and ascorbate redox state. Plant Mol. Biol. 69, 577–592.

Dixon, D.P., Cummins, L., Cole, D.J., Edwards, R., 1998. Glutathione-mediated detoxification systems in plants. Curr. Opin. Plant Biol. 1, 258–266.

Dixon, D.P., Lapthorn, A., Edwards, R., 2002. Plant glutathione transferases. Genome Biol. 3 (3), 1–10.

Dixon, D.P., McEwen, A.G., Lapthorn, A.J., Edwards, R., 2003. Forced evolution of a herbicide detoxifying glutathione transferase. J. Biol. Chem. 278 (26), 23930–23935.

Dominguez-Solis, J.R., Gutierrez-Alcala, G., Vega, J.M., Romero, L.C., Gotor, C., 2001. The cytosolic *O*-acetylserine(thiol)lyase gene is regulated by heavy metals and can function in cadmium tolerance. J. Biol. Chem. 276, 9297–9302.

Edwards, E.A., Enard, C., Creissen, G.P., Mullineaaux, P.M., 1994. Synthesis and properties of glutathione reductase in stressed peas. Planta 192, 137–143.

Edwards, R., Dixon, D.P., Walbot, V., 2000. Plant glutathione S transferase: enzymes with multiple functions in sickness and in health. Trends Plant Sci. 5 (5), 193–198.

Eshdat, Y., Holland, D., Faltin, Z., Ben-Haygim, G., 1997. Plant glutathione peroxidases. Physiol. Plant. 100, 234–240.

Foster, J.G., Hess, J.L., 1982. Oxygen effects on maize leaf superoxide dismutase and glutathione reductase. Phytochemistry 21, 1527–1532.

Foyer, C., Noctor, G., 2005a. Oxidant and antioxidant signaling in plants: a re-evaluation of the concept of oxidative stress in a physiological context. Plant Cell Environ. 28, 1056–1071.

Foyer, C.H., Halliwell, B., 1976. The presence of glutathione and glutathione reductase in chloroplasts: a proposed role in ascorbic acid metabolism. Planta 133, 21–25.

Foyer, C.H., Halliwell, B., 1977. Purification and properties of dehydroascorbate reductase from spinach leaves. Phytochemistry 16, 1347–1350.

Foyer, C.H., Lelandair, M., Kunert, K.K.J., 1994. Photooxidative stress in plants. Physiol. Plant. 92, 696–717.

Foyer, C.H., Noctor, G., 2005b. Redox homeostasis and antioxidant signaling: a metabolic interface between stress perception and physiological responses. Plant Cell 17, 1866–1875.

Foyer, C.H., Noctor, G., 2011. Ascorbate and glutathione: The heart of the redox hub. Plant Physiol. 155, 2–18.

Foyer, C.H., Souriau, N., Perret, S., Lelandais, M., Kunert, K.J., Pruvost, C., et al., 1995. Overexpression of glutathione reductase but not glutathione synthetase leads to increases in antioxidant capacity and resistance to photoinhibition in poplar trees. Plant Physiol. 109, 1047–1057.

Foyer, C.H., Theodoulou, F.L., Delrot, S., 2001. The functions of inter- and intracellular glutathione transport systems in plants. Trends Plant Sci. 6, 486–492.

Gomez, L.D., Noctor, G., Knight, M., Foyer, C.H., 2004. Regulation of calcium signaling and gene expression by glutathione. J. Exp. Bot. 55, 1851–1859.

Grill, E., Winnacker, E.L., Zenk, M.H., 1987. Phytochelatins, a class of heavy-metal-binding peptides from plants, are functional analogous to metallothioneins. Proc. Natl. Acad. Sci. USA 8, 439–443.

Gussarson, M., Asp, H., Adalsteinsson, S., Jensen, P., 1996. Enhancement of cadmium effects on growth and nutrient composition of birch (*Betula pendula*) by buthionine sulphoximine (BSO). J. Exp. Bot. 47, 211–219.

Hall, J.L., 2002. Cellular mechanisms for heavy metal detoxification and tolerance. J. Exp. Bot. 53, 1–11.

Harada, E., Kusano, T., Sano, H., 2000. Differential expression of genes encoding enzymes involved in sulfur assimilation pathways in response to wounding and jasmonate in *Arabidopsis thaliana*. J. Plant Physiol. 156, 272–276.

Heiss, S., Schafer, H.J., Haag-Kerwer, A., Rausch, T., 1999. Cloning sulfur assimilation genes of *Brassica juncea* L.: cadmium differentially affects the expression of a putative low-affinity sulfate transporter and isoforms of ATP sulfurylase and APS reductase. Plant Mol. Biol. 39, 847–857.

Hell, R., 1997. Molecular physiology of plant sulfur metabolism. Planta 202, 138–148.

Howarth, J.R., Dominguez-Solis, J.R., Gutierrez-Alcala, G., Wray, J.L., Romero, L.C., Gotor, C., 2003. The serine acetyltransferase gene family in *Arabidopsis thaliana* and the regulation of its expression by cadmium. Plant Mol. Biol. 51, 589–598.

Hyman, M., 2010. Glutathione: The mother of all antioxidants. <http://www.huffingtonpost.com/dr-mark-hyman/glutathione-the-mother-of_b_530494.html>.

Hyman, M., 2013. Glutathione, The mother of all antioxidants. <www.twitter.com/markhymanmd>.

Ishikawa, T., Li, Z.S., Lu, Y.P., Rea, P.A., 1997. The GS-X pump in plant, yeast, and animal cells: structure, function and gene expression. Biosci. Rep. 17, 189–207.

Jez, J.M., Cahoon, R.E., Chen, S., 2004. *Arabidopsis thaliana* glutamate-cysteine ligase: functional properties, kinetic mechanism, and regulation of activity. J. Biol. Chem. 279, 33463–33470.

Jozefczak, M., Remans, T., Vangronsveld, J., Cuypers, A., 2012. Glutathione is a key player in metal-induced oxidative stress defenses. Int. J. Mol. Sci. 13, 3145–3175.

Kato, Y., Urano, J., Maki, Y., Ushimaru, T., 1997. Purification and characterisation of dehydroascorbate reductase from rice. Plant Cell Physiol. 38, 173–178.

Kenyon, W.H., Duke, S.O., 1985. Effects of acifluorfen on endogenous antioxidants and protective enzymes in cucumber (*Cucumis sativus* L.). Plant Physiol. 79, 862–866.

Klapheck, S., Schlunz, S., Bergmann, L., 1995. Synthesis of phytochelatins and homophytochelatins in *Pisum sativum* L. Plant Physiol. 107, 515–521.

Kolukisaoglu, H.U., Bovet, L., Klein, M., Eggmann, T., Geisler, M., Wanke, D., et al., 2002. Family business: the multidrug-resistance related protein (MRP) ABC transporter genes in *Arabidopsis thaliana*. Planta 216, 107–119.

Korenyeyev, D., Logan, B.A., Payton, P., Allen, R.D., Holadey, A.S., 2001. Enhanced photochemical light utilization and decreased chilling-induced photoinhibition of photosystem11 in cotton overexpressing genes encoding chloroplast-targeted antioxidant enzymes. Physiol. Plant. 113, 323–331.

Kumar, M., Yadav, V., Tuteja, N., Johri, A.K., 2009. Antioxidant enzyme activities in maize plants colonized with *Piriformospora indica*. Microbiology 155, 780–790.

Kunert, K.J., Foyer, C.H., 1993. Thiol/disulfide exchange in plants. In: De Kok, L.J., Stulen, I., Rennenberg, H., Brunold, C., Rauser, W.E. (Eds.), Sulfur Nutrition and Assimilation in Higher Plants: Regulatory, Agricultural and Environmental Aspects. SPB Academic Publishers, The Hague, The Netherlands, pp. 139–151.

Lim, B., Meyer, A.J., Cobbett, C.S., 2011. Development of glutathione-deficient embryos in *Arabidopsis* is influenced by the maternal level of glutathione. Plant Biol. 13, 693–697.

Luna, C.M., Gonzalez, C.A., Trippi, V.S., 1994. Oxidative damage caused by excess of copper in oat leaves. Plant Cell Physiol. 35, 11–15.

Mannervik, B., Danielson, U.H., 1988. Glutathione transferases structure and catalytic activity. Crit. Rev. Biochem. 23 (2), 283–337.

Marí, M., Morales, A., Colell, A., García-Ruiz, C., Fernández-Checa, J.C., 2009. Mitochondrial glutathione, a key survival antioxidant. Antioxid. Redox Signal. 11 (11), 2685–2700.

Mata, A.M., Pinto, M.C., Lopez-Barea, J., 1985. Redox interconversion of glutathione reductase from *Escherichia coli*. A study with pure enzyme and cell free extracts. Mol. Cell Biochem. 67, 65–76.

Maughana, S., Foyer, C.H., 2006. Engineering and genetic approaches to modulating the glutathione network in plants. Physiol. Plant. 126, 382–397.

May, M., Vernoux, T., Leaver, C., Van Montagu, M., Inzé, D., 1998a. Glutathione homeostasis in plants: implications for environmental sensing and plant development. J. Exp. Bot. 49 (321), 649–667.

May, M.J., Vernoux, T., Sanchez-Fernandez, R., Van Montagu, M., Inzé, D., 1998b. Evidence for posttranscriptional activation of g-glutamylcysteine synthetase during plant stress responses. Proc. Natl. Acad. Sci. USA 95, 12049–12054.

Mehra, R.K., Tripathi, R.D., 1999. Phytochelatins and metal tolerance. In: Agrawal, S.B., Agarwal, M. (Eds.), Environmental Pollution and Plant Responses. CRC Press, Boca Raton, FL, pp. 367–382.

Meyer, A.J., Fricker, M.D., 2002. Control of demand-driven biosynthesis of glutathione in green Arabidopsis suspension culture cells. Plant Physiol. 130 (4), 1927–1937.

Meyer, Y., Siala, W., Bashandy, T., Riondet, C., Vignols, F., Reichheld, J.P., 2008. Glutaredoxins and thioredoxins in plants. Biochim. Biophys. Acta 1783, 589–600.

Mhamdi, A., Hager, J., Chaouch, S., Queval, G., Han, Y., Taconnat, L., et al., 2010. *Arabidopsis* GLUTATHIONE REDUCTASE1 plays a crucial role in leaf responses to intracellular hydrogen peroxide and in ensuring appropriate gene expression through both salicylic acid and jasmonic acid signaling pathways. Plant Physiol. 153 (3), 1144–1160.

Mittler, R., Vanderauwera, S., Gollery, M., van Breusegem, F., 2004. Reactive oxygen gene network of plants. Trends Plant Sci. 9, 490–498.

Morell, S., Follmann, H., De Tullio, M., Haberlein, I., 1997. Dehydroascorbate and dehydroascorbate reductase are phantom indicators of oxidative stress in plants. FEBS Lett. 414, 567–570.

Morgan, B., Ezeriņa, D., Amoako, T.N.E., Riemer, J., Seedorf, M., Dick, T.P., 2013. Multiple glutathione disulfide removal pathways mediate cytosolic redox homeostasis. Nat. Chem. Biol. 9, 119–125.

Nakano, Y., Asada, K., 1980. Spinach chloroplast scavenge hydrogen peroxide on illumination. Plant Cell Physiol. 21, 1295–1307.

Nocito, F.F., Lancilli, C., Crema, B., Fourcroy, P., Davidian, J.C., Sacchi, G.A., 2006. Heavy metal stress and sulfate uptake in maize roots. Plant Physiol. 141, 1138–1148.

Nocito, F.F., Pirovano, L., Cocucci, M., Sacchi, G.A., 2002. Cadmium-induced sulfate uptake in maize roots. Plant Physiol. 129, 1872–1879.

Noctor, G., Arisi, A.C., Jouanin, L., Kunert, K.J., Rennenberg, H., Foyer, C.H., 1998. Glutathione: biosynthesis, metabolism and relationship to stress tolerance explored in transformed plants. J. Exp. Bot. 49 (321), 623–647.

Noctor, G., Gomez, L., Vanacker, H., Foyer, C.H., 2002a. Glutathione homeostasis and signaling: the influence of biosynthesis, compartmentation and transport. J. Exp. Bot. 53, 1283–1304.

Noctor, G., Mhamdi, A., Chaouch, S., Han, Y., Neukermans, J., Marquez-Garcia, B., et al., 2012. Glutathione in plants: an integrated overview. Plant Cell Environ. 35, 454–484.

Ort, D.R., 2001. When there is too much light. Plant Physiol. 125, 29–32.

Pereira, G.J.G., Molina, S.M.G., Lea, P.J., Azevedo, R.A., 2002. Activity of antioxidant enzymes in response to cadmium in *Crotalaria juncea*. Plant Soil 239, 123–132.

Prade, L., Huber, R., Bieseler, B., 1998. Structure of herbicides in complex with their detoxifying enzyme glutathione-S-transferase-explanation for selectivity of enzyme in plants. Structure 6, 1445–1452.

Prasad, K.V., Pardha-Saradhi, P., Sharmila, P., 1999. Concerted action of antioxidant enzymes and curtailed growth under zinc toxicity in *Brassica juncea*. Environ. Exp. Bot. 42, 1–10.

Prasad, T.K., Anderson, M.D., Martin, B.A., Stewart, C.R., 1994. Evidence for chilling-induced oxidative stress in maize seedlings and a regulatory role for hydrogen peroxide. Plant Cell 6, 65–74.

Queval, G., Thominet, D., Vanacker, H., Miginiac-Maslow, M., Gakiere, B., Noctor, G., 2009. H_2O_2-activated up-regulation of glutathione in *Arabidopsis* involves induction of genes encoding enzymes involved in cysteine synthesis in the chloroplast. Mol. Plant 2, 344–356.

Ranieri, A., Castagna, A., Scebba, F., Careri, M., Zagnoni, I., Predieri, G., et al., 2005. Oxidative stress and phytochelatin characterisation in bread wheat exposed to cadmium excess. Plant Physiol. Biochem. 43, 45–54.

Rao, K.V.M., Sresty, T.V.S., 2000. Antioxidative parameters in the seedlings of pigeon pea (*Cajanus cajan* (L.) Millspaugh.) in response to Zn and Ni stress. Plant Sci. 157, 113–128.

Rappa, G., Lorico, A., Flavell, R.A., Sartorelli, A.C., 1997. Evidence that the multidrug resistance protein (MRP) functions as a co-transporter of glutathione and natural product toxins. Cancer Res. 57, 5232–5237.

Rascio, N., Navari-Izzo, F., 2011. Heavy metal hyperaccumulating plants: how and why do they do it? And what makes them so interesting? Plant Sci. 180, 169–181.

Rauser, W.E., 1995. Phytochelatins and related peptides. Plant of intracellular glutathione and resistance to melphalan. Plant Physiol. 109, 1141–1149.

Rauser, W.E., Schupp, K., Rennenberg, H., 1991. Cysteine, g-glutamylcysteine, and glutathione levels in maize seedlings. Plant Physiol. 97, 128–138.

Rea, P.A., 1999. MRP subfamily ABC transporters from plants and yeasts. J. Exp. Bot. 50, 895–913.

Rey, P., Cuiné, S., Eymery, F., Garin, J., Court, M., Jacquot, J.P., et al., 2005. Analysis of the proteins targeted by CDSP32, a plastidic thioredoxin participating in oxidative stress responses. Plant J. 41, 31–42.

Romero-Puertas, M.C., Palma, J.M., Gomez, M., del Rio, L.A., Sandalio, L.M., 2002. Cadmium causes the oxidative modification of proteins in pea plants. Plant Cell Environ. 25, 677–686.

Romero-Puertas, M.C., Rodriguez-Serrano, M., Corpas, F.J., Gomez, M., Del Rio, L.A., Sandalio, L.M., 2004. Cadmium-induced subcellular accumulation of O_2 and H_2O_2 in pea leaves. Plant Cell Environ. 27, 1122–1134.

Rüegsegger, A., Brunold, C., 1992. Effect of cadmium on glutamylcysteine synthesis in maize seedlings. Plant Physiol. 99, 428–433.

Salt, D.E., Prince, R.C., Pickering, I.J., Raskin, I., 1995a. Mechanisms of cadmium mobility and accumulation in Indian mustard. Plant Physiol. 109, 1427–1433.

Schäfer, H.J., Greiner, S., Rausch, T., Haag-Kerwer, A., 1997. In seedlings of the heavy metal accumulator *Brassica juncea* Cu^{2+} differentially affects transcript amounts for γ–glutamylcysteine synthetase (γ-ECS) and metallothionein (MT2). FEBS Lett. 404, 216–220.

Schickler, H., Caspi, H., 1999. Response of antioxidant enzymes to nickel and cadmium stress in hyperaccumulator plants of the genus *Alyssum*. Physiol. Plant. 105, 39–44.

Schneider, S., Martini, N., Rennenberg, H., 1992. Reduced glutathione (GSH) transport in cultured tobacco cells. Plant Physiol. Biochem. 30, 29–38.

Schutzendubel, A., Polle, A., 2002. Plant responses to abiotic stresses: heavy metal-induced oxidative stress and protection by mycorrhization. J. Exp. Bot. 53, 1351–1365.

Schutzendubel, A., Schwanz, P., Teichmann, T., Gross, K., Langenfeld-Heyser, R., Godbold, D.L., et al., 2001. Cadmium induced changes in antioxidative systems, H_2O_2 content and differentiation in pine (*Pinus sylvestris*) roots. Plant Physiol. 127, 887–898.

Seth, C.S., Remans, T., Keunen, E., Jozefczak, M., Gielen, H., Opdenakker, K., et al., 2012. Phytoextraction of toxic metals: a central role for glutathione. Plant Cell Environ. 35, 334–346.

Sharma, I., 2012a. Arsenic induced oxidative stress in plants. Biologia 67 (3), 447–453.

Sharma, I., 2013a. Arsenic stress in plants: An inside story. In: Khalid, R.H., Ahmad, P., Munir, O. (Eds.), Crop Improvement: New Approaches and Modern Techniques. Springer, New York Dordrecht Heidelberg London, pp. 379–400.

Sharma, I., 2013b. Arsenic induced oxidative stress and antioxidant defense system of *Pisum sativum* and *Pennisetum typhoides*: a comparative study. Res. J. Biotechnol. 8 (4), 48–56.

Sharma, I., Singh, R., Tripathi, B.N., 2007. Biochemistry of Arsenic toxicity and tolerance in plants. Biochem. Cell. Arch. 7, 165–170.

Shaw, M.L., Pither-Joyce, M.D., McCallum, J.A., 2005. Purification and cloning of a γ-glutamyl transpeptidase from onion (*Allium cepa*). Phytochemistry 66, 515–522.

Smeets, K., Opdenakker, K., Remans, T., van Sanden, S., van Belleghem, F., Semane, B., et al., 2009. Oxidative stress-related responses at transcriptional and enzymatic levels after exposure to Cd or Cu in a multipollution context. J. Plant Phys. 166, 1982–1992.

Srivastava, S., Tripathi, R.D., Dwivedi, U.N., 2004. Synthesis of phytochelatins and modulation of antioxidants in response to cadmium stress in *Cuscuta reflexa*-an angiospermic parasite. J. Plant Physiol. 161, 665–674.

Stevens, R.G., Creissen, G.P., Mullineaux, P.M., 2000. Characterisation of pea cytosolic glutathione reductase expressed in transgenic tobacco. Planta 211, 537–545.

Strasdeit, H., Duhme, A.K., Kneer, R., Zenk, M.H., Heimes, C., Molting, H.F., 1991. Evidence for discrete Cd $(Scys)_4$ units in cadmium phytochelatin complexes from EXAFS Spectroscopy. J. Chem. Soc. Chem. Comm. 16, 1129–1130.

Tanaka, K., Otsubo, T., Kondo, N., 1982. Participation of hydrogen peroxide in the inactivation of Calvin cycle SH enzymes in SO_2-fumigated spinach leaves. Plant Cell Physiol. 23, 1009–1018.

Tandogan, B., Ulusu, N.N, 2006. Kinetic mechanism and molecular properties of glutathione reductase. FABAD J. Pharm. Sci. 31, 230–237.

Thom, R., Cummins, I., Dixon, D.P., Edwards, R., Cole, D.J., Lapthorn, A.J., 2002. Structure of a Tau class glutathione S-transferase from wheat active in herbicide detoxification. Biochemistry 41 (22), 7008–7020.

Tran, T.A., Popova, L.P., 2013. Functions and toxicity of cadmium in plants: recent advances and future prospects. Turk. J. Bot. 37, 1–13.

Trümper, S., Follmann, H., Häberlein, I., 1994. A novel dehydroascorbate reductase from spinach chloroplasts homologous to trypsin inhibitor. FEBS Lett. 352, 159–162.

Tyystjarvi, E., Riikonen, M., Arisi, A.C.M., Kettunen, R., Jouanin, L., Foyer, C.H., 1995. Photoinhibition of photosystem11 in tobacco plants overexpressing glutathione reductase and Poplars overexpressing superoxide dismutase. Physiol. Plant. 105, 409–416.

Varma, A., Verma, S., Sudha, S.N., Butehorn, B., Franken, P., 1999. *Piriformospora indica*, a cultivable plant-growth-promoting root endophyte. Appl. Environ. Microbiol. 65, 2741–2744.

Vido, K., Spector, D., Laginel, G., Lopez, S., Tolendano, M.B., Labarre, J., 2001. A proteome analysis of the cadmium response in *Saccharomyces cerevisae*. J. Biol. Chem. 276, 8469–8474.

Waller, F., Baltruschat, H., Achatz, B., Becker, K., Fischer, M., Fodor, J., et al., 2005. The endophytic fungus Piriformospora indica reprograms barley to salt-stress tolerance, disease resistance, and higher yield. Proc. Natl. Acad. Sci. USA 102, 13386–13391.

Wells, W.W., Xu, D.P., Yang, Y., Rocque, P.A., 1990. Mammalian 26 512–17. thioltransferase (glutaredoxin) and protein disulfide isomerase have dehydroascorbate reductase activity. J. Biol. Chem. 265 (15), 361–364.

Xiang, C., Bertrand, D., 2000. Glutathione synthesis in *Arabidopsis*: multilevel controls coordinate responses to stress. In: Brunold, C., Rennenberg, H., De Kok, L.J., Stulen, I., Davidian, J.C. (Eds.), Sulfur Nutrition and Sulfur Assimilation in Higher Plants. P. Haupt, Bern, Switzerland, pp. 409–412.

Xiang, C., Oliver, D.J., 1998. Glutathione metabolic genes coordinately respond to heavy metals and jasmonic acid in *Arabidopsis*. Plant Cell 10, 1539–1550.

Xiang, C., Werner, B.L., Christensen, E.M., Oliver, D.J., 2001. The biological functions of glutathione revisited in *Arabidopsis* transgenic plants with altered glutathione levels. Plant Physiol. 126, 564–574.

Yannarelli, G.G., Fernandez-Alvarez, A.J., Santa-Cruz, D.M., Tomaro, M.L., 2007. Glutathione reductase activity and isoforms in leaves and roots of wheat plants subjected to cadmium stress. Phytochemistry 68, 505–512.

Yousuf, P.Y., Hakeem, K.R., Chandna, R., Ahmad, P., 2012. Role of glutathione reductase in plant abiotic stress. In: Ahmad, P., Prasad, M.N.V. (Eds.), Abiotic Stress Responses in Plants: Metabolism, Productivity and Sustainability. Springer, Science+business media, NY, USA, pp. 149–158.

Zenk, M.H., 1996. Heavy metal detoxification in higher plants – a review. Gene 179, 21–30.

Zhu, Y.L., Pilon-Smits, E.A.H., Jouanin, L., Terry, N., 1999a. Overexpression of glutathione synthetase in Indian Mustard enhances cadmium accumulation and tolerance. Plant Physiol. 119, 73–79.

Zhu, Y.L., Pilon-Smits, E.A.H., Jouanin, L., Terry, N., 1999b. Cadmium tolerance and accumulation in Indian mustard is enhanced by overexpressing γ–glutamylcysteine synthase. Plant Physiol. 121, 1169–1177.

Ziglari, T., Allameh, A., 2013. The significance of glutathione conjugation in aflatoxin metabolism. In: Razzaghi-Abyaneh, M. (Ed.), Aflatoxins: Recent Advances and Future Prospects. INTECH, pp. 264–286.

Chapter 7

Nonenzymatic Antioxidants in Plants

Agnieszka Waśkiewicz, Monika Beszterda and Piotr Goliński

7.1 INTRODUCTION

Plants during the vegetation process are exposed to a variety of biotic and abiotic stress conditions, including pathogens, adverse temperature, drought, high salinity, high ozone levels, heavy metals (cadmium toxicity), strong light and UV radiation (Ratkevicius et al., 2003; Giacomelli et al., 2006). Abiotic stresses are the major adverse environmental factors determining plant productivity, influencing a series of morphological, physiological, biochemical and molecular changes that unprofitably influence plant growth and development. Moreover, water deficiency, salinity, highly differentiated temperatures and oxidative stress are often correlated and these conditions singularly or in combination cause cellular damage. To cope with environmental abiotic stresses, it is of paramount significance to understand plant responses towards abiotic stresses that disturb the homeostatic equilibrium at cellular and molecular levels in order to identify a common mechanism for multiple stress tolerance. Agriculturally, the most damaging abiotic stress is water deficit leading to drought. Exposure to water or osmotic stress triggers the production of active oxygen species that can be extremely harmful to plant cells, causing oxidative damage and inactivation of enzymes (Loewus and Murthy, 2000; Li et al., 2008; Atkinson et al., 2011).

Under these stress conditions, reactive oxygen species (ROS) originating from molecular oxygen can accumulate in plant tissues, causing oxidation of cellular components, involving proteins, lipids, carbohydrates, nucleic acids and chlorophyll, also causing enzyme inhibition, activation of the programmed cell death (PCD) pathway and ultimately leading to death of the cells (Zurbriggen et al., 2009). Sublethal concentrations of reactive oxygen species acclimate in plants to different abiotic stresses and decrease plant development presumably as a part of the adaptation mechanism. To cope with oxidative stress, plants have evolved two general functionally

P. Ahmad (Ed): Oxidative Damage to Plants. DOI: http://dx.doi.org/10.1016/B978-0-12-799963-0.00007-1

interlocked protective mechanisms, enzymatic and nonenzymatic detoxification (Foyer and Noctor, 2005; Ahmad et al., 2009). Nonenzymatic antioxidant defense systems consist of low molecular mass compounds, such as ascorbic acid and glutathione, which are active in the aqueous phase, whereas the lipophilic antioxidants (such as α-tocopherol and β-carotene) are active in the membrane environment. Elevation of nonenzymatic antioxidant concentrations may be accomplished in two ways. The upregulation of the synthesis of these antioxidants is a general response during abiotic stress. On the other hand, the redox recycling of these antioxidants markedly enhances their biological efficacy by decreasing the need for *de novo* synthesis.

Nonenzymatic antioxidants can affect plant growth and development by modulating processes from mitosis and cell elongation to senescence and death, likewise gene expression associated with abiotic stresses, altering acclimation responses (De Pinto and De Gara, 2004; Tokunaga et al., 2005). On the other hand, these antioxidants function as redox buffers that interact with ROS and act as a metabolic interface that modulates the appropriate induction of acclimation responses (Foyer and Noctor, 2005).

Plant acclimation to unfavorable environmental conditions is associated with significant changes in proteome composition. Since proteins are engaged in cell stress response, proteomics research might meaningfully contribute to identification of potential relationships between protein content and plant stress acclimation. Also substantial progress has been made in improving stress-induced oxidative stress tolerance in crop plants by generation of transgenic lines with modified concentrations of antioxidants (Lee et al., 2007; Zaefyzadeh et al., 2009; Kosová et al., 2011).

Preharvest abiotic stress that causes fluctuations in the concentrations of the antioxidant compounds may be used as a mechanism for improving the nutritional quality of plants (Atkinson et al., 2011). Postharvest abiotic stressors can induce numerous quality defects in various plants, including scald, core and flesh browning of fruits, sweetening, pitting, water-soaking appearance, abnormal ripening, russeting and tissue softening. Stress may also lead to losses of nutrient constituents in fruits or vegetables, with ascorbic acid loss being the most sensitive indicator of this stress (Baldwin, 2003; DeLong et al., 2004; East et al., 2008).

7.2 FORMATION OF REACTIVE OXYGEN SPECIES (ROS) IN ABIOTIC STRESS IN PLANTS

As a result of aerobic metabolism, stepwise reduction of molecular oxygen (O_2) occurs by high-energy exhibition or electron-transfer reactions, leading to the formation of highly reactive ROS such as superoxide ($O_2^{\bullet -}$), hydrogen peroxide (H_2O_2), hydroxyl radicals ($^{\bullet}OH$) and singlet oxygen ($^1O_2$). The cell needs to maintain a balance between the concentrations of radicals required to elicit a proper response to a changing condition, while at the same time maintaining sufficiently low levels of reactive oxygen species to counteract large-scale cellular destruction. Under optimal growth conditions reactive oxygen species are primarily formed at low levels in the organelles, including chloroplasts, mitochondria, peroxisomes, the plasma membrane and the apoplastic space (del Río et al., 2002; Rhoads et al., 2006; Tripathy and Oelmüller, 2012). During stress conditions their generation rates are highly elevated (Kanofsky and Sima, 1991; Ahmad et al., 2009, 2010; Sharma et al., 2012).

Light-dependent electron transport processes render the chloroplast thylakoids to be the main source of radicals in plant cells (Doyle et al., 2010; Khanna-Chopra, 2011; O'Brien et al., 2012). These organelles are also considered to be the target of age-associated oxidative stress. The photosynthetic fixation of CO_2 may control the generation of reactive oxygen species – reduced CO_2 fixation accompanies decreased ATP and NADPH consumption, resulting in an excess of NADPH, especially under strong light. The decreased utilization of NADPH results in a drop in the concentration of $NADP^+$. While $NADP^+$ is a principal electron acceptor in photosystem I, depletion of $NADP^+$ exacerbates the passage of electrons from photosystem I to molecular oxygen causing the formation of H_2O_2 via $O_2^{\bullet-}$. The increased level of reactive oxygen species inhibits the restoration of the disturbed photosystem II and induces photoinhibition.

Synthesis of mitochondrial radicals is markedly lower when compared to chloroplast or peroxisome formation, because of the high photosynthesis and photorespiration activity in sunlight. Instead, in the dark or in nongreen tissues, mitochondria are the main sources of radicals. The sites of mitochondrial reactive oxygen species (mtROS) formation in the mitochondrial electron transport chain (mtETC) include complexes I and III, where the superoxide anion ($O_2^{\bullet-}$) is created and in turn is reduced by dismutation to hydrogen peroxide (H_2O_2) (Cvetkovska and Vanlerberghe, 2012; He et al., 2012; Agrawal et al., 2013; Vanlerberghe, 2013). Hydrogen peroxide, a compound of a comparatively lower toxicity, may react with reduced Fe^{2+} and Cu^+ to form highly toxic hydroxyl radicals and, being uncharged, may be able to penetrate membranes and leave the mitochondrion (Sweetlove and Foyer, 2004, Rhoads et al., 2006). For added separation of mitochondrial and submitochondrial particles, reactive oxygen species rates vary depending on the conditions, but generally they fall within the range of 0.2 to 30.0 nmol min^{-1} mg $protein^{-1}$ for hydrogen peroxide and superoxide (Popov et al., 2003).

As a result of essentially oxidative type of metabolism, peroxisomes are the third main sites of intracellular H_2O_2, $O_2^{\bullet-}$ and NO formation in plant cells. Among metabolic processes the glycolate oxidase reaction, fatty acid β-oxidation, the enzymatic reaction of flavin oxidases, and the disproportionation of the superoxide anion are mainly responsible for the formation of H_2O_2 in different types of peroxisomes. The sites of $O_2^{\bullet-}$ generation were identified in the organelle matrix associated with xanthine oxidase (XOD) processes and the peroxisomal membranes dependent on NADPH. XOD catalyzes the oxidation of xanthine and hypoxanthine to uric acid and is a generator of reactive oxygen species (Fridovich, 1986). Respectively, three integral polypeptides (PMPs) with molecular masses of 18, 29 and 32 kDa embedded in the membrane have been shown to form $O_2^{\bullet -}$ radicals (del Río et al., 2002). The 18 kDa PMP, which was proposed to be a cytochrome possibly belonging to the b-type group, and using NADH as the electron donor is specified as a major producer of radicals (López-Huertas et al., 1997).

7.3 NONENZYMATIC ANTIOXIDANTS IN PLANTS

7.3.1 Ascorbic Acid (Vitamin C)

Ascorbate is a ubiquitous molecule in eukaryotes; it occurs in a reduced form (ascorbic acid) and two oxidized forms (mono- and dehydroascorbic acid). The ratio between reduced and oxidized ascorbate is a primary factor influencing plant resistance to oxidative stress (Conklin et al., 2000; Zechmann, 2011; Cruz-Rus et al., 2012). It operates as an antioxidant, an enzyme cofactor, and also as a precursor for oxalate and tartrate biosynthesis. Ascorbate cooperates in a diversity of biological reactions, including photosynthesis, photoprotection, cell wall growth and cell development (Wolucka and Van Montagu, 2003). Resistance to environmental stresses and biosynthesis of ethylene, gibberellins, anthocyanins, and hydroxyproline are also some of the roles played by ascorbates (Mellidou et al., 2012). The detoxification of reactive oxygen species proceeds directly or through the glutathione-ascorbate cycle and participates in redox signaling, modulation of gene expression and the control of enzymatic activities (Zechmann, 2011). Ascorbate reacts with all the above reactive oxygen species, the product being monodehydroascorbate (MDHA). MDHA disproportionates to form dehydroascorbate (DHA) and originally ascorbate. Dehydroascorbate is labile at physiological pH, thus, in order to maintain the ascorbate pool, the reduction process of monodehydroascorbate and dehydroascorbate to ascorbate occurs (Smirnoff et al., 2001). Stabilization is obtained by enzyme complexes including the ascorbate-glutathione (GSH) cycle: monodehydroascorbate reductase, GSH-dependent dehydroascorbate reductase and glutathione reductase. Note that the reduction of dehydroascorbate by glutathione is carried out without limitation at pH 8 (i.e. in the stroma of illuminated chloroplasts) without participation of enzyme catalysis (Millar et al., 2003; Gallie, 2013a).

Ascorbate concentration fluctuates depending on many factors such as light intensity, hour of the day, age, plant tissue and cell compartment (Bartoli et al., 2000; Tabata et al., 2002; Tamaoki et al., 2003). Ascorbate amounts are quantitatively determined in individual plant organs or tissues and the recorded concentrations represent a combination of the ascorbate status of all organelles separately. Numerous abiotic conditions cause highly specific stress responses with fluctuations of ascorbate contents in individual organelles; hence this biomarker could not be detected during determinations in whole organs or tissues (Barth et al., 2006; Zechmann, 2011; Gallie, 2013b). In biochemical analyses ascorbate was primarily detected in isolated chloroplasts, mitochondria, peroxisomes and vacuoles. In addition, ascorbate was detected in the apoplast of barley, birch and poplar leaves and at the nuclear membrane, in nucleoli and along the plasma membrane in roots of *Cucurbita* plants (Jiménez et al., 1997, 1998; Kollist et al., 2001; Van Hove et al., 2001; Noctor et al., 2002; Liso et al., 2004; Takahama, 2004). Vacuoles demonstrate lower amounts of ascorbate, whereas mitochondria and chloroplasts contained intermediate concentrations. Within chloroplasts and mitochondria, ascorbate was alone identified in the stroma and matrix, accordingly, but not within the lumen of thylakoids or cristae. This picture changed under high light activity conditions (Valpuesta and Botella, 2004; Zechmann, 2011). With electron microscopy approaches ascorbate might be identified in the apoplast of vascular parenchyma cells facing xylem vessels, although not in other cell types within the leaf (Zechmann et al., 2011; Khan et al., 2012). Ascorbate commonly occurs at contents of over 20 mM in chloroplasts and is frequent in all cell compartments including the cell wall. Leaf structures in all compartments demonstrate 2 to 5 mM ascorbate contents. Overall, supervision of the ascorbate steady-state concentration in plants potentially involves regulation of biosynthesis, catabolism, recycling and transport of this molecule (Smirnoff and Wheeler, 2000).

7.3.2 Tocopherols

Tocopherols – lipid soluble antioxidants – are only produced by photosynthetic organisms, including all plants, algae and most cyanobacteria (Munné-Bosch, 2005; Horvath et al., 2006; Maeda and DellaPenna, 2007; Quadrana et al., 2013). In plant tissue tocopherol contents vary from extremely low <1 μg/g dry weight (DW) to very high levels in leaves and seeds (>1 mg/g DW) (Munné-Bosch and Alegre, 2002). Tocopherol composition differs between species and between tissues within one species (Grusak and DellaPenna, 1999; Badrhadad et al., 2013). As a rule, leaves commonly accumulate α-tocopherol, whereas seeds are rich in γ-tocopherol (DellaPenna, 2005; Lichtenthaler, 2007; Szymanska and Kruk, 2008; Velasco et al., 2013). β- and δ-tocopherols are not very abundant in most plant species. α-tocopherol is synthesized from γ-tocopherol in chloroplasts by γ-tocopherolmethyltransferase (γ-TMT, VTE4) in the envelope of plastids and is stored in plastoglobuli of the chloroplast stroma, and in thylakoid membranes (Gill and Tuteja, 2010; Szarka et al., 2012). Their antioxidant function is associated with the inhibition of membrane lipid peroxidation and the scavenging of reactive oxygen species such as singlet oxygen ($^1O_2$) (Munné-Bosch and Alegre, 2002; Foyer and Noctor, 2003; Kruk et al., 2005; Munné-Bosch, 2005; Kruk and Trebst, 2008). Tocopherols also play other functions in plant metabolism, such as sugar export from leaves to phloem (Hofius et al., 2004; Soltani et al., 2012). Among tocopherols, α-tocopherol is the most abundant form and shows the highest biological activity due to the presence of three methyl groups in its molecular structure (Kamal-Eldin and Appelqvist, 1996; Hirschberf, 1999). Tocopherol synthesis is regulated by environmental stresses and stress sensitive plant hormones such as jasmonic acid, salicylic acid and abscisic acid (Szarka et al., 2012). The α-tocopherol changes during plant response to environmental stress occur in two phases. In the first phase, the α-tocopherol content increases under stress and plays a protective role by reducing ROS levels. In the second phase, when the stress is too severe, the α-tocopherol degradation is higher than its synthesis. In stress-tolerant plants only the first phase occurs, while in stress-sensitive plants the second phase is usually observed (Munné-Bosch, 2005).

In photosynthetic organisms the changes in tocopherol contents were demonstrated in response to unfavorable environmental conditions such as drought, heavy metals, salinity or high light (Collakova and DellaPenna, 2003; Ledford et al., 2004; Tounekti et al., 2011a,b).

7.3.3 Glutathione

Glutathione (GSH) is a low-molecular-weight tripeptide γ-glutamylcysteinyl-glycine (γ-glu-cys-gly), which is one of the crucial metabolites in all aerobic organisms (Gill and Tuteja, 2010; Dubreuil-Maurizi and Poinssot, 2012; Szarka et al., 2012; Ramírez et al., 2013). In relation to other cellular thiols, GSH level is very high and it accumulates to millimolar concentrations, with the highest concentrations in chloroplasts (1–4 mM) (Szarka et al., 2012). Glutathione is localized in the cytosol, endoplasmic reticulum, vacuoles, mitochondria, chloroplasts, peroxisomes and apoplasts (Jiménez et al., 1998; Noctor and Foyer, 1998). The presence of cysteine in this molecule confers its biological properties mainly as an antioxidant functioning through its involvement in cell redox homeostasis (Noctor, 2006). Glutathione is one of the most important antioxidants that detoxify ROS and protect plants from oxidative damage (Noctor et al., 1998; Shiu and Lee, 2005; Shan et al., 2011). In plants this antioxidant exists in two different forms: the reduced form (GSH) and the oxidized disulphide form (GSSG) (Shu et al., 2011). Under optimal conditions, in leaves the GSH/GSSG ratio is not less than 20:1. It is necessary to maintain a high proportion of these forms (GSH/GSSG ratio) to counteract the inhibitory effects of ROS-induced oxidative stress (Noctor et al., 1998, 2011; Mhamdi et al., 2010). It is a potential scavenger of $^1O_2$, H_2O_2 and the most dangerous ROS, such as $^{\bullet}OH$ (Briviba et al., 1997). GSH can be synthesized in chloroplasts and in the cytosol of plant leaves or through glutathione reductase (GR) with NADPH as a donor electron, and can catalyze the reduction of GSSG to GSH, as in the ascorbate-glutathione cycle (Noctor and Foyer, 1998). Biosynthesis of glutathione occurs in two adenosine triphosphate

(ATP)-dependent steps catalyzed by the enzymes γ-glutamylcysteine ligase (γ-ECS, GSH1) and glutathione synthetase (GSH-S, GSH2) (Mittova et al., 2003). GSH1 catalyses the formation of a peptide bond between the γ-carbonyl group of glutamate and the amino group of cysteine to form γ-glutamylcysteine. Next, in the second reaction, GSH2 ligates a glycine residue with γ-glutamylcysteine to form GHS (Shu et al., 2011; Szarka et al., 2012). Glutathione plays multiple functions including cell differentiation, cell death and senescence, pathogen resistance and enzymatic regulation (Ogawa, 2005). Other functions of GSH concern the regulation of sulfate transport, signal transduction, the formation of phytochelatins and its participation in detoxification of xenobiotics (Xiang et al., 2001; Sharma and Dietz, 2006).

7.3.4 Carotenoids

Carotenoids in plants are biosynthesized *de novo* in most types of plastids, but are distributed in high quantities in chloroplasts and chromoplasts in all photosynthetic organisms. In chloroplasts carotenoids are integrated in photosynthetic membranes and are related with proteins binding the chlorophyll to form pigment-protein complexes. For chromoplasts, carotenoids are related to polar lipids and are linked to proteins to form carotenoid-lipoprotein interlocking substructures to effectively sequester and preserve a high amount of carotenoids (Lopez et al., 2008). Carotenoids and their apocarotenoid derivatives can also accumulate in amyloplasts, elaioplasts, leucoplasts and etioplasts (Mascio et al., 1991; Auldridge et al., 2006; Howitt and Pogson, 2006; Horner et al., 2007).

Plant carotenoids belong to the category of isoprenoid-derived compounds and their synthesis is driven by the nuclear-encoded enzyme system. Isoprenoids are formed by the cytosolic mevalonate and plastidial 2-C-methyl-D-erythritol 4-phosphate pathways, to cause the C5 structures creating scaffolding for the biosynthesis of carotenoids and the other compounds (Fraser et al., 2007). Plants use multistage regulatory processes to control metabolism and accumulation of carotenoids. While the qualitative and quantitative composition of carotenoids are exceptionally conserved in green tissues, in nongreen tissues or organs the carotenoid accumulation differs significantly. The important functions of carotenoids in photosynthesis, photomorphogenesis and plant development, indicate that their metabolism is regulated together with other mechanisms such as plastid biogenesis, flowering and fruit growth (Wurbs et al., 2007; Giuliano et al., 2008). Over flower development and fruit maturation, transcriptional regulation of carotenoid gene expression seems to be a main process, by which the synthesis and accumulation of specific carotenoids are organized. Simultaneously, metabolic turnover of carotenoids by carotenoid cleavage dioxygenases (CCDs) not only synthesizes relevant signaling and additional apocarotenoid components, but also supports maintenance of a constant level of carotenoids in plants (Taylor and Ramsay, 2005; Cuttriss et al., 2007).

Carotenoids are necessary for primary and secondary metabolisms of plants; they are required for photosynthesis processes and determine pigmentation (red, orange and yellow colors) of flower organs, fruits and seeds in many important crops (Botella-Pavia and Rodriguez-Concepcion, 2006). Through the cyclization of lycopene, the β,β branch leads to β-carotene and its derivatives, while the β, ε branch leads to α-carotene and its derivatives, which are relevant biological molecules that may inactivate electronically excited compounds, a process defined as quenching (DellaPenna and Pogson, 2006; Galpaz et al., 2008). Singlet molecular oxygen is an excited molecule and is formed by photochemical or enzymatic processes or by biomembrane lipid peroxidation by photoexcitation or by chemiexcitation (Sandmann et al., 2006). For instance, carotenoids were identified as accumulated in seeds under precipitate aging and in the course of germination at high temperatures. Seed carotenoids include primarily lutein and in lower amounts β-carotene. The ratio of carotenoid to chlorophyll amounts is suggested to be an indicator of seed tolerance to stress factors (Smolikova et al., 2011).

7.4 ROLE OF NONENZYMATIC ANTIOXIDANTS UNDER VARIOUS STRESSES

Stress-induced ROS accumulation is counteracted by enzymatic and nonenzymatic antioxidants. Among low molecular nonenzymatic antioxidants a key role is played by ascorbate, tocopherols, glutathione and carotenoids to cope with oxidative stress (Ahmad et al., 2010; Gill and Tuteja, 2010; Miller et al., 2010; Mostafa and Ibrahim, 2012). The levels of these antioxidants are elevated during the fight against ROS under different stresses. Some studies have focused on the analysis of the level of just one compound, while others refer to all or most of them (Dempsey et al., 2011; Kumar et al., 2012; Sánchez-Rodríguez et al., 2012; Ao et al., 2013).

7.4.1 Drought

As the world population still increases and water resources for crop production decrease, the drought-tolerant cultivars occurrence and water-use-efficient crops are global concerns. Water deficiency during the initial stage of plant growth results in delayed and irregular seedling emergence and stand establishment. Water deficit also disturbs plant development owing to the loss of turgor (Tohidi-Moghadam et al., 2009; Wang et al., 2012b), as water supply from the xylem to the surrounding elongating cells is interrupted (Chahal et al., 2007). Increasing crop tolerance to drought stress would be the most economical approach to improve agricultural productivity and to reduce agricultural use of fresh water resources. As a result, understanding of the mechanism of drought tolerance is a major goal for plant biologists and crop breeders (Chimenti et al., 2006; Drame et al., 2007; Cattivelli et al., 2008; Farooq et al., 2009a,b; Zhang et al., 2013).

Drought is an intense abiotic stress and the main constraint on wheat (*Triticum aestivum* L.) vegetation and productivity worldwide. The response of resistant and susceptible wheat genotypes to drought-induced oxidative stress and antioxidant management at a particular growth stage has been reported in many literature sources. As an example, two wheat (*Triticum aestivum* L.) genotypes: drought sensitive (PBW 343) and drought tolerant (C 306) were analyzed under water-deficiency conditions in India (Singh et al., 2012). Ascorbate concentration in roots and leaves in the two genotypes increased with age of plants, but decreased highly in response to water deficit, with an observed about 2-fold higher ascorbic content in vegetative tissues of tolerant plants. In the same experiment different antioxidative enzymes showed a positive correlation with ascorbate and negative with hydrogen peroxide concentration levels. Similarly, in a study by Al-Ghamdi (2009) exposure to water-deficiency conditions interrupted by a rewatering period induced radical-caused stress in Veery (drought tolerant) and Sids (drought sensitive) wheat (*Triticum aestivum* L.) cultivars, which exhibited high hydrogen peroxide and oxidized ascorbate contents. In well-watered seedlings the dehydroascorbate concentration was around 16% of the reduced form. After the completion of the first water-deficit period, the dehydroascorbate content increased considerably in the sensitive cv. Sids in comparison with the tolerant cv. Veery, amounting to 95% and 159% of the reduced form. In this study, sensitivity of Veery leaves to drought conditions was evident from the strong oxidation of the ascorbate-glutathione pool and a significant reduction in total ascorbate and dehydroascorbate proportions. In studies by Herbinger et al., (2002) on two wheat cultivars: sensitive *T. aestivum* Desf. cv. Nandu and tolerant *T. durum* L. cv. Extradur, water deficit reduced total ascorbate and chlorophyll amounts, while it contributed to an increase of carotenoid contents and total glutathione contents. An increase of α-tocopherol concentrations with maximum amounts in Nandu tissues was also reported at intermediary leaf water potential values. Efficiency of the nonenzymatic antioxidant system under drought in wheat was also studied in field experiments and *in vitro* combinations (Lascano et al., 2001). Under field conditions using tolerant cultivars Elite and La Paz, and sensitive cv. Oasis and Cruz Alta, no certain correlation between drought stress resistance and antioxidant contents was observed. Under *in vitro* conditions the tolerant cultivars demonstrated high ascorbate and glutathione amounts with a simultaneous low oxidative destruction in comparison to sensitive cultivars.

Bartoli et al. (1999) evaluated both oxidative exposure and the antioxidant protection system in leaves of wheat (*T. aestivum* cv. Buck Poncho), grown for a one-month period in a culture room. Stress conditions were created by the soil water potential of −2.0 MPa and were supported for 24 h. Under water deficit the concentration of lipophilic antioxidants, i.e. α-tocopherol and β-carotene, increased 2.4-fold and 2.6-fold, respectively. Simultaneously, the amount of reduced ascorbic acid decreased to 28.5%.

Alternately, the metabolism of glutathione and its functions in redox and energy balances under water-deficit conditions, the activities of key enzymes of antioxidant synthesis, e.g. γ-glutamylcysteine and glutathione synthetase, were also studied in the leaves of spring wheat (Chen et al., 2004). Plants were grown under irrigated and drought conditions. Compared to the control cultivation, water deficit induced a 26.9 and 50.0% decrease in reduced glutathione and a 181 to 200% and 142 to 181% increase in cysteine and γ-glutamylcysteine concentrations, respectively. Simultaneously, the glutathione/glutathione disulfide ratio increased, indicating that the redox status of glutathione was modified to a more reduced state by water deficit. Moreover, the activity of γ-glutamylcysteine synthetase was significantly higher than that of glutathione synthetase in the control and stressed plants, respectively.

Based on the assumption that exogenous ascorbate application plays a significant role during the initial stages of germination under various environmental conditions, seed priming is a common testing method. Farooq et al. (2013) used seeds of Mairaj and Lasani wheat cultivars in osmo- (ascorbic acid) and hydropriming (water) treatments to evaluate drought resistance. Ascorbic acid applied at high rates improved leaf emergence and elongation, leaf area, specific leaf area, chlorophyll levels, root length and seedling dry weight. Due to an increase in proline, phenolic and ascorbate concentrations, seedling water status was enhanced, while oxidative damage was reduced. In investigations conducted by Malik and Ashraf (2012) in order to assess the effect of drought two wheat (*Triticum aestivum* L.) genotypes were used: a water-deficiency tolerant cultivar Chakwal-86 and a water-deficiency sensitive strain 6544-6. Under osmotic stress conditions (−0.6 MPa) three different methods of exogenous ascorbic acid application were applied: through rooting medium, as a foliar spray and the seed soaking treatment. In both genotypes in seedlings supplemented with ascorbic acid the net photosynthesis, growth rate and chlorophyll concentration increased in comparison with the nontreated set and the rooting medium was a more efficient way to aid plants under drought stress conditions.

In three rice cultivars with various drought resistance levels (Gangyou 527, Yixiangyou 9 and Gangyou 188) growing under water stress conditions (water potential: −0.029, −0.039, −0.048 and −0.059 MPa) ascorbic acid and glutathione concentrations were analyzed (Wang et al., 2010). The most water-deficit tolerant cultivars demonstrated higher ascorbic acid contents compared with plants from the control and sensitive groups. The reduced glutathione amount in leaves was also higher in more tolerant cultivars than the less tolerant cultivars with the largest decrease after heading amounting to 3.0, 3.4 and 3.5% for Gangyou 527, Yixiangyou 9 and Gangyou 188, accordingly. In rice (*Oryza sativa* L.) seedling roots treated with ascorbic acid and its expected precursors: D-glucose and L-galactono-γ-lactone, electrolyte leakage from cellular structures, hydrogen peroxide and lipid peroxidation were reduced (Guo et al., 2005). In sunflower (*Helianthus annuus* L. cv. Blizar) under water deficit conditions (−0.03 at control culture, −0.6 and −1.2 MPa) with foliar application of ascorbic acid (0, 50, 100 and 150 mM), Ebrahimian and Bybordi (2012) observed reduced contents of flavonoids, anthocyanin and soluble sugars.

The effect of drought conditions on lipid peroxidation and ascorbic acid levels was investigated in cowpea (*Vigna unguiculata* L.) a major vegetable crop planted in tropical areas (Nair et al., 2008). In the two analyzed cultivars, leaves of *Kanakamony* demonstrated 1.13 times higher ascorbic acid concentration than *Pusakomal* cultivars under increasing water-deficit conditions. Simultaneously, comparing to the control culture, *Pusakomal* demonstrated a 2.7-fold increase in tiobarbituric acid reactive substance (TBARS) but a 1.2-fold increase as compared to *Kanakamony*.

The effect of drought stress was also studied on leaflets of soybean (*Glycine max*) from two cultivars: *Essex* and *Forre*st, and leaves of spinach (*Spinacia oleracea*) cv. *Nordic* (Robinson and Bunce, 2000). For the soybean samples, low leaf water potential values (−3.00 to −3.95 MPa) were associated with reduced ascorbic acid contents and increased dehydroascorbate contents per gram dry mass. Also the mole fraction of ascorbic acid remained at 93–99 mol% of the total ascorbic acid and dehydroascorbate, showing that most of the total ascorbate remained in the reduced form even at a low water potential. In spinach plants influenced by water stress (−1.8 to −2.6 MPa), leaf ascorbic acid content decreased by as much as 38%, but ascorbic acid remained at 96–99 mol% of the total pool. Alternately, Sánchez-Rodríguez et al. (2012) studied grafted plants of two tomato (*Lycopersicon esculentum* Mill) cultivars: Zarina and Josefina, used as the scion and rootstock. When shoots were of the water-deficit tolerant genotype Zarina, changes in total antioxidant activities were extensive and consistent during moderate drought. In this sequence reduced ascorbic acid level increased, while that of dehydroascorbate decreased. In the case when shoots were derived from the water-deficit sensitive genotype Josefina, the antioxidant enzyme activities were more restricted and the symptoms of free radical stress were obvious. No important differences in the total glutathione leaf amount were noted in cv. Zarina and its self-graft under drought stress. Meanwhile, in relation to the control conditions in cv. Josefina and *Zar* × *Jos* the total glutathione content decreased. Oxidized glutathione levels were reduced only in cv. Josefina.

In order to examine the influence of polyethylene glycol-induced drought stress and paraquat on cucumber (*Cucumis sativus* cv. Yuexiu) leaves in hydroponic culture, reduced ascorbate and reduced glutathione levels were quantified (Liu et al., 2009). The ascorbate and glutathione amounts significantly increased in the set of paraquat and polyethylene glycol treatment compared to the single application of both compounds.

To explain the importance of the ascorbate-glutathione cycle under water-deficit stress, the leaves of two-year-old apple (*Malus domestica*

Borkh. cv. Gala) plants grafted onto rootstock *Malus sieversii* (Ledeb.) Roem were used to investigate fluctuations of the ascorbate-glutathione cycle components under water-deficit conditions (Ma et al., 2011). The amounts of total ascorbate, reduced ascorbic acid, total glutathione and glutathione were evidently upregulated in apple leaves when the soil water content was about 40–45%. With a further increase of the water-deficit level, antioxidant concentrations and the redox state of reduced ascorbic acid and glutathione decreased. What is important, their amounts increased again after rewatering. Besides, water deficiency caused a significant elevation in the activities of enzymes participating in regeneration of reduced ascorbic acid and glutathione, in particular when the soil water content was above 40–45%. The test results suggest that the ascorbate-glutathione cycle is upregulated in response to water-deficit stress, but cannot be regulated at difficult environmental stress conditions. In *Malus prunifolia* and *Malus hupehensis* apple rootstock with different sensitivity to drought stress the antioxidant defense systems were evaluated (Wang et al., 2012b). Also in this case, the concentration of ascorbic acid and glutathione increased to a greater extent in tolerant *M. prunifolia* cultivars than in sensitive *M. hupehensis* in reaction to water deficiency.

The impact of lipophilic antioxidants in drought stress in *Rosmarinus officinalis* L. was evaluated (Munné-Bosch and Alegre, 2000). Young plants in pots were supported in a greenhouse with controlled temperature conditions (24°C and 18°C for day and night, respectively). After two years plants were transplanted to a field in the Mediterranean, receiving water purely from rainfall. The α-tocopherol amount accumulated in leaves was 9-fold higher during the summer with a relative water content reduction. After the autumn rainfalls the α-tocopherol concentration in leaves decreased, while during the winter it increased, presumably due to cold temperatures. The differentiation in the contents of β-carotene was similar to that reported for the chlorophylls, which content was reduced during the summer water deficit and increased after the autumn rainfalls.

Drought also induced changes in low molecular weight antioxidants in leaves of *Cistus clusii* grown under Mediterranean field conditions (Hernández et al., 2004). Ascorbic acid contents increased 2.1-fold after 15 days of stress conditions and the highest concentration was observed after 50 days of experiment, whereas dehydroascorbic acid content was reduced significantly after 15 days of water deficit and remained unchanged for the residual stress period. The α-tocopherol value was constant at the beginning of the experiment, but increased 3.3-fold after 50 days of drought.

In a study by Ratnayaka et al. (2012) two *Anoda cristata* varieties from Mississippi and New Mexico and two cotton types (*Gossypium hirsutum* and *Gossypium barbadense*) were rated under unstressed conditions and under a changing water deficiency environment. In the absence of stress factors cotton demonstrated at least 92% higher α-tocopherol amounts than *A. cristata*.

Under increasing water-deficit conditions α-tocopherol amounts grew only in cotton. A similar conclusion was reported by Jaleel et al. (2008), who found an increase in the contents of ascorbic acid, reduced glutathione and α-tocopherol in all examined parts of *Catharanthus roseus* (L.) G. Don. in the early stages of stress conditions. In *Populus cathayana* cuttings exposed to progressive water deficiency conditions for 12 weeks in a greenhouse, under moderate stress conditions, the carotenoid amounts significantly increased in all analyzed populations, while a significant decrease of carotenoid levels occurred in a wet climate population subjected to water deficit. The ratio of carotenoids/chlorophyll significantly rose under stress conditions in the dry climate population, whereas no significant variation was recorded in the wet climate population (Xiao et al., 2008).

The effect of ascorbic acid action on okra (*Hibiscus esculents* L.) germination and seedling growth was investigated under drought (water potential range −0.2 and −0.4 MPa) and normal development conditions (Baghizadeh and Hajmohammadrezaei, 2011). Ascorbate acid induced a decrease of average time necessary for germination under stress conditions and finally caused an extension of the germination process. In the seeds supplemented with 1 mM ascorbic acid the radical and plumule length and fresh and dry weights were also increased. In another experiment, *Satureja hortensis* plants were treated with ascorbic acid (0, 1 and 3 mM) in a greenhouse experiment with three levels of irrigation combinations (Yazdanpanah et al., 2011). During the mid-water-deficit stress 1 mM concentration of ascorbic acid (with 3 mM of salicylic acid) increased the content of proline and during the severe water-deficit stress, 1 mM of these compounds caused high resistance of the plant against the applied stress.

7.4.2 Extreme Temperatures (Cold and Heat Stress)

Each plant has its own temperature requirements, which are optimal for its growth and development. Plants in too low and too high temperature conditions in comparison to the optimum are exposed to stress. Both chilling stress (1−10°C) and heat stress (10−15°C above ambient) adversely affect the growth and development of plants (Wahid et al., 2007; Li et al., 2011; Lukatkin et al., 2012). Under low and high temperatures an overproduction of reactive oxygen species (ROS) occurs, which leads to damage to proteins, lipids, nucleic acids and carbohydrates (Mittler et al., 2004; Foyer and Noctor, 2005; Rodriguez and Redman, 2005). To alleviate the effects of ROS plants have evolved enzymatic and nonenzymatic antioxidants (Xie et al., 2008; Liu et al., 2009; Shu et al., 2011).

Chilling stress affects a multitude of physiological and biochemical processes. It causes poor germination, slow growth, discoloration or yellowing, withering, necrotic lesions, and membrane disintegration that are manifested as oxidative stress mediated by ROS (Guo et al., 2012; Santini et al., 2013).

The effect of low temperature (8°C, for 1 and 3 d) on metabolism of reactive oxygen species (ROS) and reactive nitrogen species (RNS) in leaves of pepper (*Capsicum annum* L.) was analyzed (Airaki et al., 2012). After 24 h of exposure the induction of ROS and RNS metabolism was observed and levels of ascorbate and glutathione were significantly higher, which suggests the role of antioxidants in cold acclimation through their impact on the redox state of the cell. In another study conducted by Singh and Singh, (2013) the impact of storage temperature (0°C and 5°C for 6 weeks) on antioxidant levels was investigated in Japanese plums. Glutathione contents were dependent on stress level and an increase of GSH was higher at 5°C than 0°C. Changes in ascorbate levels indicated a lack of protection at 0°C during storage.

To increase plant tolerance to chilling, cold acclimation or chill hardening methods were used (Zhou et al., 2012; Ao et al., 2013). In the first experiment GSH and GSSG contents and the GSH/GSSG ratio after cold acclimation (at 12/10°C) and then after chilling (7/4°C) were measured in tomato plants (Zhou et al., 2012). Under cold acclimation GSH content and the GSH/GSSG ratio significantly increased, while under chilling GSSG content significantly increased, resulting in a reduction of the GSH/GSSG ratio. In other studies glutathione and ascorbate levels were investigated in *Jatropha curcas* seedlings after chill hardening (12°C for 2 days) and then under chilling stress (1°C for 1–7 days) (Ao et al., 2013). In chill-hardened seedlings the activities of antioxidant enzymes and GSH and AsA contents were higher than in the group without chill hardening, which indicates an enhanced chilling tolerance in these plants.

Application of exogenous natural or synthetic molecules can also alleviate chilling stress (Li et al., 2011; Erdal, 2012; Guo et al., 2012). In studies conducted by Li et al. (2011) cucumber was pretreated with exogenous cinnamic acid (CA) (50 μM CA for 2 days) and then was cultivated under chilling stress (15/8°C and 25/18°C for 1 day). Results indicated that CA pretreatment enhances tolerance to chilling stress in cucumber leaves. Pepper seedlings under chilling stress (10/6°C, day/night) were investigated to explain how exogenous abscisic acid (ABA) (0.57 μM ABA for 72 h) can regulate physiological and biochemical mechanisms (Guo et al., 2012). GSH and AsA contents increased under chilling, whereas exogenous ABA pretreatment decreased these antioxidant levels. However, in the combination of ABA and chilling stress the AsA and GSH levels increased, but were lower than in the chilled group, which demonstrates that levels of these antioxidants were correlated with both ABA and chilling stress. In other studies androsterone (10^{-9} mol L^{-1}) was used as an alleviating factor in maize seedlings under chilling (10/7°C) (Erdal, 2012). Androsterone application enhanced ascorbic acid, glutathione and carotenoid levels and antioxidant activities, which may indicate the effect of this hormone on resistance to chilling stress in plants.

Similarly to chilling, high temperature results in impaired growth and development of plants and is an agricultural problem in many areas worldwide (Wahid et al., 2007; Dai et al., 2012). Most studies investigated the effects of heat stress on glutathione and ascorbic acid levels (Kocsy et al., 2002, 2004; Dai et al., 2012; Kumar et al., 2012). In an experiment conducted by Kumar et al., (2012) two genotypes of maize and two genotypes of rice were compared in terms of their response to different temperature stresses (35/30°C, 40/35°C, 45/40°C and control 30/25°C). Levels of nonenzymatic antioxidants such as ascorbic acid and glutathione were higher in maize genotypes than in rice ones at both 40/35 and 45/40°C, which suggests that maize genotypes have a greater ability to cope with high-temperature conditions. To verify the relationship between GSH levels in wheat genotypes with different frost sensitivity, i.e. frost-tolerant, moderately sensitive and frost-sensitive genotypes, were tested under high temperature (37°C for 3 days) (Kocsy et al., 2002). The best effects resulting in an increase in glutathione and hydroxymethylglutathione contents were demonstrated in frost-sensitive genotypes (Kocsy et al., 2004). As in chilling stress, cinnamic acid (CA) pretreatment (50 μM CA) was used to increase antioxidant activities in cucumber leaves under heat stress (42/38°C, day/night for 3 days) (Dai et al., 2011). After 2 days GSH and AsA contents and antioxidant activities increased in CA-pretreated cucumber leaves exposed to high temperature. In the light of this observation we can conclude that application of cinnamic acid enhances heat tolerance of seedlings.

7.4.3 Salinity

Salt stress is a major environmental stress, which affects seed germination, plant growth and development (Oh et al., 2009; Khan et al., 2011; Queiros et al., 2011). High salinity is known to cause both hyperionic and hyperosmotic effects in plants, leading to membrane disorganization or metabolic toxicity (Wang et al., 2012a). In addition, an important consequence of salinity stress is the excessive generation of several reactive oxygen species (ROS), which are continuously produced in plants as by-products of aerobic metabolism under high salt conditions, causing oxidative stress in plants (Chaparzadeh et al., 2004; Parida and Das, 2005; Ashraf, 2009; Noreen et al., 2009; Wang et al., 2012a,b).

Recently, several studies have shown that salinity affects the quality and nutritional value of plants, changes enzyme activities and nonenzymatic antioxidant contents, or causes imbalance in levels of plant growth regulators (Borghesi et al., 2011; Zhu et al., 2011, Lim et al., 2012; Van Meulebroek et al., 2012; Ellouzi et al., 2013; Tuna et al., 2013).

In studies conducted by Zhu et al. (2011) the effects of different NaCl levels (34, 68, 102, 136, 170 and 204 mM) on seed germination, early seedling growth and physiological properties of cauliflower (*Brassica oleracea* L.) were explained. The results indicated that seedling growth was significantly reduced by all salinity levels, particularly at 136 and 170 mM NaCl. Carotenoid content in leaves – as an important pigment for photosynthesis – was increased at lower salt stress (up to 120 mM NaCl) and decreased at higher salinity (136–102 mM NaCl). Similar results were demonstrated in maize plants under salt stress (Sheng et al., 2008, 2009). Lim et al. (2012) investigated the carotenoid content in sprouts treated with 10, 50, 100 and 200 mM NaCl and harvested following different growth periods (1, 3, 5 and 7 days). Carotenoid levels in sprouts increased continuously at all NaCl concentrations, with the greatest difference between carotenoid content and the control (2-fold higher) at 50 and 100 mM NaCl. It is worth noting that salinity induces abscisic acid biosynthesis from carotenoids via the mevalonic acid pathway. Thus, the increase in carotenoid levels under salinity may stimulate the ABA induction.

In tomatoes, among carotenoids mainly lycopene and β-carotene are of interest. Lycopene has the highest antioxidant activity among all dietary antioxidants and represents about 80–90% of all carotenoids in tomato fruits. The other carotenoid, β-carotene, is of interest due to its provitamin A activity (Borghesi et al., 2011; Van Meulebroek et al., 2012). Under different salinity levels (3.5 and 5.5 dS cm^{-1}) four tomato genotypes (one nonanthocyanin-accumulating and three anthocyanin-accumulating) were analyzed to investigate the variability in their carotenoids and anthocyanin levels (Borghesi et al., 2012a). A high level salinity was connected with an increase in the concentrations of all colored tomato carotenoids (even 2–3 fold). Van Meulebroek et al. (2012) tested the possibility of improving antioxidant concentrations (lycopene, β-carotene and α-tocopherol) in tomato fruits in a more saline growth environment (2.4, 4.0, 5.3 and 8.0 dS m^{-1}). In this experiment the application of different NaCl levels clearly counteracted the effect of salinity on antioxidant concentrations. In view of these results, there is a prospect of exploiting saline soils to obtain tomatoes with increased levels of health-promoting compounds.

The impact of salinity on tocopherol content has been widely studied in various plants such as tomato, rosemary, medicinal plants (*Salvia officinalis*), pea or sunflower (Noreen and Ashraf, 2009, 2010; Skłodowska et al., 2009; Tounekti et al., 2011a,b). Noreen and Ashraf (2009) tested whether changes in antioxidant enzyme activities and the levels of tocopherols (α-, γ- and δ-tocopherol) could be used as markers of salt tolerance (40, 80 and 120 mM NaCl) in nine genetically diverse pea cultivars. It is assumed that salt-tolerant genotypes are characterized by higher levels of antioxidant enzymes than salt-sensitive ones (Logan, 2005). Contrary to what most authors believe, the high contents of antioxidants should also be found in salt-sensitive cultivars (Munns and Tester, 2008). It has been shown that all the three forms of tocopherols detected in pea cultivars were significantly dependent on the levels of salt stress. The contents of γ- and δ-tocopherols increased in all genotypes, while the α-tocopherol level showed no visible differences under salinity. Similar results concerning α-tocopherol, lutein and β-carotene contents under salinity (50, 100 and 150 mM for 6 weeks) in rosemary plants were demonstrated in studies conducted by Tounekti et al. (2011b). α-Tocopherol content increased, while lutein and β-carotene concentrations decreased with the salt stress progress.

Skłodowska et al. (2009) compared the role of tocopherol under moderate (50 mM NaCl) and severe (150 mM NaCl) salinity in tomato plants. Tocopherol contents were measured at 1 and 3 h, and 1, 2, 5 and 7 days after NaCl treatment. Up to 1 day after NaCl application tocopherol levels decreased compared to the control under both stresses, while an increase in tocopherol contents was observed around the second day after treatment. There are significant differences between the concentrations of tocopherols at moderate (193% of the control on the second day) and severe (from 438 to 355% on the fifth and seventh days, respectively) salt stresses. It is considered that an increase of tocopherol content refers to plant stress tolerance and its decrease is associated with susceptibility to stress. Plants respond to stress through changes in tocopherol levels taking place in two phases: in the first stage the tocopherol content increases and in the second one it decreases (Munné-Bosch, 2005). In stress-tolerant cultivars only the first phase occurs, while in stress-sensitive ones it is often observed in the second phase.

To alleviate the effect of salinity, plant growth regulators were used as foliar fertilizers (Loreto et al., 2005; Tounekti et al., 2011a). The impact of kinetin application on α-tocopherol contents in sage (*Salvia officinalis*) under salinity was investigated (Tounekti et al., 2011a). This experiment was carried out in four groups: I – control, II – control + 10 μM kinetin, III – 100 mM NaCl, IV – 10 μM kinetin + 100 mM NaCl for 2, 4 and 6 weeks. Tocopherol contents increased both in group II and IV, which may indicate a reduction of the negative impact of salinity by kinetin application.

The negative impact of salt stress on oil content and fatty acid composition in oilseed crops was also reported (Bassil and Kaffka, 2002; Heuer et al., 2002). In many cases oil yield and unsaturated fatty acid contents were significantly lower under salt stress. Noreen and Ashraf (2010) found the effect of different levels of NaCl (0 and 120 mmol L^{-1}) and exogenous salicylic acid (SA) application (0, 100, 200 and 300 mg L^{-1}) on the fatty acids profile in sunflower (*Helianthus annuus* L.). The content of oil, linoleic acid and δ-tocopherol decreased, while that of linolenic acid, palmitic acid, stearic acid and α- and γ-tocopherols increased under salinity. After the application of exogenous SA the levels of stearic acid and α- and γ-tocopherols were significantly decreased.

Glutathione plays an important role in the response of plants to environmental stresses, including salinity. The effects of low and high salinity, short- and long-term salt treated, or cultivars with different salt tolerance on the glutathione levels were investigated in various plants (Ruiz and

Blumwald, 2002; Mittova et al., 2003; Chaparzadeh et al., 2004; Salama and Al-Mutawa, 2009; Wang et al., 2012a,b).

In studies conducted by Wang et al. (2012a) the impact of various levels of NaCl (1, 3, 5, 10 and 15 g NaCl/kg dry soil) on the gluthatione content in leaves and roots of switchgrass (*Panicum virgatum* L.) was observed. The GSH concentration was higher in leaves than in roots, both in the control and in salt stress treated samples. At a lower salinity (3 and 5 g NaCl/kg dry soil) only the leaf glutathione levels were significantly higher in comparison to the control. In the case of roots, the lowest GSH content was at the highest NaCl concentration (15 g NaCl/kg dry soil).

The effects of salinity (75 and 150 mM NaCl for 15 days) on the biosynthesis of glutathione and cysteine in wild-type and salt-tolerant transgenic *Brassica napus* L. plants were investigated (Ruiz and Blumwald, 2002). Only in the wild-type canola a 3-fold increase was observed in glutathione and cysteine contents in comparison to the control under salt stress. In contrast to previous results, the glutathione and ascorbate levels were decreased in marigold (*Calendula officinalis* L.) under salinity (50 and 100 mM NaCl for 3 weeks) (Chaparzadeh et al., 2004). Similar studies concerning glutathione biosynthesis under salinity (successively increasing the NaCl concentration by 25 mM per day over a period of 4 days, to 100 mM) in two tomato species with different salt tress tolerance were demonstrated (Mittova et al., 2003). Finally, the GSH content and the GSH/GSSG ratio increased only in salt-tolerant tomato plants, which was associated with glutathione peroxidase and glutathione-S-transferase activities.

Several studies were related to the response of the plasma membrane to salinity in genotypes with different salt tolerance (Mansour et al., 2005; Salama et al., 2007). The plasma membrane may be the primary site of salt injury and changes in plasma membrane lipid alteration induced by saline conditions may have a critical role in plant response to salinity (Mansour et al., 2002; Mansour and Salama, 2004; Salama et al., 2007). The influence of exogenous glutathione (0.5 mM) in the alleviation of salinity effects (150 mM NaCl for 3 h) on the plasma membrane in onion bulbs was investigated (Salama and Al-Mutawa, 2009). Salt stress significantly increased membrane permeability, while the exogenous glutathione application significantly reduced it. In similar studies, the addition of exogenous H_2O_2 alleviated salinity effects in wheat seedlings through glutathione production (Devi et al., 2008). As in previous studies, it was demonstrated that exogenous nitric oxide (NO) plays an important role in the inhibition of salt stress effects (100 mM NaCl) in cucumber hypocotyls and radicles (Lin et al., 2012).

7.4.4 Heavy Metal Stress

Agricultural soils over extensive regions of the world often display contamination with toxic heavy metals such as Cd, Cu, Zn, Ni, Cr, Pb, Se, As and

Co. This may result from natural sources, including volcanoes and continental dusts, and long-term anthropogenic activities such as the use of phosphate fertilizers, inappropriate watering practices, sewage sludge deposition, dust from smelters and industrial waste disposal in land used for agriculture (Anjum et al., 2008a). Under these conditions the solubility and distribution of toxic metals are dependent on adsorption, desorption, and the complexity of processes, which in turn are dependent on the soil type.

At high concentrations of heavy metals, plants primarily generate reactive oxygen species directly by the Haber-Weiss reaction or overproduction. Indirect mechanisms facilitate interactions with enzymatic and nonenzymatic antioxidants, disrupting the electron transport chain or membrane lipids and cellular processes in plants (Caregnato et al., 2008). Among all of these, lipid peroxidation yields the most noxious results, which may indicate biomembrane deterioration. Malondialdehyde, which is a decomposition product of polyunsaturated fatty acids in the membrane, is considered to be an indicator of free radical stress. What is important during heavy metal stress is that plants synthesize low molecular weight thiols, which exhibit a strong relation to toxic metals. Glutathione and cysteine belong to the most effective low molecular thiols. Glutathione plays a decisive role in detoxification of such metals as nickel and cadmium; it is also a substrate for phytochelatin biosynthesis, cysteine-abundant polypeptides with heavy metal-binding properties (Aina et al., 2007; Anjum et al., 2010; Huang et al., 2010). Plants appear to preserve a high ratio of reduced glutathione to its oxidized form in cells, but as a component of the ascorbate-glutathione pathway, reduced glutathione reacts nonenzymatically with $^1O_2$, $O_2^{\bullet -}$ and $^{\bullet}OH$ and is transformed into oxidized glutathione (Basantani and Srivastava, 2007).

Among toxic metals and metalloids, cadmium (Cd) is frequently found in plants; it is a common trace pollutant supplied into the soil by industrial processes or by the application of fertilizers and pesticides, wherein it is mostly a “guest” metal in Pb/Zn mineralization. Plants treated with high contents of cadmium (Cd) experience a decrease in photosynthesis, respiration and nitrogen metabolism, water and nutrient uptake, inhibited development, chlorosis processes, browning of root tips and death of plants. Cd^{2+} ions have also been found to be the most efficient inducers of phytochelatins in plants (Chen et al., 2010). Summarizing, stress-tolerant plants are required for the reclamation of polluted soils and the reclamation of industrial areas.

The review of literature revealed differential amounts of toxic metals and metalloid-induced oxidative stress, which may presumably be related not only to the contents of the metals supplied, but also to the plant species, the stage of plant development and duration of the treatment. In a study by Bi et al. (2009) the morphology and function of organelles and free radical processes were examined in *Arabidopsis thaliana*. After five days of exogenous Cd application in 20 μM rates, leaves demonstrated a phenotype of leaf roll and chlorosis symptoms in comparison to the nontreated plants.

After 15 days of Cd application all seedlings demonstrated yellow spots and wilting signs. The three-dimensional reconstructed images indicated that during 1 hour of toxic metal application there were no significant changes in the chloroplasts and cell morphology. After 4 hours of Cd application the chloroplasts and cells started to show irregular structures. A 70% cell survival rate was recorded when examining the influence of protoplast pre-incubation with ascorbic acid. Moreover, after 18 hours of cadmium application, 55% of protoplasts were detected to be living in the presence of ascorbic acid and catalase, but merely 2% in their absence.

The antioxidant system response in durum wheat (*Triticum durum* cv. Creso) plants was examined at 0–40 μM cadmium application rate in hydroponic cultivation (Paradiso et al., 2008). In wheat roots cadmium induces growth of ascorbate and primarily dehydroascorbate concentration (total ascorbate pool). Likewise, glutathione and oxidized glutathione amounts increased in roots under cadmium treatments depending on rates, similarly to ascorbate-glutathione cycle enzymes. During Cd treatment, root cytosolic ascorbate peroxidase (APX) levels were about 50–80% higher when compared to the control. The highest intensity of cycle enzyme activities were recorded for 10 and 20 μM, with the exception of glutathione reductase, which demonstrated the highest activity at 40 μM cadmium. Simultaneously, cadmium treatment caused decrease of root and leaf lengths in a rate-dependent system. The development reduction involved cell proliferation, as indicated by the cadmium-dependent inhibition of the mitotic index appearing in the root apical meristem. Similarly, Sun et al., (2005) evaluated the individual and combined application of cadmium and lead on the development of *T. aestivum* L. cv. Yangmai 10. In that experiment the amounts of cadmium tested alone were from 0 to 54 μmol/L; for lead they ranged from 0 to 54 μmol/L and for a simultaneous Cd/Pb application they were 2 + 2 μmol/L, 2 + 18 μmol/L, 18 + 2 μmol/L and 18 + 18 μmol/L, respectively. Toxic metals were rated into the nutrient solution as chlorate salts. Glutathione and phytochelatin contents increased considerably under exposure to cadmium at increasing rates, in the sequence of roots > stems > leaves, while lead application had no evident effect on glutathione and phytochelatin concentrations. What is important, the amounts of both toxic metals increased in wheat tissues in proportion to the rate applied. At cadmium and lead exposure conditions, the first one is present in subsoil elevated lead content in wheat roots, while the second, reduced cadmium amounts in plant tissues.

The effects of cadmium toxicity and glutathione treatment in barley (*Hordeum vulgare* L.) were also investigated (Chen et al., 2010; Wang et al., 2011). Using two barley genotypes, i.e. cv. Weisuobuzhi (cadmium-resistant) and Dong 17 (cadmium-susceptible), in hydroponic culture $CdCl_2$ and reduced glutathione were applied either in individual rates or simultaneously. The addition of 20 mg L^{-1} reduced glutathione in 5 μM cadmium nutrient solution mitigated the effects of plant development inhibition and decreased cadmium amounts in leaves and roots, particularly in the Cd-sensitive cv. Dong 17. Glutathione treatment in comparison to individual cadmium applications caused a relatively correct structure of chloroplasts with properly structured thylakoid membranes and a parallel pattern of lamellae.

In a study by Anjum et al. (2008b) the effect of cadmium and sulfur treatments on mustard (*Brassica campestris* L.) cv. Pusa Gold was examined. The soil was initially supplemented with $CdCl_2$ in rates from 0 mg/kg to 100 mg/kg and sulfur 0 mg/kg and 40 mg per kg. When no sulfur application was applied, ascorbic acid amounts were reduced by 16.16%, 36.36% and 48.98% under the influence of 25, 50 and 100 mg cadmium per kg, respectively. In contrast, under the same Cd treatments the concentration of leaf ascorbic acid increased under sulfur treatment by 26.54%, 14.28% and 8.18%, respectively. Simultaneously, glutathione amounts decreased significantly by 40.57%, 32.17% and 22.60%, at cadmium applied at 100, 50 and 20 mg/kg, respectively. What is more, with treatments of 40 mg/kg sulfur glutathione level significantly increased and the growth rate varied depending on cadmium content. What is essential, all the cadmium rates reduced plant dry mass and leaf area.

7.4.5 High Light Stress

The supposition that light is a key environmental factor controlling leaf antioxidant amounts is widely accepted. Excess light imposes an exceptional stress onto the plant that primarily affects metabolic changes in chloroplasts before other cell compartments are affected (Asada, 2006; Rastogi et al., 2013). On the subcellular level ascorbic acid was detected at considerably increased levels in most cell compartments during high light exposure. For instance, ascorbate is generally detected inside the lumen of thylakoids of chloroplasts subjected to high light stress. Consequently, in *Arabidopsis* leaves higher ascorbate amounts were identified after acclimatization to this stress. A significant increase in ascorbate contents in *Arabidopsis thaliana* was also detected in chloroplasts (104%) after excess light conditions of about 700 μmol m^{-2} s^{-1} for two weeks compared to plants grown at 150 μmol m^{-2} s^{-1} (Zechmann et al., 2011). While the light-dependent stimulation of ascorbate biosynthesis appears to demand photosynthetic electron transport activity, ascorbate biosynthesis and ascorbate regeneration processes are dependent on light quality as well as quantity.

The influence of increasing irradiance on ascorbic acid metabolism was *inter alia* examined in the leaves of wild-type and transformed *Arabidopsis thaliana* with changed quantities of the mitochondrial alternative oxidase protein (Bartoli et al., 2006). Seedlings were developing under low, intermediate and high light conditions (50, 100 and 250 μmol photons m^{-2} s^{-1}, respectively). Increasing irradiance positively improved leaf ascorbic acid concentrations; thereby the dark-induced decline of leaf ascorbic acid amounted to a loss of 11, 55 and 89 nmol g^{-1}fresh weight h^{-1}, from low, intermediate and high irradiance-grown leaves, accordingly. Additionally, supplementation with L-galactone-1,4-lactone induced a 2-fold higher concentration of ascorbic acid in leaves at high irradiance. In a study by Müller-Moulé et al., (2004) acclimation to changing light conditions was tested on *vtc*2 (ascorbate-deficient mutants), *vtc*2npq4 and *vtc*2npq1 (nonphotochemical quenching of chlorophyll fluorescence-deficiency mutants) *Arabidopsis* in comparison to the wild type. The *vtc*2 mutant demonstrated a 10 to 30% wild-type concentration of the ascorbate pool. The listed mutants were able to develop under high light conditions and had wild-type concentrations of Lhcb1 (light harvesting antenna protein of PSII), cytochrome f, PsaF proteins, and 2-cysteine peroxiredoxin. They also demonstrated higher glutathione amounts than the wild type in high light, implying a probable compensation for the lower ascorbate content. The obtained results also indicate the significance of ascorbate for the long-term adaptation of plants to high light conditions.

In another study Bartoli et al. (2009) examined *Phaseolus vulgaris* L. (cv. TUC 500), which was cultivated under two red/far red light systems amounting to 1.1 (simulated "sun" conditions) and 0.2 (simulated "shade" conditions), while supporting an identical photosynthetic photon flux density (PPFD, 300 μmol photon m^{-2} s^{-1}). Leaves under simulated "shade" conditions had on average a 38% reduction in ascorbic acid concentration than leaves under simulated "sun" conditions. The amounts of ascorbic acid were decreased to about 50% at the end of the dark period (3.3 and 2.5 μmol g^{-1} FW for "sun" and "shade" light conditions) and the contents of oxidized ascorbic acid were comparable under considered light treatments. Also, leaves growing under light fortified in far red had lower chlorophyll concentrations and lower specific leaf weights.

Burritt and Mackenzie (2003) examined the impact of high light activity on *Begonia* × *erythrophylla* grown in a glasshouse under shade-cloth. Throughout the experiment photosynthetically active radiation amounted to 155 μmol m^{-2} s^{-1}. No important differences in the total ascorbate pool were observed when compared to light-transferred and shade-grown leaves. Also, comparing the total ascorbate pool of sun-grown leaves and shade and light-transferred leaves, in the first combination it was 50% higher than in the others. Additionally, sun-grown leaves contained a greater pool of ascorbate and a reduced ascorbate:oxidized ascorbate ratio than shade-grown leaves.

7.5 CONCLUSIONS

Drought, extreme temperatures and salinity are major abiotic stresses that reduce yields of food crops worldwide. In the last decade significant progress has been made in understanding mechanisms regulating abiotic stress tolerance in crop plants. Exposure of plants to environmental stresses results in the generation of reactive oxygen species (ROS). To prevent the accumulation of ROS and to ensure their own survival, plants have developed an antioxidant defense system, which comprises various enzymatic and nonenzymatic molecules, produced to counter the adverse effect of environmental stresses. Among nonenzymatic molecules an important role is played by lipophilic (tocopherols and carotenoids) and hydrophilic (glutathione and ascorbic acid) antioxidant compounds. There are different methods, techniques and technologies to induce abiotic stress tolerance in high-yielding cultivars. Application of genomics, proteomics and trascriptomics approaches is necessary to provide a better understanding of the molecular basis of plant response to stresses. Understanding the multiple roles of antioxidants in detail will provide us with the necessary tools to improve agricultural practice and production.

REFERENCES

Agrawal, B., Czymmek, K.J., Sparks, D.L., Bais, H.P., 2013. Transient influx of nickel in root mitochondria modulates organic acid and reactive oxygen species production in nickel hyperaccumulator *Alyssum murale*. J. Biol. Chem. 288, 7351–7362.

Ahmad, P., Jaleel, C.A., Azooz, M.M., Nabi, G., 2009. Generation of ROS and non-enzymatic antioxidants during abiotic stress in plants. Bot. Res. Int. 2, 11–20.

Ahmad, P., Jaleel, C.A., Salem, M.A., Nabi, G., Sharma, S., 2010. Roles of enzymatic and non-enzymatic antioxidants in plants during abiotic stress. Crit. Rev. Biotechnol. 30, 161–175.

Aina, R., Labra, M., Fumagalli, P., Vannini, C., Marsoni, M., Cucchi, U., et al., 2007. Thiol-peptide level and proteomic changes in response to cadmium toxicity in *Oryza sativa* L. roots. Environ. Exp. Bot. 59, 381–392.

Airaki, M., Leterrier, M., Mateos, R.M., Valderrama, R., Chaki, M., Barroso, J.B., et al., 2012. Metabolism of reactive oxygen species and reactive nitrogen species in pepper (*Capsicum annuum* L.) plants under low temperature stress. Plant Cell Environ. 35, 281–295.

Al-Ghamdi, A., 2009. Evaluation of oxidative stress tolerance in two wheat (*Triticum aestivum*) cultivars in response to drought. Int. J. Agric. Biol. 11, 7–12.

Anjum, N.A., Umar, S., Ahmad, A., Iqbal, M., 2008a. Responses of components of antioxidant system in moongbean [*Vigna radiata* (L.) Wilczek] genotypes to cadmium stress. Comm. Soil Sci. Plant Anal. 39, 2469–2483.

Anjum, N.A., Umar, S., Ahmad, A., Iqbal, M., Khan, N.A., 2008b. Sulphur protects mustard (*Brassica campestris* L.) from cadmium toxicity by improving leaf ascorbate and glutathione. Plant Growth Regul. 54, 271–279.

Anjum, N.A., Umar, S., Chan, M.-T. (Eds), 2010. Ascorbate-Glutathione Pathway and Stress Tolerance in Plants. Springer, Dordrecht, The Netherlands.

Ao, P.-X., Li, Z.-G., Fan, D.-M., Gong, M., 2013. Involvement of antioxidant defense system in chill hardening-induced chilling tolerance in *Jatropha curcas* seedlings. Acta Physiol. Plant. 35, 153–160.

Asada, K., 2006. Production and scavenging of reactive oxygen species in chloroplasts and their functions. Plant Physiol. 141, 391–396.

Ashraf, M., 2009. Biotechnological approach of improving plant salt tolerance using antioxidants as markers. Biotechnol. Adv. 27, 84–93.

Atkinson, N.J., Dew, T.P., Orfila, C., Urwin, P.E., 2011. Influence of combined biotic and abiotic stress on nutritional quality parameters in tomato (*Solanum lycopersicum*). J. Agric. Food Chem. 59, 9673–9682.

Auldridge, M.E., McCarty, D.R., Klee, H.J., 2006. Plant carotenoid cleavage oxygenases and their apocarotenoid products. Curr. Opin. Plant Biol. 9, 315–321.

Badrhadad, A., Piri, K., Ghiasvand, T., 2013. Increasing of alpha-tocopherol in cell suspension cultures *Elaeagnus angustifolia* L. Int. J. Agr. Crop Sci. 5, 1328–1331.

Baghizadeh, A., Hajmohammadrezaei, M., 2011. Effect of drought stress and its interaction with ascorbate and salicylic acid on okra (*Hibiscus esculents* L.) germination and seedling growth. J. Stress Physiol. Biochem. 7, 55–65.

Baldwin, E.A., 2003. Coating and other supplemental treatments to maintain vegetable quality. In: Bartz, J.A., Brecht, J.K. (Eds.), Postharvest Physiology and Pathology of Vegetables, 2nd ed. Marcel Dekker, New York, pp. 413–456.

Barth, C., De Tullio, M., Conklin, P., 2006. The role of ascorbic acid in the control of flowering time and the onset of senescence. J. Exp. Bot. 57, 1657–1665.

Bartoli, C.G., Simontacchi, M., Tambussi, E., Beltrano, J., Montaldi, E., Puntarulo, S., 1999. Drought and watering-dependent oxidative stress: effect on antioxidant content in *Triticum aestivum* L. leaves. J. Exp. Bot. 50, 375–383.

Bartoli, C.G., Pastori, G.M., Foyer, C.H., 2000. Ascorbate biosynthesis in mitochondria is linked to the electron transport chain between complexes III and IV. Plant Physiol. 123, 335–343.

Bartoli, C.G., Yu, J., Gómez, F., Fernandez, L., McIntosh, L., Foyer, C.H., 2006. Inter-relationships between light and respiration in the control of ascorbic acid synthesis and accumulation in *Arabidopsis thaliana* leaves. J. Exp. Bot. 57, 1621–1631.

Bartoli, C.G., Tambussi, E.A., Diego, F., Foyer, C.H., 2009. Control of ascorbic acid synthesis and accumulation and glutathione by the incident light red/far red ratio in *Phaseolus vulgaris* leaves. FEBS Lett. 583, 118–122.

Basantani, M., Srivastava, A., 2007. Plant glutathione transferases – a decade falls short. Can. J. Bot. 85, 443–456.

Bassil, E.S., Kaffka, S.R., 2002. Response of safflower (*Carthamus tinctorius* L.) to saline soils and irrigation: II. Crop response to salinity. Agr. Water Manag. 54, 81–92.

Bi, Y., Chen, W., Zhang, W., Zhou, Q., Yun, L., Xing, D., 2009. Production of reactive oxygen species, impairment of photosynthetic function and dynamic changes in mitochondria are early events in cadmium-induced cell death in *Arabidopsis thaliana*. Biol. Cell 101, 629–643.

Borghesi, E., Gonzalez-Miret, M.L., Escudero-Gilete, M.L., Malorgio, F., Heredia, F.J., Melendez-Martínez, A.J., 2011. Effects of salinity stress on carotenoids, anthocyanins, and color of diverse tomato genotypes. J. Agr. Food Chem. 59, 11676–11682.

Botella-Pavia, P., Rodriguez-Concepcion, M., 2006. Carotenoid biotechnology in plants for nutritionally improved foods. Physiol. Plant 126, 369–381.

Briviba, K., Klotz, L.O., Sies, H., 1997. Toxic and signaling effects of photochemically or chemically generated singlet oxygen in biological systems. J. Biol. Chem. 378, 1259–1265.

Burritt, D.J., Mackenzie, S., 2003. Antioxidant metabolism during acclimation of *Begonia* × *erythrophylla* to high light levels. Ann. Bot. 91, 783–794.

Caregnato, F.F., Koller, C.E., MacFarlane, G.R., Moreira, J.C.F., 2008. The glutathione antioxidant system as a biomarker suite for the assessment of heavy metal exposure and effect in the grey mangrove, *Avicenniamarina* (Forsk.)Vierh Mar. Pollut. Bull. 56, 1119–1127.

Cattivelli, L., Rizza, F., Badeck, F.W., Mazzucotelli, E., Mastrangelo, A.M., Francia, E., et al., 2008. Drought tolerance improvement in crop plants: an integrated view from breeding to genomics. Field Crops Res. 105, 1–14.

Chahal, G.B.S., Sood, A., Jalota, S.K., Choudhury, B.U., Sharma, P.K., 2007. Yield, evapotranspiration and water productivity of rice (*Oryza sativa* L.) – wheat (*Triticum aestivum* L.) system in Punjab (India) as influenced by transplanting date of rice and weather parameters. Agr. Water Manag. 88, 14–22.

Chaparzadeh, N., D'Amico, M.L., Khavari-Nejad, R.-A., Izzo, R., Navari-Izzo, F., 2004. Antioxidative responses of *Calendula officinalis* under salinity conditions. Plant Physiol. Biochem. 42, 695–701.

Chen, F., Wang, F., Wu, F., Mao, W., Zhang, G., Zhou, M., 2010. Modulation of exogenous glutathione in antioxidant defense system against Cd stress in the two barley genotypes differing in Cd tolerance. Plant Physiol. Biochem. 48, 663–672.

Chen, K.-M., Gong, H.-J., Chen, G.-C., Wang, S.-M., Zhang, C.-L., 2004. Gradual drought under field conditions influences the glutathione metabolism, redox balance and energy supply in spring wheat. J. Plant Growth Regul. 23, 20–28.

Chimenti, C.A., Marcantonio, M., Hall, A.J., 2006. Divergent selection for osmotic adjustment results in improved drought tolerance in maize (*Zea mays* L.) in both early growth and flowering phases. Field Crops Res. 95, 305–315.

Collakova, E., DellaPenna, D., 2003. The role of homogentisate phytyltransferase and other tocopherol pathway enzymes in the regulation of tocopherol synthesis during abiotic stress. Plant Physiol. 133, 930–940.

Conklin, P.L., Saracco, S.A., Norris, S.R., Last, R.L., 2000. Identification of ascorbic acid-deficient *Arabidopsis thaliana* mutants. Genetics 154, 847–856.

Cruz-Rus, E., Amaya, I., Valpuesta, V., 2012. The challenge of increasing vitamin C content in plant foods. Biotechnol. J. 7, 1110–1121.

Cuttriss, A.J., Chubb, A.C., Alawady, A., Grimm, B., Pogson, B.J., 2007. Regulation of lutein biosynthesis and prolamellar body formation in *Arabidopsis*. Funct. Plant Biol. 34, 663–672.

Cvetkovska, M., Vanlerberghe, G.C., 2012. Alternative oxidase impacts the plant response to biotic stress by influencing the mitochondrial generation of reactive oxygen species. Plant Cell Environ. 36, 721–732.

Dai, A.-H., Nie, Y.-X., Yu, B., Li, Q., Lu, L.-Y., Bai, J.-G., 2012. Cinnamic acid pretreatment enhances heat tolerance of cucumber leaves through modulating antioxidant enzyme activity. Environ. Exp. Bot. 79, 1–10.

De Pinto, M.C., De Gara, L., 2004. Changes in the ascorbate metabolism of apoplastic and symplastic spaces are associated with cell differentiation. J. Exp. Bot. 55, 2559–2569.

del Río, L.A., Corpas, F.J., Sandalio, L.M., Palma, J.M., Gómez, M., Barroso, J.B., 2002. Reactive oxygen species, antioxidant systems and nitric oxide in peroxisomes. J. Exp. Bot. 53, 1255–1272.

DeLong, J.M., Prange, R.K., Harrison, P.A., 2004. The influence of pre-storage delayed cooling on quality and disorder incidence in 'Honeycrisp' apple fruit. Postharv. Biol. Technol. 34, 353–358.

DellaPenna, D., 2005. A decade of progress in understanding vitamin E synthesis in plants. J. Plant Physiol. 162, 729–737.

DellaPenna, D., Pogson, B.J., 2006. Vitamin synthesis in plants: tocopherols and carotenoids. Ann. Rev. Plant Biol. 57, 711–738.

Dempsey, R.W., Merchant, A., Tausz, M., 2011. Increasing leaf glutathione through stem feeding does not acclimate *Eucalyptus camaldulensis* seedlings towards high-light stress. Acta Physiol. Plant 33, 221–225.

Devi, S., Angrish, R., Datta, K.S., Kumar, B., 2008. Antioxidant defense system in wheat seedlings under sodium chloride stress: an inductive role of hydrogen peroxide. Indian J. Plant Physiol. 13, 22–26.

Doyle, S.M., Diamond, M., McCabe, P.F., 2010. Chloroplast and reactive oxygen species involvement in apoptotic-like programmed cell death in *Arabidopsis* suspension cultures. J. Exp. Bot. 61, 473–482.

Drame, K.N., Clavel, D., Repellin, A., Passaquet, C., Zuily-Fodil, Y., 2007. Water deficit induces variation in expression of stress-responsive genes in two peanut (*Arachis hypogaea* L.) cultivars with different tolerance to drought. Plant Physiol. Biochem. 45, 236–243.

Dubreuil-Maurizi, C., Poinssot, B., 2012. Role of glutathione in plant signaling under biotic stress. Plant Signal. Behav. 7, 210–212.

East, A.R., Tanner, D.J., Maguire, K.M., Mawson, A.J., 2008. The influence of breaks in storage temperature on 'Cripps Pink' (Pink Lady) apple physiology and quality. Hort. Sci. 43, 818–824.

Ebrahimian, E., Bybordi, A., 2012. Influence of ascorbic acid foliar application on chlorophyll, flavonoids, anthocyanin and soluble sugar contents of sunflower under conditions of water deficit stress. J. Food Agr. Environ. 10, 1026–1030.

Ellouzi, H., Hamed, K.B., Cela, J., Müller, M., Abdelly, C., Munné-Bosch, S., 2013. Increased sensitivity to salt stress in tocopherol-deficient *Arabidopsis* mutants growing in a hydroponic system. Plant Signal. Behav. 8, e23136.

Erdal, S., 2012. Androsterone-induced molecular and physiological changes in maize seedlings in response to chilling stress. Plant Physiol. Biochem. 57, 1–7.

Farooq, M., Wahid, A., Ito, O., Lee, D.J., Siddique, K.H.M., 2009a. Advances in drought resistance of rice. Crit. Rev. Plant Sci. 28, 199–217.

Farooq, M., Wahid, A., Kobayashi, N., Fujita, D., Basra, S.M.A., 2009b. Plant drought stress: effects, mechanisms and management. Agron. Sustain. Dev. 29, 185–212.

Farooq, M., Irfan, M., Aziz, T., Ahmad, I., Cheema, S.A., 2013. Seed priming with ascorbic acid improves drought resistance of wheat. J. Agron. Crop Sci. 199, 12–22.

Foyer, C.H., Noctor, G., 2003. Redox sensing and signaling associated with reactive oxygen in chloroplasts, peroxisomes and mitochondria. Physiol. Plant 119, 355–364.

Foyer, C.H., Noctor, G., 2005. Redox homeostasis and antioxidant signaling: a metabolic interface between stress perception and physiological responses. Plant Cell 17, 1866–1875.

Fraser, P.D., Enfissi, E.M.A., Halket, J.M., Truesdale, M.R., Yu, D., Gerrish, C., 2007. Manipulation of phytoene levels in tomato fruit: effects on isoprenoids, plastids, and intermediary metabolism. Plant Cell 19, 3194–3211.

Fridovich, I., 1986. Superoxide dismutases. Adv. Enzymol. Relat. Areas Mol. Biol. 58, 61–97.

Gallie, D.R., 2013a. The role of L-ascorbic acid recycling in responding to environmental stress and in promoting plant growth. J. Exp. Bot. 64, 433–443.

Gallie, D.R., 2013b. L-ascorbic acid: a multifunctional molecule supporting plant growth and development. Scientifica 2013, 795964.

Galpaz, N., Wang, Q., Menda, N., Zamir, D., Hirschberg, J., 2008. Abscisic acid deficiency in the tomato mutant *high-pigment* 3 leading to increased plastid number and higher fruit lycopene content. Plant J. 53, 717–730.

Giacomelli, L., Rudella, A., van Wijk, K.J., 2006. High light response of the thylakoid proteome in *Arabidopsis* wild type and the ascorbate-deficient mutant *vtc2-2*. A comparative proteomics study. Plant Physiol. 141, 685–701.

Gill, S.S., Tuteja, N., 2010. Reactive oxygen species and antioxidant machinery in abiotic stress tolerance in crop plants. Plant Physiol. Biochem. 48, 909–930.

Giuliano, G., Tavazza, R., Diretto, G., Beyer, P., Taylor, M.A., 2008. Metabolic engineering of carotenoid biosynthesis in plants. Trends Biotechnol. 26, 139–145.

Grusak, M.A., DellaPenna, D., 1999. Improving the nutrient composition of plants to enhance human nutrition and health. Plant Biol. 50, 133–161.

Guo, W.L., Chen, R.G., Gong, Z.H., Yin, Y.X., Ahmed, S.S., He, Y.M., 2012. Exogenous abscisic acid increases antioxidant enzymes and related gene expression in pepper (*Capsicum annuum*) leaves subjected to chilling stress. Gen. Mol. Res. 11, 4063–4080.

Guo, Z., Tan, H., Zhu, Z., Lu, S., Zhou, B., 2005. Effect of intermediates on ascorbic acid and oxalate biosynthesis of rice and in relation to its stress resistance. Plant Physiol. Biochem. 43, 955–962.

He, J., Duan, Y., Hua, D., Fan, G., Wang, L., Liu, Y., et al., 2012. DEXH box RNA helicase-mediated mitochondrial reactive oxygen species production in *Arabidopsis* mediates crosstalk between abscisic acid and auxin signaling. Plant Cell 24, 1815–1833.

Herbinger, K., Tausz, M., Wonisch, A., Soja, G., Sorger, A., Grill, D., 2002. Complex interactive effects of drought and ozone stress on the antioxidant defence systems of two wheat cultivars. Plant Physiol. Biochem. 40 (6), 691–696.

Hernández, I., Alegre, L., Munné-Bosch, S., 2004. Drought-induced changes in flavonoids and other low molecular weight antioxidants in *Cistus clusii* grown under Mediterranean field conditions. Tree Physiol. 24, 1303–1311.

Heuer, B., Yaniv, Z., Ravina, I., 2002. Effect of late salinization of chia (*Salvia hispanica*), stock (*Matthiola tricuspidata*) and evening primrose (*Oenothera biennis*) on their oil content and quality. Ind. Crop Prod. 15, 163–167.

Hirschberf, J., 1999. Production of high-value compounds: carotenoids and vitamin E. Curr. Opin. Biotechnol. 10, 186–191.

Hofius, D., Hajirezaei, M.R., Geiger, M., Tschiersch, H., Melzer, M., Sonnewald, U., 2004. RNAi-mediated tocopherol deficiency impairs photoassimilate export in transgenic potato plants. Plant Physiol. 135, 1256–1268.

Horner, H.T., Healy, R.A., Ren, G., Fritz, D., Klyne, A., Seames, C., 2007. Amyloplast to chromoplast conversion in developing ornamental tobacco floral nectaries provides sugar for nectar and antioxidants for protection. Am. J. Bot. 94, 12–24.

Horvath, G., Wessjohann, L., Bigirimana, J., Jansen, M., Guisez, Y., Caubergs, R., et al., 2006. Differential distribution of tocopherols and tocotrienols in photosynthetic and nonphotosynthetic tissues. Phytochem 67, 1185–1195.

Howitt, C.A., Pogson, B.J., 2006. Carotenoid accumulation and function in seeds and non-green tissues. Plant Cell Environ. 29, 435–445.

Huang, G.-Y., Wang, Y.-S., Sun, C.-C., Dong, J.-D., Sun, Z.-X., 2010. The effect of multiple heavy metals on ascorbate, glutathione and related enzymes in two mangrove plant seedlings (*Kandelia candel* and *Bruguiera gymnorrhiza*). Int. J. Oceanogr. Hydrobiol. 1, 11–25.

Jaleel, C.A., Sankar, B., Murali, P.V., Gomathinayagam, M., Lakshmanan, G.M., Panneerselvam, R., 2008. Water deficit stress effects on reactive oxygen metabolism in *Catharanthus roseus*; impacts on ajmalicine accumulation. Colloids Surf. B. Biointerfaces 62, 105–111.

Jiménez, A., Hernández, J.A., del Río, L.A., Sevilla, F., 1997. Evidence for the presence of the ascorbate-glutathione cycle in mitochondria and peroxisomes of pea leaves. Plant Physiol. 114, 275–284.

Jiménez, A., Hernández, J.A., Pastori, G., del Río, L.A., Sevilla, F., 1998. Role of the ascorbate-glutathione cycle of mitochondria and peroxisomes in the senescence of pea leaves. Plant Physiol. 118, 1327–1335.

Kamal-Eldin, A., Appelqvist, L.A., 1996. The chemistry and antioxidant properties of tocopherols and tocotrienols. Lipids 31, 671–701.

Kanofsky, J.R., Sima, P., 1991. Singlet oxygen production from the reactions of ozone with biological molecules. J. Biol. Chem. 266, 9039–9042.

Khan, A.L., Hamayun, M., Kim, Y.-H., Kang, S.-M., Lee, I.-J., 2011. Ameliorative symbiosis of endophyte (*Penicillium funiculosum* LHL06) under salt stress elevated plant growth of *Glycine* max L. Plant Physiol. Biochem. 49, 852–861.

Khan, T.A., Mazid, M., Mohammad, F., 2012. Potential of ascorbic acid against oxidative burst in plants under biotic stress: A review. J. Ind. Res. Technol. 2, 72–80.

Khanna-Chopra, R., 2011. Leaf senescence and abiotic stresses share reactive oxygen species-mediated chloroplast degradation. Protoplasma 249, 469–481.

Kocsy, G., Szalai, G., Galiba, G., 2002. Effect of heat stress on glutathione biosynthesis in wheat. Acta Biol. Szeged. 46, 71–72.

Kocsy, G., Szalai, G., Sutka, J., Paldi, E., Galiba, G., 2004. Heat tolerance together with heat stress-induced changes in glutatione and hydroxymethylglutathione levels is affected by chromosome 5A of wheat. Plant Sci. 166, 451–458.

Kollist, H., Moldau, H., Oksanen, E., Vapaavuori, E., 2001. Ascorbate transport from the apoplast to the symplast in intact leaves. Physiol. Plant 113, 377–383.

Kosová, K., Vítámvás, P., Prášil, I.T., Renaut, J., 2011. Plant proteome changes under abiotic stress-contribution of proteomics studies to understanding plant stress response. J. Proteom. 74, 1301–1322.

Kruk, J., Trebst, A., 2008. Plastoquinol as a singlet oxygen scavenger in photosystem II. Biochim. Biophys. Acta 1777, 154–162.

Kruk, J., Hollander-Czytko, H., Oettmeier, W., Trebst, A., 2005. Tocopherol as singlet oxygen scavenger in photosystem II. J. Plant Physiol. 162, 749–757.

Kumar, S., Gupta, D., Nayyar, H., 2012. Comparative response of maize and rice genotypes to heat stress: status of oxidative stress and antioxidants. Acta Physiol. Plant 34, 75–86.

Lascano, H.R., Antonicelli, G.E., Luna, C.M., Melchiorre, M.N., Gomez, L.D., Racca, R.W., et al., 2001. Antioxidant system response of different wheat cultivars under drought: field and *in vitro* studies. Austr. J. Plant Physiol. 28, 1095–1102.

Ledford, H.K., Baroli, I., Shin, J.W., Fischer, B.B., Eggen, R.I.L., Niyogi, K.K., 2004. Comparative profiling of lipid-soluble antioxidants and transcripts reveals two phases of photo-oxidative stress in a xanthophyll-deficient mutant of *Chlamydomonas reinhardtii*. Mol. Genet. Genomics 272, 470–479.

Lee, Y.P., Kim, S.H., Bang, J.W., Lee, H.S., Kwak, S.S., Kwon, S.Y., 2007. Enhanced tolerance to oxidative stress in transgenic tobacco plants expressing three antioxidant enzymes in chloroplasts. Plant Cell Rep. 26, 591–598.

Li, Q., Yu, B., Gao, Y., Dai, A.-H., Bai, J.-G., 2011. Cinnamic acid pretreatment mitigates chilling stress of cucumber leaves through altering antioxidant enzyme activity. J. Plant Physiol. 168, 927–934.

Li, S., Xue, X., Guo, W., Li, X., Chen, F., 2008. Study and application of negative pressure water supplying, controlling pot device and irrigation system. J. Shanghai Jiaotong Univ. 26, 478–482.

Lichtenthaler, H.K., 2007. Biosynthesis, accumulation and emission of carotenoids, alpha-tocopherol, plastoquinone, and isoprene in leaves under high photosynthetic irradiance. Photosynth. Res. 92, 163–179.

Lim, J.-H., Park, K.-J., Kim, B.-K., Jeong, J.-W., Kim, H.-J., 2012. Effect of salinity stress on phenolic compounds and carotenoids in buckwheat (*Fagopyrum esculentum* M.) sprout. Food Chem. 135, 1065–1070.

Lin, Y., Liu, Z., Shi, Q., Wang, X., Wei, M., Yang, F., 2012. Exogenous nitric oxide (NO) increased antioxidant capacity of cucumber hypocotyl and radicle under salt stress. Sci. Hortic. 142, 118–127.

Liso, R., de Tullio, M.C., Ciraci, S., Balestrini, R., La Rocca, N., Bruno, L., 2004. Localization of ascorbic acid, ascorbic acid oxidase and glutathione in roots of *Cucurbita maxima* L. J. Exp. Bot. 55, 2589–2597.

Liu, Z.-J., Zhang, X.-L., Bai, J.-G., Suo, B.-X., Xu, P.-L., Wang, L., 2009. Exogenous paraquat changes antioxidant enzyme activities and lipid peroxidation in drought-stressed cucumber leaves. Sci. Hort. 121, 138–143.

Loewus, F.A., Murthy, P.P.N., 2000. *myo*-Inositol metabolism in plants. Plant Sci. 150, 1–19.

Logan, B.A., 2005. Reactive oxygen species and photosynthesis. In: Smirnoff, N. (Ed.), Antioxidants and Reactive Oxygen Species in Plants. Blackwell, Oxford, pp. 250–267.

Lopez, A.B., Van Eck, J., Conlin, B.J., Paolillo, D.J., O'Neill, J., Li, L., 2008. Effect of the cauliflower Or transgene on carotenoid accumulation and chromoplast formation in transgenic potato tubers. J. Exp. Bot. 59, 213–223.

López-Huertas, E., Sandalio, L.M., Gómez, M., del Rio, L.A., 1997. Superoxide radical generation in peroxisomal membranes: evidence for the participation of the 18 kDa integral membrane polypeptide. Free Radic. Res. 26, 497–506.

Loreto, F., Centritto, M., Chartzoulakis, K., 2005. Photosynthetic limitations in olive cultivars with different sensitivity to salt stress. Plant Cell Environ. 26, 595–601.

Lukatkin, A.S., Brazaityte, A., Bobinas, C., Duchovski, P., 2012. Chilling injury in chilling-sensitive plants: a review. Agriculture 99, 111–124.

Ma, Y.-H., Ma, F.-W., Wang, Y.-H., Zhang, J.-K., 2011. The responses of the enzymes related with ascorbate-glutathione cycle during drought stress in apple leaves. Acta Physiol. Plant. 33, 173–180.

Maeda, H., DellaPenna, D., 2007. Tocopherol functions in photosynthetic organisms. Curr. Opin. Plant Biol. 10, 260–265.

Malik, S., Ashraf, M., 2012. Exogenous application of ascorbic acid stimulates growth and photosynthesis of wheat (*Triticum aestivum* L.) under drought. Soil Environ. 31, 72–77.

Mansour, M.M.F., Salama, K.H.A., 2004. Cellular basis of salinity tolerance in plants. Environ. Exp. Bot. 52, 113–122.

Mansour, M.M.F., Salama, K.H.A., Al-Mutawa, M.M., Abou Hadid, A.F., 2002. Effect of NaCl and polyamines on plasma membrane lipids of wheat roots. Biol. Plant. 45, 235–239.

Mansour, M.M.F., Salama, K.H.A., Ali, F.Z.M., Abou Hadid, A.F., 2005. Cell and plant responses to NaCl in (*Zea mays* L.) cultivars differing in salt tolerance. Gen. Appl. Plant Physiol. 31, 29–41.

Mascio, P.D., Murphy, M.E., Sies, H., 1991. Antioxidant defense systems: the role of carotenoids, tocopherols, and thiols. Am. J. Clin. Nutr. 53, 194–200.

Mellidou, I., Keulemans, J., Kanellis, A.K., Davey, M.W., 2012. Regulation of fruit ascorbic acid concentrations during ripening in high and low vitamin C tomato cultivars. BMC Plant Biol. 12, 239. 10.1186/1471–2229-12-239.

Mhamdi, A., Hager, J., Chaouch, S., Queval, G., Han, Y., Taconnat, L., et al., 2010. Arabidopsis glutathione reductase1 plays a crucial role in leaf responses to intracellular hydrogen peroxide and in ensuring appropriate gene expression through both salicylic acid and jasmonic acid signaling pathways. J. Plant Physiol. 153, 1144–1160.

Millar, A.H., Mittova, V., Kiddle, G., Heazlewood, J.L., Bartoli, C.G., Theodoulou, F.L., et al., 2003. Control of ascorbate acid synthesis by respiration and its implications for stress responses. Plant Physiol. 133, 443–447.

Miller, G., Suzuki, N., Ciftci-Yilmaz, S., Mittler, R., 2010. Reactive oxygen species homeostasis and signalling during drought and salinity stresses. Plant Cell Environ. 33, 453–467.

Mittler, R., Vanderauwera, S., Gollery, M., van Breusegem, F., 2004. The reactive oxygen gene network of plants. Trends Plant Sci. 9, 490–498.

Mittova, V., Theodoulou, F.L., Kiddle, G., Gomez, L., Volokita, M., Tal, M., et al., 2003. Coordinate induction of glutathione biosynthesis and glutathionemetabolizing enzymes is correlated with salt tolerance in tomato. FEBS Lett. 554, 417–421.

Mostafa, E.M., Ibrahim, M.M., 2012. HPLC analysis of non-enzymatic antioxidants in *Azolla caroliniana* (Pteridopsida) subjected to UV-B. Egypt. Acad. J. Biol. Sci. 3, 19–30.

Munné-Bosch, S., 2005. The role of α-tocopherol in plant stress tolerance. J. Plant Physiol. 162, 743–748.

Munné-Bosch, S., Alegre, L., 2000. Changes in carotenoids, tocopherols and diterpenes during drought and recovery, and the biological significance of chlorophyll loss in *Rosmarinus officinalis* plants. Planta 210, 925–931.

Munné-Bosch, S., Alegre, L., 2002. The function of tocopherols and tocotrienols in plants. Crit. Rev. Plant Sci. 21, 31–57.

Munns, R., Tester, M., 2008. Mechanisms of salinity tolerance. Ann. Rev. Plant Biol. 59, 651–681.

Müller-Moulé, P., Golan, T., Niyogi, K.K., 2004. Ascorbate-deficient mutants of *Arabidopsis* grow in high light despite chronic photooxidative stress. Plant Physiol. 134, 1163–1172.

Nair, A.S., Abraham, T.K., Jaya, D.S., 2008. Studies on the changes in lipid peroxidation and antioxidants in drought stress induced cowpea (*Vigna unguiculata* L.) varieties. J. Environ. Biol. 29, 689–691.

Noctor, G., 2006. Metabolic signalling in defence and stress: the central roles of soluble redox couples. Plant Physiol. 82, 1030–1035.

Noctor, G., Foyer, C.H., 1998. Ascorbate and glutathione: keeping active oxygen under control. Ann. Rev. Plant Physiol. Plant Mol. Biol. 49, 249–279.

Noctor, G., Arisi, A.M., Jouanin, L., Kunert, K.J., Rennenberg, H., Foyer, C.H., 1998. Glutathione: biosynthesis, metabolism and relationship to stress tolerance explored in transformed plants. J. Exp. Bot. 49, 623–647.

Noctor, G., Gomez, L., Vanacker, H., Foyer, C.H., 2002. Interactions between biosynthesis, compartmentation and transport in the control of glutathione homeostasis and signaling. J. Exp. Bot. 53, 1283–1304.

Noctor, G., Mhamdi, A., Chaouch, S., Han, Y., Neukermans, J., Marquez-Garcia, B., et al., 2011. Glutathione in plants: an integrated overview. Plant Cell Environ. 2, 454–484.

Noreen, S., Ashraf, M., 2009. Assessment of variation in antioxidative defense system in salt-treated pea (*Pisum sativum*) cultivars and its putative use as salinity tolerance markers. J. Plant Physiol. 166, 1764–1774.

Noreen, S., Ashraf, M., 2010. Modulation of salt (NaCl)-induced effects on oil composition and fatty acid profile of sunflower (*Helianthus annuus* L.) by exogenous application of salicylic acid. J. Sci. Food Agr. 90, 2608–2616.

Noreen, S., Ashraf, M., Hussain, M., Jamil, A., 2009. Exogenous application of salicylic acid enhances antioxidative capacity in salt stressed sunflower (*Helianthus annuus* L.) plants. Pak. J. Bot. 41, 473–479.

Ogawa, K., 2005. Gluthathione-associated regulation of plant growth and stress responses. Antiox. Red. Signal. 7, 973–981.

Oh, M.-M., Trick, H.N., Rajashekar, C.B., 2009. Secondary metabolism and antioxidants are involved in environmental adaptation and stress tolerance in lettuce. J. Plant Physiol. 166, 180–191.

O'Brien, J.A., Daudi, A., Butt, V.S., Bolwell, G.P., 2012. Reactive oxygen species and their role in plant defence and cell wall metabolism. Planta 236, 765–779.

Paradiso, A., Berardino, R., De Pinto, M.C., Sanità Di Toppi, L., Storelli, M.M., Tommasi, F., et al., 2008. Increase in ascorbate-glutathione metabolism as local and precocious systemic responses induced by cadmium in durum wheat plants. Plant Cell Physiol. 49, 362–374.

Parida, A.K., Das, A.B., 2005. Salt tolerance and salinity effects on plants: a review. Ecotoxicol. Environ. Saf. 60, 324–349.

Popov, V.N., Ruuge, E.K., Starkov, A.A., 2003. Effect of electron-transport inhibitors on the generation of reactive oxygen species by pea mitochondria during succinate oxidation. Biochem. (Mosc) 68, 727–751.

Quadrana, L., Almeida, J., Otaiza, S.N., Duffy, T., Corrêa da Silva, J.V., de Godoy, F., et al., 2013. Transcriptional regulation of tocopherol biosynthesis in tomato. Plant Mol. Biol. 81, 309–325.

Queiros, F., Rodrigues, J.A., Almeida, J.M., Almeida, D.P.F., Fidalgo, F., 2011. Differential responses of the antioxidant defence system and ultrastructure in a salt adapted potato cell line. Plant Physiol. Biochem. 49, 1410–1419.

Ramírez, L., Bartoli, C.G., Lamattina, L., 2013. Glutathione and ascorbic acid protect *Arabidopsis* plants against detrimental effects of iron deficiency. J. Exp. Bot. 64, 3169–3178.

Rastogi, A., Yadav, D.K., Szymańska, R., Kruk, J., Sedlářová, M., Pospíšil, P., 2013. Singlet oxygen scavenging activity of tocopherol and plastochromanol in *Arabidopsis thaliana*: relevance to photooxidative stress. Plant Cell Environ. Available from: http://doi.10.1111/pce.12161.

Ratkevicius, N., Correa, J.A., Moenne, A., 2003. Copper accumulation, synthesis of ascorbate and activation of ascorbate peroxidase in *Enteromorpha compressa* (L.) Grev. (Chlorophyta) from heavy metal-enriched environments in northern Chile. Plant Cell Environ. 26, 1599–1608.

Ratnayaka, H., Molin, W., Sterling, T., 2012. Comparison of physiological and antioxidant responses of *Anoda cristata* and cotton to progressive drought. Weed Res. 52, 358–366.

Rhoads, D.M., Umbach, A.L., Subbaiah, C.C., Siedow, J.N., 2006. Mitochondrial reactive oxygen species. Contribution to oxidative stress and interorganellar signaling. Plant Physiol. 141, 357–366.

Robinson, J.M., Bunce, J.A., 2000. Influence of drought-induced water stress on soybean and spinach leaf ascorbate-dehydroascorbate level and redox status. Int. J. Plant Sci. 161, 271–279.

Rodriguez, R., Redman, R., 2005. Balancing the generation and elimination of reactive oxygen species. Proc. Natl. Acad. Sci. USA 102, 3175–3176.

Ruiz, J.M., Blumwald, E., 2002. Salinity-induced glutathione synthesis in *Brassica napus*. Planta 214, 965–969.

Salama, K.H.A., Al-Mutawa, M.M., 2009. Glutathione-triggered mitigation in salt-induced alterations in plasmalemma of onion epidermal cells. Int. J. Agr. Biol. 11, 639–642.

Salama, K.H.A., Mansour, M.M.F., Ali, F.Z.M., Abou Hadid, A.F., 2007. NaCl-induced changes in plasma membrane lipids and proteins of (*Zea mays* L.) cultivars differing in their response to salinity. Acta Physiol. Plant. 29, 351–359.

Sánchez-Rodríguez, E., del Mar Rubio-Wilhelmi, M., Blasco, B., Leyva, R., Romero, L., Ruiz, J.M., 2012. Antioxidant response resides in the shoot in reciprocal grafts of drought-tolerant and drought-sensitive cultivars in tomato under water stress. Plant Sci. 188–189, 89–96.

Sandmann, G., Romer, S., Fraser, P.D., 2006. Understanding carotenoid metabolism as a necessity for genetic engineering of crop plants. Metabol. Eng. 8, 291–302.

Santini, J., Giannettini, J., Pailly, O., Herbette, S., Ollitrault, P., Berti, L., et al., 2013. Comparison of photosynthesis and antioxidant performance of several *Citrus* and *Fortunella* species (Rutaceae) under natural chilling stress. Trees 27, 71–83.

Shan, C.-J., Zhang, S.-L., Li, D.-F., Zhao, Y.-Z., Tian, X.-L., Zhao, X.-L., et al., 2011. Effects of exogenous hydrogen sulfide on the ascorbate and glutathione metabolism in wheat seedlings leaves under water stress. Acta Physiol. Plant. 33, 2533–2540.

Sharma, P., Jha, A.B., Dubey, R.S., Pessarakli, M., 2012. Reactive oxygen species, oxidative damage, and antioxidative defense mechanism in plants under stressful conditions. J. Bot. 2012, Available from: http://doi:10.1155/2012/217037.

Sharma, S.S., Dietz, K.-J., 2006. The significance of amino acids and amino acid-derived molecules in plant responses and adaptation to heavy metal stress. J. Exp. Bot. 57, 711–726.

Sheng, M., Tang, M., Chen, H., Yang, B.W., Zhang, F.F., Huang, Y.H., 2008. Influence of arbuscular mycorrhizae on photosynthesis and water status of maize plants under salt stress. Mycorrhiza 18, 287–296.

Sheng, M., Tang, M., Chen, H., Yang, B.W., Zhang, F.F., Huang, Y.H., 2009. Influence of arbuscular mycorrhizae on the root system of maize plants under salt stress. Can. J. Microbiol. 55, 879–886.

Shiu, C.T., Lee, T.M., 2005. Ultraviolet-B-induced oxidative stress and responses of the ascorbate-glutathione cycle in a marine macroalga *Ulva fasciata*. J. Exp. Bot. 56, 2851–2865.

Shu, D.-F., Wang, L.-Y., Duan, M., Deng, Y.-S., Meng, Q.-W., 2011. Antisense-mediated depletion of tomato chloroplast glutathione reductase enhances susceptibility to chilling stress. Plant Physiol. Biochem. 49, 1228–1237.

Singh, S., Gupta, A.K., Kaur, N., 2012. Differential responses of antioxidative defence system to long-term field drought in wheat (*Triticum aestivum* L.) genotypes differing in drought tolerance. J. Agron. Crop Sci. 198, 185–195.

Singh, S.P., Singh, Z., 2013. Dynamics of enzymatic and non-enzymatic antioxidants in Japanese plums during storage at safe and lethal temperatures. LWT – Food Sci. Technol. 50, 562–568.

Skłodowska, M., Gapińska, M., Gajewska, E., Gabara, B., 2009. Tocopherol content and enzymatic antioxidant activities in chloroplasts from NaCl-stressed tomato plants. Acta Physiol. Plant. 31, 393–400.

Smirnoff, N., Wheeler, G.L., 2000. Ascorbic acid in plants: biosynthesis and function. Crit. Rev. Biochem. Mol. Biol. 35, 291–314.

Smirnoff, N., Conklin, P.L., Loewus, F.A., 2001. Biosynthesis of ascorbic acid in plants: a renaissance. Ann. Rev. Plant Physiol. Plant Mol. Biol. 52, 437–467.

Smolikova, G.N., Laman, N.A., Boriskevich, O.V., 2011. Role of chlorophylls and carotenoids in seed tolerance to abiotic stressors. Rus. J. Plant Physiol. 58, 965–973.

Soltani, Y., Saffari, V.R., Moud, A.A.M., Mehrabani, M., 2012. Effect of foliar application of α-tocopherol and pyridoxine on vegetative growth, flowering, and some biochemical constituents of *Calendula officinalis* L. plants. Afr. J. Biotechnol. 11, 11931–11935.

Sun, Q., Wang, X.-R., Ding, S.-M., Yuan, X.-F., 2005. Effects of interaction between cadmium and plumbum on phytochelatins and glutathione production in wheat (*Triticum aestivum* L.). J. Integr. Plant Biol. 47, 435–442.

Sweetlove, L.J., Foyer, C.H., 2004. Roles for reactive oxygen species and antioxidants in plant mitochondria. In: Day, D.A., Millar, A.H., Whelan, J. (Eds.), Plant Mitochondria: From

Genome to Function, Vol. 1, Advances in Photosynthesis and Respiration. Kluwer Academic Press, Dordrecht, The Netherlands, pp. 307–320.

Szarka, A., Tomasskovics, B., Banhegyi, G., 2012. The ascorbate-glutathione-α-tocopherol triad in abiotic stress response. Int. J. Mol. Sci. 13, 4458–4483.

Szymanska, R., Kruk, J., 2008. Tocopherol content and isomers' composition in selected plant species. Plant Physiol. Biochem. 46, 29–33.

Tabata, K., Takaoka, T., Esaka, M., 2002. Gene expression of ascorbic acid-related enzymes in tobacco. Phytochem 61, 631–635.

Takahama, U., 2004. Oxidation of vacuolar and apoplastic phenolic substrates by peroxidase: physiological significance of the oxidation reactions. Phytochem. Rev. 3, 207–219.

Tamaoki, M., Mukai, F., Asai, N., Nakajima, N., Kubo, A., Aono, M., et al., 2003. Light-controlled expression of a gene encoding L-galactono-γ-lactone dehydrogenase which affects ascorbate pool size in *Arabidopsis thaliana*. Plant Sci. 164, 1111–1117.

Taylor, M., Ramsay, G., 2005. Carotenoid biosynthesis in plant storage organs: recent advances and prospects for improving plant food quality. Physiol. Plant 124, 143–151.

Tohidi-Moghadam, H.R., Shirani-Rad, A.H., Nour-Mohammadi, G., Habibi, D., Mashhadi-Akbar-Boojar, M., 2009. Effect of super absorbent application on antioxidant enzyme activities in canola (*Brassica napus* L.) cultivars under water stress conditions. Am. J. Agr. Biol. Sci. 4, 215–223.

Tokunaga, T., Miyahara, K., Tabata, K., Esaka, M., 2005. Generation and properties of ascorbic acid-overproducing transgenic tobacco cells expressing sense RNA for L-galactono-1,4-lactone dehydrogenase. Planta 220, 854–863.

Tounekti, T., Hernández, I., Müller, M., Khemira, H., Munné-Bosch, S., 2011a. Kinetin applications alleviate salt stress and improve the antioxidant composition of leaf extracts in *Salvia officinalis*. Plant Physiol. Biochem. 49, 1165–1176.

Tounekti, T., Vadel, A.M., Onate, M., Khemira, H., Munné-Bosch, S., 2011b. Salt-induced oxidative stress in rosemary plants: damage or protection? Environ. Exp. Bot. 71, 298–305.

Tripathy, B.C., Oelmüller, R., 2012. Reactive oxygen species generation and signaling in plants. Plant Signal. Behav. 7, 1621–1633.

Tuna, A.L., Kaya, C., Altunlu, H., Ashraf, M., 2013. Mitigation effects of non-enzymatic antioxidants in maize (*Zea mays* L.) plants under salinity stress. AJCS 7, 1181–1188.

Valpuesta, V., Botella, M.A., 2004. Biosynthesis of L-ascorbic acid in plants: new pathways for an old antioxidant. Trends Plant Sci. 9, 573–577.

Van Hove, L.W.A., Bossen, M.E., San Gabino, B.G., Sgreva, C., 2001. The ability of apoplastic ascorbate to protect poplar leaves against ambient ozone concentrations: a quantitative approach. Environ. Pollut. 114, 371–382.

Van Meulebroek, L., Vanhaecke, L., De Swaef, T., Steppe, K., De Brabander, H., 2012. U-HPLC-MS/MS to quantify liposoluble antioxidants in red-ripe tomatoes, grown under different salt stress levels. J. Agr. Food Chem. 60, 566–573.

Vanlerberghe, G.C., 2013. Alternative oxidase: a mitochondrial respiratory pathway to maintain metabolic and signaling homeostasis during abiotic and biotic stress in plants. Int. J. Mol. Sci. 14, 6805–6847.

Velasco, L., García-Navarro, E., Pérez-Vich, B., Fernández-Martínez, J.M., 2013. Selection for contrasting tocopherol content and profile in Ethiopian mustard. Plant Breed. 132, 694–700.

Wahid, A., Gelani, S., Ashraf, M., Foolad, M.R., 2007. Heat tolerance in plants: an overview. Environ. Exp. Bot. 61, 199–223.

Wang, F., Chen, F., Cai, Y., Zhang, G., Wu, F., 2011. Modulation of exogenous glutathione in ultrastructure and photosynthetic performance against Cd stress in the two barley genotypes differing in Cd tolerance. Biol. Trace Elem. Res. 144, 1275–1288.

Wang, H., Zhang, L., Ma, J., Li, X., Li, Y., Zhang, R., et al., 2010. Effects of water stress on reactive oxygen species generation and protection system in rice during grain-filling stage. Agr. Sci. China 9, 633–641.

Wang, Q., Wu, C., Xie, B., Liu, Y., Cui, J., Chen, G., et al., 2012a. Model analysing the antioxidant responses of leaves and roots of switchgrass to NaCl-salinity stress. Plant Physiol. Biochem. 58, 288–296.

Wang, S., Liang, D., Li, C., Hao, Y., Ma, F., Shu, H., 2012b. Influence of drought stress on the cellular ultrastructure and antioxidant system in leaves of drought-tolerant and drought-sensitive apple rootstocks. Plant Physiol. Biochem. 51, 81–89.

Wolucka, B.A., Van Montagu, M., 2003. GDP-Mannose 3′,5′-epimerase forms GDP-L-gulose, a putative intermediate for the novo biosynthesis of vitamin C in plants. J. Biol. Chem. 278, 47483–47490.

Wurbs, D., Ruf, S., Bock, R., 2007. Contained metabolic engineering in tomatoes by expression of carotenoid biosynthesis genes from the plastid genome. Plant J. 49, 276–288.

Xiang, C., Werner, B.L., Christensen, E.M., Oliver, D.J., 2001. The biological functions of glutathione revisited in *Arabidopsis* transgenic plants with altered glutathione levels. Plant Physiol. 126, 564–574.

Xiao, X., Xu, X., Yang, F., 2008. Adaptive responses to progressive drought stress in two *Populus cathayana* populations. Silva Fennica 42, 705–719.

Xie, Z., Duan, L., Tian, X., Wang, B., Egrinya-Eneji, A., Li, Z., 2008. Coronatine alleviates salinity stress in cotton by improving the antioxidative defense system sand radical-scavenging activity. J. Plant Physiol. 165, 375–384.

Yazdanpanah, S., Baghizadeh, A., Abbassi, F., 2011. The interaction between drought stress and salicylic and ascorbic acids on some biochemical characteristics of *Satureja hortensis*. Afr. J. Agr. Res. 6 (4), 798–807.

Zaefyzadeh, M., Quliyev, R.A., Babayeva, S.M., Abbasov, M.A., 2009. The effect of the interaction between genotypes and drought stress on the superoxide dismutase and chlorophyll content in durum wheat landraces. Turk. J. Biol. 33, 1–7.

Zechmann, B., 2011. Subcellular distribution of ascorbate in plants. Plant Signal. Behav. 6, 360–363.

Zechmann, B., Stumpe, M., Mauch, F., 2011. Immunocytochemical determination of the subcellular distribution of ascorbate in plants. Planta 233, 1–12.

Zhang, W., Tian, Z., Pan, X., Zhao, X., Wang, F., 2013. Oxidative stress and non-enzymatic antioxidants in leaves of three edible canna cultivars under drought stress. Hort. Environ. Biotechnol. 54, 1–8.

Zhou, J., Wang, J., Shi, K., Xia, X.J., Zhou, Y.H., Yu, J.Q., 2012. Hydrogen peroxide is involved in the cold acclimation-induced chilling tolerance of tomato plants. Plant Physiol. Biochem. 60, 141–149.

Zhu, S., Zhang, X., Luo, T., Liu, Q., Tang, Z., Jing, Z., 2011. Effects of NaCl stress on seed germination, early seedling growth and physiological characteristics of cauliflower (*Brassica oleracea* L. var. *botrytis* L.). Afr. J. Biotechnol. 10, 17940–17947.

Zurbriggen, M.D., Carrillo, N., Tognetti, V.B., Melzer, M., Peisker, M., Hause, B., et al., 2009. Chloroplast-generated reactive oxygen species play a major role in localized cell death during the non-host interaction between tobacco and *Xanthomonas campestris* pv. *vesicatoria*. Plant J. 60, 962–973.

Chapter 8

Ascorbic Acid

A Potent Defender Against Environmental Stresses

Ramanpreet Kaur and Harsh Nayyar

8.1 INTRODUCTION

Ascorbic acid (AsA), a powerful antioxidant in plants, is a reduced form of vitamin C having a molecular mass of 176.12 g mol^{-1}. It is water soluble and appears in the form of white to light yellow crystals or powder. Since, the vitamin is water-soluble, it is commonly found in fresh fruits, mainly in the citrus family, which is dominated by oranges, lemons, limes, and tangerines, and is also abundant in green leafy vegetables. Due to its strong reducing ability, L-AsA is used as an additive in foods and beverages (Mahmoud et al., 2013). The ancient Greeks, Egyptians and Romans reported that a deficiency of vitamin C causes the disease scurvy among humans. In the 1750s, James Lind discovered that daily consumption of fresh fruits and vegetables helps in curing the disease (Davey et al., 2000). Vitamin C is perhaps the most popular vitamin among the common nutrients and biochemicals. In 1932, Haworth confirmed the structure of vitamin C and it was first synthesized by Reichstein and co-workers in 1933 (Reichstein et al., 1933). Chemically, the vitamin occurs in three isomeric forms: L-ascorbic acid, D-arabo-ascorbic acid and L-arabo-ascorbic acid. Among these, the commercially used one is L-ascorbic acid (L-AsA).

8.2 CHEMISTRY

The structure of L-ascorbic acid is simple: a six carbon ring (aldono-1,4-lactone of a hexonic acid) (Davey et al., 2000). L-ascorbic acid can be reversibly converted to D-ascorbic acid, which is known as dehydroascorbic acid. Ascorbic acid is synthesized by almost all higher plant species, whereas only animals having the capacity to oxidize L-gulono-1,4-lactone can synthesize ascorbic acid. Ascorbic acid is a strong antioxidant and is known mainly for its antioxidant properties, i.e. it protects the plants from oxidative damage by

P. Ahmad (Ed): Oxidative Damage to Plants. DOI: http://dx.doi.org/10.1016/B978-0-12-799963-0.00008-3

acting as a substrate in the reactions catalyzed by ascorbate peroxidase enzymes. An efficient biological antioxidant is that which not only reacts with free radicals but which also possesses some additional properties: it should be present in sufficient amount, and should be able to regenerate (Arrigoni and Tullio, 2002). Ascorbic acid possesses these properties, hence it is considered to be a perfect antioxidant for all aerobic organisms. At the same time, recent reports suggest that the free radicals and reactive oxygen species play an important role in signal transduction, so the main function of an antioxidant is fine regulation rather than complete destruction of all free radicals (Schreck et al., 1991 and Uhlin and Eklund, 1994). The content of ascorbate is high in meristematic tissues (Loewus and Loewus, 1987; Smirnoff and Pallanca, 1996), flowers or young fruits (Abhagdow et al., 2007; Hancock et al., 2007); root tips (Arrigoni, 1994) and apices of stolons or tubers (Tedone et al., 2004). Factors responsible for accumulation of AsA in staple crops are only partially understood (Sanahuja et al., 2013). The fruit of *Myrciaria dubia* (*Family*: Myrtaceae) is considered to be one of the richest in ascorbate content (Justr et al., 2000).

8.3 BIOSYNTHESIS OF ASCORBIC ACID

8.3.1 Biosynthesis of Ascorbic Acid in Plants

In the case of plants, three different biosynthetic pathways have been proposed for ascorbic acid synthesis as shown in Fig. 8.1. Fluxes through different pathways depend on the specialization and metabolic regulation of the cell. One of the sequences for the biosynthesis of ascorbic acid is via the *Smirnoff-Wheeler pathway*, which is an energy dependent pathway. The various steps for this pathway involve the conversion of GDP-D-mannose to GDP-L-galactose by the enzyme GDP-mannose-3',5'-epimerase (Wheeler et al., 1998). GDP-L-galactose gets converted to L-galactose, which is an immediate precursor of L-galactono-1,4-lactone and this L-galactono-1,4-lactone is converted to L-ascorbic acid by galactono-1,4-lactone dehydrogenase. The enzyme galactono-1,4-lactone dehydrogenase is localized on the inner membrane of the mitochondria, whereas all others are found in cytoplasm (Smirnoff et al., 2001). Synthesis through this pathway does not involve inversion of the carbon chain, as shown experimentally in strawberry fruits (Loewus et al., 1956). Instead of forming GDP-L-galactose by GDP-mannose-3',5'-epimerase it may form GDP-L-gulose, depending upon the molecular form of the enzyme (Wolucka and Van Montagu, 2003; Valpuesta and Botella, 2004). There is another branch of the Smirnoff-Wheeler pathway which is known as the Wolucka-Van Montagu (WVM) pathway. Thus, GDP-L-gulose is a new intermediate in L-ascorbic acid synthesis, and it is mainly present during biosynthesis of ascorbic acid in the case of animals, which indicates the establishment of the animal pathway in plants too.

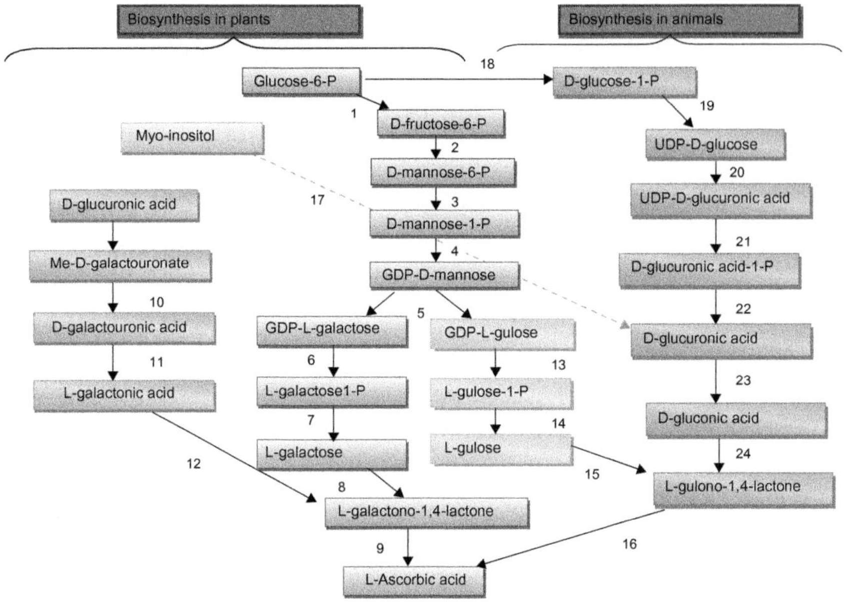

FIGURE 8.1 Multiple biosynthetic pathways in plant and a single biosynthetic pathway in animal; this also shows that the animal pathway might be functional in plants too. The various enzymes catalyzing the reaction are 1) glucose -6-phosphate isomerase; 2) mannose-6-phosphate isomerase; 3) phosphomannosemutase; 4) GDP-mannose pyrophosphorylase; 5) GDP-mannose 3,5 epimerase; 6) phosphodiesterase; 7) sugar phosphatase; 8) L-galactose dehydrogenase; 9) L-galactono-1,4-lactone dehydrogenase; 10) methyl esterase; 11) D-galactouronate reductase; 12) aldono-lactonase; 13) phosphodiesterases; 14) sugar phosphatase; 15) L-gulose dehydrogenase; 16) gulono-1,4-lactone dehydrogenase; 17) myo-inositol oxygenase; 18) phosphoglucomutase; 19) UDP-glucose pyrophosphorylase; 20) UDP-glucose dehydrogenase; 21) glucuronate-1-phosphate uridylyl transferase; 22) glucurono kinase; 23) glucuronate reductase; 24) aldono-lactonase.

In lettuce and tobacco plants, there was a 4- to 7-fold increase in ascorbic acid content when the gene from rat encoding the L-gulono-1,4-lactone (L-GulL) oxidase was expressed in these plants. This enzyme leads to the conversion of the penultimate step into ascorbic acid (Jain and Nessler, 2000). However, it is still unknown whether the action of this enzyme is to synthesize L-GulL or to act on plant L-GalL substrate.

A second pathway is the *uronic acid pathway* or the *salvage pathway* (Isherwood et al., 1954), which starts from the methyl ester of D-galacturonic acid that gets converted to D-galacturonic acid. This is reduced to L-galactonic acid, which leads to ascorbic acid through L-galactono-1,4-lactone (Agius et al., 2003). This pathway shows the inversion of the carbon chain as in the case of detached ripening strawberry fruits (Agius et al., 2003). Moreover, large amounts of L-ascorbic acid are synthesized from methyl ester of D-galacturonic acid, as in the case of cress (Isherwood et al., 1954) and *Arabidopsis* cultured cells (Davey et al., 1999). The biosynthesis through this pathway depends upon the availability of D-galacturonic acid, which is the end product of cell wall pectins in senescent cells. This pathway was proposed after the identification of the enzyme galactouronate acid reductase. For example, in ripening fruit tissue of strawberry, the enzyme D-galacturonic reductase is found (Agius et al., 2003).

Molecular and biochemical evidence has suggested a third pathway for biosynthesis of ascorbic acid, which starts from myo-inositol (Loewus and Kelly, 1961) followed by D-glucuronic acid, L-gulonic acid, L-gulono-1,4-lactone to form ascorbic acid. This acts as an alternate pathway for biosynthesis of ascorbic acid in plants. This pathway was proposed after the identification and cloning of myo-inositol oxygenase from *Arabidopsis thaliana* (Lorence et al., 2004). Out of these three pathways the most important AsA biosynthetic pathway is via the D-Man/L-Galactose pathway (Zhang et al., 2012). In plants, through various pathways, L-AsA is synthesized; however, the participation of different pathways for the synthesis of AsA depends upon the species, organs and developmental stages. In the early stage of development of achene, i.e. strawberry fruit, AsA plays an important role (Araguez et al., 2013). In strawberry (*Fragaria* spp) during fruit ripening the content of AsA was increased, due to which the transcription of genes encoding biosynthetic enzymes such as D-galacturonate reductase (FaGalUR) and myo-inositol oxygenase (FaMIOX), and the AsA recycling enzyme monodehydroascorbate reductase (FaMDHAR) were also increased (Cruz-Rus et al., 2011). The levels of AsA content and gene expression were higher in skin than in pulp in fully ripe berries (*Vaccinium myrtililus* L.) (Spinard et al., 2013). Sweet pepper (*Capsicum annuum*) is considered to have high content of ascorbic acid; however, earlier the metabolic pathway of AsA in pepper was not known. Alos and his co-workers in 2013 have made a contribution in the discovery of AsA biosynthesis, degradation and recycling pathways of pepper fruits (Alós et al., 2013).

8.3.2 Biosynthesis of Ascorbic Acid in Animals

In contrast to plants, only a single pathway exists for the biosynthesis of ascorbic acid in animals (Fig. 8.1). Burns and Mosbach in 1956 were the first to report on the biosynthesis of L-ascorbic acid from D-glucose, the various intermediates of D-glucuronic acid, L-gulonic acid, L-gulono-1,4-lactone, and the penultimate step being conversion of L-gulono-1,4-lactone to ascorbic acid by L-gulono-1,4-lactone oxidase (Chatterjee, 1973). In this pathway glucose, the primary substrate, is inverted in the end product (Burns, 1967). *De novo* synthesis of L-ascorbic acid in liver and kidney having L-gulono-1,4-lactone oxidase occurs via the hexauronic acid pathway. Humans, nonhuman primates and some other mammals such as guinea pig cannot synthesize ascorbic acid as the gene encoding the enzyme L-gulono-1,4-lactone oxidase, the last enzyme of the biosynthesis pathway, is mutated and has lost its functional form (Chatterjee, 1973).

8.3.3 Biosynthesis of Ascorbic Acid in Algae and Fungi

It has been described that green algae (Chlorophyta) and some nongreen algae follow the same Smirnoff-Wheeler pathway as followed by most of the plants for the synthesis of ascorbic acid (Renstrøm et al., 1982/1983) whereas other small photosynthetic organisms such as Bacillariophyta, Chrysophyta, protists do not show Smirnoff-Wheeler synthesis of ascorbic acid (Helsper et al., 1982; Grün and Loewus, 1984). In Ascomycota fungi such as *Candida*, *Saccharomyces* and *Sclerotia* an analogue of ascorbic acid is present, such as D-erythroascorbic acid instead of ascorbic acid (Hancock et al., 2000, Spickett et al., 2000) and the various steps for the synthesis of D-erythroascorbic acid include the conversion of D- arabinose to D-arabinose-1,5-lactone, followed by D-arabinose-1,4-lactone and finally D-erythroascorbic acid.

The ascorbate content varies among lower and higher plants as indicated in Table 8.1 (Gest et al., 2013). During the course of evolution, the concentration of AsA increased; in cyanobacteria no AsA have been found or only low content is found, whereas in higher plants a high content of AsA is observed in different compartments of the cell and in different organs as well (Tel-Or et al., 1986; Obinger et al., 1998).

TABLE 8.1 Ascorbate Content in Different Organisms Reported from the Study of Gest et al., 2013 which Showed that Concentration of Ascorbate was Several Times Lower in Cyanobacteria, Mosses, Liverworts, Algae as Compared to Higher Plants

Organisms		Conc. of Ascorbate g/FW	References
Cyanobacteria		Either no or very low	Tel-Or et al., 1986; Obinger et al., 1998
Mosses	a) *Hypnum plumaeforme*	0.1–0.6 μmol	Sun et al., 2010a
	b) *Brachythecum velutinum*	0.25–0.5 μmol	Paciolla and Tommasi, 2003
Liverworts		0.3 μmol	Paciolla and Tommasi, 2003
Algae	*Ulva compressa*	0.5 μmol	Shiu and Lee, 2005
	U. fasciata	0.5 μmol	Mellado et al., 2012
Arabidopsis thaliana		5 μmol	Kotchoni et al., 2009

As compared to lowland plants the chloroplasts of alpine plants contain up to 10 times more ascorbate, which might be due to high light, UV and low temperature of the alpine environment (Streb et al., 1997).

8.4 REGULATORY MECHANISM OF SYNTHESIS

In plants, AsA is synthesized in mitochondria and consequently transported to other compartments (Horemans et al., 2000). It is reported that most of the AsA is localized in the cytoplasm (Pignocchi et al., 2003), up to 10% in apoplast (Noctor and Foyer, 1998) and 12–13% found in chloroplasts (Horemans et al., 2000). In spite of this high and wide-ranged value, the multiple roles played by this molecule in crucial physiological processes mean that the steady-state values are tightly controlled (Smirnoff, 2000). The biosynthesis, degradation, recycling and transport of L-ascorbic acid is responsible for maintaining the steady-state level in plants.

In plants, the AsA level depends upon genotype (Cocetta et al., 2012) and on some of the environmental factors like light (Badejo et al., 2009; Li et al., 2009), harvesting date (Felicetti and Mattheis, 2010) and on certain biotic (Davey et al., 2007) and abiotic stresses (Gauter et al., 2010). The AsA levels showed progressive increase during fruit development in tomato, grapes and strawberry (Ioannnidi et al., 2009; Cruz-Rus et al., 2010, 2011). Ascorbic acid plays an important role in various processes so its level needs to be maintained by some mechanisms (Ishikawa et al., 2006) such as:

1. Modification of the enzyme activity by feedback inhibition (Pallanca and Smirnoff, 2000).
2. Expression of the gene varies according to the developmental and environment state such as light level (Smirnoff, 2000, Tamaoli, 2003), senescence (Bartoli et al., 2000), germination (Pallanca and Smirnoff, 1999) and fruit ripening (Aguis et al., 2003, Pateraki et al., 2004).
3. Regeneration of the oxidized state.
4. Transport and compartmentation of the molecule (Davey et al., 2000).

The effect of light and the feedback inhibition by L-ascorbic acid are the two main factors that are analyzed in order to control the level of L-ascorbic acid. Tabata and his co-workers (2002) worked on tobacco suspension cells and noticed that the exogenous application of L-ascorbic acid decreases the encoding capacity of genes for two enzymes such as GDP-mannose pyrophosphorylase and L-galactono-1,4-lactone dehydrogenase. Moreover, similar responses of decrease in encoding capacity of genes for enzymes GDP-mannose pyrophosphorylase and L-galactono-1,4-lactone dehydrogenase have been reported by Tabata et al., 2002 in non-irradiated tobacco leaves. In dark, the size of the ascorbate pool decreases and it can be reversed by sucrose or glucose feeding, as in the case of barley and *Arabidopsis thaliana* (Smirnoff and Pallanca, 1996; Conklin et al., 1997, 2000). The deficiency of boron reduces the ascorbate pool as in *Cucurbita* roots (Lukaszewski and Blevins, 1996). It is either due to its effects on ascorbate synthesis or turnover. Atkinson and his co-workers (2013) have demonstrated a link between the source of AsA, i.e. leaves, and the sink, i.e. fruits. The accumulation of L-AsA changes with the change in leaf and fruit number. They have reported that with the removal of leaf or girdling of phloem, the fruit weight gets reduced.

8.5 CATABOLISM OF ASCORBIC ACID

The turnover rate of ascorbate is significant (Smirnoff 2000). This is supported by the evidence that the rate of loss of the ascorbate pool from the leaves of *Arabidopsis thaliana* is 2.5% per hour (Conklin et al., 1997). The turnover rate of ascorbate in germinating pea seedlings is 13% per hour (Pallanca and Smirnoff, 2000). L-tartrate and oxalate are the metabolized products of ascorbate (Loewus, 1999) and their formation differs among species. In the grape family (*vitaceae*), tartrate is formed from the ascorbate by cleavage at C4/5 carbon skeletons whereas in the rest of the species by C2/C3 cleavage, oxalate and L-threonate is formed. The L-threonate can either be converted to L-tartrate or decarboxylate to L-glycerate. It is stated that cleavage at C2/C3 for oxalate formation involves oxygenase and hydrolase enzyme activity, whereas cleavage at C4/C5 for tartrate formation involves only a hydrolytic reaction (Saito et al., 1997). Various plant species store oxalate in the form of calcium oxalate crystals in specialized cells (Idioblasts). This oxalate may also be formed from glycolate by the photorespiratory mechanism (Loewus, 1999; Smirnoff and Wheeler., 2000). Keates et al. (2000) reported that ascorbate is the pioneer of oxalate crystals by studying ^{14}C-L-galactose in the oxalate accumulating plant *Pistia stratiotes*.

8.6 ABIOTIC STRESSES, OXIDATIVE DAMAGE AND ANTIOXIDANTS INCLUDING ASCORBIC ACID

Stress in any form whether biotic or abiotic affects the regular functioning of the plant. Under normal conditions, the plant produces various reactive oxygen species (ROS) such as hydrogen peroxide (H_2O_2), singlet oxygen ($^1O_2$), superoxide anion ($O_2^{\bullet -}$), and hydroxyl radical ($^{\bullet}OH$) (Mittler, 2002), in different organelles of the cell as shown in Table 8.2. The level of these ROS is elevated during stress. The production of ROS affects the normal functioning of cells by destroying the macromolecules such as DNA, proteins and lipids (Foyer and Fletcher, 2001). The free radicals and reaction oxygen species also play an important role in signal transduction and thus activate defense pathways (Desikin, 2001; Knight and Knight, 2001).

Plants have developed various antioxidant mechanisms to cope with ROS which involve enzymatic and nonenzymatic antioxidants (Bartoli et al., 2004),

TABLE 8.2 Production of Reactive Oxygen Species in Different Compartments

Organelle	Different Pathways for ROS Production	References
Chloroplast	PSI, PSII Electron transport of photosynthesis, Excited chlorophyll	Asada and Takahashi, 1987 Asada, 1999
Mitochondria	Electron transport during respiration	Maxwell et al., 1999 Dat et al., 2000
Peroxisomes	Glycolate oxidase, Fatty acid β-Oxidation,Xanthine oxidase	Corpas, 2001
Plasma membrane	NADPH oxidase	Hammond-Kosack and Jones, 1996 Grant and Loake, 2000
Apoplast	Oxalate oxidase Amine oxidase	Allan and Fluhr, 1997 Dat et al., 2000
Cell wall	Peroxidases, $Mn^{2+}$and NADH	Hammond-Kosack Jones, 1996 Grant and Loake, 2000

along with functions and locations of each one as given in Table 8.3 and Table 8.4. ROS and antioxidants such as ascorbate, glutathione, tocopherols, carotenoids gather information from many pathways, e.g. hormones and sugars, and form a part of a much larger stress signaling network that regulates plant growth and defense responses (Bosch et al., 2013).

Out of various nonenzymatic antioxidants, ascorbic acid is a vital component and plays a central role in the removal of hydrogen peroxide (H_2O_2) (Foyer, 1993). It has been found to have many other functions that may assist plants under stress conditions, as explained in the following sections.

8.7 FUNCTIONS OF ASCORBIC ACID

Ascorbic acid has been implicated in several functions and protects the plant from stress in both photosynthetic and nonphotosynthetic tissues (Smirnoff, 1996; Noctor and Foyer, 1998) as elaborated in the following sections.

8.7.1 As an Enzyme Cofactor

The enzymes such as 2-oxoglutarate and Fe (II)-dependent oxygenases require ascorbate for their function, as it maintains the iron as Fe (II) by acting as a reductant (Prescott and John, 1996). In the case of animals, the enzymes participating in collagen formation, such as Fe-dioxygenases

TABLE 8.3 Function of Various Enzymatic Antioxidants and their Localization

Antioxidant Enzymes	Localization	Function	References
Superoxide dismutase (SOD)	Found in almost all organelles such as Chl, Cyt, Mit, Per, Apoplast	It helps in scavenging of superoxide $O_2^{\bullet -}$ and converts into H_2O_2 thus reduces the formation of hydroxyl radical ($^{\bullet}OH$) which is highly toxic.	Scandalios, 1993; Bowler, 1992
Ascorbate peroxidase	Chl, Cyt, Mit, Per, Apoplast	It helps in scavenging H_2O_2	Asada K and Takahashi, 1987; Asada K, 1999
Catalase	Per, Gly	It also helps in removing H_2O_2	Mittler, 2002
Glutathione reductase (GR)	Chl, Cyt, Mit	It helps in reducing the glutathione	Edwards et al., 1990; Creissen et al., 1994
Dehydroascorbate reductase (DHAR)	Chl, Mit, Per	Regeneration of ascorbate from dehydroascorbate	Asada, 1994
Monodehydro-ascorbate reductase (MDHAR)	Chl, Mit, Cyt	Reduction of MDHA to give rise to ascorbate	Jimenez et al., 1997

(Padh, 1990; Jung and Hand Wells, 1997), also require ascorbic acid for maximal activity. The study of Ramirez and his co-workers (2013) reports that both AsA and GSH provide protection to *Arabidopsis* plant against the harmful effects of iron deficiency. AsA and GSH increased the activity of heme protein ascorbate peroxidase without increasing the internal iron concentration. For proper functioning of fibroblast and osteoblast, the body also needs L-ascorbic acid. It is required for the synthesis of hormones such as folate, carnitine, adrenal and it is an effective wound-healing vitamin. In the case of plants, it plays an important role in photosynthesis as an enzyme cofactor (including synthesis of ethylene, gibberellins, flavonoids and anthocyanins). The content of ascorbate and anthocyanin of plant tissues are correlated, e.g. low concentration of ascorbate reduces the accumulation of anthocyanin (Page et al., 2012). The enzymes needed for hormone synthesis like 1-aminocyclopropane-1-carboxylate oxidase and gibberellins-20-oxidase for ethylene and gibberellins synthesis, respectively, require 2-oxoglutarate dependent oxygenases (Smirnoff, 2000). In chloroplasts, a high concentration of ascorbate is found as compared to other compartments. H_2O_2 formed in

TABLE 8.4 Function of Various Nonenzymatic Antioxidants and their Localization

Nonenzymatic Antioxidant	Localization	Function	References
α-tocopherol	Membranes	It removes $^{\bullet}OH$ radicals, superoxide radical, singlet oxygen.	Asada and Takahashi, 1987
Carotenoids	Chl	It scavenges singlet oxygen and protects chlorophyll from photoxidation	Neubauer and Yamamoto 1992; Demmig-Adams and Adams, 1992
Glutathione	Chl, Cyt, Mit, Per, Apo	It scavenges peroxides	Asada, 1999; Noctor and Foyer, 1998
Ascorbic acid	Chl, Cyt, Mit, Per, Apo	It plays a central role in the removal of hydrogen peroxide (H_2O_2) and singlet oxygen	Asada, 1999; Noctor and Foyer, 1998

Abbreviations: Apo- Apoplast; Chl- Chloroplast; Cyt-Cytosol; Mit-Mitochondria; Per- Peroxisome; Gly-Glyoxisome.

chloroplast by Mehler's reaction is removed by ascorbate peroxidase (Miyake and Asada, 1992). Ascorbate also acts as a cofactor for the enzyme violaxanthin de-epoxidase that is found on the lumen side of the thylakoid membrane (Neubauer and Yamamoto, 1992). It helps in the formation of zeaxanthin (a photoprotectant) by participating in electron flow. L-ascorbic acid also participates in the synthesis of hydroxyproline-rich proteins, which are glycoproteins containing extensins and are located in the cell wall (Fry, 1986). In response to injury and pathogen attack these extensins participate in cross-linking of the cell wall (Carpita and Gibeaut, 1993; Reiter, 1998). The proteins rich in hydroxyproline help in the expansion of the cell and its division (Arrigoni et al., 1997). Ascorbic acid is also a cofactor for prolyl hydroxylases, which are involved in the synthesis of hydroxylysine (Davies et al., 1991).

8.7.1.1 Role of Ascorbate in the Thylakoid Lumen as a Cofactor

Ascorbate serves in the following roles in the thylakoid lumen:

1. It functions as a cofactor for violaxanthin–de-epoxidase (VDE), the enzyme involved in the formation of nonphotochemical quenching. Thus, it is necessary for photoprotection (Toth et al., 2013).
2. Once the oxygen-evolving complex is inactivated in the thylakoid lumen, the ascorbate acts as an alternate electron donor of PSII and protects the photosynthetic machinery by slowing down the donor side induced

photoinactivation, i.e. in short, as electron donor to PSII and to a lesser extent PSI (Toth et al., 2013).

The oxygen-evolving complex (OEC) of PSII is particularly sensitive to heat stress. The OEC comprises four oxygen bridged Mn atoms, Ca^{+2} and putatively Cl^{-}, which is liganded to the protein matrix formed by the core proteins, viz. D1, D2, CP43, CP47, and extrinsic proteins, viz. PsbO, PsbP and PsbQ (Umena et al., 2011; Grundmeier and Dau, 2012). Upon heat stress, OEC becomes functionally inactive mainly because of release of extrinsic protein PsbO and at least two manganese ions (Nash et al., 1985 and Enami et al., 1994). This in turn leads to loss of Cl^{-} (Popelkova and Yoeum, 2011), which ultimately affects the binding of Ca^{+2} (Bricker and Frankel, 2008). Once the Mn-cluster of OEC is destroyed, it has been found that AsA acts as an efficient e^{-} donor.

8.7.2 Cell growth and expansion

Ascorbate and ascorbate oxidase (AO) (the enzyme that oxidizes ascorbic acid) are found in apoplast and have a strong connection between cell metabolism and cell expansion (Smirnoff, 1996). Ascorbate is oxidized by a member of blue-colored copper oxidase glycoprotein, such as AO, into water and monodehydroascorbate (MDHA) (Esaka et al., 1990; O'Malley et al., 1993). It was reported that during early germination of pea seedling there was very low AO activity found in embryonic axis but as the growth proceeds the activity of AO enhanced (Pallanca and Smirnoff, 1999). As the plant approaches towards its final period of expansion, the activity of AO and the level of protein and mRNA decrease (Lin and Varner, 1991; Esaka et al., 1992). This action of AO is influenced by the light and the auxin. Auxin enhances the activity of AO and its mRNA level (Esaka et al., 1992; Takahama, 1994). Far red light in case of dark grown mustard seedling also enhances AO function (Hayashi and Morohashi, 1993). Smirnoff (1996) reported that the oxidation products of ascorbate such as MDHA and dehydroascorbate (DHA) are directly involved in expansion of the cell. In Onion roots, expansion of the cell and the growth is stimulated by MDHA (Hidalgo et al., 1989; Gonzalez-Reyes, 1994, 1995). As shown in Fig. 8.2, MDHA acts as an electron acceptor from the cytochrome b system, which in turn activates the H^{+}ATPase by depolarizing the plasma membrane, which in turn activates the cell wall loosening enzymes. Ascorbate also inhibits the activity of peroxidases either by detoxification or by scavenging. This in turn decreases the cross-linking activity of wall polymers by peroxidases and thus results in cell expansion (Cordoba-Pedregosa et al., 1996).

8.7.3 Cell division

The exogenous supply of ascorbic acid increased the number of cells passing from G_1 to S phase in the meristem and pericycle of the onion root

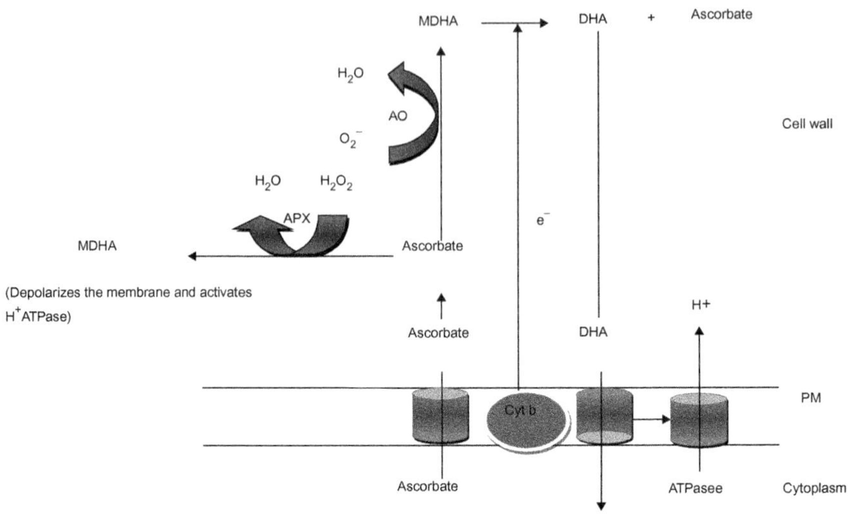

FIGURE 8.2 This model shows the involvement of ascorbic acid in cell expansion. Ascorbate is transported into cell wall via a carrier on the plasma membrane. Ascorbate oxidase present in the cell wall oxidizes the ascorbate to monodehydroascorbate. Membrane bound cytochrome b gives electrons to MDHA. The transmembrane electron transport results in activation of H^+ ATPase and causes loosening of cell wall. DHA inhibits cross linking to matrix polysaccharides. DHA is transported to cytoplasm via DHA carrier and is reduced to ascorbate by dehydroascorbate reductase.
Abbreviations: AO, Ascorbate oxidase; APX, Ascorbate peroxidase; PM, Plasma membrane; DHA, Dehydroascorbate; MDHA, Monodehydroascorbate.

(Arrigoni, 1994; Citterio et al., 1994). However, the number of cells in the quiescent center of onion cells decreased on ascorbic acid application (Innocenti et al., 1990). In the quiescent center of maize, a similar role of AsA was noticed in its involvement in controlling the transition from G_1 to S. In both plants and animals, histochemical staining with silver nitrate exposed the role of AsA in cell division and its high production in meristems. Higher levels of ascorbate oxidase are noticed in quiescent centers of cells than in the meristematic cells (Kerk and Feldman, 1995). Application of lycorine, which is a toxic crystalline alkaloid, decreases the amount of ascorbic acid and this in turn inhibits the cell division and elongation as in the case of *Avena* coleoptiles and pea internodes reported by de Leo and coworkers (1973). In onion roots, lycorine stimulates the disappearance of cells in S phase (Arrigoni, 1994). Exogenously applied ascorbic acid activates the cells of the quiescent center to re-enter the cell division as in *allium* (Liso et al., 1988). The enzyme deoxyribonucleotide reductase needs iron for its activity. In the case of a limited supply of iron, ascorbate acts as a reductant and reduces the iron from the stored phytoferritin proteins (Citterio et al., 1994).

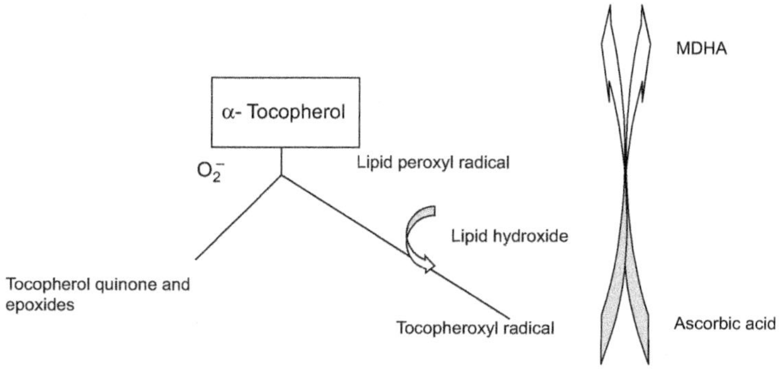

FIGURE 8.3 Regeneration of tocopherol. Tocopherol removes singlet oxygen to form tocopherol quinone and epoxides. Tocopherol also leads to formation of tocopheroxyl radical by scavenging lipid peroxyl radical. This tocopheroxyl radical is regenerated back to tocopherol by oxidation of Ascorbic acid.

8.7.4 Involvement in Tocopherol Regeneration

α-Tocopherol is located in the plastid or thylakoid membranes of the plant. Along with other antioxidants α-Tocopherol helps in the removal of ROS, especially singlet oxygen ($^1O_2$) and hydrogen peroxide from the photosynthetic membranes by converting lipid peroxyl radicals to the corresponding hydroperoxides and in this way it prevents the lipid peroxidation. It deactivates the $^1O_2$ in chloroplasts. It scavenges singlet oxygen and lipid peroxides or radicals to form tocopherol quinone, epoxides and tocopheroxyl radical, respectively (Munné-Bosch and Alegre, 2002a). This α-tocopheroxyl radical gets reduced back to α-tocopherol by ascorbic acid (Asada, 1994) as indicated in Fig. 8.3. One molecule of α-tocopherol deactivates up to 120 $^1O_2$ molecules before being degraded (Fahrenholzt et al., 1974). Here, ascorbate acts as a cofactor and is oxidized and in this way helps in regeneration of tocopherol (Smirnoff, 2000). During stress αz-tocopherol maintains the structure of the thylakoid membrane, the redox state in chloroplast and function of the plant (Munné-Bosch and Alegre, 2002a; Sattler et al., 2004). The level of α-tocopherol decreases during severe stress as larger amounts of ROS are produced in chloroplasts. Irreversible degradation of α-tocopherol occurs when α-tocopherol radicals are not recycled back to α-tocopherol by ascorbate. This occurs in *vtc*1 mutants in *Arabidopsis* as ascorbate availability is less in their chloroplasts. Under stress net tocopherol loss occurs mainly due to severe deficiency of ascorbate in chloroplasts (Munné-Bosch and Alegre, 2002b). To overcome the ROS production in chloroplast the molecules ascorbate, tocopherol and glutathione are interdependent. In the *vte*1 mutant of *Arabidopsis* (tocopherol deficient), increase in ascorbate and glutathione levels occurs whereas VTE1 overexpressing plants show high levels of tocopherol that result in decrease in ascorbate and glutathione (Kanwischer et al., 2005).

8.7.5 Floral induction

The flowering plant undergoes a shift from the vegetative to the reproductive and finally senescence phase. Various factors, such as age of the plant, and environmental factors such as quality of light, water and nutrient accessibility, photoperiod and vernalization regulate the mechanism of flowering (Bernier et al., 1993). Four genetic pathways for flowering induction have been proposed (Komeda, 2004; Corbesier and Coupland, 2005).

1. Strongly inductive long day pathway
2. Gibberellin (GA) pathway under short day conditions
3. Autonomous pathway under both short and long day conditions
4. Vernalization pathway

AsA regulates flowering via the network of the phytohormone signalling pathway, as described in Fig. 8.4. It was reported that the levels of floral meristem genes, for instance LEAFY genes, were elevated and are responsible for floral induction (Komeda, 2004; Corbesier and Coupland, 2005). Ascorbic acid-deficient mutants *vtc1* when grown under long day conditions showed earlier flowering than wild type (Conklin and Barth, 2004), whereas when *vtc1* mutants were grown under short day conditions, the occurrence of flowering is delayed (Veljovic-Jovanovic et al., 2001; Pavet et al., 2005). The gibberellin mediated pathway induced flowering under short days; however, in *vtc1* mutant the GA-mediated flowering pathway was impaired under short days (Wilson et al., 1992). This is because GA20-oxidase, gibberellins -3-β-hydroxylases requires ascorbic acid for its activity.

Moreover, increase in levels of abscisic acid ABA in *vtc1* occurs because of the upregulation of NCED dioxygenase enzyme (Pastori et al., 2003). This enzyme requires ascorbic acid for its activity. Under a limited supply of ascorbic acid as in *vtc1*, flux through the dioxygenase reaction is restricted and in such cases to compensate for the limited flux the amount of enzyme is increased. Hence, the transcription of NCED is upregulated in order to compensate the limit availability of cofactor and to increase maximal enzymatic capacity. However, ABA leads to downregulation of the LEAFY gene. Under short day conditions, the increase in ABA in *vtc1* mutant leads to the formation of FCA-ABA complexes and in turn creates interference in FCA-FY complexes. FCA is a protein participating in the regulation of flowering. This FCA-FY (negative regulator of flowering repressor FLC) is a part of the autonomous pathway and promotes flowering by inhibited FLC (Flowering Locus C) mRNA accumulation. Elevated ABA does not form the FCA-FY complex and leads to increase in FLC transcripts and thus delays in flowering (Razem et al., 2006).

Ascorbic acid deficiency leads to early flowering under long days. This might be due to no involvement of the GA pathway under long days; *vtc1* is strongly inductive light-dependent and the autonomous pathway might be

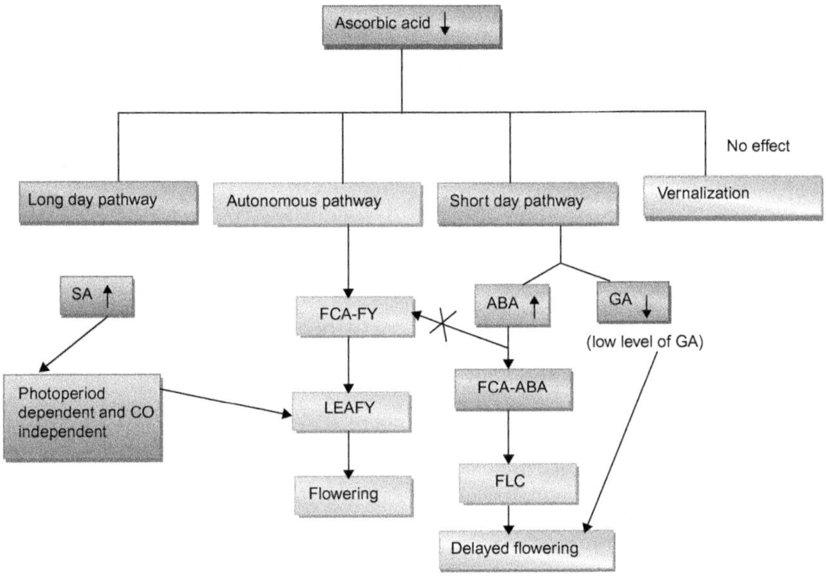

FIGURE 8.4 Involvement of ascorbic acid in floral induction under both long and short day plants. AsA has no effect on the vernalization pathway for floral induction. Low levels of AsA under short day conditions result in decrease in the GA which results in delayed flowering. Under short day conditions, low AsA leads to increase in ABA; ABA forms FCA-ABA complex which activates the repressor FLC and hence results in flowering delay. Autonomous pathway operates via FCA-FY complex, LEAFY gene induction and finally flowering. This pathway operates under both long and short day and shows no effect in response to lower AsA level. During deficiency of AsA the strongly inductive pathway shows the involvement of SA under long day conditions and promotes flowering via a photoperiod dependent but CO independent pathway. However, the involvement of SA in the strongly inductive pathway under short days is not clear.
Abbreviations: SA, Salicyclic acid; CO, Constans; ABA, Abscisic acid; GA, Gibberellic acid; FCA-FY, negative regulator of flowering repressor FLC; FLC, Flowering Locus C.

altered. Martinez et al. (2004) reported the role of salicylic acid as a regulator of flowering by CO-independent but photoperiod-dependent pathway under long days. It also regulates flowering under short day conditions by a photoperiod independent but FLC dependent pathway and delayed flowering. However, the role of SA on delayed flowering under short day conditions is not clear. It appears that control of flowering time in mutant *vtc1* is predominately determined by GA and ABA rather than any other floral induction pathway, as shown in Fig. 8.4.

8.7.6 Protection against Oxidative Stress as an Antioxidant

Genetic evidence has been reported about the role of ascorbate as an antioxidant in defense against ozone, sulfur dioxide, and UV-B radiation (Conklin et al., 1996). The reactive oxygen species and free radicals are generated during aerobic metabolism and by interaction of light. In response to various abiotic (heat, salt, drought, etc.) and biotic (pathogen attack) stresses, the generation of ROS becomes intensified. These radicals cause great damage to the proper functioning of the cell by readily oxidizing the proteins, unsaturated fatty acids and DNA. The main function of ascorbate is to react with these ROS, and form monodehydroascorbate (MDHA) catalyzed by ascorbate peroxidase (APX) (Fig. 8.5). APX not only required ascorbate as a cofactor but also is inactivated by low concentrations of ascorbate. Two molecules of MDHA are converted to Dehyroascorbate (DHA) and ascorbic acid. In order to maintain the ascorbate pool it is necessary that both DHA and MDHA gets rapidly converted back to ascorbate otherwise DHA gets converted to 2,3-diketogulonic acid (Smirnoff, 2000). MDHA is reduced to AsA by MDHA reductase either in the presence NADPH or reduced ferredoxin. DHA is converted to AsA, which comprises the following two reactions.

1. The product DHA is reduced to ascorbic acid with the help of glutathione (GSH) dependent dehydroascorbate reductase enzyme.
2. GSSG is reduced to GSH by the glutathione reductase enzyme, which is NADPH dependent.

In plants, L-AsA is the most abundant water soluble antioxidant (Gallie, 2013a). During stress conditions, the ROS increase which in turn increases the demand for antioxidant. Thus, under critical conditions the AsA content and redox state of AsA is maintained through recycling by the AsA-GSH cycle (Gallie, 2013b). The molecules AsA and GSH differ from many other primary and secondary molecules because of some important features such as:

1. Both AsA and GSH are specific molecules that link with H_2O_2 metabolism.
2. The stability of the oxidized forms of AsA and GSH.
3. Their ability to recycle to reduced form via the enzymatic system through electron transport molecule NADPH (Foyer and Noctor, 2011).

However, as compared to MDHAR, the contribution of DHAR to give rise to ascorbate is very small (Asada, 1994). Glutathione reductase catalyzes the reduction of glutathione in this cycle.

8.7.7 Ascorbic Acid and Nodulation in Leguminous Plants

It was estimated that about 20% of leguminous plants produce root nodules (Jordan, 1984). The rhizobial root nodules of leguminous plants have been studied by various scientists in order to understand the mechanism of nitrogen fixation but now the main focus is to study the production of regulating molecules and its transportation to the host (Hunter, 1989). Recently,

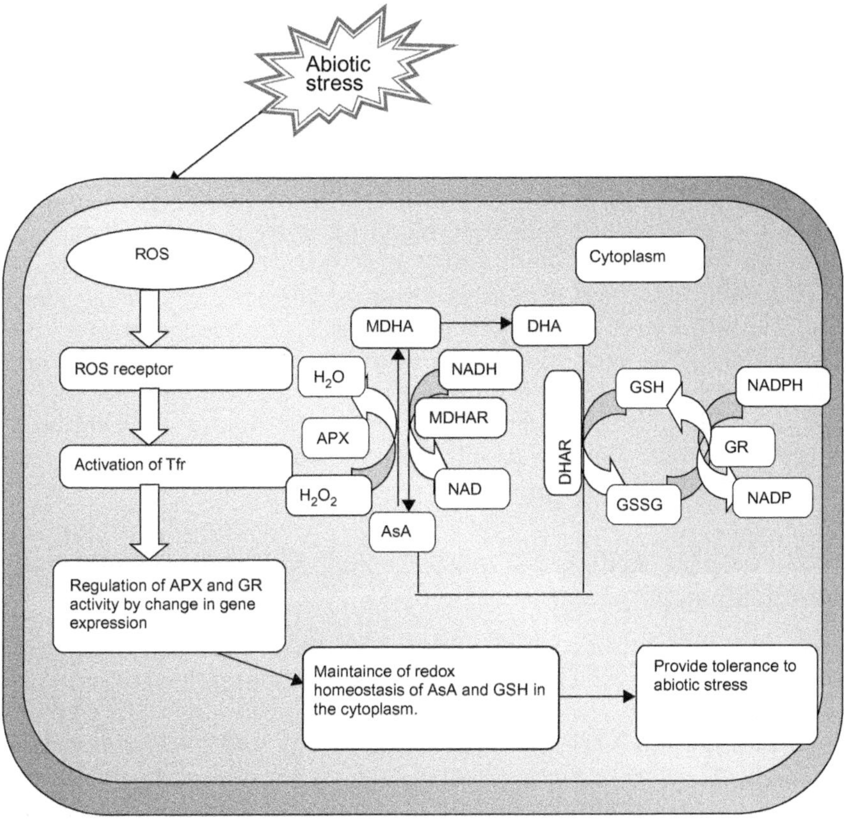

FIGURE 8.5 During abiotic stress conditions a possible regulatory mechanism for removing reactive oxygen species involves signal transduction pathway of ascorbate peroxidase and glutathione reductase. The ROS are sensed by ROS receptors. These receptors in turn send a signal which in turn leads to activation of various transcriptional factors. The transcriptional factors modify the expression of ascorbate peroxidase and glutathione reductase. These enzymes maintain the cellular homeostasis of the plant cell by maintaining the ascorbate-glutathione cycle within the plant cell. In the ascorbate-glutathione cycle, four enzymes viz. APX, MDHAR, DHAR, GR, are known to take part in a series of coupled redox reactions. This cycle plays an important role in overcoming oxidative stress in plants. Ascorbate is oxidized by APX to form monodehydroascorbate (MDHA). Regeneration of ascorbate from MDHA is necessary in order to maintain the function of the H_2O_2 scavenging system in chloroplasts. MDHAR helps in reduction of the stromal MDHA. If MDHA is not converted to AsA by MDHAR it gives rise to DHA. AsA is formed from DHA by DHAR.
Abbreviations: ROS- Reactive oxygen species; Tfr- Transcriptional factor; APX- Ascorbate peroxidase; GR- Glutathione reductase; AsA- Ascorbic acid; MDHA- Monodehydroascorbate; MDHAR- Monodehydroascorbate reductase; NAD- Nicotinamide adenine dinucleotide; NADPH- Nicotinamide adenine dinucleotide phosphate; DHA- Dehydroascorbate; DHAR-Dehydroascorbate reductase; GSH- Reduced glutathione; GSSH- Oxidized glutathione.

scientists have gained interest in the biosynthesis of ascorbic acid in the root nodules of the leguminous plants (Matamoros et al., 2006). The leguminous herb *Phaselous mungo* produces a large amount of small, pinkish root nodules. Previously, it was reported that root nodules of leguminous plants were not able to synthesize ascorbic acid but obtain it from the shoot or root. In the root nodules of *P. vulgaris,*Matamoros et al. (2006) reported the presence of ascorbic acid. AsA helps in the regulation of the senescence of the nodule. At the apex of the root nodules, AsA was found in large amounts and helps in the protection of the plant cells from peroxide damage (Matamoros et al., 2006). AsA handles various other functions during symbiosis (Matamoros et al., 2006) and it also acts as a signalling molecule to regulate various other processes (Puppo et al., 2005). In mature nodules and young roots, a glucose pool is present which acts as a precursor for synthesizing AsA. So it has been stated that production of AsA is also a beneficial aspect of symbiosis rather than only hormone production and nitrogen fixation (Matamoros et al., 2006).

8.7.8 Ascorbic Acid as a Central Defense Molecule against Abiotic Stresses

Exposure of plants to severe abiotic stress conditions results in high level production of ROS, which causes lipid peroxidation, oxidation of proteins, inhibition of enzyme activity and ultimately leads to cell death (Tanou et al., 2009). The enzymatic and nonenzymatic antioxidant molecules help in maintaining the detoxification mechanism in the plant cell. Moreover, the antioxidant also helps in proper functioning of the signalling cascades by keeping the production of ROS at optimum levels (Foyer and Noctor, 2005a). Under abiotic stress, molecules such as ascorbate and glutathione have a regulatory function and play a crucial role in maintaining the redox signaling of the plant cell (Khan et al., 2009; Szalai et al., 2009). The reduced glutathione (GSH) acts as the reducing agent of dehydroascorbate (DHA) and hence plays an essential role in the ascorbate-glutathione cycle (Garnczarska, 2005; Paradiso et al., 2008). Besides this, it can also directly remove ROS or it may act as substrate for glutathione peroxidase. The reduced (GSH) and oxidized form (GSSG) together form a redox couple (GSH/GSSG) and help in maintaining the signaling system in plants, which in turn maintains the cellular homeostasis (Clemens, 2006). During stressful conditions, the concentration of GSH is increased, which in turn helps to overcome the negative effects of oxidative stress (Anjum et al., 2010).

In plant cells, ascorbate (AsA) is considered to be the most important and powerful antioxidant. It is found in various compartments of plant cell, viz. cytosol, chloroplast, vacuoles, mitochondria and apoplast (Foyer and Noctor, 2005b; Cheng et al., 2007). Ascorbate is found in reduced form under nonstress conditions (Garnczarska, 2005). Ascorbate acts as an electron donor in a number of enzymatic and nonenzymatic reactions and thus provides protection to the cell against oxidative stress (Noctor and Foyer, 1998). As discussed earlier, ascorbate and glutathione are related through the ascorbate-glutathione cycle (outlined in Fig. 8.5).Various enzymes such as ascorbate peroxidase (APX) and dehydroascorbate reductase (DHAR) help in recycling of ascorbate (Ushimaru et al., 1997) while glutathione peroxidase helps in the recycling of glutathione (Noctor and Foyer, 1998).

8.8 ROLE OF ASA IN DEFENDING AGAINST THE ABIOTIC STRESSES

While plants use several diverse mechanisms to defend themselves against various environmental stresses, the dependence on powerful antioxidants is considered vital. AsA helps in the regulation of programmed cell death, flower senescence, tolerance against pathogen attack, protection of plants against UV, heat and high light intensity (Foyer and Noctor, 2011). In this context, ascorbic acid appears to have been involved in response to all the stresses and its role in defense has been strongly advocated. Following, some specific examples are given for various stresses where ascorbic acid is implicated in stress response.

8.8.1 Drought Stress

Drought stress is an abiotic stress, which arises because of deficiency of water availability. Different species or genotypes show varying sensitivity towards water deficit stress (Nayyar and Walia, 2004; Türkan et al., 2005). Various factors such as intensity, duration of stress and the developmental stage of plants are responsible for the diverse responses shown by plants at various organizational levels (Chaves et al., 2003). During drought stress, diverse morphological and biochemical changes occur such as reduction in leaf size, stomatal closing, and enhanced production of stress hormone such as ABA (Shinozaki and Shinozaki, 2007). It was reported by various workers that drought stress decreases the leaf area, fresh weight and dry weight of leaf, amount of starch and it also affects the amount of protein by increasing the degradation of proteins or by inhibiting protein synthesis. It also inhibits photosynthesis by affecting the photosynthetic pigments. Thus water deficit results in reduction of growth and primary production of crops (Tambussi et al., 2000). The reduction in photosynthesis is mainly because of less availability of CO_2 to chloroplast due to stomatal closure (Muller and Whitsitt, 1996). The processes such as the function of photosystems and photophosphorylation also decrease (He et al., 1995). Under normal conditions, the plant produces various reactive oxygen species (Mittler, 2002); the level of ROS is elevated during stress due to their production in the photosynthetic electron transport system (Menconi et al., 1995; Sgherri et al., 1996).

ROS are generated in chloroplast through different mechanisms, which are listed here:

1. By Mehler's reaction, which involves the production of H_2O_2 by dismutation of $O_2{\bullet}^-$ (Foyer et al., 1994).
2. Interaction of excited chlorophyll molecule with oxygen to form singlet oxygen $^1O_2$ (Foyer et al., 1994).
3. Production of superoxide by PSII (Ananyev et al., 1994).

The production of ROS affects the normal functioning of cells by destroying the macromolecules such as DNA, proteins and lipids (Foyer and Fletcher, 2001). Increase in the formation of $O_2^{\bullet -}$ causes lipid peroxidation, fatty acid saturation and damage to membranes (Menconi et al., 1995; Sgherri et al., 1996). Thylakoid proteins are damaged and are the strong indicator of oxidative damage (Berlett and Stadtman, 1997). Hence, oxidized proteins are more sensitive indicators of oxidative damage in chloroplast than of lipid peroxides and ROS production.

8.8.1.1 Role of Ascorbic Acid during Drought Stress

Drought stress decreases the GSH/GSSG ratio as found in barley and pea leaves, hence the plant must have some strong antioxidant system to maintain the oxidized state of glutathione (Tausz et al., 2001). Wheat flag leaves exposed to drought stress demonstrate elevation of total glutathione level mainly in drought susceptible cultivar (Herbinger et al., 2002). There was a significant decrease in the ratio of AsA/DHA and GSH/GSSG in *Triticum aestivum* cultivars, Moti (drought susceptible) leaves (Khanna-Chopra and Selote, 2007). Similar decrease in AsA content was reported in rice (Sharma and Dubey, 2005). Due to oxidation of AsA, the activity of APX and GR also decline, hence the antioxidant system is less effective in removing ROS (Shigeoka et al., 2002). During mild drought stress, the activity of GR and DHAR are enhanced, which results in increase in the level of glutathione concentration to impart tolerance to stress. Moreover, the oxidative damage was minimized due to induction of ascorbate-glutathione enzymes along with increase in AsA content. However, under severe water stress, the level of AsA declined, and the activity of GR and DHAR was also less, which resulted in severe oxidative damage (Menconi et al., 1995). The content of AsA decreased in maize and wheat in water-stressed plants that correlated with increase in oxidative stress (Nayyar and Gupta, 2006). The tolerant wheat cultivar shows lesser damage due to oxidative stress because of its high AsA content and increase in induction of AsA-GSH cycle enzymes (Lascano et al., 2001). The activity of APX increases in drought-stressed plants of *Euphorbia escula* (Davis and Swanson, 2001), *Zea mays* (Jiang and Zhang, 2002), soybean (Riekert van Heerden and Kruger, 2002), wheat (Dalmia and Sawhney, 2004) and *Populus accutifolius* (Türkan et al., 2005).

Resistance of wheat against drought was improved by osmopriming with AsA as this increased the accumulation of proline and antioxidant action of AsA and phenolics which in turn helps in the maintance of tissue water content, stability of membrane, better and uniform seedling growth (Farooq et al., 2013).

The involvement of ascorbic acid in drought stresses can be substantiated by its exogenous application on stressed plants. We present here a few such examples for each stress. According to the study of Amin et al. (2009) application of ascorbic acid antagonizes the effects of drought stress by decreasing the damage of stress on leaf area, fresh and dry weight. It also reduced the effect of stress on all other parameters such as reducing the malonialdehyde content, increasing the chlorophyll content, maintaining the sugars, starch, proline and protein content. This shows that application of ascorbic acid enhances resistance of plants to cope with water stress (Miguel et al., 2006). Exogenous application of L-galactono-1,4-lactone (GL) in wheat increases the pool of ascorbic acid. Formation of ascorbic acid from GL is the final step in AsA biosynthesis in plants (Arrigoni et al., 1997; Wheeler et al., 1998). During drought stress the transpiration rate increased in wheat (Malik and Ashraf, 2012) and okra (Amin et al., 2009) on ascorbic acid application as compared to nontreated plants. This is because AsA application improved the stomatal conductance of plants. AsA also provides protection to thylakoid proteins from oxidative damage during drought stress and it saves the photosynthetic membrane through various ways as shown in Fig. 8.6.

It has been found that L-galactono-1,4-lactone (GL) helps in reducing the overall thylakoid protein oxidation; however, full recovery of electron

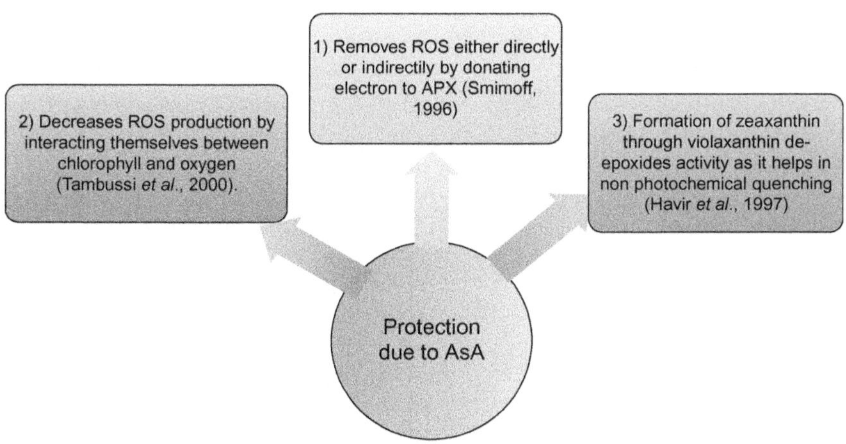

FIGURE 8.6 Role of ascorbic acid in overcoming drought stress effects through different means.

transport activity is not possible in drought-stressed plants. Thus GL helps in providing protection to the photosynthetic apparatus yet it is not able to improve the balance of carbon in drought plants. This is mainly because of low stomatal conductance for CO_2. Overexpression of APX in the chloroplast helps in providing protection against water stress (Mc Kersie et al., 1996; Yan et al., 2003). Thus, it improves the stress tolerance and increases the survival capacity of plants under stress.

8.8.2 Role of Ascorbic Acid in ABA Mediated Signaling of Stomatal Closure

Under drought stress, H_2O_2 is an important intermediatory signaling molecule that promotes closing of stomata (Chen and Gallie, 2004). This is mainly because of elevation of ABA in plants during drought conditions. As shown in Fig. 8.7 the rise in ABA level results in activation of H_2O_2 mediated cytosolic Ca^{+2} concentration via activating Ca^{+2} channels on plasma membrane as well as releasing of Ca^{+2} from intracellular spaces (Zhang et al., 2001; Kohler and Blatt, 2002). Anion channels located on plasma membrane are also activated and cause depolarization of guard cells. The efflux of K^+ occurs and the cell loses turgor and decreases volume of guard cells, which ultimately leads to stomatal closing (Blatt, 2000; Schroeder et al., 2001a, 2001b). Because AsA scavenges H_2O_2, the exogenous application of ascorbic acid reversed the H_2O_2 induced stomatal closing (Zhang et al., 2001). Plants with a high level of ascorbate are less responsive to ABA and H_2O_2 signaling. Water deficit in plants leads to stomatal closure mainly because of decrease in the AsA pool and increase in ABA (Pastori and Foyer, 2002). However, plants having overexpressed DHAR activity have more AsA redox state in whole leaf, in apoplast and in guard cells, which causes higher numbers of open stomata, more conductance and transpiration. Along with this, the DHAR overexpressed plants are also less responsive to ABA or H_2O_2 mediating signaling. During the afternoon, the production of H_2O_2 rises and the level of AsA decreases, which results in stomatal closing. This is mainly because of reduced activity of DHAR to regenerate AsA in guard cells that is insufficient to cope with the high rate of H_2O_2 production during daytime. Higher DHAR activity and more reduced AsA level results in increased tolerance to drought stress in terms of higher net photosynthesis.

8.8.3 Heat Stress and Ascorbic Acid

Heat stress is generally defined as the rise in temperature beyond the optimum level for a particular period of time and enough to cause irreversible damage to growth and development of plants. On the other side, heat tolerance is defined as the ability of the plant to grow and develop normally under high temperatures (Wahid et al., 2007). High temperature poses a great

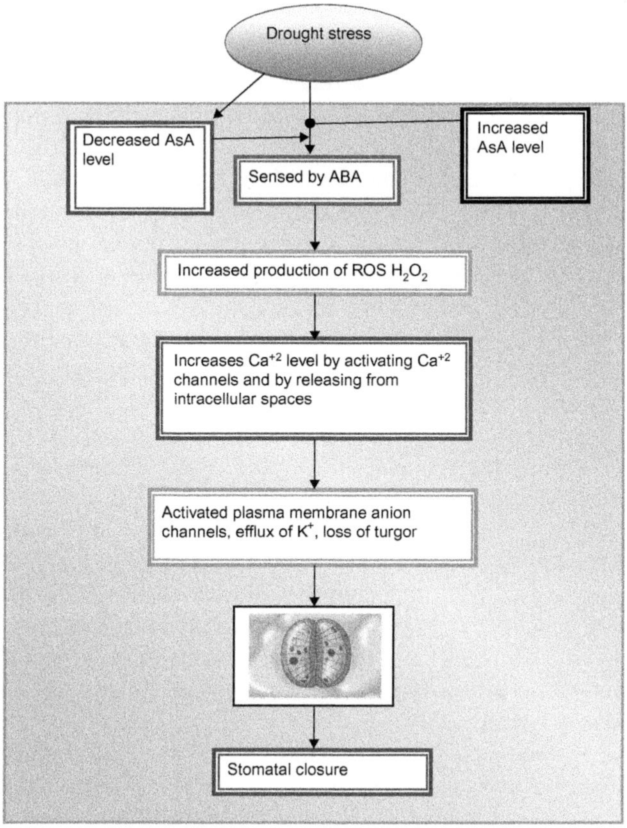

FIGURE 8.7 Role of ascorbic acid in ABA mediated signaling of stomatal closure. During drought stress, the concentration of ABA rises, which results in an increase in ROS level. The rise in ROS results in activation of H_2O_2 mediated calcium channels located on plasma membrane or on tonoplast of the cell. The anion channels are also activated and cause depolarization of plasma membrane. Potassium ions move outside from the cell and result in loss of turgor of guard cells and ultimately closing of stomata. However, when reduced AsA is present in high amounts, it inhibits the production of ROS by directly scavenging ROS or indirectly through APX and inhibits the production of ABA. Hence AsA inhibits the ABA or H_2O_2 mediated signaling and results in greater numbers of open stomata during drought stress.

threat to the worldwide crop production (Hall, 2001). High temperatures can cause significant damage to the plant, which includes scorching of leaves and twigs, sunburns on leaves, branches and stems, senescence and abscission of leaf, inhibition of shoot and root growth, fruit discoloration and yield reduction (Vollenweider and Gunthardt-Goerg, 2005). As the global temperature is likely to elevate 1.5–4.5°C by 2050 (Houghton et al., 2001), that is expected to decrease the productivity and growth of plants. High temperature causes injury to cell membrane, photosynthesis inhibition, senescence and death of the cell (Xu et al., 2006). High temperatures lead to the cell overproducing ROS. The elevated ROS poses great damage to the nucleic acid, proteins, lipids and to the normal functioning of the cell (Mittler, 2002). However, plants are able to cope with ROS by enzymatic and nonenzymatic antioxidants (Jaleel et al., 2007; Esfandiari et al., 2007). In lily plants, the enzymes of the ascorbate-glutathione cycle such as APX and GR were upregulated under heat stress, suggesting that they play an important role in alleviating the accumulation of H_2O_2. At the transcriptional level the upregulation of APX and the production of new thermostable isoforms of APX in transgenic *Arabidopsis thaliana* are in agreement with the above statement (Panchuk et al., 2002). In lily, AsA regeneration can be maintained at normal rates at 37°C and 42°C. However, at 47°C, the AsA concentration decreases significantly, which might be due to an increase in oxidation of AsA and decrease in regeneration of AsA (Yang et al., 2007). Reports on pea and wheat also stated that AsA and GSH were found at high concentrations and were essential components for the plants to defend against heat stress (Kurganova et al., 1999; Almeselmani et al., 2006). The activity of APX and foliar concentration of AsA increases when the *Arabidopsis* plants are cultivated at higher nonstressful temperature (Panchuk et al., 2002). The level of GSH rises when maize roots are treated to high temperature shock of 40°C (Nieto-Sotelo and Ho, 1986) and this helps the cells to cope with heat stress. Similar finding was observed in tomato seedlings when they were exposed to heat stress (Rivero et al., 2004). A study on wheat has reported the accumulation of GSH, enhanced GSH synthesis activity, and higher ratio of GSH/GSSH during heat stress (Kocsy et al., 2001). The heat-tolerant cabbage genotype showed a high GSH content compared to sensitive ones. In apple leaves, the content of AsA, total glutathione, GSH, GR rises when they were exposed to heat stress (40°C) for 2 h (Ma et al., 2008) because the gene expression of enzymes such as APX, DHAR, GR rises. The enzymes APX and GR prevent the oxidation of enzymes and membranes and hence play an important role in the protection of plants from high temperature (Almeselmani et al., 2006). Exogenous application of AsA provides tolerance to heat stress, as was observed on mungbean seedlings where application of AsA (50 μM) resulted in improved germination and seedling growth at 40/30°C and 40°C compared to control (Kumar et al., 2011a).

8.8.4 Chilling Stress

Chilling is one of the most important stresses that affect plant development and yield (Lang et al., 2005). It causes damage to sensitive crops like maize, rice and chickpea (Thakur et al., 2010). Chilling stress is experienced at nonfreezing temperatures by the plants; the tropical species, especially, are more sensitive. Chilling stress can induce damage at various organisational levels to reduce overall growth and yield of the plants (Thakur et al., 2010).

In stored fruits, chilling stress arises during storage below 10°C to which fleshy fruits like tomato are mostly sensitive. It is generally considered that low temperatures can be used to extend shelf life of fruits; however they may also trigger physiological disorders such as loss of quality, which ultimately results in chilling injury (Lyons, 1973). Various symptoms are developed by fruits in response to chilling injury such as rough texture, watery flesh and irregular ripening. The quality of the post-harvest fruit storage can be improved by improving the antioxidant content of sensitive fruits (Hodges et al., 2004), as in the case of apple (Hodges et al., 2004; Davey et al., 2007). During cold stress, the stomatal conductance reduces, pigment complexes changes, photochemical efficiency reduces, the biophysical properties of thylakoid lipids undergo changes, and there is inhibition in the metabolism of proteins and reduction of enzyme activity (Ensminger et al., 2006). At low temperatures during the electron transport chain, H_2O_2 is formed from superoxide dismutation. It was reported that there was accumulation of H_2O_2 in maize (Prasad et al., 1994). Physical and biochemical processes change in cold-acclimated wheat plants, which allows them to function at low temperature. Some of these changes are listed here:

1. Formation of antifreeze proteins (Yeh et al., 2000).
2. Membrane composition changes (Huner et al.,1987; Szalai et al., 2001).
3. Accumulation of osmolytes (Konstantinova et al., 2002).
4. Polyamines (Rácz et al., 1996).
5. Activation of antioxidants (Dai et al., 2009).

All these changes help in improving the antioxidant capacity of plants (Janda et al., 2003). There was an increased level of AsA-GSH in cold acclimated cereals (Dai et al., 2009; Szalai et al., 2009a). Enhancement of the antioxidant system helps in improving the chilling stress tolerance. AsA-GSH cycle scavenges H_2O_2 and helps in preventing the accumulation of H_2O_2 to toxic levels (Kocsy et al., 2001). AsA is an important antioxidant for the removal of H_2O_2. It reacts with H_2O_2 either directly or indirectly through APX (Noctor and Foyer, 1998), which has already been discussed in Fig 8.5. Studies on some other fruits also indicate that fruits that could maintain their reduced ascorbate and glutathione pools had better storage properties (Davey & Keulemans, 2004). Some of the cold tolerant genotypes of different crops show tolerance to chilling stress because of increased accumulation of GSH (glutathione) and AsA (ascorbate) as shown in Table 8.5. The loss of fruit firmness decreased under chilling stress mainly because of increased MDHAR activity and a lower oxidation level of the ascorbate pool of the fruit (Stevens et al., 2008). The increase in the function of enzymes of ascorbate-glutathione cycle by genetic engineering has resulted in enhanced tolerance to chilling stress in cotton plants (Payton et al., 2001).

To ensure proper functioning of AsA-GSH cycle, the role of GSH is very important hence during chilling the ratio of GSH/GSSG rises. It has been

TABLE 8.5 List of Some Cold Tolerant Genotypes of Different Crops which Show Tolerance to Chilling Stress Because of Increased Accumulation of GSH (Glutathione) and AsA (Ascorbate)

Tolerant Genotypes	Tolerance to Chilling Stress	References
Tomato	Increase in GSH levels	Walker and McKersie, 1993
Sorghum	Increase in GSH levels	Badiani et al., 1997
Wheat	Increase in GSH levels	Kocsy et al., 2001
Jack pine	Increase in GSH levels	Zhao and Blumwald, 1998
Poplar	Increase in GSH levels	Foyer et al., 1995
Rice	Increase in AsA levels	Huang and Guo, 2005
Barley	Increase in AsA levels	Dai et al., 2009

studied that tolerant genotypes of tomato contain more GSH/GSSG and GSH content during chilling stress as compared to sensitive ones (Walker and McKersie, 1993). Similar observations were made in cold tolerant genotypes of maize (Hodges et al., 1996). During chilling, the high ratio of GSH/GSSG in tolerant genotypes prevents the formation of disulphide bridges by keeping the sulphydryl groups of proteins in reduced form (Levitt, 1962). In strawberry, it was noticed that with increase in the chilling acclimation periods, the content of H_2O_2 and MDA (Malondialdehyde) first increased to a certain degree and then decreased. The activity of various antioxidant enzymes like SOD, DHAR, APX, GR, CAT, MDHAR increased gradually and then reached a higher level. This in turn increased the levels of AsA, DHA and GSH. Thus, it was concluded that during chilling acclimation, the chilling resistance is increased due to the role played by the AsA-GSH cycle (Zhang et al., 2008).

Seedlings of soybean (*Glycine max*) cultivars were exposed to low temperature stress in the range of 5–25°C and it was found that the antioxidant activity was induced in all the cultivars and the nature of the response depended upon the exposure time of stress, temperature and on the cultivars. During stress the activity of GR and SOD were overexpressed in all the cultivars, whereas the activity of POX and APX were more in the tolerant cultivar (Chakraborty and Pradhan, 2009). In our studies on chickpea, the cold tolerant genotypes possessed higher activity of SOD, catalase, APX and more content of proline and AsA as compared to cold-sensitive genotypes, thus showing that AsA played an important role in providing tolerance against cold (Kumar et al., 2011b). Similarly, the tolerant species of rice and barley showed increased levels of AsA content as compared to sensitive ones during chilling (Huang and Guo, 2005; Dai et al., 2009). Besides increased level of AsA there was also a report about the increased activities of APX and GR in these cultivars. A similar finding was reported by Guo et al.

(2006) in rice; as per the observations, the increased activity of APX and GR and higher level of AsA help the plant to tolerate chilling effects. The cold tolerant coffea sp (variety *Icatu*) shows an elevated level of antioxidant enzymes and molecules such as Cu/Zn-SOD, APX, ascorbate, and tocopherol which helps the plant to control oxidative stress generated during chilling (Fortunato et al., 2010). The enzymes GR and DHAR are considered to be involved in cold acclimation of the cold tolerant variety, as reported during the gene expression studied. During low temperature stress, it was demonstrated that exogenous application of reduced GSH in the concentration of 100 mg/L on young loquat fruit increased the content of GSH, AsA and activity of APX, GR and MDHAR were increased, which thus increased resistance against chilling stress (Cheng et al., 2009). In the leaves of wheat cultivars (Yunong 949 and Yanzhan 4110), which were at a later spike development state, it was found that during freezing stress, the soluble sugar, soluble proteins and AsA contents were increased and the plant suffered less damage, which indicated the involvement of these molecules in freezing tolerance (Kang et al., 2011). A recent study indicates that in the ascorbate-glutathione cycle, the recycling of ascorbate from dehydroascorbate in the choloroplast protects the PSII from chilling stress (Ding et al., 2012).

Some growth-regulating molecules, when applied exogenously, resulted in protection from chilling stress involving ascorbic acid. The application of 24-epibrassinolide (EBR) on cultured cells of *Chorispora bungeana* increased the contents of AsA and GSH and thus improved resistance against chilling stress (Liu et al., 2009). The mitochondrial antioxidant capacity in young loquat fruits during chilling stress was increased by exogenous application of nitric oxide (NO) concentration mainly due to increased concentration of GSH and AsA and this also increased the activity of APX, GR, DHAR, MDHAR (Wu et al., 2012).

8.8.5 Salt Stress and Ascorbic Acid

Salt stress arises in plants because of the presence of excess sodium chloride (NaCl) in soil that eventually gets absorbed by the plants. For most of the plants, Na^+ is not an essential mineral. Various physiological processes such as seed germination and plant development are affected by salt stress and thus limit the production potential of the plant (Bohnert et al., 1995). It was proposed that tolerance to salt stress is a stage-specific regulatory phenomenon, as it is not necessary that tolerance at one stage of plant development also provides tolerance at some other stage (Johnson et al., 1992). The physiological and biochemical function of the plant cell is altered due to salt stress, as it leads to water loss which causes drought stress and it also leads to ionic toxicity. The main target of salt stress is photosynthetic sites and its components. In response to salt stress, ABA is produced and it decreases the turgor pressure of guard cells, which ultimately results in less availability of carbon dioxide (CO_2) for photosynthesis (Leung et al., 1994). Salt stress results in an increase in the production of lignin that occurs due to the generation of ROS, which help in the lignification process and act as a signal against salt stress (Lee et al., 2007). GSH along with AsA plays a crucial role in maintaining the redox state of the cell, which ultimately provides tolerance to salinity (Gossett et al., 1996; Shalata et al., 2001). During salt stress, the oxidation of GSH occurs, mainly because of increased foliar GR activity, and this in turn decreases the GSH/GSSG ratio, as evident from the study of Chaparzadeh et al. (2004). The increase in GR function was reported in various crops, viz. pea (Hernandez et al., 2000), rice (Demiral and Türkan, 2005; Tsai et al., 2005), tomato (Molina et al., 2002; Mittova et al., 2003), *Arabidopsis* (Huang et al., 2005), wheat (Sairam et al., 2005), and mungbean (Sumithra et al., 2006). Plants having overexpressed GR were more tolerant to oxidative stress due to increase in the antioxidant capacity (Kocsy et al., 2001). As compared to salt tolerant cultivars, the salt-sensitive cultivars show a decrease in the ascorbate-glutathione cycle (Mittova et al., 2003, 2004). The oxidative damage was higher in apoplast, which results in necrotic lesions on pea leaves (Hernández et al., 2001). The salt sensitive species lack the ability to upregulate the synthesis of GSH and their utilization, which is present among salt tolerant cultivars (Mittova et al., 2003). The deficiency of NADPH instead of less activity of GR may also be responsible for decreasing the GSH/GSSG ratio, as observed in *Solanum lycopersicum*. It was found that the ascorbate content decreases in salt sensitive cultivars (Mittova et al., 2003, 2004). Salt stress results in a decrease in the biosynthesis of AsA (Song et al., 2005). The catabolism of AsA enhances as compared to its synthesis and regeneration during salinity (Shalata et al., 2001; Amor et al., 2006). Some plants acclimated to salt stress showed increased content of ascorbate (Shalata et al., 2001). During salinity, the drastic reduction of AsA/DHA ratio occurs because of involvement of APX in ROS removal. It was found that the activity of APX drops considerably at low concentrations of AsA; hence in order to maintain the activity of APX the amount of AsA should increase (Asada, 1999). By manipulation of gene expression of various antioxidant enzymes participating in the ascorbate-glutathione cycle, transgenic plants have been produced, which have acquired some tolerance to salinity (Ashraf, 2009).

Ascorbic acid along with other antioxidants assists to minimize the oxidative damage by removing H_2O_2 and other ROS and helps in stabilizing membranes (Shao et al., 2008). The content of indole acetic acid (IAA) increases due to the ascorbic acid application that promotes cell division and enlargement, which ultimately results in improving plant growth (Hassanein et al., 2009). However, application of ascorbic acid significantly decreases the lignin (Mazid et al., 2011). Application of ascorbic acid into the NaCl or mannitol media leads to an increase in ascorbate and glutathione content during the germination period (Asada, 1997). El-Hariri et al. (2010) found that exogenous application of ascorbic acid at 200 mg/L on flax cultivars helps to overcome the adverse effect of salinity and results in significant increase in growth of plant. Under salt stress ascorbic acid plays an important role in growth regulation and plant metabolism and increasing the accessibility of water and nutrients (Barakat, 2003). Exogenous supply of ascorbic acid on tomato seedlings made them more tolerant of salt stress from 0% to 50% after 9 h of NaCl exposure. Hence AsA application reduces the seed mortality and accumulation of thiobarbituric acid reactive substances (TBARS) in tomato, as reported by Shalata and Neumann (2001). AsA application also results in promotion of growth in wheat (Hamada and Al-Hakimi, 2001). The effectiveness of AsA depends upon the mode of application, i.e. whether it is applied through a rooting system or as foliar spray. From the study it was observed that foliar spray was effective in improving the growth of salt-stressed plants in two wheat cultivars, as compared to the rooting system of application.

8.9 TRANSGENICS FOR ASCORBATE METABOLISM UNDER ABIOTIC STRESS

8.9.1 Salt Stress

Increase in tolerance to salt stress due to overexpression of APX in *Tobacco* chloroplasts has been reported (Badawi et al., 2004a,b). Salt tolerant *Raphanus sativus* also shows overexpressed activity of APX (Lopez et al., 1996). It has been reported that H_2O_2 is mainly produced in peroxisomes by flavin oxidase and glycolate oxidase (Lisenbee et al., 2003). pAPX (peroxisomal type ascorbate peroxidase) plays an important role in scavenging H_2O_2 in peroxisomes (Shi et al., 2001). In order to understand the role of pAPX against oxidative stress, Shi et al. (2001) developed transgenic *Arabidiopsis thaliana* plant having a gene pAPX (HvAPX1) from *Hordeum vulgare*. Both catalase (CAT) and APX enzymes act on H_2O_2 in order to detoxify this compound to yield water and oxygen. Under severe oxidative stress, the activity of CAT is decreased in the peroxisomes of plant cells (Shi et al., 2001). The predominant mechanism for H_2O_2 detoxification is the pAPX. It has been reported that the wild types show a higher degree of H_2O_2 and MDA accumulation under salt stress than transgenic pAPX3. This difference might be due to salt tolerance in pAPX3. Tsugane et al. (1999) and Azevedo et al. (2006) while investigating the role of ascorbate peroxidase in transgenic for HvAPX1 against salt tolerance found that, under normal conditions, the total APX enzyme activity does not vary much, probably due to overexpression of HvAPX1 which only influences the single APX isoenzyme rather than improving the total APX enzyme activity.

8.9.2 Heavy Metal Stress

Human activities such as mining, smelting, electroplating, energy and fuel production, power transmission, intensive agriculture, sludge dumping and military operations can lead to the production of heavy metals, which contribute to major environmental pollution (Nedelkoska and Doran, 2000). Increased concentrations of both essential and nonessential heavy metals in the soil results in reduction of growth and toxicity symptoms in most plants (Hall, 2002). In addition, heavy metals cause oxidative stress due to increase in the production of ROS (Dietz et al., 1999). The antioxidant system comprises various enzymes and low molecular weight compounds which provide protection against oxidative stress (Noctor and Foyer, 1998).

In response to cadmium (Cd), the leaves of *Arabidopsis thaliana* show enhanced gene expression related to GSH synthesis and phytochelatin synthase (Semane et al., 2007). It was observed by various authors that there was an increase in the AsA-GSH cycle both at transcriptional and the enzymatic level in response to Cd, which showed the involvement of various antioxidant enzymes in scavenging ROS. Tolerance to Cd was observed in transgenic *Brassica juncea* where the gene expression for GSH biosynthetic enzymes was enhanced (Zhu et al., 1999). Cd-treated plants showed a decrease in GSH/GSSH ratio; GR via the ascorbate-glutathione cycle helps in the detoxification of Cd induced stress.

Exposure to Cd leads to increase in the ascorbate concentration which shows the involvement of this compound in maintaining the redox cellular homeostasis during stress (Sobrino-Plata et al., 2009). The content of AsA reduced when the plant was exposed to high Cd concentration of 100 mM (Hatata and Abdel-Aal, 2008). Exogenous application of AsA on leaves of rice seedlings causes increase in the content of AsA, GSH, ratio of AsA/DHA and GSH/GSSG rises, the activity of APX and GR increases in response to Cd treatment (Chao et al., 2010).

Plants exposed to toxic levels of copper (Cu) showed decreased levels of cellular GSH and GR activity (Tewari et al., 2006). During short-term exposure of plants to Cu the GSH/GSSG ratio played a major role (Drazkiewicz et al., 2003). Increased supply of Cu leads to an increased in the level of the total ascorbate pool. GST (Glutathione S-transferase) and GPX (glutathione peroxidase) plays a regulatory role in maintaining the redox state of the cell in *Arabidopsis* when they are exposed to Cu (Smith et al., 2004).

Aluminum (Al^{+3}) toxicity results in inhibition of root growth as observed in potato clones (Tabaldi et al., 2009). Observation of the tolerant lines showed that tolerance to Al^{+3} was due to increased activity of GST (Darko et al., 2004). Al^{+3} stressed rice seedlings show enhanced activity of ascorbate-glutathione enzymes which helps the plant to cope with oxidative stress (Sharma and Dubey, 2007). Transgenic tobacco plants overexpressing *Arabidopsis* cytosolic MDHAR or DHAR were able to tolerate Al^{+3} stress as the two enzymes MDHAR and DHAR help in the regeneration of reduced AsA (Yin et al., 2010). It was found that DHAR instead of MDHAR overexpressed transplanted plants show better root growth than wild type plants.

Addition of AsA to nickel (Ni) stressed plants showed better growth as compared to Ni-treated plants growing without ascorbic acid as AsA reduces the translocation of Ni from roots to shoots (Saeidi-Sar et al., 2007). Exposure of soybean seedlings to 0.25 mM nickel chloride (NiCl) for 5 days leads to toxicity. However, toxicity effects were reversed by exogenous application of 1mM AsA (Saeidi-Sar et al., 2007). Regeneration of AsA after Ni exposure was considered possible after the activation of MDHAR in maize as was proved because of no effect on DHAR and GR except MDHAR by Ni treatment (Baccouch et al., 2001).

8.9.3 Ozone Stress

Ozone is one of the major forms of air pollution and it affects plant development negatively (Rao and Davis, 1999; Pellinen, Palva and Kangasjarvi, 1999). The cellular damage caused by ozone can be overcome by the application of L-ascorbic acid (Conklin and Barth, 2004), suggesting its role in prevention of cellular damage. Freebairn (1960) illustrated that foliar application of ascorbic acid acts as an ozone protectant. In the plant cell, APX isoenzymes occur in subcellular compartments and cause dismutation of H_2O_2 to water along with oxidation of AsA to MDHA (Shigeoka et al., 2002). The level of cytosolic APX mRNA increases in response to ozone exposure (Kubo et al., 1995; Örvar et al., 1997) which indicated that this isoenzyme helps in detoxification of ROS generated by ozone. This was further confirmed by reduced tolerance of tobacco plants to ozone when it had an antisense version of cytosolic APX (Qrvar and Ellis, 1997). Pasqualini et al. (2002) investigated that in both control and ozone-treated leaves, the ozone resistant Bel B tobacco had an AsA pool approximately twice as high as that of ozone sensitive Bel W3. A similar situation existed in soybean leaflets, which have more ascorbic acid content in ozone tolerant cultivar Essex than in ozone sensitive Forrest cv (Robinson and Britz, 2000). Exposure to ozone can also result in premature senescence (Miller et al., 1999). This is mainly because of induction of several senescence-associated genes (SAGs) and decrease in the transcript levels of photosynthetic genes. The deficiency of AsA in *vtc1* results in the activation of many SAGs, viz. SAG13, SAG15, SAG27 (Barth et al., 2004). This is illustrated by the fact that placing of wild type and *vtc1* leaves in the dark leads to faster senescence of *vtc1* leaves than that of wild type leaves. Moreover, expression of SAG13 decreases in both wild type and *vtc1* under the application of AsA (Barth et al., 2004). Hence it shows that low AsA promotes senescence and high AsA delays senescence (Navabpour et al., 2003).

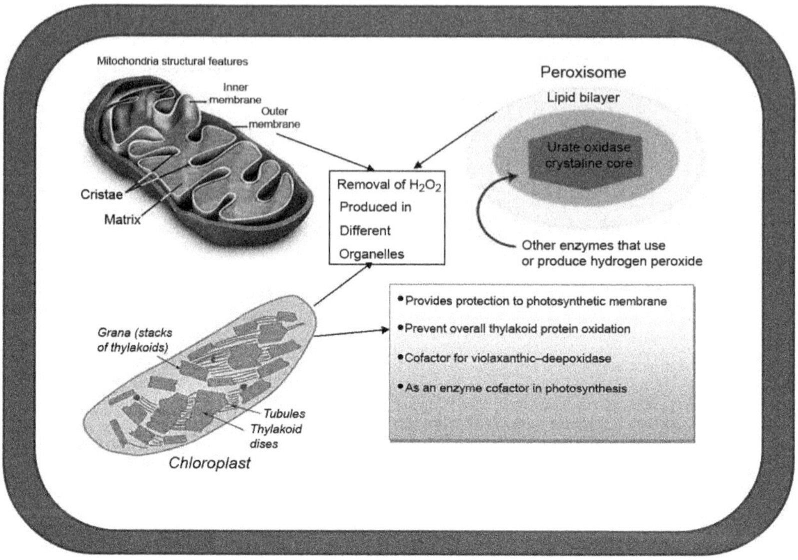

FIGURE 8.8 This figure shows the protective action of ascorbic acid at different cellular sites. During stress, H_2O_2 is produced in different organelles viz. mitochondria, chloroplast, peroxisome and is removed by ascorbate peroxidase with the help of ascorbate. In chloroplast, ascorbic acid helps in various functions.

It is evident from these findings that ascorbic acid plays an important role in overcoming adverse affects of stresses and some of the protective action of ascorbic acid at different cellular sites is discussed in Fig. 8.8.

8.10 TRANSGENICS FOR ASCORBIC ACID

Abiotic stresses such as drought, temperature extreme, salt, or ozone affect plant growth and productivity and result in a series of various morphological, physiological, biochemical and molecular changes in plants. AsA is the most abundant and important antioxidant (Smirnoff, 2000, 1996; Arrigoni and De Tullio, 2000). It is also considered to be the most powerful ROS detoxifying antioxidant due to its ability to give electrons in a number of enzymatic and nonenzymatic reactions.

The highest concentration of ascorbate was found in the fruits and flowers of *Acerola* (Badejo et al., 2007) and *Arabidopsis,* respectively, and the lowest was found in the roots of *Arabidopsis* (Lorence et al., 2004). During biosynthesis of ascorbate, the substrate GDP-D-mannose and L-galactose are of great importance in controlling the ascorbate concentration (Wheeler et al., 1998). The genes involved in its biosynthesis have been manipulated to alter the expression of ascorbic acid to make the plants tolerant to stresses.

8.10.1 Genes Targeted for Improving Ascorbate Content

Various experiments related to overexpression and virus induced genes silencing on *Arabidopsis* and tobacco reported that phosphomannosemutase (PMM) plays an important role in ascorbic acid biosynthesis (Hoeberichts et al., 2008). Transgenic tobacco having overexpressed acerola PMM was found to have three-fold more ascorbate content than that of wild type plant (Badejo et al., 2009b). In *Arabidopsis*, the genes involved in the D-Man/L-Gal pathway have been characterized (Dowdle et al., 2007; Linster et al., 2007). Mutations in the genes of D-Man/L-Gal pathway lead to a decrease in the AsA content, whereas overexpression of genes result in enhanced AsA levels (Dowdle, 2007; Linster et al., 2008). In tomato (*Solanum lycopersicon*), the overexpression of GDP-Man-3,5-epimerase gene (SlGME) increases the tolerance of plants to salt and cold due to increase in AsA content (Zhang et al., 2011). During development and stress conditions, the expression of AsA synthesis related genes was induced and increased AsA synthesis in tomato (Ioannidi et al., 2009). *At ERF98*, the *Arabidopsis* ERF gene regulates the AsA synthesis through the activation of AsA synthesis related genes (Zhang et al., 2012). Bulley and his co-workers in 2009 studied the gene expression of kiwifruit and report that there is a correlation between the increase in AsA content and the expression of GDP-L-galactose phosphorylase (GGP) and GDP-D-mannose,3',5'-epimerase (GME). The content of ascorbate increased twice when the transgenic tobacco was transformed with both GME and GGP over that of the individual gene. The transcript expression of acerola was 5- to 700-fold higher than that of *Arabidopsis* (Badejo et al., 2009a). Overexpressed tomato having GDP-D-mannose pyrophosphorylase (GMPase) gene of acerola shows a 2- to 3-fold increase in ascorbate content (Badejo et al., 2008). The level of GMPase altered due to the change in AsA content in light and darkness (Wang et al., 2011, 2013). Due to overexpression of the GMPase gene in transgenic plants under normal and temperature stress conditions, the activity of GMPase and AsA, DHA contents were increased; moreover, the DHAR, MDHAR and APX activities were also increased. Thus, it was concluded that the potato and tomato plants tolerated temperature stress due to overexpression of GMPase, which increased the AsA level (Li et al., 2011a,b). Table 8.6 shows an increase in ascorbate content with manipulation of gene expression for its biosynthetic enzymes.

Abiotic stress tolerance has been achieved in a number of transgenics through detoxification strategy. These include transgenic plants overexpressing enzymes such as glutathione peroxidase, superoxide dismutase, ascorbate peroxidase, and glutathione reductase, which are involved in oxidative protection (Zhu et al., 1999; Roxas et al., 1997). There are a number of reports that have shown that in response to different abiotic stresses such as drought, salt, or ozone, there is increased expression of APX in plants (Noctor and Foyer, 1998; Yoshimura et al., 2000). Moreover in tobacco

TABLE 8.6 Increase in Ascorbate Content in Different Plants Genetically Engineered for Overexpression of its Biosynthetic Enzymes

Plant	Overexpressed Gene	Increased % of Ascorbate	References
Arabidopsis	Mammalian L-gulono-lactone oxidase	4–7-fold	Jain and Nessler, 2000
Strawberry fruits	D-galacturonic acid	2–3-fold	Aguis et al., 2003
Tobacco	GDP-L-galactose guanyl transferase and GDP-mannose-3′,5′-epimerase	7-fold	Bulley et al., 2009
Tomato	GDP-L- galactose	Several-fold	Bulley et al., 2012
Potato	GDP-L- galactose	Several-fold	Bulley et al., 2012

TABLE 8.7 Overexpression of DHAR and MDHAR and its Consequences in Some Species

Transgenic Plant	Overexpressed Gene	Oxidative Stress Tolerance	References
Tobacco	human DHAR gene	tolerance to low temperature and salt stress	Kwon et al., 2003
Tobacco	wheat DHAR	protection against ozone in tobacco	Chen and Gallie, 2005
Arabidopsis	rice DHAR	increased salt tolerance	Ushimaru et al., 2006
Tobacco	*Arabidopsis* cytosolic DHAR	tolerance to drought and ozone stresses	Eltayeb et al., 2006
Tobacco	overexpression of MDHAR	tolerance to salt and osmotic stresses	Eltayeb et al., 2007
Tomato fruit	increased MDHAR	chilling tolerance	Stevens et al., 2008

chloroplasts overexpression of APX provides enhanced tolerance of the plant against salt and water deficit (Badawi et al., 2004a,b). Table 8.7 indicates some examples of plants where genetic engineering of DHAR and MDHAR expression has resulted in tolerance to some abiotic stresses. These studies suggest that the defense system of plants can be enhanced against oxidative stress by generating transgenic plants having overexpressed ascorbate content

(Foyer et al., 1994). Through the manipulation of biosynthetic or recycling pathways, various technologies have been developed to increase the content of ascorbic acid in plants (Chen and Gallie, 2012).

8.11 CONCLUSIONS

In conclusion, this review summarizes the various biosynthetic pathways for ascorbic acid that are fully supported by biochemical and molecular genetic evidence. It also helps us to understand more about the controlled mechanisms of ascorbic acid and its degradation. Along with glutathione, ascorbic acid is considered to be one of the major redox buffers of the plant cell, so its role in various plant functions is also discussed in addition to its role as an antioxidant, in cell growth and elongation, in cell division, as an enzyme cofactor, and in floral induction and nodulation in leguminous plants. The role of ascorbic acid in overcoming various types of abiotic stresses is increasingly becoming clear, especially with molecular approaches. Much progress has been made in our understanding of ascorbic acid synthesis, regulation and its functions. In future, the stress tolerance of economically important plants can be improved with the help of transgenic approaches. AsA has multi-functional roles and new roles are emerging, especially under stress situations, where the relevance of this molecule is vital. Exogenous application of AsA has resulted in improvement of plants under various stress situations. There is further need to understand the impact of increased function on other processes of stressed plants. More experimental work is required to understand the mechanisms of this molecule in some crops as well as to identify new steps during its metabolism. With the help of new transgenic approaches, plants can be raised that have high resistance to oxidative stress as well as fruits with higher nutritional value. Deficiency of vitamin C leads to early senescence of plants; therefore by synthesis of AsA via genetic engineering, the shelf life of fruits and vegetables can be improved. Crops can be genetically engineered for increasing the endogenous content of AsA to enhance their tolerance to a variety of abiotic stresses, which would be vital for agriculture.

REFERENCES

Agius, F., Gonzalez-Lamothe, R., Caballero, J.L., Munoz-Blanco, J., Botella, M.A., Valpuesta, V., 2003. Engineering increased vitamin C levels in plants by overexpression of a D-galacturonic acid reductase. Nat. Biotechnol. 21, 177–181.

Alhagdow, M., Mounet, F., Gilbert, L., Nunes-Nesi, A., Garcia, V., Just, D., et al., 2007. Silencing of the mitochondrial ascorbate synthesizing enzyme l-galactono-1,4-lactone dehydrogenase affects plant and fruit development in tomato. Plant Physiol. 145, 1408–1422.

Allan, A.C., Fluhr, R., 1997. Two distinct sources of elicited reactive oxygen species in tobacco epidermal cells. Plant Cell 9, 1559–1572.

Almeselmani, M., Deshmukh, P.S., Sairam, R.K., Kushwaha, S.R., Singh, T.P., 2006. Protective role of antioxidant enzymes under high temperature stress. Plant Sci. 171, 382–388.

Alós, E., Rodrigo, M.J., Zacarías, L., 2013. Transcriptomic analysis of genes involved in the biosynthesis, recycling and degradation of L-ascorbic acid in pepper fruits (*Capsicum annuum* L.). Plant Sci. 207, 2–11.

Amin, B., Mahleghah, G., Mahmood, H.M.R., Hossein, M., 2009. Evalution of interaction effect of drought stress with ascorbate and salicylic acid on some of physiological and biochemical parameters in okra (*Hibiscus esculentus* L.). Res. J. Biol. Sci. 4, 380–387.

Amor, N.B., Jimenez, A., Megdiche, W., Lundqvist, M., Sevilla, F., Abdelly, C., 2006. Response of anti-oxidant systems to NaCl stress in the halophyte *Cakile maritima*. J. Plant Physiol. 126, 446–457.

Ananyev, G., Renger, G., Wacker, U., Klimov, V., 1994. The photoproduction of superoxide radicals and the superoxide dismutase activity of photosystem II. The possible involvement of cytochrome b559. Photosynth. Res. 41, 327–338.

Anjum, N.A., Umar, S., Chan, M.T., 2010. Ascorbate-Glutathione Pathway and Stress Tolerance in Plants. Springer, Dordrecht, The Netherlands.

Aragüez, I., Rus, E.C., Botella, M.A., Escobar, N.M., Valpuesta, V., 2013. Proteomic analysis of strawberry achenes reveals active synthesis and recycling of l-ascorbic acid. J. Proteome 83, 160–179.

Arrigoni, O., 1994. Ascorbate system in plant development. J. Bioenerg. Biomembr. 26, 407–419.

Arrigoni, O., De Tullio, M.C., 2000. The role of ascorbic acid in cell metabolism: between gene-directed functions and unpredictable chemical reactions. J. Plant Physiol. 157, 481–488.

Arrigoni, O., De Tullio, M.C., 2002. Ascorbic acid: much more than just an antioxidant. Biochim. Biophys. Acta. 1569, 1–9.

Arrigoni, O., Calabrese, G., DeGara, L., Bitonti, M.B., Liso, R., 1997. Correlation between changes in cell ascorbate and growth of *Lupinus albus* seedlings. J. Plant Physiol. 150, 302–308.

Asada, K., 1994. Mechanisms for scavenging reactive molecules generated in chloroplasts under light stress. In: Baker, N.R., Bowyer, J.R. (Eds.), Photoinhibition of Photosynthesis: From Molecular Mechanisms to the Field. Bios Scientific Publishers, Oxford, pp. 129–142.

Asada, K., 1997. The role of ascorbate peroxidase and mono-dehydro ascorbate reductase in H_2O_2 scavenging in plants. In: Scandalios, J.G. (Ed.), Oxidative Stress and Molecular Biology of Antioxidant Defences. Cold Spring Harbor Laboratory Press, New York, 527–568.

Asada, K., 1999. The water–water cycle in chloroplasts: scavenging of active oxygen and dissipation of excess photons. Rev. Plant Phys. 50, 601–639.

Asada, K., Takahashi, M., 1987. Production and scavenging of active oxygen in photosynthesis. In: Kyle, D.J., et al., (Eds.), Photoinhibition. Elsevier, pp. 227–287.

Ashraf, M., 2009. Biotechnological approach of improving plant salt tolerance using antioxidants as markers. Biotechnol. Adv. 27, 84–93.

Atkinson, C.J., Davies, M.J., Taylor, J.M., Longbottom, H., 2013. Linking ascorbic acid production in *Ribes nigrum* with fruit development and changes in sources and sinks. Ann. Bot. 111, 703–712.

Baccouch, S., Chaoui, A., Ferjani, E., 2001. Nickel toxicity induces oxidative damage in *Zea mays* roots. J. Plant Nutr. 24, 1085–1097.

Badawi, G.H., Kawano, N., Yamauchi, Y., Shimada, E., Sasaki, R., Kubo, A., et al., 2004a. Over-expression of ascorbate peroxidase in tobacco chloroplasts enhances the tolerance to salt stress and water deficit. Physiol. Plant 121, 231–238.

Badawi, G.H., Yamauchi, Y., Shimada, E., Sasaki, R., Kawano, N., Tanaka, K., 2004b. Enhanced tolerance to salt stress and water deficit by overexpressing superoxide dismutase in tobacco (*Nicotiana tabacum*) chloroplasts. Plant Sci. 66, 919–928.

Badejo, A.A., Jeong, S.T., Goto-Yamamoto, N., Esaka, M., 2007. Cloning and expression of GDP-D-mannose pyrophosphorylase gene and ascorbic acid content of acerola (*Malpighia glabra* L.) fruit at ripening stages. Plant Physiol. Bioch. 45, 665–672.

Badejo, A.A., Tanaka, N., Esaka, M., 2008. Analysis of GDP-D-mannose pyrophosphorylase gene promoter from acerola (*Malpighia glabra*) and increase in ascorbate content of transgenic tobacco expressing the acerola gene. Plant Cell Physiol. 49, 126–132.

Badejo, A.A., Fujikawa, Y., Esaka, M., 2009a. Gene expression of ascorbic acid biosynthesis related enzymes of the Smirnoff-Wheeler pathway in acerola (*Malpighia glabra*). J. Plant Physiol. 166, 652–660.

Badejo, A.A., Eltelib, H.A., Fukunaga, K., Fujikawa, Y., Esaka, M., 2009b. Increase in ascorbate content of transgenic tobacco plants overexpressing the acerola (*Malpighia glabra*) phosphomanno-mutase gene. Plant Cell Physiol. 50, 423–428.

Badiani, M., Paolacci, A.R., Fusari, A., Dovidio, R., Scandalios, J.G., Porceddu, E., et al., 1997. Non-optimal growth temperatures and antioxidants in the leaves of *Sorghum bicolor* (L.) Moench. II. Short-term acclimation. J. Plant Physiol. 151, 409–421.

Barakat, H., 2003. Interactive effects of salinity and certain vitamins on gene expression and cell division. Int. J. Agric. Biol. 3, 219–225.

Barth, C., Moeder, W., Klessig, D.F., Conklin, P.L., 2004. The timing of senescence and response to pathogens is altered in the ascorbate-deficient *Arabidopsis* mutant vitamin c-1. Plant Physiol. 134, 1784–1792.

Bartoli, C.G., Pastori, G.M., Foyer, G.M., 2000. Ascorbate biosynthesis in mitochondria is linked to the electron transport chain between complexes III and IV. Plant Physiol. 123, 335–343.

Bartoli, C.G., Gómez, F., Martínez, D.E., Guiamet, J.J., 2004. Mitochondria are the main target for oxidative damage in leaves of wheat (*Triticum aestivum* L.). J. Exp. Bot. 55, 1663–1669.

Berlett, B.S., Stadtman, E.R., 1997. Protein oxidation in aging, disease, and oxidative stress. J. Biol. Chem. 272, 20313–20316.

Bernier, G., Havelange, A., Houssa, C., Petitjean, A., Lejeune, P., 1993. Physiological signals that induce flowering. Plant Cell 5, 1147–1155.

Blatt, M.R., 2000. Cellular signaling and volume control in stomatal movements in plants. Annu. Rev. Cell. Dev. Bio. 16, 221–241.

Bohnert, H.J., Nelson, D.E., Jensen, R.G., 1995. Adaptations to environmental stresses. Plant Cell 7, 1099–1111.

Bosch, S.M., Queval, G., Foyer, C.H., 2013. The impact of global change factors on redox signaling underpinning stress tolerance. Plant Physiol. 161, 15–19.

Bowler, C., 1992. Superoxide dismutases and stress tolerance. Annu. Rev. Plant Phys. Plant Mol. Biol. 43, 83–116.

Bricker, T.M., Frankel, L.K., 2008. The psbo1 mutant of *arabidopsis* cannot efficiently use calcium in support of oxygen evolution by photosystem II. J. Biol Chem. 283, 29022–29027.

Bulley, S., Wright, M., Rommens, C., Yan, H., Rassam, M., Wang, K.L., et al., 2012. Enhancing ascorbate in fruits and tubers through over-expression of the L-galactose pathway gene GDP-L-galactose phosphorylase. Plant Biotech. J. 10, 390–397.

Bulley, S.M., Rassam, M., Hoser, D., Otto, W., Schunemann, N., Wright, M., et al., 2009. Gene expression studies in kiwifruit and gene over-expression in *Arabidopsis* indicates that GDP-L-galactose guanyltransferase is a major control point of vitamin C biosynthesis. J. Exp. Bot. 60, 765–778.

Burns, J.J., 1967. Ascorbic acid. In: Greenberg, D.M. (Ed.), Metabolic Pathways, Vol. I, Third edn. Academic Press, New York, pp. 394–411.

Burns, J.J., Mosbach, E.H., 1956. Further observations in the biosynthesis of L-ascorbic acid from D-glucose in the rat. J. Biol. Chem. 221, 107–111.

Carpita, N.C., Gibeaut, D.M., 1993. Structural models of primary cell walls in flowering plants: consistency of molecular structure with the physical properties of the walls during growth. Plant J. 3, 1–30.

Chakraborty, U., Pradhan, D., 2009. Antioxidant responses of four *Glycine max* (L.) Merr. cultivars to low temperature induced stress. J. Plant Biol. 36, 23–32.

Chao, Y.Y., Hong, C.Y., Kao, C.H., 2010. The decline in ascorbic acid content is associated with cadmium toxicity of rice seedlings. Plant Physiol. Bioch. 48, 374–381.

Chaparzadeh, N., D'Amico, M.L., Khavari-Nejad, R.A., Izzo, R., Navari-Izzo, F., 2004. Antioxidative responses of *Calendula officinalis* under salinity conditions. Plant Physiol. Bioch. 42, 695–701.

Chatterjee, I.B., 1973. Evolution and the biosynthesis of ascorbic acid. Sci. 182, 1271–1272.

Chaves, M.M., Maroco, J.P., Pereira, J.S., 2003. Understanding plant responses to drought from genes to the whole plant. Funct. Plant Biol. 30, 239–264.

Chen, Z., Gallie, D.R., 2004. The ascorbic acid redox state controls guard cell signaling and stomatal movement. Plant Cell 16, 1143–1162.

Chen, Z., Gallie, D.R., 2005. Increasing tolerance to ozone by elevating foliar ascorbic acid confers greater protection against ozone than increasing avoidance. Plant Physiol. 138, 1673–1689.

Chen, Z., Gallie, D.R., 2012. Induction of monozygotic twinning by ascorbic acid in tobacco. PLoS One. 7, e39147.

Cheng, F.Y., Burkey, K.O., Robinson, J.M., Booker, F.L., 2007. Leaf extracellular ascorbate in relation to O_3 tolerance of two soybean cultivars. Environ. Pollut. 150, 355–362.

Citterio, S., Sgorbati, S., Scippa, S., Sparvoli, E., 1994. Ascorbic acid effect on the onset of cell proliferation in pea root. Physiol Plant. 92, 601–607.

Clemens, S., 2006. Evolution and function of phytochelatin synthase. J. Plant Physiol. 163, 319–332.

Cocetta, G., Karppinen, K., Suokas, M., Hohtola, A., Häggman, H., Spinardi, A., et al., 2012. Ascorbic acid metabolism during bilberry (*Vaccinium myrtillus* L.) fruit development. J. Plant Physiol. 169, 1059–1065.

Conklin, P.L., Barth, C., 2004. Ascorbic acid, a familiar small molecule intertwined in the response of plants to ozone, pathogens, and the onset of senescence. Plant Cell Physiol. 27, 959–970.

Conklin, P.L., Williams, E.H., Last, R.L., 1996. Environmental stress tolerance of an ascorbic acid-deficient *Arabidopsis* mutant. J. Nutr. Sci. Vitaminol. USA. 93, 9970–9974.

Conklin, P.L., Pallanca, J.E., Last, R.L., Smirnoff, N., 1997. L-Ascorbic acid metabolism in the ascorbate deficient *Arabidopsis* mutant vtc1. Plant Physiol. 115, 1277–1285.

Conklin, P.L., Saracco, S.A., Norris, S.R., Last, R.L., 2000. Identification of acid-deficient *Arabidopsis thaliana* mutants. Genet. 154, 847–856.

Corbesier, L., Coupland, G., 2005. Photoperiodic flowering of *Arabidopsis*: integrating genetic and physiological approaches to characterization of the floral stimulus. Plant Cell Environ. 28, 54–66.

Cordoba-Pedregosa, M.D., Gonzalez-Reyes, J.A., Canadillas, M.D., Navas, P., Cordoba, F., 1996. Role of apoplastic and cell-wall peroxidases on the stimulation of root elongation by ascorbate. Plant Physiol. 112, 1119–1125.

Corpas, F.J., 2001. Peroxisomes as a source of reactive oxygen species and nitric oxide signal molecules in plant cells. Trends Plant Sci. 6, 145–150.

Creissen, G.P., Broadbent, P., Kular, B., Reynolds, H., Wellburn, A.R., Mullineaux, P.M., 1994. Manipulation of glutathione reductase in transgenic plants: implication for plant responses to environmental stress. Proc. R. Soc. Edin. 102B, 167–175.

Cruz-Rus, E., Botella, M.A., Gomez-Jimenez, M.C., 2010. Analysis of genes involved in l-ascorbic acid biosynthesis during growth and ripening of grape berries. J. Plant Physiol. 167, 739–748.

Cruz-Rus, E., Amaya, I., Sánchez-Sevilla, J.F., Botella, M.A., Valpuesta, V., 2011. Regulation of l-ascorbic acid content in strawberry fruits. J. Exp. Bot. 62, 4191–4201.

Dai, F., Huang, Y., Zhou, M., Zhang, G., 2009. The influence of cold acclimation on antioxidative enzymes and antioxidants in sensitive and tolerant barley cultivars. Plant Biol. 53, 257–262.

Dalmia, A., Sawhney, V., 2004. Antioxidant defence mechanism under drought stress in wheat seedlings. Plant Physiol. Plant Mol. Biol. 10, 109–114.

Darkó, E., Ambrus, H., Stefanovits-Bányai, E., Fodor, J., Bakos, F., Barnabás, B., 2004. Aluminium toxicity, Al tolerance and oxidative stress in an Al-sensitive wheat genotype and in Al-tolerant lines developed by *in vitro* microspore selection. Plant Sci. 166, 583–591.

Dat, J., Vandenabeele, S., Vranova, E., Van Montagu, M., Inze, D., Breusegem, F.V., 2000. Dual action of the active oxygen species during plant stress responses. Cell Mol. Biol. 57, 779–795.

Davey, M.W., Keulemans, J., 2004. Determining the potential to breed for enhanced antioxidant status in *Malus*: mean inter- and intravarietal fruit vitamin C and glutathione contents at harvest and their evolution during storage. J. Agric. Food Chem. 52, 8031–8038.

Davey, M.W., Gilot, C., Persiau, G., Østergaard, J., Han, Y., 1999. Ascorbate biosynthesis in *arabidopsis* cell suspension culture. Plant Physiol. 121, 535–543.

Davey, M.W., Montagu, M.V., Inzé, D., Sanmartin, M., Kanellis, A., Smirnoff, N., et al., 2000. Plant L-ascorbic acid: chemistry, function, metabolism, bioavailability and effects of processing. J. Sci. Food Agric. 80, 825–860.

Davey, M.W., Auwerkerken, A., Keulemans, J., 2007. Relationship of apple vitamin C and antioxidant contents to harvest date and postharvest pathogen infection. J. Sci. Food Agric. 87, 802–813.

Davies, M.B., Austin, J., Partridge, D.A., 1991. Vitamin C: Its chemistry and biochemistry. Cambridge: Royal Society of Chemistry.

Davis, D.G., Swanson, H.R., 2001. Activity of stress-related enzymes in the perennial weed leafy spurge (*Euphorbia esula* L.). Environ. Expt. Bot. 46, 95–108.

De Leo, P., Dalessandro, G., De Santis, A., Arrigoni, O., 1973. Inhibitory effect of lycorine on cell division and cell elongation. Plant Cell Physiol. 14, 481–486.

Demiral, T., Turkan, I., 2005. Comparative lipid peroxidation, antioxidant defence systems and praline content in roots of two rice cultivars differing in salt tolerance. Environ. Exp. Bot. 53, 247–257.

Demmig-Adams, B., Adams III, W.W., 1992. Photoprotection and other responses of plants to high light stress. Annu. Rev. Plant Physiol. Plant Mol. Biol. 43, 599–626.

Desikin, R., 2001. Regulation of the *Arabidopsis* transcriptosome by oxidative stress. Plant Physiol. 127, 159–172.

Dietzk, J., Baier, M., Krämer, U., 1999. Free Radicals and Reactive Oxygen Species as Mediators of Heavy Metal Toxicity in Plants: From Molecules to Ecosystems. Springer-Verlag, Berlin, pp. 73–97.

Ding, S., Lei, M., Lu, Q., Zhang, A., Yin, Y., Wen, X., et al., 2012. Enhanced sensitivity and characterization of photosystem II in transgenic tobacco plants with decreased chloroplast glutathione reductase under chilling stress. BBA – Bioenerg. 1817, 1979–1991.

Dowdle, J., Ishikawa, T., Gatzek, S., Rolinski, S., Smirnoff, N., 2007. Two genes in *Arabidopsis thaliana* encoding GDP-L-galactose phosphorylase are required for ascorbate biosynthesis and seedling viability. Plant J. 52, 673–689.

Drazkiewicz, M., Skorzynska-Polit, E., Krupa, Z., 2003. Response of the ascorbate-glutathione cycle to excess copper in *Arabidopsis thaliana* (L.). Plant Sci. 164, 195–202.

Edwards, E.A., Rawsthorne, S., Mullineaux, P.M., 1990. Subcellular distribution of multiple forms of glutathione reductase in leaves of pea (*Pisum sativum* L.). Planta 180, 278–284.

El Hariri Sh., M., Sadak, D.M., El-Bassiouny, H.M.S., 2010. Response of flax cultivars to ascorbic acid and α-tocopherol under salinity stress conditions. IJAR 2, 101–109.

Eltayeb, A.E., Kawano, N., Badawi, G., Kaminaka, H., Sanekata, T., Morishima, I., 2006. Enhanced tolerance to ozone and drought stresses in transgenic tobacco overexpressing dehydroascorbate reductase in cytosol. Physiol. Plant 127, 57–65.

Eltayeb, A.E.K., Badawi, G.H., Kaminaka, H., Sanekata, T., Shibahara, T., Inanaga, S., 2007. Overexpression of monodehydroascorbate reductase in transgenic tobacco confers enhanced tolerance to ozone, salt and polyethylene glycol stresses. Planta 225, 1255–1264.

Enami, I., Kitamura, M., Tomo, T., Isokawa, Y., Ohta, H., Katoh, S., 1994. Is the primary cause of thermal inactivation of oxygen evolution in spinach PS II membranes release of the extrinsic 33 kDa protein or of Mn?. Biochim. Biophys. Acta. 1186, 52–58.

Ensminger, I., Busch, F., Huner, N.P.A., 2006. Photostasis and cold acclimation: sensing low temperature through photosynthesis. Plant Physiol. 126, 28–44.

Esaka, M., Hattori, T., Fujisawa, K., Sakajo, S., Asahi, T., 1990. Molecular cloning and nucleotide sequence of full-length cDNA for ascorbate oxidase from cultured pumpkin cells. Eur. J. Biochem. 191, 537–541.

Esaka, M., Fujisawa, K., Goto, M., Kisu, Y., 1992. Regulation of ascorbate oxidase expression in pumpkin by auxin and copper. Plant Physiol. 100, 231–237.

Esfandiari, E., Shekari, F., Shekari, F., Esfandiari, M., 2007. The effect of salt stress on antioxidant enzymes activity and lipid peroxidation on the wheat seedling. Not. Bot. Hort. Agrobot. Cluj. 35, 48–56.

Fahrenholzt, S.R., Doleiden, F.H., Tozzolo, A.M., Lamola, A.A., 1974. On the quenching of singlet oxygen by a–tocopherol. Photochem. Photobiol. 20, 505–509.

Farooq, M., Irfan, M., Aziz, T., Ahmad, I., Cheema, S.A., 2013. Seed priming with ascorbic acid improves drought resistance of wheat. Agron. Crop Sci. 199, 12–22.

Felicetti, E., Mattheis, J.P., 2010. Quantification and histochemical localization of ascorbic acid in 'Delicious,' 'Golden Delicious' and 'Fuji' apple fruit during on-tree development and cold storage. Postharvest Biol. Technol. 56, 56–63.

Fortunato, A., Lidon, F., Batista-Santos, A.P., Leitão, E., Pais, I., Ribeiro, A., et al., 2010. Biochemical and molecular characterization of the antioxidant system of *Coffea* sp. under cold conditions in genotypes with contrasting tolerance. J. Plant Physiol. 167, 333–342.

Foyer, C.H., 1993. Ascorbic acid. In: Alscher, R.G., Hess, J.L. (Eds.), Antioxidants in Higher Plants. CRC Press, Boca Raton. FL, pp. 31–58.

Foyer, C.H., Fletcher, J.M., 2001. Plant antioxidants: colour me healthy. Biologist 48, 115–120.

Foyer, C.H., Noctor, G., 2005a. Oxidant and antioxidant signalling in plants: a re-evaluation of the concept of oxidative stress in a physiological context. Plant Cell Environ. 28, 1056–1071.

Foyer, C.H., Noctor, G., 2005b. Redox homeostasis and antioxidant signalling: a metabolic interface between stress perception and physiological responses. Plant Cell 17, 1866–1875.

Foyer, C.H., Noctor, G., 2011. Ascorbate and glutathione: the heart of the redox hub. Plant Physiol. 155, 2–18.

Foyer, C.H., Lelandais, M., Kunert, K.H., 1994. Photooxidative stress in plants. Plant Physiol. 92, 696–717.

Foyer, C.H., Souriau, N., Perret, S., Lelandais, M., Kunert, K., Pruvost, C., et al., 1995. Overexpression of glutathione reductase but not glutathione synthetase leads to increases in antioxidant capacity and resistance to photoinhibition in poplar trees. Plant Physiol. 109, 1047–1057.

Freebairn, H.T., 1960. The prevention of air pollution damage to plants by the use of Vitamin C sprays. JAPCA 10, 314–317.

Fry, S.C., 1986. Cross-linking of matrix polymers in the growing cell walls of angiosperms. Annu. Rev. Plant Physiol. 37, 165–186.

Gallie, D.R., 2013a. L-Ascorbic Acid: a multifunctional molecule supporting plant growth and development. Scientifica 2013, 795964.

Gallie, D.R., 2013b. The role of l-ascorbic acid recycling in responding to environmental stress and in promoting plant growth. J. Exp. Bot. 64, 433–443.

Garnczarska, M., 2005. Response of the ascorbate-glutathione cycle to re-aeration following hypoxia in lupine roots. Plant Physiol. Bioch. 43, 583–590.

Gautier, H., Lopez-Lauri, F., Massot, C., Murshed, R., Marty, I., Sallanon, H., et al., 2010. Impact of ripening and salinity on tomato fruit ascorbate content and enzymatic activities related to ascorbate recycling. Funct. Plant Sci. Biotechnol. 4, 66–75.

Gest, N., Gautier, H., Stevens, R., 2013. Ascorbate as seen through plant evolution: the rise of a successful molecule? J. Exp. Bot. 64, 33–53.

Gonzalez-Reyes, J.A., Hidalgo, A., Caler, J.A., Palos, R., Navas, P., 1994. Nutrient uptake changes in ascorbate free radical-stimulated roots. Plant Physiol. 104, 271–276.

Gonzalez-Reyes, J.A., Alcain, F.J., Caler, J.A., Serrano, A., Cordoba, F., Navas, P., 1995. Stimulation of onion root elongation by ascorbate and ascorbate free radical in *Allium cepa* L. Protoplasma 184, 31–35.

Gossett, D.R., Banks, S.W., Millhollon, E.P., Lucas, M.C., 1996. Antioxidant response to NaCl stress in a control and a NaCl-tolerant cotton cell line grown in the presence of paraquat, buthionine sulfoximine and exogenous glutathione. Plant Physiol. 112, 803–809.

Grant, J.J., Loake, G.J., 2000. Role of reactive oxygen intermediates and cognate redox signalling in disease resistance. Plant Physiol. 124, 21–29.

Grün, M., Loewus, F.A., 1984. L-Ascorbic acid biosynthesis in the euryhaline diatom *Cyclotellacriptica*. Planta 160, 6–11.

Grundmeier, A., Dau, H., 2012. Structural models of the manganese complex of photosystem II and mechanistic implications. Biochim. Biophys. Acta. 1817, 88–105.

Guo, Z., Ou, W., Lu, S., Zhong, Q., 2006. Differential responses of antioxidative system to chilling and drought in four rice cultivars differing in sensitivity. Plant Physiol. Bioch. 44, 828–836.

Hall, A.E., 2001. Crop Responses to Environment. CRC Press LLC, Boca Raton, Florida, p. 232.

Hall, J.L., 2002. Cellular mechanisms for heavy metal detoxification and tolerance. J. Exp. Bot. 53, 1–11.

Hamada, A.M., Al-Hakimi, A.M.A., 2001. Salicylic acid versus salinity-drought induced stress on wheat seedlings. Rostl. Vyr. 47, 444–450.

Hammond-Kosack, K.E., Jones, J.D.G., 1996. Resistance gene-dependent plant defense responses. Plant Cell 8, 1773–1791.

Hancock, R.D., Galpin, J.R., Viola, R., 2000. Biosynthesis of L-ascorbic acid (vitamin C) by *Sacchararomyces cerevisiae*. FEMS Lett. 186, 245–250.

Hancock, R.D., Walker, P.G., Pont, S.D.A., Marquis, N., Vivera, S., Gordon, S.L., et al., 2007. L-Ascorbic acid accumulation in fruit of *Ribes nigrum* occurs by in situ biosynthesis via the l-galactose pathway. Funct. Plant Biol. 34, 1080–1091.

Hassanein, R.A., Bassony, F.M., Barakat, D.M., Khalil, R.R., 2009. Physiological effects of nicotinamide and ascorbic acid on *Zea mays* plant grown under salinity stress: Changes in growth, some relevant metabolic activities and oxidative defense systems. Res. J. Agr. Biol. Sci. 5, 72–81.

Hatata, M.M., Abdel-Aal, E.A., 2008. Oxidative stress and antioxidant defence mechanisms in response to cadmium treatments. American-Eurasian J. Agric. & Environ. Sci. 4, 655–669.

Havir, E.A., Tausta, S.L., Peterson, R.B., 1997. Purification and properties of violaxanthin de-epoxidase from spinach. Plant Sci. 123, 57–66.

Hayashi, R., Morohashi, Y., 1993. Phytochrome control of ascorbate oxidase activity in mustard (*Sinapis alba* L.) seedlings. Plant Physiol. 102, 1237–1241.

He, J.X., Wang, J., Liang, H.G., 1995. Effects of water stress on photochemical function and protein metabolism of photosystem II in wheat leaves. Plant Physiol. 93, 771–777.

Helsper, J.P., Kagan, L., Hilby, C.L., Maynard, T.M., Loewus, F.A., 1982. L-Ascorbic acid biosynthesis in *Ochromonasdanica*. Plant Physiol. 69, 465–468.

Herbinger, K., Tausz, M., Wonisch, A., Soja, G., Sorger, A., Grill, D., 2002. Complex interactive effects of drought and ozone stress on the antioxidant defence systems of two wheat cultivars. Plant Physiol. Bioch. 40, 691–696.

Hernández, J.A., Jiménez, A., Mullineaux, P.M., Sevilla, F., 2000. Tolerance of pea (*Pisum sativum* L.) to long term salt stress is associated with induction of antioxidant defenses. Plant Cell Environ. 23, 853–862.

Hernández, J.A., Ferrer, M.A., Jiménez, A., Barcelo, A.R., Sevilla, F., 2001. Antioxidant systems and O_2-/H_2O_2 production in the apoplast of pea leaves. Its relation with salt-induced necrotic lesions in minor veins. Plant Physiol. 127, 817–831.

Hidalgo, A., Gonzalez-Reyes, J.A., Navas, P., 1989. Ascorbate free radical enhances vacuolarization in onion root meristems. Plant Cell Environ. 12, 455–460.

Hodges, D.M., Andrews, C.J., Johnson, D.A., Hamilton, R.I., 1996. Antioxidant compound responses to chilling stress in differentially sensitive inbred maize lines. Plant Physiol. 98, 685–692.

Hodges, D.M., Lester, G.E., Munro, K.D., Toivonen, P.M.A., 2004. Oxidative stress: importance for postharvest quality. Hortic. Sci. 39, 924–929.

Hoeberichts, F.A., Vaeck, E., Kiddle, G., Coppens, E., van de Cotte, B., Adamantidis, A., et al., 2008. A temperature-sensitive mutation in the *Arabidopsis thaliana* phosphomannomutase gene disrupts protein glycosylation and triggers cell death. J. Biol. Chem. 283, 5708–57018.

Horemans, N., Foyer, C.H., Asard, H., 2000. Transport and action of ascorbate at the plant plasma membrane. Trends Plant Sci. 5, 263–267.

Houghton, J.T., Ding, Y., Griggs, D.J., Noguer, M., Linden, P.J., Xiaosu, D., 2001. Climate Change: the scientific basis contribution of working group first to third assessment report of the intergovernmental panel on climate change. Cambridge University Press, UK.

Huang, C., He, W., Guo, J., Chang, X., Su, P., Zhang, L., 2005. Increased sensitivity to salt stress in an ascorbate-deficient *Arabidopsis* mutant. J. Exp. Bot. 56, 3041–3049.

Huang, M., Guo, Z., 2005. Responses of antioxidant system to chilling stress in two rice cultivars differing in sensitivity. Plant Biol. 49, 81–84.

Huner, N.P.A., Krol, M., Williams, J.P., Maissan, E., Low, P.S., Roberts, D., et al., 1987. Low temperature development induces a specific decrease in trans-D3- hexadecenoic acid content which influences LHCII organization. Plant Physiol. 84, 12–18.

Hunter, W.J., 1989. Indole-3-acetic acid production by bacteroids from soybean root nodules. Plant Physiol. 76, 31–36.

Innocenti, A.M., Bitonti, M.B., Arrigoni, O., Liso, R., 1990. The size of the quiescent center in roots of *Allium cepa* L. grown with ascorbic acid. New Phytol. 114, 507–509.

Ioannidi, E., Kalamaki, M.S., Engineer, C., Pateraki, I., Alexandrou, D., Melli-dou, I., et al., 2009. Expression profiling of ascorbic acid-related genes during tomato fruit development and ripening and in response to stress conditions. J. Exp. Bot. 60, 663–678.

Isherwood, F.A., Chen, Y.T., Mapson, L.W., 1954. Synthesis of L-ascorbic acid in plants and animals. Biochem J. 56, 1–15.

Ishikawa, T., Dowdle, J., Smirnoff, N., 2006. Progress in manipulating ascorbic acid biosynthesis and accumulation in plants. Physiol. Plant 126, 343–355.

Jain, A.K., Nessler, C.L., 2000. Metabolic engineering of an alternative pathway for ascorbic acid biosynthesis in plants. Mol. Breed. 6, 73–78.

Jaleel, C.A., Gopi, R., Manivannan, P., Panneerselvam, R., 2007. Antioxidative potentials as a protective mechanism in *Catharanthusroseus* (L.) G.Don. Plants under salinity stress. Turk. J. Bot. 31, 245–251.

Janda, T., Szalai, G., Rios-Gonzalez, K., Veisz, O., Páldi, E., 2003. Comparative study of frost tolerance and antioxidant activity in cereals. Plant Sci. 164, 301–306.

Jiang, M.Y., Zhang, J.H., 2002. Water stress induced abscisic acid accumulation triggers the increased generation of reactive oxygen species and up regulates the activities of antioxidant enzymes in maize leaves. J. Exp. Bot. 53, 2401–2410.

Jimenez, A., Hernandez, J.A., Rio, L.A., Sevilla, F., 1997. Evidence for the presence of the ascorbate–glutathione cycle in mitochondria and peroxisomes of pea leaves. Plant Physiol. 114, 275–284.

JinCheng, W., Jie, L., JianQin, C., QiaoBin, D., LianHuang, C., Xin, X., et al., 2009. Effects of GSH on AsA-GSH circulation metabolism in chloroplasts of young loquat fruits under low temperature stress. Sci. Silvae. Sin. 45, 15–19.

Johnson, D.W., Smith, S.E., Dobrenz, A.K., 1992. Genetic and phenotypic relationships in response to NaCl at different developmental stages in alfalfa. Theor. Appl. Genet. 83, 833–838.

Jordan, D.C., 1984. Rhizobiaceae. In: Krieg, N.R., Holt, J.G. (Eds.), Bergey's Manual of Systematic Bacteriology, Vol. 1. Williams and Wilkins, Baltimore, MD, pp. 235–240.

Jung, C.H., Wells, W.W., 1997. Ascorbic acid is a stimulatory cofactor for mitochondrial glycerol-3-phosphate dehydrogenase. Biochem. Biophys. Res. Commun. 239, 457–462.

Justr, K.C., Visentainer, J.V., Evelazio de Souza, N., Matsushita, M., 2000. Nutritional composition and vitamin C stability in stored camu-camu (*Myrciaria dubia*) pulp. Arch. Latinoam. Nutr. 50, 405–408.

Kang, G., Yue, C., Peng, H., Han, Q., Li, G., Guo, T., 2011. Effects of freezing stress on anti-chilling contents in wheat leaves. J. Hen. Agri. Sci. 40, 56–60.

Kanwischer, M., Porfirova, S., Bergmüller, E., Dörmann, P., 2005. Alterations in tocopherol cyclase activity in transgenic and mutant plants of *Arabidopsis* affect tocopherol content, tocopherol composition, and oxidative stress. Plant Physiol. 137, 713–723.

Keates, S.E., Tarlyn, N.M., Loewus, F.A., Franceschi, V.R., 2000. L-Ascorbic acid and L-galactose: source of oxalic acid in calcium oxalate deposition in *Pistia stratiotes*. Phytochem. 53, 433–440.

Kerk, N.M., Feldman, L.J., 1995. A biochemical model for the initiation and maintenance of the quiescent center implication for organisation of root meristems. Dev. 121, 2825–2833.

Khan, N.A., Anjum, N.A., Nazar, R., Iqbal, N., 2009. Increased activity of ATP-sulfurylase and increased contents of cysteine and glutathione reduce high cadmium-induced oxidative stress in mustard cultivar with high photosynthetic potential. *Russ. J.* Plant Physiol. 56, 743–750.

Khanna-Chopra, R., Selote, D.S., 2007. Acclimation to drought stress generates oxidative stress tolerance in drought-resistant than -susceptible wheat cultivar under field conditions. Env. Exp. Bot. 60, 276–283.

Knight, H., Knight, M.R., 2001. Abiotic stress signalling pathways: specificity and cross-talk. Trends Plant Sci. 6, 262–267.

Kocsy, G., Galiba, G., Brunold, C., 2001. Role of glutathione in adaptation and signalling during chilling and cold acclimation in plants. Plant Physiol. 113, 158–164.

Kohler, B., Blatt, M.R., 2002. Protein phosphorylation activates the guard cell Ca21 channel and is a prerequisite for gating by abscisic acid. Plant J. 32, 185–194.

Komeda, Y., 2004. Genetic regulation of time to flower in *Arabidopsis thaliana*. Annu. Rev. Plant Biol. 55, 521–535.

Konstantinova, T., Parvanova, D., Atanassov, A., Djilianov, D., 2002. Freezing tolerant tobacco, transformed to accumulate osmoprotectants. Plant Sci. 163, 157–164.

Kotchoni, S.O., Larrimore, K.E., Mukherjee, M., Kempinski, C.F., Barth, C., 2009. Alterations in the endogenous ascorbic acid content affect flowering time in *Arabidopsis*. Plant Physiol. 149, 803–815.

Kubo, A., Saji, H., Tanaka, K., Kondo, N., 1995. Expression of *Arabidopsis* cytosolic ascorbate peroxidase in response to ozone or sulfur dioxide. Plant Mol. Biol. 29, 479–489.

Kumar, S., Kaur, R., Kaur, N., Bhandhari, K., Kaushal, N., Gupta, K., et al., 2011a. Heat-stress induced inhibition in growth and chlorosis in mungbean (*Phaseolus aureus* Roxb.) is partly mitigated by ascorbic acid application and is related to reduction in oxidative stress. *Acta Physiol.* Plant 33, 2091–2101.

Kumar, S., Malik, J., Thakur, P., Kaistha, S., Sharma, K.D., Upadhyaya, H.D., et al., 2011b. Growth and metabolic responses of contrasting chickpea *(Cicer arietinum* L.) genotypes to chilling stress at reproductive phase. Acta Physiol. Plant. 33, 779–787.

Kurganova, L.N., Veselov, A.P., Sinitsina, Y.V., Elikova, E.A., Kulaeva, O.N., 1999. Lipid peroxidation products as possible mediator of heat stress response in plants. Russ. J. Plant Physiol. 16, 181–185.

Kwon, S.Y., Choi, S.M., Ahn, Y.O., Lee, H.S., Lee, H.B., Park, Y.M., et al., 2003. Enhanced stress-tolerance of transgenic tobacco plants expressing a human dehydroascorbate reductase gene. J. Plant Physiol. 160, 347–353.

Lascano, H.R., Antonicelli, G.E., Luna, C.M., Melchiorre, M.N., Gomez, L.D., Racca, R.W., et al., 2001. Antioxidant system response of different wheat cultivars under drought: field and in vitro studies. Aust. J. Plant Physiol. 28, 1095–1102.

Lee, B.R., Kim, K.Y., Jung, W.D., Avice, D.C., Ourry, A., Kim, T.H., 2007. Peroxidases and lignification in relation to the intensity of water deficit stress in white clover (*Trifolium repens* L.). J. Exp. Bot. 58, 1271–1279.

Leung, J., Bouvier-Durand, M., Morris, P.C., Guerrier, D., Chedfor, F., Giraudat, J., 1994. *Arabidopsis* ABA-response gene ABI1: features of a calcium-modulated protein phosphatase. Sci. 264, 1448–1452.

Levitt, J., 1962. A sulfhydryl-disulfide hypothesis of frost injury and resistance in plants. J. Theor. Biol. 3, 355–391.

Li, C., Zhang, L., Shi, Q., Li, Q., Guo, X., Li, X., et al., 2011a. Effect of tomato GMPase overexpression on tolerance of potato plants to temperature stress. Sci. Agri. Sin. 44, 4952–4961.

Li, C., Zhang, L., Shi, Q., Li, Q., Guo, X., Li, X., et al., 2011b. Effects of GMPase overexpression on ascorbic acid content and relative index to low-temperature tolerance in tomato plants. Acta. Hort. Sin. 38, 692–700.

Li, M., Ma, F., Shang, P., Zhang, M., Hou, C., Liang, D., 2009. Influence of light on ascorbate formation and metabolism in apple fruits. Planta 230, 39–51.

Lin, L.S., Varner, J.E., 1991. Expression of ascorbic acid oxidase in zucchini squash (*Cucurbita pepo* L.). Plant Physiol. 96, 159–165.

Linster, C.L., Gomez, T.A., Christensen, K.C., Adler, L.N., Young, B.D., Brenner, C., et al., 2007. *Arabidopsis* VTC2 encodes a GDP-L-galactose phosphorylase, the last unknown enzyme in the smirnoff-wheeler pathway to ascorbic acid in plants. J. Biol. Chem. 282, 18879–18885.

Linster, C.L., Adler, L.N., Webb, K., Christensen, K.C., Brenner, C., Clarke, S.G., 2008. A second GDP-L-galactose phosphorylase in *Arabidopsis* enroute to vitamin C. J. Biol. Chem. 283, 18483–18492.

Lisenbee, C.S., Heinze, M., Trelease, R.N., 2003. Peroxisomal ascorbate peroxidase resides within a sub domain of rough endoplasmic reticulum in wild-type *arabidopsis* cells. Plant Physiol. 132, 870–882.

Liso, R., Innocenti, A.M., Bitonti, M.B., Arrigoni, O., 1988. Ascorbic acid induced progression of quiescent centre cells from G1 to S phase. New Phytol. 110, 469–471.

Liu, Y., Zhao, Z., Si, J., Di, C., Han, J., An, L., 2009. Brassinosteroids alleviate chilling-induced oxidative damage by enhancing antioxidant defense system in suspension cultured cells of *chorispora bungeana*. Plant Growth Regu.l 59, 207–214.

Loewus, F.A., 1999. Biosynthesis and metabolism of ascorbic acid in plants and of analogs of ascorbic acid in fungi. Phytochem. 52, 193–210.

Loewus, F.A., Kelly, S., 1961. The metabolism of D-galacturonic acid and its methyl ester in the detached ripening strawberry. Arch. Biochem. Biophys. 95, 483–493.

Loewus, F.A., Loewus, M.W., 1987. Biosynthesis and metabolism of ascorbic-acid in plants. Crit. Rev. Plant Sci. 5, 101–119.

Loewus, F.A., Jang, R., Seegmiller, C.G., 1956. The conversion of 14C-labeled sugars to L-ascorbic acid in ripening strawberries. J. Biol. Chem. 222, 649–664.

Lopez, F., Vansuyt, G., Case-Delbart, F., Fourcroy, P., 1996. Ascorbate peroxidase activity, not the mRNA level, is enhanced in salt-stressed *Raphanus sativas* plants. Plant Physiol. 97, 13–20.

Lorence, A., Chevone, B.I., Mendes, P., Nessler, C., 2004. Myo-Inositol oxygenase offers a possible entry point into plant ascorbate biosynthesis. Plant Physiol. 134, 1200–1205.

Lukaszewski, K.M., Blevins, D.G., 1996. Root growth inhibition in boron-deficient or aluminum-stressed squash may be a result of impaired ascorbate metabolism. Plant Physiol. 112, 1135–1140.

Lyons, J., 1973. Chilling injury in plants. Annu. Rev. Plant Physiol. 24, 445–466.

Ma, Y.H., Ma, F.W., Zhang, J.K., Li, M.J., Wang, Y.H., Liang, D., 2008. Effects of high temperature on activities and gene expression of enzymes involved in ascorbate-glutathione cycle in apple leaves. Plant Sci. 175, 761–766.

Mahmoud, M.A.A., Chedea, V.S., Detsi, A., Kefalas, P., 2013. Ascorbic acid modifies the free radical scavenging behaviour of catechin: An insight into the mechanism. Food Res. Int. 51, 907–913.

Malik, S., Ashraf, M., 2012. Exogenous application of ascorbic acid stimulates growth and photosynthesis of wheat (*Triticum aestivum* L.) under drought. Soil Environ. 31, 72–77.

Martinez, C., Pons, E., Prats, G., Leon, J., 2004. Salicylic acid regulates flowering time and links defence responses and reproductive development. Plant J. 37, 209–217.

Matamoros, M.A., Loscos, J., Coronado, M.J., Ramos, J., Sato, S., Testillano, P.S., 2006. Biosynthesis of ascorbic acid in legume root nodules. Plant Physiol. 141, 1068–1077.

Maxwell, D.P., Wang, Y., McIntosh, L., 1999. The alternative oxidase lowers mitochondrial reactive oxygen production in plant cells. Proc. Nat. Acad. Sci. USA. 96, 8271–8276.

Mazid, M., Khan, T.A., Khan, Z.H., Quddusi, S., Mohammad, F., 2011. Occurrence, biosynthesis and potentialities of ascorbic acid in plants. Int. J. Plant, Animal and Environmental Sciences 1, 167–184.

McKersie, B.D., Bowley, S.R., Harjanto, E., Leprince, O., 1996. Water deficit tolerance and field performance of transgenic alfalfa overexpressing superoxide dismutase. Plant Physiol. 111, 1177–1181.

Mellado, M., Contreras, R.A., Gonzalez, A., Dennett, G., Moenne, A., 2012. Copper-induced synthesis of ascorbate, glutathione and phytochelatins in the marine alga *Ulva compressa* (Chlorophyta). Plant Physiol. Biochem. 51, 102–108.

Menconi, M., Sgherri, C.L.M., Pinzino, C., Navari-Izzo, F., 1995. Activated oxygen production and detoxification in wheat plants subjected to a water deficit programme. J. Exp. Bot. 46, 1123–1130.

Miguel, A., Rosales, Ruiz, J.M., Hernandez, J., Soriano, T., Castilla, N., Romero, L., 2006. Antioxidant content and ascorbate metabolism in cherry tomato exocarp in relation to temperature and solar radiation. J. Sci. Food Agric. 86, 1545–1551.

Miller, J.D., Arteca, R.N., Pell, E.J., 1999. Senescence-associated gene expression during ozone-induced leaf senescence in *Arabidopsis*. Plant Physiol. 120, 1015–1024.

Mittler, R., 2002. Oxidative stress; antioxidants and stress tolerance. Trends Plant Sci. 7, 405–410.

Mittova, V., Theodoulou, F.L., Kiddle, G., Gómez, L., Volokita, M., Tal, M., et al., 2003. Coordinate induction of glutathione biosynthesis and glutathione-metabolizing enzymes is correlated with salt tolerance in tomato. FEBS (Fed. Eur. Biochem. Soc.) Lett. 554, 417–421.

Mittova, V., Guy, M., Tal, M., Volokita, M., 2004. Salinity up-regulates the antioxidative system in root mitochondria and peroxisomes of the wild salt-tolerant tomato species *Lycopersicon pennellii*. J. Exp. Bot. 55, 1105–1113.

Miyake, C., Asada, K., 1992. Thylakoid bound ascorbate peroxidase in spinach chloroplasts and photoreduction of its primary oxidation product, Monodehydroascorbate radicals in the thylakoids. Plant Cell Physiol. 33, 541–553.

Molina, A., Bueno, P., Marín, M.C., Rosales, M.P.R., Belver, A., Venema, K., et al., 2002. Involvement of endogenous salicylic acid content, lipoxygenase and antioxidant enzyme activities in the response of tomato cell suspension cultures to NaCl. New Phytol. 156, 409–415.

Muller, J.E., Whitsitt, M.S., 1996. Plant cellular responses to water deficit. Plant Growth Regul. 20, 41–46.

Munné-Bosch, S., Alegre, L., 2002a. The function of tocopherols and tocotrienols in plants. Crit. Rev. Plant Sci. 21, 31–57.

Munné-Bosch, S., Alegre, L., 2002b. Plant aging increases oxidative stress in chloroplasts. Planta 214, 608–615.

Nash, D., Miyao, M., Murata, N., 1985. Heat inactivation of oxygen evolution in photosystem II particles and its acceleration by chloride depletion and exogenous manganese. Biochim. Biophys. Acta. 807, 127–133.

Navabpour, S., Morris, K., Allen, R., Harrison, E., Buchanan-Wollaston, V., 2003. Expression of senescence enhanced genes in response to oxidative stress. J. Exp. Bot. 54, 2285–2292.

Nayyar, H., Gupta, D., 2006. Differential sensitivity of C3 and C4 plants to water deficit stress: association with oxidative stress and antioxidants. Environ. Exp. Bot. 58, 106–113.

Nayyar, H., Walia, D.P., 2004. Genotypic variation in response to water stress and abscisic acid induced accumulation of osmolytes in flag leaf and developing wheat grains. J. Agron. Crop Sci. 190, 39–45.

Nedelkoska, T.V., Doran, P.M., 2000. Characteristics of heavy metal uptake by plants species with potential for phytoremediation and phytomining. Miner. Eng. 13, 549–561.

Neubauer, C., Yamamoto, H.Y., 1992. Mehler-peroxidase reaction mediates zeaxanthin formation and zeaxanthin-related fluorescence quenching in intact chloroplasts. Plant Physiol. 99, 1354–1361.

Nieto-Sotelo, J., Ho, T.H.D., 1986. Effect of heat shock on the metabolism of glutathione in maize roots. Plant Physiol. 82, 1031–1035.

Noctor, G., Foyer, C.H., 1998. Ascorbate and glutathione: keeping active oxygen under control. Annu. Rev. Plant. Physiol. Plant. Mol. Biol. 49, 249–279.

Obinger, C., Regelsberger, G., Pircher, A., Strasser, G., Peschek, G.A., 1998. Scavenging of superoxide and hydrogen peroxide in blue-green algae (cyanobacteria). Physiol. Plant 104, 693–698.

Örvar, B.L., McPherson, J., Ellis, B.E., 1997. Pre-activating wounding response in tobacco prior to high-level ozone exposure prevents necrotic injury. Plant J. 11, 203–212.

O'Malley, D.M., Whetten, R., Bao, W., Chen, C.L., Sederoff, R.R., 1993. The role of laccase in lignification. Plant J. 4, 751–757.

Paciolla, C., Tommasi, F., 2003. The ascorbate system in two Bryophytes: *brachythecium velutinum* and *marchantia polymorpha*. Biol. Plant. 47, 387–393.

Padh, H., 1990. Cellular functions of ascorbic acid. J. Biochem.Cell Biol. 68, 1166–1173.

Page, M., Sultana, N., Paszkiewicz, K., Florance, H., Smirnoff, N., 2012. The influence of ascorbate on anthocyanin accumulation during high light acclimation in *arabidopsis thaliana:* further evidence for redox control of anthocyanin synthesis. Plant Cell Environ. 35, 388–404.

Pallanca, J.E., Smirnoff, N., 1999. Ascorbic acid metabolism in pea seedlings. A comparison of d-glucosone, l-sorbosone, and l-galactono- 1,4-lactone as ascorbate precursors. Plant Physiol. 20, 453–462.

Pallanca, J.E., Smirnoff, N., 2000. The control of ascorbic acid synthesis and turnover in pea seedlings. J. Exp. Bot. 51, 669–674.

Panchuk, I.I., Volkov, R.A., Schoffl, F., 2002. Heat stress and heat shock transcription factor dependent expression and activity of ascorbate peroxidase. Plant Physiol. 129, 838–853.

Paradiso, A., Berardino, R., De Pinto, M.C., Sanitá Di Toppi, L., Storelli, M.M., Tommasi, F., et al., 2008. Increase in ascorbate-glutathione metabolism as local and precocious systemic responses induced by cadmium in durum wheat plants. Plant Cell Physiol. 49, 362–374.

Pasqualini, S., Della Torre, G., Ferranti, F., Ederli, L., Piccioni, C., Reale, L., et al., 2002. Salicylic acid modulates ozone induced hypersensitive cell death in tobacco plants. Physiol. Plant 115, 204–212.

Pastori, G.M., Foyer, C.H., 2002. Common components, net-works, and pathways of cross-tolerance to stress. The central role of "redox" and abscisic acid-mediated controls. Plant Physiol. 129, 460–468.

Pastori, G.M., Kiddle, G., Antoniw, J., Bernard, S., Veljovic-Jovanovic, S., Verrier, P.J., et al., 2003. Leaf vitamin C contents modulate plant defense transcripts and regulate genes that control development through hormone signaling. Plant Cell 15, 939–951.

Pateraki, I., Sanmartin, I.A.M., Kalamaki, M.S., Gerasopoulos, D., Kanellis, A.K., 2004. Modecular characterization and expression studies during melon fruit development and ripening of L-galactono-1,4-lactone dehydrogenase. J. Exp. Bot. 55, 1623–1633.

Pavet, V., Olmos, E., Kiddle, G., Mowla, S., Kumar, S., Antoniw, J., et al., 2005. Ascorbic acid deficiency activates cell death and disease resistance responses in *arabidopsis*. Plant Physiol. 139, 1291–1303.

Payton, P., Webb, R., Kornyeyev, D., Allen, R., Holaday, A., 2001. Protecting cotton photosynthesis during moderate chilling at high light intensity by increasing chloroplastic antioxidant enzyme activity. J. Exp. Bot. 52, 2345–2354.

Pellinen, R., Palva, T., Kangasjarvi, J., 1999. Short communication subcellular localization of ozone-induced hydrogen peroxide production in birch (*Betula pendula*) leaf cells. Plant J. 20, 349–356.

Pignocchi, C., Fletcher, J.M., Wilkinson, J.E., Barnes, J.D., Foyer, C.H., 2003. The function of ascorbate oxidase in tobacco. Plant Physiol. 132, 1631–1641.

Popelkova, H., Yocum, C.F., 2011. PsbO, the manganese-stabilizing protein: analysis of the structure-function relations that provide insights into its role in photosystem II. J. Photochem. Photobiol. B. 104, 179–190.

Prasad, T.K., Anderson, M.D., Martin, B.A., Stewart, C.R., 1994. Evidence for chilling-induced oxidative stress in maize seedlings and a regulatory role for hydrogen peroxide. Plant Cell 6, 65–74.

Prescott, A.G., John, P., 1996. Dioxygenases: molecular structure and role in plant metabolism. Annu. Rev. Plant Physiol. Plant Mol. Biol. 47, 245–271.

Puppo, A., Groten, K., Bastian, F., Carzaniga, R., Soussi, M., Lucas, M.M., 2005. Legume nodule senescence: roles for redox and hormone signaling in the orchestration of the natural aging process. New Phytol. 165, 683–701.

Rácz, I., Kovács, M., Lásztity, D., Veisz, O., Szalai, G., Páldi, E., 1996. Effect of short-term and long-term low temperature stress on polyamine biosynthesis in wheat genotypes with varying degrees of frost tolerance. J. Plant Physiol. 148, 368–373.

Ramírez, L., Bartoli, C.G., Lamattina, L., 2013. Glutathione and ascorbic acid protect *arabidopsis* plants against detrimental effects of iron deficiency. J. Exp. Bot. 64, 3169–3178.

Rao, M.V., Davis, K.R., 1999. Ozone-induced cell death occurs via two distinct mechanisms in arabidopsis: the role of salicylic acid. Plant J. 17, 603–614.

Razem, F.A., El-Kereamy, A., Abrams, S.R., Hill, R.D., 2006. The RNAbinding protein FCA is an abscisic acid receptor. Nat. 439, 290–294.

Reichstein, T., Grüssner, A., Oppenauer, R., 1933. Synthesis of D- and L-ascorbic acids (vitamin C). Helv. Chim. Acta. 16, 1019–1033.

Reiter, W.D., 1998. The molecular analysis of cell wall components. Trends Plant Sci. 3, 27–32.

Renstrøm, B., Grün, M., Loewus, F.A., 1982. Biosynthesis of L-ascorbic acid in *chlorella pyrenoidosa*. Plant Sci. Lett. 28, 299–305.

Riekert van Heerden, P.D., Kruger, G.H.J., 2002. Separately and simultaneously induced dark chilling and drought stress effects on photosynthesis, proline accumulation and antioxidant metabolism in soybean. J. Plant Physiol. 159, 1077–1086.

Rivero, R.M., Ruiz, J.M., Romero, L., 2004. Oxidative metabolism in tomato plants subjected to heat stress. J. Hortic. Sci. Biotech. 79, 560–564.

Robinson, J.M., Britz, S.J., 2000. Tolerance of a field grown soybean cultivar to elevated ozone level is concurrent with higher leaflet ascorbic acid level, higher ascorbate-dehydroascorbate redox status, and long term photosynthetic productivity. Photosynth. Res. 64, 77–87.

Roxas, V.P., Smith Jr, R.K., Allen, E.R., Allen, R.D., 1997. Overexpression of glutathione S- growth of transgenic tobacco seedlings during stress. Nat. Biotechnol. 15, 988–991.

Saeidi-Sar, S., Khavari-Nejad, R.A., Fahimi, H., Ghorbanli, M., Majd, A., 2007. Interactive effects of gibberellin A3 and ascorbic acid on lipid peroxidation and antioxidant enzyme activities in *glycine max* seedlings under nickel stress. Russ. J. Plant Physiol. 54, 74–79.

Sairam, R.K., Srivastava, G.C., Agarwal, S., Meena, R.C., 2005. Differences in antioxidant activity in response to salinity stress in tolerant and susceptible wheat genotypes. Plant Biol. 49, 85–91.

Saito, K., Ohmoto, J., Kuriha, N., 1997. Incorporation of 18 O into oxalic, L-threonic and L-tartaric acids during cleavage of L-ascorbic acid and 5-keto-D-gluconic acids in plants. Phytochem. 44, 805–809.

Sanahuja, G., Farré, G., Bassie, L., Zhu, C., Christou, P., Capell, T., 2013. Ascorbic acid synthesis and metabolism in maize are subject to complex and genotype-dependent feedback regulation during endosperm development. Biotechnol. J. 8, 1221–1230.

Sattler, S.E., Gilliland, L.U., Magallanes-Lundback, M., Pollard, M., DellaPenna, D., 2004. Vitamin E is essential for seed longevity and for preventing lipid peroxidation during germination. Plant Cell 16, 1419–1432.

Scandalios, J.G., 1993. Oxygen stress and superoxide dismutase. Plant Physiol. 101, 7–12.

Schreck, R., Rieber, P., Baeuerle, P.A., 1991. Reactive oxygen intermediates as apparently widely used messengers in the activation of the NF-κB transcription factor and HIV-1. EMBO J 10, 2247–2258.

Schroeder, J.I., Allen, G.J., Hugou vieux, V., Kwak, J.M., Waner, D., 2001a. Guard cell signal transduction. Annu. Rev. Plant Physiol. Plant Mol. Biol. 52, 627–658.

Schroeder, J.I., Kwak, J.M., Allen, G.J., 2001b. Guard cell abscisic acid signaling and engineering drought hardiness in plants. Nat. 410, 327–330.

Semane, B., Cuypers, A., Smeets, K., Van Belleghem, F., Horemans, N., Schat, H., et al., 2007. Cadmium responses in *arabidopsis thaliana*: glutathione metabolism and antioxidant defence system. Plant Physiol. 129, 519–528.

Sgherri, C.L.M., Pinzino, C., Navari-Izzo, F., 1996. Sunflower seedlings subjected to increasing water stress by water deficit: Changes in O_2^- production related to the composition of thylakoid membranes. Plant Physiol. 96, 446–452.

Shalata, A., Neumann, P.M., 2001. Exogenous ascorbic acid (vitamin C) increases resistance to salt stress and reduces lipid peroxidation. J. Exp. Bot. 52, 2207–2211.

Shalata, A., Mittova, V., Volokita, M., Guy, M., Tal, M., 2001. Response of the cultivated tomato and its wild salt-tolerant relative *Lycopersicon pennellii* to salt-dependent oxidative stress: the root antioxidant system. Plant Physiol. 112, 487–494.

Shao, H.B., Chu, L.Y., Zhao, H.L., Kang, C., 2008. Primary antioxidant free radical scavenging and redox signalling pathways in higher plant cells. Int. J. Biochem. Sci. 4, 8–14.

Sharma, P., Dubey, R.S., 2005. Drought induces oxidative stress and enhances the activities of anti-oxidant enzymes in growing rice seedlings. Plant Growth Regul. 46, 209–221.

Sharma, P., Dubey, R.S., 2007. Involvement of oxidative stress and role of antioxidative defense system in growing rice seedlings exposed to toxic concentrations of aluminum. Plant Cell Rep. 26, 2027–2038.

Shi, W.M., Muramoto, Y., Ueda, A., Takabe, T., 2001. Cloning of peroxisomal ascorbate peroxidase gene from barley and enhanced thermotolerance by overexpressing in *Arabidopsis thaliana*. Gene 273, 23–27.

Shigeoka, S., Ishikawa, T., Tamoi, M., Miyagawa, Y., Takeda, T., Yabuta, Y., et al., 2002. Regulation and function of ascorbate peroxidase isoenzymes. J. Exp. Bot. 53, 1305–1319.

Shinozaki, K., Shinozaki, K.Y., 2007. Gene networks involved in drought stress response and tolerance. J.Exp.Bot. 58, 221–227.

Shiu, C.T., Lee, T.M., 2005. Ultraviolet-B-induced oxidative stress and responses of the ascorbate–glutathione cycle in a marine macroalga *ulva fasciata*. J. Exp. Bot. 56, 2851–2865.

Smirnoff, N., 1996. The function and metabolism of ascorbic acid in plants. Ann. Bot. 78, 661–669.

Smirnoff, N., 2000. Ascorbic acid: metabolism and functions of a multifaceted molecule. Curr. Opin. Plant Biol. 3, 229–235.

Smirnoff, N., Pallanca, J.E., 1996. Ascorbate metabolism in relation to oxidative stress. Biochem. Soc. Trans. 24, 472–478.

Smirnoff, N., Wheeler, G.L., 2000. Ascorbic acid in plants: biosynthesis and function. Crit. Rev. Biochem. Mol. Biol. 35, 291–314.

Smirnoff, N., Conklin, P.L., Loewus, F.A., 2001. Biosynthesis of ascorbic acid in plants: a renaissance. Annu. Rev. Plant Physiol. Plant Mol. Biol. 52, 437–467.

Smith, A.P., De Ridder, B.P., Guo, W.J., Seeley, E.H., Regnier, F.E., Goldsbrough, P.B., 2004. Proteomic analysis of *arabidopsis* glutathione s-transferases from benoxacor- and copper-treated seed-lings. J. Biol. Chem. 279, 26098–26104.

Sobrino-Plata, J., Ortega-Villasante, C., Flores-Cáceres, M.L., Escobar, C., Del Campo, F.F., Hernández, L.E., 2009. Differential alterations of antioxidant defences as bioindicators of mercury and cadmium toxicity in alfalfa. Chemosphere 77, 946–954.

Song, X.S., Hu, W.H., Mao, W.H., Ogweno, J.O., Zhou, Y.H., Yu, J.Q., 2005. Response of ascorbate peroxidase isoenzymes and ascorbate regeneration system to abiotic stresses in *Cucumis sativus* L. Plant Physiol. Biochem. 43, 1082–1088.

Spickett, C.M., Smirnoff, N., Pitt, A.R., 2000. The biosynthesis of erythroascorbate in *saccharomyces cerevisiae* and its role as an antioxidant. Free Radical Biol. Med. 28, 183–192.

Stevens, R., Page, D., Gouble, B., Garchery, C., Zamir, D., Causse, M., 2008. Tomato fruit ascorbic acid content is linked with monodehydroascorbate reductase activity and tolerance to chilling stress. Plant Cell Environ. 31, 1086–1096.

Streb, P., Feierabend, J., Bligny, R., 1997. Resistance of photoinhibition of photosystem II and catalase and antioxidative protection in high mountain plants. Plant Cell Environ. 20, 1030–1040.

Sumithra, K., Jutur, P.P., Dalton, C.B., Reddy, A.R., 2006. Salinity-induced changes in two cultivars of *vigna radiata*: responses of antioxidative and proline metabolism. J. Plant Growth Regul. 50, 11–22.

Sun, S.Q., He, M., Cao, T., Yusuyin, Y., Han, W., Li, J.L., 2010a. Antioxidative responses related to H_2O_2 depletion in *hypnum plumaeforme* under the combined stress induced by Pb and Ni. Environ. Monit. Assess. 163, 303–312.

Szalai, G., Janda, T., Páldi, E., Dubacq, J.P., 2001. Changes in the fatty acid unsaturation after hardening in wheat chromosome substitution lines with different cold tolerance. J. Plant Physiol. 158, 663–666.

Szalai, G., Kellos, T., Galiba, G., Kocsy, G., 2009. Glutathione as an antioxidant and regulatory molecule in plants under abiotic stress conditions. J. Plant Growth Regul. 28, 66–80.

Tabaldi, L.A., Cargnelutti, D., Gonçalves, J.F., Pereira, L.B., Castro, G.Y., Maldaner, J., et al., 2009. Oxidative stress is an early symptom triggered by aluminum in al-sensitive potato plantlets. Chemosphere 76, 1402–1409.

Tabata, K., Takoka, T.T., Esaka, M.M., 2002. Gene expression of ascorbic acid-related enzymes in tobacco. Phytochem. 61, 631–635.

Takahama, U., 1994. Changes induced by abscisic acid and light in the redox state of ascorbate in the apoplast of epicotyls of *vigna angularis*. Plant Cell Physiol. 35, 975–978.

Tamaoli, M., 2003. Light-controlled expression of a gene encoding L-galactono-g-lactone dehydrogenase which affects ascorbate pool size in *arabidopsis thaliana*. Plant Sci. 164, 1111–1117.

Tambussi, E.A., Bartoli, C.G., Beltrano, J., Guiamet, J.J., Araus, J.L., 2000. Oxidative damage to thylakoid proteins in water-stressed leaves of wheat (*Triticum aestium*). Physiol. Plant 108, 398–404.

Tanou, G., Molassiotis, A., Diamantidis, G., 2009. Induction of reactive oxygen species and necrotic death-like destruction in strawberry leaves by salinity. Environ. Exp. Bot. 65, 270–281.

Tausz, M., Bytnerowicz, A., Arbaugh, M.J., Wonisch, A., Grill, D., 2001. Biochemical response patterns in Pinus ponderosa trees at field plots in the San Bernardino Mountains (*Southern California*). Tree Physiol. 21, 329–336.

Tedone, L., Hancock, R.D., Alberino, S., Haupt, S., Viola, R., 2004. Long-distance transport of l-ascorbic acid in potato. BMC Plant Biol. 4, 16.

Tel-Or, E., Huflejt, M.E., Packer, L., 1986. Hydroperoxide metabolism in cyanobacteria. Arch. Biochem. Biophys. 246, 396–402.

Tewari, R.K., Parma, P.K., Sharma, N., 2006. Antioxidant responses to enhanced generation of super-oxide anion radical and hydrogen peroxide in the copper-stressed mulberry plants. Planta 223, 1145–1153.

Thakur, P., Kumàr, S., Malik, J.A., Berger, J.D., Nayyar, H., 2010. Cold stress effects on reproductive development in grain crops: an overview. Environ. Exp. Bot. 67, 429–443.

Tóth, S.Z., Schansker, G., Garab, G., 2013. The physiological roles and metabolism of ascorbate in chloroplasts. Physiol. Plant 148, 161–175.

Tsai, Y.C., Hong, C.Y., Liu, L.F., Kao, C.H., 2005. Expression of ascorbate peroxidase and glutathione reductase in roots of rice seedlings in response to NaCl and H_2O_2. J. Plant Physiol. 162, 291–299.

Tsugane, K., Kobayashi, K., Niwa, Y., Ohba, Y., Wada, K., Kobayashi, H., 1999. A recessive *arabidopsis* mutant that grows photoautotrophically under salt stress shows enhanced active oxygen detoxification. Plant Cell 11, 1195–1206.

Türkan, I., Bor, M., Özdemir, F., Koca, H., 2005. Differential responses of lipid peroxidation and antioxidants in the leaves of drought tolerant *P. Acutifolius* Gray and drought-sensitive *P. vulgaris* L. subjected to polyethylene glycol mediated water stress. Plant Sci. 168, 223–231.

Uhlin, U., Eklund, H., 1994. Structure of ribonucleotide reductase protein R1. Nat. 370, 533–539.

Umena, Y., Kawakami, K., Shen, J.R., Kamiya, N., 2011. Crystal structure of oxygen-evolving photosystem II at a resolution of 1.9 Å. Nat. 473, 55–61.

Ushimaru, T., Maki, Y., Sano, S., Koshiba, K., Asada, K., Tsuji, H., 1997. Induction of enzymes involved in the ascorbate-dependent antioxidant system, namely ascorbate peroxidase, mono dehydroascorbate reductase and dehydroascorbate reductase, after exposure to air of rice (*Oryza sativa*) seedlings germinated under water. Plant Cell Physiol. 38, 541–549.

Ushimaru, T., Nakagawa, T., Fujioka, Y., Daicho, K., Naito, M., Yamauchi, Y., et al., 2006. Transgenic *arabidopsis* plants expressing the rice dehydroascorbate reductase gene are resistant to salt stress. J. Plant Physiol. 163, 1179–1184.

Valpuesta, V., Botella, M.A., 2004. Biosynthesis of L-ascorbic acid in plants: new pathways for an old antioxidant. Trends Plant Sci. 9, 573–577.

Veljovic-Jovanovic, S.D., Pignocchi, C., Noctor, G., Foyer, C.H., 2001. Low ascorbic acid in the *vtc-1* mutant of *arabidopsis* is associated with decreased growth and intracellular redistribution of the antioxidant system. Plant Physiol. 127, 426–435.

Vollenweider, P., Gunthardt-Goerg, M.S., 2005. Diagnosis of abiotic and biotic stress factors using the visible symptoms in foliage. Environ. Pollut. 137, 455–465.

Wahid, A., Gelani, S., Ashraf, M., Foolad, M.R., 2007. Heat tolerance in plants: an overview. Environ. Exp. Bot. 61, 99–223.

Walker, M.A., McKersie, B.D., 1993. Role of the ascorbate-glutathione antioxidant system in chilling resistance of tomato. J. Plant Physiol. 141, 234–239.

Wang, H.S., Yu, C., Zhu, Z.J., Yu, X.C., 2011. Overexpression in tobacco of a tomato GMPase gene improves tolerance to both low and high temperature stress by enhancing antioxidation capacity. Plant Cell Rep. 30, 1029–1040.

Wang, J., Yu, Y., Zhang, Z., Quan, R., Zhang, H., Ma, L., et al., 2013. *arabidopsis* CSN5B interacts with VTC1 and modulates ascorbic acid synthesis. Plant Cell 25, 625–636.

Wheeler, G.L., Jones, M.A., Smirnoff, N., 1998. The biosynthetic pathway of vitamin C in higher plants. Nat. 393, 365–369.

Wilson, R.N., Heckman, J.W., Somerville, C.R., 1992. Gibberellin is required for flowering in *arabidopsis thaliana* under short days. Plant Physiol. 100, 403–408.

Wolucka, B.A., Van Montagu, M., 2003. GDP-mannose 3,5-epimerase forms GDP-L-gulose, a putative intermediate for the novo biosynthesis of vitamin C in plants. J. Biol. Chem. 278, 47483–47490.

Wu, J., Wu, J., Liang, J., Chen, J., Cai, L., 2012. Effects of exogenous no on AsA-GSH circulation metabolism in young loquat fruit mitochondria under low temperature stress. Pak. J. Bot. 44, 847–851.

Xu, S., Li, J., Zhang, X., Wei, H., Cui, L., 2006. Effects of heat acclimation pretreatment on changes of membrane lipid peroxidation, antioxidant metabolites, and ultrastructure of chloroplasts in two cool-season turfgrass species under heat stress. Environ. Exp. Bot. 56, 274–285.

Yan, J., Wang, Y., Allen, R., Holaday, A.S., Tissue, D., Zhang, H., 2003. Protection of photosynthesis and seed protection under water deficit conditions in transgenic tobacco plants that over-express *arabidopsis* ascorbate peroxidase. Plant Physiol. 43, 1477–1483.

Yang, S.H., Wang, L.J., Li, S.H., 2007. Ultraviolet-B irradiation induced freeing tolerance in relation to antioxidant system in winter wheat (*Triticum aestivum* L.) leaves. Environ. Exp. Bot. 60, 300–307.

Yeh, S., Moffatt, B.A., Griffith, M., Xiong, F., Yang, D.S.C., Wiseman, S.B., et al., 2000. Chitinase genes responsive to cold encode antifreeze proteins in winter cereals. Plant Physiol. 124, 1251–1263.

Yin, L., Wang, S., Eltayeb, A.E., Uddin, M.I., Yamamoto, Y., Tsuji, W., et al., 2010. Overexpression of dehydroascorbate reductase, but not monodehydroascorbate reductase, confers tolerance to aluminum stress in transgenic tobacco. Planta 231, 609–621.

Yoshimura, K., Yabuta, Y., Ishikawa, T., Shigeoka, S., 2000. Expression of spinach ascorbate peroxidase isoenzymes in response to oxidative stresses. Plant Physiol. 123, 223–234.

Zhang, C., Liu, J., Zhang, Y., Cai, X., Gong, P., Zhang, J., et al., 2011. Overexpression of SlGMEs leads to ascorbate accumulation with enhanced oxidative stress, cold, and salt tolerance in tomato. Plant Cell Rep. 30, 389–398.

Zhang, X., Zhang, L., Dong, F., Gao, J., Galbraith, D.W., Song, C.P., 2001. Hydrogen peroxide is involved in abscisic acid-induced stomatal closure in *vicia faba*. Plant Physiol. 126, 1438–1448.

Zhang, Y., Luo, Y., Hou, Y.X., Jiang, H., Chen, Q., Tang, H.R., 2008. Chilling acclimation induced changes in the distribution of H_2O_2 and antioxidant system of strawberry leaves. Agri. J. 3, 286–291.

Zhang, Z., Wang, J., Zhang, R., Huang, R., 2012. The ethylene response factor AtERF98 enhances tolerance to salt through the transcriptional activation of ascorbic acid synthesis in *arabidopsis*. Plant J. 71, 273–287.

Zhao, S., Blumwald, E., 1998. Changes in oxidation reduction state and antioxidant enzymes in the roots of jack pine seedlings during cold acclimation. Plant Physiol. 104, 134–142.

Zhu, O.Y.L., Pilon-Smits, E.A.H., Jouanin, L., Terry, N., 1999. Overexpression of glutathione synthetase in Indian mustard enhances cadmium accumulation and tolerance. Plant Physiol. 11901, 73–79.

Chapter 9

Carotenoids Involved in Antioxidant System of Chloroplasts

Dariusz Latowski, Renata Szymanska and Kazimierz Strzałka

9.1 INTRODUCTION

Carotenoids are natural pigments synthesized by plants and other photosynthetic organisms such as algae, some bacteria, and some fungi. These pigments, however, cannot be synthesized by even the most colorful species in the animal kingdom, so animals obtain carotenoids in their diets, and may employ them in various ways in metabolic processes. In photosynthetic organisms carotenoids serve two key roles: they absorb light energy for use in photosynthesis, and they protect chlorophylls and other elements of the photosynthetic apparatus from photodamage (Isaacs, 1995; Armstrong and Hearst, 1996; Ramel et al., 2012; Christaki et al., 2013). In mammals α-carotene, β-carotene, γ-carotene, and β-cryptoxanthin can be transformed to retinal, a form of vitamin A (Eroglu et al., 2012). Some other carotenoids such as astaxanthin, lutein or zeaxanthin protect the macula of the retina, which is the part of the eye with the sharpest vision. These carotenoids absorb near-ultraviolet and blue light. Generally carotenoids are considered to be widespread antioxidants (Kidd, 2011; Abdel-Aal et al., 2013; Bojorquez et al., 2013).

9.2 CHEMICAL STRUCTURE OF CAROTENOIDS AND THEIR ANTIOXIDANT PROPERTIES

9.2.1 Chemical Structure and Types of Carotenoids

Structurally, all carotenoids are tetraterpenoid pigments, which means that they are biosynthesized from four 10-carbon terpene units, which in turn are formed from eight 5-carbon isoprene units (Fig. 9.1).

P. Ahmad (Ed): Oxidative Damage to Plants. DOI: http://dx.doi.org/10.1016/B978-0-12-799963-0.00009-5

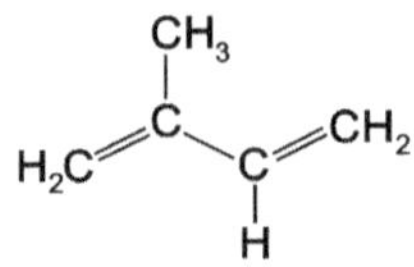

FIGURE. 9.1 Chemical structure of isoprene (2-methyl-1,3-butadiene).

Hence, carotenoids are polyunsaturated hydrocarbons containing 40 carbon atoms per molecule and variable numbers of hydrogen atoms and oxygen atoms in the case of xanthophylls. Due to the presence of the conjugated double bond system, they can absorb blue light in the range of 350–500 nm and they are colored to the human eye. There are over 600 known carotenoids.

All known and naturally occurring carotenoids can be divided by their chemical structure into the following subgroups:

- Acids and acid esters (e.g. torularhodin)
- Alcohols (e.g. zeaxanthin)
- Aldehydes (e.g. rhodopinal)
- Apo carotenoids (e.g. bixin)
- Esters of alcohols (e.g. fucoxanthin)
- Epoxides (e.g. diadinoxanthin)
- Ethers (e.g. spheroidene)
- Glycosides (e.g. oscillaxanthin)
- Higher carotenoids (e.g. bacterioruberin)
- Hydrocarbons (e.g. phytofluene)
- Ketones (e.g. astaxanthin)
- Nor- and seco-carotenoids (e.g. β-carotenone)

A different and better-known division of carotenoids includes two classes:

1. Carotenes
2. Xanthophylls

9.2.1.1 Chemical Characteristics of Carotenes

Carotenes contain only carbon and hydrogen atoms and no other elements. Some carotenes are terminated by hydrocarbon rings, on one or both ends of the molecule. Carotenes are found in plants in two primary forms designated by characters from the Greek alphabet: α-carotene and β-carotene (Fig. 9.2).

The two primary isomers of carotene, α-carotene and β-carotene, differ in the position of a double bond (and thus a hydrogen) in the cyclic group at one end. β-Carotene is the more common form and can be found in yellow, orange, and green leafy fruits and vegetables. Generally, the greater the intensity of the orange color of the fruit or vegetable, the more β-carotene it contains. The molecule of β-carotene possesses two rings called β-rings

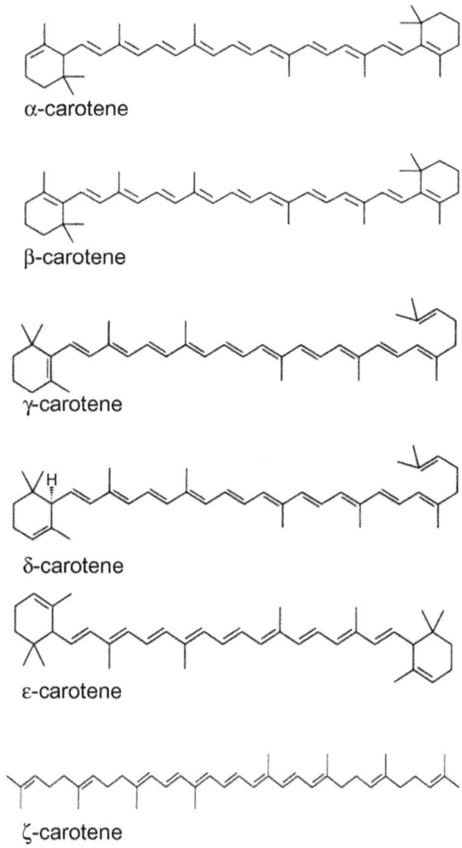

FIGURE 9.2 Chemical structure of carotenes.

which are formed by nine carbon atoms, while α-carotene has a β-ring at one end of the molecule and a ε-ring on the other end. Based on the type of molecule chain termination, the following nomenclature of carotenoids is used:

- α-carotene possessing one β and one ε ring;
- β-carotene possessing two β rings;
- γ-carotene possessing only one β ring, while the other end is uncyclized and labelled *psi* (ψ);
- δ-carotene possessing only one ε ring while the other end is uncyclized (ψ-end);
- ε-carotene possessing two ε rings
- ζ-carotene is the biosynthetic precursor of neurosporene, which is the precursor of lycopene, which, in turn, is the precursor of the carotenes α through ε. Lycopene such as ζ-carotene is ψ,ψ-carotene (Fig. 9.2).

Carotenes are lipophilic because they are hydrocarbons containing no oxygen and thus are insoluble in water.

9.2.1.2 Chemical Characteristics of Xanthophylls

The second group of carotenoids are xanthophylls, originally called phylloxanthins. The name of this carotenoid group originates from Greek xanthos (yellow) + phyllon (leaf), due to their formation of the yellow band seen in early chromatography of leaf pigments. Their molecular structure is similar to carotenes, but xanthophylls contain oxygen atoms. Xanthophylls contain their oxygen either as hydroxyl groups (zeaxanthin and lutein) and/or as pairs of hydrogen atoms that are substituted by oxygen atoms acting as a bridge forming epoxide group (antheraxanthin or violaxanthin). Some xanthophylls contain oxygen as methoxy (spirilloxanthin) or oxo (echinenone) groups (Goodwin 1980). For this reason, they are more polar than the purely hydrocarbon carotenes, and it is this difference that allows their separations from carotenes in many types of chromatography. Typically, xanthophylls are less orange in color than carotenes. The only xanthophyll known to contain a β-ionone ring is β-cryptoxanthin and thus it is the only xanthophyll that is known to possess pro-vitamin A activity for mammals. Even then, it is a vitamin only for plant-eating mammals that possess the enzyme to make retinal from carotenoids that contain β-ionone. In species other than mammals, certain xanthophylls may be converted to hydroxylated retinal-analogues that function directly in vision.

9.2.2 Antioxidant Properties of Carotenoids in Photosynthesis

Carotenoids were intensively studied during the last 50 years and many functions and associations of these pigments were documented. Among the first studied were the antenna and photoprotective functions of carotenoids in photosynthesis. Under low light conditions, carotenoids may act as energetic antennae, harvesting solar radiation at the wavelengths not absorbed by chlorophylls and transmitting the energy by energy transfer to the chlorophylls or bacteriochlorophylls in the reaction centers (RCs). In this way, they broaden the range of light used in photosynthesis. Carotenoids also function as structural backbones holding the structure of the light-harvesting complexes, LH1, LH2 as well as higher plant LHCs located peripherally with respect to the reaction centers (Liu et al., 2004; McDermott et al., 2007; Pan et al., 2011).

On the other hand, it has been shown that carotenoids act as quenchers of the potentially damaging chlorophyll (or bacteriochlorophyll) triplet states and singlet molecular oxygen, i.e. they play a crucial role as photoprotectors in photosynthetic units preventing self-oxidation of the photosystems (PS)

(Siefermann-Harms, 1985; Koyama, 1991; Frank and Cogdell, 1993; Frank, 1999; Pospíšil, 2012). In excess light conditions, reaction centers, where light dependent charge separation and the first steps of electron transport occur, cannot efficiently quench Chl excited states, thus increasing the probability of intersystem crossing to triplet states (Apel and Hirt, 2004). Chl triplets readily react with molecular oxygen, yielding harmful singlet oxygen ($^1O_2$) (Krieger-Liszkay, 2005; Fischer et al., 2012). LHC proteins catalyze thermal dissipation of Chl singlet excited states, thus limiting the formation of triplet states (Triantaphylides, 2009).The deactivation of $^1O_2$ in the PSII reaction center is provided mainly by β-carotene (Ewart, 1986), while thermal dissipation of Chl singlet ($^1P^*_{680}$) excited states is operated by xanthophylls in the process called nonphotochemical quenching (NPQ) (Barber and Rivas, 1993; Nishiyama et al., 2001, Jahns and Holzwarth, 2012, Johnson et al., 2012). The reason that β-carotene cannot quench triplet states of Chl in RC of PSII ($^3P_{680}$) is probably the extremely high oxidizing power of PSII necessary to oxidize water (Van Gorkom and Schelvis, 1993). The midpoint potential of $P^+{}_{680}/P_{680}$ must be higher than the value of about 1.0 V of β-carotene (Edge et al., 2000). In this situation, if the distance from β-carotene to $^1P_{680}$ was close enough to allow quenching of $^3P_{680}$, electron transfer between these molecules would be exceedingly fast and would prevent electron transfer from tyrosine molecule bound to the D1 protein (YZ) to $^1P_{680}$. This problem is unique for PSII; the redox potentials of the primary electron donors in other photosystems are less oxidizing than the other pigments in the complexes and consequently the cationic forms are not dangerous (Thompson and Brudvig, 1988).

Recently, a novel photoprotection mechanism to be operative in intact closed PSII RCs (i.e. with reduced quinone Q_A) was proposed, whereas a consequence of the negative charge on Q_A, the RC triplet, is formed by the so-called radical pair (RP) mechanism. The observed triplet state was interpreted as a triplet state of accessory D2-chlorophyll ($^3Chl_{accD2}$) on the normally inactive D2 branch. It has been observed that this triplet state was quenched by β-carotene in D2 protein (Braslavsky and Holzwarth, 2012). Among different redox states of oxygen, the fully reduced form H_2O and the fully oxidized form O_2, differing by 4 e^-, are relatively stable. Intermediate states, called reactive oxygen species (ROS), produced by univalent electron transport within photosystems, are toxic and need to be scavenged.

Accumulation of ROS in plants at the cellular level is observed not only during high-light stress or under ultraviolet radiation but also during the exposure to stresses such as drought and desiccation, salt stress, chilling, heat shock, heavy metals, air pollutants, mechanical stress, nutrient deprivation, and pathogen attack (Alscher et al., 2002). LHC-bound xanthophylls are excellent quenchers of ROS. They can carry out this process with far greater efficiency in comparison to lipid free carotenoids (Telfer, 2002;

Horton et al., 2005; Foyer and Noctor, 2005, 2011). This mechanism is essential for photoprotection even at low light intensity (Jansson, 1994; Tounekti et al., 2012). Xanthophyll biosynthesis mutants lacking β-xanthophylls have been shown to be extremely photosensitive (Jansson, 1994) and produced singlet oxygen even at low light intensity. Xanthophylls bound to peripheral antenna proteins, specified as Lhcb proteins, are thought to be crucial for preventing ROS formation by quenching the triplet chlorophyll excited state (Triantaphylides and Havaux, 2009; Foyer and Noctor, 2011) and finally by scavenging the ROS produced. In higher plants, six different peripheral antenna proteins, designated as Lhcb1–6, form the light-harvesting antenna of PSII. The light-harvesting antenna of PSI contains at least four different antenna proteins marked as Lhca1–4 (Jansson, 1994; Morosinotto et al., 2003; Dekker and Boekema, 2005). The structure of the trimeric LHCII complex is composed of monomers called Lhcb1–3 and it binds 8 molecules of chlorophyll *a*, 6 molecules of chlorophyll *b* and 4 xanthophylls (2 luteins (Lut), 1 neoxanthin (Nx) and one of the pigments engaged in the xanthophyll cycle). The second group of Lhcb proteins includes monomeric Lhcb4, Lhcb5, and Lhcb6, which bind Lut and Vx or Zx (Croce et al., 2000; Jahns et al., 2009).

The analysis of photoprotection in xanthophyll biosynthesis mutants has shown that photoprotection activity of all xanthophylls is enhanced by binding to LHC proteins (Foyer and Noctor, 2005; Demmig-Adams et al., 1989) and scavenging within LHC proteins is indispensable to normal operation of photosynthesis (Johnson et al., 2010). Xanthophyll binding to LHC proteins appears to be stabilized through multiple weak interactions (Pogson et al., 1998).

Four types of xanthophyll-binding sites have been reported in plant Lhcb proteins, called respectively L1, L2, N1 and V1 (Siefermann and Yamamoto, 1975; Woodall et al., 1997a,b; Davison et al., 2002; Triantaphylides and Havaux, 2009). Site L1 is reserved to host Lut in all Lhc proteins. The L2 site displays a different nature in several aspects:

1. Its occupancy is different depending on protein, i.e. in Lhcb1–3 in L2 site Lut is bounded whereas in Lhcb4–6 this site is occupied by Vx ;
2. In Lhcb4–6 proteins Vx can be substituted by Zx upon Zx accumulation in high light conditions, when Vx is de-epoxidased to Zx (Horton et al., 1991; McNulty et al., 2007; Müller et al., 2010).This L2 site has an allosteric nature and binding of Zx induces a conformational change in LHCII, causing shortening of its fluorescence lifetime (Lambrev et al., 2010).

The next type of xanthophyll-binding site is site V1. It is occupied by Vx and it is present only in LHCII (Lhcb1–3 – in proteins Lhcb4–6 there is no V1 site) (Kühlbrandt et al., 1994; Ruban et al., 2007). The occupancy of this site by Vx does not induce conformational changes in the protein

(Siefermann and Yamamoto, 1975; Kühlbrandt et al., 1994; Gruszecki and Strzałka, 2005).

Site N1 is a peripheral site in LHCII and binds Nx. It was previously suggested that Nx in Lhcb4 and Lhcb5 shared the site L2 with Vx; but later it was found that Nx has its own binding site (Pogson et al., 1998; Ruban et al., 2007). Nx seems to be selective for scavenging superoxide (Adams et al., 1999) and Vx, or even more, Zx appears to be specific for $^1O_2$ quenching. Recently it was postulated that Zx is the major player in the deactivation of excited singlet Chl, while the photoprotective role of Lut is predominantly restricted to its function in the deactivation of triplet chlorophyll. Additionally, Zx serves important functions as an antioxidant in the lipid phase of the membrane and is likely to act as a key component in the memory of the chloroplast with respect to preceding photo-oxidative stress (Jahns and Holzwarth, 2012).

The antioxidant actions of carotenoids are based on their singlet oxygen quenching properties and their ability to trap peroxyl radicals (Holzwarth et al., 2006). The ability of carotenoids to quench singlet oxygen is the best documented antioxidant activity of these molecules. The quenching of singlet oxygen leads to excitation of carotenoids. The original unexcited state in which these pigments can be used again as singlet oxygen quenchers is obtained by a series of rotational and vibrational interactions of carotenoids with the solvent, resulting in dissipation of excess energy. In this way, the efficiency of the carotenoid antioxidant activity is dependent on the speed of recovery of their basic state, which strongly depends on the number of conjugated double bonds of the molecule. The cyclic or acyclic end groups or the nature of substituents in carotenoids containing cyclic end groups play a much less significant role in this mechanism. This explains why lycopene, having 11 conjugated double bonds, is one of the most efficient singlet oxygen quenchers among the natural carotenoids (Kamlowski et al., 1996). The prevention of lipid peroxidation by carotenoids has been suggested to be mainly via singlet oxygen quenching (Holzwarth et al., 2006; Ramel et al., 2012).

β-carotene is also a scavenger of peroxyl radicals, especially at low oxygen tension (Yruela et al., 1994). This activity may also be exhibited by other carotenoids. The interaction of carotenoids with peroxyl radicals may proceed via an unstable β-carotene radical adduct (Yruela et al., 1994; Losi et al., 2003). These carotenoid adduct radicals are highly resonance stabilized and therefore relatively unreactive. They may both terminate radical reactions by binding to the attacking free radicals and undergo a decay to generate nonradical products (Losi et al., 2003). The reaction of carotenoids with peroxyl radicals is more rapid than analogical reaction of unsaturated acyl chains; therefore these pigments can act as antioxidants. In this process, carotenoids are sacrificed for the benefit of cells and are destroyed (Braslavsky and Holzwarth, 2012).

9.3 PHOTOPROTECTIVE ROLE OF β-CAROTENE IN PHOTOSYNTHETIC REACTION CENTER

The PSII RC contains two all-trans β-Car molecules, which are clearly distinguishable by their spectroscopic properties. One of these molecules, called β-Car507, has maxima at 507, 473 and 443 nm and the other, called β-Car489, has maxima at 489, 458 and 429 nm (Van Dorssen et al., 1987; Breton et al., 1998; Kwa et al., 1992; Tomo et al., 1997). Spectroscopic measurements indicate that both of the β-Car molecules in the PSII RC are bound in different environments, one to the D1 and the other to the D2 protein. It has been shown that these two molecules of carotene are unable to quench chlorophyll triplets in PSII (Telfer, 2002) but that they can scavenge $^1O_2$ directly and thus provide significant protection against oxidative damage (Van Gorkom and Schelvis, 1993). However, this protection can only be partial because the β-Car molecules are bound some distance away from the source of the $^1O_2$. Therefore, the other target molecules that could be oxidized are always in competition with the ability of the β-Car to scavenge $^1O_2$ (Telfer, 2002). Further examination of the oxidation reactions in the PSII RC showed that initially β-Car is oxidized by oxidized P_{680} ($P^+{}_{680}$) and that the carotenoid cation created in this way is unstable and is irreversibly bleached in an oxygen-independent reaction (Telfer and Barber, 1989; de Las Rivas et al., 1993). It has been calculated that β-Car must be at a distance of 18–20 Å from the nearest oxidized chlorophyll (Telfer and Barber, 1995). As already reported a β-Car molecule located in the D2 protein (β-CarD2) is the first electron donor for $P^+{}_{680}$ when the electron transport from water to $P^+{}_{680}$ becomes limited. In such a situation electrons can be cycled around PSII via Q_B, Cyt b_{559} (HP), β-CarD2 and $P^+{}_{680}$. The β-CarD2 acts as an intermediate between $P^+{}_{680}$ and cyt b_{559} in its high potential (HP) form (Hanley et al., 1999; Faller et al., 2001). The β-CarD2 forms a link between $P^+{}_{680}$ and one of the peripheral chlorophyll molecules bound near the edge of the PSII (Chl *z*) when the electron supply from the cytochrome is limited (Barber and Rivas, 1993; Stewart et al., 1998; Shigemori et al., 1998; Ruffle et al., 2001; Wang et al., 2002). The β-CarD2 is thought to oxidize the Chl *z*, which in turn can quench fluorescence and reduce the photochemical pressure on PSII (Stewart et al., 1998; Shigemori et al., 1998). It is known that when the lifetime of $P^+{}_{680}$ is extended, it oxidizes other components of the RC. Thus β-CarD2, as a donor of electrons, protects the PS II against photoinhibition (Thompson and Brudvig, 1988). On the other hand, there is no evidence that the D1 protein is engaged in cyclic electron transfer. Thus oxidation of β-Car on that side would seem to be undesirable. Under high light both a bleaching of β-Car and the destabilization of the RC structure is observed. It has been suggested that a low quantum yield of β-Car oxidation leads to its degradation (Edge et al., 1997).

When the electron supply from water to P^{+}_{680} is limited the β-CarD2 catalyzes an efficient cycle of electrons around PSII via Cyt b_{559}, whereas the β-Car located in the D1 polypeptide (β-CarD1) is oxidized. The lack of an efficient electron donor to β-CarD1 would serve as an important signal for protease attack, which is able to selectively destroy the D1 protein and prevent further charge separations. These findings explain why D1 is much more labile than D2 and offers a mechanism by which PSII is deactivated without any of its other proteins being too drastically affected (Telfer, 2002). Recently, a novel photoprotection mechanism operating in intact closed PSII RCs (i.e. with reduced quinone Q_A) was proposed. As a consequence of the negative charge on Q_A, the RC triplet is formed by the so-called radical pair (RP) mechanism on the normally inactive D2 protein, most likely the $^3Chl_{accD2}$, where it can be subsequently quenched by the relatively close D2 β-carotene (D2-carotene is only 13.2 Å away from Chl_{accD2}). In isolated RCs of PSII, i.e. the D1−D2−cyt b_{559} complex, the quinone Q_A is absent. Thus, the forward electron transport cannot proceed and the charge-separated singlet state P^{+}_{680}-$Pheo^{-}$ (Holzwarth et al., 2006) forms a triplet RP ($^3RP^{\pm}$) in high yield through spin dephasing, which then recombines to the RC ^3Chl state located at the accessory Chl on the D1 protein (Chl_{accD1}) (Kamlowski et al., 1996) with a lifetime up to 1 ms. β-carotene in the D1-branch, located at a relatively long distance from the Chl_{accD1} (17 Å), cannot quench the RC triplet and this molecule is assigned to play a structural role, necessary to keep the functional organization of the RC. The D2 branch becomes active in electron transfer most likely by the influence of the negative charge on the reduced Q_A,Q^{-}_{A}, which exerts an electrostatic repulsion on the potential charge of the D1 pheophytin. Thus, the normally formed (in open RCs and in the isolated D1−D2−cytb_{559} RC (Holzwarth et al., 2006) secondary RP P^{+}_{680}-$Pheo^{-}$ cannot be created in a closed intact RC and the charge separation and subsequent triplet formation are switched to the D2 side. Therefore, the D2-carotene is enabled to play its crucial photoprotective role by quenching the $^3Chl_{accD2}$. This interpretation explains why RC triplet quenching is not observed in isolated D1−D2−cytb_{559} RCs. If Q_A is either not present at all (as in the isolated RC) or is not charged (as in open RCs or with doubly reduced Q_A), then the RC ^3Chl is formed on the D1 branch. The D1 branch ^3Chl cannot be quenched by β-carotene due to the large distance. This D1 β-carotene is necessary for holding the structure.

9.4 PHOTOPROTECTIVE ROLE OF XANTHOPHYLL CYCLE PIGMENTS

In living cells the level of ROS has to be kept under tight control because they play a role both as cellular indicators of stress and as secondary

TABLE 9.1 The Enzymatic and Nonenzymatic Scavengers that Enable Plants to Regulate ROS Level

Enzymatic Antioxidants	Nonenzymatic Antioxidants
– catalase (Cat); – glutathione reductase (GR); – peroxidases (POD), e.g. ascorbate peroxidase (APX), glutathione peroxidase (GPX); – superoxide dismutase (SOD);	– ascorbic acid (Asc); – glutathione; – products of light phase of the xanthophyll cycles; – tocopherols

messengers engaged in the signal transduction pathway generated by stress. The mechanisms controlling the level of ROS can be based on enzymatic or nonenzymatic antioxidants (Table 9.1). One group of nonenzymatic antioxidants that exists in plants are the products of de-epoxidation occurring in the processes generally called the xanthophyll cycle.

The xanthophyll cycle is one of the most efficient mechanisms protecting plants and other photosynthesizing organisms against light stress under overexcitation conditions. At the molecular level some pigments of this cycle act as quenchers of $^{1}P^{*}_{680}$ thus preventing the formation of ROS (Latowski et al., 2004a; Gruszecki et al., 2006). Six types of the xanthophyll cycle are listed today. Four of them engage β-xanthophylls and two α-xanthophylls (Table 9.2) (Garcia-Plazaola et al., 2007). Five of the six described xanthophyll cycles have in common the de-epoxidation of mono- or diepoxidized xanthophylls under high light condition. The de-epoxidized products facilitate the dissipation of excitation energy, and then when high light stress declines they are again epoxidized (Latowski et al., 2004a; Garcia-Plazaola et al., 2007).

The dissipation of excitation energy by de-epoxidized xanthophylls is due to mechanisms collectively referred to as nonphotochemical quenching (NPQ). The predominant NPQ component is induced as a result of the acidification of the thylakoid lumen associated with the formation of the chloroplast proton motive force and defined as energy quenching (qE) (Niyogi, 1999). In addition to qE, relaxing within 2–5 minutes, a slowly relaxing component of the NPQ process is known as qI (photoinhibitory quenching), with a half-time of approximately 30 minutes and longer (depending on the degree of photoinhibition) (Horton et al., 1996). A third quenching component (qT), relaxing within 15–20 minutes, also has been reported (Horton et al., 1996; de Bianchia et al., 2010). The most commonly occurring and most intensively studied type of xanthophyll cycle is the *violaxanthin cycle* (Vx-cycle), also called *xanthophyll cycle*, in which the main product of strong light-stimulated de-epoxidation is Zx. An increase in NPQ after high light treatment and its correlation with an increase both in Zx and Ax concentrations have been observed in leaves of spinach (*Spinacia oleracea*), isolated chloroplasts and

TABLE 9.2 Types of Xanthophyll Cycles (Latowski et al., 2011)

Type of Xanthophyll Cycle		Products of Light Phase of the Xanthophyll Cycles	Description
Violaxanthin cycle (xanthophyll cycle)	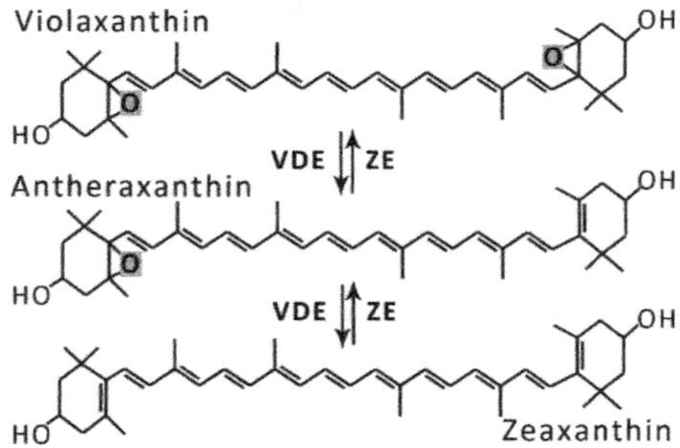	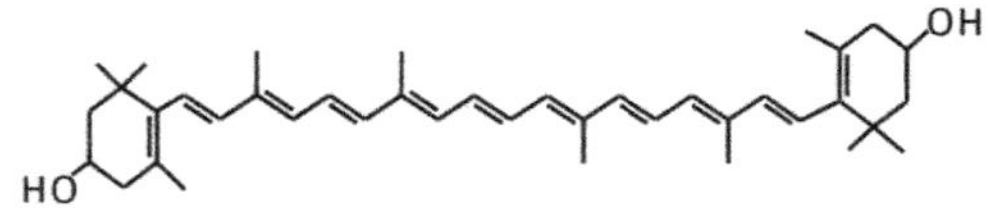	Enzymes: – violaxanthin de-epoxidase (VDE) – zeaxanthin epoxidase (ZE) Occurrence: – higher plants – mosses – lichens – algae (Phaeophyta, Chlorophyta, Rhodophyta)
Diadinoxanthin cycle	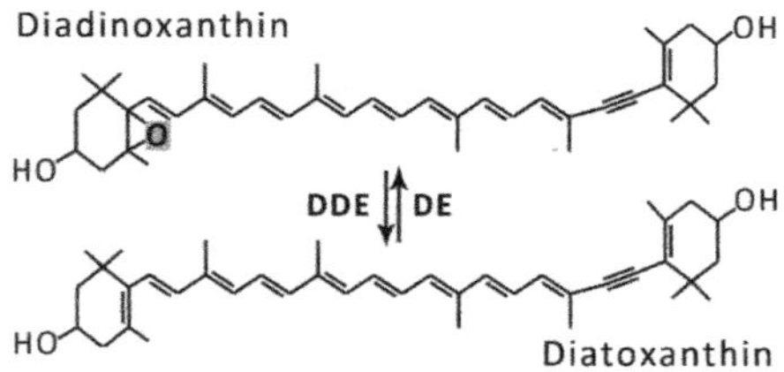	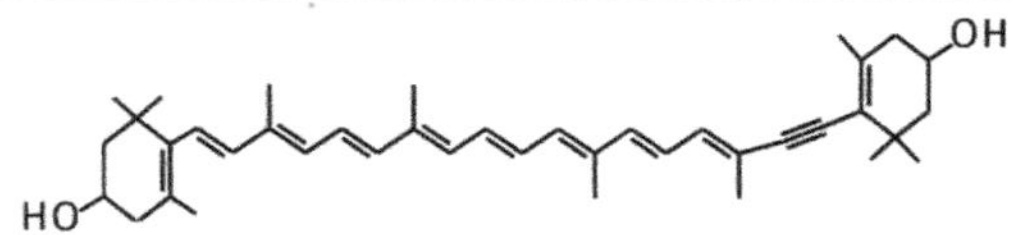	Enzymes: – diadinoxanthin de-epoxidase (DDE) – diatoxanthin epoxidase (DE) Occurrence: – algae (diatoms, phaeophytes, dinophytes, haptophytes)
Antheraxanthin cycle	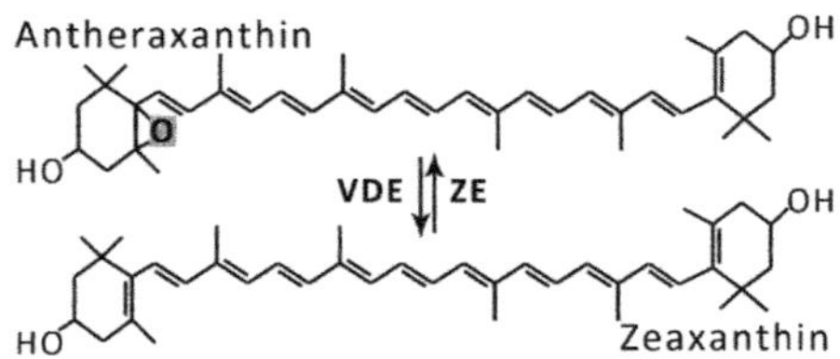	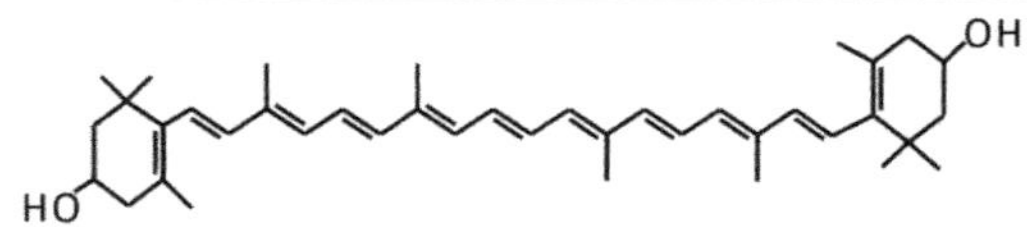	Enzymes: – violaxanthin de-epoxidase (VDE) – zeaxanthin epoxidase (ZE) Occurrence: – *Gracilaria gracilis* – *Gracilaria multipartite*

(*Continued*)

TABLE 9.2 (Continued)

Type of Xanthophyll Cycle		Products of Light Phase of the Xanthophyll Cycles	Description
Xanthophyll cycle in *mantoniella squamata*	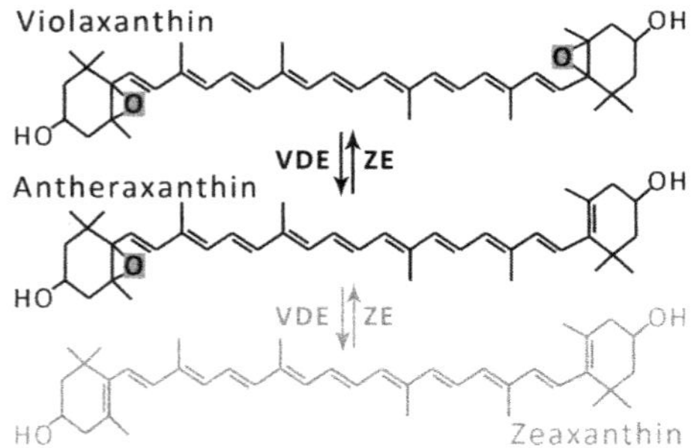	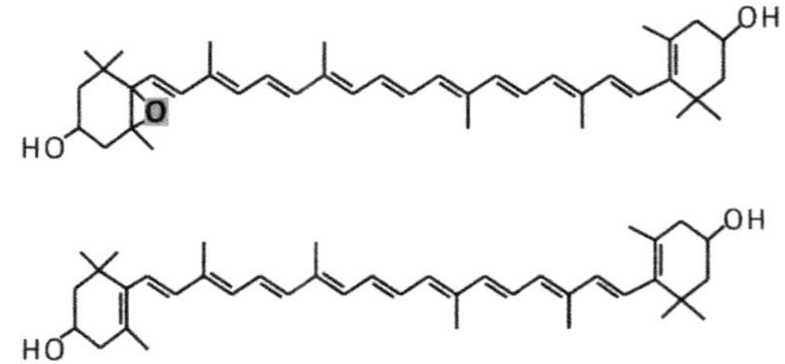	Enzymes: – violaxanthin de-epoxidase (VDE) – zeaxanthin epoxidase (ZE) Occurrence: – *Mantoniella squamata*
Lutein epoxide cycle	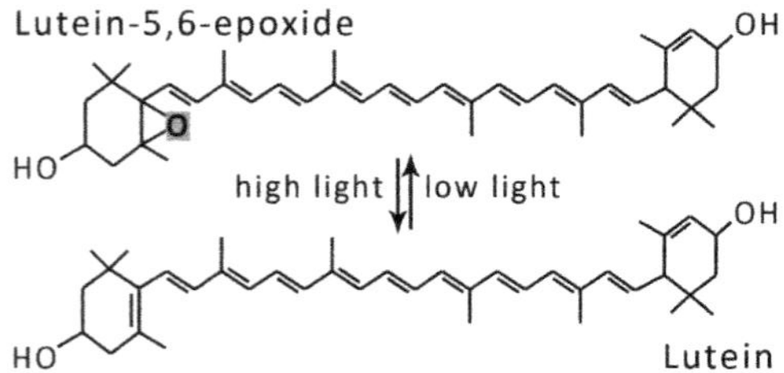	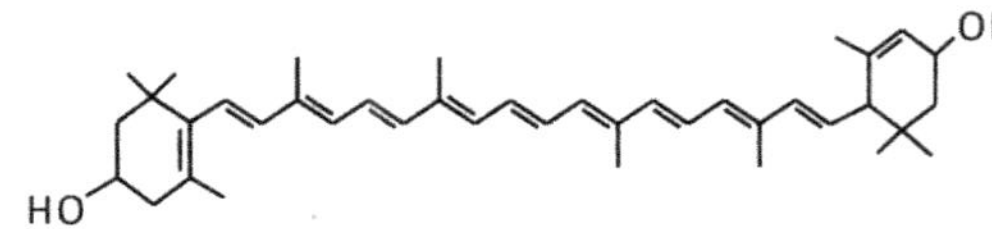	Enzymes: – violaxanthin de-epoxidase (VDE) – zeaxanthin epoxidase (ZE) Occurrence: – leaves of 62% of the 188 tested species
Siphonaxanthin cycle	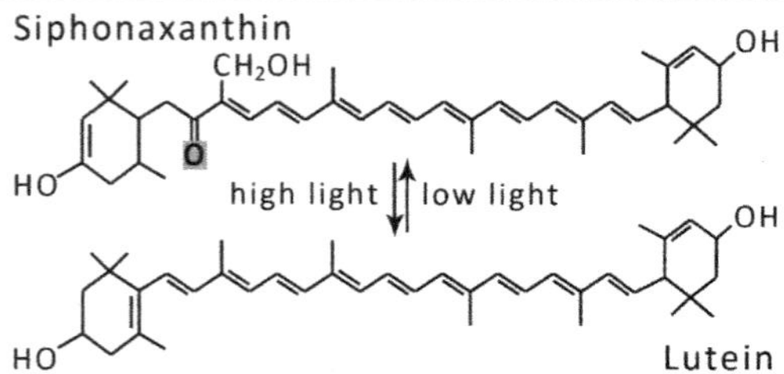	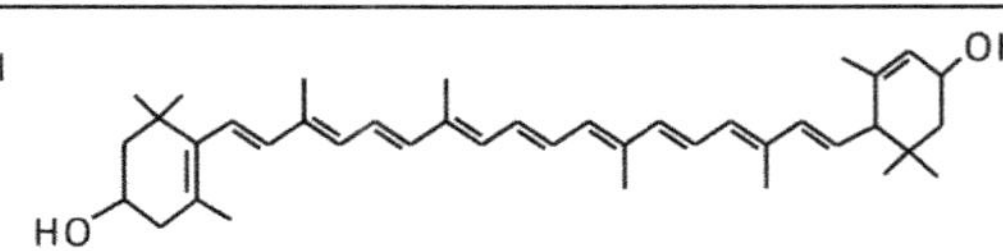	Enzymes: no data available Occurrence: – *Caulerpa racemosa*

purified LHC complexes (Adams et al., 1995; Ruban and Horton, 1999). These observations resulted in the introduction of an additional component of NPQ, known as qZ, Zx-dependent quenching. It is suggested that qZ develops both at moderate and high light intensities in a time range from 10 to 30 min (Nilkens et al., 2010; Lavaud et al., 2002). Similarly, in diatoms a good correlation was also observed between the level of NPQ and level of diatoxanthin (Dtx), which is a product of diadinoxanthin de-epoxidation (Table 9.2) (Lavaud et al., 2002; Lavaud et al., 2002b). Additionally, during prolonged UV-A and UV-B illumination of one of the diatom species (*Thalassiosira weissflogii*) a photoprotective action of Dtx has been observed. The increased activity of the Dtx cycle as a response to light stress was observed also in UV-insensitive diatoms (Lavaud et al., 2002b).

It has also been observed that the de-epoxidation of lutein-epoxide (Lx) to Lut (Table 9.2) facilitated the rapid engagement of NPQ, and that this process may be fine-tuned by concurrent Zx accumulation inducing strong energy dissipation in plants having both an Lx-cycle and Vx-cycle (Garcia-Plazaola et al., 2007). On the other hand, photoconversion of siphonaxanthin (Sx) to Lut (Table 9.2) was detected in the green alga *Caulerpa racemosa*, showing significant similarity to Vx–Ax interconversion. This transformation seems to be activated by the same mechanism as the first step of Vx de-epoxidation. In particular, both Ax and L achieved their highest level not only under strong light intensity, but also at sunrise when light intensity was lower (Young and Frank, 1996). This last feature suggests a photoprotective role of the conversion between Sx and Lut and a similar and very high sensitivity of the two cycles to light. Additionally, there are the biochemical similarities between Lut and Ax. The energetic state analysis also revealed similar S1 values for Ax and Lut (Young and Frank, 1996). All these observations show that the products of the xanthophyll cycles that are created under light conditions are effective quenchers of ROS. Several mechanisms explaining the role of xanthophylls cycle pigments in NPQ have been suggested. These can be grouped in two categories. One of these engages direct overexcitation quenching by products of the light phase of the xanthophyll cycles. The other covers models based on the indirect participation of the xanthophyll cycle carotenoids in photoprotection (Latowski et al., 2011). In the former, i.e. direct overexcitation quenching, a key factor in evaluating the efficiency of NPQ is the energy of the lowest excited singlet states (S1) of the epoxy- and de-epoxy-xanthophylls. The xanthophylls which are created as the result of light-induced enzymatic conversion, Ax, Zx, L, Dtx (Table 9.2), possess a longer conjugated double bond system, as compared to their oxidized derivatives (11 conjugated double bonds for Zx versus 9 for Vx). This fact implies that the S1 of these pigments is located at a lower level on the energy scale with respect to the Qy level of chlorophyll a (Chl a), and suggests a possible quenching mechanism of the excessive singlet excitation state of Chl a by these xanthophylls but not by their oxidized derivatives (Frank et al., 2000).

9.4.1 Direct Quenching of Overexcitation By-products of the Light Phase of the Xanthophyll Cycles

One of the hypotheses explaining the role of the xanthophyll cycles in NPQ assumes a direct quenching of overexcitation by the de-epoxy-xanthophylls created under high light conditions. This model proposes that a pH-generated structural change in the pigment–protein complex facilitates a downhill energy transfer from Chl a to Zx, Ax, Ddx, or Lut (Demmig-Adams, 1990; Frank et al., 1994). This idea was supported by estimation of the energies of the lowest-lying singlet (S1) states of the xanthophylls from either the dynamics of these states or the fluorescence from molecules of carotenoids contains less than 10 carbon–carbon double bonds and extrapolation of the energies to the longer molecules, including also the pigments engaged in the xanthophyll cycles (Frank et al., 1994). Spectroscopic and kinetic investigations have shown that the energies of the S1 states of the de-epoxidized xanthophylls are low enough to quench Chl excited states. It was previously published that a clear relationship exists between the xanthophyll S1 energy level and its ability to quench Chl fluorescence (Young and Frank, 1996). The use of xanthophylls with S1 energies above that of Chl had little, if any, effect on NPQ. A xanthophyll molecule may be considered as a "quencher" or "non-quencher" depending on whether the S1 energy is below or above that of Chl a, respectively. Model system studies using liposomes with embedded LHCII and xanthophyll cycle pigments revealed that the xanthophylls decrease the relative quantum yield of Chl fluorescence, quenching the Chl via singlet excitation transfer. Zx was a better quencher than Vx (Gruszecki et al., 1997). Moreover, it was also demonstrated that Zx quenching ability was related to Zx dimer formation immediately on the addition of this pigment to the Chl containing micelles. It has been postulated that this dimerization may play a role in Zx functioning in the photosynthetic apparatus (Avital et al., 2006). It also should be mentioned that some of the recent determinations of the energy levels of the xanthophyll cycle pigments show that the difference between the energies of the S1 state of Vx and Zx is too small in order to obtain effective Chl excitation quenching efficiencies by these two compounds (Polivka and Sundstrom, 2004). In the antenna complexes isolated from the L mutants of *Arabidopsis thaliana*, in which L was replaced with Zx, no exceptional Chl singlet excitation quenching was observed (Gruszecki et al., 2006; Lokstein et al., 2002). Moreover, it was also reported that the S1 level of not only Zx but also Vx lies below the Q_y level of Chl (Polivka et al., 1999). In light of this, a mechanism involving direct quenching via singlet–singlet energy transfer seems to be invalid. According to these data, Vx could be an even more efficient quencher than Zx, because its S1 level lies below but closer to the Chl Q_y transition than the S1 level of Zx. It has recently been suggested that quenching may also be caused by Chl–Chl interactions. Some authors have postulated that in the quenched state an increase in interaction of both Chl–Chl and xanthophylls–Chl occurs. Liao et al. (2010) using two-photon excitation spectroscopy, demonstrated that in quenched LHCII the excitation of xanthophyll S1 always leads to an excitation of Chl and simultaneously the excitation of Chl also always leads to an excitation of xanthophyll S1. These results support the hypothesis that excitonic interactions between xanthophylls and chlorophylls occur, resulting in low-lying states which serve as energy traps and dissipative valves for excess excitation energy (Bode et al., 2009; Liao et al., 2010). If excitation energy can be coherently transferred between two pigment molecules then they can be seen as excitonically coupled. These two molecules create a molecular dimer that functions as a single quantum-mechanical individual and the excitation is scavenged by both molecules. Thus, the immediate appearance of the xanthophyll S1 signal when excited chlorophyll molecules are present could be explained by excitonic coupling (Ruban et al., 2012). Indirect participation of xanthophyll cycle pigments in excess energy dissipation is the basis of another group of mechanisms. This has led to several models involving xanthophyll cycle products created under high light conditions (Latowski et al., 2011).

9.4.2 Role of the Xanthophylls in LHCII Aggregation-Dependent Indirect Quenching of Overexcitation

According to models of LHCII aggregation-dependent indirect quenching of overexcitation, qE occurs upon aggregation of the major, trimeric LHCII complex of PSII. This produces a conformational change within the protein and promotes energy transfer from Chl *a* to S1 excited state of Lut bound to the LHCII (Berera et al., 2007; Ruban et al., 2007). Recently, an aggregation-dependent LHCII quenching model was supported by the observation of a red shifted fluorescence lifetime component both in aggregated LHCII trimers binding Zx and in quenched leaves (Miloslavina et al., 2008). In this model, Zx bound at site V1 of LHCII acts as an allosteric modulator of L-dependent quenching, whereas aggregation *in vitro* has been proposed to entrain an intrinsic conformational transition in the LHCII complex, responsible for the establishment of the quenching reaction (Ilioaia et al., 2008). Xanthophyll S1–Chl excited state coupling was recently measured in isolated LHCII complexes and correlated with the NPQ amplitude *in vivo* in different mutants such as *npq1*, *npq2*, *lut2*, and *PsbS* overaccumulating lines (Bode et al., 2009).

Indirect spectroscopic evidence suggesting that upon the formation of qE, part of the major LHCII undergoes separation from the PSII supercomplex is presented by Holzwarth's group (Miloslavina et al., 2008; Holzwarth et al., 2009). Subsequently, biochemical evidence showing that under NPQ conditions PsbS controls the dissociation of a part of the PSII–LHCII supercomplex containing LHCII, CP24 and CP29 was obtained by the group of Bassi (Betterle et al., 2009). Ruban et al. (2012) have more recently obtained further structural evidence for a clustering of PSII core units as well as LHCII antenna aggregation upon the formation of qE.

These results support the role of ΔpH and the de-epoxidation of Vx to Zx in LHCII aggregation. It has been proposed that, in the qE state, the part of the PSII–LHCII supercomplex that contains trimers of LHCII is dissociated and that this reorganization leads to the aggregation and partial segregation of LHCII from PSII (Ruban et al., 2012). In this model Zx is seen as favoring LHC aggregates, while the more polar Vx favors fluorescent LHCII conformations (Ruban et al., 2012).

9.4.3 Light-Driven Reactions in LHCII as a Mechanism of the Indirect Quenching of Overexcitation by the Products of the Light Phase of the Xanthophyll Cycles

It was observed that under strong light photo-*all-trans* Vx bound to LHCII undergoes isomerization to the *cis* form (Grudziński et al., 2001). Under the same conditions light-induced trimer to monomer transition in LHCII was detected (Garab et al., 2002; Hill et al., 2012). Fluorescence lifetime imaging microscopy revealed that *all-trans* Vx stabilizes the trimeric organization of the LHCII, whereas Zx, product of Vx de-epoxidation, promotes a monomeric state of LHCII. Because the rate of excitation energy transfer from Vx to Chl in LHCII is extremely low (Caffarri et al., 2001) the light energy absorbed by Vx is postulated to be utilized by acceleration of the isomerization of *all-trans* Vx to the *cis* isomer. Decreasing of *all-trans* Vx and increasing of *cis* isomer of this pigment causes a trimer to monomer transition in LHCII, which then leads to a reduction in the Chl fluorescence lifetime. The shortening of the Chl fluorescence lifetime reflects a more efficient thermal dissipation of singlet excitation and therefore the light-dependent process is discussed in terms of photoprotective activity within LHCII.

Moreover, the operation of the xanthophyll cycle in the photosynthetic apparatus requires Vx to be freely available within the lipid phase of the thylakoid membrane for de-epoxidation to Zx (Jahns et al., 2009). Vx is a xanthophyll relatively weakly bound to the protein environment of LHCII, and the process of the light-driven change of this pigment's molecular configuration can result in its uncoupling from the protein and its transfer to the lipid environment of the membrane. Certainly, light dependent LHCII monomerization makes it possible, or easier, for Vx to migrate from the protein to the lipid environment. Vx in an *all-trans*, fully relaxed configuration is a specific substrate of the de-epoxidase enzyme (Yamamoto and Higashi, 1978) and the pigment tends to adopt such a configuration after light-driven transformation, due to the energy minimization process (Niedzwiedzki et al., 2005). Recently lamellar aggregates were observed in Zx enriched LHCII which were isolated from leaves preilluminated with high-intensity light. It was postulated that these aggregates are well suited to dissipate excess energy upon overexcitation (Janik et al., 2013).

9.4.4 Model of Charge Transfer Quenching between Chl a and Zx

The basis of another model, which presents the role of de-epoxidized xanthophylls in NPQ, is charge-transfer (CT) quenching between Chl *a* and Zx. The suggestion that NPQ is connected with the formation of the CT state between Chl *a* and Zx was proposed on the basis of quantum chemical calculations and ultra-fast pump-probe experiments on isolated thylakoid membranes (Holt et al., 2005). The CT mechanism involves energy transfer from a majority of Chl molecules to a Chl–Zx heterodimer that undergoes charge separation followed by recombination, thereby transiently producing a Zx radical cation ($Zx^{\bullet+}$) with a very short relaxation time between 50–200 ps, as expected of an efficient quencher. The formation of $Zx^{\bullet+}$ in thylakoids depends on the three components needed for qE *in vivo*: lumen acidification, PsbS activation and Zx production (Avenson et al., 2008). Signals indicating $Zx^{\bullet+}$ formation have been found in isolated monomeric forms of Lhcb, but not in LHCII (Avenson et al., 2008; Ahn et al., 2008; Avenson et al., 2009).

On the basis of the mutation analysis of Chl binding sites in Lhcb458 it is thought that for CT quenching a Chl pair is critical rather than a single Chl *a* chromophore. These Chl molecules were called Chl A5 and Chl B5. The involvement of a Chl pair is reasonable since charge delocalization over the Chl pair would stabilize the CT state. Chl A5–B5 are located in the proximity of the L2 domain. Zx binding to this site induces a conformational change bringing Chl A5 into excitonic interaction with Chl B5 and switching the protein to a dissipative state by $Zx^{\bullet+}$ formation. Also, Lhcb6 antenna complexes show $Zx^{\bullet+}$ formation. Two distinct CT quenching sites were detected in the Lhcb5 complex. They involve Zx and Lut radical cation species, depending on Zx binding to the L2 binding site. Thus, Zx in the L2 acts both as a quencher and as an allosteric modulator of Lut CT efficiency into site L1 (Avenson et al., 2009). Recently, L radical cation was also detected in Lhcb6 and Lhcb4 complexes binding L as the only xanthophyll (Li et al., 2009). The interaction of Zx with LHCII via PsbS was also postulated as an explanation of the fact that LHCII is involved in the catalysis of NPQ, despite the fact that LHCII trimers containing $Zx^{\bullet+}$ could not be found in solution (Avenson et al., 2008; Li et al., 2009). However, PsbS was not found to bind Zx to LHCII either in *in vivo* or *in vitro* studies (Bonente et al., 2008).

9.4.5 Protection of Plants Against Oxidative Stress by the Xanthophylls Localized Outside the Pigment–Protein Complexes

A very important aspect of the xanthophyll function as molecules protecting plants against oxidative stress is the presence of these pigments in the lipid phase of the thylakoid membrane in free form, not assembled into pigment–protein complexes (Gruszecki et al., 2006; Gruszecki and Strzałka, 1991; Latowski et al., 2004b). The protective effect of Zx on lipid degradation under strong light conditions was observed in pea leaves (Havaux et al., 1991). Under strong illumination in leaf cells of this plant the content of lipids decreased and the ratio of saturated to unsaturated lipids increased. When Zx formation was inhibited by DTT more significant lipid degradation was observed (Havaux and Niyogi, 1999). Similar results were obtained from experiments with *Arabidopsis npq1* mutant deficient in the production of Zx. Lipid content was measured in this mutant in response to strong illumination. In comparison to the wild *Arabidopsis* form, the *npq1* mutant had a significantly higher level of photooxidized lipids (Havaux and Niyogi, 1999). The same effect was also observed in tomato leaves; the Zx level and lipid degradation (measured as ethylene formation) were also well correlated, although this degradation appeared to be temperature dependent. At 3°C and under high light (low level of Zx), ethylene production was intensive but at 23°C and in high light, ethylene secretion was lower and the Zx content increased (Sarry et al., 1994).

The antioxidant properties of Zx were also tested in model systems. It was observed that Zx was the most effective of all tested carotenoids such as β-cryptoxanthin, β-carotene, astaxanthin, canthaxanthin, and lycopene, against oxidation initiated both in the aqueous and lipid phases. In a homogeneous organic solution, all tested carotenoids ameliorated lipid peroxidation. Zx, as well as β-carotene, reacted with ROS at similar rates, giving a similar degree of protection in an organic solution. The reactivity and protective efficiency of the astaxanthin and canthaxanthin were lower (Woodall et al., 1997a,b). Also lutein in model systems reacted rapidly with oxidizing agents and this xanthophyll was recognized as an important antioxidant factor (Woodall et al., 1997a,b). These results point to the significance of the xanthophyll pigments in direct protection of the photosynthetic apparatus against ROS. Although Zx and Vx are normally bound to the antenna proteins, they must be liberated from their binding sites to the lipid domains surrounding the antenna complexes so as to be accessible as substrates for the xanthophyll cycle enzymes (Schaller et al., 2010). Thus, significant proportions of Zx and Vx are transiently present in the lipid phase, where they may directly quench ROS.

It was also postulated that Vx-cycle and particularly Zx play a role in senescence, as a photoprotectant against lipid photooxidation (Munné-Bosch and Alegre, 2002). In addition it was observed that the level of Lut increased gradually during the aging of primary cabbage leaves while the level of Lx was decreased, although the correlation of this phenomenon with lipid peroxidation was not tested (Misra et al., 2011). Notwithstanding differences between authors, the Vx-cycle is recognized as one of the main adaptation mechanisms responsible for a fast response to peroxidation and for the creation of antioxidant substances in thylakoid membranes that can quench singlet oxygen (Krinsky, 1979) and other reactive oxygen species (Lim et al., 1992).

It was also observed that some stress factors like drought (Peguero-Pina et al., 2007), salt stress (Lu et al., 2003), or chilling stimulate the production of Zx (Sui et al., 2007). It is commonly suggested that this is due to xanthophyll cycle activity which plays an important role in protecting the photosynthetic apparatus from photoinhibitory damage under a variety of stressors. Another aspect of the xanthophyll cycle in protection against oxidative stress was noticed when the effect on the xanthophyll cycle of short-term ozone pollution at high doses under photoinhibitory conditions was studied. It was found that xanthophyll cycle appeared to counteract the effects of O_3 on the PSII reaction center, preventing a decline in D1 content (Ciompi et al., 1997).

In other experiments plants were subjected to direct treatment with H_2O_2, $O_2^{\bullet -}$, and to paraquat as a herbicide. Decrease in the Vx content was observed in these experiments, which was not compensated by the sum of Ax + Zx. It was hypothesized that, under photoinhibitory conditions combined with strong oxidative stress, Vx is used in large part not for the xanthophyll cycle reaction but for the synthesis of growth inhibitory substances such as abscisic acid (ABA) (Pasqualini et al., 1999; Ederli et al., 1997). It was documented that Vx is one of the intermediate products in ABA synthesis (Audran et al., 1998). One may suppose that conditions causing an increase in Vx de-epoxidase activity would result in a decrease in ABA production. Exogenously added ABA, which inhibits its synthesis, resulted in a higher concentration of Zx and greater photoprotection of PSII (Ivanov et al., 1995).

9.4.6 The Xanthophyll Cycle Activity and Ascorbate-Glutathione Cycle

Although there are different xanthophyll cycles in different taxonomic groups, enzymes participating in them have very similar properties and requirements. The factors influencing the activity of de-epoxidases are: temperature and pH, redox potential, availability of substrates and presence of specific lipids. In living cells exposed to low light intensities, when lumenal pH is around 7, de-epoxidases are water soluble. Shift to high light conditions causes acidification of lumen, which triggers de-epoxidation. De-epoxidation of Vx starts when lumenal pH drops below 6.5 and optimal pH for this reaction is 5, while de-epoxidation of Ddx to Dtx starts at pH 7 and the optimal pH is at 5.5. The lower the pH, the higher the de-epoxidation state of the xanthophyll cycle pigments (Jakob et al., 2001). Low pH is needed for binding of de-epoxidase to the membrane, as well as for protonation of ascorbate (pKa 4.1), which leads to creation of acid form AscH

TABLE 9.3 Dependence on the pH-Optimum for Activity of Violaxanthin De-epoxidase (VDE) and Diadinoxanthin De-epoxidase (DDE) on the Asc Concentration

Asc Concentration [mM]	*Optimum pH for Activity of		**Optimum pH for Activity of Partially Purified DDE
	Partially Purified VDE	VDE in Intact Thylakoids	
0.5	4.88	4.70	5.00
3.0	5.15	4.8–4.9	5.30
30.0	5.25	5.00	5.58

**Bratt et al., 1995,*
***Grouneva et al., 2006*

(Bratt et al., 1995). All known de-epoxidases use ascorbate (Asc) as a reductant in de-epoxidation (Hager, 1969; Yamamoto, 1979). Availability of Asc is important for regulation of de-epoxidase activity. This phenomenon was discovered in experiments, where VDE was inhibited by ascorbate peroxidase (APX) after addition of hydrogen peroxide. It was interpreted to be a result of competition of APX with VDE for Asc (Neubauer and Yamamoto, 1994). It was shown that optimal concentration of Asc for VDE activity strongly depends on pH. At pH 4.5–5.5, saturation of the enzyme is observed at 10–20 mM Asc (Table 9.3), whereas at pH 6.0 de-epoxidase remains not saturated even at 100 mM (Bratt et al., 1995). Proposed range of Asc concentration in chloroplasts is 10–50 mM (Gillham and Dodge, 1986, Schöner and Krause, 1990, Foyer, 1993), but increase of Asc level is often observed during response to stress, such as chilling or high light irradiance (Gillham and Dodge, 1987, Schöner and Krause, 1990). The Asc-dependent pH optimum for partially purified VDE is similar to the optimum measured for enzyme located in thylakoids (Yamamoto, 1979), although for VDE in thylakoids more narrow pH optimum and a shift for lower pH compared with partially purified enzyme were observed (Table 9.3). Supposedly, these differences are the result of different passage of Asc across the thylakoid membranes at a certain pH. At higher pH Asc is negatively charged (Asc^-), which limits membrane permeability for it. It was suggested that for VDE K_M for Asc is pH-dependent (Table 9.4) and Asc is a co-substate for VDE rather than simply a cofactor (Brat et al., 1995; Eskling et al., 1997).

Interestingly, the range of K_M for Asc was 0.10 ± 0.02 mM for all determined values in the cases when concentrations of the reduced acid form of Asc (AscH) were taken into consideration and assuming that pKa value was 4.1. Therefore, it was suggested that not Asc^- but rather AscH acts as a substrate for VDE and DDE.

TABLE 9.4 Obtained K_M Values (mM) of Violaxanthin De-epoxidase (VDE) and Diadinoxanthin De-epoxidase (DDE) for Asc at Different pH.

pH	K_M VDE [mM]		K_M DDE**
	*	**	
6.5	–	ND	8.7 ± 0.79
6.0	10	13.8 ± 0.91	6.6 ± 0.34
5.5	2.5	5.0 ± 0.56	1.2 ± 0.05
5.0	1.0	2.3 ± 0.09	0.7 ± 0.04
4.5	0.3	ND	0.6 ± 0.05

**Bratt et al., 1995,*
***Grouneva et al., 2006*
ND – not determined;

Being a protonated form of Asc, AscH is an endogenous proton and electron donor during de-epoxidation; it activates both DDE and VDE (Yamamoto, 1979; Bratt et al., 1995; Neubauer and Yamamoto, 1994; Eskling et al., 1997). These enzymes catalyze electron and proton transfer from AscH to one or two epoxide groups in epoxy-xanthophylls. The products of the reaction are completely de-epoxidized xanthophylls (i.e. Zx is formed from Vx and Dtx from Ddx), dehydroascorbate (DHA) and water (Hager, 1969). As DHA is an oxidized form of AscH, there is a need for a turnover mechanism, although, up to now, no mechanism of DHA re-reduction is known to occur in thylakoid lumen. Therefore, it was suggested that in thylakoid membrane there is a system exporting DHA outside (Bratt et al., 1995), but this hypothesis has not been empirically confirmed. The presence of an Asc carrier in the plasma membrane and chloroplast envelope was suggested by Foyer and Lelandais (1996), although there is no evidence confirming this postulate. Negatively charged Asc^- cannot simply diffuse across the lipid bilayer, so it is possible that antiport transport for Asc^- and DHA occurs in thylakoid membrane (Eskling et al., 1997). If such system really operates, re-reduction of Asc could be carried out in the stroma via the GSH cycle, using NADPH and GSH. Then, the resulting Asc^- could be protonated to AscH in thylakoid lumen acidified during high light exposure. Therefore, AscH concentration would depend on light-driven acidification of the lumen. It was observed that levels of Asc and enzymes scavenging free radicals are enhanced in plants grown at high light compared to low-light grown plants (Gillham and Dodge, 1987; Schöner and Krause, 1990). Simultaneous increase of V, A, Z pool pigments and Asc level was observed in plants exposed to high light intensities for longer time (Demmig-Adams et al., 1995; Logan et al., 1996).

Further studies on isolated chloroplasts seem to confirm the requirement for active transport of Asc, as well as possible limitation of de-epoxidation by Asc consuming reactions occurring in the stroma (Neubauer and Yamamoto, 1994). It was shown that the affinity of partially purified DDE for the AscH is 3–4 times higher than the affinity of VDE (Grouneva et al., 2006). Both enzymes show similar strong dependence of the ascorbate affinity on pH (Bratt et al., 1995), although determined K_M value of DDE for Asc at pH 5 was 0.7 mM, while for VDE it was 2.3 mM (Table 9.4) (Grouneva et al., 2006). High concentrations of Asc cause significant shift of the pH optimum towards higher pH values. It was shown that in such a case DDE remains active at almost neutral pH. It was also observed that this enzyme displays strong and very fast change of activity in response to small changes of lumenal pH. When intact diatoms are incubated in the dark for a longer time, chlororespiration leads to slight pH changes. Supposedly, DDE sensitivity for such small changes is responsible for significant induction of de-epoxidation of Ddx to Dtx in those dark-incubated cells (Jakob et al., 2001). Interestingly, the affinity for ascorbate does not depend on the de-epoxidized substrate, but is an intrinsic feature of enzyme. While DDE has high affinity for Asc, it can efficiently catalyze de-epoxidation at low pH, even in cases when availability of reduced Asc is limited. Presumably, different affinities of DDE and VDE for Asc are caused by differences in their amino acid sequences. It was discovered that four highly conserved histidines are crucial for binding of Asc by VDE. When those histidines (H 121, H124, H167, H173) were substituted by alanine or arginine, the modified enzyme displayed lowered activity and affinity to Asc, compared to native protein (Emanuelsson et al., 2003). Lower affinity of de-epoxidase for Asc results in a decreased activity of the enzyme. This indicates that histidine residues as mentioned previously may be responsible for enzymatic activity and binding of Asc. It was hypothesized that modification in the de-epoxidase region rich in histidines could occur during evolution of diatoms and lead to enhanced affinity for Asc, displayed by DDE. The higher affinity for Asc may also be a consequence of physiological differences between higher plants and diatoms. At low pH values, high activity of de-epoxidase is possible in spite of limited availability of AscH. Such a situation can occur in cells exposed to high light, where the Mehler reaction is strongly stimulated, which leads to enhanced consumption of Asc by APX (Claquin et al., 2004). Thus, high Asc affinity of DDE enables efficient de-epoxidation of Ddx, even if large amounts of Asc are used by photoprotective processes. It is also worthy of notice that, in contrast to higher plants and green algae, the Asc pool in diatoms exposed to high light is highly oxidized (Grouneva et al., 2006).

As shown, the main mechanism of regulation of the xanthophyll cycles by ascorbate-glutathione cycle depends on the AscH level. This makes AscH availability an important factor influencing protective functions of all types of xanthophyll cycles.

9.5 CONCLUSIONS

Carotenoids play a crucial role in both photoautotrophic and heterotrophic organisms. In photosynthesis two important functions for these pigments are imputed. The first is light harvesting and transfer of singlet excitation energy to chlorophylls. The second function is protection of chlorophylls and other elements of the photosynthetic apparatus from photodamage. The antioxidant actions of carotenoids are based on their singlet oxygen quenching properties and their ability to trap peroxyl radicals (Holzwarth et al., 2006). The quenching of singlet oxygen leads to excitation of carotenoids. The unexcited state of these pigments is obtained again by a series of rotational and vibrational interactions of carotenoids with the solvent resulting in dissipation of excess energy. The interaction of carotenoids with peroxyl radicals may proceed via an unstable β-carotene radical adduct (Yruela et al., 1994; Losi et al., 2003). These carotenoid adduct radicals are highly resonance stabilized and therefore relatively unreactive. They may both terminate radical reactions by binding to the attacking free radicals and undergo a decay to generate nonradical products (Losi et al., 2003; Braslavsky and Holzwarth, 2012). The deactivation of $^1O_2$ in the PSII reaction center is provided mainly by β-carotene (Ewart, 1986), while thermal dissipation of Chl singlet ($^1P^*_{680}$) excited states is operated by xanthophylls in the NPQ process (Barber and Rivas, 1993; Nishiyama et al., 2001).

Several mechanisms explaining the role of xanthophylls in NPQ have been suggested. They can be grouped into two categories. One of these engages direct overexcitation quenching by products of the light phase of the xanthophyll cycles. The other covers models based on the indirect participation of the xanthophylls in photoprotection (Latowski et al., 2011).

Another group of mechanisms postulated for excess energy dissipation by xanthophylls is based on the indirect participation of these pigments in excess energy dissipation. Which of the postulated mechanisms for photoprotection of carotenoids in photosynthesis is the closest to the truth must be shown in future research.

REFERENCES

Abdel-Aal, E.-S.M., Akhtar, H., Zaheer, K., Ali, R., 2013. Dietary Sources of lutein and zeaxanthin carotenoids and their role in eye health. Nutrients 5, 1169–1185.

Adams, W.W., Demmig-Adams, B., Verhoeven, A.S., Baker, D.H., 1995. Photoinhibition during winter stress: involvement of sustained xanthophylls cycle-dependent energy dissipation. Plant Physiol. 22, 261–276.

Adams, W.W., Demmig-Adams, B., Logan, B.A., Barker, D.H., Osmond, C.B., 1999. Rapid changes in xanthophyll cycle-dependent energy dissipation and photosystem II efficiency in two vines, *Stephania japonica* and *Smilax australis*, growing in the understory of an open eucalyptus forest. Plant Cell Environ. 22, 125–136.

Ahn, T.K., Avenson, T.J., Ballottari, M., Cheng, Y.C., Niyogi, K.K., Bassi, R., et al., 2008. Architecture of a charge transfer state regulating light harvesting in a plant antenna protein. Science 320, 794–797.

Alscher, R.G., Erturk, N., Heath, L.S., 2002. Role of superoxide dismutases (SODs) in controlling oxidative stress in plants. J. Exp. Bot. 53, 1331–1341.

Apel, K., Hirt, H., 2004. Reactive oxygen species: metabolism, oxidative stress, and signal transduction. Ann. Rev. Plant. Biol. 55, 373–399.

Armstrong, G.A., Hearst, J.E., 1996. Carotenoids 2: genetics and molecular biology of carotenoid pigment biosynthesis. FASEB J. 10 (2), 228–237.

Audran, C., Borel, C., Frey, A., Sotta, B., Meyer, C., Simonneau, T., et al., 1998. Expression studies of the zeaxanthin epoxidase gene in *Nicotiana plumbaginifolia*. Plant Physiol. 118, 1021–1028.

Avenson, T.J., Ahn, T.K., Zigmantas, D., Niyogi, K.K., Li, Z., Ballottari, M., et al., 2008. Zeaxanthin radical cation formation in minor light-harvesting complexes of higher plant antenna. J. Biol. Chem. 283, 3550–3558.

Avenson, T.J., Ahn, T.K., Niyogi, K.K., Ballottari, M., Bassi, R., Fleming, G.R., 2009. Lutein can act as a switchable charge transfer quencher in the CP26 light-harvesting complex. J. Biol. Chem. 284, 2830–2835.

Avital, S., Brumfeld, V., Malkin, S., 2006. A micellar model system for the role of zeaxanthin in the non-photochemical quenching process of photosynthesis – chlorophyll fluorescence quenching by the xanthophylls. Biochim. Biophys. Acta 1757, 798–810.

Barber, J., Rivas, J.D., 1993. A functional model for the role of cytochrome b-559 in the protection against donor and acceptor side photoinhibition. Proc. Natl. Acad. Sci. USA 90, 10942–10946.

Berera, R., van Stokkum, I.H., Kodis, G., Keirstead, A.E., Pillai, S., Herrero, C., et al., 2007. Energy transfer, excited-state deactivation, and exciplex formation in artificial caroteno-phthalocyanine light-harvesting antennas. J. Phys. Chem. B 111, 6868–6877.

Betterle, N., Ballottari, M., Zorzan, S., de Bianchi, S., Cazzaniga, S., Dall'Osto, L., et al., 2009. Light-induced dissociation of an antenna hetero-oligomer is needed for non-photochemical quenching induction. J. Biol. Chem. 284, 15255–15266.

Bode, S., Quentmeier, C.C., Liao, P.N., Hafi, N., Barros, T., Wilk, L., et al., 2009. On the regulation of photosynthesis by excitonic interactions between carotenoids and chlorophylls. Proc. Nat. Acad. Sci. USA 106, 12311–12316.

Bojorquez, R.M.C., Gallego, J.G., Collado, P.S., 2013. Functional properties and health benefits of lycopene. Nutr. Hosp. 28, 6–15.

Bonente, G., Howes, B.D., Caffarri, S., Smulevich, G., Bassi, R., 2008. Interactions between the photosystem II subunit PsbS and xanthophylls studied in vivo and in vitro. J. Biol. Chem. 283, 8434–8445.

Braslavsky, S.E., Holzwarth, A.R., 2012. Role of carotenoids in Photosystem II (PSII) reaction centers. Int. J. Thermophys. 33, 2021–2025.

Bratt, C.E., Arvidsson, P.E., Carlsson, M., Åkerlund, H.E., 1995. Regulation of violaxanthin de-epoxidase activity by pH and ascorbate concentration. Photosyth. Res. 45, 165–175.

Breton, J., Duranton, J., Satoh, K., 1998. Orientation of pigments in the reaction center and the core antenna of photosystem II. In: Scheer, H., Schneider, S. (Eds.), Photosynthetic Light-Harvesting Systems: Organization and Function. Walter de Gruyter, Berlin, pp. 375–386.

Caffarri, S., Croce, R., Breton, J., Bassi, R., 2001. The major antenna complex of photosystem II has a xanthophyll binding site not involved in light harvesting. J. Biol. Chem. 276, 35924–35933.

Christaki, E., Bonos, E., Giannenas, I., Florou-Paneri, P., 2013. Functional properties of carotenoids originating from algae. J. Sci. Food Agr. 93, 5–11.

Ciompi, S., Castagna, A., Ranieri, A., Nali, C., Lorenzini, G., Soldatini, F.G., 1997. CO_2 assimilation, xanthophyll cycle pigments and PSII efficiency in pumpkin plants as affected by ozone fumigation. Physiol. Plant. 101, 881–889.

Claquin, P., Kromkamp, J.C., Martin-Jezquel, V., 2004. Relationship between photosynthetic metabolism and cell cycle in a synchronized culture of the marine alga *Cylindrotheca fusiformis* (Bacillariophyceae). Eur. J. Phycol. 39, 33–41.

Croce, R., Cinque, G., Holzwarth, A.R., Bassi, R., 2000. The Soret absorption properties of carotenoids and chlorophylls in antenna complexes of higher plants. Photosynth. Res. 64, 221–231.

Davison, P.A., Hunter, C.N., Horton, P., 2002. Overexpression of beta-carotene hydroxylase enhances stress tolerance in *Arabidopsis*. Nature 418, 203–206.

de Bianchia, S., Ballottaria, M., Dall'Ostoa, L., Bassi, R., 2010. Regulation of plant light harvesting by thermal dissipation of excess energy. Biochem. Soc. Trans. 38, 651–660.

de Las Rivas, J., Telfer, A., Barber, J., 1993. Two coupled β-carotene molecules protect P_{680} from photodamage in isolated photosystem II reaction centers. Biochim. Biophys. Acta 1142, 155–164.

Dekker, J.P., Boekema, E.J., 2005. Supramolecular organization of thylakoid membrane proteins in green plants. Biochim. Biophys. Acta. 1706, 12–39.

Demmig-Adams, B., 1990. Carotenoids and photoprotection in plants: a role for the xanthophyll zeaxanthin. Biochim. Biophys. Acta – Bioenerg. 1020 (1), 1–24.

Demmig-Adams, B., Winter, K., Krüger, A., Czygan, F.C., 1989. Zeaxanthin and the induction and relaxation kinetics of the dissipation of excess excitation energy in leaves in 2% O_2, 0% CO_2. Plant. Physiol. 90, 887–893.

Demmig-Adams, B., Adams III, W.W., Laogan, B.A., Verhoeven, A.S., 1995. Xanthophyll cycle-dependent energy dissipation and flexible photosystem II efficiency in plants acclimated to light stress. Aust. J. Plant. Physiol. 22, 249–260.

Ederli, L., Pasqualini, S., Batini, P., Antonielli, M., 1997. Photoinhibition and oxidative stress: effects on xanthophyll cycle, scavenger enzymes and ABA content in tobacco plants. J. Plant. Physiol. 151 (4), 422–428.

Edge, R., McGarvey, D.J., Truscott, T.G., 1997. The carotenoids as antioxidants. J. Photochem. Photobiol. B: Biol. 41, 189–200.

Emanuelsson, A., Eskling, M., Åkerlund, H.E., 2003. Chemical and mutational modification of histidines in violaxanthin de-epoxidase from *Spinacia oleracea*. Physiol. Plant. 84, 218–224.

Eroglu, A., Hruszkewycz, D.P., Dela Sena, C., Narayanasamy, S., Riedl, K.M., Kopec, R.E., et al., 2012. Naturally-occurring eccentric cleavage products of provitamin A β-carotene function as antagonists of retinoic acid receptors. J. Biol. Chem. 287, 15886–15895.

Eskling, M., Arvidsson, P.O., Åkerlund, H.E., 1997. The xanthophyll cycle, its regulation and components. Physiol. Plant. 100, 806–816.

Ewart, A.J., 1986. On assimilatory inhibition in plants. J. Linnean. Soc. Bot. 31, 364–461.

Faller, P., Maly, T., Rutherford, A.W., MacMillan, F., 2001. Chlorophyll and carotenoid radicals in photosystem II studied by pulsed ENDOR. Biochem 40, 320–326.

Fischer, B.B., Ledford, H.K., Wakao, S., Huang, S.G., Casero, D., Pellegrini, M., et al., 2012. Singlet oxygen resistant 1 links reactive electrophile signalling to singlet oxygen acclimation in *Chlamydomonas reinhardtii*. Proc. Nat. Acad. Sci. USA 109, 1302–1311.

Foyer, C., Lelandais, M., 1996. A comparision of the relative rates of transport of ascorbate and glucose across the thylakoid, chloroplast and plasmalemma membranes of pea leaf mosophyll cells. J. Plant. Physiol. 148, 391–398.

Foyer, C.H., 1993. Ascorbic acid. In: Alscher, R.G., Hess, J.L. (Eds.), Antioxidants in Higher Plants. CRC Press, Boca Raton, pp. 312–358.

Foyer, C.H., Noctor, G., 2005. Oxidant and antioxidant signalling in plants: a reevaluation of the concept of oxidative stress in a physiological context. Plant Cell Environ. 28, 1056–1071.

Foyer, C.H., Noctor, G., 2011. Ascorbate and glutathione: the heart of the redox hub. Plant Physiol. 155, 2–18.

Frank, H., Cogdell, R.J., 1993. Photochemistry and function of carotenoids. in photosynthesis. In: Young, A., Britton, G. (Eds.), Carotenoids in Photosynthesis. Chapman and Hall, London, pp. 253–315.

Frank, H.A., 1999. Incorporation of carotenoids into reaction center and light-harvesting pigment–protein complexes. In: Frank, H.A., Young, A.J., Britton, G., Cogdell, R.J. (Eds.), Photochemistry of Carotenoids. Kluwer, Dordrecht, pp. 235–244.

Frank, H.A., Cua, A., Chynwat, V., Young, A., Gosztola, D., Wasielewski, M.R., 1994. Photophysics of the carotenoids associated with the xanthophyll cycle in photosynthesis. Photosynth. Res. 41, 389–395.

Frank, H.A., Bautista, J.A., Josue, S.J., Young, A.J., 2000. Mechanism of nonphotochemical quenching in green plants: energies of the lowest excited singlet states of violaxanthin and zeaxanthin. Biochemistry 39, 2831–2837.

Garab, G., Cseh, Z., Kovacs, L., Rajagopal, S., Varkonyi, Z., Wentworth, M., et al., 2002. Light-induced trimer to monomer transition in the main light-harvesting antenna complex of plants: thermo-optic mechanism. Biochemistry 41, 15121–15129.

Garcia-Plazaola, J.I., Matsubara, S., Osmond, C.B., 2007. The lutein epoxide cycle in higher plants: its relationships to other xanthophylls cycles and possible functions. Funct. Plant. Biol. 34, 759–773.

Gillham, D.J., Dodge, A.D., 1986. Hydrogen-peroxide scavenging systems within pea chloroplasts. A quantitative study. Planta 167, 246–251.

Goodwin, T.W., 1980. The Biochemistry of the Carotenoids, Vol. 1: Plants. Chapman and Hall, New York, pp. 203.

Grouneva, I., Jakob, T., Wilhelm, C., Goss, R., 2006. Influence of ascorbate and pH on the activity of the diatom xanthophyll cycle-enzyme diadinoxanthin de-epoxidase. Physiol. Plant 126, 205–211.

Grudziński, W., Matula, M., Sielewiesiuk, J., Kernem, P., Krupa, Z., Gruszecki, W.I., 2001. Effect of 13-cis violaxanthin on organization of light harvesting complex II in monomolecular layers. Biochim. Biophys. Acta 1503, 291–302.

Gruszecki, W.,I., Matuła, M., Myśliwa-Kurdziel, B., Kernem, P., Krupa, Z., Strzałka, K., 1997. Effect of xanthophyll pigments on fluorescence of chlorophyll a in LHC II embedded to liposomes. J. Photochem. Photobiol. B: Biol. 37, 84–90.

Gruszecki, W.I., Strzałka, K., 1991. Does the xanthophyll cycle take part in the regulation of fluidity of the thylakoid membrane? Biochim. Biophys. Acta 1060, 310–314.

Gruszecki, W.I., Strzałka, K., 2005. Carotenoids as modulators of lipid membrane physical properties. Biochim. Biophys. Acta 1740, 108–115.

Gruszecki, W.I., Grudziński, W., Gospodarek, M., Patyra, M., Maksymiec, W., 2006. Xanthophyll-induced aggregation of LHCII as a switch between light-harvesting and energy dissipation systems. Biochim. Biophys. Acta 1757, 1504–1511.

Hager, A., 1969. Lichtbedingte pH-Erniedrigung in einem Chloroplastenkompartiment als Ursache der enzymatischen Violaxanthin- Zeaxanthin-Umwandlung; Beziehungen zur Photophosphorylierung. Planta 89, 224–243.

Hanley, J., Deligiannakis, Y., Pascal, A., Faller, P., Rutherford, A.W., 1999. Carotenoid oxidation in photosystem II. Biochemistry 38, 8189–8195.

Havaux, M., Niyogi, K.K., 1999. The violaxanthin cycle protects plants from photooxidative damage by more than one mechanism. Proc. Nat. Acad. Sci. USA 96, 8762–8767.

Havaux, M., Gruszecki, W.I., Dupont, I., Leblanc, R.M., 1991. Increased heat emission and its relationship to the xanthophyll cycle in pea leaves exposed to strong light stress. J. Photochem. Photobiol. B Biol. 8, 361–370.

Hill, R., Larkum, A.W.D., Prášil, O., Kramer, D.M., Szabó, M., Kumar, V., et al., 2012. Light-induced dissociation of antenna complexes in the symbionts of scleractinian corals correlates with sensitivity to coral bleaching. Coral Reefs 31 (4), 963–975.

Holt, N.E., Zigmantas, D., Valkunas, L., Li, X.P., Niyogi, K.K., Fleming, G.R., 2005. Carotenoid cation formation and the regulation of photosynthetic light harvesting. Science 307, 433–436.

Holzwarth, A.R., Müller, M.G., Reus, M., Nowaczyk, M., Sander, J., Rögner, M., 2006. Kinetics and mechanism of electron transfer in intact photosystem II and in the isolated reaction center: pheophytin is the primary electron acceptor. Proc. Natl. Acad. Sci. USA 103, 6895–6900.

Holzwarth, A.R., Miloslavina, Y., Nilkens, M., Jahns, P., 2009. Identification of two quenching sites active in the regulation of photosynthetic light-harvesting studied by time-resolved fluorescence. Chem. Phys. Lett. 483, 262–267.

Horton, P., Ruban, A.V., Rees, D., Pascal, A.A., Noctor, G., Young, A.J., 1991. Control of the light-harvesting function of chloroplast membranes by aggregation of the LHCII chlorophyll protein complex. FEBS Lett. 292, 1–4.

Horton, P., Ruban, A.V., Walters, R.G., 1996. Regulation of light harvesting in green plants. Ann. Rev. Plant. Physiol. Plant. Mol. Biol. 47, 655–684.

Horton, P., Wentworth, M., Ruban, A.V., 2005. A control of the light harvesting function of chloroplast membranes: the LHCII-aggregation model for non-photochemical quenching. FEBS Lett. 579, 4201–4206.

Ilioaia, C., Johnson, M.P., Horton, P., Ruban, A.V., 2008. Induction of efficient energy dissipation in the isolated light-harvesting complex of photosystem II in the absence of protein aggregation. J. Biol. Chem. 283, 29505–29512.

Isaacs, N.W., 1995. Crystal structure of an integral membrane light-harvesting complex from photosynthetic bacteria. Nature 374, 517–521.

Ivanov, A.G., Krol, M., Maxwell, D., Huner, N.P., 1995. Abscisic acid induced protection against photoinhibition of PSII correlates with enhanced activity of the xanthophyll cycle. FEBS Lett. 371, 61–64.

Jahns, P., Holzwarth, A.R., 2012. The role of the xanthophyll cycle and of lutein in photoprotection of photosystem II. Biochim. Biophys. Acta 1817, 182–193.

Jahns, P., Latowski, D., Strzałka, K., 2009. Mechanism and regulation of the violaxanthin cycle: the role of antenna proteins and membrane lipids. Biochim. Biophys. Acta 1787, 3–14.

Jakob, T., Goss, R., Wilhelm, C., 2001. Unusual pH-dependence of diadinoxanthin de-epoxidase activation causes chlororespiratory induced accumulation of diatoxanthin in the diatom Phaeodactylum tricornutum. J. Plant. Physiol 158, 383–390.

Janik, E., Bednarska, J., Zubik, M., Puzio, M., Luchowski, M., Grudziński, W., et al., 2013. Molecular architecture of plant thylakoids under physiological and light stress conditions: a study of lipid–light-harvesting complex II model membranes. Plant Cell 25, 2155–2170.

Jansson, S., 1994. The light-harvesting chlorophyll a/b-binding proteins. Biochim. Biophys. Acta 1184, 1–19.

Johnson, M.P., Zia, A., Horton, P., Ruban, A.V., 2010. Effect of xanthophyll composition on the chlorophyll excited state lifetime in plant leaves and isolated LHCII. Chem. Phys. 373, 23–32.

Johnson, M.P., Zia, A., Ruban, A.V., 2012. Elevated ΔpH restores rapidly reversible photoprotective energy dissipation in *Arabidopsis* chloroplasts deficient in lutein and xanthophyll cycle activity. Planta 235 (1), 193–204.

Kamlowski, A., Frankemöller, L., van der Est, A.J., Stehlik, D., Holzwarth, A.R., 1996. Evidence for delocalization of the triplet state 3P680 in the D1D2cytb559-complex of Photosystem II. Ber. Bunsen-Ges. Phys. Chem. 100, 2045–2051.

Kidd, P.M., 2011. Astaxanthin, cell membrane nutrient with diverse clinical benefits and anti-aging potential. Altern. Med. Rev. 16 (4), 335–364.

Koyama, Y., 1991. Structures and functions of carotenoids in photosynthetic systems. J. Photochem. Photobiol. Ser. Biol. 9, 265–280.

Krieger-Liszkay, A., 2005. A singlet oxygen production in photosynthesis. J. Exp. Bot. 56, 337–346.

Krinsky, N.I., 1979. Carotenoid protection against oxidation. Pure Appl. Chem. 51, 649–660.

Kwa, S.L.S., Newell, W.R., Van Grondelle, R., Dekker, J.P., 1992. The reaction center of photosystem II studied by polarized fluorescence spectroscopy. Biochim. Biophys. Acta 1099, 193–202.

Kühlbrandt, W., Wang, D.N., Fujiyoshi, Y., 1994. Atomic model of plant light-harvesting complex by electron crystallography. Nature 367, 614–621.

Lambrev, P.H., Nilkens, M., Miloslavina, Y., Jahns, P., Holzwarth, A.R., 2010. Kinetic and spectral resolution of multiple non-photochemical quenching components in *Arabidopsis* leaves. Plant Physiol. 152, 1611–1624.

Latowski, D., Grzyb, J., Strzałka, K., 2004a. The xanthophyll cycle – molecular mechanism and physiological significance. Acta Physiol. Plant 26, 197–212.

Latowski, D., Akerlund, H.E., Strzałka, K., 2004b. Violaxanthin de-epoxidase, the xanthophyll cycle enzyme, requires lipid inverted hexagonal structures for its activity. Biochemistry 43, 4417–4420.

Latowski, D., Kuczynska, P., Strzałka, K., 2011. Xanthophyll cycle – a mechanism protecting plants against oxidative stress. Redox Rep. 16, 78–90.

Lavaud, J., Rosseau, B., Etienne, A.L., 2002. In diatoms, a transthylakoid proton gradient alone is not sufficient to induce a nonphotochemical fluorescence quenching. FEBS Lett. 523, 163–166.

Lavaud, J., Rosseau, B., van Gorkom, H.J., Etienne, A.L., 2002b. Influence of the diadinoxanthin pool size on photoprotection in the marine planktonic diatom *Phaeodactylum tricornutum*. Plant Physiol. 129, 1398–1406.

Li, Z., Ahn, T.K., Avenson, T.J., Ballottari, M., Cruz, J.A., Kramer, D.M., et al., 2009. Lutein accumulation in the absence of zeaxanthin restores nonphotochemical quenching in the *Arabidopsis thaliana* npq1 mutant. Plant Cell 21, 1798–1812.

Liao, P.N., Holleboom, C.P., Wilk, L., Kuehlbrandt, W., Walla, P.J., 2010. Correlation of Car S1 → Chl with Chl → Car S1 energy transfer supports the excitonic model in quenched light harvesting Complex II. J. Phys. Chem. B. 114, 15650–15655.

Lim, B.P., Nagao, A., Terao, J., Tanaka, T., Takama, K., 1992. Antioxidant activity of xanthophylls on peroxyl radical-mediated phospholipid membrane. Biochem. Biophys. Acta 1126, 178–184.

Liu, Z., Yan, H., Wang, K., Kuang, T., Zhang, J., Gui, L., et al., 2004. Crystal structure of spinach major light-harvesting complex at 2.72 Å resolution. Nature 428, 287–292.

Logan, B.A., Barker, D.H., Demmig-Adamas, B., Adams III, W.W., 1996. Acclimation of leaf carotenoid composition and ascorbate levels to gradients in the light environment within an Australian rainforest. Plant Cell Environ. 19, 1083–1090.

Lokstein, H., Tian, L., Polle, J.E.W., DellaPenna, D., 2002. Xanthophyll biosynthetic mutants of *Arabidopsis thaliana*: altered nonphotochemical quenching of chlorophyll fluorescence is due to changes in Photosystem II antenna size and stability. Biochim. Biophys. Acta 1553, 309–319.

Losi, A., Yruela, I., Reus, M., Holzwarth, A.R., Braslavsky, S.E., 2003. Structural changes upon excitation of D1-D2-Cyt b(559) photosystem II reaction centers depend on the beta-carotene content. Photochem. Photobiol. Sci. 2, 722–729.

Lu, C.M., Qiu, N.W., Lu, Q.T., 2003. Photoinhibition and the xanthophylls cycle are not enhanced in the salt-acclimated halophyte *Artimisia anethifolia*. Physiol. Planta 118 (4), 532–537.

McDermott, G., Prince, S.M., Freer, A.A., Hawthornthwalte-Lawless, A.M., Papiz, M.Z., Cogdell, R.J., et al., 2007. Differential effects of carotenoids on lipid peroxidation due to membrane interactions: X-ray diffraction analysis. Biochim. Biophys. Acta 1768, 167–174.

Miloslavina, Y., Wehner, A., Lambrev, P.H., Wientjes, E., Reus, M., Garab, G., et al., 2008. Far-red fluorescence: a direct spectroscopic marker for LHCII oligomer formation in non-photochemical quenching. FEBS Lett. 582, 3625–3631.

Misra, A.N., Latowski, D., Strzałka, K., 2011. Photosynthetic excitation pressure causes violaxanthin de-epoxidation in aging cabbage (*Brassica oleracea* L.) leaves. J. Life Sci. 5, 182–191.

Morosinotto, T., Caffarri, S., Dall'Osto, L., Bassi, R., 2003. Mechanistic aspects of the xanthophyll dynamics in higher plant thylakoids. Physiol. Plant. 119, 347–354.

Müller, M.G., Lambrev, P., Reus, M., Wientjes, E., Croce, R., Holzwarth, A.R., 2010. Singlet energy dissipation in the photosystem II light-harvesting complex does not involve energy transfer to carotenoids. Chemphyschem 11, 1289–1296.

Munné-Bosch, S., Alegre, L., 2002. Plant aging increases oxidative stress in chloroplasts. Planta 214, 608–615.

Neubauer, C., Yamamoto, H.Y., 1994. Membrane barriers and Mehler-peroxidase reaction limit the ascorbate availability for violaxanthin de-epoxidase activity in intact chloroplasts. Photosynth. Res. 39, 137–147.

Niedzwiedzki, D., Krupa, Z., Gruszecki, W.I., 2005. Temperature-induced isomerization of violaxanthin in organic solvents and in lightharvesting complex II. J. Photochem. Photobiol. B Biol. 78, 109–114.

Nilkens, M., Kress, E., Lambrev, P.H., Miloslavina, Y., Müller, M., Holzwarth, A.R., et al., 2010. Identification of a slowly inducible zeaxanthin-dependent component of nonphotochemical quenching of chlorophyll fluorescence generated under steady state conditions in *Arabidopsis*. Biochim. Biophys. Acta 1797, 466–475.

Nishiyama, Y., Yamamoto, H., Allakhverdiev, S.I., Inaba, M., Yokota, A., Murata, N., 2001. Oxidative stress inhibits the repair of photodamage to the photosynthetic machinery. EMBO J. 20, 5587–5594.

Niyogi, K.K., 1999. Photoprotection revisited: genetic and molecular approaches. Ann. Rev. Plant. Physiol. Plant. Mol. Biol. 50, 333–359.

Pan, X.W., Li, M., Wan, T., Wang, L.F., Jia, C.J., Hou, Z.Q., et al., 2011. Structural insights into energy regulation of light-harvesting complex CP29 from spinach. Nat. Struct. Mol. Biol. 18, 309–315.

Pasqualini, S., Batini, P., Ederli, L., Antonielli, M., 1999. Responses of the xanthophyll cycle pool and ascorbate-glutathione cycle to ozone stress in two tobacco cultivars. Free Rad. Res. 31 (Suppl), 67–73.

Peguero-Pina, J.J., Morales, F., Flexas, J., Gil-Pelegrín, E., Moya, I., 2007. Photochemistry, remotely sensed physiological reflectance index and de-epoxidation state of the xanthophyll cycle in *Quercus coccifera* under intense drought. Oecologia 156, 1–11.

Pogson, B.J., Niyogi, K.K., Björkman, O., DellaPenna, D., 1998. Altered xanthophylls compositions adversely affect chlorophyll accumulation and nonphotochemical quenching in *Arabidopsis* mutants. Proc. Natl. Acad. Sci. USA 95, 13324–13329.

Polivka, T., Sundstrom, V., 2004. Ultrafast dynamics of carotenoid excited states – from solution to natural and artificial systems. Chem. Rev. 104, 2021–2071.

Polivka, T., Herek, J.L., Zigmantas, D., Akerlund, H., Sundstrom, V., 1999. Direct observation of the (forbidden) S1 state in carotenoids. Proc. Natl. Acad. Sci. USA 96, 4914–4917.

Pospíšil, P., 2012. Molecular mechanisms of production and scavenging of reactive oxygen species by photosystem II. Biochim. Biophys. Acta 1817, 218–231.

Ramel, F., Birtic, S., Cuiné, S., Triantaphylidčs, C., Ravanat, J.L., Havaux, M., 2012. Chemical quenching of singlet oxygen by carotenoids in plants. Plant Physiol. 158, 1267–1278.

Ruban, A.V., Horton, P., 1999. The xanthophyll cycle modulates the kinetics of non-photochemical energy dissipation in isolated light harvesting complexes, intact chloroplasts, and leaves of spinach. Plant Physiol. 119, 531–542.

Ruban, A.V., Berera, R., Ilioaia, C., van Stokkum, I.H.M., Kennis, J.T.M., Pascal, A.A., et al., 2007. Identification of a mechanism of photoprotective energy dissipation in higher plants. Nature 450, 575–578.

Ruban, A.V., Johnson, M.P., Duffy, C.D.P., 2012. The photoprotective molecular switch in the photosystem II antenna. Biochim. Biophys. Acta 1817, 167–181.

Ruffle, S.V., Wang, J., Johnston, H.G., Gustafson, T.L., Hutchison, R.S., Minagawa, J., et al., 2001. Photosystem II peripheral chlorophyll mutants in *Chlamydomonas reinhardtii*. biochemical characterization and sensitivity to photo-inhibition. Plant Physiol. 127, 633–644.

Sarry, J.E., Montillet, J.L., Sauvaire, Y., Havaux, M., 1994. The protective function of the xanthophyll cycle in photosynthesis. FEBS Lett. 353, 147–150.

Schaller, S., Latowski, D., Jemioła-Rzemińska, M., Wilhelm, C., Strzałka, K., Goss, R., 2010. The main thylakoid membrane lipid monogalactosyldiacylglycerol (MGDG) promotes the de-epoxidation of violaxanthin associated with the light-harvesting complex of photosystem II (LHCII). Biochim. Biophys. Acta-Bioenerg. 1797, 414–424.

Schöner, S., Krause, G.H., 1990. Protective systems against active oxygen species in spinach. Response to cold-acclimation in excess light. Planta 180, 383–389.

Shigemori, K., Hara, H., Kawamori, A., Akabori, K., 1998. Determination of the distances from tyrosine D to Q_A and chlorophyll Z in photosystem II studied by '2 + 1' pulsed EPR. Biochim. Biophys. Acta 1363, 187–198.

Siefermann, D., Yamamoto, H.Y., 1975. NADPH and oxygen-dependent epoxidation of zeaxanthin in isolated chloroplasts. Biochem. Biophys. Res. Commun. 62, 456–461.

Siefermann-Harms, D., 1985. Carotenoids in photosynthesis: 1. Location in photosynthetic membranes and light-harvesting function. Biochim. Biophys. Acta 811, 325–355.

Stewart, D.H., Cua, A., Chisholm, D.A., Diner, B.A., Bocian, D.F., Brudvig, G.W., 1998. Identification of histidine 118 in the D1 polypeptide of photosystem II as the axial ligand to chlorophyll Z. Biochem 37, 10040–10046.

Sui, N., Li, M., Liu, X.Y., Wang, N., Fang, W., Meng, Q.W., 2007. Response of xanthophyll cycle and chloroplastic antioxidant enzymes to chilling stress in tomato over-expressing glycerol-3-phosphate acyltransferase gene. Photosynthetica 45, 447–454.

Telfer, A., 2002. What is beta-carotene doing in the photosystem II reaction centre?. Phil. Trans. R. Soc. London. Ser. B 357, 1431–1439.

Telfer, A., Barber, J., 1989. Evidence for the photo-induced oxidation of the primary electron donor P_{680} in the isolated photosystem II reaction centre. FEBS Lett. 246, 223–228.

Telfer, A., Barber, J., 1995. Role of carotenoid bound to the photosystem II reaction centre. In: Mathis, P. (Ed.), Photosynthesis: From Light to the Biosphere, vol. IV. Kluwer, Dordrecht, The Netherlands, pp. 15–20.

Thompson, L.M., Brudvig, G.W., 1988. Cytochrome b559 may function to protect photosystem II from photoinhibition. Biochemistry 27, 6653–6658.

Tomo, T., Mimuro, M., Iwaki, M., Kobayashi, M., Itoh, S., Satoh, K., 1997. Topology of pigments in the isolated photosystem II reaction center studied by selective extraction. Biochim. Biophys. Acta 1321, 21–30.

Tounekti, T., Abreu, M.E., Khemira, H., Munné-Bosch, S., 2012. Canopy position determines the photoprotective demand and antioxidant protection of leaves in salt-stressed *Salvia officinalis* L. plants. Environ. Exper. Bot. 78, 146–156.

Triantaphylides, C., Havaux, M., 2009. Singlet oxygen in plants: production, detoxification and signaling. Trends Plant Sci. 14, 219–228.

Van Dorssen, R.J., Breton, J., Plijter, J.J., Satoh, K., Van Gorkom, H.J., Amesz, J., 1987. Spectroscopic properties of the reaction center and of the 47 kDa chlorophyll protein of photosystem II. Biochim. Biophys. Acta 893, 267–274.

Van Gorkom, H.J., Schelvis, J.P.M., 1993. Kok's oxygen clock: what makes it tick? The structure of P680 and consequences of its oxidizing power. Photosynth. Res. 38, 297–301.

Wang, J., Gosztola, D., Ruffle, S.V., Hemann, C., Seibert, M., Wasielewski, M.R., et al., 2002. Functional asymmetry of photosystem II D1 and D2 peripheral chlorophyll mutants of *Chlamydomonas reinhardtii*. Proc. Natl. Acad. Sci. USA 99, 4091–4096.

Woodall, A.A., Britton, G., Jackson, M.J., 1997a. Carotenoids and protection of phospholipids in solution or in liposomes against oxidation by peroxyl radicals: relationship between carotenoid structure and protective ability. Biochim. Biophys. Acta 1336, 575–586.

Woodall, A.A., Lee, S.W.-M., Weesie, R.J., Jackson, M.J., Britton, G., 1997b. Oxidation of carotenoids by free radicals: relationship between structure and reactivity. Biochim. Biophys. Acta 1336, 33–42.

Yamamoto, H.Y., 1979. Biochemistry of the violaxanthin cycle in higher-plants. Pure Appl. Chem. 51, 639–648.

Yamamoto, H.Y., Higashi, R.M., 1978. Violaxanthin de-epoxidase. Lipid composition and substrate specificity. Arch. Biochem. Biophys. 190, 514–522.

Young, A.J., Frank, H.A., 1996. Energy transfer reactions involving carotenoids: quenching of chlorophyll fluorescence. J. Photochem. Photobiol B: Biol. 36, 3–15.

Yruela, I., Churio, M.S., Gensch, T., Braslavsky, S.E., Holzwarth, A.R., 1994. Optoacoustic and singlet oxygen near-IR emission study of the isolated D1-D2-cyt b-559 reaction center complex of photosystem II. Protein movement associated with charge separation. J. Phys. Chem. 98, 12789–12795.

Chapter 10

Lipophilic Molecules as a Part of Antioxidant System in Plants

Renata Szymanska, Dariusz Latowski, Beatrycze Nowicka and Kazimierz Strzałka

10.1 INTRODUCTION

Oxygen-containing free radicals such as superoxide ($O_2^{\bullet -}$), hydroperoxide ($HO_2^{\bullet}$) or hydroxyl radical ($^{\bullet}OH$), as well as nonradical forms: hydrogen peroxide (H_2O_2) and singlet oxygen ($^1O_2$), are generally known as reactive oxygen species (ROS). Various ROS can be formed as by-products during oxygen-related metabolic pathways. Both harmful and beneficial effects of ROS have been observed (Gechev et al., 2006). When generated in high levels, ROS are damaging to the cells. The state when an excessive amount of ROS is present in the cells is called oxidative stress. In order to suppress deleterious effects of ROS, organisms evolved mechanisms of ROS detoxification. These processes are governed by specific enzymes or less bulky molecules with ROS scavenging properties (Halliwell, 2006).

As plants perform oxygenic photosynthesis, which means a production of molecular oxygen in chloroplasts, they are particularly exposed to ROS. High amounts of ROS are formed especially during stresses, such as high light stress, when absorbed light energy cannot be efficiently used for photosynthetic reactions. Therefore, plants possess a wide range of mechanisms to detoxify ROS, which includes several antioxidant compounds and enzymes (Gechev et al. 2006). An important group of lipophilic antioxidants are prenyllipids. Among these, chromanols, such as tocopherols, tocotrienols, and plastochromanol, as well as plastoquinone belonging to the prenylquinones are important compounds known to scavenge oxygen and organic radicals, as well as quench and scavenge $^1O_2$. They are crucial for protection of membrane lipids containing polyunsaturated fatty acid chains from lipid peroxidation. Recently, it was discovered that prenyllipids are also involved in cellular signaling pathways (Mene-Saffrane and DellaPenna, 2010; Nowicka and Kruk, 2010).

P. Ahmad (Ed): Oxidative Damage to Plants. DOI: http://dx.doi.org/10.1016/B978-0-12-799963-0.00010-1

10.2 ROS PRODUCTION SITES IN CHLOROPLASTS

During normal metabolic processes plants produce a variety of reactive oxygen species (ROS), including $O_2^{\bullet -}$, H_2O_2, hydroxyl ($^{\bullet}OH$), $^1O_2$ and others. ROS production is elevated under biotic and abiotic stress (Apel and Hirt, 2004). Due to their high reactivity, ROS can react with proteins, pigments, nucleic acids, carbohydrates and membrane lipids, leading to cell and tissue destruction, reduction of photosynthesis yield, increase of electrolyte leakage, accelerating cell death and senescence (O'Brien et al., 2012). Increased level of ROS is tightly connected with increased biosynthesis of low molecular weight antioxidants and higher activity of antioxidant enzymes (Munne-Bosch and Alegre, 2002; Apel and Hirt, 2004; Asada 2006). These compounds can efficiently scavenge ROS and protect cell molecules from ROS-dependent destruction.

There are many cellular compartments where ROS are generated. In plants, ROS are formed in: chloroplasts, mitochondria, peroxisomes, cell wall and apoplast (Galvez-Vadivieso and Mullineaux, 2010). Moreover, some enzymes such as NADPH oxidases, cell wall peroxidases, amine oxidases, and flavin-containing oxidases can also produce ROS (Galvez-Vadivieso and Mullineaux, 2010). Although in plants the significant part of ROS is produced in apoplast (Kreslavsky et al., 2012), the largest amount of ROS is generated in chloroplasts (Fig. 10.1). This is connected with electron transport occurring in thylakoid membranes.

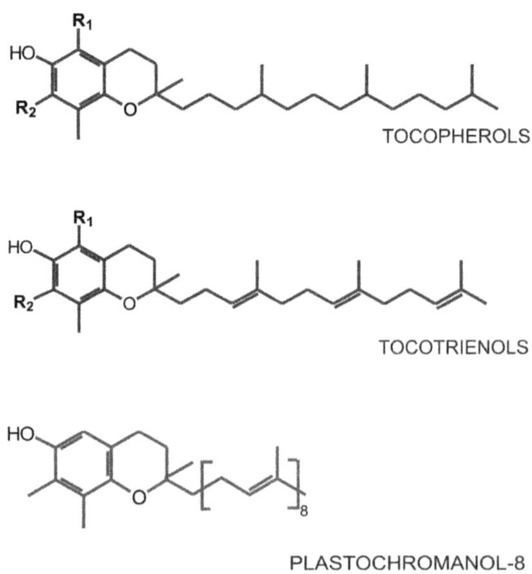

Tocopherol/Tocotrienol	R_1	R_2
α	CH_3	CH_3
β	CH_3	H
γ	H	CH_3
δ	H	H

FIGURE 10.1 Chemical structure of tocochromanols.

Although $O_2^{\bullet -}$-generating systems encompass NADPH oxidases, leakage from electron transport chains and xanthine oxidase (O'Brien et al., 2012), this reactive oxygen form is generated mostly in PSI, on its acceptor site. A low amount of $O_2^{\bullet -}$ is also produced within the plastoquinone pool (Pospisil, 2009). Exposure to light of high intensity and decreased CO_2 availability causes enhanced reduction of molecular oxygen to $O_2^{\bullet -}$ at the PSI (in the Mehler reaction) (Fig. 10.1). After that, $O_2^{\bullet -}$ is rapidly dismutated to H_2O_2 – the most long-living ROS (Foyer and Shigeoka, 2011). Dismutation reaction is catalyzed by membrane-bound Cu/Zn dismutases (Asada, 1999). Further detoxification of hydrogen peroxide (H_2O_2) is catalyzed in the chloroplasts by ascorbate peroxidases (APXs). This mechanism, called the *water–water cycle*, is a very sufficient mechanism of excessive energy dissipation (Asada, 2006; Trianthaphylides et al., 2008). It was shown that knockout of APX1 in *Arabidopsis* caused an increase in the level of heat shock transcription factors (HSfA4a and HSfA8), but this effect was not observed in CAT2 mutant (Davletova et al., 2005; Vanderauwera et al., 2005). Tobacco with deficiency in single cytoplasmic APX1 or peroxisomal CAT shows constant symptoms of oxidative stress (Rizhsky et al., 2002). In contrast to single mutation, double tobacco mutant (APX1/CAT) is more resistant to oxidative stress and even more tolerant than wild-type (Rizhsky et al., 2002).

Furthermore, a decreased level of CO_2 affects the first step in the CO_2 fixation pathway catalyzed by Rubisco by shifting the carboxylation toward oxygenation – a process called photorespiration (Trianthaphylides et al., 2008). The effect of this process is H_2O_2 production in peroxisomes. In physiological conditions the concentration of H_2O_2 is less than 1 μM (Foyer and Shigeoka, 2011; Kreslavsky et al., 2012). Pospisil (2009) claims that H_2O_2 can be produced at the donor side of PSII, in the process of water photooxidation, but this reaction is unlikely to occur in nature. More probably, H_2O_2 is created on the acceptor side of PSII at O_2 uptake by the reduced pool of plastoquinone when this pool is highly reduced (Pospisil, 2009).

Photosynthetic electron transport chain is also the source of singlet oxygen ($^1O_2$) (Fig. 10.1). This ROS species is produced by light-excited chlorophyll and its precursors. Its occurrence is connected with formation of chlorophyll P680 triplet states in the reaction centers of PSII and/or in LHC when the electron transport chain is overreduced due to absorption of excess light or under action of other stress factors (Kriger-Liszkay, 2005; Szymanska et al., 2012). In such conditions, PSII can be damaged in the process called photoinhibition (Szymanska et al., 2012). When the energy absorbed by chlorophyll is not used sufficiently, the singlet excited state of its molecules can be transformed into the triplet excited state. The chlorophyll triplet excited state has a relatively longer lifetime, which enables reaction with oxygen (triplet ground-state), leading to formation of reactive $^1O_2$ (Kriger-Liszkay, 2005; Fischer et al., 2013). Because half-life of $^1O_2$ in cells is about 200 ns this kind of ROS can react with molecules in its neighborhood (Kriger-Liszkay, 2005). The possible diffusion distance of $^1O_2$ was calculated to be ~10 nm (Sies and Menck, 1992). Singlet oxygen is able to react with proteins, pigments and fatty acids and causes loss of PSII activity due to the degradation of D1 protein (Aro et al., 1993; Szymanska et al., 2012). The dangerous triplet excited state of chlorophyll can be quenched directly by the carotenoids, and this situation was observed in the antenna system (Telfer, 2002), but not in the reaction center. Although, it was suggested that two β-carotene molecules present in the reaction center of PSII probably quench $^1O_2$ (Telfer, 2002). Apart from carotenoids, α-tocopherol is also an efficient $^1O_2$ scavenger (Trebst, 2003). Singlet oxygen formed in the PSII reaction center reacts with D1 protein causing its degradation. It is thought that the degradation of D1 protein is a physiological defense system, which prevents uncontrolled damage of PSII. This process serves as a safety valve to detoxify $^1O_2$ quickly and directly at the place of its production (Trebst, 2002; 2003). Damaged D1 protein is degraded and PSII is repaired by the newly synthesized D1 (D1 protein turnover) (Aro et al., 1993; Szymanska et al., 2012). Despite the fact that PSII is the main source of $^1O_2$ formation, some literature data showed evidence that $^1O_2$ is also formed during illumination of isolated cytochrome b_6f (Suh et al., 2000). It is more probable that the Fe-S cluster is responsible for this production, not the cytochrome itself (Suh et al., 2000), although participation of chlorophyll *a* molecule bound to cyt b_6f complex was also postulated (Sang et al., 2009).

Singlet oxygen can also be formed in the course of a greening process when etioplasts are transformed into chloroplasts. During formation of thylakoid membranes disorganized chlorophyll can act as a photosensitizer – the triplet excited state of chlorophyll is formed under illumination and $^1O_2$ is produced (Marder et al., 1998). Moreover, it was proved that protochlorophyllide is also involved in $^1O_2$ production (Op den Camp et al., 2003). There is strong evidence that $^1O_2$ can also be produced by free chlorophyll and its degradation products, which are strong photosensitizers (Kriger-Liszkay, 2005). Free chlorophyll is bound with the ELIPs (Early Light Induced Proteins) or WSCPs (Water Soluble Chlorophyll Proteins) (Schmidt et al., 2003). The binding of free chlorophyll by those molecules reduces $^1O_2$ production, but the mechanism of this action is still unknown.

10.3 EFFECT OF DIFFERENT ABIOTIC FACTORS ON ROS PRODUCTION AND OXIDATIVE STRESS

Most types of abiotic stress such as high light, drought, salinity, flood, heat or cold disrupt the metabolic balance of cells, resulting in enhanced production of ROS (Mittler, 2002; Sharma et al., 2012; Suzuki et al., 2012). In recent years it was clearly recognized that ROS play a dual role – not only as toxic molecules but also as key regulators of many biological processes such as cell cycle, growth, development, programmed cell death, hormone signaling, and cell responses to biotic and abiotic factors (Foyer and Noctor, 2003; Miller et al., 2008). Identification of the chemical nature and cellular source of ROS under abiotic stress conditions can explain certain defense and acclimation mechanisms (Mittler et al., 2004). While most of the studies treat the impact of different environmental factors on physiology and biochemistry of plants as a separate case, in nature plants are usually subjected to the combination of stress factors, both biotic and abiotic. Moreover there are big differences between laboratory and field conditions. For example, Semchuk et al. (2009) have shown that *Arabidopsis* plants growing outdoors over a long period had up to 8- to 12-fold higher total tocopherol content than those grown under laboratory conditions.

As described above, one consequence of ROS presence in chloroplast is photoinhibition. Another and more pronounced deleterious effect is lipid peroxidation. Lipids can be oxidized by two nonenzymatic reactions: type I is initiated by free radicals, i.e. $^{\bullet}OH$, hydroperoxide ($HO_2^{\bullet}$) or peroxyl radical, whereas type II is a direct oxidation by $^1O_2$ (Trianthaphylides et al., 2008). Both reaction types lead to respective oxygenated fatty acids. Using HPLC-electrospray ionization-tandem mass spectrometry based method Trianthaphylides et al. (2008) confirmed that nonenzymatic lipid peroxidation in *Arabidopsis* leaves is almost exclusively mediated by $^1O_2$ and that photo-oxidative stress-dependent cell death involves $^1O_2$ generation in the final step.

Plants have evolved various strategies to acclimate to high light conditions. The mechanisms of that acclimation work not only at the gene level, but also at other levels of organization. Under strong illumination plants synthesize antioxidant enzymes and low molecular weight antioxidants (Krause et al., 2012). A number of studies focused on the tocochromanol response to various stress conditions. It was shown that the level of tocochromanols is related to stress intensity. The most studied member of tocochromanol family is α-tocopherol, due to its essential role in plant and animal metabolism. However, it was proved that the other tocochromanols can fulfill the protective functions as effectively as α-tocopherol. Szymanska and Kruk (2010a) have shown for the first time that under high light conditions, plastoquinol is the main lipid-soluble antioxidant and it plays a crucial role in plant acclimation to high light in addition to α-tocopherol. Experiments with tocopherol biosynthetic mutants have shown that in the high light-treated plants the level of plastoquinol increased 10-fold and this increase was independent of tocopherol level. This increase was mainly attributed to the photochemically nonactive fraction of this compound localized in plastoglobules, which are the storage site of prenyllipids (Szymanska and Kruk, 2010a).

Under high light conditions production of ROS is elevated and they attack cellular molecules, including polyunsaturated fatty acids (PUFA), proteins and pigments. Tocochromanols in these conditions are not restricted to the scavenging of $^1O_2$, but they are able to protect all molecules localized in the lipid phase (Havaux et al., 2005). Under intensive illumination the total level of tocochromanols increases, with the predominance of α-tococopherol and plastoquinol (Szymanska and Kruk, 2010a). The same tendency is observed when plants are exposed to high light in combination with nutrient deficiency and low temperature (Collakova and DellaPenna, 2003). Under water deficiency ROS are produced as a result of stomatal closure and over-reduction of the electron transport chain. Increased production of ROS ($^1O_2$, $^\bullet OH$ and H_2O_2) leads to lipid peroxidation and photosynthetic apparatus damage (Cela et al., 2009). Drought stress induced the increase of α-tocopherol level in *Arabidopsis*, tobacco (Liu et al., 2008), field-grown rosemary, sage, lemon balm leaves and holm oak plants (Munné-Bosch, 2005). Recently, it was shown that tocopherols play a crucial role in the tolerance to oxidative stress induced by heavy metals (zinc, cadmium, lead, nickel). Exposure to heavy metals leads to increase of tocopherol level (Lushchak and Semchuk, 2012). The exposure of *Arabidopsis* plants to copper and cadmium resulted in 6- and 5-fold increases in tocopherol concentration, respectively (Collin et al., 2008). This increase was connected to the upregulation of genes encoding tocopherol biosynthesis enzymes. The cadmium enhanced levels of HPPD and VTE2 (in the first stage of stress) and VTE5 (in a later stage) while copper induced upregulation of VTE2 transcript level (Collin et al., 2008). Salinity is another stress factor, which has a great impact on tocopherol level. In *Arabidopsis*, salt treatment leads to decrease of α-tocopherol and increase of the level of γ-tocopherol (Ellouzi et al., 2011). In tobacco plants, salt treatment resulted in 4-fold increase of total tocopherol content, mainly represented by the α-form (Abbasi et al., 2007). Increase of tocopherol content during response to salinity was also observed in basil (Tarchoune et al., 2013). Transgenic tobacco plant – deficient in α-tocopherol, accumulated 70-fold higher amounts of γ-tocopherol than α-tocopherol in salt-stressed plants as compared to wild and unstressed plants (Abbasi et al., 2007). The observed changes in tocochromanol concentration are the result of altered expression of genes involved in biosynthesis of these compounds. On the other hand they are caused by tocochromanol action in the cells – reducing ROS levels and inhibition of lipid peroxidation (Munné-Bosch, 2005).

10.4 CHARACTERIZATION OF STRUCTURE, BIOSYNTHESIS AND FUNCTION OF PRENYLLIPIDS OCCURRING IN CHLOROPLASTS (TOCOCHROMANOLS, PLASTOQUINOL)

Chloroplasts are equipped with a number of antioxidant molecules and mechanisms. Tocochromanols, very efficient antioxidants, are considered to be the major antioxidants in the lipid phase of membranes (Munne-Bosch and Alegre, 2002). Tocochromanols are also essential nutrients for humans and animals. They are lipid-soluble, amphipathic compounds, exclusively synthesized by

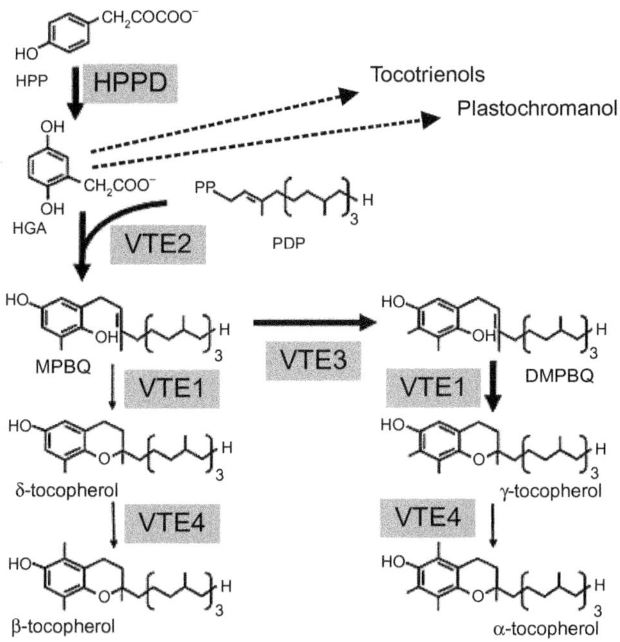

FIGURE 10.2 Biosynthesis of tocopherols and plastochromanol.

plants and some cyanobacteria (Szymanska and Kruk, 2010a). Tocochromanols are divided into two groups: tocopherols and tocotrienols (Fig. 10.2). Tocopherols have a fully saturated isoprenoid side chain (derived from phythyl diphosphate) opposite to tocotrienols, which have an unsaturated chain (derived from geranylgeranyl diphosphate). Each of two groups is composed of four homologues differing in the number and position of methyl groups in the aromatic ring (position 5, 7 and 8). According to the distribution of these methyl groups they are named: α-, β-, γ- and δ-homologue (Fig. 10.2).

Another compound belonging to the group of tocochromanols is plastochromanol – a naturally occurring homologue of γ-tocotrienol with a longer side chain composed of eight subunits. Hydroxyl-derivative of plastochromanol, a hydroxy-plastochromanol, was also found in plants (Fig. 10.2) (Szymanska and Kruk, 2010b). Plastochromanol is synthesized from plastoquinol by tocopherol cyclase (VTE1) (Szymanska and Kruk, 2010a). All these molecules have antioxidant properties, but α-tocopherol is the most biologically and chemically active (Schneider, 2005). Recently, it was shown that antioxidant properties of plastochromanol are similar to these of tocopherols; moreover it was suggested that due to the longer chain this compound could be very effective in the hydrophobic core of the lipid bilayer (Nowicka et al., 2013). The importance of plastochromanol for $^1O_2$ detoxification was also confirmed in *Arabidopsis* (Rastogi et al., 2013).

The main sources of tocopherols are plant oils, vegetables, fruits and seeds. In plants they are present in all tissues, but the highest amount can be found in green leaves and seeds (Munne-Bosch and Alegre, 2002) (Table 10.1).

In leaves the predominant form is α-tocopherol, whereas in seeds, the γ-form. β- and δ-tocopherols, as well as tocotrienols are less abundant (Schneider, 2005). In general, tocochromanol content in leaves is ~50 μg/g FW (Szymanska and Kruk, 2008a) (Table 10.1). Plastochromanol was found in great amounts in flax and rape seed oils (Gruszka and Kruk, 2007). In contrast to tocotrienol, whose occurrence is limited to the nonphotosynthetic organs, PC was also found in leaves of many plant species (Strzałka et al., 2009, 2011; Szymanska and Kruk, 2010b). Its function in leaves and seeds seems to be the same as tocopherols. In plants, tocochromanols occur mainly in plastids, but their presence in other organelles has been reported. Tocopherols have also been found in vacuoles and nuclei of barley leaves (Rautenkranz et al., 1994) and in both chloroplasts and mitochondria of green algae (Shigeoka et al., 1986; Kusmic et al., 1999). In seeds, apart from plastids, tocopherols were found to be associated with cytoplasmic lipid bodies (White et al., 2006). In chloroplasts tocochromanols occur in thylakoid membranes, plastoglobules and the chloroplast envelope (Munne-Bosch and Alegre, 2002).

TABLE 10.1 Tocopherol Content and Composition in Selected Plant Species

Species	α-Toc [μg/g FW]	γ-Toc/α-Toc [%]	δ-Toc/α-Toc [%]	Total Tocs [μg/g FW]
Spinach	16,2 ± 1,7	5,0 ± 1,2	2,0 ± 1,2	17,3
Parsley	47,1 ± 7,5	1,7 ± 0,2	0,5 ± 0,3	48,1
Tobacco	180 ± 12	0,5 ± 0,2	0,7 ± 0,1	182,2
Ficus elastica (leaf)	301 ± 18	1,1 ± 0,3	0,03 ± 0,002	304,4
Kalanchoe daigremontiana	16,7 ± 1,5	85,2 ± 8,6	12,0 ± 2,0	32,9
Lettuce	16,2 ± 0,1	37 ± 1,7	0,1 ± 0,1	22,2
Cuscuta europea	10 ± 0,8	20,8 ± 2,2	1,2 ± 0,1	12,1
Cuscuta epithymum	10 ± 3	160 ± 20	96 ± 14	35,6
Viscum album	56	3	0,7	57,7
Betula verrucosa	293 ± 108	2,5 ± 2	2,15 ± 0,8	307
Acer psudoplatanus	435 ± 138	25 ± 14	10 ± 7	587
Alnus incana	519 ± 90	6,5 ± 3,7	0,2 ± 0,2	553
Pinus sylvestris	118 ± 26	0,65 ± 0,15	0,06 ± 0,02	119
Populus tremula	637 ± 98	3,0 ± 0,3	0,1 ± 0,1	656
Juniperus communis	145 ± 37	1,5 ± 0,5	0,4 ± 0,2	147

From Szymanska and Kruk, 2008.

Another very important plant prenyllipid is plastoquinone (PQ) and its reduced form, plastoquinol (PQH_2). Plastoquinone is the key isoprenoid quinone in the oxygenic photosynthesis of cyanobacteria and all plants. It carries electrons between membrane protein complexes and shuttles protons across the membrane. Within chloroplasts PQ occurs in thylakoids, as well as in plastoglobules, which are the storage site of this compound and the inner chloroplast envelope, where its synthesis proceeds. Plastoquinone was found not only in photosynthetic plant organs, but also in minor amounts in all other examined organs, e.g. roots, bulbs, flowers, fruits and etiolated leaves (Nowicka and Kruk, 2010).

The biosynthesis pathway of tocochromanol has been identified in the 1990s using *Arabidopsis thaliana* and *Synechocystis* sp. PCC6803 as model organisms. Tocochromanol biosynthesis enzymes are mostly localized in the chloroplast envelope, except for tocopherol cyclase, which was found in plastoglobules (Szymanska and Kruk, 2010a, b; Piller et al., 2012). In tocochromanol biosynthesis the precursors of the chromanol ring and isoprenoid chain are formed in two different metabolic pathways (Fig. 10.3).

Homogentisic acid (HGA) derived from cytosolic shikimate pathway is a precursor of head group, while phytyl diphosphate (PDP) or geranylgeranyl diphosphate (GGDP) derived from the plastidial methylerythriol phosphate pathway are precursors of side chains of all tocochromanols (DellaPenna, 2005; DellaPenna and Pogson, 2006) (Fig. 10.3).

The tocochromanol biosynthetic pathway encompasses at least five different enzymes (HPPD and VTE1–VTE4). Homogentisic acid is produced from tyrosine derived catabolite – *p*-hydroxyphenylpyruvate (HPP). This step is catalyzed by 4-hydroxyphenylpyruvate dioxygenase (HPPD) (Fig. 10.3).

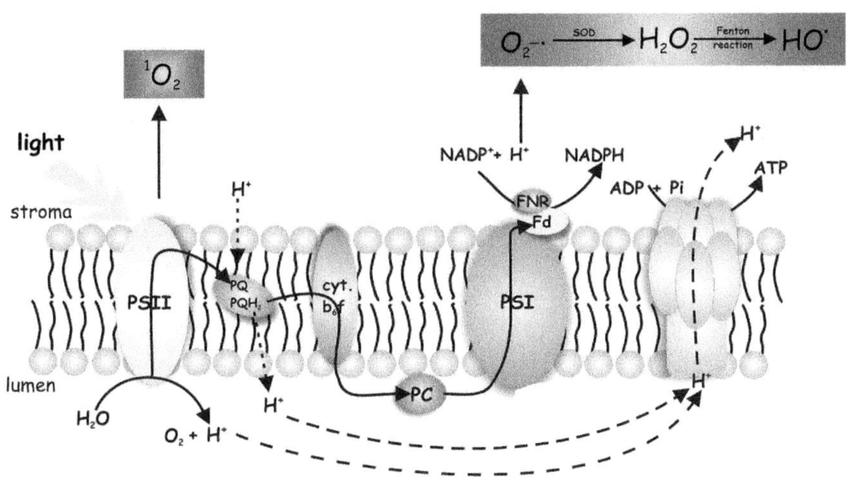

FIGURE 10.3 ROS generation in photosynthethic transport chain.

In *Arabidopsis* this enzyme is also involved in PQ biosynthesis (Norris et al, 1998). Presumably, HPPD is also involved in degradation of tyrosine (Lushchak and Semchuk, 2012). This enzyme was purified from several mammals, plants and bacteria (Lushchak and Semchuk, 2012). It is supposed to be a regulatory enzyme in tocopherol biosynthesis. HPPD engaged in prenyllipid biosynthesis in plants is localized in cytosol (Garcia et al., 1997). Its inhibition leads to direct inhibition of PQ and tocopherol biosynthesis and indirect inhibition of carotenoids desaturation which causes phytoene accumulation and photo-oxidation of the plastids (Norris et al., 1998). In *Arabidopsis* disruption of HPPD activity results in inability to synthesize both PQ and tocopherols (Norris et al., 1998). However, in *Synechocystis*, disruption in *hpd* gene (homologue of plant HPPD gene) leads to impaired tocopherol biosynthesis, without affecting PQ level, which suggests that in *Synechocystis* the HPPD gene product is not required for PQ synthesis (Dahnhardt et al., 2002). Recently, it was shown that the cyanobacterial PQ biosynthetic pathway differs from the pathway of higher plants and resembles the ubiquinone biosynthetic pathway of *E. coli*. A prenyltransferase needed for PQ synthesis turned out to be a close homologue of plenyltransferase present in *E. coli* (Sadre et al., 2012). Transgenic rice seeds with overexpression of *Arabidopsis* gene HPPD, accumulate slightly higher levels of total tocochromanols than the controls, but opposite to the control, they exhibit higher levels of α- than γ-isoform (Farre et al., 2012).

In the next step, precursors of α-tocochromanol side chain – PDP and GGDP condensate with HGA – to yield 2-methyl-6-phytylbenzoquinol (MPBQ) (Fig. 10.3) and 2-methyl-6-geranylgeranylbenzoquinol (MGGBQ) for tocopherol and tocotrienol, respectively. The prenylation reaction is catalyzed by homogentisate prenyltransferase (HPT, VTE2). In *Arabidopsis* this enzyme is able to attach only PDP to HGA to yield MPBQ (DellaPenna, 2005). Monocots possess also another prenyltransferase, a geranylgeranyl transferase (HGGT), which can use GGDP as a substrate and catalyze formation of MGGBQ (Cahoon et al., 2003). A null mutation in *vte2 Arabidopsis* mutant results in the lack of all tocopherols and their biosynthetic intermediates, both in leaves and seeds (Sattler et al., 2004). Investigation of substrate specificity of VTE2 isolated from *Arabidopsis* and *Synechocystis* has shown that VTE2 could use both, PDP and GGDP, as substrates with preferences for PDP (Collakova and DellaPenna, 2001).

Another enzyme – MPBQ methyltransferese (MPBQ MT; VTE3) – catalyzes methylation of MPBQ and MGGBQ to yield 2,3-dimethyl-5-phytyl-1,4-benzoquinone (DMPBQ) (Fig. 10.3) and 2,3-dimethyl-5-geranylgeranylbenzoquinol (DMGGBQ), respectively. This enzyme was originally discovered in *Synechocystis* sp. PCC6803 (Shintani et al., 2002). Disruption of VTE3 in plants leads to the lack of PQ and α-tocopherol in leaves.

Subsequently, tocopherol cyclase (VTE1) converts MPBQ and DMPBQ to δ- and γ-tocopherol, respectively, as well as MGGBQ and DMGGBQ to yield δ- and γ-tocotrienol, respectively (Li et al., 2008). For the first time VTE1 was purified from *Anabaena variabilis* (Stocker et al., 1993) and stepwise the gene encoding VTE1 was cloned from several species: *Synechocystis sp.* PCC6803, *Arabidopsis*, and maize (Porfirova et al., 2002; Sattler et al., 2003). Using mutants of tocopherol biosynthesis genes Szymanska and Kruk (2010a) revealed that VTE1 is also engaged in plastochromanol biosynthesis, where it catalyzes cyclization of plastoquinol. Mutation in gene encoding VTE1 in maize causes inhibition of sugar transport (Provencher et al., 2001). On the other hand, overexpression of VTE1 in *Arabidopsis* leads to decrease of the amount of other antioxidants (Li et al., 2008). Overexpression of VTE1 gene in *Arabidopsis* leads to a 7-fold increase in total tocopherol content (Kanwischer et al., 2005). The same effect was observed in tobacco (Liu et al., 2008). Moreover, Zbierzak et al. (2010) have shown that overexpression of VTE1 gene in *Arabidopsis* leaves caused a 7-fold increase in plastochromanol content, whereas γ-tocopherol content was similar to the wild-type. Mene-Saffrane et al. (2010) have demonstrated that double *Arabidopsis* mutant, *vte2vte1*, which lacks PC-8, DMPBQ and tocopherols, exhibits the most severe physiological and biochemical phenotype among all available tocopherol-deficient mutants.

The last step of tocochromanol biosynthesis involves methylation of δ- and γ-tocopherol to β- and α-tocopherol, as well as δ- and γ-tocotrienol to the corresponding tocotrienols (Munne-Bosch and Alegre, 2002; DellaPenna, 2005). This last stage is due to activity of γ-tocopherol methyltransferase (γ-TMT; VTE4). Together with VTE1, γ-TMT is essential for genetic engineering and manipulation of tocochromanol levels, as well as composition of its homologues (Li et al., 2008). Disruption of VTE4 gene (and *slr0089* in *Synechocystis sp.* PCC6803) leads to lack of α-tocopherol and accumulation of γ-tocopherol (Shintani and DellaPenna, 1998). Overexpression of the gene encoding γ-TMT in *Arabidopsis* resulted in 80-fold increase of α-tocopherol content in seeds (Shintani and DellaPenna, 1998). In double overexpressing transgenic lines *vte1vte4* PC-8 was converted into 5-methyl-plastochromanol-8, suggesting that γ-TMT can use not only γ-tocopherol, but also plastochromanol as a substrate, although γ-tocopherol was the preferred substrate of γ-TMT (Zbierzak et al., 2010). Experiments with overexpression of *vte4* gene were also carried out using other plant species important from an agricultural point of view, e.g. soybean (Chen et al., 2012).

It is clearly established that enzymes described above (VTE1-VTE4) play a crucial role in tocochromanol biosynthesis; however, other enzymes also seem to have an impact on their production. One of these enzymes is a phytyl kinase (VTE5), which can use free phytol as a substrate and condensates it with HGA (Valentin et al., 2006). Another enzyme indirectly engaged in tocochromanol synthesis is a tyrosine aminotransferase (TAT), involved in tyrosine biosynthesis (Riewe et al., 2012). When two T-DNA knockout plant mutants for the gene encoding TAT were analyzed, it turned out that TAT activity is severely reduced and total tyrosine content is strongly elevated (Riewe et al., 2012). In addition, these mutants displayed significantly reduced levels of tocopherols, suggesting a major role of this enzyme in tocopherol biosynthesis (Riewe et al., 2012). The isolation of genes involved in tocochromanol biosynthesis allows development of the genetic manipulations that will enhance the tocochromanols level. This will change the ratio of each homologue, which explains the role of those compounds in plants' metabolism (Schneider, 2005). It also helps breeders to improve their crop and its nutrition value.

10.5 ANTIOXIDANT FUNCTION OF PRENYLLIPIDS IN CHLOROPLASTS

The best studied function of tocochromanols is their antioxidant function. Tocochromanols are able to scavenge and quench ROS, and in that way they protect membrane lipids from peroxidation (Munne-Bosch and Alegre, 2002; Schneider, 2005). Furthermore, it was established that tocochromanols are engaged in several signal transduction pathways and regulation of transcription (Munné-Bosch, 2005). In some experiments mutations in tocopherol biosynthesis genes caused differences in phenotype in comparison to the wild type under normal and stress conditions, but most of the studies did not show distinguishable differences between these lines (Sattler et al., 2004; Szymanska and Kruk, 2010a). Results gained from the work of Kanwischer et al. (2005) and Collakova and DellaPenna showed two different effects of manipulation of tocopherol biosynthetic genes. Overexpression of *vte1* gene in *Arabidopsis* leads to the increase of γ-tocopherol content (Kanwischer et al., 2005). On the other hand, overexpression of *vte2* or *vte4* genes individually yields an enhanced level of α-tocopherol (Collakova and DellaPenna, 2001). Interestingly, correlation between tocopherol deficiency and the levels of other antioxidants (i.e. ascorbate, glutathione) has also been observed. In most cases this correlation is negative – when tocopherol content decreases, the content of other antioxidants increases (Li et al., 2008). In some cases this correlation in positive – high tocopherol content is coupled with increased content of ascorbate and glutathione (Caretto et al., 2002). Moreover it was clearly demonstrated that tocopherol biosynthetic intermediates, such as DMPBQ and MPBQ, also have antioxidant properties, and these compounds were accumulated in *vte1 Arabidopsis* mutant (Zbierzak et al., 2010; Szymanska et al., *unpublished data*). This phenomenon explains the lack of differences between *vte1* mutants and wild type. The cited data clearly show that tocopherols have multiple functions and their role in plants is much more complex than it was thought in the past.

Reactive oxygen species are produced in plant cells continuously under normal and stress conditions. Enzymatic and nonenzymatic antioxidants allow plants to maintain a certain level of ROS adequate to the current conditions and overall plant physiology. The plant's goal is to achieve a balance between antioxidant production and ROS scavenging. Plants need protection from excessive levels of ROS, while low amounts of ROS are needed for cellular signaling. If the concentration of ROS exceeds a certain level, the oxidative stress develops and the cell gets the signal for programmed cell death (Evans et al., 2005).

The common function of tocochromanols in plants is scavenging of ROS in photosynthetic membranes and lowering the extent of lipid peroxidation by reducing lipid peroxyl radicals to corresponding hydroperoxides (Munne-Bosch and Alegre, 2002). Lipid peroxidation is initiated by $^{\bullet}OH$ or $HO_2^{\bullet}$, which can abstract a hydrogen atom from a lipid molecule leading to formation of lipid peroxyl radicals. Tocopherols inhibit propagation of peroxidation by scavenging of lipid radicals. In this reaction tocopherols "sacrifice" hydrogen from their hydroxyl group and are converted into tocopheroxyl radicals. Moreover, tocopherols can protect hydrosulfide groups of proteins from oxidation by direct reaction with protein sulfur radical or indirectly by the mediating enhanced synthesis of other antioxidants (Li et al., 2008). Tocopheroxyl radical formed in the reactions of scavenging of other radicals can be re-reduced to tocopherol by the ascorbate-gluthatione system (Munné-Bosch, 2005). When reducing tocopheroxyl radicals to tocopherol, ascorbate is being oxidized to monodehydroascorbate and later to ascorbate by monodehydroascorbate reductase (MDHAR), an enzyme using NADH as a cofactor. Reaction of tocopherol with $^1O_2$ leads to formation of hydroxyperoxide, which is hydrolyzed irreversibly to tocopherol quinone (TQ) (Kruk and Strzałka, 1995). Tocopherol quinone can be enzymatically converted to tocopherol quinol in NADPH-dependent reaction (Kruk et al., 2005). Both TQ and its reduced form have antioxidant properties (Kruk and Strzałka, 1995; Kruk et al., 2005). Moreover, Szymanska and Kruk (2008a) have shown that α-tocopherol quinone (α-TQ) accumulates in leaves of runner beans exposed to high light and low temperature. These authors also proved that the α-TQ level is higher in evergreen plants (Szymanska and Kruk, 2008a). Tocopherols, together with β-carotene and plastoquinol can quench $^1O_2$ in the PSII (Trebst, 2003; Kruk et al., 2005, Havaux et al., 2005). Singlet oxygen leads to PSII photo-inhibition and D1 protein degradation. In *Chlamydomonas* cells, D1 turnover is correlated with the turnover of α-tocopherol (Kruk et al., 2005). It was also found that $^1O_2$ inhibits elongation of translation of D1 protein (Nishiyama et al., 2006). Based on the results obtained it was assumed that tocopherols protect the resynthesis of PSII, not PSII center itself (Inoue et al., 2011). Deficiency of vitamin E in *Arabidopsis* leads to strong zeaxanthin accumulation and vice versa (Havaux et al., 2005), suggesting that the two compounds have overlapping roles. Double mutant *vte1npq1* displays strong photosensitivity and $^1O_2$-mediated lipid peroxidation in high light (Havaux et al., 2005). Similar results were obtained by Kanwischer et al. (2005) who found that *vte1cad2* double mutants of *Arabidopsis* are more sensitive to photo-oxidative stress than *vte1* or *cad2* single mutants, thus indicating that the simultaneous loss of tocopherols and glutathione strongly affects the stability and the efficiency of the photosynthetic apparatus. Recently Valentin et al. (2006) identified a new enzyme responsible for condensation of free phytol with HGA. This enzyme – VTE5 – has a phytol kinase activity and seems to be a great explanation of linkage between changes of tocochromanol and chlorophyll levels during senescence of leaves (Valentin et al., 2006).

Taking into account that plants accumulate tocopherols mainly in seeds, it is clear that these compounds fulfill a special role in these plant organs. Seeds of *vte2 Arabidopsis* mutant had greatly reduced longevity in comparison with wild type (Sattler et al., 2004). This mutant also exhibited a range of developmental disorders in early seedlings growth, i.e. impaired cotyledon expansion, limited root growth, defects in storage lipids metabolism and dramatic increase of nonenzymatic lipid peroxidation (Sattler et al., 2004, 2006). Interestingly, these defects are completely attenuated in another mutant with impaired tocopherol biosynthesis, a *vte1* mutant. Although, *vte1* mutant accumulated DMPBQ, an intermediate of tocopherol biosynthetic pathway, which takes over and compensates tocopherol functions in these plants (Maeda and DellaPenna, 2007). Results obtained with *vte2* mutant strongly confirm that tocopherol functions in controlling nonenzymatic lipid peroxidation during seed storage, germination and seed development (Sattler et al., 2006). Analysis of overall gene expression patterns during seed germination in *vte2* mutants shows that in *vte2* seeds more than 160 genes are upregulated, when compared with the wild type, whereas in *vte1* mutant the number of such genes is 12 only (Sattler et al., 2006). This indicates that DMPBQ can fulfill many, but not all functions of tocopherol (Sattler et al., 2006).

In experiments on unicellular green alga *Chlamydomonas reinhardtii*, it was observed that not only chromanols, but also PQH_2 fulfills significant antioxidant role *in vivo* (Kruk and Trebst, 2008; Nowicka and Kruk, 2012). In another study on *Chlamydomonas* overlapping of protective functions of tocopherols and carotenoids was noticed, which clearly shows that in photosynthetic organisms different lipophilic antioxidants act in synergy (Li et al., 2012).

10.6 INVOLVEMENT OF ROS AND LIPOPHILIC ANTIOXIDANTS IN SIGNALING NETWORK

The subtle balance between ROS production and scavenging seems to have an influence on a large network of genes called the "ROS gene network" (Petrov et al., 2013). In *Arabidopsis* this network includes more than 152 genes sufficiently regulating ROS metabolism in plants (Miller et al., 2008). On the other hand, different ROS-detoxifying enzymes, as well as low molecular weight antioxidants such as tocochromanols, ascorbate and glutathione play important roles in the regulation of cellular ROS homeostasis. These antioxidants act as redox buffers and can influence gene expression connected with biotic and abiotic stresses (Foyer and Noctor, 2005; Miller et al., 2008, Sierla et al., 2013). Plants constantly sense the changes in ROS level and then adjust their gene expression to optimally react and acclimate to the fluctuating environment. The last decade has brought a large amount of evidence, supporting the thesis that ROS are crucial for signaling, as well as for inter- or intracellular communication (Choudhury et al., 2013; Scheler et al., 2013). Use of lack-of-function mutants and other genetic tools allowed characterization of key regulatory proteins involved in sensing of ROS and regulation of transcript expression by different ROS signals (Vanderauwera et al., 2005; Miller et al., 2008). Nowadays much attention is given to superoxide-generating plant homologue of respiratory burst NADPH oxidase (Rboh). Most likely this enzyme is engaged in many signaling pathways, in which it is responsible for the amplification of ROS production (Torres and Dangl, 2005; Bell et al., 2009).

Reactive oxygen species can modulate gene expression directly or serve as a signals molecules. The presence of $^1O_2$ upregulates genes responsible for molecular defense response to photo-oxidative stress (Op den Camp et al., 2003). Taking into account the short lifetime of $^1O_2$, it is unlikely that $^1O_2$ directly takes part in a signal transduction chain. More likely, $^1O_2$ oxidizes some components (i.e. pigments or fatty acids) and the resulting oxidation products are involved in signaling pathways (Weber, 2002).

Superoxide and hydrogen peroxide act directly as second messengers in the regulation and expression of the genes encoding proteins responsible for oxidative stress response, such as glutathione peroxidases, glutathione-*S*-transferases and ascorbate peroxidase (Mullineaux et al., 2000, Vranova et al., 2002).

Some data show that tocopherols can also play a role in signal transduction (Munné-Bosch, 2007). First of all, it was clearly demonstrated that tocopherol biosynthesis pathway is regulated by ROS derived as by-products of photosynthesis. Taking into account that tocopherols share a biosynthetic pathway with PQ, the mutation in genes involved in tocopherol biosynthesis may affect PQ production. One example is *hppd* gene, which is needed for both tocopherol and PQ synthesis. The expression of *hppd* gene during senescence is most likely connected with oxidative stress (Falk et al., 2002). During senescence, chlorophyll degradation occurs, which results in increase of free phytol. Phytol can be used for tocopherol synthesis and in that way its availability may regulate tocopherol level in plants (Li et al., 2008). Munné-Bosch et al. (2007) revealed that α-tocopherol deficiency in *Arabidopsis* can reduce plants growth and enhance anthocyanin accumulation by transiently increasing jasmonic acid (JA) levels. Therefore, it was suggested that tocopherols play a role in cellular signaling by altering phytohormone levels in plants, rather than by directly regulating gene expression (Munné-Bosch et al., 2007). By controlling the ROS level and thus the extent of lipid peroxidation and hydroperoxide content in chloroplasts, tocopherols may regulate JA concentration in leaves, and indirectly influence JA-dependent gene expression in the nucleus (Munné-Bosch et al., 2007).

Tocopherols are also involved in carbohydrate transport in plants. The experiments with *sdx1* (*sucrose export defective 1*) mutant of maize showed that this gene encodes protein involved in chloroplast-to-nucleus signaling necessary for proper differentiation of bundle sheet cells and plasmodesmata in C4 plants (Porfirova et al., 2002; Provencher et al., 2004). Mutation in *sdx1* gene causes a characteristic phenotype with the symptoms of defects in photoassimilate transport (callose deposition, death of vascular parenchyma cells, occlusion of plasmodesmata and lack of symplastic connection between vascular parenchyma and bundle sheet cells) (Hofius et al., 2004). As it turned out, *sdx1* gene has a great homology with *vte1* and mutation in *sdx1* gene causes tocopherol deficiency in maize (Provencher et al., 2001). These results were confirmed by experiments with *Arabidopsis* mutants *vte1* and *vte2*, which also displayed a characteristic phenotype caused by disorders of photoassimilate transport. On the contrary, in *Arabidopsis* this phenotype occurred only under low temperature (Maeda et al., 2006). Some recent reports suggested that deficiency of tocopherols in tobacco (with silenced *vte2* and *vte4* genes) increases the sensitivity to methyl viologen, severe salt and sorbitol stresses (Abbasi et al., 2007). *Vte4* plants, which accumulate γ-tocopherol instead of α-form, are more sensitive to salt stress, but have improved tolerance to sorbitol-induced osmotic stress (Abbasi et al., 2007). Recently, Szymanska and Kruk (2008b) have shown that γ-tocopherol accumulates in etiolated and young seedlings of runner bean. It was proposed that γ-tocopherol is responsible for drought and osmotic stress resistance (Szymanska and Kruk, 2008b). Analysis of the expression level of genes involved in ethylene biosynthesis, signaling and perception in *vte1* and *vte4 Arabidopsis* mutants treated with salt or water stress revealed that the biggest differences occurred in *vte4* mutant, but not in *vte1* (Cela et al., 2011). This change in gene expression was the most evident in mature leaves under salt stress. Accumulation of γ-tocopherol instead of α-tocopherol in *vte4* mutant led to elevated levels of transcripts of genes connected with ethylene signaling pathway (mainly *ctr1, ein2, ein3, erf1*). Moreover, leaf-induced JA accumulated in both *vte1* mutant and wild type, but not in the *vte4* mutant, suggesting that JA and ethylene signaling pathways are downregulated in mature leaves of salt-stressed *vte4* mutants (Cela et al., 2011). Therefore, it was suggested that different tocopherol homologues have specific functions in plants. Cela et al. (2011) postulated that the function of γ-tocopherol may be particularly relevant in desiccated tissues, such as the seeds, where this isoform is known to accumulate in significant amounts. Furthermore, it has been shown that a water deficit specifically induces γ-tocopherol accumulation in leaves of the CAM plant *Aptenia cordifolia* (Cela et al., 2009). Taken together, all these results suggest that, aside from its potential as an antioxidant and its important role as a precursor of α-tocopherol, γ-tocopherol could exert additional functions in plant.

Recently it has been shown that exogenous treatment with ethylene induced transcription of the HPD gene (Falk et al., 2002). Experiments carried out with water-stressed *Arabidopsis* mutant *ein3-1*, which shows impaired ethylene signaling, revealed that tocopherol levels increased progressively during water stress in wild type plants, but the water-stress-induced increase of tocopherol content was delayed and reduced in *ein3-1* mutants, thus suggesting that ethylene signaling may be involved in tocopherol biosynthesis (Cela et al., 2009).

An important signaling function can also be asserted for plastoquinone. In that case, the redox state of PQ pool is the factor that triggers response of the cells (Nowicka and Kruk, 2010). The impact of the redox state of PQ pool on signaling was shown in experiments in which photosystems were selectively excited using light with certain wavelengths or specific inhibitors were added to block reduction of PQ or oxidation of PQH_2 (Nowicka and Kruk, 2010). It is known that the redox state of the PQ pool regulates expression of chloroplast genes, such as *psbA*, *psaAB*, encoding peptides in reaction centers of PSI and PSII, as well as nuclear genes: *i.e. lhcb* encoding antenna complexes, ascorbate peroxidase, superoxide dismutase, nitrate reductase (Karpinski et al., 1999; Pfannschmidt et al., 1999, Fey et al., 2005; Giordano et al., 2005; Lepetit et al., 2013). The redox state of the PQ pool also regulates activity of kinase STN7, responsible for antenna phosphorylation during state transitions (Rochaix, et al., 2012; Rochaix, 2013).

10.7 CONCLUSIONS

Up to now a significant amount of data concerning antioxidant functions of tocopherols has been obtained. The reactions of tocopherol with different ROS and lipid radicals was examined in several *in vitro* experiments, which enabled determination of rate constants and identification of oxidation products. After years of studies, the tocopherol biosynthetic pathway in plants and cyanobacteria is well known. Due to detailed knowledge of tocopherol biosynthetic enzymes, it is possible to obtain mutants and transgenic plants with impaired or enhanced tocopherol biosynthesis. The former were used to determine tocopherol functions *in vivo*, whereas the latter are potentially important for the food industry. Although there were several attempts to obtain transgenic plants with enhanced production of α-tocopherol, they have not been used for commercial scale yet. Introducing varieties with high α-tocopherol content to the market is still a question for the future.

The participation of chromanols in response to different kind of stress factors was widely examined, especially in case of high light or temperature stress. Some other stress factors, such as exposure to heavy metals and other environmental pollutants, still need more detailed study. Apart from stress response studies, recently discovered chromanol – a plastochromanol – is less examined than tocopherols. The main tocopherol biosynthetic enzymes are known, but there are some enzymes connected with metabolism of these compounds, such as tocopherol oxidase present in young seedlings of *Phaseolus*, which are only poorly known.

The participation of tocopherols in signal transduction is only superficially known and needs further study. It is crucial to develop reliable and specific methods for determination of ROS inside living cells. Such methods are already known (Kalyanaraman, et al., 2012) but they have to be improved to enable understanding of signaling function of ROS, and therefore participation of ROS scavenging molecules in that process. Also, the signaling function of tocopherols, which are not connected to its antioxidant action, are still obscure. In particular, the molecular mechanisms of tocopherol-mediated signaling are unknown.

The participation of PQ in photosynthetic electron transport chain is well known, whereas antioxidant functions of this compound are known to a much lesser extent. There are also some questions concerning PQ biosynthesis. Recently, it was discovered that in cyanobacteria the PQ biosynthetic pathway is different from the pathway present in higher plants. The cyanobacterial pathway seems to be more similar to ubiquinone biosynthesis. Up to now only two enzymes needed for PQ biosynthesis in blue-green algae are known. The signaling function of plastoquinone is now a subject of intensive studies.

It is clear that nowadays we know a lot about prenyllipid antioxidants, but there are many questions to be answered in the future, especially concerning participation of different ROS and lipophilic antioxidants in signal transduction.

REFERENCES

Abbasi, A.R., Hajirezaei, M., Hofius, D., Sonnevald, U., Voll, L.M., 2007. Specific roles of αα-and γ-tocopherol in biotic stress responses of transgenic tobacco. Plant Physiol. 143, 1720–1738.

Apel, K., Hirt, H., 2004. Reactive oxygen species: metabolism, oxidative stress and signal transduction. Annu. Rev. Plant Biol. 55, 373–399.

Aro, E.M., Virgin, I., Andersson, B., 1993. Photoinhibition of photosystem II: inactivation, protein damage and turnover. Biochim. Biophys. Acta. 1143, 113–134.

Asada, K., 1999. The water-water cycle in chloroplasts: scavenging of active oxygens and dissipation of excess photons. Annu. Rev. Plant Physiol. Plant Mol. Biol. 50, 601–639.

Asada, K., 2006. Production and scavenging of reactive oxygen species in chloroplasts and their functions. Plant Physiol. 141, 391–396.

Bell, E., Takeda, S., Dolan, L., 2009. Reactive oxygen species in growth and development. In: Rio, L.A., Puppo, A. (Eds.), Signaling and Communication in Plants. Springer-Verlag, Berlin, pp. 43–53.

Cahoon, E.B., Hall, S.H., Ripp, K.G., Ganzke, T.S., Hitz, W.D., Coughlan, S.J., 2003. Metabolic redesign of vitamin E biosynthesis in plants for tocotrienol production and increased antioxidant content. Nat. Biotechnol. 21, 1082–1087.

Caretto, S., Paradiso, A., D'Amico, L., Gara, L.D., 2002. Ascorbate and gluthatione metabolism in two sunflower cell lines of differing α-tocopherol biosynthetic capability. Plant Physiol. Biochem. 40, 509–513.

Cela, J., Falk, J., Munné-Bosch, S., 2009. Ethylene signaling may be involved in the regulation of tocopherol biosynthesis in *Arabidopsis thaliana*. FEBS Lett. 583, 992–996.

Cela, J., Chang, C., Munné-Bosch, S., 2011. Accumulation of γ- rather than α-tocopherol alters ethylene signaling gene expression in the vte4 mutant of *Arabidopsis thaliana*. Plant Cell Physiol. 52, 1389–1400.

Chen, D.F., Zhang, M., Wang, Y.Q., Chen, X.W., 2012. Expression of γ-tocopherol methyltransferase gene from Brassica napus increased α-tocopherol content in soybean seed. Biol Plant. 56, 131–134.

Choudhury, S., Panda, P., Sahoo, L., Panda, S.K., 2013. Reactive oxygen species signaling in plants under abiotic stress. Plant Signal. Behav. 8, e23681.

Collakova, E., DellaPenna, D., 2001. Isolation and functional analysis of homogentisite phytyltransferase from *Synechocystis* sp. PC6803 and *Arabidopsis*. Plant Physiol. 127, 1113–1124.

Collakova, E., DellaPenna, D., 2003. Homogentisate phytyltransferase activity is limiting for tocopherol biosynthesis in *Arabidopsis thaliana*. Plant Physiol. 131, 632–642.

Collin, V.C., Eymery, F., Genty, B., Rey, P., Havaux, M., 2008. Vitamin E is essential for the tolerance of *Arabidopsis thaliana* to metal induced oxidative stress. Plant Cell Environ. 31, 99–111.

Dahnhardt, D., Falk, J., Appel, J., van der Kooij, T.A.W., Schulz-Friedrich, R., Krupinska, K., 2002. The hydroxyphenylpyruvate dioxygenase from *Synechocystis sp.* PCC 6803 is not required for plastoquinone biosynthesis. FEBS Lett. 523, 177–181.

Davletova, S., Rizhsky, L., Liang, H., Shengqiang, Z., Oliver, D.J., Coutu, J., et al., 2005. Cytosolic ascorbate peroxidase 1 is a crucial component of the reactive oxyen gene network of Arabidopsis. Plant Cell 17, 268–281.

DellaPenna, D., 2005. A decade of progress in understanding vitamin E synthesis. J. Plant Physiol. 162, 729–737.

DellaPenna, D., Pogson, B.J., 2006. Vitamin synthesis in plants: tocopherols and carotenoids. Annu. Rev. Plant Biol. 57, 711–738.

Ellouzi, H., Hamed, K.B., Cela, J., Munne-Bosch, S., Abdelly, C., 2011. Early effects of salt stress on the physiological and oxidative status of *Cakile maritima* (halophyte) and *Arabidopsis thaliana* (glycophyte). Physiol. Plant. 142, 128–143.

Evans, N.H., McAinsh, M.R., Hetherington, A.M., Knight, M.R., 2005. ROS perception in *Arabidopsis thaliana*: the ozone-induced calcium response. Plant J. 41, 615–626.

Falk, J., Krauss, N., Dahnhardt, D., Krupinska, K., 2002. The senescence associated gene of barley encoding 4-hydroxyphenylpyruvate dioxygenase is expressed during oxidative stress. J. Plant Physiol. 159, 1245–1253.

Farre, G., Sudhakar, D., Naqvi, S., Sandmann, G., Christou, P., Capell, T., et al., 2012. Transgenic rice grains expressing a heterologous p-hydroxyphenylpyruvate dioxygenase shift tocopherol synthesis from the γ- to the α-isoform without increasing absolute tocopherols levels. Transgenic Res. 21, 1093–1097.

Fey, V., Wagner, R., Brautigam, K., Pfannschmidt, T., 2005. Photosynthetic redox control of nuclear gene expression. J. Exp. Bot. 56, 1491–1498.

Fischer, B.B., Hideg, É., Krieger-Liszkay, A., 2013. Production, detection, and signaling of singlet oxygen in photosynthetic organisms. Antiox. Redox Signal. 18, 2145–2162.

Foyer, C.H., Noctor, G., 2003. Redox sensing and signaling associated with reactive oxygen in chloroplasts, peroxisomes and mitochondria. Physiol. Plant. 119, 355–364.

Foyer, C.H., Noctor, G., 2005. Redox homeostasis and antioxidant signaling: a metabolic interface between stress perception and physiological responses. Plant Cell 17, 1866–1875.

Foyer, C.H., Shigeoka, S., 2011. Understanding oxidative stress and antioxidant function to enhance photosynthesis. Plant Physiol. 155, 93–100.

Galvez-Vadivieso, G., Mullineaux, P.M., 2010. The role of reactive oxygen species in signaling from chloroplasts to the nucleus. Plant Physiol. 138, 430–439.

Garcia, I., Rodgers, M., Lenne, C., Rolland, A., Sailland, A., Matringe, M., 1997. Subcellular localization and purification of a *p*-hydroxyphenylpyruvate dioxygenase from cultured carrot cells and characterization of the corresponding cDNA. Biochem. J. 325, 761–769.

Gechev, T.S., van Breusegem, F., Stone, J.M., Denev, I., Laloi, C., 2006. Reactive oxygen species as signals that modulate plant stress responses and programmed cell death. BioEssay 28, 1091–1101.

Giordano, M., Chen, Y.B., Koblizek, M., Falkowski, P.G., 2005. Regulation of nitrate reductase in *Chlamydomonas reinhardtii* by the redox state of the plastoquinone pool. Eur. J. Phycol. 40, 345–352.

Gruszka, J., Kruk, J., 2007. RP-LC for determination of plastochromanol, tocotrienols and tocopherols in plant oils. Chromatographia 66, 909–913.

Halliwell, B., 2006. Reactive species and antioxidants. Redox biology is a fundamental theme of aerobic life. Plant Physiol. 141, 312–322.

Havaux, M., Eymery, F., Porfirova, S., Rey, P., Dormann, P., 2005. Vitamin E protects against photo-inhibition and photooxidative stress in *Arabidopsis thaliana*. Plant Cell 17, 3451–3469.

Hofius, D., Hajirezaei, M., Geiger, M., Tschiersch, H., Meizer, M., Sonnewald, U., 2004. RNAi-mediated tocopherol deficiency impairs photoassimilate export in transgenic potato plants. Plant Physiol. 135, 1256–1268.

Inoue, S., Ejima, K., Iwai, E., Hayashi, H., Appel, J., Tyystjärvi, E., et al., 2011. Protection by α-tocopherol of the repair of photosystem II during photoinhibition in *Synechocystis* sp. PCC 6803. Biochim. Biophys. Acta 1807, 236–241.

Kalyanaraman, B., Darley-Usmar, V., Davies, K.J., Dennery, P.A., Forman, H.J., Grisham, M.B., et al., 2012. Measuring reactive oxygen and nitrogen species with fluorescent probes: challenges and limitations. Free Radic. Biol. Med. 52, 1–6.

Kanwischer, M., Porforova, S., Bergmuller, E., Dormann, P., 2005. Alternation of tocopherol cyclase activity in transgenic and mutant plants of *Arabidopsis* affect tocopherol content, tocopherol composition and oxidative stress. Plant Physiol. 137, 713–723.

Karpinski, S., Reynolds, H., Karpinska, B., Wingsle, G., Creissen, G., Mullineaux, P.M., 1999. Systemic signaling and acclimation in response to excess excitation energy in Arabidopsis. Science 284, 654–657.

Krause, G.H., Winter, K., Matsubara, S., Krause, B., Jahns, P., Virgo, A., et al., 2012. Photosynthesis, photoprotection, and growth of shade-tolerant tropical tree seedlings under full sunlight. Photosynth. Res. 113, 273–285.

Kreslavsky, V.D., Los, D.A., Allakhverdiev, S.I., Kuznetsov, V.I.V., 2012. Signaling role of reactive oxygen species in plants under stress. Russ. J. Plant Physiol. 59, 141–154.

Kriger-Liszkay, A., 2005. Singlet oxygen production in photosynthesis. J. Exp. Bot. 56, 337–346.

Kruk, J., Strzałka, K., 1995. Occurrence and function of α-tocopherol quinone in plants. J. Plant Physiol. 145, 405–409.

Kruk, J., Trebst, A., 2008. Plastoquinol as a singlet oxygen scavenger in photosystem II. Biochim. Biophys. Acta 1777, 154–162.

Kruk, J., Hollander-Czytko, H., Oettmeier, W., Trebst, A., 2005. Tocopherol as a singlet oxygen scavenger in photosystem II. J. Plant Physiol. 162, 749–757.

Kusmic, C., Barsacchi, R., Barsanti, L., Gualtieri, P., Passarelli, V., 1999. *Euglena gracilis* as source of the antioxidant vitamin E. Effects of culture conditions in the wild strain and in the natural mutant WZSL. J. App. Phycol. 10, 555–559.

Lepetit, B., Sturm, S., Rogato, A., Gruber, A., Sachse, M., Falciatore, A., et al., 2013. High light acclimation in the secondary plastids containing diatom *Phaeodactylum tricornutum* is triggered by the redox state of the plastoquinone pool. Plant Physiol. 161, 853–865.

Li, Y., Wang, Z., Sun, X., Tang, K., 2008. Current opinions on the functions of tocopherol based on the genetic manipulation of tocopherol biosynthesis in plants. J. Integr. Plant Biol. 50, 1057–1069.

Li, Z., Keasling, J.D., Niyogi, K.K., 2012. Overlapping photoprotective function of vitamin E and carotenoids in *Chlamydomonas*. Plant Physiol. 158, 313–323.

Liu, X., Hua, X., Guo, J., Qi, D., Wang, L., Liu, Z., et al., 2008. Enhanced tolerance to drought stress in transgenic tobacco plants overexpressing VTE1 for increased tocopherol production from *Arabidopsis thaliana*. Biotechnol. Lett. 30, 1275–1280.

Lushchak, V.I., Semchuk, N.M., 2012. Tocopherol biosynthesis: chemistry, regulation and effects of environmental factors. Acta Physiol. Plant 34, 1607–1628.

Maeda, H., DellaPenna, D., 2007. Tocopherol functions in photosynthetic organisms. Curr. Opin. Plant Biol. 10, 260–265.

Maeda, H., Song, W., Sage, T.L., DellaPenna, D., 2006. Tocopherols play a crucial role in low-temperature adaptation and phloem loading in *Arabidopsis*. Plant Cell 18, 2710–2732.

Marder, J.B., Droppa, M., Caspi, V., Raskin, V.I., Horvath, G., 1998. Light-dependent thermoluminescence from thylakoids from greening barely leaves: evidence for involvement of oxygen radicals and free chlorophyll. Physiol. Plant. 104, 713–719.

Mene-Saffrane, L., DellaPenna, D., 2010. Biosynthesis, regulation and functions of tocochromanols in plants. Plant Physiol. Biochem. 48, 301–309.

Mene-Saffrane, L., Jones, A.D., DellaPenna, D., 2010. Plastochromanol-8 and tocopherols are essential lipid-soluble antioxidants during seed desiccation and quiescence in *Arabidopsis*. Proc. Natl. Acad. Sci. USA 107, 17815–17820.

Miller, G., Shulaev, V., Mittler, R., 2008. Reactive oxygen signaling and abiotic stress. Physiol. Plant. 133, 481–489.

Mittler, R., 2002. Oxidative stress, antioxidants and stress tolerance. Trends Plant Sci. 7, 405–410.

Mittler, R., Vanerauwera, S., Gollery, M., Van Breusegem, F., 2004. Reactive oxygen network of plants. Trends Plant Sci. 9, 490–498.

Mullineaux, P., Ball, L., Escobar, C., Karpinska, B., Creissen, G., Karpinski, S., 2000. Are diverse signaling pathways integrated in the regulation of *Arabidopsis* antioxidant defense gene expression in response to excess excitation energy?. Phil. Trans. R. Soc. Lond. B 355, 1531–1540.

Munné-Bosch, S., 2005. The role of α-tocopherol in plant stress tolerance. J. Plant Physiol. 162, 743–748.

Munné-Bosch, S., 2007. Alpha-tocopherol: a multifaceted molecule in plants. Vitam. Horm. 76, 375–392.

Munne-Bosch, S., Alegre, L., 2002. The function of tocopherols and tocotrienols in plants. Crit. Rev. Plant Sci. 21, 31–57.

Munné-Bosch, S., Weiler, E., Alegre, L., Muller, M., Duchting, P., Falk, J., 2007. α-Tocopherol may influence cellular signaling by modulating jasmonic acid levels in plants. Planta 225, 681–691.

Nishiyama, Y., Allakhverdiev, S., Murata, N., 2006. A new paradigm for the action of reactive oxygen species in the photo-inhibition of photosystem II. Biochim. Biophys. Acta 1757, 742–749.

Norris, S.R., Shen, X., DellaPenna, D., 1998. Complementation of the *Arabiopsis pds1* mutation with the gene encoding p-hydroxyphenylpyruvate dioxygenase. Plant Physiol. 117, 1317–1323.

Nowicka, B., Kruk, J., 2010. Occurrence, biosynthesis and function of isoprenoid quinones. Biochim. Biophys. Acta 1797, 1587–1605.

Nowicka, B., Kruk, J., 2012. Plastoquinol is more active than α-tocopherol in singlet oxygen scavenging during high light stress of *Chlamydomonas reinhardtii*. Biochim. Biophys. Acta 1817, 389–394.

Nowicka, B., Gruszka, J., Kruk, J., 2013. Function of plastochromanol and other biological prenyllipids in the inhibition of lipid peroxidation – a comparative study in model systems. Biochim. Biophys. Acta 1828, 233–240.

Op den Camp, R.G.L., Przybyla, D., Ochsenbein, C., Laloi, C., Kim, C., Danon, A., et al., 2003. Rapid induction of distinct stress responses after the release of singlet oxygen in *Arabidopsis*. Plant Cell 15, 2320–2332.

O'Brien, J., Daudi, A., Butt, V.S., Bolwell, G.P., 2012. Reactive oxygen species and their role in plant defense and cell wall metabolism. Planta 236, 765–779.

Petrov, V., Schippers, J., Benina, M., Minkov, I., Mueller-Roeber, B., Gechev, T., 2013. In search for new players of the oxidative stress network by phenotyping an *Arabidopsis* T-DNA mutant collection on reactive oxygen species-eliciting chemicals. Plant Omics J. 6, 46–54.

Pfannschmidt, T., Nilsson, A., Allen, J.F., 1999. Photosynthetic control of chloroplast gene expression. Nature 397, 625–628.

Piller, L.E., Abraham, M., Dörmann, P., Kessler, F., Besagni, C., 2012. Plastid lipid droplets at the crossroads of prenylquinone metabolism. J. Exp. Bot. 63, 1609–1618.

Porfirova, S., Bergmuller, E., Tropf, S., Lemke, R., Dormann, P., 2002. Isolation of an *Arabidopsis* mutant lacking vitamin E and identification of cyclase essential for tocopherol biosynthesis. Proc. Natl. Acad. Sci. USA 99, 12495–12500.

Pospisil, P., 2009. Production of reactive oxygen species by Photosystem II. Biochim. Biophys. Acta 1787, 1151–1160.

Provencher, L.M., Miao, L., Sinha, N., Lucas, W.J., 2001. Sucrose export defective1 encodes a novel protein implicated in chloroplast-to-nucleus signaling. Plant Cell 13, 1127–1141.

Rastogi, A., Yadav, D.K., Szymanska, R., Kruk, J., Sedlarova, M., Pospisil, P., 2013. Singlet oxygen scavenging activity of tocopherol and plastochromanol in *Arabidopsis thaliana*: relevance to photooxidative stress. Plant Cell Environ.. 10.1111/pce.12161.

Rautenkranz, A.A.F., Li, L., Mauchler, F., Mautinoia, E., Oertli, J.J., 1994. Transport of ascorbic acid and dehydroascorbic acids across protoplast and vacuole membranes isolated from barley (*Hordeum vulgare* L. cv Gerbel) leaves. Plant Physiol. 106, 187–193.

Riewe, D., Koohi, M., Lisec, J., Pgeiffer, M., Lippmann, R., Schmeiche, J., et al., 2012. A tyrosine aminotransferase involved in tocopherol synthesis in *Arabidopis*. Plant J. 71, 850–859.

Rizhsky, L., Hallak-Herr, E., van Breusegem, F., Rachmilevitch, S., Barr, J.E., Rodermel, S., et al., 2002. Double antisense plants lacking ascorbate peroxidase and catalase are less sensitive to oxidative stress than single antisense plants lacking ascorbate peroxidase or catalase. Plant J. 32, 329–342.

Rochaix, J.D., 2013. Redox regulation of thylakoid protein kinases and photosynthetic gene expression. Antiox. Redox Signal. 18, 2184–2201.

Rochaix, J.D., Lemeille, S., Shapiguzov, A., Samol, I., Fucile, G., Willig, A., et al., 2012. Protein kinases and phosphatases involved in the acclimation of the photosynthetic apparatus to a changing light environment. Phil. Trans. R. Soc. Lond. B 367, 3466–3474.

Sadre, R., Pfaff, C., Buchkremer, S., 2012. Plastoquinone-9 biosynthesis in cyanobacteria differs from that in plants and involves a novel 4-hydroxybenzoate solanesyltransferase. Biochem. J. 442, 621–629.

Sang, M., Ma, F., Xie, J., Chen, X.-B., Wang, K.-B., Qin, X.-C., et al., 2009. High-light induced singlet oxygen formation in cytochrome b_6f complex from *Bryopsis corticulans* as detected by EPR spectroscopy. Biophys. Chem. 146, 7–12.

Sattler, S.E., Cahoon, E.B., Coughlan, S.J., DellaPenna, D., 2003. Characterization of tocopherol cyclase from higher plants and cyanobacteria. Evolutionary implications for tocopherol synthesis and function. Plant Physiol. 132, 2184–2195.

Sattler, S.E., Gilliland, L.U., Magallanes-Lundback, M., Pollard, M., DellaPenna, D., 2004. Vitamin E is essential for seed longevity and for preventing lipid peroxidation during germination. Plant Cell 16, 1419–1432.

Sattler, S.E., Mene-Saffrane, L., Farmer, E.E., Krischke, M., Mueller, M.J., DellaPenna, D., 2006. Non-enyzmatic lipid peroxidation reprograms gene expression and activates defense markers in *Arabidopsis* tocopherol-deficient mutans. Plant Cell 18, 3706–3720.

Scheler, C., Durner, J., Astier, J., 2013. Nitric oxide and reactive oxygen species in plant biotic interactions. Curr. Opin. Plant Biol. doi:org/10.1016/j.pbi.2013.06.020.

Schmidt, K., Fufezan, C., Kriger-Liszkay, A., Satoh, H., Paulsen, H., 2003. Recombinant water-soluble chlorophyll protein from *Brassica oleracea* var. *botrys* binds various chlorophyll derivatives. Biochemistry 42, 7427–7433.

Schneider, C., 2005. Chemistry and biology of vitamin E. Mol. Nutr. Food Res. 49, 7–30.

Semchuk, N.M., Lushchak, O.V., Falk, J., Krupinska, K., Lushchak, V.I., 2009. Inactivation of genes, encoding tocopherol biosynthetic pathway enzymes, results in oxidative stress in outdoor grown *Arabidopsis thaliana*. Plant Physiol. Biochem. 47, 384–390.

Sharma, P., Jha, A.B., Dubey, R.S., Pessarakli, M., 2012. Reactive oxygen species, oxidative damage, and antioxidative defense mechanism in plants under stressful conditions. J. Bot. 2012, http://dx.doi.org/10.1155/2012/217037.

Shigeoka, S., Onishi, T., Nakano, Y., Kitaoka, S., 1986. The contents and subcellular distribution of tocopherols in *Euglena gracilis*. Agric. Biol. Chem. 50, 1063–1065.

Shintani, D., DellaPenna, D., 1998. Elevating the vitamin E content of plants through metabolic engineering. Science 282, 2098–2100.

Shintani, D.K., Cheng, Z., DellaPenna, D., 2002. The role of 2-methyl-6-phytylbenzoquinone methyltransferase in determining tocopherol composition in *Synechocystis* sp. PCC6803. FEBS Lett. 511, 1–5.

Sierla, M., Rahikainen, M., Salojärvi, J., Kangasjärvi, J., Kangasjärvi, S., 2013. Apoplastic and chloroplastic redox signaling networks in plant stress responses. Antiox. Redox Signal. 18, 2220–2239.

Sies, H., Menck, C.F., 1992. Singlet oxygen induced DNA damage. Mutat. Res. 275, 367–375.

Stocker, A., Rüttimann, A., Woggon, W.-D., 1993. Identification of the tocopherol cyclase in the blue-green algae *Anabaena variabilis* Kutzing (cyanobacteria). Helv. Chim. Acta 76, 1729–1738.

Strzałka, K., Szymanska, R., Swiezewska, E., Skorupinska-Tudek, K., Suwalsky, M., 2009. Tocochromanols, plastoquinone and polyprenols in selected plant species from Chilean Patagonia. Acta Biol. Crac. Bot. 51, 39–44.

Strzałka, K., Szymanska, R., Suwalsky, M., 2011. Prenyllipids and pigments content in selected Antarctic lichens and mosses. J. Chilean Chem. Soc. 56, 808–811.

Suh, H.J., Kim, C.S., Jung, J., 2000. Cytochrome b_6f complex as an indigenous photodynamic generator of singlet oxygen in thylakoid membranes. Photochem. Photobiol. 71, 10–109.

Suzuki, N., Koussevitzky, S., Mittler, R., Miller, G., 2012. ROS and redox signalling in the response of plants to abiotic stress. Plant Cell Environ. 35, 259–270.

Szymanska, R., Kruk, J., 2008a. Tocopherol content and isomer's composition in selected plant species. Plant Physiol. Biochem. 46, 29–33.

Szymanska, R., Kruk, J., 2008b. γ-tocopherol dominates in young leaves of runner bean (*Phaseolus coccineus*) under a variety of growing conditions: the possible functions of γ-tocopherol. Phytochemistry 69, 2142–2148.

Szymanska, R., Kruk, J., 2010a. Plastoquinol is the main prenyllipid synthesized during acclimation to high-light conditions in *Arabidopsis* and converted to plastochromanol by tocopherol cyclase. Plant Cell Physiol. 51, 537–545.

Szymanska, R., Kruk, J., 2010b. Identification of hydroxy-plastochromanol in *Arabidopis* leaves. Acta Biochim. Pol. 57, 105–107.

Szymanska, R., Latowski, D., Strzalka, K., 2012. Chloroplasts and strong photoprotective mechanisms. Curr. Chem. Biol. 6, 254–264.

Tarchoune, I., Sgherri, C., Baâtour, O., Izzo, R., Lachaâl, M., Navari-Izzo, F., et al., 2013. Effects of oxidative stress caused by NaCl or Na_2SO_4 excess on lipoic acid and tocopherols in Genovese and Fine basil (*Ocimum basilicum*). Ann. Appl. Biol. 163, 23–32.

Telfer, A., 2002. What is β-carotene doing in the photosystem II reaction centre? Phil. Trans. R. Soc. Lond. B 357, 1431–1440.

Torres, M.A., Dangl, J.L., 2005. Functions of the respiratory burst oxidase in biotic interactions, abiotic stress and development. Curr. Opin. Plant Biol. 8, 397–403.

Trebst, A., 2002. A specific role for tocopherol and of chemical singlet oxygen quenchers in the maintenance of photosystem II structure and function in *Chlamydomonas reinhardii*. FEBS Lett. 516, 156–160.

Trebst, A., 2003. Function of β-carotene and tocopherol in photosystem II. Z Naturforsch. C 58, 609–620.

Trianthaphylides, C., Krischke, M., Hoeberichts, A., Ksas, B., Gresser, G., Havaux, M., et al., 2008. Singlet oxygen is a major reactive oxygen species involved in photooxidative damage in plants. Plant Physiol. 148, 960–968.

Valentin, H.E., Lincoln, K., Moshiri, F., Jensen, P.K., Qi, Q., Venkatesh, T.V., 2006. The *Arabidopsis* vitamin E pathway gene 5–1 mutant reveals a critical role for phytol kinase in seed tocopherol biosynthesis. Plant Cell 18, 212–224.

Vanderauwera, S., Zimmermann, P., Rombauts, S., Vandenabeele, S., Langebartels, C., Gruissem, W., et al., 2005. Genome-wide analysis of hydrogen peroxide-regulated gene expression in *Arabidopsis* reveals a high-light induced transcriptional cluster involved in anthocyanins biosynthesis. Plant Physiol. 139, 806–821.

Vranova, E., Inze, D., van Breusegem, F., 2002. Signal transduction during oxidative stress. J. Exp. Bot. 53, 1227–1236.

Weber, H., 2002. Fatty acid-derived signals in plants. Trend Plant Sci. 7, 217–224.

White, D.A., Fisk, I.D., Gray, D.A., 2006. Characterization of oat (*Avena sativa* L.) oil bodies and intrinsically associated E-vitamers. J. Cer. Sci. 43, 244–249.

Zbierzak, A.M., Kanwischer, M., Wille, C., Vidi, P.A., Giavalisco, P., Lohmann, A., et al., 2010. Intersection of tocopherol and plastoquinol metabolic pathways at the plastoglobule. Biochem. J. 425, 389–399.

Chapter 11

Drought Stress Induced Oxidative Damage and Antioxidants in Plants

Parvaiz Ahmad, Sumiya Jamsheed, Asiya Hameed, Saima Rasool, Iti Sharma, MM Azooz and Mirza Hasanuzzaman

11.1 INTRODUCTION

Physiological responses of plants in different situations constitute stress (Jaleel et al., 2008a,b,c,d). All metabolic activities are affected regularly by environmental changes over daily and seasonal cycles (Heerden and Krüger, 2002). However, each deviation from a normal factor doesn't always result in stress. Drought stress when accompanied with very high constraint or fluctuations results in injury, abnormal physiological changes, etc. (Siemens and Zwiazek, 2003). The changed physiological condition tends to change the equilibrium and results in stress and strain (physical and chemical) (Jaleel et al., 2007a,b,c,d). Environmental stresses induce a number of responses in plants such as changed gene expression, cell metabolism, growth rate and production (Ashraf and Foolad, 2007; Ahmad and Prasad, 2012a,b). Plant growth and productivity face many types of abiotic stresses; of these, drought stress has the most adverse effects (Chartzoulakis et al., 2002). The condition in which the water potential of the plant and turgor are reduced to such an extent that it will interfere with normal metabolic activities is called *water deficit stress* (Tsialtas et al., 2001). A situation in which we observe stomatal closure which leads to limitation of gaseous exchange is water stress (Edward and Wright, 2008). A very extensive loss of water which leads to disturbances in cell structure and metabolism by termination of enzymatic reactions is known as *desiccation* (Zhang et al., 2004). The ability of plants to tolerate water scarcity is of great value in agriculture as this stress causes major problems in agriculture (Chaves et al., 2002, 2003). Water stress tolerance is the result of changes in cellular biochemistry of plants (Sivamani et al., 2000). It results in the accumulation of different solutes and proteins induced by

P. Ahmad (Ed): Oxidative Damage to Plants. DOI: http://dx.doi.org/10.1016/B978-0-12-799963-0.00011-3

osmotic stress (Chimenti et al., 2002; Jaleel et al., 2007e). Both the duration for which the plant is being exposed to water stress and its severity affect the various physiological responses of plant to water deficit (Colom and Vazzana, 2001; Davis and Swanson, 2001).

Severe water stress affects almost all the metabolic functions and ultimately leads to death of the cell (Zhang and Schmidt, 2000). Since plant growth and development are affected by interacting effects of stress factors and various molecular, biochemical and physiological phenomena, tolerance to environmental stress is very complex (Ashraf and Harris, 2004). The mechanism of drought tolerance includes ion homeostasis, biosynthesis of osmolytes, scavenging of harmful radicals, water circulation and coordination of a long distance response system (Reddy et al., 2004). Apart from osmotic adjustment, a very noteworthy effect of abiotic stress is production of reactive oxygen species such as hydrogen peroxide, etc. (Mittler, 2002; Ahmad, 2010; Ahmad et al., 2006, 2008a,b, 2010a,b,c, 2011a,b, 2012a,b, 2013; Ahmad and Umar, 2011; Ahmad and Prasad, 2012a,b). The generation of reactive oxygen species is responsible for the damage of the biomolecules of the cell (Ahmad et al., 2010a,b,c, 2011a,b; Ahmad and Umar, 2011; Ahmad and Prasad, 2012a,b). In order to withstand stress conditions, plants use their antioxidant enzymes, e.g. catalase, ascorbate peroxidase, superoxide dismutase, glutathione reductase, etc. (Prochazkova et al., 2001; Ahmad et al., 2006, 2008a,b, 2010a,b,c, 2012a,b, 2013; 2011; Ahmad and Umar, 2011; Ahmad and Prasad 2012a,b). A combined action of non-enzymatic antioxidants like ascorbic acid, glutathione, tocopherols, etc. also help in scavenging the ROS in plants (Srivalli et al., 2003; Dalmia and Sawhney, 2004; Ahmad et al., 2006, 2008a,b, 2010a,b,c, 2012a,b, 2013; 2011; Ahmad and Umar 2011; Ahmad and Prasad 2012a,b). The antioxidant defense system is responsible for elevating tolerance in plants exposed to various environmental stress conditions (Jaleel et al., 2007b; Ahmad et al., 2008b; Ahmad, 2010). In order to understand drought tolerance better, a careful reconsideration is needed to understand the limited success of the physiological and molecular approaches (Fig. 11.1). The responses of plants to stress are due to a variety of changes in biochemical and physiological activities and are governed by multiple signaling pathways (Knight and Knight, 2001). It is known that the gene expressions that are induced in response to stress overlap between the signaling pathways (Seki et al., 2001; Chen et al., 2002). Thus, decoding the mechanism of perceiving environmental signals and then transmitting to the cellular machinery to activate the responses for adaptation. Based on this multiplicity of signaling, there are manifold increases in primary sensors that recognize early signals of stress and modify the expression of genes that are involved in it. These metabolic reactions to stress that involve the insight of signal transduction alterations result in stress tolerance. "Omics" technologies that include transcriptomics, proteomics, fluxomics and metabolomics have provided

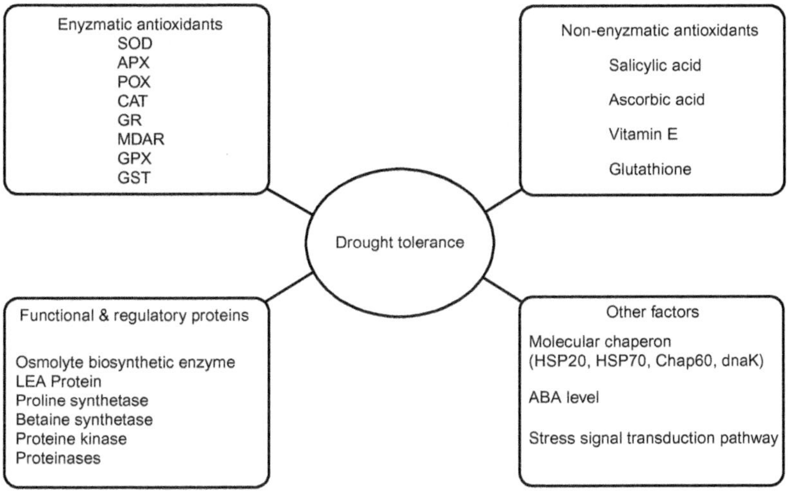

FIGURE 11.1 Factors contributing to drought tolerance in plants.

integrated genome pathway recognition. Analysis using transcriptome microarray technology (Zhu et al., 2001) reveals that stress-induced genes are of two types depending on their end products: one type is mainly composed of functional proteins like membrane proteins, osmolyte biosynthetic enzymes, detoxifying enzymes and proteins for protection of macromolecules, while the other group consists of the regulatory proteins, protein kinases and transcription-regulation protein kinase, etc., and proteinases (phosphoesterases and phospholipases C, etc.) involved in the regulation of signal transduction and gene expression. Thus, it can be said that, using transcriptomics, drought tolerance may be developed by incorporation of LEA proteins, proline synthetase or betaine synthetase encoding genes. The expression of tolerant genes is regulated by interaction of cis-elements located in the promoter region of abiotic stress-related genes and transcription activation factors. Several transcriptional regulatory mechanisms that participate in the field of stress response induction of genes are displayed by molecular and genomic analysis. The tools of genomics produce inclusive data sets on alteration in gene expression, protein profiles, and metabolites that result from exposure to drought. Since the stress response has numerous complexities due to the involvement of several genes, biotechnology and its special insights bring about useful effects. Transgenic plants with a single regulatory gene (such as transcription factor) have been developed to tolerate drought stress. These genes regulate the expression of downstream genes involved in the stress response. Knowing specific target traits, marker-assisted selection can also play a role in drought tolerance (Tuberosa and Salvi, 2006) (Fig. 11.1). The present efforts toward elucidation of metabolic/genomic profiles of plants to drought stress indicate that omics-assisted technologies will soon become routine.

11.2 ROS GENERATION IN CELL

Oxygen is a very important constituent for the sustainability of all the organisms. The reduction of O_2 to H_2O provides energy to the organism but when it is incompletely reduced ROS are generated which are highly reactive and oxidize biological molecules (Ahmad and Umar, 2011). ROS generated in different subcellular locations are actually the by-products of various metabolic activities. Here rises the query of whether the specificity and selectivity of ROS signals are dependent on the subcellular site of ROS generation. ROS are always formed by electron seepage during electron transport activities of chloroplasts, mitochondria, and plasma membranes or as a derivative of different metabolic pathways running in various cellular compartments (Sharma et al., 2012). Many important processes such as growth and development of plants, cell death and responses to stress are regulated by ROS (Gapper and Dolan, 2006; Dietz, 2008). The production of ROS is detrimental to proteins, lipids and nucleic acids as they induce the production of various oxidative radicals which cause oxidation. Although antioxidants help in keeping the levels of ROS low they are not able to remove them completely. ROS contributes to the aging process in plants; thus, repair systems are important to repair the cellular damage. The fact that drought stress is responsible for enhanced production of ROS is a well-recognized fact in aerial parts of the plants (Noctor and Foyer, 1998; Noctor et al., 2002). The proteomics study in watermelon carried out by Yoshimura et al. (2008) showed that the root tissue accumulated several ROS metabolism-related proteins on either day 1 or day 3 after drought stress. Drought stress induces the production of reductases and glutamylcysteine synthetases. Chloroplasts, mitochondria, and peroxisomes are main sites for ROS production. Both endogenous and exogenous substances can produce ROS. Mitochondria, cytochrome P450 metabolism, peroxisomes, and inflammatory cell activation are potential endogenous sources (Inoue et al., 2003).

11.2.1 ROS Generation in Mitochondria

Mitochondria are also a potential site for ROS production, primarily generating oxygen and hydrogen peroxide molecules. This is because of the overreduction of electrons in the electron transport chain. Direct reduction of O_2 to O_2^- occurs in the flavoprotein region of the enzyme NADH dehydrogenase, which is an enzyme of the respiratory system in plants. When electrons pass through the mitochondrial electron transport chain the oxygen radicals are enhanced in the presence of antimycin −A, resulting in the accumulation of ubiquinone that is capable of autooxidation, thus producing

O_2^- (Loschen et al., 1973; Forman and Boveris, 1982). Under *in vitro* conditions, ubiquinone of the mitochondrial electron transport chain appears as a prominent location for H_2O_2 generation and it seems that O_2^- is a major precursor of H_2O_2 (Boveris and Cadenas, 1973; Winston, 1990).

11.2.2 ROS Generation in Chloroplasts

Because of their oxygenic conditions and the abundance of the photosynthesizers, photosynthetic plants are at a high risk of oxidative damage. In the Mehler reaction (Mehler, 1951) during photosynthetic electron transport, the photoreduction of O_2 results in the production of O_2^-. In spite of acting as a substitute for the intake of surplus energy, photoreduction of O_2 is associated with the generation of toxic ROS. The rate of production of ROS is activated when photon intensity is in excess of that required for the assimilation of CO_2. Photoreduction of ROS in chloroplast is suppressed by the relaxation systems under excess photon production. Examples of these include photorespiration, cyclic electron flow either through PSI or PSII, downregulation of PSII quantum yield regulated by the xanthophyll cycle and photon gradient across thylakoid membrane (Asada, 2006).

11.2.3 ROS Generation in Microsomes and Peroxisomes

In plants NADPH dependent microsomal electron transport produces O_2^- (Winston, 1990). The NADH activity generates H_2O_2 through chain reaction of peroxidases in the presence of NADPH or related compounds (Chen and Schopfer, 1999). The production of H_2O_2 is distinguished from phagocyte-type NADPH oxidase as it has different K_m value for oxygen, NADH and NADPH and response towards enzyme inhibitors (cyanide, azide and diphenyleneiodonium (DPI) (Vera-Estrella et al., 1992). In addition to its NAD(P)H-oxidase activity, horseradish peroxidase can reduce H_2O_2 to OH (Chen and Schopfer, 1999). During the infection of barley by *Erysiphe graminis* a Germin like oxalate oxidase got induced which is another alternative source of ROS (Dumas et al., 1995). Allan and Fluhr (1997) reported that addition of fungal elicitor gives rise to two distinct ROS producing mechanisms in epidermal cells of tobacco. The two sources of ROS production are (1) NADPH oxidase and/or a xanthine oxidase and (2) peroxidase and/or amine oxidase.

11.2.4 Other Sources of ROS Generation in Plants

ROS may also be generated during the detoxification reaction catalyzed by cytochrome P450 in cytoplasm and endoplasmic reticulum (Dybing et al., 1976). Plasma membrane and extracellular apoplast are also the sites of ROS generation. The source of H_2O_2 in apoplast of plant cells is cell wall peroxidases and germin like oxidases (Bolwell and Woftastek, 1997). Alkaline pH activates cell-wall peroxidases which results in the production of H_2O_2. The cell wall peroxidases catalyze the pH-dependent alkalization of apoplast upon elicitor recognition causing H_2O_2 production, which leads to generation of ROS and oxidative burst during biotic stress conditions (Bolwell and Woftastek, 1997). Gadjev et al. (2006) reported that ROS alters the gene expression to select the specificity of ROS-induced transcript expression. The experiments were conducted on transgenic plants having alterations in ion and catalase activity due to treatment with oxidative stress agents. Many experiments included disruption in antioxidant enzymes, which resulted in reduction or complete abolishing of enzyme activity. This experiment showed that only one species of ROS was able to alter a majority of transcripts. These transcripts were characterized by place of ROS production and their chemical characteristics (Gadjev et al., 2006). Many genes with changed expression reacted in such a way that it highlighted the ROS type and its location to determine the gene response. Usually in flu mutant ethylene production may partially block during their movement from dark to light. Danon et al. (2005) observed that ROS can block the ethylene production, which may alter the early response and downstream transcripts expression in flu mutant. ROS mediated plant response and disorders in pathways have been documented in genome-wide expression inventories. The selectivity and specificity of ROS signaling due to ROS sensors within the cell is still a major challenge.

11.3 NONENZYMATIC ANTIOXIDANTS

11.3.1 Ascorbic Acid

Nonenzymatic antioxidant defense components act against drought-induced oxidative stress in plants (Wang et al., 2005). Drought stress has been reported to result in a significant increase in antioxidant concentration in turf grass (Zhang and Schmidt, 2000; Vranova et al., 2002). In maize and wheat, ascorbic acid is reduced under oxidative stress indicating its involvement in deciding the oxidative response (Vertovec et al., 2001; Nayyar and Gupta, 2006). Ascorbic acid acts as a reducing substrate for H_2O_2 and assists in detoxification of plant cells exposed to oxidative stress (Fujita et al., 2003; Türkan et al., 2005). In the majority of the plant cell types, organelles and apoplast, ascorbate has been detected as one of the extensively studied antioxidants (Smirnoff, 2000). It is being synthesized in mitochondria and moved to other cell organelles through facilitated diffusion or through a proton electrochemical gradient (Ogbonnaya et al., 2003). Ascorbic acid also plays a vital role in cell elongation (Tahkokorpi et al., 2007). Monodehydroascorbate in ascorbate-glutathione cycle is generated by the reduction of H_2O_2 to water requiring two molecules of ascorbate

(Pastori et al., 2000; Pan et al., 2003). Monodehydroascorbate is a radical with a short lifetime and can break into dehydroascorbate and ascorbic acid (Nam et al., 2001). In the water cycle in chloroplast, NADPH acts as the electron donor and the reaction is catalyzed by monodehydroascorbate reductase or ferredoxin (Noctor et al., 2002). Yue et al. (2013) have reported that small amounts of natural ascorbate and damaged ascorbate-glutathione cycle induces a decrease in the reduced form of ascorbate, which leads to enhanced sensitivity to drought stress in the ascorbate-deficient mutant of *Arabidopsis*.

11.3.2 α-Tocopherol

α-Tocopherols (Vitamin E) are lipophillic antioxidants synthesized in green parts of all plants. They scavenge lipid-peroxy radicals during the intensive activities of other antioxidants (Munné-Bosch, 2005; Massacci et al., 2008). α-Tocopherols shield lipids and other membrane structures and functioning of PSII (Lopez-Huertas et al., 2000; Nordberg and Arner, 2001). By interacting with acyl groups of lipids, α-tocopherols scavenge many ROS and lipid soluble compounds produced during oxidative stress (Munné-Bosch, 2005). Drought-stressed plants have been reported to synthesize α-tocopherols (Lawlor and Cornic, 2002; Munné-Bosch, 2005; Martinez et al., 2007). Oxidative stress activates the expression of genes responsible for the synthesis of tocopherols in plants (Falk et al., 2002). Manivannan et al. (2007a) reported an enhancement in α-tocopherol content in drought-stressed *Vigna* plants. Zhang and Schmidt (2000) reported a two-fold increase in α-tocopherol in turf grass under water stress. Synthesis of α-tocopherol is under genetic control activated by oxidative stress in plants (Falk et al., 2002; Gong et al., 2012).

11.3.3 Reduced Glutathione (GSH)

Glutathione (L-glutamyl-L-cysteinylglycine, GSH) is composed of L-glutamic acid, L-cysteine and glycine. It is a thiol compound present in all plants, animals and microorganisms, and behaves as an important compound for free radical scavenging (Monneveux et al., 2006). GSH participates in many cellular activities and protects the cells from toxic effects of many reactive species (Petropoulos et al., 2008). GSH is involved in many biological functions such as maintainance of membrane integrity, protein activities and DNA synthesis (Cabuslay et al., 2002). Glutathione also participates in the control of H_2O_2 levels (Meyer et al., 2005). The change in the ratio of reduced GSH to oxidized (GSSG) form during the degradation of H_2O_2 is important in certain redox signaling pathways (Noctor et al., 2002; Pastori and Foyer, 2002). It is presumed that this ratio (GSH/GSSG) is an indicator of the cellular redox balance and is involved in ROS perception

(Prochazkova et al., 2001; Foyer and Noctor, 2003). Reduced glutathione (GSH) participates in the reactive oxygen radical reduction by stress and thus acts as a strong antioxidant. Chen et al. (2012) reported that glutathione helps to withstand drought stress in transgenic *Arabidopsis* plants. Glutathione is a tripeptide (α-glutamyl cysteinylglycine), present in all cell compartments such as cytosol, chloroplasts, endoplasmic reticulum, vacuoles and mitochondria (Sankar et al., 2007a,b). In most plant cells it is the main source of nonprotein thiols. Nonprotein thiols are mostly the product of glutathione in plants. In all organisms, glutathione is suitable to serve a broad range of biochemical functions because of the chemical reactivity of the thiol groups in it (Heerden and Krüger, 2002). The nucleophilic nature of the thiol group makes it capable of mercaptide bond formation with metals and for reacting with electrophiles (Rodríguez et al., 2005). All these properties therefore make it a very suitable biochemical to protect plants against stress of different kinds such as oxidative stress, heavy metals and organic chemicals (Noctor et al., 2002; Samarah, 2005).

11.4 ENZYMATIC ANTIOXIDANTS

Stress results in the production of ROS species which promote oxidative damage in biological systems (Reddy et al., 2004; Samarah et al., 2006). The detoxification of ROS exist in all plants and can be divided into enzymatic (SOD, APX, POX, GR, etc.) and nonenzymatic (ascorbic acid, flavanones, anthocyanins, etc.). The levels of response depend on various factors such as species involved, the metabolic state of the plant and intensity and duration of stress. Increase in foliar antioxidant activity has been reported in many stress situations (Pastori et al., 2000; Sharp and LeNoble, 2002). The degree and quantity of increase in antioxidants during oxidative stress are varied in different species and even between the two cultivars of the same species (Manivannan et al., 2007a,b). Plants can resist the stress-induced production of reactive oxygen species (ROS) by increasing the ingredients of their defense systems (Ahmad et al. 2008a,b, 2010a,b, 2011). However, various enzymatic and nonenzymatic antioxidant systems protect the plant cells against such effects (Ahmad et al., 2008a,b, 2010a,b,c, 2011). A lot of research has been done on drought stress on plants and it describes how exposed plants adapt to the detoxifying enzyme activities (Djibril et al., 2005; Manivannan et al., 2008). Analysis of ultrastructural changes in cells by Dong et al. (2013) suggests that drought-mediated damage to cucumber cells is prevented through stabilization of cell structure and alleviation of lipid peroxidation by increased activity of antioxidant enzymes and contents of antioxidant metabolites. Superoxide dismutase (SOD), catalase (CAT), peroxidase (POD), ascorbate peroxidase (APX) and other enzymes of ascorbate-glutathione pathway (Halliwell and Asada cycle) are the main enzymes involved in the detoxification of ROS. All these enzymes show enhanced activity during stress conditions (Jaleel et al., 2007a,b). Enzymes SOD, CAT and POD form important antioxidant systems and SOD catalyzes the dismutation of O_2 to H_2O_2 and O_2, while catalase and peroxidase scavenge H_2O_2 (Egert and Tevini, 2002). Feng et al. (2003) have reported that oxidative stress can be induced by exposing the cucumber seedlings to chilling treatment which weakens the antioxidant system and increases H_2O_2 levels. Similar effects have been observed in rice (Huang, 2000). Mohammad et al. (2013) suggests that tolerance against drought-induced oxidative stress may be achieved by modulation in antioxidative and glyoxalase systems in mustard.

11.4.1 Peroxidase

Peroxidase (POX) helps in decreasing H_2O_2 content and cell peroxidation of membrane lipids by decreasing MDA content. Irrespective of the genotypes, *Radix astragali* plants showed enhanced peroxidase activity under water deficit (Sankar et al., 2007a,b, 2008). Increased peroxidase activity has been reported in cucumber seedlings subjected to chilling stress (Feng et al., 2003). Similarly, increased peroxidase activity has also been observed in low temperature stressed wheat seedlings (Zhao et al., 2008). Extensive research on oxidative stress has shown that adverse environmental conditions induce overproduction of reactive oxygen species (ROS) in plants, such as $O_2^{\bullet -}$, H_2O_2 and hydroxyl radical ($^{\bullet}OH$) in plants (Shao et al., 2008). Increase in the activity of SOD, POX, and CAT under salinity on tolerant chickpea genotypes has been reported (Rasool et al., 2013). Increased peroxidase activity was also reported previously in wheat genotypes by Shao et al., (2008). Soybean plants showed enhanced peroxidase activity under water deficit (Zhang et al., 2006). Peroxidase when coordinated with superoxide dismutase (SOD) appears to play an important role in the scavenging process (Farooq et al., 2008). They are chloroplastic enzymes, which scavenge H_2O_2 generated primarily through superoxide dismutase action (Chaitanya et al., 2002).

11.4.2 Superoxide Dismutase

In biological evolution and processes such as signal transduction, gene transcription and regulation of enzyme performance, molecules with unpaired electrons are essential (Fang et al., 2002). These can act as important mediators in immune response. For example, if phagocytes are activated, O_2 will be converted to O by NADPH oxidase. Excessive production of free radicals causes mutations due to nucleotide oxidation and dimerization during the replication for their unstable and very reactive properties. Due to lipid oxidation, these reactive species may bring about lysis of cells and membrane dysfunction by associating with membrane lipids, especially the glycolipids.

During the past decades, different dietary antioxidants were developed and served the purpose of supporting nutrition to lessen hazardous effects of free radicals through enzymatic and nonenzymatic protection. Superoxide anionic radical ($O_2^{\bullet-}$), which is a precursor of reactive species, is generated *in vivo* in the epithelial cells, macrophages and neutrophils in the electro-rich aerobic environment (Fang et al., 2002). $O_2^{\bullet-}$ can be converted into H_2O_2 by superoxide dismutase (SOD), which is a defense enzyme, protecting the cells from cellular damage caused by ROS in the living cells (Wu et al., 2008). SOD has a protective effect against oxygen toxicity by regulating concentration of superoxide anionic radical (Nordberg and Arner, 2001). Hence SOD attains an extensive potential in the pharmaceutical and food industry (Meyer et al., 2005). That overproduction of chloroplastic SOD enhances stress tolerance has been shown by several studies (Pastori and Foyer, 2002). Enhanced activity of SOD in *Radix astragali* varied in three different genotypes under water-deficit stress (Tan et al., 2006). There is a significant activation of SOD in cucumber seedlings under chilling stress (Feng et al., 2003). Zhang et al. (2006) reported 25% enhancement under drought stress in SOD activity in soybean. In higher plants drought stress induces increased SOD activity (Reddy et al., 2004; Salekjalali et al., 2012; Chakraborty and Pradhan, 2012; Rai et al., 2013). Double the amount of SOD activity was observed in drought-stressed citrus plants when compared to well-watered control plants at the seedling stage (Wu et al., 2006). The activity of SOD has been reported to increase under drought stress in *Zea mays* (Pastori et al., 2000), *Euphorbia esula* (Davis and Swanson, 2001), *Cassia angustifolia* (Agarwal and Pandey, 2003), *Triticum aestivum* (Singh and Usha, 2003), *Oryza sativa* (Wang et al., 2005), *P. acutifolius* (Türkan et al., 2005). In aerobic organisms, SOD is one of the most important enzymes and plays a very important role in the cellular defense mechanism against reactive oxygen species as it modulates the reactive amount of O_2 and H_2O_2 and decreases the production of $^{\bullet}OH$ radical which may cause severe damage to membrane, proteins and DNA (Kage et al., 2004).

11.4.3 Catalase

Plants produce H_2O_2 in the metabolic process and cause damage to cell function, while catalase (CAT) eliminates H_2O_2 and plays a key role in the elimination of O_2 (Gaspar et al., 2002). Water stress caused an enhancement of CAT activity in both wild and cultivated species of *Radix astragli* at the seedling stage (Tan et al., 2006) and *Arachis hypogaea* (Sankar et al., 2007a). A decrease in CAT activity was reported by Feng et al. (2003) in cucumber seedlings under chilling stress. A two-fold increase in CAT activity has been observed in *Poncirus trifoliata* under water stress (Wu et al., 2006). Nayyar and Gupta (2006) also have reported foliar CAT activity enhancement under water stress in wheat and maize. Increased activity of catalase under drought stress has been reported in *Zea mays* (Pastori et al., 2000), *Allium schoenoprasum* (Egert and Tevini, 2002) and *Triticum aestivum* (Dalmia and Sawhney, 2004; Chakraborty and Pradhan, 2012), *P. acutifolius* (Türkan et al., 2005). Increased CAT activity has been observed in higher plants during drought stress (Reddy et al., 2004, 2005). Similar results have been reported in *Lotus corniculatus* (Borsani et al., 2001) and *Oryza sativa* (Wang et al., 2005). Catalases are heme-containing enzymes that catalyze the dismutation of hydrogen peroxide into water and oxygen (Ahmad et al., 2010a,b, 2011a,b). Glyoxisomes and peroxisomes have abundant quantities of catalases where it is involved in decomposition of H_2O_2 generated during the oxidation of fatty acids in germinating barley (Jiang and Zhang, 2001) and for the removal of H_2O_2 produced by photorespiration by converting glycolate into glyoxalate in C3 plants (Li et al., 2000; Kiani et al., 2008). This is due to the proliferation of peroxisomes during stress, which helps in scavenging of H_2O_2 diffused from the cytosol (Lopez-Huertas et al., 2000; Kusaka et al., 2005).

11.4.4 Ascorbate Peroxidase

Increased activity of ascorbate peroxidase (APX) has been reported in *Euphorbia escula* under drought stress (Davis and Swanson, 2001), *Zea mays* (Jiang and Zhang, 2001), *Triticum aestivum* (Dalmia and Sawhney, 2004), *P. accutifolius* (Türkan et al., 2005), *G. max* (Heerden and Krüger, 2002), and *Lycopersicon esculantum* (Rai et al., 2013). Enhanced activity of APX under drought stress was observed in *Vigna* (Manivannan et al., 2007a). The APX activity also increased in *G. max* and *Catharanthus roseus* under water stress (Heerden and Krüger, 2002; Jaleel et al., 2007b). Enhanced activity of APX in the leaves of olive trees experiencing severe drought stress has been reported by Sofo et al. (2005). APX is one of the most important antioxidant enzymes of plants involved in detoxification of H_2O_2 and ascorbate for reduction. APX reduces H_2O_2 to water with ascorbate as the specific electron donor (Lawlor and Cornic, 2002). In trifoliate orange an increased APX activity was observed under water stress, but at mild water deficit no significant variation in APX activity was recorded in maize and wheat (Lawson et al., 2003; Nayyar and Gupta, 2006).

11.5 PROTEOMICS UNDER DROUGHT STRESS

Finally, metallomics considers that metal cofactors in proteins help in their functional placement and put it in the known cellular pathways; biomolecules that bind to the metals and metalloids constitute a major part of molecules participating in metabolism and behavior. The use of chromatography and highly sensitive atomic detectors or inductively coupled plasma-mass spectrometer (ICP-MS) for heterotram tags species in the metallomic analytical approach, combined with MS, are the central components. With the ICP-MS detector the molecular tracing has become more sensitive, selective and reliable. Therefore, metallomics is now applied in traditional biomarker analysis and assists in quantitative analytical approaches that change this "omic" technology to a very beneficial alternative for biomolecule analysis at massive levels. Therefore, massive identification of proteins by proteomic analysis can help in general appraisal of affected pollutants. Thus, without requiring any previous knowledge, a comprehensive assessment of the toxic and defensive mechanisms triggered by pollutants is provided by environmental proteomics. The identification of proteins with high throughput proteomics methods and de novo sequencing is used. The plant responses to high salinity/drought stress have been greatly advanced by new proteomic technology and mass spectrometry-based protein recognition (Parker et al., 2006; Qureshi et al., 2007; Caruso et al., 2008). Drought-responsive proteins in several plants have been identified by many research groups using the proteomic approach. Several drought-responsive proteins in root proteome of drought-tolerant and drought sensitive rice varieties, including ABA- and stress-responsive proteins, ascorbate peroxidase and many others have been identified (Salekdeh et al., 2002). Several proteins were found to be modulated in expression by salt concentration in a coordinated manner (Nohzadeh et al., 2007). These proteins were found to be involved in protein folding, regulation of metabolic processes, biomass production, photorespiration, antioxidant defense system and operation of ion channels. It is reported that application of exogenous salycilic acid can also decrease drought-induced oxidative stress in plants (Kabiri et al., 2012).

The deficiency of water around the root zone leads to development of drought stress in plants. As a harsh abiotic stress factor, drought hinders growth, development and production of plants. A vast range of reports has been accumulated on plant response to drought stress. Although site specific alterations in activity of antioxidant enzymes have been reported by Devi et al. (2012) as CAT stimulates in root and shoot, and APX in endosperm, GR and POD are upregulated in shoots of drought-tolerant wheat genotypes.

However, very little information is available on drought sensing and tolerance mechanisms at the organelle proteome level in plants. On the basis of proteomic analysis of modern and wild wheat Budak et al. (2013) reported that wild wheat holds a genetic potential to stress and advancement for drought tolerance can be accomplished by the introduction of drought tolerance-related genes and QTLs to modern wheat cultivars.

The expression of the drought-responsive gene is always triggered by a temporary increase in the endogenous ABA level of plants (Hirayama and Shinozaki, 2007). Water deficiency also leads to H_2O_2 production in photosynthetic organisms. Interestingly, ABA itself stimulates H_2O_2 formation through superoxide radical production from reaction of oxygen and NADPH, catalyzed by membrane bound NADPH oxidase. However, the latest contradictory reports suggest that drought stress-induced ABA may control the increasing collection of H_2O_2, by inducing CATB gene expression (Hu et al., 2005). In the plant cell, H_2O_2 acts as a signaling molecule and activates mitogen-activated protein kinase (MAPK) cascade for stress responsive gene expression (Ye et al., 2011). The components of the stress signal transduction pathway, namely serine/threonine protein kinase, histidine kinase, and tyrosine phosphatase, have been recognized by nuclear proteome analyses. The process of transcription is assisted by transcription factor (TF) proteins. Maruyama et al. (2004) have reported that drought stress leads to upregulation of TFs, specifically RF2b, bZIP, HB3, and homeobox-leucine zipper protein. Usually TFs for delayed stress response genes are activated by early stress response genes. ROS such as superoxide radicals and hydrogen peroxide are formed by NADPH oxidase activity and their rapid scavenging is essential to protect oxidative damage of DNA (Ahmad et al., 2011a,b). Antioxidant defense enzyme SOD is capable of scavenging superoxide radical whereas hydrogen peroxide is scavenged by APX or GPX. The upregulation of SOD, APX, GPX and GST protein indicates a competent antioxidant defense system in plants exposed to environmental stress (Ahmad et al., 2011a,b). Along with antioxidants, molecular chaperon (HSP20, HSP70, Chap60, dnaK) aided refolding of incorrectly folded proteins, also facilitating protection of plants exposed to drought stress. Koag et al. (2009) also suggested that drought-induced synthesis of DHNs proteins have an important role in safety of plant cell membranes.

At subcellular levels drought-mediated overreduction of electron transport chain leads to superoxide radicals and singlet oxygen production in the mitochondria and chloroplast of plants. A large quantity of Cu-Zn SOD was observed in the mitochondrial proteome of water deficit pea plant. Drought-induced upregulation of APX, SOD and MDAR was also noted in cell walls of plant cells. Moreover, the correct folding of proteins may confirm from excess amount of chaperonin in mitochondrial and cell wall proteomes of drought-stressed plants (Taylor et al., 2005). Reports suggest that a detailed analysis of chloroplastic proteome is essential to fulfill knowledge gaps about the role of stress-responsive proteins during drought stress.

11.6 CONCLUSIONS

Inducing cellular damage by free radicals and other reactive oxygen species, drought stress causes plants to respond through the change of physiological and biochemical properties, including growth, protein degradation, and lipid peroxidation. To minimize the damaging influence of activated oxygen species, plants have excreted some metabolites, such as soluble protein and antioxidants as well. To overcome the limiting of crop production, plant cells were examined extensively in terms of an antioxidant defense system, such as nonenzymatic and enzymatic compounds. Ascorbic acid, α-tocopherol, and reduced glutathione are considered to be dominant compounds for drought tolerance, which markedly contribute to their antioxidant capability. On the other hand, peroxidase, superoxide dismutase, catalase, as well as ascorbate peroxidase are the principal enzymatic defenses used by the plants against oxidative damage under drought stress. Thus it is well documented that the antioxidant defense as a protective mechanism in higher plants plays an essential role in the removal of reactive oxygen species under water deficit stress. However, the existence of metabolic limitations is only partially investigated and the precise mechanisms are still poorly understood. Therefore, the adverse relationship between drought stress and the response of the many biological macromolecules and micromolecules involved needs much more study. Examining the mechanism of stress tolerance concomitant with the involved genes in terms of the signal network is very important for crop enhancement. If some drought-inducible genes associated with representative performance can be further identified and exerted at a higher level of transcription, then the effect of investigation can carry out a positive regulation of the cellular milieu and plant tolerance. It is very intriguing to understand the possible regulatory mechanism or genes involved on the down or up expression of the transgenic plants under stress resistance. It will be a rather useful guide for the illumination of a sophisticated metabolic network and a feasible strategy for improving stress tolerance toward crop productivity increase. A number of internal factors and environmental signals coordinate and affect the growth and development of plants. The higher plants have developed a complex signaling network for sensing these environmental signals. The extracellular signals are perceived through various receptors. In plants, the abiotic stress signaling involves changes in calcium, MAP kinase and growth regulators. The road to engineering such tolerance into sensitive species is a long one, with much effort yet required to uncover each product of genes induced by abiotic stress and signal transduction pathways. The biologists have to look for particular sets of markers to predict tolerance toward a particular stress. MicroRNAs emerge as an important tool in abiotic stress tolerance and should be studied extensively in the future. For transcriptional profiling, microarrays make the most commonly used platform and the combination of their technical and practical advantages explains their continued use until some major improvements are made in the proteomics technology. Because of extensive posttranscriptional regulation particularly relevant in stress response, all transcriptome profiling techniques have limited relevance in describing the physiology of a whole cell or organism and, thus, the microarrays describe the transcriptome accurately, but not all transcriptomes. The relevance of microarray data to physiology can be increased by the use of polysomal m-RNA fractions. Other advancements like higher spatial resolution in tissue sampling and efforts to make treatment conditions relevant to plant growth in field conditions will definitely help to ensure the transcriptomic profiles from drought and salinity stresses of an increasing number of species, which will be useful in the systems biology, comparative genomics and gene discovery. Efforts for the detection and characterization of stress responsive organelle proteins are essential to develop genetically drought-tolerant crops for human welfare.

REFERENCES

Agarwal, S., Pandey, V., 2003. Stimulation of stress-related antioxidative enzymes in combating oxidative stress in Cassia seedlings. Indian J. Plant Physiol. 8, 264–269.

Ahmad, P., 2010. Growth and antioxidant responses in mustard (*Brassica juncea* L.) plants subjected to combined effect of gibberellic acid and salinity. Arch. Agro. Soil. Sci. 56, 575–588.

Ahmad, P., Prasad, M.N.V., 2012a. Environmental Adaptations and Stress Tolerance in Plants in the Era of Climate Change. Springer Science + Business Media, New York.

Ahmad, P., Prasad, M.N.V., 2012b. Abiotic Stress Responses in Plants: Metabolism, Productivity and Sustainability. Springer Science + Business Media, New York.

Ahmad, P., Umar, S., 2011. Antioxidants: Oxidative Stress Management in Plants. Studium Press Pvt. Ltd, New Delhi, India.

Ahmad, P., Sharma, S., Srivastava, P.S., 2006. Differential physio-biochemical responses of high yielding varieties of Mulberry (*Morus alba*) under alkalinity (Na_2CO_3) stress *in vitro*. Physiol. Mol. Biol. Plant 12, 59–66.

Ahmad, P., Sarwat, M., Sharma, S., 2008a. Reactive oxygen species, antioxidants and signaling in plants. J. Plant Biol. 51, 167–173.

Ahmad, P., Jhon, R., Sarwat, M., Umar, S., 2008b. Responses of proline, lipid peroxidation and antioxidative enzymes in two varieties of *Pisum sativum* L. under salt stress. Int. J. Plant Prod. 2, 353–366.

Ahmad, P., Jaleel, C.A., Salem, M.A., Nabi, G., Sharma, S., 2010a. Roles of enzymatic and non-enzymatic antioxidants in plants during abiotic stress. Crit. Rev. Biotechnol. 30, 161–175.

Ahmad, P., Umar, S., Sharma, S., 2010b. Mechanism of free radical scavenging and role of phytohormones during abiotic stress in plants. In: Ashraf, M., Ozturk, M., Ahmad, M.S.A. (Eds.), Plant Adaptation and Phytoremediation. Springer, Dordrecht, pp. 99–108.

Ahmad, P., Jaleel, C.A., Sharma, S., 2010c. Antioxidative defence system, lipid peroxidation, proline metabolizing enzymes and biochemical activity in two genotypes of *Morus alba* L. subjected to NaCl stress. Russ. J. Plant Physiol. 57, 509–517.

Ahmad, P., Nabi, G., Jaleel, C.A., Umar, S., 2011a. Free radical production, oxidative damage and antioxidant defense mechanisms in plants under abiotic stress. In: Ahmad, P., Umar, S. (Eds.), Oxidative Stress: Role of Antioxidants in Plants. Studium Press Pvt. Ltd, New Delhi, India, pp. 19–53.

Ahmad, P., Nabi, G., Ashraf, M., 2011b. Cadmium-induced oxidative damage in mustard [*Brassica juncea* (L.) Czern. & Coss.] plants can be alleviated by salicylic acid. S. Afr. J. Bot. 77, 36–44.

Ahmad, P., Ozturk, M., Gucel, S., 2012a. Oxidative damage and antioxidants induced by heavy metal stress in two cultivars of mustard (L) plants. Fresenius Environ. Bull. 21, 2953–2961.

Ahmad, P., Hakeem, K.R., Kumar, A., Ashraf, M., Akram, N.A., 2012b. Salt-induced changes in photosynthetic activity and oxidative defense system of three cultivars of mustard (*Brassica juncea* L.). Afr. J. Biotechnol. 11, 2694–2703.

Ahmad, P., Ashraf, M., Azooz, M.M., Rasool, S., Akram, N.A., 2013. Potassium starvation-induced oxidative stress and antioxidant defense responses in *Brassica juncea.* J. Plant Interact.. 10.1080/17429145.2012.747629) In Press.

Allan, A.C., Fluhr, R., 1997. Two distinct sources of elicited reactive oxygen species in tobacco epidermal cells. Plant Cell 9, 1559–1572.

Asada, K., 2006. Production and scavenging of reactive oxygen species in chloroplasts and their functions. Plant Physiol. 141, 391–396.

Ashraf, M., Foolad, M.R., 2007. Roles of glycine betaine and proline in improving plant abiotic stress resistance. Environ. Exp. Bot. 59, 206–216.

Ashraf, M., Harris, P.J.C., 2004. Potential biochemical indicators of salinity tolerance in plants. Plant Sci. 166, 3–16.

Bolwell, G.P., Woftastek, P., 1997. Mechanism for the generation of reactive oxygen species in plant defense: broad perspective. Physiol. Mol. Plant Pathol. 51, 347–349.

Borsani, O., Díaz, P., Agius, F., Valpuesta, V., Monza, J., 2001. Water stress generates an oxidative stress through the induction of a specific Cu/Zn superoxide dismutase in *Lotus corniculatus* leaves. Plant Sci. 161, 757–763.

Boveris, A.E., Cadenas, E., 1973. Mitochondrial generation of H_2O_2. Biochem. J. 138, 707–716.

Budak, H., Kantar, M., Kurtoglu, K.Y., 2013. Drought tolerance in modern and wild wheat. Sci. World J. 548246.

Cabuslay, G.S., Ito, O., Alejal, A.A., 2002. Physiological evaluation of responses of rice (*Oryza sativa L.*) to water deficit. Plant Sci. 163, 815–827.

Caruso, G., Cavaliere, C., Guarino, C., Gubbiotti, R., Foglia, P., Lagana, A., 2008. Identification of changes in *Triticum durum* L. leaf proteome in response to salt stress by two-dimensional electrophoresis and MALDI-TOF mass spectrometry. Anal. Bio. Chem. 391, 381–390.

Chaitanya, K.V., Sundar, D., Masilamani, S., Reddy, A.R., 2002. Variation in heat stress-induced antioxidant enzyme activities among three mulberry cultivars. Plant Growth Regul. 36, 175–180.

Chakraborty, U., Pradhan, B., 2012. Drought stress-induced oxidative stress and antioxidative responses in four wheat (*Triticum aestivum* L.) varieties. Arch. Agron. Soil Sci. 58, 617–630.

Chartzoulakis, K., Patakas, A., Kofidis, G., Bosabalidis, A., Nastou, A., 2002. Water stress affects leaf anatomy, gas exchange, water relations and growth of two avocado cultivars. Sci. Hortic. 95, 39–50.

Chaves, M.M., Pereira, J.S., Maroco, J., Rodrigues, M.L., et al., 2002. How plants cope with water stress in the field: photosynthesis and growth. Ann. Bot. 89, 907–916.

Chaves, M.M., Pereira, J.S., Maroco, J., 2003. Understanding plant response to drought – from genes to the whole plant. Funct. Plant Biol. 30, 239–264.

Chen, J.H., Jiang, H.W., Hsieh, E.J., Chen, H.Y., Chien, C.T., Hsieh, H.L., et al., 2012. Drought and salt stress tolerance of an *Arabidopsis* glutathione *S*-transferase U17 knockout mutant are attributed to the combined effect of glutathione and abscisic acid. Plant Physiol. 158, 340–351.

Chen, S.X., Schopfer, P., 1999. Hydroxylradical production in physiological reactions. A novel function of peroxidase. Eur. J. Biochem. 260, 726–735.

Chen, W., Provart, N.J., Glazebrook, J., Katagiri, F., Chang, H.S., Eulgem, T., et al., 2002. Expression profile matrix of *Arabidopsis* transcription factor genes suggests their putative functions in response to environmental stresses. Plant Cell 14, 559–574.

Chimenti, C.A., Pearson, J., Hall, A.J., 2002. Osmotic adjustment and yield maintenance under drought in sunflower. Field Crop Res. 75, 235–246.

Colom, M.R., Vazzana, C., 2001. Drought stress effects on three cultivars of *eragrostis curvula*: photosynthesis and water relations. Plant Growth Regul. 34, 195–202.

Dalmia, A., Sawhney, V., 2004. Antioxidant defense mechanism under drought stress in wheat seedlings. Physiol. Mol. Biol. Plant 10, 109–114.

Davis, D.G., Swanson, H.R., 2001. Activity of stress-related enzymes in the perennial weed leafy spurge (*Euphorbia esula* L.). Environ. Exp. Bot. 46, 95–108.

Devi, R., Kaur, N., Gupta, A.K., 2012. Potential of antioxidant enzymes in depicting drought tolerance of wheat (*Triticum aestivum L.*). Indian J. Biochem. Biophy. 49, 257–265.

Dietz, K., 2008. Redox signal integration: from stimulus to networks and genes. Physiol. Plant. 133, 459–468.

Djibril, S., Mohamed, O.K., Diaga, D., Diégane, D., Abaye, B.F., Maurice, S., et al., 2005. Growth and development of date palm (*Phœnix dactylifera* L.) seedlings under drought and salinity stresses. Afr. J. Biotechnol. 4, 968–972.

Dong, X., Bi, H., Wu, G., Ai, X., 2013. Drought-induced chilling tolerance in cucumber involves membrane stabilisation improved by antioxidant system. Int. J. Plant Prod. 7, 67–80.

Dumas, B., Freyssinet, G., Pallett, K.E., 1995. Tissue-specific expression of germin-like oxalate oxidase during development and fungal infection of barley seedlings. Plant Physiol. 107, 1091–1096.

Dybing, E., Nelson, J.R., Mitchell, J.R., Sesame, H.A., Gillette, J.R., 1976. Oxidation of a methyldopa and other catechols by chytochromes R450-generated superoxide anion: possible mechanism of methyldopa hepatitis. Mol. Pharmacol. 12, 911–920.

Edward, D., Wright, D., 2008. The effects of winter water-logging and summer drought on the growth and yield of winter wheat (*Triticum aestivum* L.). Eur. J. Agron. 28, 234–244.

Egert, M., Tevini, M., 2002. Influence of drought on some physiological parameters symptomatic for oxidative stress in leaves of chives (*Allium schoenoprasum*). Environ. Exp. Bot. 48, 43–49.

Falk, J., Krauss, N., Dahnhardt, D., Krupinska, K., 2002. The senescence associated gene of barley encoding 4-hydroxyphenylpyruvate dioxygenase is expressed during oxidative stress. J. Plant Physiol. 159, 1245–1253.

Fang, Y., Yang, S., Wu, G., 2002. Free radicals, antioxidants and nutrition. Nutrition 18, 872–879.

Farooq, M., Basra, S.M.A., Wahid, A., Cheema, Z.A., Cheema, M.A., Khaliq, A., 2008. Physiological role of exogenously applied glycinebetaine in improving drought tolerance of fine grain aromatic rice (*Oryza sativa* L.). J. Agron. Crop Sci. 194, 325–333.

Feng, Z., Guo, A., Feng, Z., 2003. Amelioration of chilling stress by triadimefon in cucumber seedlings. Plant Growth Regul. 39, 277–283.

Forman, H.J., Boveris, A., 1982. Superoxide radical and hydrogen peroxide in mitochondria. In: Pryor, W.A. (Ed.), Free Radicals in Biology and Medicine, vol. 5. Academic Press, New York, pp. 65–90.

Foyer, C.H., Noctor, G., 2003. Redox sensing and signaling associated with reactive oxygen in chloroplasts, peroxisome and mitochondria. Physiol. Plant. 119, 355–364.

Fujita, T., Maggio, A., Rios, M.G., Stauffacher, C., Bressan, R.A., Csonka, L.N., 2003. Identification of regions of the tomato γ-glutamyl kinase that are involved in allosteric regulation by proline. J. Biochem. 278, 14203–14210.

Gadjev, I., Vanderauwera, S., Gechev, T.S., Laloi, C., Minkov, I.N., Shulaev, V., et al., 2006. Transcriptomic footprints disclose specificity of reactive oxygen species signaling in *Arabidopsis*. Plant Physiol. 141, 436–445.

Gapper, C., Dolan, L., 2006. Control of plant development by reactive oxygen species. Plant Physiol. 141, 341–345.

Gaspar, T., Franck, T., Bisbis, B., Kevers, C., Jouve, L., Hausman, J.F., et al., 2002. Concepts in plant stress physiology. Application to plant tissue cultures. Plant Growth Regul. 37, 263–285.

Gong, C., Ning, P., Bai, J., 2012. Responses of antioxidative protection to varying drought stresses induced by micro-ecological fields on desert C4 and C3 plants in northwest China. Life Sci. J. 9, 2006–2013.

Heerden, P.D.R., Krüger, G.H.J., 2002. Separately and simultaneously induced dark chilling and drought stress effects on photosynthesis, proline accumulation and antioxidant metabolism in soybean. J. Plant Physiol. 159, 1077–1086.

Hirayama, T., Shinozaki, K., 2007. Perception and transduction of abscisic acid signals: keys to the function of the versatile plant hormone ABA. Trend Plant Sci. 12, 343–351.

Hu, X., Jiang, M., Zhang, A., Lu, J., 2005. Abscisic acid-induced apoplastic H_2O_2 accumulation up-regulates the activities of chloroplastic and cytosolic antioxidant enzymes in maize leaves. Planta 223, 57–68.

Huang, J., Hariji, R., Adam, L., Rozwadowski, K.L., Hammerlineli, J.L., Keller, W.A., et al., 2000. Genetic engineering of glycinebetaine production towards enhancing stress tolerance in plants. Plant Physiol. 122, 747–756.

Inoue, M., Sato, E.F., Nishikawa, M., Park, A.M., Kira, Y., Imada, I., et al., 2003. Mitochondrial generation of reactive oxygen species and its role in aerobic life. Curr. Med. Chem. 10, 2495–2505.

Jaleel, C.A., Gopi, R., Panneerselvam, R., 2007a. Alterations in lipid peroxidation, electrolyte leakage, and proline metabolism in *Catharanthus roseus* under treatment with triadimefon, a systemic fungicide. C. R. Biol. 330 (12), 905–912.

Jaleel, C.A., Manivannan, Paramasivam, Sankar, B., Kishorekumar, A., Gopi, Ragupathi, Somasundaram, R., et al., 2007b. Water deficit stress mitigation by calcium chloride in *Catharanthus roseus*; effects on oxidative stress, proline metabolism and indole alkaloid accumulation. Colloids Surf. B: Biointerfaces 60, 110–116.

Jaleel, C.A., Manivannan, Paramasivam, Sankar, B., Kishorekumar, A., Gopi, Ragupathi, Somasundaram, R., et al., 2007c. *Pseudomonas fluorescens* enhances biomass yield and ajmalicine production in *Catharanthus roseus* under water deficit stress. Colloids Surf. B: Biointerfaces 60, 7–11.

Jaleel, C.A., Manivannan, Paramasivam, Lakshmanan, G.M.A., Sridharan, Ramalingam, Panneerselvam, R., 2007d. NaCl as a physiological modulator of proline metabolism and antioxidant potential in *Phyllanthus amarus*. C. R. Biol. 330, 806–813.

Jaleel, C.A., Manivannan, Paramasivam, Sankar, B., Kishorekumar, A., Gopi, Ragupathi, Somasundaram, R., et al., 2007e. Alterations in osmoregulation, antioxidant enzymes and indole alkaloid levels in *Catharanthus roseus* exposed to water deficit. Colloids Surf. B: Biointerfaces 59, 150–157.

Jaleel, C.A., Manivannan, P., Lakshmanan, G.M.A., Gomathinayagam, M., Panneerselvam, R., 2008a. Alterations in morphological parameters and photosynthetic pigment responses of *Catharanthus roseus* under soil water deficits. Colloids Surf. B: Biointerfaces 61 (2), 298–303.

Jaleel, C.A., Sankar, B., Murali, P.V., Gomathinayagam, M., Lakshmanan, G.M.A., Panneerselvam, R., 2008b. Water deficit stress effects on reactive oxygen metabolism in *Catharanthus roseus*; impacts on ajmalicine accumulation. Colloids Surf. B: Biointerfaces 62, 105–111.

Jaleel, C.A., Manivannan, P., Murali, P.V., Gomathinayagam, M., Panneerselvam, R., 2008c. Antioxidant potential and indole alkaloid profile variations with water deficits along different parts of two varieties of *Catharanthus roseus*. Colloids Surf. B: Biointerfaces 62, 312–318.

Jaleel, C.A., Gopi, R., Sankar, B., Gomathinayagam, M., Panneerselvam, R., 2008d. Differential responses in water use efficiency in two varieties of *Catharanthus roseus* under drought stress. C. R. Biol. 331, 42–47.

Jiang, M., Zhang, J., 2001. Effect of abscisic acid on active oxygen species, antioxidative defense system and oxidative damage in leaves of maize seedlings. Plant Cell Physiol. 42, 1265–1273.

Kabiri, R., Farahbakhsh, H., Nasibi, F., 2012. Salicylic acid ameliorates the effects of oxidative stress Induced by water deficit in hydroponic culture of *Nigella sativa*. Am-Euras. J. Agric. Environ. Sci. 12, 1420–1425.

Kage, H., Kochler, M., Stützel, H., 2004. Root growth and dry matter partitioning of cauliflower under drought stress conditions: measurement and simulation. Eur. J. Agron. 20, 379–394.

Kiani, S.P., Maury, P., Sarrafi, A., Grieu, P., 2008. QTL analysis of chlorophyll fluorescence parameters in sunflower (*Helianthus annuus* L.) under well-watered and water-stressed conditions. Plant Sci. 175, 565–573.

Knight, H., Knight, M.R., 2001. Abiotic stress signaling pathway: specificity and cross-talk. Trends Plant Sci. 6, 262–267.

Koag, M.C., Wilkens, S., Fenton, R.D., Resnik, J., Vo, E., Close, T.J., 2009. The K-segment of maize DHN1 mediates binding to anionic phospholipid vesicles and concomitant structural changes. Plant Physiol. 150, 1503–1514.

Kusaka, M., Ohta, M., Fujimura, T., 2005. Contribution of inorganic components to osmotic adjustment and leaf folding for drought tolerance in pearl millet. Physiol. Plant 125, 474–489.

Lawlor, D.W., Cornic, G., 2002. Photosynthetic carbon assimilation and associated metabolism in relation to water deficits in higher plants. Plant Cell Environ. 25, 275–294.

Lawson, T., Oxborough, K., Morison, J.I.L., Baker, N.R., 2003. The responses of guard and mesophyll cell photosynthesis to CO2, O2, light and water stress in a range of species are similar. J. Exp. Bot. 54, 1743–1752.

Li, B.S., Peng, X.X., Li, M.Q., 2000. Relationship between oxalate accumulation and photorespiratory glycolate metabolism in plants. Acta Phytophysiol. Sin. 26, 148–152.

Lopez-Huertas, E., Charlton, W.L., Johnson, B., Graham, I.A., Baker, A., 2000. Stress induces peroxisome biogenesis genes. EMBO J. 19, 6770–6777.

Loschen, G., Azzi, A., Richtler, C., Flohebp, L., 1973. Mitochondrial H_2O_2 formation: relationship with energy conservation. FEBS Lett. 33, 84–88.

Manivannan, P., Jaleel, A.C., Kishorekumar, A., Sankar, B., Somasundaram, R., Sridharan, R., et al., 2007a. Changes in antioxidant metabolism of *Vigna unguiculata* (L.) Walp. by propiconazole under water deficit stress. Colloids Surf. B: Biointerfaces 57, 69–74.

Manivannan, P., Jaleel, C.A., Sankar, B., Kishorekumar, A., Somasundaram, R., Alagu-Lakshmanan, G.M., et al., 2007b. Growth, biochemical modifications and proline metabolism in *Helianthus annuus* L. as induced by drought stress. Colloids Surf. B: Biointerfaces 59, 141–149.

Manivannan, P., Jaleel, C.A., Somasundaram, R., Panneerselvam, R., 2008. Osmoregulation and antioxidant metabolism in drought stressed *Helianthus annuus* under triadimefon drenching. C. R. Biol. 331, 418–425.

Martinez, J.P., Silva, H., Ledent, J.F., Pinto, M., 2007. Effect of drought stress on the osmotic adjustment, cell wall elasticity and cell volume of six cultivars of common beans (*Phaseolus vulgaris* L.). Eur. J. Agron. 26, 30–38.

Maruyama, K., Sakuma, Y., Kasuga, M., Ito, Y., Seki, M., Goda, H., et al., 2004. Identification of cold-inducible downstream genes of the *Arabidopsis* DREB1A/CBF3 transcriptional factor using two microarray systems. Plant J. 38, 982–993.

Massacci, A., Nabiev, S.M., Pietrosanti, L., Nematov, S.K., Chernikova, T.N., Thor, K., et al., 2008. Response of the photosynthetic apparatus of cotton (*Gossypium hirsutum*) to the onset of drought stress under field conditions studied by gas-exchange analysis and chlorophyll fluorescence imaging. Plant Physiol. Biochem. 46, 189–195.

Mehler, A.H., 1951. Studies of reactions of illuminated chloroplasts. I. Mechanics of reduction of oxygen and other Hill reagents. Arch. Biochem. Biophys. 33, 65–77.

Meyer, A., Hansen, D.B., Gomes, C.S.G., Hobley, T.J., Thomas, O.R.T., Franzreb, M., 2005. Demonstration of a strategy for product purification by high-gradient magnetic fishing: recovery of superoxide dismutase from unconditioned whey. Biotechnol. Prog. 21, 244–254.

Mittler, R., 2002. Oxidative stress, antioxidants and stress tolerance. Trends Plant Sci. 7, 405–410.

Mohammad, A.H., Mohammad, G.M., Masayuki, F., 2013. Cross protection by cold-shock to salinity and drought stress-induced oxidative stress in mustard (*Brassica campestris* L.) seedlings. Mol. Plant Breed. 4, 50–70.

Monneveux, P., Sánchez, C., Beck, D., Edmeades, G.O., 2006. Drought tolerance improvement in tropical maize source populations: evidence of progress. Crop Sci. 46, 180–191.

Munné-Bosch, S., 2005. The role of α-tocopherol in plant stress tolerance. J. Plant Physiol. 162, 743–748.

Nam, N.H., Chauhan, Y.S., Johansen, C., 2001. Effect of timing of drought stress on growth and grain yield of extra-short-duration pigeonpea lines. J. Agric. Sci. 136, 179–189.

Nayyar, H., Gupta, D., 2006. Differential sensitivity of C3 and C4 plants to water deficit stress: association with oxidative stress and antioxidants. Environ. Exp. Bot. 58, 106–113.

Noctor, G., Foyer, C.H., 1998. Ascorbate and glutathione: keeping active oxygen under control. Annu. Rev. Plant Physiol. Plant Mol. Biol. 49, 249–279.

Noctor, G., Gomez, L., Vanacker, H., Foyer, C.H., 2002. Interaction between biosynthesis, compartmentation and transport in the control of glutathione homeostasis and signilling. J. Exp. Bot. 53, 1283–1308.

Nohzadeh, M.S., Habibi, R.M., Heidari, M., Salekdeh, G.H., 2007. Proteomics reveals new salt responsive proteins associated with rice plasma membrane. Biosci. Biotechnol. Biochem. 71, 2144–2154.

Nordberg, J., Arner, E.S.J., 2001. Reactive oxygen species, antioxidants, and the mammalian thioredoxin system. Free Radic. Biol. Med. 31, 1287–1312.

Ogbonnaya, C.I., Sarr, B., Brou, C., Diouf, O., Diop, N.N., Roy-Macauley, H., 2003. Selection of cowpea genotypes in hydroponics, pots, and field for drought tolerance. Crop Sci. 43, 1114–1120.

Pan, X.Y., Wang, G.X., Yang, H.M., Wei, X.P., 2003. Effect of water deficits on within-plot variability in growth and grain yield of spring wheat in northwest china. Field Crop Res. 80, 195–205.

Parker, R., Flowers, T.J., Moore, A.L., Harpham, N.V.J., 2006. An accurate and reproducible method for proteome profiling of the effects of salt stress in the rice leaf lamina. J. Exp. Bot. 57, 1109–1118.

Pastori, G., Mullineaux, P., Foyer, C.H., 2000. Post-transcriptional regulation prevents accumulation of glutathione reductase protein and activity in the bundle sheath cells of maize. Plant Physiol. 122, 667–675.

Pastori, G.M., Foyer, C.H., 2002. Common components, networks, and pathways of cross-tolerance to stress. The central role of "redox" and abscisic acid-mediated controls. Plant Physiol. 129, 460–468.

Petropoulos, S.A., Daferera, D., Polissiou, M.G., Passam, H.C., 2008. The effect of water deficit stress on the growth, yield and composition of essential oils of parsley. Sci. Hort. 115, 393–397.

Prochazkova, D., Sairam, R.K., Srivastava, G.C., Singh, D.V., 2001. Oxidative stress and antioxidant activity as the basis of senescence in maize leaves. Plant Sci. 161, 765–771.

Qureshi, M.I., Qadir, S., Zolla, L., 2007. Proteomics-based dissection of stress-responsive pathways in plants. J. Plant Physiol. 164, 1239–1260.

Rai, G.K., Rai, N.P., Rathaur, S., Kumar, S., Singh, M., 2013. Expression of rd29A::AtDREB1A/CBF3 in tomato alleviates drought-induced oxidative stress by regulating key enzymatic and non-enzymatic antioxidants. Plant Physiol. Biochem. 69, 90–100.

Rasool, S., Ahmad, A., Siddiqi, T.O., Ahmad, P., 2013. Changes in growth, lipid peroxidation and some key antioxidant enzymes in chickpea genotypes under salt stress. Acta Physiol. Plant 35, 1039–1050.

Reddy, A.R., Chiatanya, K.V., Vivekanandan, M., 2004. Drought induced responses of photosynthesis and antioxidant metabolism in higher plants. J. Plant Physiol. 161, 1189–1202.

Reddy, K.J., Theriappan, P., Sreenivasulu, N., 2005. Regulation of proline biosynthesis, degradation, uptake and transport in higher plants: its implications in plant growth and abiotic stress tolerance. Curr. Sci. 88, 424–438.

Rodríguez, M., Canales, E., Hidalgo, O.B., 2005. Molecular aspects of abiotic stress in plants. Biotecnol. Apl. 22, 1–10.

Salekdeh, G.H., Siopongco, J., Wade, L.J., Ghareyazie, B., Bennett, J., 2002. Proteomic analysis of rice leaves during drought stress and recovery. Proteomics 2, 1131–1145.

Salekjalali, M., Haddad, R., Jafari, B., 2012. Effects of soil water shortages on the activity of antioxidant enzymes and the contents of chlorophylls and proteins in barley. Am-Euras. J. Agric. Environ. Sci. 12, 57–63.

Samarah, N.H., 2005. Effects of drought stress on growth and yield of barley. Agron. Sustain. Dev. 25, 145–149.

Samarah, N.H., Mullen, R.E., Cianzio, S.R., Scott, P., 2006. Dehydrin-like proteins in soybean seeds in response to drought stress during seed filling. Crop Sci. 46, 2141–2150.

Sankar, B., Jaleel, A.C., Manivannan, P., Kishorekumar, A., Somasundaram, R., Panneerselvam, R., 2007a. Effect of paclobutrazol on water stress amelioration through antioxidants and free radical scavenging enzymes in *Arachis hypogaea* L. Colloids Surf. B: Biointerfaces 60, 229–235.

Sankar, B., Jaleel, C.A., Manivannan, P., Kishorekumar, A., Somasundaram, R., Panneerselvam, R., 2007b. Drought induced biochemical modifications and praline metabolism in *Abelmoschus esculentus* (L.) Moench. Acta Bot. Croa. 66, 43–56.

Sankar, B., Jaleel, C.A., Manivannan, P., Kishorekumar, A., Somasundaram, R., Panneerselvam, R., 2008. Relative efficacy of water use in five varieties of *Abelmoschus esculentus* (L.) Moench. under water-limited conditions. Colloids Surf. B: Biointerfaces 62, 125–129.

Seki, M., Narusaka, M., Abe, H., Kasuga, M., Yamaguchi-Shinozaki, K., Carnici, P., et al., 2001. Monitoring the expression pattern of 1300 *Arabidopsis* genes under drought and cold stresses by using a full-length cDNA microarray. Plant Cell 13, 61–72.

Shao, H.B., Chu, L.Y., Shao, M.A., Jaleel, C.A., Mi, H.M., 2008. Higher plant antioxidants and redox signaling under environmental stresses. C. R. Biol. 331, 433–441.

Sharma, P., Jha, A.B., Dubey, R.S., Pessarakli, M., 2012. Reactive oxygen species, oxidative damage, and antioxidative defense mechanism in plants under stressful conditions. J. Bot. 217037, 1–26.

Sharp, R.E., LeNoble, M.E., 2002. ABA, ethylene and the control of shoot and root growth under water stress. J. Exp. Bot. 53, 33–37.

Siemens, J.A., Zwiazek, J.J., 2003. Effects of water deficit stress and recovery on the root water relations of trembling aspen (*Populus tremuloides*) seedlings. Plant Sci. 165, 113–120.

Singh, B., Usha, K., 2003. Salicylic acid induced physiological and biochemical changes in wheat seedlings under water stress. Plant Growth Regul. 39, 137–141.

Sivamani, E., Bahieldin, A., Wraith, J.M., Al-Niemi, T., Dyer, W.E., Ho, T.D., et al., 2000. Improved biomass productivity and water use efficiency under water deficit conditions in transgenic wheat constitutively expressing the barley HVA1 gene. Plant Sci. 155, 1–9.

Smirnoff, N., 2000. Ascorbic acid: metabolism and functions of a multifacetted molecule. Curr. Opin. Plant Biol. 3, 229–235.

Sofo, A., Dichio, B., Xiloyannis, C., Masia, A., 2005. Antioxidant defences in olive trees during drought stress: changes in activity of some antioxidant enzymes. Funct. Plant Biol. 32, 45–53.

Tahkokorpi, M., Taulavuori, K., Laine, K., Taulavuori, E., 2007. After-effects of drought-related winter stress in previous and current year stems of *Vaccinium myrtillus* L. Environ. Exp. Bot. 61, 85–93.

Tan, Y., Liang, Z.S., Shao, H.B., Du, F., 2006. Effect of water deficits on the activity of anti-oxidative enzymes and osmoregulation among 3 different genotypes of *Radix Astragali* at seeding stage. Biointerfaces 48, 60–65.

Taylor, N.L., Heazlewood, J.L., Day, D.A., Millar, A.H., 2005. Differential impact of environmental stresses on the pea mitochondrial proteome. Mol. Cell Prot. 4, 1122–1133.

Tsialtas, J.T., Handley, L.L., Kassioumi, M.T., Veresoglou, D.S., Gagianas, A.A., 2001. Interspecific variation in potential water use efficiency and its relation to plant species abundance in a water limited grass land. Funct. Ecol. 15, 605–614.

Tuberosa, R., Salvi, S., 2006. Genomics-based approaches to improve drought tolerance of crops. Trends Plant Sci. 11, 405–412.

Türkan, I., Bor, M., Özdemir, F., Koca, H., 2005. Differential responses of lipid peroxidation and antioxidants in the leaves of drought-tolerant *P. acutifolius* Gray and drought-sensitive *P. vulgaris* L. subjected to polyethylene glycol mediated water stress. Plant Sci. 168, 223–231.

Vera-Estrella, R., Blumwald, E., Higgins, V.J., 1992. Effect of specific elicitors of *Cladosporium fulvum* on tomato suspension cells. Evidence for the involvement of active oxygen species. Plant Physiol. 99, 1208–1215.

Vertovec, M., Sakcali, S., Ozturk, M., Salleo, S., Giacomich, P., Feoli, E., et al., 2001. Diagonosing plant water status as a tool for quantifying water stress on a regional basis in Mediterranean drylands. Ann. For. Sci. 58, 13–125.

Vranova, E., Inzé, D., Van Breusegem, F., 2002. Signal transduction during oxidative stress. J. Exp. Bot. 53, 1227–1236.

Wang, F.Z., Wang, Q.B., Kwon, S.Y., Kwak, S.S., Su, W.A., 2005. Enhanced drought tolerance of transgenic rice plants expressing a pea manganese superoxide dismutase. J. Plant Physiol. 162, 465–472.

Winston, G.W., 1990. Free radicals in cells. In: Alscher, R.G., Cumming, J.R. (Eds.), Stress Responses in Plants: Adaptation and Acclimation Mechanisms. Wiley-Liss Inc, pp. 57–86.

Wu, Q.S., Xia, R.X., Zou, Y.N., 2006. Reactive oxygen metabolism in mycorrhizal and non-mycorrhizal citrus (*Poncirus trifoliata*) seedlings subjected to water stress. J. Plant Physiol. 163, 1101–1110.

Wu, Q.S., Xia, R.X., Zou, Y.N., 2008. Improved soil structure and citrus growth after inoculation with three arbuscular mycorrhizal fungi under drought stress. Eur. J. Soil Biol. 44, 122–128.

Ye, N., Zhu, G., Liu, Y., Li, Y., Zhang, J., 2011. ABA controls H_2O_2 accumulation through the induction of *OsCATB* in rice leaves under water stress. Plant Cell Physiol. 52, 689–698.

Yoshimura, K., Masuda, A., Kuwano, M., Yokota, A., Akashi, K., 2008. Programmed proteome response for drought avoidance/tolerance in the root of a C3 xerophyte (wild watermelon) under water deficits. Plant Cell Physiol. 49, 226–241.

Yue, N., Yuping, W., Ping, L.i., Feng, Z., Heng, L., Guochang, Z., 2013. Drought stress induces oxidative stress and the antioxidant defense system in ascorbate-deficient mutants of *Arabidopsis thaliana*. Acta Physiol. Plant. 35 (4), 1189–1200.

Zhang, M., Duan, L., Zhai, Z., Li, J., Tian, X., Wang, B., et al., 2004. Effects of plant growth regulators on water deficit-induced yield loss in soybean. In: Proceedings of the Fourth International Crop Science Congress, Brisbane, Australia, 26 Sep–1 Oct 2004.

Zhang, M., Duan, L., Tian, X., He, Z., Li, J., Wang, B., et al., 2006. Unicanazole-induced tolerance of soybean to water deficit stress in relation to changes in photosynthesis, hormones and antioxidant system. J. Plant Physiol. 164, 709–717.

Zhang, X., Schmidt, R.E., 2000. Hormone-containing products' impact on antioxidant status of tall fescue and creeping bentgrass subjected to drought. Crop Sci. 40, 1344–1349.

Zhao, C.X., Yu, G.L., Jaleel, C.A., Shao, H.B., Yang, H.B., 2008. Prospects for dissecting plant-adaptive molecular mechanisms to improve wheat cultivars in drought environments. C. R. Biol. 331, 579–586.

Zhu, T., Budworth, P., Han, B., Brown, D., Chang, H.S., Zou, Z., et al., 2001. Towards elucidating global gene expression in developing *Arabidopsis*: parallel analysis of 8,300 genes. Plant Physiol. Biochem. 39, 221–242.

Chapter 12

Antioxidant Enzymes: Defense against High Temperature Stress

Rohit Joshi and Vishwanathan Chinnusamy

12.1 INTRODUCTION

Plant growth and yield are adversely affected by various abiotic stresses such as drought, salinity, extreme temperature, heavy metals and toxic chemicals (Kochhar and Kochhar, 2005). Abiotic stresses such as temperature, drought and salinity are the primary cause of crop loss worldwide, reducing average yields for major crops by more than 50% (Bray et al., 2000). High temperature is more pervasive and economically damaging, causing reduction in shoot dry mass, growth and net assimilation rates in a number of plants (Wahid et al., 2007). Globally high temperature stress (HTS) affects 7 million hectares, while terminal heat stress affects almost 40% of the irrigated wheat growing areas (Almeselmani et al., 2009). Further, the expected rise in global temperatures due to global warming will exacerbate heat stress impact on several crops in the near future. In addition, plant growth and productivity is also adversely affected due to sequential changes in morphological, physiological, biochemical and molecular properties because of HTS (Wang et al., 2001).

The optimum temperature for growth of every plant species is specific and is determined by the original temperature zone in which the plant species was evolved. Extreme temperature limits plant growth throughout its ontogeny, but the threshold level varies significantly with the developmental stages (Wahid et al., 2007). The potentially serious effects of global warming by the latter half of the 21st century are of a major concern. It is projected that global warming resulting from doubling of atmospheric concentrations of CO_2 and other greenhouse gases could jeopardize agriculture and forestry as early as 2040 AD (Grover et al., 2000). Although a lot of research has already been carried out to evaluate the effects of global warming on agriculturally important plants (Iba, 2002), practical approaches for improved HTS tolerant cultivars have only begun recently (Sharkey, 2000).

P. Ahmad (Ed): Oxidative Damage to Plants. DOI: http://dx.doi.org/10.1016/B978-0-12-799963-0.00012-5

All the organisms have an inherent ability to survive exposure to temperatures above the optimal for growth. This is called *basal thermotolerance*. Organisms acquire enhanced ability to tolerate otherwise lethal temperature if they are exposed to sublethal temperature before exposing them to lethal temperatures. This process is called *acquired thermotolerance* (Hong and Vierling, 2000). Some prokaryotic and eukaryotic thermophiles can exhibit acquired thermotolerance, tolerating even temperatures above 60°C (Hong and Vierling, 2000). Pretreatment with moderate or sublethal temperatures allows plants to acquire thermotolerance (Massie et al., 2003; Larkindale et al., 2005) to withstand supraoptimal temperatures. These plants are termed *thermotolerant* plants (Sharkey et al., 2001). Prevention of heat damage and repair of heat-sensitive components, in combination, increase the ability of plants to withstand and acclimate to supraoptimal temperatures.

The majority of lab experiments related to plant stress response to changes in environmental conditions have focused on a single stress. In contrast, a number of different stresses, e.g. drought, salinity and high temperature stresses, occur simultaneously in the field and often induce similar cellular damage (Zhu, 2002; Mittler et al., 2004). However, several studies have examined the impacts of different combinations of stresses on the growth and productivity of plants, and found higher detrimental effect of combined stresses as compared to individual stress (Rizhsky et al., 2004). For example, oxidative stress, which frequently accompanies high temperature, salinity, or drought stress, activates similar cell signaling pathways (Shinozaki and Yamaguchi-Shinozaki, 2000; Zhu, 2001, 2002) and cellular responses, e.g. production of heat-shock proteins (HSPs) (Qu et al., 2013), upregulation of antioxidants to detoxify reactive oxygen species (ROS), and accumulation of compatible solutes (Kochhar and Kochhar, 2005; Nahakpam and Shah, 2012). This chapter will focus on the role of the active oxygen scavenging systems in plants, and the physiological and biochemical processes and mechanisms that govern protective, repair and acclimation processes.

12.2 ACCLIMATIVE RESPONSE TO HEAT STRESS

High temperature stress alters membrane fluidity and permeability resulting in dysfunction of membrane-linked processes (Howarth, 2005; Larkindale et al., 2005) and can cause complete denaturation and inactivation of enzymes that lead to imbalance in metabolic pathways (Wahid et al., 2007). These adverse effects on membrane and proteins further reduce photosynthesis, impaired translocation of assimilates and reduce carbon gain, leading to altered growth and reproduction (Rizhsky et al., 2004). In addition this results in the production of reactive oxygen species that cause heat-induced oxidative stress (Panchuk et al., 2002) and promotes programmed cell death (Vacca et al., 2004).

In both monocotyledon and dicotyledon plants, HTS critically affects the floral development process and quality of seeds. In *B. napus*, 27/17°C light/dark throughout the life cycle causes total sterility, while HTS treatment, i.e. 35/15°C light/dark for a week during early flowering, reduces seed production (Young et al., 2004). Reduced gametophyte fertility or function is the major cause for reduction in seed set under HTS. Pollen viability was shown to be reduced due to HTS in *Sorghum bicolor* (L.) Moench (Prasad et al., 2006), *Lycopersicon esculentum* Mill (Singh et al., 2010), *Cicer arietinum* (Devasirvatham et al., 2012) and *Glycine* max L. Merr. (Djanaguiraman et al., 2012), whereas in *T. aestivum* (Rang et al., 2011) and *B. napus* (Hua et al., 2012), HTS leads to abnormal megagametophyte formation. The adverse effects of HTS during the reproductive stage point toward some common mechanisms involved in HTS induced reduction in seed development. Although rice is considered tolerant to high temperature, HTS causes pollen and spikelet sterility in rice. Even 1°C rise in ambient temperature for just one hour at the time of anthesis can lead to a high level of spikelet sterility (Jagadish et al., 2007).

Similarly the optimal temperature for pollen tube growth in cotton is 27 to 32°C (Kakani et al., 2005), for photosynthesis 33°C (Bibi et al., 2008) and for boll formation and fruit development 30°C (Snider et al., 2008). However, every 1°C increase in daily temperature reduces 110 kg/ha of cotton fiber (Singh et al., 2007). In the Mississippi river delta in the USA, maximum day temperatures often exceed 38°C during peak bloom, causing reduction in the number of fertilized ovules. Decreased pollen tube growth as evidenced by decline in pollen tube/ovule ratio may be a major cause of decline in yield (Rahman et al., 2004; Hua et al., 2012). Unlike drought and salinity stresses, ambient temperature change occur within hours. Therefore, plants need to respond in a very short time and the gradual increase in temperature during the day causes alterations in antioxidant metabolism and physiological responses.

12.3 REACTIVE OXYGEN SPECIES AND OXIDATIVE STRESS

Abiotic stresses including heat stress enhance the production and accumulation of reactive oxygen species leading to oxidative stress. Compounds that induce thermotolerance also induce oxidative bursts, and very short heat pulses can induce bursts of superoxide and hydrogen peroxide (Panchuk et al., 2002). Both HTS and ROS can induce heat shock proteins (HSP) in plants (Lee et al., 2000; Panchuk et al., 2002). Enhancement of the ROS detoxification system is typical of plant responses to biotic and abiotic stresses (Munné-Bosch et al., 2004).

Cellular homeostasis is disrupted by heat stress. This results in reduction of molecular oxygen by high energy electrons to form ROS (Mittler, 2002; Rizhsky et al., 2002). In a living cell ROS, e.g. $^1O_2$, H_2O_2, $O_2^{\bullet -}$ and $^{\bullet}OH$, are able to damage nearly every organic component (Snider et al., 2008). Under normal conditions, ROS are produced at low levels in chloroplasts, mitochondria and peroxisomes. However, their rate of production is dramatically increased during stress (Mittler et al., 2011). Under stress conditions ROS react directly with lipids, proteins and nucleic acids, which leads to membrane injury because of lipid peroxidation, degradation of proteins, inactivation of enzymes, bleaching of pigments and disruption of DNA strands (Apel and Hirt, 2004; Gao et al., 2008), eventually resulting in cell death (Bencze and Veisz, 2005). Limitation of CO_2 fixation along with reduction of electron transport chain is the primary cause of ROS production in chloroplast. In peroxisomes, during photorespiration, oxidation of glycolate to glyoxylic acid causes production of H_2O_2 (Mittler et al., 2004). Heat stress disrupts mitochondrial functions, which results in ROS accumulation and oxidative damage to membrane lipids (Vacca et al., 2004; Suzuki and Mittler, 2006). During heat shock, electrons from NADH produced by Krebs cycle enzymes are directed towards the reduction of O_2 to reactive oxygen intermediates (ROI) by various components of uncoupled electron transport chain (Davidson and Schiestl, 2001). This process involves suppression of pathways for production of ROI during stress and induction of various defense genes, e.g. ROI-scavenging enzymes and HSPs (Mittler, 2002).

ROS-scavenging mechanisms have an important role in protecting plants against HTS as evidenced by previous studies (Yabuta et al., 2002; Yoshimura et al., 2004). Recently, it was identified that ROS play an important signaling role in plants during growth, development, response to biotic and abiotic stresses and programmed cell death. In recent years, it became apparent that rapid production of ROS, referred to as *oxidative burst*, is an essential process, and molecular studies have shown that respiratory burst oxidase homolog (Rboh) genes, encoding NADPH oxidases, are the key producers of signal transduction-associated ROS in cells (Miller et al., 2009). The localized and temporal production and scavenging of ROS are likely to be extremely critical in the cellular and intracellular transduction of ROS signals. Accumulation of H_2O_2 was also found to be involved in stress signaling and mediating the cellular redox status (Panchuk et al., 2002). Thus two, somewhat opposing, roles of ROS underscore the need to control the steady-state level of ROS in cells, therefore providing an additional powerful strategy to enhance the tolerance of crops to these environmental stress conditions (Suzuki and Mittler, 2006).

12.3.1 Photoinhibition

Photosynthesis is one of the most susceptible physiological processes to heat stress in plants (Wahid et al., 2007). Photochemical reactions (PSII) in thylakoid lamellae are the most heat sensitive, whereas PSI activity, stromal enzymes and chloroplast envelope are relatively more thermostable (Sairam et al., 2000; Larkindale and Huang, 2004). Heat stress primarily targets the reaction center of PSII (Wang et al., 2010). Thermal denaturation of PSII was found to be linked with physical changes that occur in the lipid matrix of thylakoid membranes during HTS (such as formation of cylindrical inverted lipid micelles and increased fluidity), which can modify lipid-protein interactions (Wahid et al., 2007) resulting in conformational changes in thylakoid proteins (Mathur et al., 2011). Heat inactivation of PS II is reported to be accompanied by aggregation and subsequent dissociation of LHC II (Li et al., 2009). The lipid phase of the thylakoid membranes and the thermostability of PSII is altered by mutation in the activity of a chloroplast fatty acid desaturase and catalytic hydrogenation of thylakoids (Havaux, 1992). It was also observed that increase in temperature initially blocks the PSII reaction centers, and then dissociates antennae pigment protein complexes from the core of PSII light-harvesting apparatus (Mathur et al., 2011). It was reported previously that gradual elevation in leaf temperature (1°C min^{-1}) causes progressive rise in Chl a fluorescence intensity even with low excitation light, reaching a peak due to the dissociation between light-harvesting Chl-protein complexes and the PSII reaction centers (Havaux, 1992). Heat stress also dissociates the oxygen evolving complex (OEC), resulting in an imbalance of electron flow from OEC toward the acceptor side of PSII (Ronde et al., 2004). Heat stress also impairs other parts of the reaction center, e.g. D1 and D2 proteins (De las Rivas and Barber, 1997). Other studies also have hypothesized that heat stress inhibits electron transport at the acceptor side of PSII (Wen et al., 2005). In heated leaves, chl fluorescence increase is associated with a shift in the redox state of Q_A (Mathur et al., 2011). Direct measurements of the redox potential of Q_A provides an evidence that heat stress induces an increase in the redox potential of the Q_A/Q_A^- couple and thus electron transfer from Q_A to the secondary quinone electron acceptor of PSII (Q_B) is inhibited (Wen et al., 2005).

On the other hand, a few workers have reported that decrease in photosynthesis is linked with Calvin cycle reaction, i.e. inactivation of Rubisco activity and RuBP or Pi regeneration capacity (Salvucci and Crafts-Brandner, 2004; Mengutay et al., 2013). The thermal tolerance of photosynthesis is also influenced by changes in the leaf water potential and osmotic potential (Havaux, 1992). Taking the apparent correlation between heat tolerance of PSII and chloroplast lipid compositions together, it can be proposed that stability of PSII to HTS could be achieved via alteration in the lipid composition and degree of saturation, changes in the levels of antioxidants (Snider et al., 2010), membrane permeability, cyclic electron flow (Sharkey and Zhang, 2010) and trienoic fatty acids (Mathur et al., 2011). Several *in vitro* studies reported the occurrence of water soluble protective compounds in the chloroplast stroma under stress conditions that stabilize thylakoid membranes. Santarius and Muller (1979) found that after removing stroma from chloroplasts, the thermostability of the thylakoids from high temperature stressed leaves decreased as compared to normal plants. In addition, thylakoid membranes showed high stability when the isolated chloroplasts were heated in presence of these water soluble compounds, suggesting the importance of these compounds in leaves during heat stress (Shakeel et al., 2011).

The chloroplasts are especially prone to oxyradical production, as they have a high rate of oxygen turnover. First, activated oxygen forming superoxide is produced as a consequence of electron leakage from the normal pathway in the photosynthesis electron transport chain. Superoxide then initiates the chain reaction, generating other free radicles. This oxygenase reaction forming superoxide in a pseudocyclic electron flow is a major oxygen-consuming reaction (Asada, 1999). The Mehler-peroxidase reaction was found to increase in *Cyperus* during drought stress when the availability of CO_2 is limited (Farage et al., 2006). This suggested that the Mehler-peroxidase reaction is an important sink for excess electrons (Kramer and Evans, 2011).

12.4 ANTIOXIDANT ENZYMES: ACTIVE OXYGEN SPECIES DEFENSE SYSTEMS

Several reports showed that heat induces oxidative stress and expression of antioxidative enzymes in bacteria (Morgan et al., 1986), yeast (Davidson et al., 1996) and plants (Jaleel et al., 2007; Esfandiari et al., 2007). ROS are produced as by-products after exposure of plants to these stresses, which can damage cellular components (Xiaozhong and Huang, 2000; Polle, 2001). To protect cells from oxidative damage, plants have developed several enzymatic and nonenzymatic detoxification systems to counteract ROS (Sairam and Tyagi, 2004). Plants protect their cellular and subcellular systems from the cytotoxic effects of the ROS with protective molecules, e.g. isoprene (Young et al., 2004), glycinebetaine (Sakamoto and Murata, 2001), α-tocopherol (Vitamin E) and carotenoids, antioxidant metabolites such as ascorbic acid and glutathione (Markovska et al., 2009), and antioxidant enzymes such as dehydroascorbate reductase (Kubo et al., 1999), superoxide dismutase (SOD), ascorbate peroxidase (APX), catalase (CAT), glutathione peroxidase (GPX), glutathione reductase, and peroxiredoxin (PrxR) (Sairam et al., 2000; Shah et al., 2001; Iba, 2002; Mittler et al., 2004).

12.4.1 Nonenzymatic Defense

One of the most common responses of plants to various abiotic stresses is the accumulation of compatible organic solutes, e.g. choline O-sulfate, proline and glycine betaine (Ashraf and Fooland, 2007). Proline accumulates during several stresses and acts as a compatible solute that buffers cellular redox potential (Wahid and Close, 2007). Proline is suggested to protect plants by acting as a cellular osmotic regulating, ROS detoxifying molecule, thereby maintaining membrane integrity and stabilizing antioxidant enzymes (Joshi et al., 2013). However, overaccumulation of proline was also found to be toxic for plant cells (Rizhsky et al., 2004).

Some growth regulators have been reported to enhance heat stress tolerance (Horvath et al., 2007). Salicylic acid (SA) is reported to stabilize the trimers of heat shock transcription factors and assist them in binding to the heat shock element in the promoter of *HSP* genes (Wang et al., 2010). Changes in endogenous SA and antioxidants may be involved in heat acclimation and exogenous application of SA or acetylsalicylate was shown to enhance antioxidant capacity in rice (Sakata et al., 2010) and *Arabidopsis* (Clark et al., 2004). Photosynthetic efficiency in spring wheat, barley (El-Tayeb, 2005; Arfana et al., 2007) and *Phillyrea angustifolia* can be improved under drought stress through SA application (Munné-Bosch and Penuelas, 2003; Singh and Usha, 2003). Thiobarbituric acid reactive substances and relative electrolyte leakage was found to be reduced after exogenous application of SA in young grape leaves under heat stress, indicating that SA can induce intrinsic heat tolerance in grapevines (Wang and Li, 2006, 2007). These results suggested that SA enhances heat tolerance by reducing oxidative stress. SA also affects photochemical aspects of PSII (such as absorption, utilization as well as dissipation of excess energy) under heat stress. SA pretreatment alleviated the decrease of Pn under heat stress, mainly because of elevating rubisco activation state and PSII efficiency. In addition rapid recovery of PSII function by SA accelerated the increase of Pn (Wahid et al., 2007).

12.4.2 Enzymatic Defense

Antioxidant enzymes play a major role in detoxification of ROS during abiotic stress (Jiang and Huang, 2001; Vacca et al., 2004). Antioxidant enzymes such as POX, CAT, SOD and GR are involved in scavenging of superoxide and H_2O_2 (Mittler, 2002; Ceylan et al., 2013) in conjunction with MDAR and GR, which regenerate ascorbate (Nagesh-Babu and Devaraj, 2008). Accumulation of the ROS such as H_2O_2 under different environmental stresses results in enhanced activity of CAT and POX. High temperature stress enhanced guaiacol-POX but deactivated CAT in French bean and other crop species (Gür et al., 2010). In contrast, heat stress enhanced the activities of all the antioxidative enzymes viz. CAT, APX, SOD, POX and GR in thermotolerant genotypes of *B. juncea*. Thus, tolerant genotypes have greater expression of genes encoding antioxidant enzymes viz. SOD, CAT as well as Halliwell-Asada Pathway enzymes (DHAR, MDHAR, GR and APX). However, heat stress reduces activity of catalase and SOD in a wide range of species (Iba, 2002). Catalases are not the decisive enzyme during oxidative defense since it is not found in mitochondria and chloroplasts where most of the $O_2^{\bullet-}$ is generated (Halliwell and Gutteridge, 2006). Catalases detoxify

H_2O_2 produced by oxidase enzymes using substrates such as glycollate, urate, and D amino acids in peroxisomes (Schrader and Fahimi, 2004).

Superoxide dismutase converts $O_2^{\bullet-}$ to equally toxic H_2O_2 (Gür et al., 2010). In plants, SOD isoforms localized in different compartments within an apoplast of the cell and are differentially expressed (Wingsle et al., 1999). Animals contain MnSOD in the mitochondrial matrix and CuZnSOD in the rest of the cell, including mitochondrial intermembrane space (Halliwell, 2006). Prominent induction of cytosolic CuZnSOD was also reported under high-temperature treatment in both light and dark conditions. Plants contain almost similar enzymes, but a few of them also contain FeSOD in the chloroplast (Alscher et al., 2002). Bacteria contain MnSOD, CuZnSOD, FeSOD and sometimes NiSOD (Halliwell and Gutteridge, 2006). Some anaerobic bacteria can cope with $O_2^{\bullet-}$ using superoxide reductase, which catalyzes the overall reaction (Niviere and Fontecave, 2004). As soon as SOD converts superoxide radical to H_2O_2, either APX, CAT or POD reduces it to water and oxygen through the ascorbate-glutathione cycle, cytoplasm and other organelles, respectively (Howarth, 2005).

Plants are rich in peroxidases, which remove H_2O_2 by using it to oxidize different cosubstrates. Many plant peroxidases are nonspecific, and can use multiple cosubstrates (Mano et al., 2001). Ascorbate peroxidase (APX) is an important antioxidant enzyme of plants with different isoforms in chloroplasts, cytosol and microsomes that detoxifies H_2O_2. APX catalyzes the conversion of H_2O_2 to monodihydroascorbate (MDHA) using ascorbic acid as substrate and thus removes toxic H_2O_2 (Payton et al., 2001; Blokhina et al., 2003). The ascorbic acid is recovered by MDAR, DHAR and GR that was oxidized by APX. Heat, drought and combined stress treatments increased the transcript level of *APX* gene in tobacco plants (Rizhsky et al., 2002). Increase in APX throughout the stress period and recovery in tolerant genotypes suggests that the glutathione-ascorbate cycle is more efficient in tolerant genotype, thereby imparting tolerance to the seedlings (Pang and Wang, 2010). Heat shock transcription-factor-dependent expression of *APX* in *Arabidopsis* suggested that HSFs are involved not only in expression of *HSPs* but also in oxidative stress regulation of antioxidant gene expression (Iba, 2002).

Earlier it was thought that most important H_2O_2 scavenging enzymes in animals are glutathione peroxidases, a selenium-containing enzyme. GPX removes H_2O_2 by coupling its reduction to water with oxidation of reduced glutathione (GSH) forming oxidized glutathione (GSSG) (Brigelius-Flohe, 1999) and the reaction is reversed by the glutathione reductase enzyme. Four types of GPXs are currently reported; the first one is the classical enzyme, called GPX1. Another, GPX2, is found in the cell lining of the gastrointestinal tract where it helps to metabolize peroxides in ingested food lipids. The third form, GPX3, is found at low levels in mammalian body fluids. The fourth member is phospholipid hydroperoxide glutathione peroxidase (PHGPX or GPX4), which reduces H_2O_2 and fatty acid hydroperoxide. In contrast, due to lack of selenoprotein GPX activity in the chloroplast, APX removes entire

H_2O_2 (Halliwell, 2006). Plant GPXs contain Cys rather than seleno-Cys at the active sites, leading to the decrease in their catalytic activity. Indeed, few plant enzymes prefer thioredoxin to GSH as a substrate (Rodriguez-Milla et al., 2003). Glutathione reductase (GR) prevents the oxidation of enzymes and membranes from HTS, thereby playing an important role in the protection of plants (Almeselmani et al., 2006). Peroxiredoxins are homodimers without any prosthetic group which are realized to be important H_2O_2-scavenging systems in plants, animals and bacteria (Rhee et al., 2005). H_2O_2 oxidizes an –SH group on the peroxiredoxin to a sulfenic acid, cys-SOH, in all the cases. They are classified into three classes: typical 2-cys, atypical 2-cys and 1-cys peroxiredoxins. The 2-cys enzymes react further with another –SH on the protein to form a disulphide, which is reduced by thioredoxin. An unknown cellular reductant regenerates the –SH group in 1-cys peroxiredoxins. Peroxiredoxins were found to be weaker in H_2O_2 scavenging than GPX, although large amounts are present (up to 0.8% of total soluble protein) in some animal cells, yet they are readily inactivated by H_2O_2 (Georgiou and Masip, 2003).

12.5 ANTIOXIDANT SIGNALING: UNRAVELING THE TAPESTRY OF NETWORKS

Acquired thermotolerance is in part conferred by cellular mechanisms that prevent oxidative damage under heat stress (Bergmüller et al., 2003; Larkindale and Huang, 2004). Many ROS-scavenging enzymes showed elevated levels of transcript and protein during heat stress (Rizhsky et al., 2002; Vacca et al., 2004).

12.5.1 Second Messengers and Secondary Signals

Various second messengers involved in HTS signaling include H_2O_2 (Zhang et al., 2011), GSH and GSSG (Foyer and Shigeoka, 2011), plastoquinone (Bhattacharjee, 2012), ethylene, SA, ABA, Ca^{2+} (Larkindale and Knight, 2002; Larkindale and Huang, 2004) and ROI (de Carvalho, 2008). ROIs regulate diverse processes, such as chilling responses (Prasad et al., 1994), cross talk between different abiotic stresses (Bowler et al., 1992) and photosynthesis (Hormann et al., 1993). HTS triggers specific transient Ca^{2+} influx across the plasma membrane (Wu and Jinn, 2010). Calcium regulates enzymatic ROI-scavengers and the oxidative stress responses (Knight et al., 1996; Gong et al., 1998). Calmodulin inhibitors and calcium channel blockers induced oxidative damage to membrane and reduced survival rate after heat treatment (Saidi et al., 2009). Application of membrane fluidizers, such as benzyl alcohol, triggers an isothermal Ca^{2+} influx, *HSP* expression and thermotolerance (Suri and Dhindsa, 2008). Moreover, reducing the extracellular Ca^{2+} reversibly inhibited the heat-induced upregulation of *HSPs* and accordingly thermotolerance in *Arabidopsis*, tobacco, soybean and rice (Saidi et al., 2009; Wu and Jinn, 2010). This suggests the role of specific calcium-permeable channels in response to heat stress.

SA is also known as an important signaling component in calcium signaling, oxidative responses and systemic acquired resistance (Wahid et al., 2007). In maize, *CuZn-SOD4* and *CuZn-SOD4A* transcript levels increase with H_2O_2 pretreatment (Kernodle and Scandalios, 1996). Changes in glutathione levels regulate the expression of genes encoding enzymatic ROI-scavenging system in plants (Wingsle and Karpinski, 1996; Karpinski et al., 1997). The GR activity was increased in response to higher GSSG levels, signifying *in vivo* redox intraconversion of the enzyme. However, GSH reduces the transcript levels of cytosolic and chloroplastic CuZn-SOD. The signaling pathway is a complex network regulating expression of genes encoding enzymatic ROI-scavenging system in plant cells (Mittler et al., 2011).

ABA is a universal plant hormone that regulates plant development and environmental stress responses. ABA signaling is most critical for acquired thermotolerance, while ethylene signaling and antioxidant protection are critical for basal thermotolerance (Larkindale and Knight, 2002). Studies with the mutants of ABA-, ethylene-, ROS-, and SA-signaling pathways, and knockout mutants of respiratory burst oxidase (RbohD), revealed that control of oxidative damage is more critical for stress tolerance (Table 12.1; Miller et al., 2008). Together, these observations indicate that heat tolerance involves multiple processes, either critical during basal thermotolerance or during acquired thermotolerance. ABA plays an important role in signaling of HTS. ABA also has been shown to increase both GR and APX activities in *Arabidopsis* (Dat et al., 1998). Chimeric genes with a small HSP promoter, which acts synergistically with heat shock transcription factor 3 in *Arabidopsis*, were also induced by ABA (Larkindale et al., 2005). Heat stress-responsive signal transduction pathways and defense mechanisms, involving HSFs and HSPs, are reported to be involved in the sensing of ROS (Qu et al., 2013). Earlier studies also reported the presence of HSF-binding sequence at the promoter region of gene encoding APX1. Similarly transgenic *Arabidopsis* overexpressing HSF3 showed higher activity of APX during HTS than wild type plants (Panchuk et al., 2002). Recently, a plasma membrane ATP-binding cassette (ABC) transporter, AtABCG25, was identified in *Arabidopsis*, which exhibits ATP-dependent ABA efflux transport activity. It was shown that plants over-expressing AtABCG25 tolerate higher leaf temperatures, suggesting its role in stomatal regulation through intercellular ABA signaling pathway (Kuromori et al., 2010).

Increased heat shock tolerance can be obtained by heat-induced metabolites such as nitric oxide (NO) and hydrogen peroxide (H_2O_2) (Hua, 2009). Increase in H_2O_2 levels after heat shock has been associated with HTS-induced *HSP* expression reported in tobacco and *Arabidopsis* suspension cells. When peroxide scavengers or inhibitors of H_2O_2 generation are added during HTS, *HSP* expression is reduced in both *Arabidopsis* and tobacco (Volkov et al., 2006; Konigshofer et al., 2008; Banti et al., 2010). It was found that treatment with membrane fluidizers triggers rapid and transient elevation of cellular H_2O_2 (Konigshofer et al., 2008). H_2O_2 then might activate one or

TABLE 12.1 Heat-Stress Responsive Genes and their Involvement During Heat Stress in Mutant and Transgenic Plants

Plant	Mutant/ Transgenic	Response During Heat Stress	Reference
Arabidopsis	*apx1/ apx2*	Plants lacking APX2 produced more seeds under prolonged heat stress	Suzuki et al., 2012
	HSF3/APX2	Enhanced ROS accumulation in plants	Panchuk et al., 2002
	ACS7	Mutants exhibited enhanced tolerance heat stress	Dong et al., 2011
	CAM3	Thermotolerance significantly reduced in CAM3 mutants and rescued / enhanced in CAM3 over-expression	Zhang et al., 2009
	CAM3	AtCaM3 is involved in NO signal transduction as a downstream factor and is dependent on increased HS transcription factor DNA-binding activity and HSP accumulation	Xuan et al., 2010
	CBK3	T-DNA insertion *AtCBK3* knockout lines showed impaired basal thermotolerance, while overexpression of *AtCBK3* resulted in increased basal thermotolerance	Liu et al., 2008
	Apx1/ Apx2	H_2O_2 is an essential component in the heat stress signaling pathway and is involved in HSF activation during the early phase of heat stress	Volkov et al., 2006
	HsfA2/ HsfA1a/ HsfA1b	Heat-dependent acclimation is lost in knockout mutant as well as in double mutant HsfA1a/HsfA1b (hsfA1a/1b), indicating role of these three HSFs to confer anoxia tolerance	Banti et al., 2010
	HsfA3	HsfA3 behaves as a typical Hsf and ectopic overexpression conferred increased thermotolerance	Li et al., 2013
	HSFA2	Knockout and overexpression lines showed that HSFA2 is a regulatory component of cytoplasmic protein response	Sugio et al., 2009
	HSP90	Heat shock reduces cytosolic HSP90 activity *in vivo* and HSF is involved in the response to HSP90 inhibition	Yamada et al., 2007

(*Continued*)

TABLE 12.1 (Continued)

Plant	Mutant/ Transgenic	Response During Heat Stress	Reference
	ROF1	Exposure to heat stress induces nuclear localization of ROF1-HSP90.1 complex, which is dependent upon the presence of the transcription factor HsfA2.	Meiri and Breiman, 2009
	BOB1	BOB1 is cytoplasmic at basal temperatures, but at high temperatures it forms heat shock granules containing canonical small HSPs.	Perez et al., 2009
Alfa alfa	HAMK	Activation of HAMK by temperature is inhibited by blocking the influx of extracellular calcium and is also prevented by an antagonist of CDPKs.	Sangwan, et al., 2002
Potato	StMPK1	StMPK1 transcript levels were found to increase after a heat-shock treatment.	Blanco et al., 2006
Rice	OsMSRMK2	High temperatures differentially regulate the OsMSRMK2 mRNA accumulation in leaves of intact plants.	Agrawal, et al., 2002
	PME	Pectin methylesterase was activated in response to HS and its elevated activity was related to the recovery of the HS-released Ca^{2+} concentration.	Wu and Jinn, 2010
	AZRE	L-azetidine-2-carboxylic acid-responsive element is an alternative form of HSE, and its interaction with canonical HSEs through HSFs is required to respond to HS.	Guan et al., 2010
	SBPase	High temperature stress led to the association of SBPase with the thylakoid membranes from the stroma fractions which maintains the activation of Rubisco by preventing the sequestration of Rubisco activase to the thylakoid membranes from the soluble stroma fraction.	Feng et al., 2007
	CAM1-1	CaM1-1 isoform can interpret a Ca^{2+} signature regulating transcription of nucleoplasmic small HSP gene during heat stress.	Wu and Jinn, 2012

(*Continued*)

TABLE 12.1 (Continued)

Plant	Mutant/ Transgenic	Response During Heat Stress	Reference
Tobacco	HAMK	Heat-triggered HAMK cascade play an essential role in launching of heat shock response and hsp gene expression.	Suri and Dhindsa, 2008
	sHSPs	H_2O_2 is an early component of the heat signaling pathway and responds rapidly to changes in membrane fluidity and required for activation of sHSP synthesis.	Konigshofer et al., 2008
	HSP101	Overexpression of HSP101 showed increased thermotolerance.	Chang et al., 2007
	BADH	Overexpression led to increased thermotolerance induced by accumulation of glycinebetaine *in vivo* and is associated with the enhancement of the repair of PSII from heat-enhanced photo inhibition.	Yang et al., 2007
	GPAT	Overexpression of glycerol-3-phosphate acyltransferase enhances the stability of photosynthetic apparatus under HTS by increasing saturated extent of thylakoid membrane lipids.	Yan et al., 2008
	PanD	Expression of L-aspartate-alpha-decarboxylase resulted in increased thermotolerance.	Fouad and Rathinasabapathi, 2006
	HSF1	BhHsf1 play dual roles in providing heat stress tolerance and growth retardation by regulating target genes related to stress protection and mitotic cell cycle.	Zhu et al., 2009
Tomato	CodA	Overexpression of choline oxidase shows higher levels MT-sHSP, HSP70 and HSC70 during heat stress.	Li et al., 2011
	cAPX	Overexpression of cAPX shows enhanced resistance to heat (40°C) in transgenic plants	Wang et al., 2006
	ClpB	Antisense Lehsp100 / ClpB lines exhibited extreme repression of heat-induced expression of Lehsp100/ClpB and were greatly impaired and withered after exposing to lethal heat shock.	Yang et al., 2006

more calcium channels (Saidi et al., 2011). NO is also a secondary messenger found to be increased in tobacco and *Arabidopsis* after HTS (Xuan et al., 2010). The mutants of NO synthesis (*noa1*) display reduced basal thermotolerance (Xuan et al., 2010). However, pretreatment of sodium nitroprusside (NO donor) restored optimal thermotolerance (Saidi et al., 2011).

12.5.2 Signal Transduction

Calmodulins (CaMs) have been implicated in signaling in response to HTS (Liu et al., 2005; Zhang et al., 2009). *In vitro* addition of CaM antagonists at higher temperatures reduced the DNA-binding ability of HSFs (Li et al., 2004). This shows the downstream action of CaMs to heat-sensitive Ca^{2+} channels. Additionally, Ca^{2+} has been shown to be essential for heat activation of mitogen-activated protein (MAP) kinase (HAMK) (Suri and Dhindsa, 2008), suggesting downstream regulation of kinases through the Ca^{2+} dependent heat signaling pathway. Ca^{2+} dependent protein kinase inhibitor reduces HAMK activity and inhibitors of MAPKK repress thermotolerance (Suri and Dhindsa, 2008; Saidi et al., 2009). High temperature was shown to elevate the activity of MAPK in rice and *Solanum tuberosum* (Blanco et al., 2006; Rao et al., 2010). Various MAPKs cascades can be activated following ROS accumulation including ROS-responsive MEKK1, MPK4, MAPKKK and MPK6 (Mittler et al., 2011). The guard cells highly expressing MPK9 and MPK12 were recently found to be involved in ABA-induced stomata closure functioning downstream to ROS signals and necessary for the activation of anion channels (Jammes et al., 2009). SLAC1, a protein essential in the functioning of the guard cell anion channel, was found to be essential for stomatal closure in response to CO_2, O_3, ABA, light–dark transitions, humidity, Ca^{2+}, NO and H_2O_2 (Vahisalu et al., 2010). Thus HTS-induced calcium second messenger is perceived by calcium sensor proteins to regulate the MAPK cascade, gene expression and regulation of stomata, which is important to control the plant tissue temperatures (Fig. 12.1).

Phospholipid-based signaling is also involved in the response to heat stress (Munnik and Nielsen, 2011). The levels of PIP_2 and PA were found to be increased under heat stress (Mishkind et al., 2009). IP_3 is produced from PIP_2 and is rapidly converted to IP_6 (Qu et al., 2013), which further act to release Ca^{2+}. G protein-based signaling has also been identified as a factor for heat stress response (Tuteja and Sopory, 2008). All the three subunits of G-proteins were found to be involved in high temperature stress response (Asakura and Kurosaki, 2007; Tuteja and Sopory, 2008).

12.5.3 Transcriptional Regulation

Synthesis of HSPs in response to heat stress responses has been well studied in a wide range of organisms (Gurley, 2000; Wang et al., 2004; Kotak et al.,

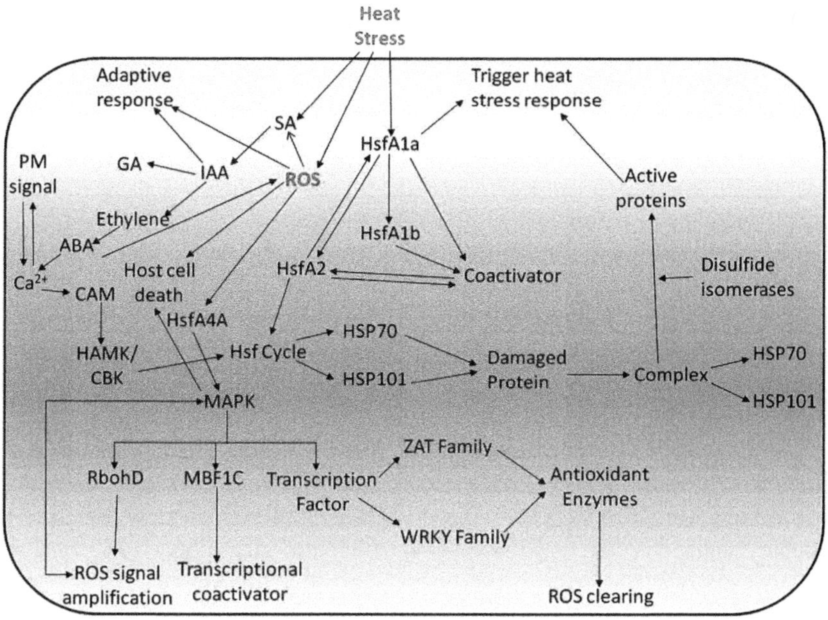

FIGURE 12.1 Schematic representation for acclimation and regulation to oxidative stress of ROS scavenging mechanism providing heat-stress and thermotolerance in plants. Heat-stress transcription factors (Hsfs) act as central regulators of the heat-stress response.

2007; Waters, 2013). A major breakthrough for HTS response was the pioneering work on *Drosophila melanogaster*. An unexpected rise in the incubator temperature led to the discovery of changing activity of gene patterns of the polytene chromosomes in the salivary glands of larva (Ritossa, 1962). After 10 years newly formed heat stress proteins were identified (Tissieres et al., 1974). At that point the field of HTS biology was rapidly linked with bacteria and other eukaryotic organisms (Forreiter and Nover, 1998; Scharf et al., 1998; Bharti and Nover, 2002). Through transcriptional profiling using microarray, a large number of genes that are induced or repressed by heat stress have been identified recently (Yeh et al., 2012).

During acclimation to heat stress, plants induce massive transcription and translations of HSPs (Wahid et al., 2012). Synthesis of HSPs is associated with acquired thermotolerance in plants and other organisms (Waters, 2013). They have now been classified into different families based on their molecular mass, and most of them act as chaperones to protect cellular proteins against irreversible heat-induced denaturation and facilitate reorganization of heat-damaged proteins (Kotak et al., 2007). All organisms produce HSPs, which can be classified into the following major families: small HSPs (Park and Bong, 2002), HSP90 (Ludwig-Muller et al., 2000) and HSP101 (Queitsch et al., 2000; Hong and Vierling, 2001), but small HSPs are abundant in plants (Wang et al., 2004). Hsp100 family proteins were found to be essential for the acquisition of thermotolerance in plants as evidenced by genetic studies. Loss-of-function mutants of HSP101 in maize (Larkindale et al., 2005) and *Arabidopsis* (Hong and Vierling, 2000, 2001) are unable to attain thermotolerance at different growth stages. A link between HSP101, mitochondria and oxidative stress in heat tolerance was unraveled by the identification of s suppressor mutant of *hot1-4* (*hsp101*) mutant. The *SUPPRESSOR OF HOT1-4* encodes a mitochondrial transcription termination factor (mTERF)-related protein. Microarray analysis revealed that expression of *HSPs* and redox-related genes increased in *shot1-2* through retrograde signaling. Further evidence suggested that HSP101 may repair proteins damaged by oxidative stress (Kim et al., 2012).

Heat shock transcription factors (HSF) are mainly involved in stress-dependent and developmental expression of *HSPs* in plants (Figure 12.1) (Huang and Xu, 2008; Mittal et al., 2011; Singh et al., 2012). Plants express more than 20 different *HSF* genes as compared to four different *HSFs* in human cells (Panchuk et al., 2002). Differences between different classes occur due to structural and functional implications. In addition to HSPs, non-HSP genes are also essential for thermotolerance. Transgenic and mutant analyses revealed that, in addition to HSPs, transcription factors such as DREB2A, bZIP28 and WRKY proteins play a key role in stress responsive transcription and heat tolerance (Todaka et al., 2012). Further, genetic engineering for compatible osmolytes, ROS detoxification, membrane-lipids saturation and polyamines also enhanced heat tolerance of plants, suggesting the involvement of diverse stress protection mechanisms in response to heat stress (Grover et al., 2013).

Cytosolic APX is induced by high temperatures and *APX1* gene induction is regulated by the *HSE* cis-elements in its promoter (Iba, 2002). *APX1* is a member of a multigene family and found to be induced by excessive light, sulfur dioxide and ozone (Panchuk et al., 2002) and heat stress (Rizhsky et al., 2002). This family also includes cytosolic isoforms, APX1 and APX2, microsomal enzyme APX3, chloroplastic stromal sAPX, and thylakoid-bound tAPX (Zhang et al., 1997). The activation of *APX2* gene expression was linked with hydrogen peroxide signaling in a number of abiotic and biotic stresses (Noctor et al., 2000). *apx1* knockout plants show higher transcript levels of a heat shock factor *HSF21*. *HSF21* was also found to be upregulated on H_2O_2 treatment in wild-type plants (Pnueli et al., 2003; Davletova et al., 2005a,b). In contrast, dominant negative variant of *HSF21* showed reduced expression of *ZAT12*, a H_2O_2 responsive zinc finger protein required for *APX1 and APX2* expression (Qu et al., 2013). This evidence indicates that HSFs are involved in the sensing and detoxification of ROS. Another stress-responsive gene, SP1, which is a homo-oligomeric protein from *Populus tremula*, has also been reported to provide stability under extreme conditions, such as boiling. SP1 was demonstrated as a molecular chaperone in protecting and repairing various heat-labile enzymes (Wang et al., 2003) and correlated with salt stress tolerance. The pace of understanding biochemical and molecular mechanisms essential for heat stress regulation is expected to increase with the dawn of functional genomics approaches.

12.6 CONCLUSIONS

Various stresses from their surrounding environment affect plants in a complex manner. In previous studies, molecular breeding approaches altered only a single character for stress tolerance by introducing a single gene in plants. Now we know that multiple factors are involved in providing tolerance over a wide range of temperatures. Thus, emphasis now should be placed on an effective approach for creating plants that can tolerate a wide range of stresses. The rapid development of functional genomics with elaborate study on transcriptomics and proteomics will facilitate the understanding of signaling pathways providing resistance to various stresses.

Most of the work discussed here clarifies the interlinkage of temperature stress and ROS signaling. The timing and localization of ROS production and scavenging during HTS play a key role in these processes. Further work is required to identify and characterize the coordination of signaling events between ROS and temperature stress. Also there is a need to clarify the role of ROS in regulating plant hormones during temperature stress.

REFERENCES

Agrawal, G.K., Rakwal, R., Iwahashi, H., 2002. Isolation of novel rice (*Oryza sativa* L.) multiple stress responsive MAP kinase gene, OsMSRMK2, whose mRNA accumulates rapidly in response to environmental cues. Biochem. Biophys. Res. Commun. 294, 1009–1016.

Almeselmani, M., Deshmukh, P.S., Sairam, R.K., Kushwaha, S.R., Singh, T.P., 2006. Protective role of antioxidant enzymes under high temperature stress. Plant Sci. 171, 382–388.

Almeselmani, M., Deshmukh, P.S., Sairam, R.K., 2009. High temperature stress tolerance in wheat genotypes: role of antioxidant defence enzymes. Acta Agron. Hung. 57, 1–14.

Alscher, R.G., Erturk, N., Heath, L.S., 2002. Role of superoxide dismutases (SODs) in controlling oxidative stress in plants. J. Exp. Bot. 53, 1331–1341.

Apel, K., Hirt, H., 2004. Reactive oxygen species: metabolism, oxidative stress, and signal transduction. Annu. Rev. Plant Biol. 55, 373–399.

Arfana, M., Atharb, H.R., Ashraf, M., 2007. Does exogenous application of salicylic acid through the rooting medium modulate growth and photosynthetic capacity in two differently adapted spring wheat cultivars under salt stress? J. Plant Physiol. 164, 685–694.

Asada, K., 1999. The water-water cycle in chloroplasts: scavenging of active oxygens and dissipation of excess photons. Annu. Rev. Plant Physiol. Plant Mol. Biol. 50, 601–639.

Asakura, Y., Kurosaki, F., 2007. Cloning and expression of Dcga encoding alpha subunit of GTP binding protein in carrot seedlings. Biol. Pharm. Bull. 30, 1800–1804.

Ashraf, M., Fooland, M.R., 2007. Roles of glycine betaine and proline in improving plant abiotic stress resistance. Environ. Exp. Bot. 59, 206–216.

Banti, V., Mafessoni, F., Loreti, E., Alpi, A., Perata, P., 2010. The heat inducible transcription factor HsfA2 enhances anoxia tolerance in *Arabidopsis*. Plant Physiol. 152, 1471–1483.

Bencze, S., Veisz, O., 2005. Resistance to environmental stress in wheat at elevated levels of CO_2. Növénytermelés 54, 467–475.

Bergmuller, E., Porfirova, E., Dormann, P., 2003. Characterization of an *Arabidopsis* mutant deficient in tocopherol methyltransferase. Plant Mol. Biol. 52, 1181–1190.

Bharti, K., Nover, L., 2002. Heat stress-induced signaling. In: Scheel, D., Wasternack, D.C. (Eds.), Plant Signal Transduction: Frontiers in Molecular Biology. Oxford University Press, Oxford, pp. 74–115.

Bhattacharjee, S., 2012. The language of reactive oxygen species signaling in plants. J. Bot., 2012, 10.1155/2012/985298.

Bibi, A.C., Oosterhuis, D.M., Gonias, E.D., 2008. Photosynthesis, quantum yield of photosystem II, and membrane leakage as affected by high temperatures in cotton genotypes. J. Cotton Sci. 1, 1–10.

Blanco, F.A., Zanetti, M.E., Casalongué, C.A., Daleo, G.R., 2006. Molecular characterization of a potato MAP kinase transcriptionally regulated by multiple environmental stresses. Plant Physiol. Biochem. 44, 315–322.

Blokhina, O., Virolainen, E., Fagerstedt, K.V., 2003. Antioxidants: oxidative damage and oxygen deprivation stress: a review. Annal. Bot. 91, 179–194.

Bowler, C., Van Montagu, M., Inzé, D., 1992. Superoxide dismutases and stress tolerance. Annu. Rev. Plant Physiol. Plant Mol. Biol. 43, 83–116.

Bray, E.A., Bailey-Serres, J., Weretilnyk, E., 2000. Responses to abiotic stresses. In: Gruissem, W., Buchannan, B.W., Jones, B.R. (Eds.), Biochemistry and Molecular Biology of Plants. American Society of Plant Biology, Rockville, MD, pp. 1158–1249.

Brigelius-Flohe, R., 1999. Tissue-specific functions of individual glutathione peroxidases. Free Radic. Biol. Med. 27, 951–965.

Ceylan, H.A., Türkan, I., Sekmen, A.H., 2013. Effect of coronatine on antioxidant enzyme response of chickpea roots to combination of PEG-induced osmotic stress and heat stress. J. Plant Gr. Reg. 32, 72–82.

Chang, C.C., Huang, P.S., Lin, H.R., Lu, C.H., 2007. Transactivation of protein expression by rice HSP101 in planta and using Hsp101 as a selection marker for transformation. Plant Cell Physiol. 48, 1098–1107.

Clark, S.M., Mur, L.A.J., Wood, J.E., Scott, I.M., 2004. Salicylic acid dependent signaling promotes basal thermotolerance but is not essential for acquired thermotolerance in *Arabidopsis thaliana*. Plant J. 38, 432–437.

Dat, J.F., Foyer, C.H., Scott, I.M., 1998. Changes in salicylic acid and antioxidants during induced thermotolerance in mustard seedlings. Plant Physiol. 118, 1455–1461.

Davidson, J.F., Schiestl, R.H., 2001. Mitochondrial respiratory electron carriers are involved in oxidative stress during heat stress in *Saccharomyces cerevisiae*. Mol. Cell Biol. 21, 8483–8489.

Davidson, J.F., Whyte, B., Bissinger, P.H., Schiestl, R.H., 1996. Oxidative stress is involved in heat-induced cell death in *Saccharomyces cerevisiae*. Proc. Natl. Acad. Sci. U.S.A. 93, 5116–5121.

Davletova, S., Rizhsky, L., Liang, H., Shengqiang, Z., Oliver, D.J., Coutu, J., et al., 2005a. Cytosolic ascorbate peroxidase 1 is a central component of the reactive oxygen gene network of *Arabidopsis*. Plant Cell 17, 268–281.

Davletova, S., Schlauch, K., Coutu, J., Mittler, R., 2005b. The zinc-finger protein Zat12 plays a central role in reactive oxygen and abiotic stress signaling in *Arabidopsis*. Plant Physiol. 133, 847–856.

de Carvalho, M.H.C., 2008. Drought stress and reactive oxygen species production, scavenging and signaling. Plant Signal Behav. 3, 156–165.

De las Rivas, J., Barber, J., 1997. Structure and thermal stability of photosystem II reaction centers studied by infrared spectroscopy. Biochem. 36, 8897–8903.

Devasirvatham, V., Gaur, P.M., Mallikarjuna, N., Raju, T.N., Trethowan, R.M., Tan, D.K.Y., 2012. Effect of high temperature on the reproductive development of chickpea genotypes under controlled environments. Func. Plant Biol. 39, 1009–1018.

Djanaguiraman, M., Prasad, P.V.V., Boyle, D.L., Schapaugh, W.T., 2012. Soybean pollen anatomy, viability and pod set under high temperature stress. J. Agro. Crop. Sci. 199, 171–177.

Dong, H., Zhen, Z., Peng, J., Chang, L., Gong, Q., Wang, N.N., 2011. Loss of ACS7 confers abiotic stress tolerance by modulating ABA sensitivity and accumulation in *Arabidopsis*. J. Expt. Bot. 62, 4875–4887.

El-Tayeb, M.A., 2005. Response of barley grains to the interactive effect of salinity and salicylic acid. Plant Growth Regul. 45, 215–224.

Esfandiari, E., Shekari, F., Shekari, F., Esfandiari, M., 2007. The effect of salt stress on antioxidant enzymes activity and lipid peroxidation on the wheat seedling. Not. Bot. Hort. Agrobot. Cluj. 35, 48–56.

Farage, P.K., Blowers, D., Long, S.P., Baker, N.R., 2006. Low growth temperatures modify the efficiency of light use by photosystem II for CO_2 assimilation in leaves of two chilling-tolerant C_4 species, *Cyperus longus* L. and *Miscanthus* × *giganteus*. Plant Cell Environ. 29, 720–728.

Feng, L.L., Wang, K., Li, Y., Tan, Y.P., Kong, J., Li, H., et al., 2007. Overexpression of SBPase enhances photosynthesis against high temperature stress in transgenic rice plants. Plant Cell Rep. 26, 1635–1646.

Forreiter, C., Nover, L., 1998. Heat stress-induced proteins and the concept of molecular chaperones. J. Biosci. 23, 287–302.

Fouad, W.M., Rathinasabapathi, B., 2006. Expression of bacterial L-aspartate-alpha decarboxylase in tobacco increases beta-alanine and pantothenate levels and improves thermotolerance. Plant Mol. Biol. 60, 495–505.

Foyer, C.H., Shigeoka, S., 2011. Understanding oxidative stress and antioxidant: functions to enhance photosynthesis. Plant Physiol. 155, 93–100.

Gao, J.G., Xiao, Q., Ding, L.P., Chen, M.J., Yin, L., Li, J.Z., et al., 2008. Differential responses of lipid peroxidation and antioxidants in *Alternanthera philoxeroids* and *Oryza sativa* subjected to drought stress. Plant Growth Regul. 56, 89–95.

Georgiou, G., Masip, L., 2003. Biochemistry: an overoxidation journey with a return ticket. Science 300, 592–594.

Gong, M., van der Luit, A., Knight, M.R., Trewavas, A.J., 1998. Heat shock induced changes in intracellular Ca^{2+} level in tobacco seedlings in relation to thermotolerance. Plant Physiol. 116, 429–437.

Grover, A., Agarwal, M., Agarwal, S.K., Sahi, C., Agarwal, S., 2000. Production of high temperature tolerant transgenic plants through manipulation of membrane lipids. Cur. Sci. 79, 557–559.

Grover, A., Mittal, D., Negi, M., Lavania, D., 2013. Generating high temperature tolerant transgenic plants: achievements and challenges. Plant Sci. 205–206, 38–47.

Guan, J.C., Yeh, C.H., Lin, Y.P., Ke, Y.T., Chen, M.T., You, J.W., et al., 2010. A 9 bp cis-element in the promoters of class I small heat shock protein genes on chromosome 3 in rice mediates Lazetidine-2-carboxylic acid and heat shock responses. J. Expt. Bot. 61, 4249–4261.

Gurley, W.B., 2000. HSP101: a key component for the acquisition of thermotolerance in plants. Plant Cell 12, 457–459.

Gür, A., Demirel, U., Özden, M., Kahraman, A., Çopur, O., 2010. Diurnal gradual heat stress affects antioxidant enzymes, proline accumulation and some physiological components in cotton (*Gossypium hirsutum* L.). Afr. J. Biotech. 9, 1008–1015.

Halliwell, B., 2006. Reactive species and antioxidants. Redox biology is a fundamental theme of aerobic life. Plant Physiol. 141, 312–322.

Halliwell, B., Gutteridge, J.M.C., 2006. Free Radicals in Biology and Medicine, fourth ed. Clarendon Press, Oxford.

Havaux, M., 1992. Stress tolerance of photosystem II *in vivo* antagonistic effects of water, heat, and photoinhibition stresses. Plant Physiol. 100, 424–432.

Hong, S.W., Vierling, E., 2000. Mutants of *Arabidopsis thaliana* defective in the acquisition of tolerance to high temperature stress. Proc. Natl. Acad. Sci. U.S.A. 97, 4392–4397.

Hong, S.W., Vierling, E., 2001. Hsp101 is necessary for heat tolerance but dispensable for development and germination in the absence of stress. Plant J. 27, 25–35.

Hormann, H., Neubauer, C., Asada, K., Schraiber, U., 1993. Intact chloroplasts display pH 5 optimum of O2-reduction in the absence of methyl viologen: indirect evidence for a regulatory role of superoxide protonation. Photosyn. Res. 37, 69–80.

Horvath, E., Szalai, G., Janda, T., 2007. Induction of abiotic stress tolerance by salicyclic acid signaling. J. Plant Growth Regul. 26, 290–300.

Howarth, C.J., 2005. Genetic improvements of tolerance to high temperature. In: Ashraf, M., Harris, M.P.J.C. (Eds.), Abiotic Stresses: Plant Resistance through Breeding and Molecular Approaches. Howarth Press Inc., New York.

Hua, J., 2009. From freezing to scorching, transcriptional responses to temperature variations in plants. Curr. Opin. Plant Biol. 12, 568–573.

Hua, S., Yu, H., Zhang, Y., Lin, B., Ding, H., Zhang, D., et al., 2012. High temperature induced fertility transition and anther carbohydrate metabolism modification in a canola recessive genic male sterile line. Af. J. Agric. Res. 7, 2475–2489.

Huang, B., Xu, C., 2008. Identification and characterization of proteins associated with plant tolerance to heat stress. J. Integr. Plant Biol. 50, 1230–1237.

Iba, K., 2002. Acclimative response to temperature stress in higher plants: approaches of gene engineering for temperature tolerance. Ann. Rev. Plant Biol. 53, 225–245.

Jagadish, S.V.K., Craufurd, P.Q., Wheeler, T.R., 2007. High temperature stress and spikelet fertility in rice. J. Exp. Bot. 58, 1627–1635.

Jaleel, C.A., Gopi, R., Manivannan, P., Panneerselvam, R., 2007. Antioxidative potentials as a protective mechanism in Catharanthus roseus (L.) G. Don. Plants under salinity stress. Tur. J. Bot. 31, 245–251.

Jammes, F., Song, C., Shin, D., Munemasa, S., Takeda, K., Gu, D., et al., 2009. MAP kinases MPK9 and MPK12 are preferentially expressed in guard cells and positively regulate ROS-mediated ABA signaling. Proc. Natl. Acad. Sci. U.S.A. 106, 20520–20525.

Jiang, Y., Huang, B., 2001. Drought and heat stress injury to two cool season turfgrasses in relation to antioxidant metabolism and lipid peroxidation. Crop Sci. 41, 436–442.

Joshi, R., Ramanarao, V.M., Baisakh, N., 2013. *Arabidopsis* plants constitutively overexpressing a myo-inositol 1-phosphate synthase gene (*SaINO1*) from the halophyte smooth cordgrass exhibits enhanced level of tolerance to salt stress. Plant Physiol. Biochem. 65, 61–66.

Kakani, V.G., Reddy, K.R., Koti, S., Wallace, T.P., Prasad, P.V.V., Reddy, V.R., et al., 2005. Differences in *in vitro* pollen germination and pollen tube growth of cotton cultivars in response to high temperature. Ann. Bot. 96, 59–67.

Karpinski, S., Escobar, C., Karpinska, B., Creissen, G., Mullineaux, P.M., 1997. Photosynthetic electron transport regulates the expression of cytosolic ascorbate peroxidase genes in *Arabidopsis* during excess light stress. Plant Cell 9, 627–640.

Kernodle, S.P., Scandalios, J.G., 1996. A comparison of the structure and function of the highly homologous maize antioxidant Cu/Zn superoxide dismutase genes, SOD4 and SOD4A. Genet. 143, 317–328.

Kim, M., Lee, U., Small, I., des Francs-Small, C.C., Vierling, E., 2012. Mutations in an *Arabidopsis* mitochondrial transcription termination factor-related protein enhance thermotolerance in the absence of the major molecular chaperone HSP101. Plant Cell 24, 3349–3365.

Knight, H., Trewavas, A.J., Knight, M., 1996. Cold calcium signaling in *Arabidopsis* involves two cellular pools and change in calcium signature after acclimation. Plant Cell 8, 489–503.

Kochhar, S., Kochhar, V.K., 2005. Expression of antioxidant enzymes and heat shock proteins in relation to combined stress of cadmium and heat in *Vigna mungo* seedlings. Plant Sci. 168, 921–929.

Konigshofer, H., Tromballa, H.W., Loppert, H.G., 2008. Early events in signaling high-temperature stress in tobacco BY2 cells involve alterations in membrane fluidity and enhanced hydrogen peroxide production. Plant Cell Environ. 31, 1771–1780.

Kotak, S., Larkindale, J., Lee, U., von Koskull-Döring, P., Vierling, E., Scharf, K.D., 2007. Complexity of the heat stress response in plants. Curr. Opin. Plant Biol. 10, 310–316.

Kramer, D.M., Evans, J.R., 2011. The importance of energy balance in improving photosynthetic productivity. Plant Physiol. 155, 70–78.

Kubo, A., Aono, M., Nakajima, N., Saji, H., Tanaka, K., Kondo, N., 1999. Differential responses in activity of antioxidant enzymes to different environmental stresses in *Arabidopsis thaliana*. J. Plant Res. 112, 279–290.

Kuromori, T., Miyaji, T., Yabuuchi, H., Shimizu, H., Sugimoto, E., Kamiya, A., et al., 2010. ABC transporter AtABCG25 is involved in abscisic acid transport and responses. Proc. Natl. Acad. Sci. U.S.A. 107, 2361–2366.

Larkindale, J., Huang, B., 2004. Thermo-tolerance and antioxidant systems in *Agrostis stoloifera*: involvement of salicylic acid, abscisic acid, calcium, hydrogen peroxide, and ethylene. J. Plant Physiol. 161, 405–413.

Larkindale, J., Knight, M.R., 2002. Protection against heat stress-induced oxidative damage in *Arabidopsis* involves calcium, abscisic acid, ethylene, and salicylic acid. Plant Physiol. 128, 682–695.

Larkindale, J., Hall, J.D., Knight, M.R., Vierling, E., 2005. Heat stress phenotypes of *Arabidopsis* mutants implicate multiple signaling pathways in the acquisition of thermotolerance. Plant Physiol. 138, 882–897.

Lee, B.H., Won, S.H., Lee, H.S., Miyao, M., Chung, W.I., Kim, I.J., et al., 2000. Expression of the chloroplast-localized small heat shock protein by oxidative stress in rice. Gene. 245, 283–290.

Li, B., Liu, H.T., Sun, D.Y., Zhou, R.G., 2004. Ca^{2+} and calmodulin modulate DNA-binding activity of maize heat shock transcription factor *in vitro*. Plant Cell Physiol. 45, 627–634.

Li, S., Li, F., Wang, J., Zhang, W.E.N., Meng, Q., Chen, T.H.H., et al., 2011. Glycinebetaine enhances the tolerance of tomato plants to high temperature during germination of seeds and growth of seedlings. Plant Cell Environ. 34, 1931–1943.

Li, Z., Wakao, S., Fischer, B.B., Niyogi, K.K., 2009. Sensing and responding to excess light. Ann.u Rev. Plant Biol. 60, 239–260.

Li, Z., Zhang, L., Wang, A., Xu, X., Li, J., 2013. Ectopic overexpression of SlHsfA3, a heat stress transcription factor from tomato, confers increased thermotolerance and salt hypersensitivity in germination in transgenic *Arabidopsis*. PLoS ONE 8, e54880.

Liu, H.T., Un, D.Y., Zhou, R.G., 2005. Ca^{2+} and AtCaM3 are involved in the expression of heat shock protein gene in *Arabidopsis*. Plant Cell Environ. 28, 1276–1284.

Liu, H.T., Gao, F., Li, G.L., Han, J.L., Liu, D.L., Sun, D.Y., et al., 2008. The calmodulin-binding protein kinase 3 is part of heat-shock signal transduction in *Arabidopsis thaliana*. Plant J. 55, 760–773.

Ludwig-Muller, J., Krishna, P., Forreiter, C., 2000. A glucosinolate mutant of *Arabidopsis* is thermosensitive and defective in cytosolic Hsp90 expression after heat stress. Plant Physiol. 123, 949–958.

Mano, J., Ohno, C., Domae, Y., Asada, K., 2001. Chloroplastic ascorbate peroxidase is the primary target of methyl viologen-induced photooxidative stress in spinach leaves: its relevance to monodehydroascorbate radical detected with *in vivo* ESR. Biochem. Biophys. Acta. 504, 275–287.

Markovska, Y.K., Gorinova, N.I., Nedkovska, M.P., Miteva, K.M., 2009. Cadmium-induced oxidative damage and antioxidant responses in *Brassica juncea* plants. Biol. Plant 53, 151–154.

Massie, M.R., Lapoczka, E.M., Boggs, K.D., Stine, K.E., White, G.E., 2003. Exposure to the metabolic inhibitor sodium azide induces stress protein expression and thermotolerance in the nematode *Caenorhabditis elegans*. Cell Stress Chap. 8, 1–7.

Mathur, S., Allakhverdiev, S.I., Jajoo, A., 2011. Analysis of high temperature stress on the dynamics of antenna size and reducing side heterogeneity of Photosystem II in wheat leaves (*Triticum aestivum*). Biochim. Biophys. Acta. 1807, 22–29.

Meiri, D., Breiman, A., 2009. *Arabidopsis* ROF1 (FKBP62) modulates thermotolerance by interacting with HSP90.1 and affecting the accumulation of HsfA2-regulated sHSPs. Plant J. 59, 387–399.

Mengutay, M., Ceylan, Y., Kutman, U.B., Cakmak, I., 2013. Adequate magnesium nutrition mitigates adverse effects of heat stress on maize and wheat. Plant Soil. 368, 57–72.

Miller, G., Shulaev, V., Mittler, R., 2008. Reactive oxygen signaling and abiotic stress. Physiol. Plant 133, 481–489.

Miller, G., Schlauch, K., Tam, R., Cortes, D., Torres, M.A., Shulaev, V., et al., 2009. The plant NADPH oxidase RBOHD mediates rapid systemic signaling in response to diverse stimuli. Sci. Signal 2, ra45.

Mishkind, M., Vermeer, J.E.M., Darwish, E., Munnik, T., 2009. Heat stress activates phospholipase D and triggers PIP2 accumulation at the plasma membrane and nucleus. Plant J. 60, 10–21.

Mittal, D., Enoki, Y., Lavania, D., Singh, A., Sakurai, H., Grover, A., 2011. Binding affinities and interactions among different heat shock element types and heat shock factors in rice (*Oryza sativa* L.). FEBS J. 278, 3076–3085.

Mittler, R., 2002. Oxidative stress, antioxidants and stress tolerance. Trends Plant Sci. 7, 405–410.

Mittler, R., Vanderauwera, S., Gollery, M., Van Breusegem, F., 2004. Reactive oxygen gene network of plants. Trends Plant Sci. 9, 490–498.

Mittler, R., Vanderauwera, S., Suzuki, N., Miller, G., Tognetti, V.B., Vandepoele, K., et al., 2011. ROS signaling: the new wave? Trends Plant Sci. 16, 300–309.

Morgan, R.W., Christman, M.F., Jacobson, F.S., Stroz, G., Ames, B.N., 1986. Hydrogen peroxide-inducible proteins in *Salmonella typhimurium* overlap with heat shock and other stress proteins. Proc. Natl. Acad. Sci. U.S.A. 83, 8059–8063.

Munné-Bosch, S., Penuelas, J., 2003. Photo- and antioxidative protection, and a role for salicylic acid during drought and recovery in field-grown *Phillyrea angustifolia* plants. Planta. 217, 758–766.

Munné-Bosch, S., Peñuelas, J., Asensio, D., Llusia, J., 2004. Airborne ethylene may alter antioxidant protection and reduce tolerance of Holm oak to heat and drought stress. Plant Physiol. 136, 2937–2947.

Munnik, T., Nielsen, E., 2011. Green light for polyphosphoinositide signals in plants. Curr. Opin. Plant Biol. 14, 489–497.

Nagesh-Babu, R., Devaraj, V.R., 2008. High temperature and salt stress response in French bean (*Phaseolus vulgaris*). Aust. J. Crop. Sci. 2, 40–48.

Nahakpam, S., Shah, K., 2012. Expression of key antioxidant enzymes under combined effect of heat and cadmium toxicity in growing rice seedlings. Plant Growth Regul. 63, 23–35.

Niviere, V., Fontecave, M., 2004. Discovery of superoxide reductase: an historical perspective. J. Biol. Inorg. Chem. 9, 119–123.

Noctor, G., Veljovic-Jovanovic, S., Foyer, C.H., 2000. Peroxide processing in photosynthesis: antioxidant coupling and redox signalling. Phil. Trans. R. Soc. Lond. B. 355, 1465–1475.

Panchuk, I.I., Volkov, R.A., Schöffl, F., 2002. Heat-stress and heat-shock transcription factor-dependent expression and activity of ascorbate peroxidase in *Arabidopsis*. Plant Physiol. 129, 838–853.

Pang, C.H., Wang, B.S., 2010. Role of ascorbate peroxidase and glutathione reductase in ascorbate–glutathione cycle and stress tolerance in plants. In: Anjum, N.A., Chan, M.T., Umar, S. (Eds.), Ascorbate–Glutathione Pathway and Stress Tolerance in Plants. Springer, Dordrecht, pp. 91–112.

Park, S.M., Bong, H.C., 2002. Class I small heat-shock protein gives thermotolerance in tobacco. J. Plant Physiol. 159, 25–30.

Payton, P., Holaday, A.S., Allen, R.D., 2001. Antioxidant systems and environmental stress tolerance in cotton. In: Jenkins, J.N., Saha, S. (Eds.), Genetic Improvement of Cotton Emerging Technologies. Science Publishers, Enfield, NH, USA, pp. 227–237.

Perez, D.E., Hoyer, J.S., Johnson, A.I., Moody, Z.R., Lopez, J., Kaplinsky, N.J., 2009. BOBBER1 is a non-canonical *Arabidopsis* small heat shock protein required for both development and thermotolerance. Plant Physiol. 151, 241–252.

Pnueli, L., Liang, H., Rozenberg, M., Mittler, R., 2003. Growth suppression, altered stomatal responses, and augmented induction of heat shock proteins in cytosolic ascorbate peroxidase (Apx1) -deficient *Arabidopsis* plants. Plant J. 34, 187–203.

Polle, A., 2001. Dissecting the superoxide dismutase-ascorbate glutathione pathway in chloroplasts by metabolic modeling. Computer simulations as a step towards flux analysis. Plant Physiol. 126, 445–462.

Prasad, P.V.V., Boote, K.J., Allen, L.H., Sheehy, J.E., Thomas, J.M.G., 2006. Species, ecotype and cultivar differences in spikelet fertility and harvest index of rice in response to high temperature stress. Field Crops Res. 95, 398–411.

Prasad, T.K., Anderson, M.D., Martin, B.A., Stewart, C.R., 1994. Evidence for chilling-induced oxidative stress in maize seedlings and regulatory role of hydrogen peroxide. Plant Cell 6, 65–74.

Qu, A.L., Ding, Y.F., Jiang, Q., Zhu, C., 2013. Molecular mechanisms of the plant heat stress response. Biochem. Biophys. Res. Commun. 432, 203–207.

Queitsch, C., Hong, S., Vierling, E., Lindquist, S., Hong, S.W., 2000. Heat shock protein 101 plays a crucial role in thermotolerance in *Arabidopsis*. Plant Cell 12, 479–492.

Rahman, H., Malik, S.A., Saleem, M., 2004. Heat tolerance of upland cotton during the fruiting stage evaluated using cellular membrane thermostability. Field Crops Res. 85, 149–158.

Rang, Z.W., Jagadish, S.V.K., Zhou, Q.M., Craufurd, P.Q., Heuer, S., 2011. Effect of high temperature and water stress on pollen germination and spikelet fertility in rice. Environ. Exp. Bot. 70, 58–65.

Rao, K.P., Vani, G., Kumar, K., Wankhede, D.P., Mishra, M., Gupta, M., et al., 2010. Arsenic stress activates MAP kinase in rice roots and leaves. Arch. Biochem. Biophys. 506, 73–82.

Rhee, S.G., Chae, H.Z., Kim, K., 2005. Peroxiredoxins: a historical overview and speculative preview of novel mechanisms and emerging concepts in cell signaling. Free Radic. Biol. Med. 38, 1543–1552.

Ritossa, F., 1962. A new puffing pattern induced by heat shock and DNP in *Drosophila*. Experientia 18, 571–573.

Rizhsky, L., Hongjian, L., Mittler, R., 2002. The combined effect of drought stress and heat shock on gene expression in tobacco. Plant Physiol. 130, 1143–1151.

Rizhsky, L., Liang, H., Shuman, J., Shulaev, V., Davletova, S., Mitler, R., 2004. When defense pathways collide: the response of *Arabidopsis* to a combination of drought and heat stress. Plant Physiol. 134, 1683–1696.

Rodriguez-Milla, M.A., Maurer, A., Huete, A.R., Gustafsonm, J.P., 2003. Glutathione peroxidase genes in *Arabidopsis* are ubiquitous and regulated by abiotic stresses through diverse signaling pathways. Plant J. 36, 602–615.

Ronde, J.A.D.D., Cress, W.A., Kruger, G.H.J., Strasser, R.J., Staden, J.V., 2004. Photosynthetic response of transgenic soybean plants containing an *Arabidopsis* P5CR gene, during heat and drought stress. J. Plant Physiol. 61, 1211–1244.

Saidi, Y., Finka, A., Muriset, M., Bromberg, Z., Weiss, Y.G., Maathuis, F.J., et al., 2009. The heat shock response in moss plants is regulated by specific calcium-permeable channels in the plasma membrane. Plant Cell 21, 2829–2843.

Saidi, Y., Finka, A., Goloubinoff, P., 2011. Heat perception and signaling in plants: a tortuous path to thermotolerance. New Phytol. 190, 556–565.

Sairam, R.K., Tyagi, A., 2004. Physiological and molecular biology of salinity stress tolerance in deficient and cultivated genotypes of chickpea. Plant Growth Regul. 57, 109–114.

Sairam, R.K., Srivastava, G.C., Saxena, D.C., 2000. Increased antioxidant activity under elevated temperature: a mechanism of heat stress tolerance in wheat genotypes. Biol. Plant. 43, 245–251.

Sakamoto, A., Murata, N., 2001. The use of bacterial choline oxidase, a glycinebetaine-synthesizing enzyme, to create stress resistant transgenic plants. Plant Physiol. 125, 180–188.

Sakata, T., Oshino, T., Miura, S., Tomabechi, M., Tsunaga, Y., Higashitani, N., et al., 2010. Auxins reverse plant male sterility caused by high temperatures. Proc. Natl. Acad. Sci. U.S.A. 107, 8569–8574.

Salvucci, M.E., Crafts-Brandner, S.J., 2004. Inhibition of photosynthesis by heat stress: the activation state of Rubisco as a limiting factor in photosynthesis. Physiol. Plant 120, 179–186.

Sangwan, V., Orvar, B.L., Beyerly, J., Hirt, H., Dhindsa, R.S., 2002. Opposite changes in membrane fluidity mimic cold and heat stress activation of distinct plant MAP kinase pathways. Plant J. 31, 629–638, 2002.

Santarius, K.A., Muller, M., 1979. Investigations on heat resistance of spinach leaves. Planta 146, 529–538.

Scharf, K.-D., Höhfeld, I., Nover, L., 1998. Heat stress response and heat stress transcription factors. J. Biosci. 23, 313–329.

Schrader, M., Fahimi, H.D., 2004. Mammalian peroxisomes and reactive oxygen species. Histochem. Cell Biol. 122, 383–393.

Shah, K., Kumar, R.G., Verma, S., Dubey, R.S., 2001. Effect of cadmium on lipid peroxidation, superoxide anion generation and activities of antioxidant enzymes in growing rice seedlings. Plant Sci. 161, 1135–1144.

Shakeel, S., Haq, N.U., Heckathorn, S.A., Hamilton, E.W., Luthe, D.S., 2011. Ecotypic variation in chloroplast small heat-shock proteins and related thermotolerance in *Chenopodium album*. Plant Physiol. Biochem. 49, 898–908.

Sharkey, T.D., 2000. Some like it hot. Sci. 287, 435–437.

Sharkey, T.D., Zhang, R., 2010. High temperature effects on electron and proton circuits of photosynthesis. J. Int. Plant Biol. 52, 712–722.

Sharkey, T.D., Chen, X.Y., Yeh, S., 2001. Isoprene increases thermotolerance of fosmidomycin-fed leaves. Plant Phyisol. 125, 2001–2006.

Shinozaki, K., Yamaguchi-Shinozaki, K., 2000. Molecular responses to dehydration and low temperature: differences and cross-talk between two stress signaling pathways. Curr. Opin. Plant Biol. 3, 217–223.

Singh, A., Mittal, D., Lavania, D., Agarwal, M., Mishra, R.C., Grover, A., 2012. OsHsfA2c and OsHsfB4b are involved in the transcriptional regulation of cytoplasmic OsClpB (Hsp100) gene in rice (*Oryza sativa* L.). Cell Stress Chaperones 17, 243–254.

Singh, B., Usha, K., 2003. Salicylic acid induced physiological and biochemical changes in wheat seedlings under water stress. Plant Growth Regul. 39, 137–141.

Singh, R., Singh, S., Cheema, D.S., Dhaliwal, M.S., 2010. Effect of high temperature on pollen viability and reproductive organs of tomato (*Lycopersicon esculentum* Mill). Crop Improv. 37, 209.

Singh, R.P., Prasad, P.V.V., Sunita, K., Giri, S.N., Reddy, K.R., 2007. Influence of high temperature and breeding for heat tolerance in cotton: a review. In: Spark Ed., D.L. (Ed.), Advances in Agronomy 93. Academic Press, California, pp. 315–385.

Snider, J.L., Oosterhuis, D.M., Skulman, B.M., Kawakami, E.M., Storch, D.K., 2008. The effect of high temperature on *in vivo* pollen tube growth, calcium levels, antioxidant response, and superoxide production in the cotton pistil. In: Oosterhuis, D.M. (Ed.), Summaries of Arkansas Cotton Research. pp. 46–51.

Snider, J.L., Oosterhuis, D.M., Kawakami, E.M., 2010. Genotypic differences in thermotolerance are dependent upon pre-stress capacity for antioxidant protection of the photosynthetic apparatus in *Gossypium hirsutum*. Physiol. Plant. 138, 268–277.

Sugio, A., Dreos, R., Aparicio, F., Maule, A.J., 2009. The cytosolic protein response as a subcomponent of the wider heat shock response in *Arabidopsis*. Plant Cell 21, 642–654.

Suri, S.S., Dhindsa, R.S., 2008. A heat-activated MAP kinase (HAMK) as a mediator of heat shock response in tobacco cells. Plant Cell Environ. 31, 218–226.

Suzuki, N., Mittler, R., 2006. Reactive oxygen species and temperature stresses: a delicate balance between signaling and destruction. Physiol. Plant. 126, 45–51.

Suzuki, N., Miller, G., Sejima, H., Harper, J., Mittler, R., 2012. Enhanced seed production under prolonged heat stress conditions in *Arabidopsis thaliana* plants deficient in cytosolic ascorbate peroxidase. J. Exp. Bot. 64, 253–263.

Tissieres, A., Mitchell, H.K., Tracy, U.M., 1974. Protein synthesis in salivary glands of *D. melanogaster*. Relation to chromosome puffs. J. Mol. Biol. 84, 389–398.

Todaka, D., Nakashima, K., Shinozaki, K., Yamaguchi-Shinozaki, K., 2012. Toward understanding transcriptional regulatory networks in abiotic stress responses and tolerance in rice. Rice 5, 6.

Tuteja, N., Sopory, S.K., 2008. Plant signaling in stress: G–protein coupled receptors, heterotrimeric. G–proteins and signal coupling via phospholipases. Plant Signal Behav. 3, 79–86.

Vacca, R.A., de Pinto, M.C., Valenti, D., Passarella, S., Marra, E., De Gara, L., 2004. Production of reactive oxygen species, alteration of cytosolic ascorbate peroxidase, and impairment of mitochondrial metabolism are early events in heat shock-induced programmed cell death in tobacco Bright-Yellow 2 cells. Plant Physiol. 134, 1100–1112.

Vahisalu, T., Puzõrjova, I., Brosché, M., Valk, E., Lepiku, M., Moldau, H., et al., 2010. Ozone-triggered rapid stomatal response involves the production of reactive oxygen species, and is controlled by SLAC1 and OST1. Plant J. 62, 442–453.

Volkov, R.A., Panchuk, I.I., Mullineaux, P.M., Schöffl, F., 2006. Heat stress induced H_2O_2 is required for effective expression of heat shock genes in *Arabidopsis*. Plant Mol. Biol. 61, 733–746.

Wahid, A., Close, T.J., 2007. Expression of dehydrins under heat stress and their relationship with water by cerium based in a salt-adapted cell line of *Pisum sativum*. J. Exp. Bot. 48, 1529–1535.

Wahid, A., Gelani, S., Ashraf, M., Foolad, M.R., 2007. Heat tolerance in plants: an overview. Environ. Exp. Bot. 61, 199–223.

Wahid, A., Farooq, M., Hussain, I., Rasheed, R., Galani, S., 2012. Responses and management of heat stress in plants. In: Ahmad, P., Prasad, M.N.V. (Eds.), Environmental Adaptations and Stress Tolerance of Plants in the Era of Climate Change. Springer, NewYork, USA, pp. 135–157.

Wang, L.J., Li, S.H., 2006. Salicylic acid-induced heat or cold tolerance in relation to Ca^{2+} homeostasis and antioxidant systems in young grape plants. Plant Sci. 170, 685–694.

Wang, L.J., Li, S.H., 2007. The effects of salicylic acid on distribution of ^{14}C assimilation and photosynthesis in young grape plants under heat stress. Acta. Hort. 738, 779–7851.

Wang, L.J., Fan, L., Loescher, W., Duan, W., Liu, G.J., Cheng, J.S., et al., 2010. Salicylic acid alleviates decreases in photosynthesis under heat stress and accelerates recovery in grapevine leaves. BMC. Plant Biol. 10, 34–44.

Wang, W., Vinocur, B., Shoseyov, O., Altman, A., 2004. Role of plant heat-shock proteins and molecular chaperones in the abiotic stress response. Trends Plant Sci. 9, 244–252.

Wang, W.X., Vinocur, B., Shoseyov, O., Altman, A., 2001. Biotechnology of plant osmotic stress tolerance: physiological and molecular considerations. Acta Hort. 560, 285–292.

Wang, W.X., Barak, T., Vinocur, B., Shoseyov, O., Altman, A., 2003. Abiotic resistance and chaperones: possible physiological role of SP1, a stable and stabilizing protein from Populus. In: Vasil, I.K. (Ed.), Plant Biotechnology 2000 and Beyond. Kluwer, Dordrecht, pp. 439–443.

Wang, Y., Wisniewski, M., Meilan, R., Cui, M., Fuchigami, L., 2006. Transgenic tomato (*Lycopersicon esculentum*) overexpressing cAPX exhibits enhanced tolerance to UV-B and heat stress. J. Appl. Hort. 8, 87e90.

Waters, E.R., 2013. The evolution, function, structure, and expression of the plant sHSPs. J. Exp. Bot. 64, 391–403.

Wen, X.G., Gong, H.M., Lu, C.M., 2005. Heat stress induces a reversible inhibition of electron transport at the acceptor side of photosystem II in a cyanobacterium *Spirulina platensis*. Plant Sci. 168, 1471–1476.

Wingsle, G., Karpinski, S., 1996. Differential redox regulation of glutathione reductase and CuZn-superoxide dismutase gene expression in *Pinus sylvestris* L. needles. Planta. 198, 151–157.

Wingsle, G., Karpiński, S., Hällgren, J.E., 1999. Low temperature, high light stress and antioxidant defence mechanisms in higher plants. Phyton-Ann. Rei. Bot. 39, 253–268.

Wu, H.C., Jinn, T.L., 2010. Ethylesterase activity and cytosolic Ca^{2+} oscillation are crucial for plant thermotolerance. Plant Signal. Behav. 5, 1252–1256.

Wu, H.C., Jinn, T.L., 2012. Oscillation regulation of Ca^{2+}/calmodulin and heat-stress related genes in response to heat stress in rice (*Oryza sativa* L.) Plant Signal. Behav. 7, 1056–1057.

Xiaozhong, L., Huang, B., 2000. Heat stress injury in relation to membrane lipid peroxidation in creeping bentgrass. Crop Sci. 40, 503–510.

Xuan, Y., Zhou, S., Wang, L., Cheng, Y., Zhao, L., 2010. Nitric oxide functions as a signal and acts upstream of AtCaM3 in thermotolerance in *Arabidopsis* seedlings. Plant Physiol. 153, 1895–1906.

Yabuta, Y., Motoki, T., Yoshimura, K., Takeda, T., Ishikawa, T., Shigeoka, S., 2002. Thylakoid membrane-bound ascorbate peroxidase is a limiting factor of antioxidative systems under photo-oxidative stress. Plant J. 32, 915–925.

Yamada, K., Fukao, Y., Hayashi, M., Fukazawa, M., Suzuki, I., Nishimura, M., 2007. Cytosolic HSP90 regulates the heat shock response that is responsible for heat acclimation in *Arabidopsis thaliana*. J. Biol. Chem. 282, 37794–37804.

Yan, K., Chen, N., Qu, Y.Y., Dong, X.C., Meng, Q.W., Zhao, S.J., 2008. Overexpression of sweet pepper glycerol-3-phosphate acyltransferase gene enhanced thermotolerance of photosynthetic apparatus in transgenic tobacco. J. Integr. Plant Biol. 50, 613–621.

Yang, J.Y., Sun, Y., Sun, A.Q., Yi, S.Y., Qin, J., Li, M.H., et al., 2006. The involvement of chloroplast HSP100/ClpB in the acquired thermotolerance in tomato. Plant Mol. Biol. 62, 385–395.

Yang, X., Wen, X., Gong, H., Lu, Q., Yang, Z., Tang, Y., et al., 2007. Genetic engineering of the biosynthesis of glycinebetaine enhances thermotolerance of photosystem II in tobacco plants. Planta. 225, 719–733.

Yeh, C.H., Kaplinsky, N.J., Hu, C., Charng, Y.Y., 2012. Some like it hot, some like it warm: phenotyping to explore thermotolerance diversity. Plant Sci. 195, 10–23.

Yoshimura, K., Miyao, K., Gaber, A., Takeda, T., Kanaboshi, H., Miyasaka, H., et al., 2004. Enhancement of stress tolerance in transgenic tobacco plants overexpressing *Chlamydomonas* glutathione peroxidase in chloroplasts or cytosol. Plant J. 37, 21–33.

Young, L.W., Wilen, R.W., Bonham-Smith, P.C., 2004. High temperature stress of *Brassica napus* during flowering reduces micro- and mega gametophyte fertility, induces fruit abortion, and disrupts seed. J. Exp. Bot. 55, 485–495.

Zhang, H., Wang, J., Nickel, U., Allen, R., Goodman, H., 1997. Cloning and expression of an *Arabidopsis* gene encoding a putative peroxisomal ascorbate peroxidase. Plant Mol. Biol. 34, 967–971.

Zhang, L., Jiang, H., Gao, X., Zou, Y., Liu, M., Liang, Y., et al., 2011. Heat shock transcription factor-1 inhibits H_2O_2-induced apoptosis via down-regulation of reactive oxygen species in cardiac myocytes. Mol. Cell Biochem. 347, 21–28.

Zhang, W., Zhou, R.G., Gao, Y.J., Zheng, S.Z., Xu, P., Zhang, S.Q., et al., 2009. Molecular and genetic evidence for the key role of AtCaM3 in heat-shock signal transduction in *Arabidopsis*. Plant Physiol. 149, 1773–1784.

Zhu, J.K., 2001. Cell signaling under salt, water and cold stresses. Curr. Opin. Plant Biol. 4, 401–406.

Zhu, J.K., 2002. Salt and drought stress signal transduction in plants. Annu. Rev. Plant Biol. 53, 247–273.

Zhu, Y., Wang, Z., Jing, Y., Wang, L., Liu, X., Liu, Y., et al., 2009. Ectopic overexpression of BhHsf1, a heat shock factor from the resurrection plant *Boea hygrometrica*, leads to increased thermotolerance and retarded growth in transgenic *Arabidopsis* and tobacco. Plant Mol. Biol. 71, 451–467.

Chapter 13

Reactive Oxygen Species and Antioxidants in Response to Pathogens and Wounding

Fakiha Afzal, Rabia Khurshid, Muhammad Ashraf and Alvina Gul Kazi

13.1 INTRODUCTION

Unlike other eukaryotic organisms which are mobile and have complex locomotory organs to move from one place to another in order to find food, shelter and protection, plants are confined to their growth place and they lack intricate and sophisticated immune and locomotory systems. In order to overcome these deficiencies, plants have evolved a broad range of mechanisms which not only protect them against pathogens, but also help them fight abiotic stresses such as drought, high salinity, frost, dehydration, metal toxicity, ultraviolet type B radiation, chemicals such as herbicides, pesticides and fungicides, ozone, extremely high or low temperature, air pollutants, topography, wounding and hypoxia (Yana et al., 2013). These mechanisms help in evading pathogens and in turn help plants to grow healthier. Plants do so by producing ROS (reactive oxygen species) naturally through oxygen metabolism. Related terminologies used are ROI (reactive oxygen intermediates) or RNI (reactive nitrogen intermediates) (Sharma et al., 2012). These are chemically active molecules containing nascent oxygen and are biochemically very reactive. These are not only involved in cell defense but also in intra- and extracellular cell signaling and maintaining homeostasis (Ali and Alqurainy, 2006). During exogenous stress to the plant body, their concentration in the body dramatically rises in order to eradicate the stress. But their higher concentration can be damaging to the plant cells themselves, and therefore plants have evolved various complex mechanisms to keep the concentrations of ROS under balance by strict surveillance (Ozgur et al., 2013). Reactive oxygen intermediates

P. Ahmad (Ed): Oxidative Damage to Plants. DOI: http://dx.doi.org/10.1016/B978-0-12-799963-0.00013-7

(ROI) are by-products of electron reduction of water including hydroxyl ion (OH^-), superoxide anion radical, and hydrogen peroxide (H_2O_2). ROS includes:

1. Reactive oxygen intermediates
2. Ozone
3. Singlet oxygen species

Due to oxidation of halides, catalyzed by an enzyme called peroxidase, HOCl, HOBr and HOI are produced. All of these are also included in ROS (also sometimes referred to as ROI). Some nitrogen intermediates that are also very reactive influence the levels of ROI nitrite, nitric oxide radical, peroxynitrite and nitric oxide radical (Choudhury and Panda, 2013). Stresses to plants lead to significant crop losses. Pathogens are both crop and location specific. Research studies are ongoing in order to improve crop yield in areas that are affected more easily by such stresses (Ali and Alqurainy, 2006). Chloroplasts, mitochondria and peroxisomes are the key players in providing defense to plants against ROS by the production of several enzymatic and non-enzymatic antioxidants to scavenge ROS (Gill and Tuteja, 2010). A balance between ROS and antioxidants is created as the complete eradication of ROS means the loss of an important second messenger in intracellular signaling cascades. Whenever this balance is disturbed, it leads to oxidative stress (Tuteja, 2007; Tuteja, 2010; Ahmad et al., 2010a, b, 2011; Ahmad and Umar, 2011; Koyro et al., 2012; Khan and Singh, 2008; Dalton et al., 1999).

13.2 REACTIVE OXYGEN SPECIES

Plants have successfully evolved their defense mechanisms by the clever use of reactive oxygen species (ROS), the by-products of cell metabolism. Certain signaling pathways exponentially increase the amount of ROS that helps plants fight against infection and stressful conditions. This exponential increase in ROS is termed the *oxidative burst*. The positive power of ROS has been discovered in recent years along with the fact that ROS in low concentrations is crucial in certain vital pathways in plants. The estimates tell us that out of all the O_2 absorbed, around 1% of it is diverted to produce ROS (Sharma et al., 2012; Tuteja, 2010).

Superoxide anion ($O_2^{\bullet -}$), hydrogen peroxide (H_2O_2) and hydroxyl radical ($^{\bullet}OH$) are the major forms of ROS produced during photosynthesis and glycolysis normally and aid in certain signaling pathways. However, their excess in the cells can have devastating effects if not scavenged properly. These effects include: lipid peroxidation, protein oxidation, nucleic acid damage, and programmed cell death activation (Sharma and Dubey, 2007). Higher concentrations of ROS result from the phenomenon of oxidative burst, which is a major defense strategy for plants.

13.2.1 ROS and Signaling

ROS, as discussed earlier, is a role player in several different cell functions, including:

1. Cell growth
2. Pathogen recognition (Apel and Hirt, 2004)
3. Response to biotic and abiotic stresses (Laloi et al., 2007; Miller et al., 2007)
4. Establishment of symbiotic relationship between legume and rhizobia (Rubio et al., 2004)
5. Endo- and ectomycorrhiza formation (Fester and Hause, 2005; Baptista et al., 2007)

As a secondary messenger, Table 13.1 illustrates the further roles of ROS as described in Sharma et al., 2012.

13.2.2 ROS Gene Network

The production of ROS is governed by a large network of genes. In *Arabidopsis*, this network consists of a total of 150 genes (Mittler et al., 2004). A number of proteins are also involved in ROS production pathways as regulatory units. ROS signaling is controlled through transient production and continuous scavenging. Both environmental and developmental factors contribute to the modulation of ROS.

13.2.3 ROS Generators and Scavengers

NADPH oxidases (NOx/RBOH) generate ROS. Major ROS scavengers include glutatione peroxidase, certain phenolic compounds, peroxiredoxins and thioredoxins (Margis et al., 2008; Alkhalfioui et al., 2008; Rouhier and Jacquot, 2002).

TABLE 13.1 ROS in Lower Concentrations as Second Messenger Signaling Molecule

Hormones Activated By ROS	Plant Responses After Hormonal Activation
Auxin	Root gravitropism
Abscisic acid	Stomata closure (Pei et al., 2000)
Gibberelic acid	Programmed cell death (Gechev and Hille, 2005)
Jasmonic acid	Lignin biosynthesis
Salicylic acid	Hypersensitive response, Osmotic stress

TABLE 13.2 Modes of ROS Production in Plant Cell Organelles

Cell Organelle	Modes of ROS Production
Chloroplast	PSII: electron transport chain
	Fd, 2Fe-2S, and 4Fe-4S clusters
	PSI: electron transport chain
	QA and QB
	Chlorophyll pigments
Mitochondria	Complex I: NADH dehydrogenase segment
	Complex II: reverse electron flow to complex I
	Complex III: ubiquinone-cytochrome region
Enzymes	Aconitase, 1-galactono-γ
	lactone, dehydrogenase (GAL)
Cell wall	Cell-wall-associated peroxidase
	diamine oxidases
Peroxisomes	Matrix: xanthine oxidase (XOD)
	Membrane: electron transport chain flavoprotein
	NADH and Cyt b
	Metabolic processes: glycolate oxidase,
	fatty acid oxidation, flavin oxidases,
	disproportionation of $O_2^{\bullet-}$
	radicals
Endoplasmic Reticulum	NAD(P)H-dependent
	electron transport
	involving Cyt P450
Apoplast	Cell-wall-associated
	oxalate oxidase
	Amine oxidases

In addition, Class III peroxidases, which are specific to plants, are both producers and scavengers of ROS (Passardi et al., 2004). There are certain modes of production of ROS in plant cell organelles as tabulated in Table 13.2.

13.2.4 Plant Pathogen Interaction and the Oxidative Burst

Whenever there is pathogen invasion, there is increase in ROS production at the site of pathogen interaction. Superoxide ($O_2^{\bullet -}$) and hydrogen peroxide (H_2O_2) are produced primarily (Apel and Hirt, 2004). The generation of $O_2^{\bullet -}$ has been observed in several plant–pathogen interactions including: avirulent bacteria, fungi, and viruses (Low and Merida, 1996).

The response to avirulent pathogens includes a two-phase ROS production episode. The first phase is low powered and the second phase is high powered and includes higher ROS accumulation, which ensues before the onsets of hypersensitive response thus triggering the oxidative burst (Lamb et al., 1997; Torres et al., 2006). It leads to programmed cell death (PCD), recognition of pathogen and incompatible interface (Mehdy, 1994; Levine et al., 1996).

13.2.5 ROS Production

By-products of various metabolic processes in cells and electron leakage from electron transport chain, mitochondria, plasma membranes and chloroplast cause the production of ROS in plant cells (Foyer and Lelandais, 1996), Fig. 13.1.

13.2.5.1 Singlet Oxygen ($^1O_2$)

The singlet oxygen can be formed by:

1. Absorption of sufficient energy by the oxygen atom
2. Monovalent reduction of oxygen

O_2 is a stable molecule with two free electrons that encircle the nucleus in parallel fashion in the same direction. But, at high energy states, one of the electrons starts rotating in reverse, making both the electrons revolve in opposite directions. This $^1O_2$ form can participate in divalent reduction, or the transfer of two electrons simultaneously (Apel and Hirt, 2004).

Also, in photosystem II, through triplet chlorophyll formation, $^1O_2$ that is highly reactive is produced in the antenna and the reaction system (Krieger-Liszkay, 2005). This involves the formation of triplet state chlorophyll by the dissipation of chlorophyll from the photosystem during the process of photosynthesis (Sharma et al., 2012). The triplet chlorophyll molecule (^{3}Chl) reacts with $^3O_2$ and the $^1O_2$ ROS is liberated. The reaction thus occurs in the following way:

$$\text{Chl (in the presence of light)} \rightarrow {}^3\text{Chl} + {}^3O_2 \rightarrow \text{Chl} + {}^1O_2 \text{ (highly destructive ROS)}$$

The closing of stomata due to some abiotic stresses, such as drought and salinity, causes further formation of $^1O_2$. This molecule has 3μs or even less within the cell (Hatz et al., 2007; Hackbarth et al., 2010). Even a small fraction of this $^1O_2$ can easily diffuse up to a distance of several hundreds of nanometers. $^1O_2$ as a destructive element causes oxidation of the vital

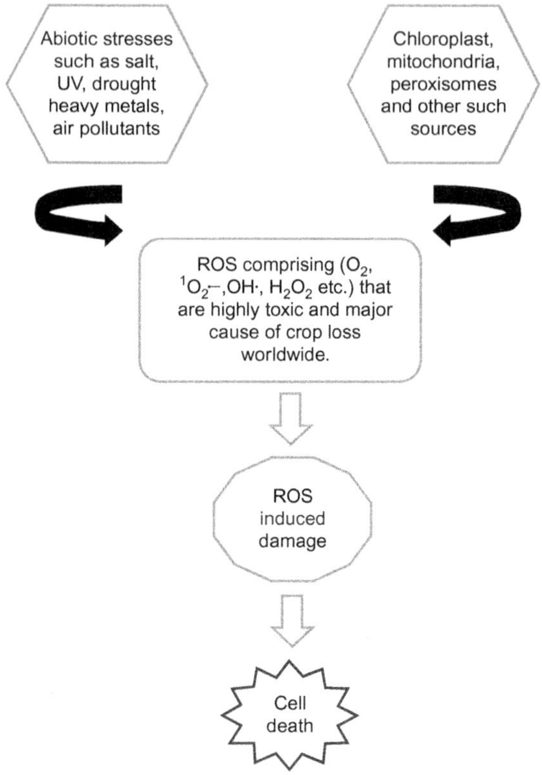

FIGURE 13.1 The causative agents include both abiotic and biotic stresses.

biomolecules such as proteins, DNA and unsaturated fatty acids (Wagner et al., 2004; Ahmad et al., 2011). In DNA, it reacts with and causes modification in the nucleic acid, deoxyguanosine (Kasai, 1997; Tuteja et al., 2009). $^1O_2$ is also considered to be the ROS that causes loss of photosystem II, which is light induced, and in turn the loss of photosystem II may ultimately cause cell death (Krieger-Liszkay et al., 2008). Scavengers of $^1O_2$ include α-tocopherol and β-carotene (Krieger-Liszkay, 2005).

13.2.5.2 The Superoxide ($O_2^{\bullet -}$)

It is formed with the input of some energy. The oxygen that is produced in the chloroplasts during normal photosynthesis can act as an electrophile, accepting the electrons from photosystem II and forming superoxide (Singh and Tuteja, 2010). O_2 is reduced in steps and the intermediates include the primary ROS molecule; the superoxide $O_2^{\bullet -}$. O_2 undergoes reduction, gaining one electron and turning to $O_2^{\bullet -}$. This superoxide then moves on to generate the secondary ROS either through catalysis by a metal or an enzyme

TABLE 13.3 Major and Minor Sources of H_2O_2

Major Sources of H_2O_2	Minor Sources of H_2O_2
Chloroplast	NADPH oxidase
Mitochondria	Photoxidation reactions
Endoplasmic Reticulum	Xanthin oxidase
Plasma membrane	
Fatty acid β-oxidation	
Photorespiration	

depending on the location inside the cell (Valko et al., 2005). This superoxide is a nucleophilic molecule with a short half-life of 1 μs and it can both oxidize and reduce and has a moderate reactive nature. The superoxide, in a slow-proceeding reaction with little yield, reacts with H_2O_2 in the Haber-Weiss reaction and in turn the most reactive ROS, the hydroxyl radical ($^\bullet OH$), is formed.

13.2.5.3 Hydrogen Peroxide (H_2O_2)

Both abiotic and biotic stress leads to the formation of H_2O_2 in cells, just like all other important ROS. Apart from that, it is naturally required and thus prepared by plant tissue. It is formed when the superoxide radical undergoes reduction (univalent). H_2O_2 has the ability to react with the thiol group of enzymes and deactivate them (Tewari et al., 2006). Its half-life is longer (1 ms) as compared to other ROS molecules (Bhattachrjee, 2005). It performs both the roles as a signaling molecule, occurring at lower concentrations, and as an important ROS in triggering cell death at higher concentrations (programmed cell death) (Quan et al., 2008). There are major and minor sources of hydrogen peroxide, as given in Table 13.3.

13.2.5.3.1 Role of H_2O_2 in Plant Defense

Hydrogen peroxide induces system-acquired resistance (SAR) that is triggered after localized exposure to pathogens (Chen et al., 1993). SAR resembles the innate immune response in animals. There are certain pathogen recognition receptors (PRR) that recognize pathogen-associated molecular patterns (PAMPS) and, thus, the defense strategies are launched:

1. Helps in hypersensitive cell-death response (HR) (Tenhaken et al., 1995; Levine et al., 1994).

2. Oxidative cross-linking of glycoproteins of the cell wall – helps them overcome the enzyme degradation by pathogens (Bradley et al., 1992).
3. Toxicity towards pathogens (Peng and Kuc, 1992).

13.2.5.4 Hydroxyl Radical (OH•)

Known to be the most reactive ROS, the hydroxyl radical is produced when superoxide and hydrogen peroxide react in the presence of Fe or some other transitional metal, pH being neutral. The reaction that is triggered by the superoxide anion is called the *Fenton reaction*:

$$H_2O_2 + O_2(\text{in the presence of } Fe^{2+} \text{and } Fe^{3+}) \rightarrow OH^- + O_2 + OH\bullet$$

Hydroxyl radical cannot be eliminated from the cells due to the lack of enzymatic action. Thus, if produced in alarming quantities, this ROS is lethal to cells (Vranova et al., 2002).

13.2.6 Vitality of ROS in Plant Defense

The initial response is the most important against pathogens. It is mostly this initial response that determines how badly the plant will be affected and how quickly it will recover from the attack (Gayoso et al., 2010). The more efficient the defense at the initial stage of the pathogen invasion, the more easily the establishment of infection can be avoided (Ferreira et al., 2006). Certain defense strategies are controlled by the pathogenesis genes: the PR genes. Pathogen recognition by the pathogen recognition receptors (PRRs) leads to PR gene upregulation (Albrecht & Bowman, 2008). The oxidative burst has been found in studies to have an important role to play during plant-pathogen interactions and wound healing. ROS has been shown to help in the formation of certain barriers of a physical nature in the plant cell wall that involve phenolic polymers, glycoproteins, callose and lignin (Lamb & Dixon, 1997; Huckelhoven and Kogel, 2003). Lignin in tissues provides a strong physical barrier in plants and thus increases resistance in plants (Vilanova et al., 2013). The phenylpropanoid pathway is an important pathway that occurs during lignification of tissues and is triggered by either biotic and abiotic stresses or stimuli (Vogt, 2010).

13.3 ANTIOXIDANTS IN RESPONSE TO PATHOGEN AND WOUNDING

Reactive oxygen species produced as an exposure to harsh conditions such as drought, extreme temperature variations, pollutants, heavy metals and deficiency of nutrients, are very toxic and destructive to the plant cells and cause oxidative damage; therefore a defense mechanism is employed by the plant body, such as the antioxidative mechanism (Ahmad et al., 2010a, b,

2011; Ahmad and Umar, 2011; Koyro et al., 2012; Choudhury and Panda, 2013). Enzymatic antioxidants in plants are:

1. Dehydroascorbate reductase (DHAR)
2. Glutathione-S-transferase (GST)
3. Superoxide dismutase (SOD)
4. Glutathione peroxidase (GPX)
5. Monodehydro ascorbate reductase (MDHAR)
6. Catalase (CAT)
7. Guaicol peroxidase (GOPX)
8. Ascorbate peroxidase (APX)
9. Glutathione reductase (GR)

Their expression is controlled at the genetic level and regulation of the genes is done according to the need for the removal of ROS in cells. Both environmental and developmental stimuli are involved in the regulation (Gara et al., 2003). There are also different isozymes which are associated with different cellular compartments. Nonenzymatic antioxidants include:

1. Ascorbic acid (AsA)
2. Glutathione (GSH)
3. Flavonoids, which are phenolic compounds
4. Carotenoids (Car)
5. Nonprotein amino acids
6. Vitamin E (α-tocopherols)

All of these compounds help plants against ROS by scavenging them. Concentration of nonenzymatic antioxidants is high as compared to the enzymatic ones (Ali and Alqurainy, 2006; Gill & Tuteja, 2010; Tuteja 2007).

There can be more than one antioxidant present in one cellular location, as in cytosol there are a minimum of three of the ROS scavengers: catalase, ascorbate peroxidase, glutathione peroxidase (Nobuhiro and Mittler, 2006). It has been found that plants possessing antioxidant activity have tolerance to different kind of stresses so that plants are able to grow in semifatal environments (Fecht-Christoffers et al., 2003). In order to elucidate and explore the positive role of antioxidants in plants under stress, several transgenic lines have been established and characterized (Sarowar et al., 2005).

In making a plant resistant to pathogenic infections, the following are the requirements: (a) pathogen identification, (b) activation of complex metabolic pathways in the infected cells, (c) propagation impedance within plant tissues (Gara et al., 2003). Whichever mechanism is opted for, the final aim is to block the penetration of the pathogen into the plant tissues, without damage. The basal response is the hypersensitivity response (HR) in which certain genes are switched on near the position of pathogen penetration which encode proteins specific for pathogenesis, especially antimicrobial phytoalexin and hydrolytic enzymes (Dixon et al., 1994). After this, certain events such as programmed cell death (PCD) are activated. A large amount of ROS production is a crucial event in HR that involves cellular antioxidative machinery of the plant.

To study the threshold levels between antioxidants and ROS, much research work is being done that suggests that both of them are strictly regulated. This has been further investigated and confirmed by transgenesis (Wang et al., 2010). It is still debatable whether activation of ROS system scavenging antioxidants is a result to evade pathogens or a reaction against oxidative stress. Some of these responses are common but some of them are species-specific.

Certain antioxidants are termed "elicitors" and can be used to make plants pathogen tolerant (Thakur and Sohal, 2013). As mitochondria is a known source of ROS, therefore evidence has showed that it is a target of biotic stress and tobacco plant inoculated with pathogenic *Pseudomonas syringae* showed a rapid burst of superoxide from mitochondria and showed increased antioxidants (Cvetkovska & Vanlerberghe, 2013). Major antioxidants and their cellular localizations are mentioned in Table 13.4.

13.4 ENZYMATIC ANTIOXIDANTS IN PLANT-PATHOGEN INTERACTION

13.4.1 Superoxide Dismutase

Superoxide dismutase (SOD) is a metalloenzyme, ubiquitous and the most effective intracellular antioxidant in all aerobic organisms. Its upregulation is directly related to fighting against oxidative stresses which are triggered by both biotic and abiotic stress. SOD gives the first line of defense for plants and plays a crucial role in plant survival under stress, making them stress tolerant (Ahmad et al., 2010a, b, 2011; Ahmad and Umar, 2011; Koyro et al., 2012; Farhoudi et al., 2012). It catalyzes the following reaction:

$$O_2{\bullet}^- + O_2{\bullet}^- + 2H^+ \longrightarrow 2H_2O_2 + O_2$$

One of the oxygen radicals is reduced to hydrogen peroxide by catalyzing dismutation and the second one is oxidized to oxygen, thus evading the ROS. In this way $O_2^{\bullet -}$ is removed, which otherwise would form $^{\bullet}OH$ (Haber-Weiss reaction). The rate of this reaction is 10,000 times quicker than spontaneous dismutation. On the basis of their metal cofactors, SOD is classified into three types: (a) iron (Fe-SOD), (b) manganese (Mn-SOD) and (c) copper/zinc SOD (Cu/Zn-SOD) and their cellular localization is different (Gupta et al., 1993; Ahmad et al., 2010a, b) and described in Table 13.5. In the genome of *Arabidopsis thaliana* there are three iron SOD genes (FSD1, FSD2 and FSD3), three copper/zinc SOD genes (CSD1, CSD2 and CSD3), and one manganese SOD gene (MSD1). Every isozyme

TABLE 13.4 Major ROS Scavengers and Their Cellular Localization

Major ROS Scavengers	Abbreviation	Cellular Localization
Superoxide dismutase	SOD	Chloroplasts,
		Cytosol,
		Mitochondria,
		Peroxisomes,
		Apoplasts
Ascorbate peroxidase	APX	Chloroplasts,
		Cytosol,
		Mitochondria,
		Peroxisomes,
		Apoplasts
Catalase	CAT	Peroxisomes
Glutathione peroxidase	GPX	Cytosol
Ascorbic acid	ASH	Chloroplasts,
		Cytosol,
		Mitochondria,
		Peroxisomes,
		Apoplast
Glutathione	GSH	Chloroplasts,
		Cytosol,
		Mitochondria,
		Peroxisomes,
		Apoplast
α-Tocopherol	–	Membranes

is nuclear encoded and after formation moves to its destined localization by a peptide leader sequence to a specific location in the cell. Many transgenic plants have been produced having upregulated SOD activities, including all three isozymes, and showed higher stress tolerance towards variable stresses (Jomova et al., 2012).

TABLE 13.5 Isozymes of SOD and Their Cellular Localization

Isozymes of SOD	Cellular Localization
Copper/Zinc SOD	Chloroplast and Cytosol
Iron SOD	Chloroplast
Manganese SOD	Mitochondria and Peroxisomes

13.4.2 Catalase

Catalase (CAT) is one of the absolute important tetrameric enzymes for detoxifying ROS. It dismutates hydrogen peroxide produced by peroxisomes by the following reaction:

$$H_2O_2 \longrightarrow H_2O + 1/2O_2$$

In maize, CAT has three isozymes located in different cellular compartments. Research studies have investigated that CAT is also involved in various other reactions with hyper peroxides and methyl hydrogen peroxide is one of them (Ali et al., 2006). It has been seen that the gene responsible for encoding CAT makes the plant resistant to various stresses such as salt stress (Nagamiya et al., 2007), metal stress (Azpilicueta et al., 2007), pathogenic stress (Mittler et al., 2004), or drought stress (Sharma and Dubey, 2005). Different transgenic plants have been produced with increased CAT activities and have showed good results against stress and postharvest physiological deterioration (PPD), which is caused by oxidative burst (Xu et al., 2013).

13.4.3 Ascorbate Peroxidase

Ascorbate peroxidase (APX) is found in all advanced plants and algae. Its role in foraging ROS and defending cells is never ignored. It uses ascorbate (ASH) as an electron donor and scavenges hydrogen peroxide in ascorbate glutathione (ASH/GSH) cycle. It catalyzes the following reaction:

$$H_2O_2 + AA \longrightarrow 2H_2O + DHA$$

There are four main isoforms of APX family. (a) Thylakoid APX (tAPX), (b) glyoxisome membrane forms (gmAPX), (c) chloroplast stromal soluble form (sAPX), (d) cytosolic form (cAPX) (Ahmad et al., 2011; Ahmad and Umar, 2011). As compared to catalase, it is more vital in stress as it has high affinity for hydrogen peroxide. Transgenic studies showed its role in different kind of stresses. Enhanced tolerance against salt, metal and drought stress tolerance has been seen in transgenic *Ceratophyllum demersum* (Arvind and Prasad, 2003), *Nicotiana tabacum* (Badawi et al., 2004), rice (Yang et al., 2009), *P. vulgaris* and *P. asperata* (Yang et al., 2008) expressing high amounts of APX. Transgenic studies have showed that APX presence does not confer resistance in the plant to many of the pathogens, especially to *Pseudomonas syringae.* Some plants developed weak resistance to certain pathogens such as *Ralstonia solanacearum.* It was also shown that plants with APX activity become more fungus resistant due to increased POD activity.

13.4.4 Glutathione Reductase (GR), Glutathione S-transferases (GST), and Glutathione Peroxidase (GPX)

GR (flavo-protein), GST and GPX are involved in defense against ROS in plants and prokaryotes. GR is mainly present in chloroplast but a small amount is present in mitochondria and cytosol also. GR is mainly involved in reduction of glutathione (GSH). Both of them are important for plant survival. GR is more important towards abiotic stress than biotic. It catalyzes the following reaction:

$$\mathrm{GSSG} + \mathrm{NAD(P)}H \longrightarrow 2\mathrm{GSH} + \mathrm{NAD(P)}^{+}$$

GST is involved in the conjugation reaction between the substrates of xenobiotic and GSH. GSH is involved in many essential plant functions such as (a) enhancing tolerance against various biotic and abiotic stresses, (b) maintaining hormonal homeostasis in plants, (c) detoxification of herbicides and hydro peroxide, (d) impounding anthocyanin in the plant vacuole, (e) tyrosine metabolism. An amazing role of GST in the plant body is to detoxify those compounds which are involved in damaging DNA, RNA or proteins. It has also been observed that GST turns out to be a negative regulator, as its silencing makes a plant resistant to pathogen (Dixon et al., 2010).

$$\mathrm{RX} + \mathrm{GSH}\{\mathrm{ReversReact}\}\mathrm{HX} + \mathrm{R} - \mathrm{S} - \mathrm{GSH}$$

where R = aliphatic, aromatic or heterocyclic group, X = sulfate, nitrite or halide group. GPXs scavenge hydrogen peroxide and hydro peroxides. It is a diverse family, having many isozymes. Its overexpression is related to stress tolerance (Noctor et al., 2002).

13.4.5 Dehydroascorbate Reductase

In the presence of glutathione (GSH), which acts as a reducing agent, dehydroascorbate (DHA) (Foyer and Mullineaux, 1998) is reduced to ascorbate (AsA) catalyzed by dehydroascorbate reductase (DHAR). Thus DHAR is directly involved in keeping AsA in its reduced form (Sharma et al., 2012).

$$2GSH + DHA\{ReversReact\}GSSG + AsA$$

Its activity in plants is related to increasing stress tolerance such as to temperature, salinity and drought (Slooten et al., 1995; Kubo et al., 1999). In *L. japonicas* gene upregulation associated with DHAR production is found and it was seen to be more salt tolerant than other leguminous plants (Rubio et al., 2009) but it has also been reported that certain stresses are not associated with increased DHAR activity in plants (Tanaka et al., 1991).

13.4.6 Monodehydroascorbate Reductase

Monodehydroascorbate reductase (MDHAR) is the major constituent of the ASH/GSH cycle. It is an FAD (falvin dinucleotide) enzyme. This enzyme is highly specific for the substrate MDHA (monodehydro ascorbate), which acts as an electron acceptor. Usually NADH acts as an electron donor but MDAR prefers NADPH for this (Gill and Tuteja, 2010).

$$NADH + 2MDHA \rightleftharpoons NAD^+ + 2AA(\text{where AA} = \text{electron donor})$$

MDHAR isozymes are located in various cellular compartments as mentioned in Table 13.2. Many transgenic plants have been produced having increased MDHAR activity and have showed tolerance against various stresses (Boo and Jung, 1999; Mittler et al., 2004; Sharma and Dubey, 2007; Maheshwari and Dubey, 2009).

13.5 NONENZYMATIC ANTIOXIDANTS

13.5.1 Ascorbic Acid

Ascorbic acid, vitamin C or simply ascorbate (AsA) is the most renowned, water soluble, powerful, low molecular weight and the most abundant ROS scavenger present in plants (Smirnoff 1996, 2000; Arrigoni and De, 2000; Horemans et al., 2000; Ahmad et al., 2010a, b, 2011; Gill and Tuteja, 2010; Ahmad and Umar, 2011; Koyro et al., 2012; Sharma et al., 2012). A convincing argument can be made for the importance of ascorbate from its involvement in many vital physiological processes in plants, such as metabolism, cell growth, cell differentiation and the antioxidant system. It is undoubtedly a very abundant molecule, as its presence has been confirmed in many plant tissues and cells including meristems, photosynthetic cells, certain fruits, cytosol, chloroplast stroma and apoplast. Under unstressed and normal physiological circumstances, ascorbate is found in a reduced form in leaves chloroplast stroma, which forms 90% of its pool (Smirnoff, 2000). In cytosol its concentration is about 20 mM while in chloroplast it can be around 20–300 mM (Foyer and Lelandais, 1996; Foyer and Noctor, 2005) showing that chloroplast and stroma has its highest concentration around

30–40% of the total ascorbate (Gill and Tuteja, 2010). Ascorbate metabolism involves the following key players:

1. Mitochondria play the major part in ASH synthesis (Shao et al., 2008). In fact they are not only involved in its synthesis but also in its regeneration. ASH is synthesized in mitochondria by L-galactono-γ-lactone dehydrogenase (Szarka et al., 2007; Shao et al., 2008). By facilitated diffusion and electrochemical gradient (positive), it is transported to other cellular localizations (Smirnoff et al., 2004).

2. Certain uronic acid intermediates are also involved in the synthesis of ascorbic acid (Isherwood et al., 1954) by the pathway depicted in Fig. 13.2.

Being a universal antioxidant, ascorbate's main functions are highlighted as:

1. It is involved in the reduction of hydrogen peroxide into water in the presence of ascorbate peroxidase (Noctor and Foyer, 1998).
2. As a ROS scavenger, it directly hunts for OH^-, OH^- and O_2^- (Noctor and Foyer, 1998).
3. For the enzyme violaxantin de-epoxidase, which is present in chloroplast, it acts as a cofactor, in a way supporting extra excitation energy dissipation (Smirnoff et al., 2004).
4. Ascorbate is also involved in membrane protection by regenerating tocopherol from tocopheroxyl radical (Thomas et al., 1992).
5. It plays a vital role in cell division (Smirnoff, 1996).
6. It regulates the cell cycle (Liso et al., 1988).
7. It also helps in cell elongation (De Tullio et al., 1999).
8. It is involved in the reduction of ferryl leghaemoglobin and ferric leghaemoglobin (Moreau et al., 1995).
9. Concentration of ascorbic acid in the nodules of legumes is approximately 1–2 mM (Dalton et al., 1986); thus it is involved in the effectiveness of legumes (Dalton et al., 1993).
10. Ascorbate also helps in hydroxylation of proline and various other vital processes in plants.

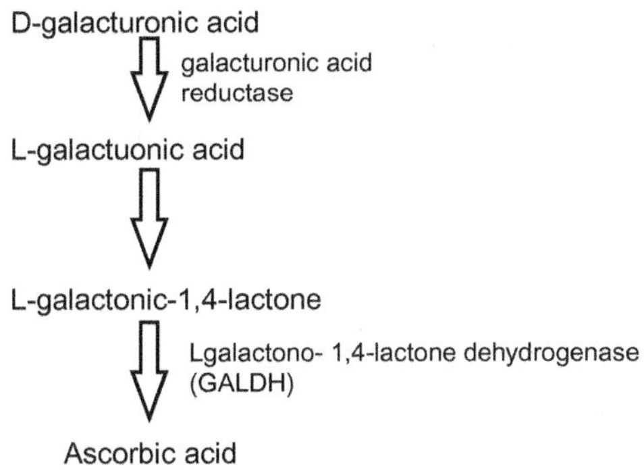

FIGURE 13.2 Synthesis of ascorbic acid by uronic acid.

11. Providing the first line of defense against ROS, it helps plants against oxidative damages, making them resistant to stress (Sharma and Dubey, 2005; Maheshwari and Dubey, 2009; Mishra et al., 2011; Srivastava and Dubey, 2011).

13.5.2 Glutathione

Glutathione (GSH) is a tripeptide, nonenzymatic antioxidant. It gains a lot of attention from scientists for the vast number of its vital roles, which most importantly includes protection of plants against various pathogens (Ali and Alqurainy, 2006). Glutathione (γ-glutamylcysteine) or its homologues play such a dominating role in plant life that, without it, survival of plants would be impossible. Its obligatory role in plants is not fully understood but the following indispensable functions of GSH make this small molecule very important:

1. It plays an important role in many biosynthetic pathways, events related to plant growth and development such as differentiation of cells, resistance against pathogenic stress, regulation of different enzymes involved in plant development, and senescence (Rausch and Wachter, 2005).
2. It is involved in detoxification of many xenobiotics (Xiang et al., 2001), redox homeostasis (Noctor et al., 2012), regulation in the transportation of sulphate, regulation of expression of stress-related genes (Mullineaux and Rausch, 2005) and signal transduction.
3. GSH is also involved in the conjugation of certain metabolites (Rausch and Wachter, 2005).
4. It is also involved in protecting thiols (Rausch and Wachter, 2005).
5. GSH plays a positive role in many protein–protein interactions by thiol-disulphide exchange.

It mainly scavenges $^1O_2$, H_2O_2 and most strikingly OH^- (Larson, 1988).

Several studies have proved that plants lacking the glutathione gene have depleted levels of this important antioxidant which in turn raises its oxidized molecule, glutathione disulphide (GSSG) and thus leads to the accretion of phytoalexins (Guo et al., 1993; Gustine, 1987; Stossel, 1984). Furthermore, its elevated levels in pathogenic attacks confer plant resistance to that pathogen. Elevated levels of glutathione with the activation of hypersensitivity response were seen in plant leaves which were attacked through avirulent biotrophic pathogens (El-Zahaby et al., 1995; Fodor et al., 1997; Vanacker et al., 1998; Vanacker et al., 1999). Table 13.6 illustrates the positive role of GSH in certain pathogen attacks. It is seen that ASC levels are raised when GSH levels are depleted in a plant providing compensation. It clearly shows working GSH to be synergic with other antioxidants. Studies were reported on *Arabidopsis* mutants with 70% lower glutathione levels than wild against fungal and bacterial pathogens (May et al., 1996).

TABLE 13.6 GSH Response Against Various Pathogenic Stresses

Disease/ Pathogen	Plant Name	Antioxidant	Plant Response	References
Powdery mildew	Barley	Raised level of oxidized GSH and hydrogen peroxide	Resistant to attack	Vanacker et al., 2000
Yeast	*Linum ussitatissimum* cv. Linola	Increased cysteine and methionine biosynthesis resulted in significant increase in glutathione	Protection against *Fusarium* infection	Czuj et al., 2009
Tomato leaves	*Botrytis cinerea*	Decrease in GSH content	Decrease in antioxidant defense promoting the spread of necrotic areas that facilitate the penetration of necrotrophic phytopathogens	Kuzniak and Sklodowska, 1999
Tomato leaves	*Avena sativa*	Decrease in GSH content	Decrease in antioxidant defense promoting the spread of necrotic areas that facilitates the penetration of necrotrophic phytopathogens	Gonnen and Schlosser, 1993
Cotyledons of tomato	*Cladosporium fulvum*	90% increase in oxidized GSH	Protection against *C. fulvum*	May et al., 1999

13.5.2.1 Ascorbate-Glutathione Cycle

This is the best studied and main pathway involved in quenching and detoxifying hydrogen peroxide in plant chloroplast, leaf peroxisomes and cytosol of nodules (Noctor and Foyer, 1998). It involves the following molecules:

a. Ascorbate (AsA)
b. Glutathione (GSH)
c. NADPH
d. Ascorbate peroxidase (APX)
e. Certain enzymes

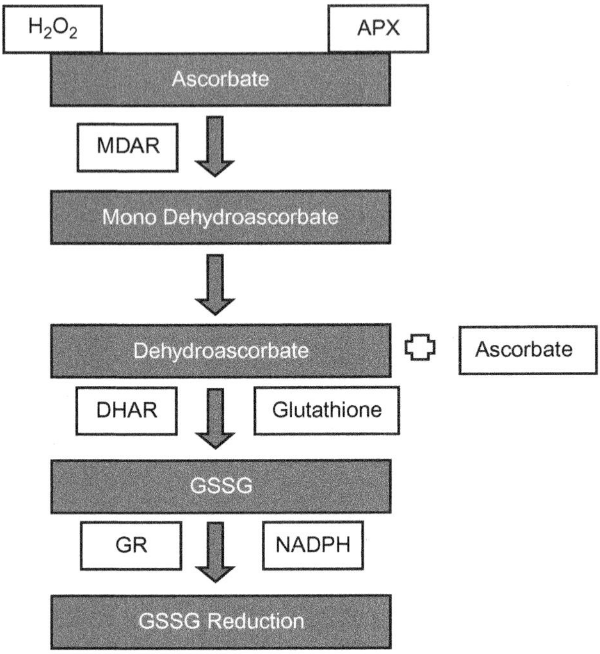

FIGURE 13.3 Glutathione/ascorbate pathway.

Following are the steps involved in this pathway:

1. Reduction of hydrogen peroxide in the presence of ascorbate, which acts as an electron donor by ascorbate peroxidase (APX).
2. Monodehydroascorbate reductase (MDAR) catalyzes the oxidation of ascorbate into monodehydroascorbate (MDA).
3. MDA rapidly fuses into ascorbate and dehydroascorbate.
4. Then dehydroascorbate reductase (DHAR) catalyzes the reduction of dehydroascorbate with glutathione, yielding GSSG which is oxidized glutathione.
5. Finally, glutathione reductase (GR) catalyzes the reduction of GSSG in the presence of NADPH which acts as an electron donor.
6. Therefore, ascorbate and glutathione stayed unconsumed while electrons flow from NADPH to hydrogen peroxide. Fig. 13.3 explains the flow of electron in this pathway (Wells and Xu, 1994; Whitbread et al., 2005; Rouhier and Jacquot, 2002).

13.5.3 Vitamin E (α-Tocopherols)

α-Tocopherol, also called vitamin E, is a lipophilic ROS scavenger. It is synthesized by many plants and algae. Concentration of α-tocopherol is highest in the seeds. On the difference of position of the methyl group, isomers of tocopherols are α, β, γ and δ (Scott et al., 2006). Of all of these four isomers, α-tocopherol gains more importance due to its higher antioxidative activity. It is because of its structure having three methyl groups (Kamal-Eldin & Appelqvist, 1996). Vitamin E prevents chain formation during lipid autooxidation. It not only scavenges ROS but also lipid radicals (Hollander-Czytko et al., 2005). Being a part of biological membranes, it mainly acts as an antioxidant in the membranes. Studies showed that in membranes it plays both parts as a nonantioxidant and antioxidant. It mainly quenches $^1O_2$ which is one of the fatal ROS (Bolkhina et al., 2003). Its major role is to provide tolerance against abiotic stresses, although transgenic studies have provided evidence for its providing tolerance to plants in biotic stress too (Szarka et al., 2012). Mutant *Arabidopsis* plants with depleted levels of vitamin E showed decreased antioxidative activity, thus making plants less defensive against various pathogens and herbivores (Demmig-Adams et al., 2013).

13.5.4 Carotenoids

Carotenoids (Car) are pigments and also act as lipid soluble antioxidants by providing tolerance against oxidative stress. They also play vital roles in plant metabolism (Tuteja, 2007; Ahmad et al., 2011; Ahmad and Umar, 2011; Koyro et al., 2012). Carotenoids include:

1. Beta carotene
2. Zeaxanthin
3. Tocopherols

Approximately 600 carotenoids are found in nature and are present in microorganisms and plants (Collins, 2001). These act as a photoprotective by one of these mechanisms:

1. By disintegrating additional excitation energy.
2. As a ROS scavenger.
3. By suppression of lipid radicals.

They are the major scavenger of $^1O_2$ in photosynthetic machinery (Tounekti et al., 2013).

13.5.5 Flavonoids

Flavonoids are phenolic compounds found abundantly in plants. Flavonoids are stored in the vacuole of plants as glycosides. They are also found in leaf exudates and floral parts of plants (Grace and Logan, 2000; Tounekti et al., 2013). It has been found that in nitrogen-deficient conditions they help to build a symbiotic relationship between plant and microbe. One study confirmed that flavonoid presence makes *A. thaliana* resistant to nitrogen

TABLE 13.7 Two Major Pathogens and Plant Association Involving Antioxidant Activity

Pathogen	Plant	Antioxidant	References
Odium lini (fungus)	*Linum usitatissimum*	GPX and CAT	Ashry & Mohamed, 2012
Bean yellow mosaic virus	*Vicia faba*	POD, CAT, APX and SOD	Radwan et al., 2010

deficiency (Peng et al., 2008). Many studies have showed that flavonoids confer tolerance to pathogens to plants (Gould & Lister, 2006). In one study isolation of prenylated flavonoids was done from *Tephrosia apollinea* L. and it was found that there are four types of prenylated flavonoids:

1. Tephroapollin-F
2. Isoglabratephrin
3. lanceolatin-A
4. (+) (−)glabratephrin.

These prenylated flavonoids have antifungal activity against four phytopathogenic fungus, namely: *Helminthosporium* sp., *Pestalotiopsis* sp., *Alternaria alternatel* and *Colletotrichum acutatum* which were reliant on dose. Specific antioxidants are produced to specific ROS after specific pathogenic attack. Table 13.7 gives the two major examples.

13.6 CONCLUSIONS

Plants, although less specialized in terms of physiology than animals, are well adapted to survive the environmental challenges around them. Reactive oxygen species (ROS) in meager quantities are the by-products of normal cell metabolism and act as second messengers in several of the cell signaling pathways. The main ROS consists of: the singlet oxygen, superoxide anion, hydrogen peroxide and the hydroxyl ion. These ROS molecules have moderate to high activity in cells. Overproduction of ROS is toxic to cells and can cause lipid peroxidation, protein oxidation, nucleic acid damage, and programmed cell death activation (Sharma and Dubey, 2007). Nevertheless, this overproduction is a marvelous defensive technique used by plants to overcome abiotic and biotic stresses. The overproduction of ROS triggers the hypersensitive response (HR) that kills the affected cells. But since the longer the ROS in such alarming quantities stays inside the cells, the more the plant cells are at the risk of dying out. So, this is taken care of by scavenger molecules that include antioxidants, which makes sure that all the excessive

ROS is quenched out in time before it starts ailing the plant too. Interestingly, it is actually the balance between the antioxidants and the ROS produced that determines the harmful or useful aspects of ROS. In pathogenesis, ROS provides the first line of defense. Furthermore, more research regarding more sophisticated mechanisms giving detailed insight into the interplay of ROS and ROS scavengers is required.

Despite much research on ROS and ROS scavengers, there is still a lack of knowledge regarding a comprehensive view of ROS: their formation, their effects on plants and practicality. Due to their short half-life and extreme reactivity, it is very difficult to study ROS. Therefore there is a need for advanced analytical studies, including the latest fields of biological science such as proteomics, metabolomics and genomics. This will help in studying deep insights of molecules, their structures, their interactions and their impact on plant genome. Genetic engineering has been used for some time to make plants resistant to pathogens. Elucidating plants' integral functions, it is possible to employ an advanced level of manipulation in the plant genome. When the antioxidant system is concerned, as many studies have proved them to be ROS scavengers, their properties can be explored to have more beneficial results. Genes encoding antioxidants can be transformed to produce transgenic plants with unregulated activities of antioxidants. In this way, not only natural mechanisms of plants are explored against ROS, but also greater resistance against pathogens can be achieved, as it is a key player. In this way harmful chemicals could be avoided which are now being used against pests and pathogens. This genetic transformation can open a new field to improve crop yield and crop tolerance against various biotic and abiotic stresses. Another approach known as gene pyramiding can be used in order to sum the useful antioxidants together.

REFERENCES

Ahmad, P., Umar, S., 2011. Antioxidants: Oxidative Stress Management in Plants. Studium Press Pvt. Ltd, New Delhi, India.

Ahmad, P., Jaleel, C.A., Salem, M.A., Nabi, G., Sharma, S., 2010a. Roles of enzymatic and non-enzymatic antioxidants in plants during abiotic stress. Crit. Rev. Biotechnol. 30, 161–175.

Ahmad, P., Umar, S., Sharma, S., 2010b. Mechanism of free radical scavenging and role of phytohormones during abiotic stress in plants. In: Ashraf, M., Ozturk, M., Ahmad, M.S.A. (Eds.), Plant Adaptation and Phytoremediation. Springer, Dordrecht Heidelberg London, NY, pp. 99–108.

Ahmad, P., Nabi, G., Jaleel, C.A., Umar, S., 2011. Free radical production, oxidative damage and antioxidant defense mechanisms in plants under abiotic stress. In: Ahmad, P., Umar, S. (Eds.), Oxidative Stress: Role of Antioxidants in Plants. Studium Press Pvt. Ltd, New Delhi, India, pp. 19–53.

Albrecht, U., Bowman, K.D., 2008. Gene expression in *Citrus sinensis* (L.) Osbeck following infection with the bacterial pathogen *Candidatus Liberibacter asiaticus* causing Huanglongbing in Florida. Plant Sci. 175, 291–306.

Ali, A.A., Alqurainy, F., 2006. Activities of antioxidants in plants under environmental stress. In: Motohashi, N. (Ed.), The Lutein-prevention and Treatment for Diseases. Transworld Research Network, India, pp. 187–256.

Alkhalfioui, F., Renard, M., Frendo, P., Keichinger, C., Meyer, Y., Gelhaye, E., et al., 2008. A novel type of thioredoxin dedicated to symbiosis in legumes. Plant Physiol. 148, 424–435.

Apel, K., Hirt, H., 2004. Reactive oxygen species: metabolism, oxidative stress, and signal transduction. Annu. Rev. Plant Biol. 55, 373–399.

Arrigoni, O., De, T.M.C., 2000. The role of ascorbic acid in cell metabolism between gene-directed functions and unpredictable chemical reactions. J. Plant Physiol. 157, 481–488.

Arvind, P., Prasad, M.N.V., 2003. Zinc alleviates cadmium-induced oxidative stress in *Ceratophyllum demersum L*: a free-floating freshwater macrophyte. Plant Physiol. Biochem. 41, 391–397.

Ashry, N.A., Mohamed, H.I., 2012. Impact of secondary metabolites and related enzymes in flax resistance and/or susceptibility to powdery mildew. Afr. J. Biotechnol. 11, 1073–1077.

Azpilicueta, C.E., Benavides, M.P., Tamaro, M.L., Gallego, S.M., 2007. Mechanism of CATA3 induction by cadmium in sunflower leaves. Plant Physiol. Biochem. 45, 589–595.

Badawi, G.H., Yamauchi, Y., Shimada, E., Sasaki, R., Kawano, N., Tanaka, K., et al., 2004. Enhanced tolerance to salt stress and water deficit by overexpressing superoxide dismutase in tobacco (*Nicotiana tabacum*) chloroplasts. Plant Sci. 166, 919–928.

Baptista, P., Martins, A., Pais, M.S., Tavares, R.M., Lino-Neto, T., 2007. Involvement of reactive oxygen species during early stages of ectomycorrhiza establishment between *Castanea sativa* and *Pisolithus tinctorius*. Mycorrhiza. 17, 185–193.

Bhattachrjee, S., 2005. Reactive oxygen species and oxidative burst: roles in stress, senescence and signal transduction in plant. Curr. Sci. 89, 1113–1121.

Bolkhina, O., Virolainen, E., Fagerstedt, K.V., 2003. Antioxidants, oxidative damage and oxygen deprivation stress: a review. Ann. Botany. 91, 179–194.

Boo, Y.C., Jung, J., 1999. Water deficit – Induced oxidative stress and antioxidative defenses in rice plants. J. Plant. Physiol. 155, 255–261.

Bradley, D.J., Kjellbom, P., Lamb, C.J., 1992. Elicitor- and wound-induced oxidative cross-linking of a proline-rich plant cell wall protein: a novel, rapid defense response. Cell 70, 21–30.

Chen, Z., Silva, H., Klessig, D.F., 1993. Active oxygen species in the induction of plant systemic acquired resistance by salicylic acid. Sci. 262, 1883–1886.

Choudhury, S., Panda, P., 2013. Reactive oxygen species signalling in plants under abiotic stress. Plant Sign.. Behav. 8, e23681.

Collins, C., 2001. Carotenoids and genomic stability. Mutat. Res. 475, 1–28.

Cvetkovska, M., Vanlerberghe, G.C., 2013. Alternative oxidase impacts the plant response to biotic stress by influencing the mitochondrial generation of reactive oxygen species. Plant Cell. Environ. 36, 721–732.

Czuj, T., Zuk, M., Starzycki, M., Amir, R., Szopa, J., 2009. Engineering increases in sulphur amino acid contents in flax by overexpressing the yeast Met25 gene. Plant Sci. 177, 584–592.

Dalton, D.A., Russell, S.A., Hanus, F.J., Pascose, G.A., Evans, H.J., 1986. Enzymatic reactions of ascorbate and glutathione that prevent peroxide damage in soybean root nodules. Proc. Natl. Acad. Sci. U.S.A. 83, 3811–3815.

Dalton, D.A., Langeberg, L., Treneman, N., 1993. Correlations between the ascorbate glutathione pathway and effectiveness in legume root nodules. Physiol. Plant. 87, 365–370.

Dalton, T.P., Shertzer, H.G., Puga, A., 1999. Regulation of gene expression by reactive oxygen. Pharmacol. Toxicol. 39, 67–101.
De Tullio, M.C., Paciolla, C., Dalla Vecchia, F., Rascio, N., D'Emerico, S., De Gara, L., et al., 1999. Changes in onion root development induced by the inhibition of peptidyl-prolyl hydroxylase and influence of the ascorbate system on cell division and elongation. Planta. 209, 424–434.
Demmig-Adams, B., Christopher, M., Cohu, V., Amiard, E., Zadelhoff, G., Gerrit, A., et al., 2013. Emerging trade-offs – impact of photoprotectants (PsbS, xanthophylls, and vitamin E) on oxylipins as regulators of development and defense. New phytol. 197, 720–729.
Dixon, D.P., Skipsey, M., Edwards, R., 2010. Roles for glutathione transferases in plant secondary metabolism, Phytochemistry. J. Phytochem. 77, 338–350.
Dixon, R.A., Harrison, M.J., Lamb, C.J., 1994. Early events in the activation of plant defense responses. Annu. Rev. Phytopathol. 32, 479–501.
El-Zahaby, H.M., Gullner, G., Kiraly, Z., 1995. Effects of powdery mildew infection of barley on the ascorbate–glutathione cycle and other antioxidant in different host–pathogen interactions. Phytopathol. 85, 1225–1230.
Farhoudi, R., Hussain, M., Lee, D.J., 2012. Modulation of enzymatic antioxidants improves the salinity resistance in canola (*Brassica napus*). Intl. J. Agri. Biol. 14, 465–468.
Fecht-Christoffers, M.M., Maier, P., Horst, W.J., 2003. Apoplastic peroxidases and ascorbate are involved in manganese toxicity and tolerance of *Vigna unguiculata*. Plant Physiol. 117, 237–244.
Ferreira, R.B., Monteiro, S., Freitas, R., Santos, C.N., Chen, Z., Batista, L.M., et al., 2006. Fungal pathogens: the battle for plant infection. Cri. Rev. Plant Sci. 25, 505–524.
Fester, T., Hause, G., 2005. Accumulation of reactive oxygen species in arbuscular mycorrhizal roots. Mycorrhiza. 15, 373–379.
Fodor, J., Gullner, G., Adam, A.L., Barna, B., Komives, T., Kiraly, Z., 1997. Local and systemic responses of antioxidant to tobacco mosaic virus infection and to salicylic acid in tobacco. Plant Physiol. 114, 1443–1451.
Foyer, C.H., Lelandais, M.A., 1996. A comparison of the relative rates of transport of ascorbate and glucose across the thylakoid, chloroplast and plasma lemma membranes of pea leaves mesophyll cells. J. Plant Physiol. 148, 391–398.
Foyer, C.H., Mullineaux, P.M., 1998. The presence of dehydroascorbate and dehydroascorbate reductase in plant tissues. FEBS Lett. 425, 528–529.
Foyer, C.H., Noctor, G., 2005. Redox homeostasis and antioxidant signaling: a metabolic interface between stress perception and physiological responses. Plant Cell 17, 1866–1875.
Gara, L.D., Pinto, M.C.D., Tommasi, F., 2003. The antioxidant systems vis-à-vis reactive oxygen species during plant–pathogen interaction. Plant Physiol. Biochem. 41, 863–870.
Gayoso, C., Pomar, F., Novo-Uzal, E., Merino, F., de Ilarduya, O.M., 2010. The Vemediated resistance response of the tomato to *Verticillium dahliae* involves H_2O_2, peroxidase and lignins and drives PAL gene expression. BMC Plant Biol. 10, 232–251.
Gechev, T.S., Hille, J., 2005. Hydrogen peroxide as a signal controlling plant programmed cell death. J. Cell Biol. 168, 17–20.
Gill, S.S., Tuteja, N., 2010. Reactive oxygen species and antioxidant machinery in abiotic stress tolerance in crop plants. Plant Physiol. Biochem. 48, 909–930.
Gonnen, M.V., Schlösser, E., 1993. Oxidative stress in interaction between *Avena sativa L.* and *Drechlera spp.*. Physiol. Mol. Plant Pathol. 42, 221–234.
Gould, K.S., Lister, C., 2006. Flavonoid functions in plants. In: Andersen, Ø.M., Markham, K.R. (Eds.), Flavonoids: Chemistry, biochemistry and applications, 2006. CRC Press, Boca Raton, pp. 397–411.

Grace, S., Logan, B.A., 2000. Energy dissipation and radical scavenging by the plant phenylpropanoid pathway. Philos. T. Roy. Soc. B. 355, 1499–1510.

Guo, Z.J., Nakagavara, S., Sumitani, K., Ohta, Y., 1993. Effect of intracellular glutathione level on the production of 6-methoxy-mellein in cultured carrot (Daucus carota) cells. Plant Physiol. 102, 45–51.

Gupta, S.A., Webb, R.B., Holaday, A.S., Allen, R.D., 1993. Overexpression of superoxide-dismutase protects plants from oxidative stress (Induction of ascorbate peroxidase in superoxide dismutase-overexpressing plants). Plant Physiol. 103, 1067–1073.

Gustine, G.L., 1987. Induction of medicarpin biosynthesis in landino clover by *p*–chloromercuribenzoic acid is reverted by dithiotreitol. Plant Physiol 84, 3–6.

Hackbarth, S., Schlothauer, J., Preub, A., Roder, B., 2010. New insights to primary photodynamic effects – singlet oxygen kinetics in living cells. J. Photochem. Photobiol. 98, 173–179.

Hatz, S., Lambert, J.D.C., Ogilby, P.R., 2007. Measuring the lifetime of singlet oxygen in a single cell: addressing the issue of cell viability. Photochem. Photobiol. Sci. 10, 1106–1116.

Hollander-Czytko, H., Grabowski, J., Sandorf, I., Weckermann, K., Weiler, E.W., 2005. Tocopherol content and activities of tyrosine aminotransferase and cysteine lyase in *Arabidopsis* under stress conditions. J. Plant Physiol. 162, 767–770.

Horemans, N., Foyer, C.H., Potters, G., Asard, H., 2000. Ascorbate functions and associated transport systems in plants. Plant Physiol. Biochem. 38, 531–540.

Huckelhoven, R., Kogel, K.H., 2003. Reactive oxygen intermediates in plant-microbe interactions: who is who in powdery mildew resistance?. Planta. 216, 891–902.

Isherwood, F.A., Chen, Y.T., Mapson, L.W., 1954. Synthesis of L-ascorbic acid in plants and animals. The Biochem. J. 56, 1–15.

Jomová, K., Feszterová, M., Morovič, M., 2012. Expression of pathogenesis-related protein genes and changes of superoxide dismutase activity induced by toxic elements in *Lupinus luteus L.* J. Microbiol. Biotechnol. Food Sci. 1, 437–445.

Kamal-Eldin, A., Appelqvist, L.A., 1996. The chemistry and antioxidant properties of tocopherols and tocotrienols. Lipids. 31, 671–701.

Kasai, H., 1997. Analysis of a form of oxidative DNA damage, 8-hydroxy-2′-deoxyguanosine, as a marker of cellular oxidative stress during carcinogenesis. Mut. Res. 387, 147–163.

Khan, N.A., Singh, S., 2008. Abiotic Stress and Plant Responses. IK International, India.

Krieger-Liszkay, A., 2005. Singlet oxygen production in photosynthesis. J. Exp. Bot. 56, 337–346.

Koyro, H.W., Ahmad, P., Nicole, G., 2012. Abiotic stress responses in plants: an overview. In: Ahmad, P., Prasad, M.N.V. (Eds.), Environmental Adaptations and Stress Tolerance of Plants in the Era of Climate Change. Springer Science + Business Media, New York, pp. 1–28.

Krieger-Liszkay, A., Fufezan, C., Trebst, A., 2008. Single oxygen production in photosystem II and related protection mechanism. Photosynth. Res. 98, 551–564.

Kubo, A., Aono, M., Nakajima, N., Saji, H., Tanaka, K., Kondo, N., 1999. Differential responses in activity of antioxidant enzymes to different environmental stresses in *Arabidopsis thaliana*. J. Plant Res. 112, 279–390.

Kuzniak, E., Sklodowska, M., 1999. The effect of *Botrytis cinerea* infection on ascorbate glutathione cycle in tomato leaves. Plant Sci. 148, 69–76.

Laloi, C., Stachowiak, M., Pers-Kamczyc, E., Warzych, E., Murgia, I., Apel, K., 2007. Crosstalk between singlet oxygen- and hydrogen peroxide-dependent signaling of stress responses in *Arabidopsis thaliana*. Proc. Natl. Acad. Sci. U.S.A. 104, 672–677.

Lamb, C., Dixon, R.A., 1997. The oxidative burst in plant disease resistance. Annu. Rev. Plant Physiol. Plant Mol. Biol. 48, 251–275.

Larson, R.A., 1988. The antioxidants of higher plants. Phytochem. 27, 969–978.

Levine, A., Tenhaken, R., Dixon, R., Lamb, C., 1994. Cell 79, 583–593.

Levine, A., Pennell, R.I., Alvarez, M.E., Palmer, R., Lamb, C., 1996. Calcium-mediated apoptosis in a plant hypersensitive disease resistance response. Curr. Biol. 6, 427–437.

Liso, R., Innocenti, A.M., Bitonti, M.B., Arrigoni, O., 1988. Ascorbic acid induced progression of quiescent centre cells from G1 to S phase. New Phytol. 110, 469–471.

Low, P.S., Merida, J.R., 1996. The oxidative burst in plant defense: unction and signal transduction. Plant physiol. 96, 533–542.

Maheshwari, R., Dubey, R.S., 2009. Nickel-induced oxidative stress and the role of antioxidant defence in rice seedlings. Plant Growth Regul. 59, 37–49.

Margis, R., Dunand, C., Teixeira, F.K., Margis-Pinheiro, M., 2008. Glutathione peroxidase family – an evolutionary overview. FEBS J. 275, 3959–3970.

May, M.J., Parker, J.E., Daniels, M.J., Leaver, C.J., Christopher, S., 1996. An *Arabidopsis* mutant depleted in glutathione shows unaltered responses to fungal and bacterial pathogens. Mol. Plant Microbiol. 9, 349–356.

May, M.J., Hammond-Kosack, K.E., Jones, J.D.G., 1999. Involvement of reactive oxygen species, glutathione metabolism, and lipid peroxidation in the Cf-gene-dependent defence response of tomato cotyledons induced by race-specific elicitors of *Cladosporium fulvum*. Plant Physiol. 110, 1367–1379.

Mehdy, M.C., 1994. Active oxygen species in plant defense against pathogens. Plant Physiol. 105, 467–472.

Miller, G., Suzuki, N., Rizhsky, L., Hegie, A., Koussevitzky, S., Mittler, R., 2007. Double mutants deficient in cytosolic and thylakoid ascorbate peroxidase reveal a complex mode of interaction between reactive oxygen species, plant development, and response to abiotic stresses. Plant Physiol. 144, 1777–1785.

Mishra, S., Jha, A.B., Dubey, R.S., 2011. Arsenite treatment induces oxidative stress, upregulates antioxidant system, and causes phytochelatin synthesis in rice seedlings. Protoplasma. 248, 565–577.

Mittler, R., Vanderauwera, S., Gollery, M., Van, B.F., 2004. Reactive oxygen gene network of plants. Trends Plant Sci. 9, 490–498.

Moreau, S., Puppo, A., Davies, M.J., 1995. The reactivity of ascorbate with different redox states of leg hemoglobin. Phytochem. 39, 1281–1286.

Mullineaux, P.M., Rausch, T., 2005. Glutathione, photosynthesis and the redox regulation of stress-responsive gene expression. Photosynthetic Res. 86, 459–474.

Nagamiya, K., Motohashi, T., Nakao, K., Prodhan, S.H., Hattori, E., Hirose, S., et al., 2007. Enhancement of salt tolerance in transgenic rice expressing an *Escherichia coli* catalase gene, katE. Plant Biotechnol. 49–55, rep1.

Nobuhiro, S., Mittler, R., 2006. Reactive oxygen species and temperature stresses: a delicate balance between signalling and destruction. Plant Physiol. 126, 45–51.

Noctor, G., Foyer, C., 1998. Ascorbate and glutathione: keeping active oxygen under control. Annu. Rev. Plant Physiol. Plant Mol. Biol. 49, 249–279.

Noctor, G., Gomez, L., Vanacker, H., Foyer, C.H., 2002. Interactions between biosynthesis, compartmentation, and transport in the control of glutathione homeostasis and signalling. J. Exp. Bot. 53, 1283–1304.

Noctor, G., Mhamd, A., Chaouch, S., Han, Y., Neukermans, J., Marquez-Garcia, B., et al., 2012. Glutathione in plants: an integrated overview. Plant Cell Environ. 35, 454–484.

Ozgur, R., Uzilday, B., Sekmen, A., Türkan, I., 2013. Reactive oxygen species regulation and antioxidant defence in halophytes. Functl. Plant Biol. 40, 832–847.

Passardi, F., Penel, C., Dunand, C., 2004. Performing the paradoxical: how plant peroxidases modify the cell wall. Trends Plant Sci. 9, 534–540.

Pei, Z.M., Murata, Y., Benning, G., Thomine, S., Klusener, B., Allen, G.J., et al., 2000. Calcium channels activated by hydrogen peroxide mediate abscisic acid signalling in guard cells. Nat. 406, 731–734.

Peng, M., Kuc, J., 1992. Peroxidase-generated hydrogen peroxide as a source of antifungal activity in vitro and on tobacco leaf disks. Phytopathol. 82, 696–699.

Peng, M., Hudson, D., Schofield, A., Tsao, R., Yang, R., Gu, H., et al., 2008. Adaptation of *Arabidopsis* to nitrogen limitation involves induction of anthocyanin synthesis which is controlled by the NLA gene. J. Exp. Bot. 59, 2933–2944.

Quan, L.J., Zhang, B., Shi, W.W., Li, H.Y., 2008. Hydrogen peroxide in plants: a versatile molecule of the reactive oxygen species network. J. Integrat. Plant Biol. 50, 2–18.

Radwan, D.E.M., Fayez, K.A., Mahmoud, S.Y., Lu, G., 2010. Modifications of antioxidant activity and protein composition of bean leaf due to *Bean yellow mosaic virus* infection and salicylic acid treatments. Acta. Phys.-Chim. Sin. 32, 891–904.

Rausch, T., Wachter, A., 2005. Sulfur metabolism: a versatile platform for launching defence operations. Trends Plant Sci. 10, 503–509.

Rouhier, N., Jacquot, J.P., 2002. Plant peroxiredoxins: alternative hydroperoxide scavenging enzymes. Photosyn. Res. 74, 259–268.

Rouhier, N., Gelhaye, E. and Jacquot, J.P. (2002). Exploring the active site of plant glutaredoxin by site-directed mutagenesis. FEBS Let 511, 145–149.

Rubio, M.C., James, E.K., Clemente, M.R., Bucciarelli, B., Fedorova, M., Vance, C.P., et al., 2004. Localization of superoxide dismutases and hydrogen peroxide in legume root nodules. Mol. Plant Microbe. In. 17, 1294–1305.

Rubio, M.C., Bustos-Sanmamed, P., Clemente, M.R., Becana, M., 2009. Effects of salt stress on the expression of antioxidant genes and proteins in the model legume *Lotus japonicas*. New Phytol. 181, 851–859.

Sarowar, S., Kimb, E.N., Kima, Y.J., Oka, S.H., Kimb, K.D., Hwangb, B.K., et al., 2005. Overexpression of a pepper ascorbate peroxidase-like 1 gene in tobacco plants enhances tolerance to oxidative stress and pathogens. Plant Sci. 169, 55–63.

Scott, E., Laurent Méne-Saffrane, S., Farmer, E.E., Krischke, M., Mueller, M.J., Pennaa, D. D., 2006. Nonenzymatic lipid peroxidation reprograms gene expression and activates defense markers in *Arabidopsis* tocopherol-deficient mutants. The Plant Cell 18, 3706–3720.

Shao, H.B., Chu, L.Y., Lu, Z.H., Kang, C.M., 2008. Primary antioxidant free radical scavenging and redox signalling pathways in higher plant cells. Intl. J. Biol. Sci. 4, 8–14.

Sharma, P., Dubey, R.S., 2005. Drought induces oxidative stress and enhances the activities of antioxidant enzymes in growing rice seedlings. Plant Growth Reg. 46, 209–221.

Sharma, P., Dubey, R.S., 2007. Involvement of oxidative stress and role of antioxidative defense system in growing rice seedlings exposed to toxic concentrations of aluminium. Plant Cell Rep. 26, 2027–2038.

Sharma, P., Dubey, R.S., 2005. Modulation of nitrate reductase activity in rice seedlings under aluminium toxicity and water stress: role of osmolytes as enzyme protectant. J. Plant Physiol. 162, 854–864.

Sharma, P., Jha, A.B., Dubey, R.A., Pessarakli, M., 2012. Reactive oxygen species, oxidative damage, and antioxidative defense mechanism in plants under stressful conditions. J. Bot. 2012, 10.1155/2012/217037.

Singh, G.S., Tuteja, N., 2010. Reactive oxygen species and antioxidant machinery in crop plants. Plant Physiol. Biochem. 48, 909–930.

Slooten, L., Capiau, K., Van Comp, W., Van Montagu, M., Sybesma, C., Inzé, D., 1995. Factors affecting the enhancement of oxidative stress tolerance in transgenic tobacco overexpressing manganese superoxide dismutase in chloroplasts. Plant Physiol. 107, 737–750.

Smirnoff, N., 1996. The function and metabolism of ascorbic acid in plants. Annu. Bot. 78, 661–669.

Smirnoff, N., 2000. Ascorbic acid: metabolism and functions of a multifacetted molecule. Curr. Opin. Plant Biol. 3, 229–235.

Smirnoff, N., Running, J.A., Gatzek, S., 2004. Ascorbate biosynthesis: a diversity of pathways. In: Asard, H., May, J.M., Smirnoff, N. (Eds.), Vitamin C: Its Functions and Biochemistry in Animals and Plants. BIOS Scientific, New York, USA, pp. 7–29.

Srivastava, S., Dubey, R.S., 2011. Manganese-excess induces oxidative stress, lowers the pool of antioxidants and elevates activities of key antioxidative enzymes in rice seedlings. Plant Growth Regul., 1–16.

Stossel, P., 1984. Regulation of sulphydryl groups of glyceollin accumulation in soybean hypocotyls. Planta. 160, 314–319.

Szarka, A., Horemans, A., Kovacs, Z., Grof, P., Mayer, M., Banhegyi, G., 2007. Dehydroascorbate reduction in plant mitochondria is coupled to the respiratory electron transfer chain. Physiol. Plant. 129, 225–232.

Szarka, A., Tomassakovics, B., Banhegyi, G., 2012. The ascorbate-glutathione-α-tocopherol triad in abiotic stress response. Intl. J. Mol. Sci. 13, 4458–4483.

Tanaka, K., Tankeuchi, E., Kubo, A., Sakaki, T., Haraguch, K., Kawamura, Y., 1991. Two immunologically different isoenzymes of ascorbate peroxidase from spinach leaves. Arch. Biochem. Biophys. 286, 371–375.

Tenhaken, R., Levine, A., Brisson, L.F., Dixon, R.A., Lamb, C., 1995. Proc. Natl. Acad. Sci. U. S. A 92, 4158–4163.

Tewari, R.K., Kumar, P., Sharma, P.N., 2006. Antioxidant responses to enhanced generation of superoxide anion radical and hydrogen peroxide in the copper-stressed mulberry plants. Planta. 223, 1145–1153.

Thakur, M., Sohal, B.S., 2013. Role of elicitors in inducing resistance in plants against pathogen infection: a review. ISRN Biochem. 20, 1155–1165.

Thomas, C.E., McLean, L.R., Parker, R.A., Ohlweiler, D.F., 1992. Ascorbate and phenolic antioxidant interactions in prevention of liposomal oxidation. Lipids. 27, 543–550.

Torres, M.A., Jones, J.D.G., Dangl, J.I., 2006. Reactive oxygen species signalling in response to pathogens. Plant Physiol. 141, 373–378.

Tounekti, T., Hernández, I., Munné-Bosch, S., 2013. Salicylic acid biosynthesis and role in modulating terpenoid and flavonoid metabolism in plant responses to abiotic stress. In: Hayat, S., Ahmad, A., Alyemeni, M. (Eds.), Salicylic Acid: Plant Growth and Development. Springer, Dordrecht, pp. 141–162.

Tuteja, N., 2007. Mechanisms of high salinity tolerance in plants. Methods Enzymol. 428, 419–438.

Tuteja, N., 2010. Cold, salt and drought stress. In: Hirt, H. (Ed.), Plant Stress Biology: From Genomics towards System Biology. Wiley-Blackwell, Weinheim, Germany, pp. 137–159.

Valko, M., Morris, H., Cronin, M.T.D., 2005. Metals, toxicity and oxidative stress. Curr. Med. Chem. 12, 1161–1208.

Vanacker, H., Carver, T.L.W., Foyer, C.H., 1998. Pathogen-induced changes in the antioxidant status of the apoplast in barley leaves. Plant Physiol. 117, 1103–1114.

Vanacker, H., Carver, T.L.W., Foyer, C.H., 1999. Changes in apoplastic antioxidants induced by powdery mildew attack in oat genotypes with race non-specific resistance. Planta. 208, 444–452.

Vanacker, H., Carver, T.L.W., Foyer, C.H., 2000. Early H_2O_2 accumulation in mesophyll cells leads to induction of glutathione during hypersensitive response in the barley-powdery mildew interaction. Plant Physiol. 123, 1289–1300.

Vilanova, L., Torresa, R., Viñasb, I., González-Candelasc, L., Usalla, J., Fiori, D.S., et al., 2013. Wound response in orange as a resistance mechanism against *Penicillium digitatum* (pathogen) and *P. expansum* (non-host pathogen). Postharvest Biol. Technol. 78, 113–122.

Vogt, T., 2010. Phenylpropanoid biosynthesis. Mol. Plant 3, 2–20.

Vranova, E., Atichartpongkul, S., Villarroel, R., Montagu, M.V., Inze, D., Camp, W.V., 2002. Comprehensive analysis of gene expression in *Nicotiana tabacum* leaves acclimated to oxidative stress. Proc. Natl. Acad. Sci. U.S.A. 99, 870–875.

Wagner, D., Przybyla, D., Camp, R.O.D., 2004. The genetic basis of singlet oxygen-induced stress response of *Arabidopsis thaliana*. Sci. 306, 1183–1185.

Wang, Y.C., Qu, G.Z., Li, H.Y., Wu, Y.J., Wang, C., Liu, G.F., et al., 2010. Enhanced salt tolerance of transgenic poplar plants expressing a manganese superoxide dismutase from *Tamarix androssowii*. Mol. Biol. 37, 1119–1124.

Wells, W.W., Xu, D.P., 1994. Dehydroascorbate reduction. J. Bioeng. Bio. Membr. 26, 369–377.

Whitbread, A.K., Masoumi, A., Tetlow, N., Schmuck, E., Coggan, M., Board, P.G., 2005. Characterization of the omega class of glutathione transferases. Methods Enzymol. 401, 78–99.

Xiang, C., Werner, B.L., Christensen, E.L., Oliver, D.J., 2001. The biological functions of glutathione revisited in *Arabidopsis* transgenic plants with altered glutathione levels. Plant Physiol. 126, 564–574.

Xu, J., Duan, X., Yang, J., Beeching, J.R., Zhang, P., 2013. Enhanced reactive oxygen species scavenging by overproduction of superoxide dismutase and catalase delays postharvest physiological deterioration of cassava storage root. Plant Physiol. 161, 1517–1528.

Yana, W., Tianhonga, Z., Junlib, W., Yixina, Z., Weia, G. and Minjiana, C. (2013). Reactive oxygen species, ascorbic acid and glutathione to ozone stress in soybean (Glycine max) leave. Third International Conference on Intelligent System Design and Engineering Applications. China.

Yang, Y., Han, C., Liu, Q., Lin, B., Wang, J., 2008. Wang Effect of drought and low light on growth and enzymatic antioxidant system of *Picea asperata* seedlings. Acta. Physiol. Plant 30, 433–440.

Yang, Z., Wu, Y., Li, Y., Ling, H.Q., Chu, C., 2009. OsMT1a, a type 1 metallothionein, plays the pivotal role in zinc homeostasis and drought tolerance in rice. Plant Mol. Biol. 70, 219–229.

Chapter 14

Role of Ascorbate Peroxidase in Postharvest Treatments of Horticultural Crops

Yoshihiro Imahori

14.1 INTRODUCTION

Horticultural crops are an important source of carbohydrates, proteins, organic acids, vitamins and minerals for human nutrition. When humans use plants or plant parts, whether for food or for aesthetic purposes, there is always a postharvest component that leads to loss (Fallik, 2004). Their losses in quantity and quality affect horticultural crops between harvest and consumption. Thus, to reduce the losses, producers and handlers must understand the biological and environmental factors involved in deterioration (Kader, 1992). Fresh horticultural crops are living tissues subject to continuous changes after harvest. While some changes are desirable, most are not. In addition all fresh horticultural crops are high in water content and thus are subject to desiccation and to mechanical injury (Kader, 1992). Their commodities are perishable products with active metabolism and subject to extensive postharvest losses through microbial decay, physical injury, and senescence during the postharvest period. However, these postharvest changes in horticultural crops cannot be stopped, but they can be slowed within certain limits. Consequently, the maintenance or improvement of the postharvest life of fresh horticultural crops is becoming increasingly important. Proper postharvest handling plays an important role in increasing food availability (Kader, 1992). Most postharvest treatments involve the alteration of the natural conditions of horticultural crops in order to prolong their postharvest life (González-Aguilar et al., 2010). Additionally, fresh horticultural crops are diverse in morphological structure (roots, stems, leaves, flowers, fruits, and so on), in composition, and in general physiology. Commodity requirements and recommendations for maximum postharvest life vary among the commodities (Kader, 1992).

P. Ahmad (Ed): Oxidative Damage to Plants. DOI: http://dx.doi.org/10.1016/B978-0-12-799963-0.00014-9

As stress is generally defined as any environmental factor potentially unfavorable to living organisms, with the exception of decay, quality losses in actual postharvest produce can be directly or indirectly attributable to a combination of abiotic stress and stress-induced senescence (Lester, 2003). During harvest or postharvest treatments horticultural crops are exposed to several abiotic stresses that often lead to the accumulation of metabolites. These abiotic stresses lead to a series of morphological, physiological, biochemical and molecular changes in horticultural crops (Wang et al., 2003). The plant can acquire resistance to abiotic stresses. The acquired resistance is often associated with enhanced mobilization of defense responses after subsequent exposure of the plants to stresses (Capanoglu, 2010). Also, abiotic stresses are used in postharvest activities as traditional tools to extend the shelf life of product, focusing mainly on color, texture, and flavor quality change. Also, these stresses can be used in preharvest activities to enhance the quality and yield of products in the field (Kader, 1992). Therefore, approaches to modulate or control abiotic stresses in plant tissues can be very important for improving shelf life and quality retention during postharvest handling of horticulture crops. In relation to these approaches, there have been many strategies using various types of postharvest treatments.

Abiotic stresses are often interconnected and lead to the production of reactive oxygen species (ROS) that are derived from molecular oxygen. In relation to this action, plants have developed several antioxidation strategies to scavenge ROS (Ahmad et al., 2010a, b, 2011; Ahmad and Umar, 2011; Koyro et al., 2012). The activation of various pathways leading to the accumulation of antioxidants has been associated with plant responses to abiotic stresses, such as drought, UV, high light, salinity, low or high temperature and wounding (Capanoglu, 2010; Imahori, 2012). Enhancement of antioxidant defense in plants can increase tolerance to abiotic stresses (Wang et al., 2003; Ahmad et al., 2010a, b, 2011; Ahmad and Umar, 2011; Koyro et al., 2012). The ROS detoxification process in plants is essential for the protection of plant cells and their organelles against the toxic effect of ROS (Caverzan et al., 2012). These ROS scavenging systems include enzymatic and nonenzymatic antioxidants. The ROS-scavenging enzymes including ascorbate peroxidase (APX) respond to environmental stress. APX is one of the most important enzymes of the ROS scavenging system and plays a central role in the ascorbate-glutathione cycle (Ishikawa and Shigeoka, 2008; Zhang, 2013). Thus, APX induction is one element of the response to the oxidative stress caused in plants. APX plays an important role in controlling the concentration of ROS that participates in the signal transduction in many physiological processes, and the management of ROS during stress in horticultural crops. The function of APX does not only have to be limited to antioxidative protection, but also there can be a more extensive connection between APX induction and regulation of metabolism. Expression of APX genes is activated by various abiotic stress factors. There are some connections between APX induction and regulation of metabolism.

This chapter focuses on the role of APX in postharvest treatments for horticultural crops. Approaches to modulate or control abiotic stresses in plant tissues can be very important for improving shelf life and quality retention during postharvest handling of horticultural crops. In relation to approaches, the physiological function and modulation of APX in response to abiotic stress conditions such as low or high temperature, atmosphere, UV and salinity are discussed.

14.2 OXIDATIVE STRESS AND ANTIOXIDANT MACHINERY

Even under optimal conditions, ROS are mainly produced at a low level in organelles such as chloroplasts, mitochondria, peroxisomes and glyoxysomes (Ahmad et al., 2010a, b). However, during biotic and abiotic stress, ROS production is dramatically elevated. ROS may be produced in response to abiotic and biotic stresses, and associated with a number of physiological responses in plants. When the generation of ROS exceeds the capacity of the plant to maintain cellular redox homeostasis, or when the production of ROS exceeds the capacity of the plant to scavenge them, the oxidative stress occurs (Hodges et al., 2004; Bhattacharjee, 2013). Oxidative stress can be induced at various subcellular sites and initiated by an increased production of ROS as a result of numerous processes (Toivonen, 2004; Ahmad et al., 2011). The three sites related to ROS generation are the apoplastic region (cell wall, apoplastic space and the external surface of the plasma membrane), the cytoplasm, and cellular organelles (chloroplasts, mitochondria and peroxisomes/glyoxysomes) (Toivonen, 2004). ROS include such compounds as superoxide radical ($O_2^{\bullet -}$), singlet oxygen ($^1O_2$), hydrogen peroxide (H_2O_2), and the hydroxyl radical ($^{\bullet}OH$). They are partially reduced forms of molecular oxygen (O_2), resulting from either the excitation of O_2 to form $^1O_2$ or the transfer of one, two or three electrons to O_2 to form, respectively, $O_2^{\bullet -}$, H_2O_2 or $^{\bullet}OH$ (Mittler, 2002; Hodges et al., 2004). Hydrogen peroxide has important useful functions in several cellular processes, and its production increases in stressed plant tissues (Purvis, 2004).

ROS are generated by all aerobic organisms. When exposed to abiotic stress conditions, such as a low temperature (Sala and Lafuente, 2004), a high temperature (Sairam et al., 2000), water-deficit stress (Jin et al., 2006), and ozone pollution (Mehlhorn et al., 1990), the production of ROS increases and causes oxidative damage to the plant itself (Shigeoka et al., 2002). The toxicity of ROS is considered to be due to their ability to initiate radical cascade reactions that lead to the production of the hydroxyl radical and other destructive species which can cause protein damage, lipid peroxidation, DNA damage and finally cell death (Davey et al., 2000; Ahmad et al., 2011; Hossain et al., 2012). ROS accumulation may cause oxidative damage to lipids, forming toxic products such as malondialdehyde (MDA), a secondary end product of polyunsaturated fatty acid oxidation. Thus, MDA is usually an indicator of the degree of plant oxidative stress (Hodges et al., 1999) and the structural integrity of the membranes of plants subjected to the abiotic stress, such as low temperatures (Posmyk et al., 2005). Therefore, oxidative stress, resulting from the deleterious effects of ROS, is an important phenomenon in many biological systems. These oxidative stresses can be caused at various subcellular sites and initiated by the accumulation of ROS (Toivonen, 2004).

Scavenging of ROS is important for maintenance of normal plant growth. Plants have evolved an efficient antioxidant defense system that would prevent the accumulation of ROS and repair oxidative damage. This system involves both lipid-soluble antioxidants (α-tocopherol and carotene), water-soluble reductants (ascorbate and glutathione), and enzymes such as catalase (CAT), APX, superoxide dismutase (SOD) and glutathione reductase (GR) (Imahori et al., 2008; Ahmad et al., 2010a,b, 2011; Ahmad and Umar, 2011; Koyro et al., 2012; Zhang, 2013). Ascorbate, glutathione and α-tocopherol have each been shown to act as antioxidants in the detoxification of ROS in aerobic cell (Szarka et al., 2012). These substances have central and interrelated roles acting both chemically and as substrates in enzyme-catalyzed detoxification reactions (Alscher et al., 1997). SOD catalyzes superoxide radical to H_2O_2 and O_2 in a reaction that is spontaneous and extremely rapid, thus protecting the cells from damage by superoxide radical reaction products. The product, which is a potentially toxic compound, is then reduced to water by a number of enzymes such as CAT and APX (Kochhar et al., 2003). CAT converts H_2O_2 to water and molecular oxygen. This enzyme has extremely high catalytic rates but low substrate affinities, since the reaction requires the simultaneous access of two hydrogen peroxide to the active site (Willekens et al., 1995; Noctor and Foyer, 1998).

Reducing substances, such as ascorbate and glutathione, seem to be involved in effectively scavenging ROS. In addition to reacting nonenzymatically with ROS, those reducing substances are known to form an ascorbate-glutathione cycle. Both ascorbate and glutathione can scavenge ROS not only by a direct nonenzymatic reaction, but also through the cycle (Larson, 1988). Hence, this cycle could play an important role in scavenging H_2O_2 in the plant (Foyer et al., 1997; del Río et al., 1998). This ascorbate-glutathione cycle is implicated in the scavenging of ROS and free radicals. Hydrogen peroxide is relatively stable and can penetrate organellar membranes, but in the presence of ferrous iron it is further reduced to the highly reactive and detrimental hydroxyl radical. The ascorbate-glutathione cycle is an efficient way for plant cells to dispose of H_2O_2 in certain cellular compartments where this metabolite is produced, and it makes use of the nonenzymatic antioxidants, ascorbate and glutathione, in a series of reactions catalyzed by four antioxidative enzymes (del Río et al., 1998). The four enzymes are

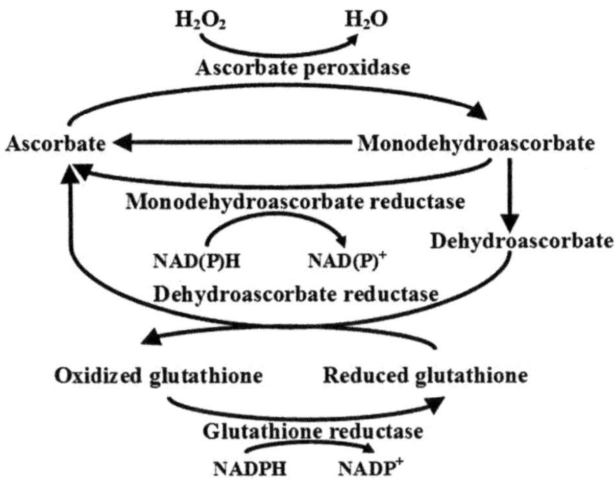

FIGURE 14.1 The ascorbate-glutathione cycle.

APX, monodehydroascorbate reductase (MDHAR), dehydroascorbate reductase (DHAR) and GR (Smirnoff, 1996). APX uses two molecules of ascorbate to reduce H_2O_2 to water, with the concomitant generation of two molecules of monodehydroascorbate (MDHA). MDHA is a radical with a short lifetime that, if not rapidly reduced, disproportionates to ascorbate and dehydroascorbate. MDHA can be reduced directly to ascorbate. The electron donor for MDHA reduction may be reduced ferredoxin or NAD(P)H. The latter redaction is catalyzed by MDHAR. Despite the possibility of enzymatic and nonenzymatic regeneration of ascorbate directly from MDHA, rapid disproportionation of the MDHA radical means that some dehydroascorbate is always produced when ascorbate is oxidized in plant tissues. Dehydroascorbate is reduced to ascorbate by the action of DHAR, using reduced glutathione. This reaction generates oxidized glutathione, which is in turn rereduced to reduced glutathione by NADPH, a reaction catalyzed by GR (Noctor and Foyer, 1998). The ascorbate-glutathione cycle is shown in Fig. 14.1.

Therefore, as horticulture crops experience many types of postharvest stress during storage and/or transport, it is important to understand how biotic and abiotic (environmental) changes imposed on the plant tissues result in oxidative stress and how plant tissues regulate oxidative stress.

14.3 CHARACTERISTICS OF ASCORBATE PEROXIDASE

Plant peroxidases are heme proteins and contain iron as the prosthetic groups which play a key role in the catalytic site. These heme-containing peroxidases have been further subdivided into three classes based on cellular location and function (González-Rábade et al., 2012). APX belongs to the Class

I peroxidases including the intracellular peroxidases of prokaryotic origin. The molecular and enzymatic properties of APX are different from those of other heme-containing peroxidases such as guaiacol peroxidase (Shigeoka et al., 2002; González-Rábade et al., 2012). APX is specific to plant, and is a H_2O_2-scavenging enzyme that utilizes ascorbate as a specific electron donor to reduce H_2O_2 to water, and has high specificity for ascorbate. This enzyme is extremely sensitive to the ascorbate concentrations. The activity of APX rapidly loses stability and declines under conditions in which the ascorbate concentration is lower than 20 μM (Dąbrowska et al., 2007; Ishikawa and Shigeoka, 2008). In the presence of ascorbate, the reaction intermediate of APX Compound I oxidizes ascorbate to produce two molecules of MDHA in the catalytic cycle. Unless APX Compound I is reduced by ascorbate, it is oxidized and degraded to the inactivated form by 10 nM H_2O_2 (Shigeoka et al., 2002). Therefore ascorbate is indispensable to protect chloroplasts and other cell constituents from damage by H_2O_2 and hydroxyl radicals produced from it (Asada, 1992). The high level of endogenous ascorbate is necessary to effectively maintain the antioxidant system that protects plants from oxidative stress.

APX is found in several cellular compartments, and consists of a family of isoenzymes with different characteristics. Its isoenzymes are classified into at least four distinct cellular compartments according to their subcellular localization (Shigeoka et al., 2002). Soluble isoenzymes are found in cytosol, mitochondria and chloroplast stroma, and membrane-bind isoenzymes are found in microbody (including glyoxisome and peroxisome) and chloroplast thylakoids (Shigeoka et al., 2002; Ishikawa and Shigeoka, 2008; Caverzan et al., 2012). APX isoenzymes differ in molecular weight, optimal pH, stability and substrate (Dąbrowska et al., 2007). Subcellular localization of APX isoenzymes is determined by the presence of organelle-specific targeting peptides and transmembrane domains in the N- and C-terminal regions of the protein (Shigeoka et al., 2002; Dąbrowska et al., 2007; Caverzan et al., 2012).

Plant chloroplastic APX isoenzymes have stromal APX (sAPX) and thylacoid APX (tAPX) isoenzymes. Both isoenzymes are involved in the water-water cycle in which $O_2^{\bullet-}$ is reduced to water in a two-step reaction catalyzed by SOD and APX. Superoxide is produced at the reducing site of photosystem I (PSI) by reduction of O_2 by electrons donated from PSI (Asada, 1999). The water-water cycle maintains a proper ATP/NADPH ratio and alleviates the overreduction of photosystems when plants are exposed to photoinhibitory conditions (Dąbrowska et al., 2007). tAPX has one major hydrophobic domain (Shigeoka et al., 2002). The tAPX sequence exhibits a C-terminal extension with a hydrophobic anchor region for binding the thylakoid membrane (Ishikawa and Shigeoka, 2008). As transcript levels of tAPX are weakly responsive to external stimuli, exposing the plant to oxidative stress, tAPX is involved in direct detoxification of H_2O_2 but not protection from excess ROS produced under environmental stress (Dąbrowska et al., 2007). mAPX has a similar C-terminal region to tAPX and its N-terminal active domain faces the cytosol (Shigeoka et al., 2002; Dąbrowska et al., 2007). The physiological function of mAPX is to scavenge H_2O_2 generated both in microbodies and the cytosol in combination with CAT and cytosolic APX (cAPX). Chloroplastic APX isoenzymes are especially sensitive to inactivation. Their half-inactivation time is less than 30 seconds, while that for cAPX and mAPX is around one hour or more (Chen and Asada, 1989; Dąbrowska et al., 2007; Ishikawa and Shigeoka, 2008; Caverzan et al., 2012). For the encoding genes of chloroplastic APX isoenzymes, they are divided into two groups. The first group comprises single genes encoding two isoenzymes through a post-transcriptional method regulated by alternative splicing. The second group includes individual genes in which different isoenzymes are individually regulated. The mechanism of alternative splicing in chloroplastic APX isoenzymes is fundamental for controlling the expression of sAPX and tAPX isoenzymes (Ishikawa and Shigeoka,, 2008; Caverzan et al., 2012).

14.4 THE ROLE OF ASCORBATE PEROXIDASE

APX isoenzymes play an important role in scavenging H_2O_2 that is generated continuously in plant cells and protecting them. In order to maintain an adequate cellar concentration of ROS, APX isoenzymes play a protective role against ROS produced in excess under environmental stress (Dąbrowska et al., 2007). Their expression level reflects the occurrence of stress conditions and is modulated by various environmental stresses. The transcriptional expression of *APX* genes is tissue and developmental stage dependent (Caverzan et al., 2012). The chloroplastic ascorbate peroxidase activity in broccoli florets strongly decreased with senescence during storage, suggesting that storage of broccoli florets might lead to a sharp decrease of APX activity in the chloroplast. On the other hand, cAPX activity decreased slightly (Shigenaga et al., 2005). The findings suggest that the hydrogen peroxide produced in broccoli florets could be associated with the enhancement of the cAPX activity and the cAPX formed effectively scavenges H_2O_2 during storage (Shigenaga et al., 2005). Morita et al. (1999) determined the oxidative stress response of rice cAPX, and suggested that H_2O_2 is involved in oxidative stress signaling, leading to the induction of cAPX. Thus, hydrogen peroxide seems to play a signaling role in the cell when the plant is placed under stress conditions. APX, which is localized in chloroplasts, mitochondria, microbodies, and cytosol, is a key enzyme to decompose H_2O_2 in plants. Because of the different affinities of APX and CAT for H_2O_2, APX might be responsible for the fine modulation of ROS for signaling, whereas CAT might be responsible for the removal of excess ROS under stress (Mittler, 2002). CAT does not require reducing power and has a high reaction rate but a low affinity for H_2O_2. On the contrary, APX has a higher affinity for H_2O_2, allowing for the scavenging of small amounts of H_2O_2 in more specific locations (Dat et al., 2000).

Plant cAPX isoenzymes are usually homodimers. Two or three isoenzymes are found in plant cytosol (Dąbrowska et al., 2007; Ishikawa and Shigeoka, 2008). Among APX isoenzymes, cAPX is highly responsive to environmental conditions. The steady-state level of the cAPX transcript significantly increased in response to high-light illumination, drought, and heat in pea and spinach leaves (Ishikawa and Shigeoka, 2008). Recently the signaling mechanisms inducing cAPX expression and its regulation have been shown (Ishikawa and Shigeoka, 2008). APX plays important roles in the regulation of cellular ROS levels and is the central components of H_2O_2-scavenging networks (Suzuki et al., 2013). In *Arabidopsis* cytosolic APX1 is constitutively expressed in many plant tissues, and its expression is significantly upregulated in response to a large number of biotic and abiotic stresses (Suzuki et al., 2013). Also, cytosolic APX1 was shown to be essential for the protection of the thylakoid and stromal/mitochondrial APXs as well as the protection of nuclear DNA during light stress (Davletova et al., 2005; Suzuki et al., 2013). The function of cytosolic APX1 is tightly linked to ROS signaling pathways in different cellular compartments and could be involved in the regulation of ROS levels in the entire cell (Suzuki et al., 2013). A second cytosolic APX isoenzyme, cytosolic APX2, is also involved in the response of plants to abiotic stress. Expression of cytosolic APX2 is significantly upregulated in roots in response to oxidative stress, and in roots and shoots in response to salinity and osmotic stress, and in roots, shoots, leaves and pollen in response to heat stress. Cytosolic APX2 is highly responsive to heat stress and might play important roles in the regulation of heat tolerance in plants (Suzuki et al., 2013).

14.5 ENVIRONMENTAL STRESS AND POSTHARVEST PRODUCE

14.5.1 Temperature Treatments

14.5.1.1 Heat Treatment

Heat treatments have been used for insect disinfestations, decay control, ripening decay and modification of horticultural crop responses to other stresses and maintenance of quality during storage (Lurie, 1998; Paull and Chen, 2000). There are three methods in use to heat commodities: hot water, vapor heat and hot air. Postharvest heat treatments are often applied for a relatively short time, at temperatures higher than heat treatments designed to kill insect pests located at the interior of a commodity, because the target pathogens are found on the surface or in the first few cell layers under the skin of the commodity (Lurie, 1998; Fallik, 2004; Imahori, 2012). Many horticultural crops tolerate exposure to water temperatures of 50–60°C for up to 10 min (Lurie, 1998). A recent extension of the hot water treatment has been the development of hot water rinsing and brushing (Fallik, 2004). Vapor heat treatment is a method of heating commodity with air saturated with water vapor at temperatures of 50–60°C. This was developed mainly to kill insect eggs and larvae as a quarantine treatment before fresh market shipment (Lurie, 1998; Fallik, 2004). Hot air treatment can be applied by placing horticultural crops in a heated chamber with a ventilating fan, or by applying forced hot air where the speed of air circulation is precisely controlled, and this has been used for both fungal and insect control (Lurie, 1998). Heat treatments can also be used to inhibit ripening processes or to induce resistance to chilling injury which is a physiological disorder caused by the exposure of crops to low temperatures above the freezing point, and external skin damage during storage, through changes in gene expression and protein synthesis (Lurie, 1998; Paull and Chen, 2000; Fallik, 2004).

Development of thermo tolerance has been associated with synthesis of heat shock protein (HSP) and loss of thermo tolerance with the disappearance of HSP (Lurie, 1998). There is a correlation between the development of thermo tolerance and the synthesis of HSP, as well as a correlation between the loss of thermo tolerance and the disappearance of HSP (Lurie, 1998). The development of thermo tolerance is dependent on protein synthesis (Liu et al., 1996). Thus it also is dependent on the incubation temperature. The exposure temperature is high enough to initiate HSP synthesis (Lurie, 1998). Temperatures in the range 35–40°C have been found to be effective, depending upon the commodity. At 42°C or higher, transcription and translation of HSP are inhibited and HSP synthesis is attenuated (Ferguson et al., 1994). Continued exposure to 42°C, though still allowing heat shock polypeptide synthesis, is the limit for induction of heat shock tolerance and enhancement of protein degradation (Ferguson et al., 1994). Consequently, its thermo tolerance decreases and heat damage may result (Paull and Chen, 2000). The disruption of transcription occurs by mRNA being released from the ribosomes (Stuger et al., 1999). Protein degradation, particularly of rate-limiting enzymes, continues at a higher rate due to the higher temperature (Paull and Chen, 2000).

Heat treatment may induce oxidative stress. Oxidative stress occurs when the generation of ROS exceeds the capacity of the plant to maintain cellular redox homeostasis or to scavenge ROS. ROS can cause membrane and protein damage (Paull and Chen, 2000; Hodges et al., 2004; Vicente et al., 2006). Decrease in antioxidant activity in stressed tissues results in higher levels of ROS that may contribute to cellular injury. As a result, heat stress often activates cell signaling pathways and cellular response, followed by the production of oxygen radical scavengers such as APX (Wang et al., 2003). Protection against oxidative stress is an important component in determining the survival of plant under heat stress. Plants have developed a series of both enzymatic and nonenzymatic detoxification systems to scavenge ROS, for protecting plant cells from oxidative stress (Wahid et al., 2007). Thereby the nondamaging heating treatment causes a moderate stress in the horticultural crops that mobilized antioxidant responses. Moderate heat treatment of certain horticultural commodities before storage can diminish the effects of applied oxidative stress (Lurie, 1998; Hodges et al., 2004). For example, in mandarin, strawberry and banana fruit, heat treatment induced the activity of APX involved in ROS detoxification (Sala and Lafuente, 2000; Vicente et al., 2006; Wang et al., 2012). cAPX activity in heat-treated broccoli florets increased during storage (Shigenaga et al., 2005). Expression and activation of APX is related to the appearance of physiological injuries caused in plants by heat stress (Wahid et al., 2007).

Heat stress can affect the capacity of biological systems to synthesize proteins, resulting in the synthesis of a new set of special proteins termed HSPs. The heat shock response is manifested in most living organisms as induction or enhanced synthesis of HSPs. These HSPs are believed to prevent irreversible protein denaturation and breakdown that would be detrimental to the cell and to confer tolerance to heat (Ferguson et al., 2000; Paull and Chen, 2000). The heat shock response is primarily regulated at the transcriptional levels (Wang et al., 2003). The accumulation of HSP is controlled under heat stress transcription factors (HSFs). HSFs play a central role in stress dependent and developmental expression of HSPs in plants (Kotak et al., 2007). The heat shock response can be completely inhibited by effective removal of H_2O_2 because the expression of genes such as those encording HSFs and cAPX is modulated by the ROS signal (Davletova et al., 2005). HSP gene expression and protein synthesis are associated with high temperature exposure of various plant parts. With postharvest heat treatments, HSP transcripts and protein levels have been shown to increase (Ferguson et al., 2000). The lag period for induction for heat shock response is slower than other stress responses, although HSP are synthesized within 30 min after heat treatment in the range 34–42°C (Paull and Chen, 2000). Plant modification for enhanced tolerance is mostly based on the manipulation of genes that protect and maintain the function and structure of cellular components (Wang et al., 2003).

14.5.1.2 Chilling Treatment

Chilling is an important abiotic stress that inhibits growth and development of plants, limits geographical distribution of many species and reduces agricultural productivity (Imahori, 2011). Generally reductions in temperature can substantially reduce the velocity of many metabolic processes, which leads to fruit senescence, deterioration and loss of crop quality (Saltveit and Morris, 1990). Common effects of chilling temperatures include an alteration in the rate of ripening, as indicated by delayed or even total failure of fruit color development and reduced softening (Imahori et al., 2008; Imahori, 2011, 2012). Thus, low-temperature storage can extend shelf life of horticultural crops. Storage temperatures for horticultural crops are selected to optimize quality attributes such as texture, flavor, and appearance over time (Hodges et al., 2004). However, crop-specific temperature thresholds exist behind which symptoms of chilling-induced oxidative injury and oxidative-related disorders will occur (Toivonen, 2003; Wismer, 2003; Hodges et al., 2004). Chilling temperatures alter the equilibrium between ROS generation and defense mechanisms such that oxidatively induced stress is observed (Hodges et al., 2004; Imahori et al., 2008; Imahori, 2011).

The metabolic status and the chemical composition of the tissue at the time of chilling can affect the tolerance of the tissue to the chilling (Wang, 2010). Chilling-tolerant plants usually have a higher proportion of unsaturated fatty acid and correspondingly a lower transition temperature in the membrane lipids (Mahajan and Tuteja, 2005). On the other hand, chilling-sensitive plants are marked by a higher proportion of saturated fatty acid and, therefore, a higher transition temperature. Chilling-resistant tissues tend to have a higher degree of unsaturation of fatty acids in membrane lipids than chilling-sensitive tissues (Wang, 2010; Singh and Singh, 2013). At chilling temperatures, the tissues weaken because they are to carry on normal metabolic processes. Various physiological and biochemical alterations and cellular dysfunctions occur in chilling-sensitive plants in response to chilling stress (Wang, 2010). In chilling-sensitive crops, low temperatures impair the energy state of cell and/or provoke alterations in membrane integrity. Chilling-sensitive plants characteristically exhibit structural injuries and may suffer from metabolic dysfunction when chilled (Mahajan and Tuteja, 2005). This may result in enhanced ROS proliferation, while also reducing scavenging efficacy through such factors as chilling-related inactivation of antioxidants and/or impeded antioxidant turnover (Hodges et al., 2004). As a result, chilling-tolerant plants or plant parts may initially contain or can generate more antioxidants during stress and/or produce fewer ROS than chilling-sensitive plants (Hodges et al., 2004).

Chilling injury (CI) is a physiological disorder that occurs in several horticultural commodities when they are exposed to or stored at temperatures below certain threshold temperatures that are characteristic for the particular species (Purvis, 2002). The disorder may be related to tissue deterioration or senescence, which leads to changes in membrane permeability. The physical properties of membranes are affected by low temperature and are associated with low temperature-induced quality defects. CI was induced by the decrease in the degree of unsaturated fatty acids and the membrane permeability (Imahori et al., 2008; Imahori, 2011). The decline in the ratio of unsaturated to saturated fatty acids may indicate an increase in lipid peroxidation during chilling. The lipid peroxidation contributes to the development of CI (Wang et al., 1992). In fact, chilling has been found to induce lipid degradation in cucumber fruit, cucumber seedling and tomato pericarp, and these results may indicate that CI enhances lipid peroxidation in the membranes (Wang, 1995). Oxidative stress has been associated with the appearance of chilling damage in horticultural crops. Oxidative damage is considered to be an early response of sensitive tissues to chilling (Hariyadi and Parkin, 1991). ROS produced in response to chilling stress contributes to membrane damage.

Membrane lipid peroxidation may be one of the first events in the manifestation of CI (Hodges et al., 2004). Lipid peroxidation can be indicated by measuring the content of MDA, because MDA is a secondary end product of polyunsaturated fatty acid oxidation. The extent of lipid peroxidation activation depends on the degree of chilling stress and is correlated to the extent of CI (Lukatkin, 2002). However, the increase of MDA indicates a distinct deterioration of membrane integrity and activation of lipid peroxidation in the plant commodities. Cold exposure can alter the structure of the membranes due to lipid peroxidation (Posmyk et al., 2005). The degradation of the polyunsaturated fatty acids due to peroxidation produces MDA at the same time, inducing membrane rigidification and death of cells. Changes in MDA level in a tissue can be a good indicator of the structural integrity of the membranes of plants subjected to low temperature (Posmyk et al., 2005). Free radical-induced lipid peroxidation is one of the main causes of membrane deterioration. Many free radicals are highly reactive chemically and can induce the oxidative breakdown of double bonds in the fatty acids of membrane lipids. For instance, free radical production in thylakoid membranes of leaves in low-temperature treated plants has been observed by Hidey and Bjorn (1996). Mitochondria in chilling-sensitive green bell peppers are a source of oxidative stress during chilling as superoxide production from the electron transport chain is increased (Hodges et al., 2004). The antioxidant activity derives from endogenous bioactive compounds. The activity of antioxidant systems to quench various oxygen radicals is considered to be inversely proportional to lipid peroxidation levels.

Sato et al. (2001) observed that chilling injury in heat-treated rice seedlings did not occur during 7 days at 5°C, whereas the control developed a severe symptom of chilling injury. APX activity was higher in heat-treated rice seedlings than in the control. cAPX mRNA expression in heat-treated rice seedlings was also higher than without heat treatment under chilling stress. They suggested that the reduction of chilling injury could be because of heat induction of the cAPX gene. Hence, the induced H_2O_2 could be involved in the production of APX (İşeri et al., 2013). Potato tubers had transient accumulation of cAPX mRNA after storage at low temperature, showing that APX expression was induced in response to low temperature (Kawakami et al., 2002). The combined action of SOD, CAT and APX that converts the potentially dangerous $O_2^{\bullet -}$ and H_2O_2 is supposed to efficiently eliminate these ROS, and, consequently, protect cellular components against the more active reactive hydroxyl radical. The balance between SOD, CAT and APX activities is critical to cell survival during cold storage (Sala and Lafuente, 2004). The increase in SOD activity could enhance the ability of the fruit to dismutate $O_2^{\bullet -}$, whereas the increases in CAT and APX activities would contribute to the elimination of H_2O_2. The damage of plant tissues has been associated with ROS arising under stress conditions such as exposure to chilling. Oxidative damage is considered to be an early response of sensitive tissues to chilling. Stimulation of some detoxifying enzymes during storage might result from an induction of these enzymes through the production of $O_2^{\bullet -}$ and H_2O_2. Chilling stress may alter the equilibrium between ROS production and defense mechanisms such that oxidatively induced CI is observed (Hodges et al., 2004). This balance between the formation and detoxification of ROS is critical to cell survival during chilling storage. It is interesting to note that the mume (*Prunus mume*) fruit stored at 1°C resulted in a greater induction in SOD, CAT and APX activities than that at 6°C (Imahori et al., 2008). These results indicate that the linked action of SOD, CAT and APX may contribute, at least to some extent, to the mitigation of CI in fruit stored at 1°C. It seems that, in fruit stored at 1°C, the APX isoforms were synthesized that reduced H_2O_2 concentration and protected the fruit from oxidative stress. APX evidently takes part in plant protection against CI during cold storage (Wang et al., 2012).

14.5.2 Atmospheric Treatments

14.5.2.1 Controlled and Modified Atmospheres

Exposing fresh horticultural crops to reduced O_2 and/or elevated CO_2 can either be beneficial or harmful, depending on the concentration of these gases, temperature, exposure duration and commodity. Controlled atmosphere (CA) or modified atmosphere (MA) storage utilizing reduced O_2 and/or elevated CO_2 are known to maintain quality and consequently extend shelf life of many fresh horticultural crops (Kader, 1986; Imahori et al., 2013). The beneficial effects of CA or MA storage include delayed ripening, reduced physiological and pathological disorders and the possibility for disinfesting fruit (Burdon et al., 2007). The gas composition of CA is monitored and deviations from the set points corrected. MA differs in that they are not actively controlled and the gas composition results from a balance between the plant gas consumption or production and gas diffusion through a permeable membrane (Chervin et al., 1996). To date, much research has been conducted to evaluate the effects of CA and MA storage on the quality and storability for a large number of horticultural crops and specific cultivars of each commodity (Weichmann, 1987). However, despite the enormous economic significance of CA or MA storage, accompanying the use of low O_2 or high CO_2 atmospheres for maintaining quality of fresh horticultural crops during CA or MA storage is the risk that very low O_2 and/or high CO_2 atmospheres may cause damage to the produce (Burdon et al., 2007).

Exposing fresh horticultural crops to low O_2 can be beneficial or harmful, depending on concentrations of these gases, temperature, and exposure duration. Exposing products to stress O_2 levels for long periods can lead to abnormal ripening, browning of tissues, and accumulation of ethanol and acetaldehyde (Imahori et al., 2007; Imahori, 2012). Oxygen levels as low as 0.2% in the plant cell may result in anaerobic respiration (Kader, 1986). Plant responses to low O_2 concentrations include induction of fermentation pathways, accumulation of fermentation products, and decreases in intracellular pH and ATP levels (Imahori et al., 2003). During fermentation, acetaldehyde which is produced through pyruvate decarboxylation by pyruvate decarboxylase (PDC) is converted to ethanol by alcohol dehydrogenase (ADH) using NADH (Imahori et al., 2012). On the other hand, lactate is formed in a single step by the reduction of pyruvate by lactate dehydrogenase (LDH) and NADH. Thus, the major function of fermentative metabolism is to use NADH and pyruvate, when electron transport and oxidative phosphorylation are inhibited, so that glycolysis can proceed. Both ethanol and lactate are produced to a varying degree by most plants under low O_2. Therefore, many plants have two simultaneous pathways competing for pyruvate and NADH under low O_2 conditions (Imahori et al., 2003). The induction of PDC, ADH, and/or LDH is one of the mechanisms for accumulations of anaerobic products. Fermentative metabolism results in the accumulation of anaerobic products by the actions of the enzymes PDC, ADH and LDH under low O_2 concentrations (Imahori et al., 2000, 2003, 2012; Imahori, 2012). However, the activities of ADH and LDH are not necessarily the rate limiting factors for the accumulations of ethanol and lactate in some plant tissues, if the activities of these enzymes are high (Xia and Saglio, 1992).

The responses of horticultural crops to elevated CO_2 levels vary considerably within or among species, cultivars, organ types and developmental stages, and include both undesirable and beneficial physiological and biochemical changes (Beaudry, 1999). Moreover, it is well known that the effect of CO_2 depends on its dosage and environmental conditions such as temperature (Smith, 1992). Carbon dioxide may act both as an inducer and a suppressor of respiration depending on its concentration in situ as well as internal oxygen concentration, duration of exposure, the commodity and temperature (Imahori et al., 2007). During storage, the physiological effects of elevated CO_2 are a decrease in respiration rate and ethylene production, and retention of chlorophyll content, textural quality and sensory attributes of horticultural commodities (Herner, 1987). The responses of horticultural crops to very high CO_2 concentrations include induction of the glycolytic pathway, fermentation pathways, accumulation of succinate and/or alanine and decreases in pH and ATP levels (Mathooko, 1996). Similarly, an atmosphere enriched with more than 20% CO_2 in the presence of atmospheric oxygen also caused ethanol accumulation in lettuce, fig fruit and strawberry fruit (Mathooko, 1996). Elevated CO_2 concentrations, above a level of about

20% or higher, depending on the commodity and the O_2 concentrations, can result in accumulation of ethanol within the tissues (Kader, 1986). The accumulation of ethanol, as a product of fermentative metabolism, indicates that some substrates of energy metabolism are passing through the fermentation pathway (Imahori et al., 2007). Accumulation of ethanol occurs typically when plant tissues are exposed to anaerobic conditions, and are usually associated with increased activities of ADH and, less consistently with pyruvate decarboxylase (Imahori et al., 2007; Imahori, 2012). The increase in ADH activity could be due to an increased rate of synthesis or activation of the enzyme by a decrease in cytoplasmic pH and/or increased concentrations of substrates such as pyruvate, acetaldehyde and NADH (Mathooko, 1996; Imahori et al., 2002; Imahori, 2012).

CA (2% O_2 and 5% CO_2) storage induces significant metabolic changes such as a decrease in ascorbate contents and transitory increase in APX and GR activities, which may partially explain the behavior of fruit during long-term storage, and may be useful in preventing the CA-induced physiological disorders in Conference pears (Larrigaudiere et al., 2001). Additionally, CA (2% O_2 and 5% CO_2) storage elicited a general increase in *APX* gene transcripts in two pear cultivars. However, a large increase in APX activity in response to CA stress was observed in nonbrowning "Doyenne du Comice," whereas the APX activity in browning "WBC" decreased (Cascia et al., 2013). In spinach leaves, declining activities of APX, CAT over the 35 d storage period regardless of the composition of the storage atmosphere suggests that regulation of H_2O_2 plays an important role in both the dynamics and severity of postharvest senescence of leaves (Hodges et al., 2004). Also, the elevated O_2 level increased the ability of the antioxidative defense mechanism including APX in Chinese bayberry in order to first control the ROS level and eventually the fruit decay severity (Yang et al., 2009).

14.5.2.2 Ethanol Fumigation

Ethanol has been found to be beneficial in either counteracting senescent processes or reducing chilling injury in horticultural crops (Toivonen, 1995; Imahori, 2012). Postharvest ethanol treatment can have beneficial effects on fruit physiology such as enhancing the sensory quality of apples, reducing astringency of persimmons and bananas, reducing postharvest decay of citrus and stone fruit and controlling scald in apples (Jamieson et al., 2003). Also, treatment with ethanol significantly inhibited the yellowing of broccoli and maintained higher activities of CAT, APX and SOD. The effect of ethanol in delaying the yellowing was correlated to enhanced antioxidant enzyme activity (Xu et al., 2012). In raspberries fruit, ethanol may induce the accumulation of antioxidant compounds, inactivating ROS by antioxidative enzymes such as APX, SOD and CAT, improving antioxidant status of crops (González-Aguilar et al., 2010).

14.5.3 UV Irradiation

Ultraviolet (UV) light causes biological stress in plants and defense mechanisms of plant tissues with the consequent production of phytoalexin compounds. Phytoalexin accumulation could be accompanied by other inducible defenses such as cell wall modifications, defense enzymes, and antioxidant activity (González-Aguilar et al., 2007; Imahori, 2012). The UV portion of the electromagnetic spectrum ranges approximately 10 to 400 nm (Shama, 2007). UV radiation has been applied to produce in long wave (UV-A: 315–400 nm), medium (UV-B: 280–315 nm) and short wave (UV-C: 100–280 nm) dosages. The shortest wavelengths of the UV spectrum are the most energetic ones and are more effective biocides for surface sterilization of some food products (Shama, 2007; Perkins-Veazie et al., 2008). Exposure to UV leads to retardation of plant growth and alteration in metabolic processes such as photosynthesis and respiration. UV radiation induced damage is related to acceleration in ROS generation, leading to oxidative stress. However, plants have developed several antioxidation strategies to scavenge ROS. Enhancement of antioxidant defense in plants can increase tolerance to UV radiation (Wang et al., 2003). Additionally the secondary metabolism in plant is activated together with the enzymatic antioxidant system (González-Aguilar et al., 2010).

UV irradiation has been used to maintain the postharvest quality and extend the shelf life of several fresh horticultural crops (Perkins-Veazie et al., 2008; Yamauchi, 2013). Low UV doses induce production of antifungal compounds, ripening delay and reduction of chilling injury (Pombo et al., 2009). The exposure to UV-C delays fruit softening which is one of the main factors determining fruit postharvest life (Pan et al., 2004). UV-C decreased the activity of enzymes involved in tomato cell wall degradation and delayed the fruit softening (Barka et al., 2000). Treatment with UV-C increases the contents of antioxidants, ascorbic acid and total phenolic compound, and improves nutritional qualities of tomato fruit (Liu et al., 2011). UV irradiation can affect physiological processes at the genetic level. In parsley, UV-B upregulates genes encoding the flavonoid biosynthetic pathway, such as chalcone synthase and phenylalanine ammonia lyases (PAL), which are key enzymes in anthocyanin formation (Perkins-Veazie et al., 2008). Although treatment with UV-A or UV-B seems to be less harmful than UV-C radiation, it has the effect of increasing antioxidants such as ascorbate, α-tocophol and polyphenol (Yamauchi, 2013).

Hormetic doses of UV-C radiation have been used as a physical treatment to extend postharvest life of several horticultural crops (Pombo et al., 2009). Low doses of UV-C radiation stimulated beneficial reactions in biological organs, a phenomenon known as hormesis (González-Aguilar et al., 2010; Shama, 2007; Beckles, 2012). Hormesis has been defined as the use of potentially harmful agents at low doses in order to induce a beneficial stress response (Shama and Alderson, 2005). Although UV light leads to the deleterious effects such as decreased protein synthesis, impaired chloroplast function, and DNA damage on plant tissues, low doses of UV could inflict repairable damage to DNA, and this slight trauma would activate repair mechanisms for radiation-induced DNA damage (Costa et al., 2006; Shama, 2007). Hormetic effects manifest themselves in treated plant tissue through the action of a variety of induced chemical species. They include phytoalexins such as scoparone in oranges, 6-methoxymellein in carrots and resveratrol in grapes (Shama, 2007). Also induced are enzymes, such as chitinases and glucanases in peaches and PAL in peaches and tomatoes (Shama, 2007).

Changes in the antioxidative defense system were shown in several horticultural crops subjected to UV radiation. UV-C radiation enhanced the activities of antioxidant enzymes including APX in strawberry fruit during storage (Erkan et al., 2008). The effect of UV-C irradiation is the induction of enzymes that played a role in oxidative stress, such as APX and guaiacol peroxidase in tomato fruit during ripening (Barka et al., 2000; Barka, 2001). Also, in UV-B treated broccoli florets during storage, the contents in these antioxidants such as ascorbate and glutathione, and APX activity increased (Yamauchi, 2013). These effects of UV radiation on the synthesis of antioxidant compounds and enzymes can vary depending on the hermetic doses, time of exposure and treated horticultural crops (González-Aguilar et al., 2010). When a hermetic dose of UV irradiation is absorbed by biological material, it can interact with atoms and molecules, mainly water, producing ROS by the univalent reduction of O_2 in a rapid and controlled manner. The primary ROS formed in the cell is $O_2^{\bullet-}$ which triggers a cascade of reactions that results in the formation of a variety of ROS and induction of antioxidative enzymes such as APX. A key ROS is H_2O_2 produced by SOD (González-Aguilar et al., 2010). Therefore, UV radiation induces production of ROS such as $O_2^{\bullet-}$ by exciting electrons by electronic transition in appropriate photosensitizers. SOD then dismutases the produced $O_2^{\bullet-}$ to H_2O_2. Accordingly H_2O_2 scavenging enzymes such as APX efficiently catalyze the breakdown of H_2O_2 (Dat et al., 2000). H_2O_2 may act more as a signal molecule than directly inducing oxidative damage at lower UV dose, and be responsible for the improvement of the antioxidant status of horticultural crops activating gene expression of enzymes related to the synthesis and accumulation of antioxidant capacity (González-Aguilar et al., 2010).

14.5.4 Water-Deficit Stresses

Water deficit stress occurs in plants when the water potential outside the plant tissue declines. The drop in water potential may be caused by reduced water availability by drought or low temperature or by increased osmolarity by salt (Hunter et al., 2011). Water deficit stress leads to severe effects in plants. Responses to water stress include loss of cell turgor, reduced leaf water potential and growth rate, and decreased stomatal conductance, photochemical efficiency, and chlorophyll content (Hunter et al., 2011). Removal of water from the membrane disrupts the normal bilayer structure and results in the membrane becoming exceptionally porous when desiccated. Stress within the lipid bilayer may also result in displacement of membrane integrity, selectivity, disruption of cellular compartmentalization and a loss of activity of enzymes, which are primarily membrane based. In addition, cytosolic and organelle protein may exhibit reduced activity or may even undergo complete denaturation when dehydrated. The high concentration of cellular electrolytes due to the dehydration of protoplasm may cause disruption of cellular metabolism (Mahajan and Tuteja, 2005). Water-deficit stress is one of the most common adverse environmental stresses. Plants can respond and adapt to the stress by altering their cellular metabolism and invoking various defense mechanisms under the stress conditions (Jin et al., 2006). This defense mechanism includes accumulating antioxidant enzymes, increasing the amounts of antioxidant metabolites and organic osmolytes, and a rise in the concentration of abscisic acid (Hunter et al., 2011).

Water-deficit stress leads to enhanced production of ROS, induces oxidative stress, and causes ROS-associated injury. In accordance, this stress also occurs to activation of enzymes involved in the production and removal of ROS (Mahajan and Tuteja, 2005; Türkan and Demiral, 2009; Miller et al., 2010; Murshed et al., 2013). Transcript profiling studies on a range of intact plants have shown that water deficit stress rapidly alters the expression of a large number of genes, encoding proteins of diverse functions (Hunter et al., 2011). In particular, the importance of APX in tolerance to water-deficit stress has been shown in some horticultural crops. In banana plantlets, APX activity was accelerated during water deficit stress in the tolerant variety, but was unaffected in sensitive variety (Chai et al., 2005). In cut rose, regulation of APX at transcript level may be involved in the response to water deficit stress, and the changing patterns of *APX* gene expression paralleled those of APX activity (Jin et al., 2006). In mulberry, APX activity was significantly high in the water-stressed leaves (Reddy et al., 2004).

Drought stress is known as the more common water-deficit stress. The stress component that defines drought is a decrease in the availability of soil water. This decreased water availability can be quantified as a decreaser in water potential (Verslues et al., 2006). Drought induced inhibition of photosynthesis leads to increased production of ROS in the chloroplasts. The accumulation of ROS during such conditions originates from the decline in CO_2 fixation, leading to higher leakage of electrons to O_2 (Dat et al., 2000). Drought stress alters the oxidative balance of the cell and acclimation to drought is generally correlated with keeping the levels of ROS relatively low through the ROS scavenging system (Dat et al., 2000). In response to drought stress, a higher increase in transcript levels of cytosolic and peroxisomal *APX* genes was observed in the drought-sensitive cowpea cultivar (Caverzan et al., 2012).

Plant responses to salt stress include drought mediated symptoms. The components of drought and salt stress cross talk with each other as both these stresses ultimately result in dehydration of the cell and osmotic imbalance (Mahajan and Tuteja, 2005). Plant is greatly affected by salinity, which causes alteration in nutrient uptake, accumulation of toxic ions, osmotic stress and oxidative stress (Verslues et al., 2006; Agarwal et al., 2013). High salinity causes ionic stress, osmotic stress and secondary stresses such as hormonal and nutritional imbalances. Salt stress causes oxidative damage resulting in the formation of ROS (Seckin et al., 2010). Several changes occur inside the plant cell, including changes in gene expression levels, synthesis of molecular chaperones, and activation of enzymes involved in the production and removal of ROS (Caverzan et al., 2012). High salt concentrations decrease the osmotic potential of soil solution creating a water stress in plants. They cause severe ion imbalance and toxicity. The interactions of salts with mineral nutrition may result in nutrient imbalances and deficiencies (Sairam and Tyagi, 2004). Higher salinity causes broad adjustments in plant metabolism and physiology. Salt stress causes oxidative damage resulting in the ROS production. For example, salt-induced responses include ROS-mediated lipid peroxidation in citrus cells (Piqueras et al., 1996; Dat et al., 2000). Mitochondrial and chloroplastic SOD and APX isozymes increase during salt stress in pea (Dat et al., 2000; Hernández et al., 2000). The expression of APX isozymes in response to salinity was tissue specific and also dependent on stress duration (Lin and Pu, 2010).

14.6 CONCLUSIONS

After harvest horticultural crops experience many types of postharvest stress during storage and and/or transport. Therefore, the maintenance or improvement of the postharvest quality and the postharvest life of horticultural crops is becoming increasingly. Different postharvest treatments are employed in order to extend the shelf life in horticultural crops of commercial interest. These treatments can be of a physical, chemical or biotechnological property. The reluctance of consumers regarding the chemical treatment of horticultural commodities has promoted the use of physical treatments, which consist mainly of heat treatments before or during the low-temperature storage period and regulation and control of the gaseous composition around the horticultural commodities during storage. These technologies can be explained by the existence in plants of cross-resistance to different stress conditions. The exposure of plant to certain conditions of moderate stress not only induces resistance to this kind of stress, but also protects against other kinds of stress (Sevillano et al., 2009). Thus, approaches to modulate or control the stresses in plant tissues are very important to improving shelf life and quality retention during postharvest handling of horticultural crop.

During harvest or postharvest treatments horticultural crops are exposed to several abiotic stresses that often lead to the accumulation of metabolites. These abiotic stresses lead to a series of morphological, physiological, biochemical and molecular changes in horticultural crops (Wang et al., 2003). Abiotic stresses are often interconnected and lead to the production of ROS. The role of ROS during abiotic stress has become a subject of considerable interest given that ROS have been implicated in processes leading to plant stress acclimation. High ROS levels can lead to phytotoxicity, whereas relatively low concentrations can be employed for acclamatory signaling (Dat et al., 2000). Some postharvest treatments could induce some mechanisms that affect the metabolic activity of the treated commodity, such as the triggering of the antioxidant mechanism of horticultural crops. The activation of the antioxidant system as a response to postharvest stress can result in improving the antioxidant status of horticultural crops (González-Aguilar et al., 2010).

APX induction is one element of the response to the oxidative stress caused in plants. Expression of APX genes is activated by various abiotic stress factors. There are some connections between APX induction and regulation of metabolism. APX is a key enzyme regulating ROS concentrations acting in plants. The function of APX does not only have to be limited to antioxidative protection, but also be more extensive connection between APX induction and regulation of metabolism. APX plays an important role in controlling the concentration of ROS that participate in the signal transduction in many occurring physiological processes, and the management of ROS during stress in horticultural crops.

The controlled abiotic stresses can be used as tools by the fresh produce industry to create healthier products by enhancing the nutraceutical content of fresh-cut or whole products. This would be the basis for designing strategies to develop novel tools that will open the possibility of tailoring fresh commodity with enhanced benefit properties for use of the fresh produce and processing industries. Therefore, there is need to understand how different plant tissues and their metabolic pathways respond to different abiotic stresses, applied alone or in combination with others. There is also a need to understand how different stresses trigger the specific enzymes involved in the synthesis of targeted compound, as well as the possible interaction between different stresses and the response of the plant tissue.

REFERENCES

Agarwal, P.K., Shukla, P.S., Gupta, K., Jha, B., 2013. Bioengineering for salinity tolerance in plants: state of the art. Mol. Biotechnol. 54, 102–123.

Ahmad, P., Umar, S., 2011. Antioxidants: Oxidative Stress Management in Plants. Studium Press Pvt. Ltd, New Delhi, India.

Ahmad, P., Jaleel, C.A., Salem, M.A., Nabi, G., Sharma, S., 2010a. Roles of enzymatic and non-enzymatic antioxidants in plants during abiotic stress. Crit. Rev. Biotechnol. 30, 161–175.

Ahmad, P., Umar, S., Sharma, S., 2010b. Mechanism of free radical scavenging and role of phytohormones during abiotic stress in plants. In: Ashraf, M., Ozturk, M., Ahmad, M.S.A. (Eds.), Plant Adaptation and Phytoremediation. Springer, New York, USA, pp. 99–108.

Ahmad, P., Nabi, G., Jeleel, C.A., Umar, S., 2011. Free radical production, oxidative damage and antioxidant defense mechanisms in plants under abiotic stress. In: Ahmad, P., Umar, S. (Eds.), Oxidative Stress: Role of Antioxidants in Plants. Studium Press Pvt. Ltd, New Delhi, India, pp. 19–53.

Alscher, R.G., Donahue, J.L., Cramer, C.L., 1997. Reactive oxygen species and antioxidants: relationships in green cells. Physiol. Plant 100, 224–233.

Asada, K., 1992. Ascorbate peroxidase – a hydrogen peroxide-scavenging enzyme in plants. Physiol. Plant 85, 235–241.

Asada, K., 1999. The water–water cycle in chloroplasts: scavenging of active oxygens and dissipation of excess photons. Annu. Rev. Plant Physiol. Plant Mol. Biol. 50, 601–639.

Barka, E.A., 2001. Protective enzymes against reactive oxygen species during ripening of tomato (*Lycopersicon esculentum*) fruits in response to low amounts of UV-C. Aust. J. Plant Physiol. 28, 785–791.

Barka, E.A., Kalantari, S., Makhlouf, J., Arul, J., 2000. Effects of UV-C irradiation on lipid peroxidation markers during ripening of tomato (*Lycopersicon esculentum*) fruits. Aust. J. Plant Physiol. 27, 147–152.

Beaudry, R.M., 1999. Effects of O_2 and CO_2 partial pressure on selected phenomena affecting fruit and vegetable quality. Postharvest Biol. Technol. 15, 293–303.

Beckles, D.M., 2012. Factors affecting the postharvest soluble solids and sugar content of tomato (*Solanum lycopersicum* L.) fruit. Postharvest Biol. Technol. 63, 129–140.

Bhattacharjee, S., 2013. Heat and chilling induced disruption of redox homeostasis and its regulation by hydrogen peroxide in germinating rice seeds (*Oryza sativa* L., Cultivar Ratna). Physiol. Mol. Biol. Plant 19, 199–207.

Burdon, J., Lallu, N., Yearsley, C., Burmeister, D., Billing, D., 2007. The kinetics of acetaldehyde and ethanol accumulation in 'Hass' avocado fruit during induction and recovery from low oxygen and high carbon dioxide conditions. Postharvest Biol. Technol. 43, 207–214.

Capanoglu, E., 2010. The potential of priming in food production. Trends Food Sci. Technol. 21, 399–407.

Cascia, G., Bulley, S.M., Punter, M., Bowen, J., Rassam, M., Schotsmans, W.C., et al., 2013. Investigation of ascorbate metabolism during inducement of storage disorders in pear. Physiol. Plant 147, 121–134.

Caverzan, A., Passaia, G., Rosa, S.B., Ribeiro, C.W., Lazzarotto, F., Margis-Pinheiro, M., 2012. Plant responses to stresses: role of ascorbate peroxidase in the antioxidant proection. Genet. Mol. Biol. 35, 1011–1019.

Chai, T.T., Fadzillah, N.M., Kusnan, M., Mahmood, M., 2005. Water stress-induced oxidative damage and antioxidant reponses in micropropagated banana plantlets. Biol. Plant 49, 153–156.

Chen, G.X., Asada, K., 1989. Ascorbate peroxidase in tea leaves: occurrence of two isozymes and the differences in their enzymatic and molecular properties. Plant Cell Physiol. 30, 987–998.

Chervin, C., Brady, C.J., Patterson, B.D., Faragher, J.D., 1996. Could studies on cell responses to low oxygen levels provide improved options for fruit storage and disinfestations? Postharvest Biol. Technol. 7, 289–299.

Costa, L., Vicente, A.R., Civello, P.M., Chaves, A.R., Martinez, G.A., 2006. UV-C treatment delays postharvest senescence in broccoli florets. Postharvest Biol. Technol. 39, 204–210.

Dat, J., Vandenabeele, S., Vranová, E., Van Montagu, M., Inzé, D., Van Breusegem, F., 2000. Dual action of the active oxygen species during plant stress responses. Cell Mol. Life Sci. 57, 779–795.

Davey, M.W., Montagu, M.V., Inzé, D., Sanmartin, M., Kanellis, A., Smirnoff, N., et al., 2000. Plant L-ascorbic acid: chemistry, function , metabolism, bioavailability and effects of processing. J. Sci. Food Agric. 80, 825–869.

Davletova, S., Rizhsky, L., Liang, H., Shengqiang, Z., Oliver, D.J., Coutu, J., et al., 2005. Cytosolic ascorbate peroxidase 1 is a central component of the reactive oxygen gene network of *Arabidopsis*. Plant Cell 17, 268–281.

Dąbrowska, G., Kata, A., Goc, A., Szechyńska-Hebda, M., Skrzypek, E., 2007. Characteristics of the plant ascorbate peroxidase family. Acta. Biol. Cracov. Ser. Bot. 49, 7–17.

del Río, L.A., Pastori, G.M., Palma, J.M., Sandalio, L.M., Sevilla, F., Corpas, F.J., et al., 1998. The activated oxygen role of peroxisome in senescence. Plant Physiol. 116, 1195–1200.

Erkan, B., Wang, S., Wang, C.Y., 2008. Effect of UV treatment on antioxidant capacity, antioxidant enzyme activity and decay in strawberry fruit. Postharvest Biol. Technol. 48, 163–171.

Fallik, E., 2004. Prestorage hot water treatments immersion, rinsing and brushing. Postharvest Biol. Technol. 32, 125–134.

Ferguson, I.B., Lurie, S., Bowen, J.H., 1994. Protein synthesis and breakdown during heat shock of cultured pear (*Pyrus communis* L.) cells. Plant Physiol. 104, 1429–1437.

Ferguson, I.B., Ben-Yehoshua, S., Mitcham, E.J., McDonald, R.E., 2000. Postharvest heat treatments: introduction and workshop summary. Postharvest Biol. Technol. 21, 1–6.

Foyer, C.H., Lopez-Delgado, H., Dat, J.F., Scott, I.M., 1997. Hydrogen peroxide- and glutathione-associated mechanisms of acclamatory stress tolerance and signaling. Physiol. Plant 100, 241–254.

González-Aguilar, G., Villegas-Ochoa, M.A., Martínez-Téllez, M.A., Gardea, A.A., Ayala-Zavala, J.F., 2007. Improving antioxidant capacity of fresh-cut mangoes treated with UV-C. J. Food Sci. 72, S197–S202.

González-Aguilar, G., Villa-Rodriguez, J.A., Ayala-Zavala, J.F., Yahia, E.M., 2010. Improvement of the antioxidant status of tropical fruits as a secondary response to some postharvest treatment. Trends Food Sci. 21, 475–482.

González-Rábade, N., dei Carmen Oliver-Salvador, M., Salgado, E., Badillo-Corona, J.A., 2012. In vitro production of plant peroxidases – a review. Appl. Biochem. Biotechnol. 166, 1644–1660.

Hariyadi, P., Parkin, K.L., 1991. Chilling-induced oxidative stress in cucumber fruits. Postharvest Biol. Technol. 1, 33–45.

Hernández, J.A., Jiménez, A., Mullineaux, P., Sevilla, F., 2000. Tolerance of pea (*Pisum sativum* L.) to long-term salt stress is associated with induction of antioxidant defences. Plant Cell Environ. 23, 853–862.

Herner, R.C., 1987. High CO_2 effects on plant organs. In: Weichmann, J. (Ed.), Postharvest Physiology of Vegetables. Marcel Dekker, Dublin, Ireland, pp. 239–253.

Hidey, E., Bjorn, L.O., 1996. Ultraweak light emission, free radicals chilling and light sensitivity. Physiol. Plant 98, 223–228.

Hodges, D.M., Delong, J.M., Forney, C.F., Prange, R.K., 1999. Improving the thiobarbituric acid-reactive-substances assay for estimating lipid peroxidation in plant tissues containing anthocyanin and other interfering compounds. Planta 207, 604–611.

Hodges, D.M., Lester, G.E., Munro, K.I.D., Toivonen, P.M.A., 2004. Oxidative stress: importance for postharvest quality. HortScience 39, 924–929.

Hossain, Z., Nouri, M.-Z., Komatsu, S., 2012. Plant cell organelle proteomics in response to abiotic stress. J. Proteome Res. 11, 37–48.

Hunter, D.A., Pinkney, T.T., Watson, L.M., Trivellini, A., Janssen, B.J., Brummell, D.A., et al., 2011. Effect of postharvest water deficit stress on gene expression in heads of broccoli (*Brassica oleracea* Var. *italica*). Postharvest Biol. Technol. 59, 113–123.

Imahori, Y., 2011. Postharvest oxidative stress in fruits and vegetables. In: Ahmad, P., Umar, S. (Eds.), Oxidative Stress: Role of Antioxidants in Plants. Studium Press Pvt. Ltd, New Delhi, India, pp. 251–268.

Imahori, Y., 2012. Postharvest stress treatments in fruits and vegetables. In: Ahmad, P., Prasad, M.N.V. (Eds.), Abiotic Stress Responses in Plants, Metabolism, Productivity and Sustainability. Springer, Science + business media, New York, USA, pp. 347–358.

Imahori, Y., Kanetsune, Y., Ueda, Y., Chachin, K., 2000. Changes in hydrogen peroxide content and antioxidative enzyme activities during the maturation of sweet pepper (*Capsicum annuum* L.) fruit. J. Japan. Soc. Hort. Sci. 69, 690–695.

Imahori, Y., Kota, M., Ueda, Y., Ishimaru, M., Chachin, K., 2002. Regulation of ethanolic fermation in bell pepper fruit under low oxygen stress. Postharvest Biol. Technol. 25, 159–167.

Imahori, Y., Matushita, K., Kota, M., Ueda, Y., Ishimaru, M., Chachin, K., 2003. Regulation of fermentative metabolism in tomato fruit under low oxygen stress. J. Hort. Sci. Biotech. 78, 386–393.

Imahori, Y., Kishioka, I., Uemura, K., Makita, E., Fujiwara, H., Nishiyama, Y., et al., 2007. Effects of short-term exposure to low oxygen atmospheres on physiological responses of sweetpotato roots. J. Japan. Soc. Hort. Sci. 76, 258–265.

Imahori, Y., Takemura, M., Bai, J., 2008. Chilling-induced oxidative stress and antioxidant responses in mume (*Prunus mume*) fruit during low temperature storage. Postharvest Biol. Technol. 49, 54–60.

Imahori, Y., Kota, M., Tanaka, H., 2012. The changes of ethanolic fermentation in bell pepper fruit during induction and recovery from low oxygen conditions. Acta Hortic. 934, 185–190.

Imahori, Y., Yamamoto, K., Tanaka, H., Bai, J., 2013. Residual effects of low oxygen storage of mature green fruit on ripening processes and ester biosynthesis during ripening in bananas. Postharvest Biol. Technol. 77, 19–27.

İşeri, Ö., Körpe, D.A., Sahin, F.I., Haberal, M., 2013. Hydrogen peroxide pretreatment of roots enhanced oxidative stress response of tomato under cold stress. Acta Physiol. Plant 35, 1905–1913.

Ishikawa, T., Shigeoka, S., 2008. Recent advances in ascorbate biosynthesis and the physiological significance of ascorbate peroxidase in photosynthesizing organisms. Biosci. Biotechnol. Biochem. 72, 1143–1154.

Jamieson, L.E., Meier, X., Smith, K.J., Lewthwaite, S.E., Dentener, P.P., 2003. Effect of ethanol vapor treatments on lightbrown apple moth larval mortality and 'Braeburn' apple fruit characterization. Postharvest Biol. Technol. 28, 391–403.

Jin, J., Shan, N., Ma, N., Bai, J., Gao, J., 2006. Regulation of ascorbate peroxidase at the transcript level is involved in tolerance to postharvest water deficit stress in the cut rose (*Rosa hybrida* L.) cv. Samantha. Postharvest Biol. Technol. 40, 236–243.

Kader, A.A., 1986. Biochemical and physiological basis for effects of controlled and modified atmospheres on horticultural crops. Food Technol. 40, 99–104.

Kader, A.A., 1992. Postharvest biology and technology: an overview. In: Kader, A.A. (Ed.), Postharvest Technology of Horticultural Crops. Division of Agricultural and Natural Resources, University of California, Oakland, USA, pp. 15–20.

Kawakami, S., Matsumoto, Y., Matsunaga, A., Mayama, S., Mizuno, M., 2002. Molecular cloning of ascorbate peroxidase in potato tubers and its response during storage at low temperature. Plant Sci. 163, 829–836.

Kochhar, S., Watkin, C.B., Conklin, P.L., Brown, S.K., 2003. A quantitative and qualitative analysis of antioxidant enzymes in relation to susceptibility of apples to superficial scald. J. Am. Sci. Hort. Sci. 128, 910–916.

Kotak, S., Larkindale, J., Lee, U., von Koskull-Döring, R., Vierling, E., Scharf, K.-D., 2007. Complexity of the heat stress response in plants. Curr. Opin. Plant Biol. 10, 310–316.

Koyro, H.W., Ahmad, P., Geissler, N., 2012. Abiotic stress responses in plants: an Overview. In: Ahmad, P., Prasad, M.N.V. (Eds.), Environmental Adaptations and Stress Tolerance of Plants in the Era of Climate Change. Springer, Science + business media, New York, USA, pp. 1–28.

Larrigaudiere, C., Lentheric, I., Pinto, E., Vendrell, M., 2001. Short-term effects of air and controlled atmosphere storage on antioxidant metabolism in conference pears. J. Plant Physiol. 158, 1015–1022.

Larson, R.A., 1988. The antioxidants of higher plants. Phytochemistry 27, 969–978.

Lester, G.E., 2003. Oxidative stress affecting fruit senescence. In: Hodges, D.M. (Ed.), Postharvest Oxidative Stress in Horticultural Crops. Haworth Press Inc., New York, USA, pp. 113–129.

Lin, K.H., Pu, S.F., 2010. Tissue- and genotype-specific ascorbate peroxidase expression in sweet potato in response to salt stress. Biol. Plant 54, 664–670.

Liu, C., Han, X., Ca,i, L., Lu, X., Ying, T., Jiang, Z., 2011. Postharvest UV-B irradiation maintains sensory qualities and enhances antioxidant capacity in tomato fruit during storage. Postharvest Biol. Technol. 59, 232–237.

Liu, X., Anderson, J.A., Maness, N.O., Martin, B., 1996. Protein synthesis inhibitors block high temperature acclimation in bell pepper. HortScience 31, 160–161.

Lukatkin, A.S., 2002. Contribution of oxidative stress to the development of cold-induced damage to leaves of chilling-sensitive plants. 2. The activity of antioxidant enzymes during plant chilling. Russ. J. Plant Physiol. 49, 782–788.

Lurie, S., 1998. Postharvest heat treatments. Postharvest Biol. Technol. 14, 257–269.

Mahajan, S., Tuteja, N., 2005. Cold, sanlinity and drought stresses: an overview. Arch. Biochem. Bioph 444, 139–158.

Mathooko, F.M., 1996. Regulation of respiratory metabolism in fruits and vegetables by carbon dioxide. Postharvest Biol. Technol. 9, 247–264.

Mehlhorn, R., Tabner, B.J., Wellburn, A.R., 1990. Electron spin resonance evidence for the formation of free radicals in plants exposed to ozone. Physiol. Plant 79, 377–383.

Miller, G., Suzuki, N., Ciftci-Yilmaz, S., Mittler, R., 2010. Reactive oxygen species homeostasis and signaling during drought and salinity stresses. Plant Cell Environ. 33, 453–467.

Mittler, R., 2002. Oxidative stress, antioxidants and stress tolerance. Trends Plant Sci. 7, 405–410.

Morita, S., Kaminaka, H., Masumura, T., Tanaka, K., 1999. Induction of rice cytosolic ascorbate peroxidase mRNA by oxidative stress signalling. Plant Cell Physiol. 40, 417–422.

Murshed, R., Lopez-Lauri, F., Sallanon, H., 2013. Effect of water stress on antioxidant systems and oxidative parameters in fruits of tomato (*Solanum lycpopersicon* L, cv. Micro-tom). Physiol. Mol. Biol. Plant 19, 363–378.

Noctor, G., Foyer, C.H., 1998. Ascorbate and glutathione: keeping active oxygen under control. Annu. Rev. Plant Physiol. Plant Mol. Biol. 49, 249–279.

Pan, J., Vicente, A.R., Martinez, G.A., Chaves, A.R., Civello, P., 2004. Combined use UV-C irradiation and heat treatment to improve postharvest life of strawberry fruit. J. Sci. Food Agric. 84, 1831–1838.

Paull, R.E., Chen, N.J., 2000. Heat treatment and fruit ripening. Postharvest Biol. Technol. 21, 21–37.

Perkins-Veazie, P., Collins, J., Howard, L., 2008. Blueberry fruit response to postharvest application of ultraviolet radiation. Postharvest Biol. Technol. 47, 280–285.

Piqueras, A., Hernández, J.A., Olmos, E., Hellín, E., Sevilla, F., 1996. Changes in antioxidant enzymes and organic solutes associated with adaptation of citrus cells to salt stress. Plant Cell Tissue Organ. Cult. 45, 53–60.

Pombo, M., Dotto, M.C., Martinez, G.A., Civello, P.M., 2009. UV-C irradiation delays strawberry fruit softening and modifies the expression of genes involved in cell wall degradation. Postharvest Biol. Technol. 51, 141–148.

Posmyk, M.M., Bailly, C., Szafránska, K., Jana, K.M., Corbineau, F., 2005. Antioxidant enzymes and isoflavonoids in chilled soybean (*Glycine max* (L.) Merr.) seedling. J. Plant Physiol. 162, 403–412.

Purvis, A.C., 2002. Diphenylamine reduces chilling injury of green bell pepper fruit. Postharvest Biol. Technol. 25, 41–48.

Purvis, A.C., 2004. Regulation of oxidative stress in horticultural crops. HortScience 39, 930–932.

Reddy, A.R., Chaitanya, K.V., Jutur, P.P., Sumithra, K., 2004. Differential antioxidative responses to water stress among five mulberry (*Morus alba* L.) cultivars. Environ. Exp. Bot. 52, 33–42.

Sairam, R.K., Tyagi, A., 2004. Physiology and molecular biology of salinity stress tolerance in plants. Curr. Sci. 86, 407–421.

Sairam, R.K., Srivastava, G.C., Saxena, D.C., 2000. Increased antioxidant activity under elevated temperature: a mechanism of heat stress tolerance in wheat genotypes. Biol. Plant 43, 245–251.

Sala, J.M., Lafuente, M.T., 2000. Catalase enzyme activity is related to tolerance of mandarin fruits to chilling. Postharvest Biol. Technol. 20, 81–89.

Sala, J.M., Lafuente, M.T., 2004. Antioxidant enzymes activities and rind-staining in 'Naveline' oranges as affected by storage relative humidity and ethylene conditioning. Postharvest Biol. Technol. 31, 277–285.

Saltveit Jr, M.E., Morris, L.L., 1990. Overview on chilling injury of horticultural crops. In: Wang, C.Y. (Ed.), Chilling Injury of Horticultural Crops. CRC Press, Boca Raton, Florida, USA, pp. 3–15.

Sato, Y., Murakami, T., Funatsuki, H., Matsuba, S., Saruyama, H., Tanida, M., 2001. Heat shock-mediated APX gene expression and protection against chilling injury in rice seedlings. J. Exp. Bot. 52, 145–151.

Seckin, B., Türkan, I., Sekmen, A.H., Ozfidan, C., 2010. The role of antioxidant defense systems at differential salt tolerance of *Hordeum marinum* Huds. (sea barleygrass) and *Hordeum vulgare* L. (cultivated barley). Environ. Exp. Bot. 69, 76–85.

Sevillano, L., Sanchez-Ballesta, M.T., Romojaro, F., Flores, F.B., 2009. Physiological, hormonal and molecular mechanisms regulating chilling injury in horticultural species. Postharvest technologies applied to reduce its impact. J. Sci. Food Agric. 89, 555–573.

Shama, G., 2007. Process challenges in applying low doses of ultraviolet light to fresh produce for eliciting beneficial hermetic responses. Postharvest Biol. Technol. 44, 1–8.

Shama, G., Alderson, P., 2005. UV homesis in fruits: a concept ripe for commercialization. Trends Food Sci. Technol. 16, 128–136.

Shigenaga, T., Yamauchi, N., Funamoto, Y., Shigyo, M., 2005. Effects of heat treatment on an ascorbate-glutathione cycle in stored brocoli (*Brassica oleracea* L.) florets. Postharvest Biol. Technol. 38, 152–159.

Shigeoka, S., Ishikawa, T., Tamoi, M., Miyagawa, Y., Takeda, T., Yabuta, Y., et al., 2002. Regulation and function of ascorbate peroxidase isoenzymes. J. Exp. Bot. 53, 1305–1319.

Singh, S.P., Singh, Z., 2013. Dynamics of enzymatic and non-enzymatic antioxidants in Japanese plums during storage and lethal temperatures. LWT-Food Sci. Technol. 50, 562–568.

Smirnoff, N., 1996. The function and metabolism of ascorbic acid in plants. Ann. Bot. 78, 661–669.

Smith, R.B., 1992. Controlled atmosphere storage of 'Redcoat' strawberry fruit. J. Am. Soc. Hort. Sci. 117, 260–264.

Stuger, R., Ranostaj, S., Materna, T., Forreiter, C., 1999. Message RNA-binding properties of non-polysomal ribonucleoproteins from heat stressed tomato cells. Plant Physiol. 120, 23–31.

Suzuki, N., Miller, G., Sejima, H., Harper, J., Mittler, R., 2013. Enhanced seed production under prolonged heat stress conditions in Arabidopsis thaliana plants deficient in cytosolic ascorbate peroxidase 2. J. Exp. Bot. 64, 253–263.

Szarka, A., Tomasskovics, B., Bánhegyi, G., 2012. The ascorbate-glutathione-α-tocopherol triad in abiotic stress response. Int. J. Mol. Sci. 13, 4458–4483.

Toivonen, P.M.A., 1995. Non-ethylene, non-respiratory volatiles in harvested fruits and vegetables: their occurrence, biological activity and control. Postharvest Biol. Technol. 12, 109–125.

Toivonen, P.M.A., 2003. Effects of storage conditions and postharvest procedures on oxidative stress in fruits and vegetables. In: Hodges, D.M. (Ed.), Postharvest Oxidative Stress in Horticultural Crops. Haworth Press Inc., New York, USA, pp. 69–90.

Toivonen, P.M.A., 2004. Postharvest storage procedures and oxidative stress. HortScience 39, 938–942.

Türkan, I., Demiral, T., 2009. Recent developments in understanding salinity tolerance. Environ. Exp. Bot. 67, 2–9.

Verslues, P.E., Agarwal, M., Katiyar-Agarwal, S., Zhu, J., Zhu, J.K., 2006. Methods and concepts in quantifying resistance to drought, salt and freezing, abiotic stresses that affect plant water status. Plant J. 45, 523–539.

Vicente, A.R., Martínez, G.A., Chaves, A.R., Civello, P.M., 2006. Effect of heat treatment on strawberry fruit damage and oxidative metabolism during storage. Postharvest Biol. Technol. 40, 116–122.

Wahid, A., Gelani, S., Asharf, M., Foolad, M.R., 2007. Heat tolerance in plants: an overview. Environ. Exp. Bot. 61, 199–223.

Wang, C.Y., 1995. Effects of temperature preconditioning on catalase, peroxidase, and superoxide dismutase in chilled zucchini squash. Postharvest Biol. Technol. 5, 67–76.

Wang, C.Y., 2010. Alleviation of chilling injury in tropical and subtropical fruits. Acta. Hort. 864, 267–274.

Wang, C.Y., Kramer, G.F., Whitaher, B.D., Lusby, W.R., 1992. Temperature preconditioning increases tolerance to chilling injury and alters lipid composition in zucchini squash. J. Plant Physiol. 140, 229–235.

Wang, H., Zhang, Z., Xu, L., Huang, X., Pang, X., 2012. The effect of delay between heat treatment and cold storage on alleviation of chilling injury in banana fruit. J. Sci. Food Agric. 92, 2624–2629.

Wang, W., Vinocur, B., Altman, A., 2003. Plant responses to drought, salinity and extreme temperatures: towards genetic engineering for stress tolerance. Planta 218, 1–14.

Weichmann, J., 1987. Low oxygen effects. In: Weichmann, J. (Ed.), Postharvest Physiology of Vegetables. Marcel Dekker, Dublin, Ireland, pp. 231–237.

Willekens, H., Inzé, D., Van Montagu, M., Van Camp, W., 1995. Catalase in plants. Mol. Breed. 1, 207–228.

Wismer, W.V., 2003. Low temperature as a causative agent of oxidative stress in postharvest crops. In: Hodges, D.M. (Ed.), Postharvest Oxidative Stress in Horticultural Crops. Haworth Press Inc., New York, USA, pp. 55–68.

Xia, J., Saglio, P.H., 1992. Lactic acid efflux as a mechanism of hypoxic acclimation of maize root tips to anoxia. Plant Physiol. 100, 40–46.

Xu, F., Chen, X., Jin, P., Wang, X., Wang, J., Zheng, Y., 2012. Effect of ethanol treatment on quality and antioxidant activity in postharvest broccoli florets. Eur. Food Res. Technol. 235, 793–800.

Yamauchi, N., 2013. Quality maintenance of postharvest horticultural crops by stress treatments and approach for the elucidation of its mechanism. J. Japan Soc. Hort. Sci. 82, 1–10.

Yang, Z., Zheng, Y., Cao, S., 2009. Effect of high oxygen atmosphere storage on quality, antioxidant enzymes, and DPPH-radical scavenging activity of Chinese bayberry fruit. J. Agric. Food Chem. 57, 176–181.

Zhang, Y., 2013. Ascorbic Acid in Plants: Springerbriefs in Plant Science. Springer, New York, USA.

Chapter 15

Mycorrhizal Association and ROS in Plants

Qiang-Sheng Wu, Ying-Ning Zou and Elsayed Fathi Abd-Allah

15.1 INTRODUCTION

The molecular oxygen (O_2) of organisms often easily absorbs electrons from other molecules, and then forms reactive oxygen. The biological reactive oxygen species (ROS) mainly include hydroxyl radical ($^{\bullet}OH$), superoxide anion radical ($O_2^{\bullet -}$), singlet oxygen ($^1O_2^{-}$), hydrogen peroxide (H_2O_2), hypochlorous acid (HOCl), peroxyl radical ($HOO^{\bullet}$), etc. (Apel and Hirt, 2004; Ahmad et al., 2010a,b, 2011, 2012; Ahmad and Umar, 2011; Koyro et al., 2012; Rasool et al., 2013). The sites of ROS generation in plant cells are located in cell membrane, chloroplasts, mitochondria and peroxisomes (Movafegh and Hoseini, 2013). Herein, chloroplasts and mitochondria are the main parts of ROS generation in plants via electron transfer and β-oxidation (Dutilleul et al., 2003). The photoreduction reaction in chloroplasts is the main pathway to produce ROS (Gill and Tuteja, 2010). In mitochondria, electron transport, especially complex I and complex III, cannot reach the terminal oxidase normally, but directly react with oxygen, then form $O_2^{\bullet -}$ (Moller, 2001). In general, ROS generation and elimination are in dynamic balance under normal metabolic processes and are controlled by both the energy level and the redox potential (Pospíšil, 2012). However, under adverse conditions ROS is generated in large quantities, thus breaking the dynamic equilibrium. An excess of ROS can cause protoplast membrane oxidation and degradation of macromolecules, thus resulting in cell damage and loss of normal physiological function, even cell death (Ahmad et al., 2010a,b, 2011; Ahmad and Umar, 2011; Koyro et al., 2012; Rasool et al., 2013). Interestingly, ROS in a plant defense role acts as signaling molecules and also plays an important role against pathogen invasion in plants (Torres et al., 2002, 2005; Barna et al., 2012). The signal process may mainly involve the regulation of heat shock protein 70 production and accumulation, thereby triggering the defense mechanisms (Piterková et al., 2013).

P. Ahmad (Ed): Oxidative Damage to Plants. DOI: http://dx.doi.org/10.1016/B978-0-12-799963-0.00015-0

Plants also develop an intact defense system to allow the scavenging of ROS for dynamic equilibrium, and the antioxidant defense systems include both enzymatic and nonenzymatic systems. Nonenzymatic antioxidant defense systems mainly refer to the antioxidant substances in plants, which include ascorbic acid (ASC), glutathione (GSH), phenolic compounds, alkaloids, tocopherol, carotenoids, etc. (Caverzan et al., 2012). The antioxidants not only play a direct role in the elimination of ROS but also act as a substrate for the antioxidant enzymes in the removal of ROS. The main antioxidant enzymes are superoxide dismutase (SOD), guaiacol peroxidase (G-POD), glutathione peroxidase (GPX), catalase (CAT), glutathione reductase (GR), ascorbate peroxidase (APX), monodehydroascorbate reductase (MDHAR), and dehydroascorbate reductase (DHAR) (Wu et al., 2006b; Khanna-Chopra and Selote, 2007; Ahmad et al., 2010a,b, 2011; Ahmad and Umar, 2011; Koyro et al., 2012; Rasool et al., 2013). The enzymatic defense system not only scavenges ROS but also repairs the damage caused by ROS (Moller, 2001). In addition, proline is an inefficient $^1O_2$ scavenger, because the proline accumulation did not quench $^1O_2$ in neutral aqueous solutions (Signorelli et al., 2013). If the abiotic and biotic stresses are too severe, these antioxidant defense systems are not able to neutralize the excess ROS, possibly still resulting in cell death and necrosis development (Barna et al., 2012).

Recently, the relationship between the mycorrhizal association and ROS has been of wide concern and has been reviewed (Wu and Zou, 2011). Approximately 90% of terrestrial plants can form mycorrhizal associations between their roots and soil mycorrhizal fungi, which provide a variety of benefits to the host plant, including improvement of nutrient uptake under poor soil nutrient conditions, enhancement of stressed tolerance, and increments of growth and yield (Smith and Read, 2008). Seven categories of mycorrhizas, including ectomycorrhizas, arbuscular mycorrhizas (AMs), ericoid mycorrhizas, arbutoid mycorrhizas, monotropoid mycorrhizas, ectoendomycorrhizas, and orchid mycorrhizas, have been described in detail by Peterson et al. (2004). Meanwhile, AMs are by far the most prevalent of all mycorrhizal types. AMs occur in all major grain crops and horticultural plants with 80% of plant species (Brundrett, 2009). In general, AM associations are characterized by the presence of extraradical hyphae, intraradical hyphae, arbuscules, vesicles, and spores of extraradical hyphae (Figs. 15.1 and 15.2). Arbuscules are the substance exchange sites between plants and AMs. Arbuscular mycorrhizal fungi (AMF) can influence competitive interaction among plant species as well as plant community composition (Chmuraa and Gucwa-Przepiórab. 2012). Several studies indicated that inoculation with AMF under nursery conditions substantially enhanced growth of the host plant and tolerance of environmental stress (Muthukumar et al., 2001; Muthukumar and Udaiyan, 2010; Porcel et al., 2012; Tahat and Sijam, 2012).

FIGURE 15.1 Mycorrhizal infection of *Poncirus trifoliata* seedlings infected by *Glomus mosseae*. Here, lots of intercellular hyphae appear within the roots.

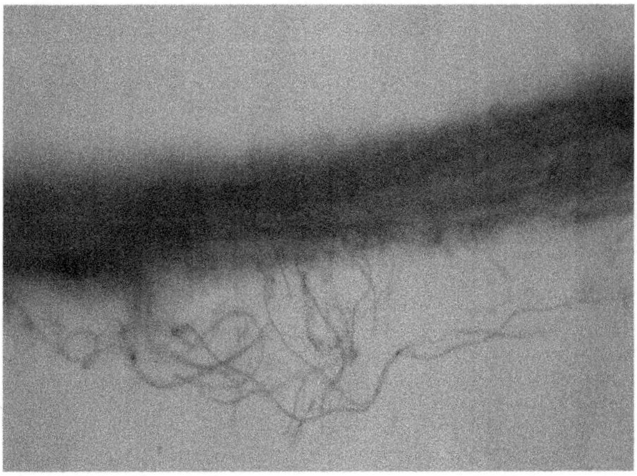

FIGURE 15.2 Mycorrhizal infection of *Trifolium repens* seedlings infected by *Glomus mosseae*. Here, lots of mycorrhizal extraradical hyphae appear on the outside of the roots.

It is well documented that AMs have certain effects on the ROS metabolism of the host plant. In particular, the arbuscule life period of AM symbiosis clearly suggested the accumulated status of H_2O_2 (Salzer et al., 1999; Fester and Hause, 2005). Furthermore, AM colonization was connected with antioxidant enzyme activities and antioxidant concentrations, thereby alleviating oxidative damage of plants grown under stress (Peterson et al., 2004). The AM-mediated antioxidant defense systems may depend on the combination of the fungal species with the host plant, as well as the micronutrient availability (Ruiz-Lozano et al., 2012). Based on the potential relation, we focus on H_2O_2 observation in AM symbiosis and then assess the relationship of accumulation of H_2O_2 with arbuscule life period. We also illustrate the effects of AM symbiosis on ROS burst, antioxidant enzyme activities and antioxidant concentrations of the host plant. Some of the potential mechanisms are described in detail.

15.2 ROS OBSERVATION IN MYCORRHIZAL ASSOCIATIONS

Although the presence of AM symbiosis can obviously alter antioxidant defense systems to decrease ROS burst in the host plant (see below), changes in biochemical variables often showed inconsistency in many research studies. Therefore, few researchers attempted to find the evidence directly of ROS in mycorrhizal associations through morphological observation. A diaminobenzidine (DAB) staining technique was used to examine the cellular localization of H_2O_2 in parts of the roots of *Medicago truncatula* inoculated with *Glomus intraradices*, in combination with bright-field and fluorescence microscopy (Salzer et al., 1999). Generally, localization of H_2O_2 is presented in cells of roots containing clumped or less-branched arbuscules. In addition, H_2O_2 also accumulated in the apoplasmic space around hyphal tips. In contrast, H_2O_2 was not accumulated in highly branched arbuscules, vesicles, appressoria, middle lamella, and adjacent cells of arbuscules. The accumulation of H_2O_2 suggests that H_2O_2 occurred in the later stages of the arbuscule life cycle, when arbuscules lose their branches and degenerate. So, accumulation of H_2O_2 may be involved in arbuscule degradation. In other words, the elimination of H_2O_2 is highly positively correlated with arbuscule richness (Salzer et al., 1999).

Fester and Hause (2005) further investigated the exact localization of H_2O_2 accumulation in the roots of *Medicago truncatula*, *Zea mays* and *Nicotiana tabacum* infected by *G. intraradices* using three independent staining techniques with DAB, dihydrorhodamine 123 (DHR 123) and $CeCl_3$. Comparative analysis showed that staining with $CeCl_3$ was less sensitive than staining with DAB or DHR 123. Fungal cytosol, surface of intercellular fungal hyphae, spore wall of *G. intraradices*, and hyphal wall detected the accumulation of H_2O_2. Accumulation of H_2O_2 could be clearly localized with the fungal cytosol in the case of intercellular fungal hyphae and in the vicinity of collapsing and of intact fungal hyphae, when stained with DHR 123 or observed under the electron imcroscope. Arbuscule observation revealed that H_2O_2 diffusing across the thin hyphal wall of arbuscule branches might induce the fungal program for senescence (Fester and Hause, 2005).

Taken together, it suggests that more accumulation of H_2O_2 in arbuscules may predict the collapse or degradation of arbuscules during the mycorrhizal development.

15.3 MYCORRHIZAS AND ROS BURST

15.3.1 Superoxide ($O_2^{\bullet -}$)

Superoxide ($O_2^{\bullet -}$) is the main source of ROS *in vivo* (Zhao, 1999) and can be formed by oxidases, peroxidases, or by photosynthetic and respiratory electron transport chain (Foyer and Noctor, 2000; Mittler et al., 2004; Bindschedler et al., 2006; Sagi and Fluhr, 2006; Kreslavski et al., 2012). In addition, $O_2^{\bullet -}$ can activate a radical chain reaction and thus induce the production of $^{\bullet}OH$, $^1O_2$, and H_2O_2 (Halliwell, 1981; Yang and Gao, 2001). Under stress conditions, the plant cell can accumulate a large amount of ROS, which trigger membrane lipid peroxide and damage the membrane system, even causing plant death. Meanwhile, drought (Sun et al., 2013), low temperature (Tartoura and Youssef, 2011), hard light (Hodgson and Rasion, 1991), and infection by germs (Barna et al., 2012) induce more accumulation of $O_2^{\bullet -}$.

It has been recognized that AMF can enhance tolerance of the host plant to environmental stresses through reduction of $O_2^{\bullet -}$ concentrations (Table 15.1). Under drought stress conditions, inoculation with *G. versiforme* significantly decreased leaf and root $O_2^{\bullet -}$ concentrations of *Citrus tangerine* (Wu et al., 2006a, 2007b). In *Poncirus trifoliata* seedlings, the $O_2^{\bullet -}$ concentration of AM roots was 30% and 26% less than that of non-AM roots under well water and drought stress conditions, respectively (Wu et al., 2006b). The result may be attributed to higher antioxidant protected systems. Currently, it is not clear whether the elimination of $O_2^{\bullet -}$ is directly by AMF or by some other mechanisms. Correlation analysis showed that mycorrhizal colonization and arbuscules, but not vesicles and entry points, were significantly negatively correlated with root $O_2^{\bullet -}$ under drought stress conditions (Table 15.2; Wu and Zou, 2009).

15.3.2 Hydrogen Peroxide (H_2O_2)

Hydrogen peroxide (H_2O_2) is generated in plant cells under normal and stressful conditions such as drought, saline, chilling, UV irradiation, wounding, and intrusion by pathogens (Ahmad et al., 2010a,b, 2011; Ahmad and Umar, 2011). The electron transport chain of chloroplast, mitochondria, endoplasmic reticulum and plasma membrane, β-oxidation of fatty acid, and photorespiration are major sources of H_2O_2 generation in plant cells (Ahmad et al., 2011; Ahmad and Umar, 2011). Photooxidation reactions, NADPH oxidase as well as xanthine oxidase also contribute to H_2O_2 production in plants (Sharma et al., 2012). H_2O_2 has no unpaired electron, unlike other ROS, and it can readily cross biological membranes and consequently cause oxidative damage far from the site of its formation. In addition, H_2O_2 is the only ROS that can diffuse through aquaporins in the membranes and over larger distances within the cell (Bienert et al., 2007). H_2O_2 receives special attention as a signal molecule, which is involved in the regulation of specific biological

TABLE 15.1 **Effects of Inoculation with Arbuscular Mycorrhizal Fungi on $O_2^{\bullet -}$ and H_2O_2 Concentrations of Host Plants Exposed to Various Abiotic Stresses**

Fungal species	Host species	Conditions	Variables		References
			O_2^-	H_2O_2	
Glomus versiforme	*Citrus tangerine*	drought stress	<leaf	< leaf	Wu et al., 2007b
		drought stress	<root	<root	Wu et al., 2006a
	Poncirus trifoliata	drought stress	<leaf; < root	<leaf; < root	Wu et al., 2006b
		salt stress		<leaf	Wu et al., 2010
	C. sinensis /P. trifoliata	drought stress	<leaf; <root	<leaf; <root	Wu and Zou, 2009
G. mosseae	*P. trifoliata*	salt stress		<leaf	Wu et al., 2010
	Glycine max	drought stress		= nodule	Porcel et al., 2003
	Lycopersicon esculentum	salt stress	>leaf		Huang et al., 2010
	Avena nuda	Charcoal filtered air SO_2 stress		>leaf < leaf	Huang et al., 2008
G. intraradices	*Solanum lycopersicum*	salt stress		<leaf	Hajiboland et al., 2010

Note: Symbols > and < indicate that the variable is significantly increased and decreased by AMF colonization.

processes, and triggering tolerance against various environmental stresses (Neill et al., 2002; Torres et al., 2002; Yan et al., 2007). It seems that H_2O_2 exhibits two functions: as a signaling molecule at a certain concentration range and also as ROS which induces oxidative damage at higher concentration.

Studies have shown that AM colonization decreased H_2O_2 concentration of the host plant (Table 15.1). The higher activities of antioxidant enzymes in mycorrhizal plants induced lower accumulation of H_2O_2 as compared to non-mycorrhizal plants, indicating lower oxidative damage in the colonized plants (Hajiboland et al., 2010). Wu et al. (2006b, 2007b, 2010) found that the leaf H_2O_2 concentration was significantly lower in the inoculated

TABLE 15.2 Pearson's Correlations (*r*) between Symbiotic Development and Biochemical Variables

Tissue	AM structure	O_2^-	H_2O_2	MDA	SOD	G-POD	CAT	ASC	GSH
Leaf	AM colonization	−0.89*	−0.85	−0.95*	0.04	−0.89*	−0.86	0.87	0.91*
	Entry points	−0.48	−0.41	−0.72	−0.09	−0.52	−0.71	0.52	0.71
	Vesicles	−0.55	−0.43	−0.72	0.00	−0.53	−0.69	0.54	0.70
	Arbuscules	−0.94*	−0.97**	−0.95*	−0.19	−0.96**	−0.87	0.99**	0.93*
Root	AM colonization	−0.96*	−0.97**	−0.92*	0.35	−0.87*	−0.94*	0.98**	0.70
	Entry points	−0.67	−0.67	−0.56	0.24	−0.51	−0.59	0.67	0.62
	Vesicles	−0.17	−0.71	−0.59	0.31	−0.52	−0.64	0.71	0.56
	Arbuscules	−0.99**	−0.99**	−0.99**	0.07	−0.94*	−0.88*	0.91*	0.74

Note: * and ** indicate that correlations were significant at $P < 0.05$ or 0.01, respectively.
Adapted from Wu and Zou (2009) with permission.

seedlings with *G. versiforme* or *G. mosseae* than in the non-AM citrus plants under salinity stress and drought stress. The decreases of root H_2O_2 caused by mycorrhization were significantly negatively correlated with AM colonization and arbuscules but without entry points and vesicles (Table 15.2; Wu and Zou, 2009). It seems that mycorrhizal colonization and arbuscules has an important role in H_2O_2 elimination. Cellular observation also confirmed the roles of arbuscules in H_2O_2 elimination (Fester and Hause, 2005).

15.4 MYCORRHIZAS AND ANTIOXIDANT ENZYMES

15.4.1 Superoxide Dismutase

Superoxide dismutases (SODs) are a family of metalloenzymes, exist widely in microorganisms, animals and plants, and also participate in many physiological processes of the organism against various adverse conditions (Huseynova and Aliyev, 2012). SODs are commonly found in chloroplasts, mitochondria, peroxisomes, and cytoplasm of higher plants in three molecular forms, namely Mn-SOD, Cu/Zn-SOD, and Fe-SOD. In general, Mn-SOD is located in mitochondria and peroxisome, Fe-SOD in chloroplasts, and Cu/Zn-SOD in chloroplast and cytoplasm. Mn-SOD and Fe-SOD seem to be high homology in both sequence and structure, whereas Cu/Zn-SOD do not have homology with either Mn-SOD or Fe-SOD. SODs play a central role in protecting from oxidative damage, because SODs can catalyze the dismutation of $O_2^{\bullet -}$ to H_2O_2.

Generally, the SOD activity of plants inoculated with AMF were significantly higher than non-AMF inoculated plants (Table 15.3). For example, a potted experiment showed that SOD activity in roots of tomato colonized by *G. mosseae* was significantly higher than the non-AM plants under salinity or nonsalinity conditions (He et al., 2007). The SOD activity of *G. mosseae*-colonized *Suaeda salsa* seedlings in salt-stressed (Wang et al., 2004), *G. mosseae*-inoculated tomato seedlings (Li et al., 2012), and *G. versiforme*-inoculated *Lactuca sativa* under water stress (Ruiz-Lozano et al., 1996) was significantly higher than that in the non-inoculated controls. In both well-watered and drought-stressed red tangerine roots, SOD activity was significantly higher in AM than in non-AM plants (Wu et al., 2006a), indicating the induction of some new SOD isozymes (Palma et al., 1993) or expression of the *Mn-SOD II* gene (Ruiz-Lozano et al., 2001). However, negative correlation ($r = -0.66$, $P < 0.05$) between AM colonization and SOD activity was in drought-stressed roots (Wu et al., 2007a). Interestingly, during 12 days of drought, mycorrhizal structures did not show any correlation with leaf and root SOD activity (Table 15.2; Wu and Zou, 2009). Lanfranco et al. (2005) cloned a SOD gene (namely *GmarCuZnSOD*) from germinated spores of an AM fungus, *Gigaspora margarita*, which encoded a functional polypeptide to increase oxidative tolerance and also was controlled during the

TABLE 15.3 Effects of Inoculation with Arbuscular Mycorrhizal Fungi on SOD Activity of Host Plants Exposed to Various Abiotic Stresses

Fungal species	Host species	Conditions	SOD	References
G. versiforme	*Citrus tangerine*	well water; drought stress	>leaf; >root	Wu et al., 2006b
		drought stress	= leaf	Wu et al., 2007a
	Poncirus trifoliata	salt stress	= leaf	Wu et al., 2010
		ample water	= leaf; = root	Wu et al., 2006c
G. mosseae	*Suaeda salsa*	salt stress	>leaf	Li et al., 2012
	Glycine max	well water drought stress	<nodule; = root <nodule; = root	Porcel et al., 2003
	Lactuca sativa	drought stress	>leaf; >root	Ruiz-Lozano et al., 1996
	C. tangerine	drought stress	>leaf	Wu et al., 2007a
	P. trifoliata	high temperature	> leaf	Wu et al., 2006b
		low temperature	= leaf	
		salt stress	>leaf	Wu et al., 2010
	Lycopersicon esculentum	salt stress	>leaf; >root	Huang et al., 2010
			>leaf	Abdel Latef and He, 2011
	Capsicum annuum	salt stress (1 mM NaCl)	= leaf	Çekic et al., 2012
		salt stress (2 mM NaCl)	= leaf	
		salt stress (4 mM NaCl)	= leaf	
		salt stress (8 mM NaCl)	= leaf	
	Avena nuda	Charcoal filtered air	= leaf > leaf	Huang et al., 2008

(*Continued*)

TABLE 15.3 (Continued)

Fungal species	Host species	Conditions	SOD	References
		SO_2 stress		
G. intraradices	*Capsicum annuum*	salt stress (1 mM NaCl)	= leaf	Çekic et al., 2012
		salt stress (2 mM NaCl)	<leaf	
		salt stress (4 mM NaCl)	<leaf	
		salt stress (8 mM NaCl)	= leaf	
G. deserticola	*Lactuca sativa*	drought stress	>leaf; >root	Ruiz-Lozano et al., 1996
G. geopsorum	*C. tangerine*	drought stress	= leaf	Wu et al., 2007a
G. diaphanum	*C. tangerine*	drought stress	>leaf	Wu et al., 2007a
G. etunicatum	*C. tangerine*	drought stress	= leaf	Wu et al., 2007a
	Glycine max	salt stress	>leaf; >root	Ghorbanli et al., 2004
Mixture of three Glomus species	*Juniperus oxycedrus*	drought stress Initial middle final recovery	<leaf < leaf < leaf > leaf	Roldán et al., 2008

Note: Symbols >, <, and = indicate that the variable is significantly increased, decreased and not affected by AMF colonization.

life cycle of AMF. Corradi et al. (2009) analyzed the diversity of the *SOD1* gene from AMF and found that the *SOD1* gene was monomorphic within each isolate of *G. intraradices*.

In addition, after 14 days of 400 mM NaCl stress, the expression of *Fe-SOD*, *Cu/Zn-SOD 1*, and *Cu/Zn-SOD 2* in leaves was 1.20, 1.19, and 1.07 times higher in the mycorrhizal plants than in the nonmycorrhizal plants, respectively (Li et al., 2012). AMF is able to induce new SOD isoforms to tolerate stresses. Compared to the non-AM inoculated plants, clover plants presented two new SOD isoforms in roots: mycCuZn-SOD and a Mn-SOD under well-watered conditions (Palma et al., 1993).

Different *Glomus* species also showed dissimilar increments in SOD, and *G. mosseae* was more effective than *G. versiforme* in micropropagated citrus plantlets (Wu et al., 2006c). He et al. (2007) reported higher SOD activity of salt-stressed soybean inoculated with *G. etunicatum* or tomato inoculated with *G. mosseae*, compared with other *Glomus* species.

Four *Glomus* species (*G. mosseae*, *G. geosporum*, *G. versiforme*, and *G. diaphanum*) significantly increased the SOD activity of red tangerine seedlings grown in well water, except for *G. etunicatum* (Wu et al., 2007a). Higher SOD activities were also observed in lettuce (*Lactuca sativa*) roots colonized by *G. mosseae* or *G. deserticola* under drought-stress. However, the SOD activity in different tissues of a plant caused by mycorrhization is different. For instance, Porcel et al. (2003) found that in nodules, SOD activity was lower in drought-stressed AM soybean plants than in the corresponding non-AM plants. However, in bean (*Phaseolus vulgaris*) colonized by *G. clarum*, SOD was induced in roots at a late stage of the symbiosis development under low P and the activity of SOD was evaluated (Lambais et al., 2003). Roldán et al. (2008) reported that AMF inoculation produced a decrease in SOD activity in shoots of drought-stressed *Juniperus oxycedrus* plants, suggesting that the AM plants may be subjected to a lower oxidative stress under drought stress and show a primary drought avoidance mechanism.

Interestingly, SOD changes by mycorrhization may be dependent on long-term stress. In tomato (*Lycopersicon esculentum*) infected by *G. mosseae*, the highest SOD activity of leaves happened after 30 days of 0.5% NaCl, and decreased at 40 days (Huang et al., 2010). Under 1% NaCl treatment, the highest SOD activity of AM plants occurred after 20 days. Similarly, compared with non-AM controls, a decrease of SOD activity was found in AM *Juniperus oxycedrus* plant during drought stress, and an increase of SOD activity after drought recovery (Roldán et al., 2008).

In short, AM symbiosis can enhance SOD activity or induce new SOD genes to protect the host plant against abiotic stresses. The AM induction may be indirectly related to the AM effects on host plant growth and N and P acquisition (Ruiz-Lozano et al., 2012).

15.4.2 Peroxidase (POD)

Peroxidase is widespread in a variety of animals, plants and microorganisms and comprises bacteria peroxidase, fungi peroxidase and plant peroxidase. POD can act as catalysts to facilitate a variety of biological processes, especially donating electrons to bind to other substances, such as ferrocyanide and ascorbate, to break them down into harmless components (Li et al., 2008). POD also degrades H_2O_2 into water and oxygen (Tian et al., 2001). According to the catalytic substrate characteristics, POD can be divided into guaiacol peroxidase (G-POD), ascorbate peroxidase (APX) and glutathione peroxidase (GPX), etc. Under abiotic stress conditions, POD also functions to interrupt the cascades of uncontrolled ROS in some organelles (Jiang and Zhang, 2001; Harinasut et al., 2003).

Many experiments showed that AMs could alter POD activities of host plants, to a certain degree. Under salt stress, soybean plants inoculated with salt pretreated *G. etunicatum* showed higher POD activities in shoots, compared with the plants inoculated with the nonpretreated fungi (Ghorbanli et al., 2004). In *Olea european* plants, treatment by *G. claroideum* significantly increased APX activities in shoots (Alguacil et al., 2003). Under well-watered conditions the *Poncirus trifoliata* seedlings infected by *G. versiforme* has higher leaf G-POD activity and root APX activity, and under water stress conditions the infected seedlings also recorded with higher leaf G-POD and leaf and root APX activity (Wu et al., 2006b). Similarly, *G. mosseae* markedly increased POD and APX activities in leaf and root of *Lycopersicon esculentum* plants subjected to salt stress (Huang et al., 2010). Inoculation with *G. etunicatum* significantly enhanced G-POD and APX activities in soybean shoots and roots under salt stress (Ghorbanli et al., 2004). During drought stress and the re-watering, POD activity in shoots of *Juniperus oxycedrus* was significantly increased by inoculation with a mixture of *G. intraradices*, *G. deserticola*, and *G. mosseae* (Roldán et al., 2008). Electrophoretic studies revealed that inoculation with *G. fasciculatum* significantly increased POD activities and also induced two extra POD isozyme bands in three *Ziziphus* species, which may be due to increased P uptake by mycorrhization and to AM colonization and arbuscules (Mathur and Vyas, 1995; Wu and Zou, 2009). However, in root and nodule of drought-stressed soybean plants, *G. mosseae* inoculation obviously decreased APX activity (Porcel et al., 2003), implying that AMs might offer avoidance mechanisms for AM plants.

15.4.3 Catalase

Catalase (CAT) is the first to be discovered and characterized among the antioxidant enzymes. According to the structural differences of catalytic center, CAT can be divided into two classes, namely Fe-CAT and Mn-CAT. CAT is located in peroxisomes/glyoxysomes, cytosol, chloroplast, and mitochondria. CAT catalyzes the dismutation of two molecules of H_2O_2 into water and oxygen without consuming cellular reducing equivalents (Havir and McHale, 1989; Boguszewska and Zagdańska, 2012).

Along with the increasing mycorrhizal studies, various antioxidant enzymes are widely concerned in combination with AM symbiosis. Many reports related to the relationship between AMF and CAT of the host plant are well documented (Table 15.4). After 14 days of 400 mM NaCl stress, CAT activity in leaves of the *Suaeda salsa* plants colonized by *G. mosseae* was significantly higher than in the non-AM plants (Li et al., 2012). Wu

TABLE 15.4 Effect of Inoculation with Arbuscular Mycorrhizal Fungi on CAT Activity of Host Plants Exposed to Various Abiotic Stresses

Fungal species	Host species	Conditions	CAT	References
Glomus mosseae	*Poncirus trifoliata*	low temperature high temperature ample water	= leaf > root > leaf; = root	Wu, 2006b Wu et al., 2006c
	Suaeda salsa	salt stress	>leaf	Li et al., 2012
	Citrus tangerine	drought stress	>leaf	Wu et al., 2007a
	Glycine max	well water drought stress	= nodule; >root < nodule; = root	Porcel et al., 2003
	P. trifoliata	salt stress	>leaf	Wu et al., 2010
	Lycopersicon esculentum	salt stress	>leaf; >root	Huang et al., 2010
			>leaf	Abdel Latef and He, 2011
	Capsicum annuum	salt stress (1 mM NaCl)	= leaf	Çekic et al., 2012
		salt stress (2 mM NaCl)	<leaf	
		salt stress (4 mM NaCl)	= leaf	
		salt stress (8 mM NaCl)	= leaf	
	Avena nuda	charcoal filtered air SO_2 stress	<leaf > leaf	Huang et al., 2008
G. versiforme	*P. trifoliata*	salt stress sample water	>leaf > leaf; = root	Wu et al., 2006b Wu et al., 2006c
	C. tangerine	drought stress	>leaf	Wu et al., 2007a
G. intraradices	*Capsicum annuum*	salt stress (1 mM NaCl)	>leaf	Çekic et al., 2012
		salt stress (2 mM NaCl)	<leaf	

(Continued)

TABLE 15.4 (Continued)

Fungal species	Host species	Conditions	CAT	References
		salt stress (4 mM NaCl)	= leaf	
		salt stress (1 mM NaCl)	= leaf	
	Phaseolus vulgaris	low P high P	= root = root	Lambais et al., 2003
G. clarum	*Phaseolus vulgaris*	low P high P	>root <root	Lambais et al., 2003
G. geosporum	*C. tangerine*	drought stress	= leaf	Wu et al., 2007a
G. diaphanum	*C. tangerine*	drought stress	= leaf	Wu et al. 2007a
G. etunicatum	*Glycine max*	salt stress	= leaf; = root	Ghorbanli et al., 2004
	C. tangerine	drought stress	= leaf	Wu et al., 2007a

Note: Symbols >, <, and = indicate that the variable is significantly increased, decreased and not affected by AMF colonization.

et al. (2010) also found that AMF infection significantly increased the CAT activity in leaves of *Poncirus trifoliata* under salt stress. AM symbiosis markedly increased the CAT activity in leaves of *Citrus tangerine* seedlings under both well-watered and water stress conditions (Wu et al., 2007b). Since H_2O_2 may be involved in signal transduction in plant pathogen interactions, it seems that degradation of H_2O_2 by CAT in AM plants is a possible mechanism for avoiding the activation of responsed genes (García-Garrido and Ocampo, 2002). However, under the conditions of 8 mM NaCl application, CAT activity was similar between the pepper (*Capsicum annuum*) plants inoculated with or without *G. intraradices* (Çekic, 2012). Similar result was also found in the soybeans inoculated by *G. etunicatum* or *G. mosseae* (Porcel et al., 2003; Ghorbanli et al., 2004).

Interestingly, mycorrhizal association did not affect CAT activity of trifoliate orange (*Poncirus trifoliata*) under low temperature conditions but significantly increased CAT activity under high temperature conditions (Wu, 2011). The changes of CAT activity appear to be dependent on P nutrition of substrates, because CAT activity in bean plant (*Phaseolus vulgaris*) roots colonized by *G. clarum* was significantly induced in low P conditions but was restricted in high P conditions (Lambais et al., 2003). In nodules of soybean (*Glycine max*), the CAT activity was lower in the drought stressed mycorrhizal plants than in the non-mycorrhizal controls (Porcel et al. 2003).

Certainly, CAT changes induced by mycorrhization might be related to mycorrhizal colonization and arbuscules (Table 15.2; Wu and Zou, 2009).

It concludes that CAT activity induced by AMF may not be important for enhancing tolerance in AM plants; that depends on host plant, AMF, and environmental factors, such as P and temperature levels.

15.5 MYCORRHIZAS AND ANTIOXIDANTS

Glutathione, first discovered and named by Hopkins in 1929, is a tripeptide thiol (L-glutamyl-L-cysteinyl-glycine) with multiple functions in plant organelles (Sies, 1999; Maher, 2005). In general, glutathione is divided into two types – oxidized glutathione (GSSG) and reduced glutathione (GSH). GSH is one of the most important antioxidants in plants and is characterized by high reduction in plant cells. In the absence of stresses, the GSH:GSSG ratio in leaf typically is at least 20:1 (Mhamdi et al., 2010; Noctor et al., 2011). In addition, high concentration is the key characteristic of GSH in comparison to other cellular thiols (Tausz et al., 2004). However, GSH is not evenly distributed in plant cells and 90% of GSH is distributed in the cytoplasm (May et al., 1998). As a strong antioxidant in the form of a cysteine residue, GSH performs its antioxidation through enzymatic or nonenzymatic reactions: participating directly in the neutralization of ROS in the ASC-GSH cycle, and chemical reactions with $^1O_2$, $^\bullet OH$ and $O_2^{\bullet -}$ directly to scavenge ROS (Cooper et al. 2011; Gill et al., 2013). Another antioxidant, ascorbate (ASC), is the most abundant water-soluble antioxidant, coping with ROS effectively in plants (Chen and Gallif, 2004; Ahmad et al., 2010a,b, 2011; Ahmad and Umar, 2011; Sharma et al., 2012). ASC is synthesized through L-galactose and is an electron donor in a number of enzymatic and nonenzymatic reactions (Wheeler et al., 1998). ASC can be synthesized in all plant tissues, particularly chloroplast, and used as a reducing agent to react directly with ROS and then scavenge ROS (Foyer and Noctor, 2005). In addition, ASC as an enzyme substrate plays a vital role in counteracting ROS and detoxifying H_2O_2 via the ASC-GSH cycle (Potters et al., 2004). In general, GSH and ASC levels appear to be coordinated. All components of the ASC-GSH cycle have their specific functionings in metabolism regulation, thereby controlling and affecting plant growth and development (Kocsy et al., 2013).

As stated above, GSH and ASC are the two essential nonenzymatic antioxidants coping with ROS. Various abiotic stresses such as drought, temperature, salinity, and heavy metal stresses disturb the equilibrium between antioxidants and ROS (Ali et al., 2005) and reports suggested the important roles of AMF on antioxidants (Table 15.5). Under drought stress conditions GSH and ASC concentrations were higher in AM than in non-AM citrus plants (Wu et al., 2006a, b; Wu and Zou, 2009), suggesting that ASC-GSH cycle of AM plants utilizes more ASC as an antioxidant and more GSH as a reductant to regenerate ASC, thereby controlling a lower level of H_2O_2 in

TABLE 15.5 Effects of inoculation with arbuscular mycorrhizal fungi on GSH and ASC concentrations of host plants exposed to various abiotic stresses

Fungal species	Host plant	Conditions	GSH	ASC	References
Glomus mosseae	*Avena nuda*	charcoal filtered air SO_2 stress	>leaf > leaf	<leaf > leaf	Huang et al., 2008
	Poncirus trifoliata	no-salt salt stress	>leaf > leaf	>leaf > leaf	Wu et al., 2010
G. versiforme	*Citrus tangerine*	well-watered water stress	>leaf; >root = leaf; = root	>leaf; = root > leaf; >root	Wu et al., 2006a, 2007b
	P. trifoliata	well-watered water stress water stress (0 day) water stress (3 day) water stress (6 day) water stress (9 day) water stress (12 day) no-salt salt stress	>leaf; >root > leaf; >root > leaf; >root = leaf; = root = leaf; = root > leaf; = root = leaf; = root > leaf = leaf	>leaf; = root > leaf; >root > leaf; = root > leaf; = root = leaf; >root = leaf; >root = leaf; = root > leaf > leaf	Wu et al., 2006b Wu and Zou, 2009 Wu et al., 2010

Note: Symbols >, <, and = indicate that the variable is significantly increased, decreased and not affected by AMF colonization.

AM plants. On the other hand, AM colonization and arbuscules were highly positively correlated with ASC and without GSH (Table 15.2; Wu and Zou, 2009). The data from an experiment on *Poncirus trifoliata* stated that GSH and ASC concentrations were significantly decreased by 100 mM NaCl, whereas inoculation with *G. mosseae* and *G. versiforme* significantly increased the concentrations of the two antioxidants (Wu et al., 2010). Huang et al. (2008) conducted an experiment to assess the effect of *G. mosseae* on *Avena nuda* seedling under sulfur dioxide (SO_2) stress conditions. They found that AM symbiosis significantly increased the concentrations of GSH and ASC, which may help AM plants to counteract SO_2-induced changes.

Another antioxidant, flavonoid, was recently researched by Abbaspour et al. (2012). They found that pistachio (*Pistacia vera* L.) seedlings colonized by *G. intraradices* presented higher flavonoid content during well water and drought stress. Higher flavonoids in AM plants will help the host plant to scavenge excess ROS, thereby contributing protection against oxidative damage.

15.6 CONCLUSIONS

In conclusion, when AM and non-AM plants are subjected to abiotic stresses, concentrations of nonenzymatic antioxidants and activities of antioxidant enzymes are usually higher in the AM than in non-AM plants. Keeping higher concentrations of non-enzymatic antioxidants and activities of antioxidant enzymes due to mycorrhization would benefit the host plant to scavenge ROS generation, thereby increasing the stress tolerance of AM plants.

Although a great deal of research has recently been involved in describing the relationship between AMs and ROS in plants, much of the data is only descriptive, without related mechanisms induced by mycorrhization. We recommend that the following aspects should be strengthened in the future:

(i) Use the biochemical dyeing agents, combined with microscopy techniques, to locate the ROS distribution in AM structures. Based on the morphological and cell observations, it may prove the mechanism of ROS burst limited by AMs.

(ii) In addition to being one of the ROS, H_2O_2, a relatively long-lived molecule, is regarded as a signal molecule to induce a range of molecular, biochemical and physiological responses of plants to regulate plant development, stress adaptation, and programmed cell death (Neill et al., 2002; Gechev and Hille, 2005). During the early stages of ectomycorrhizal formation between *Castanea sativa* and *Pisolithus tinctorius*, H_2O_2 played a role in signal transduction, thereby resulting in an inhibition of ROS-scavenging enzymes (Baptista et al., 2007). Moreover, in the arbuscule life cycle, H_2O_2 accumulation can imply the arbuscule formation and degradation (see above). Therefore, we have sufficient reason to believe that H_2O_2 might also act as a signaling molecule in either mycorrhizal establishment, particularly arbuscules, or stress tolerance, which needs to be confirmed by further experiments.

(iii) In the antioxidant defense systems of plants, the ASC-GSH cycle plays a key role in association with stress tolerance. To our knowledge, only five involved documents were reported by Wu and his colleagues (Wu et al. 2006a, b, 2007b, 2010; Wu and Zou, 2011). These are not enough to understand the role of the ASC-GSH cycle to enhance stress tolerance of host plants by mycorrhization. Further research is still necessary.

(iv) The plant responses to abiotic stress not only involve physiological but also biochemical and molecular mechanisms. Various specific genes respond to abiotic stress in various plants, and their functions have been identified (Rodríguez et al., 2005). To understand the molecular regulation of stress responses, isolation and expression of stress-responsive genes must be conducted in AM plants as compared to non-AM plants. The molecular analysis of responsive genes will provide precise information on the relationship of AMs with ROS metabolism of plants.

ACKNOWLEDGMENTS

This study was supported by the National Natural Science Foundation of China (31101513) and was the Key Project of Natural Science Foundation of Hubei Province (2012FFA001).

REFERENCES

Abbaspour, H., Saeidi-Sar, S., Afshari, H., Abdel-Wahhab, M.A., 2012. Tolerance of mycorrhizal infected pistachio (*Pistacia vera* L.) seedlings to drought stress under glasshouse conditions. J. Plant Physiol. 169, 704–709.

Abdel Latef, A.A., He, C.X., 2011. Effect of arbuscular mycorrhizal fungi on growth, mineral nutrition, antioxidant enzymes activity and fruit yield of tomato grown under salinity stress. Sci. Hortic. 127, 228–233.

Ahmad, P., Umar, S., 2011. Antioxidants: Oxidative Stress Management in Plants. Studium Press Pvt. Ltd., New Delhi, India.

Ahmad, P., Jaleel, C.A., Salem, M.A., Nabi, G., Sharma, S., 2010a. Roles of enzymatic and non-enzymatic antioxidants in plants during abiotic stress. Crit. Rev. Biotechnol. 30, 161–175.

Ahmad, P., Umar, S., Sharma, S., 2010b. Mechanism of free radical scavenging and role of phytohormones during abiotic stress in plants. In: Ashraf, M., Ozturk, M., Ahmad, M.S.A. (Eds.), Plant Adaptation and Phytoremediation. Springer, Dordrecht, pp. 99–108.

Ahmad, P., Nabi, G., Jeleel, C.A., Umar, S., 2011. Free radical production, oxidative damage and antioxidant defense mechanisms in plants under abiotic stress. In: Ahmad, P., Umar, S. (Eds.), Oxidative Stress: Role of Antioxidants in Plants. Studium Press Pvt. Ltd., New Delhi, India, pp. 19–53.

Ahmad, P., Ozturk, M., Gucel, S., 2012. Oxidative damage and antioxidants induced by heavy metal stress in two cultivars of mustard (*Brassica juncea* L.) plants. Fresenius Environ. Bull. 21, 2953–2961.

Alguacil, M.M., Hernandez, J.A., Caravaca, F., Portillo, B., Roldán, A., 2003. Antioxidant enzyme activities in shoots from three mycorrhizal shrub species afforested in a degraded semi-arid soil. Physiol. Plant. 118, 562–570.

Ali, M.B., Hahn, E.J., Peak, K.Y., 2005. Effects of temperature on oxidative stress defense systems, lipid peroxidation and lipoxygenase in *Phalaenopsis*. Plant Physiol. Biochem. 43, 13–23.

Apel, K., Hirt, H., 2004. Reactive oxygen species: metabolism, oxidative stress, and signal transduction. Annu. Rev. Plant Biol. 55, 373–399.

Baptista, P., Martins, A., Pais, M.S., Tavares, R.M., Lino-Neto, T., 2007. Involvement of reactive oxygen species during early stages of ectomycorrhizal establishment between *Castanea sativa* and *Pisolithus tinctorius*. Mycorrhiza. 17, 185–193.

Barna, B., Fodor, J., Harrach, B.D., Pogany, M., Kiraly, Z., 2012. The Janus face of reactive oxygen species in resistance and susceptibility of plants to necrotrophic and biotrophic pathogens. Plant Physiol. Biochem. 59, 37–43.

Bienert, G.P., Møller, A.L.B., Kristiansen, K.A., Schulz, A., Møller, I.M., Schjorring, J.K., et al., 2007. Specific aquaporins facilitate the diffusion of hydrogen peroxide across membranes. J. Biol. Chem. 282, 1183–1192.

Bindschedler, L.V., Dewdney, J., Blee, K.A., Stone, J.M., Asai, T., Plotnikov, J., et al., 2006. Peroxidase-dependent apoplastic oxidative burst in *Arabidopsis* required for pathogen resistance. Plant J. 47, 851–863.

Boguszewska, D., Zagdańska, B., 2012. ROS as signaling molecules and enzymes of plant response to unfavorable environmental conditions. In: Lushchak, V. (Ed.), Oxidative Stress – Molecular Mechanisms and Biological Effects. InTech, Rijeka, Croatia, pp. 341–362.

Brundrett, M.C., 2009. Mycorrhizal associations and other means of nutrition of vascular plants, understanding the global diversity of host plants by resolving conflicting information and developing reliable means of diagnosis. Plant Soil. 320, 37–77.

Caverzan, A., Passaia, G., Rosa, S.B., Ribeiro, C.W., Lazzarotto, F., Margis-Pinheiro, M., 2012. Plant responses to stresses: role of ascorbate peroxidase in the antioxidant protection. Genet. Mol. Biol. 35, 1011–1019.

Çekic, F.O., Unyayar, S., Ortas, I., 2012. Effects of arbuscular mycorrhizal inoculation on biochemical parameters in *Capsicum annuum* grown under long term salt stress. Turk. J. Bot. 36, 63–72.

Chen, Z., Gallif, D.R., 2004. The ascorbic acid redox state controls guard cell signaling and stomatal movement. Plant Cell 16, 1143–1162.

Chmuraa, D., Gucwa-Przepiórab, E., 2012. Interactions between arbuscular mycorrhiza and the growth of the invasive alien annual *Impatiens parviflora* DC: A study of forest type and soil properties in nature reserves (S Poland). Appl. Soil Ecol. 62, 71–80.

Cooper, A.J.L., Pinto, J.T., Callery, P.S., 2011. Reversible and irreversible protein glutathionylation: biological and clinical aspects. Expert Opin. Drug Metabol. Toxicol. 7, 891–910.

Corradi, N., Buffner, B., Croll, D., Colard, A., Horak, A., Sanders, I.R., 2009. High-level molecular diversity of copper-zinc superoxide dismutase genes among and within species of arbuscular mycorrhizal fungi. Appl. Environ. Microbiol. 75, 1970–1978.

Dutilleul, C., Garmier, M., Noctor, G., Mathieu, C., Chetrit, P., Foyer, C.H., et al., 2003. Leaf mitochondria modulate whole cell redox homeostasis, set antioxidant capacity, and determine stress resistance through altered signaling and diurnal regulation. Plant Cell 15, 1212–1226.

Fester, T., Hause, G., 2005. Accumulation of reactive oxygen species in arbuscular mycorrhizal roots. Mycorrhiza. 15, 373–379.

Foyer, C.H., Noctor, G., 2000. Oxygen processing in photosynthesis: regulation and signalling. New Phytol. 146, 359–388.

Foyer, C.H., Noctor, G., 2005. Redox homeostasis and antioxidant signaling: a metabolic interface between stress perception and physiological responses. Plant Cell 17, 1866–1875.

García-Garrido, J.M., Ocampo, J.A., 2002. Regulation of the plant defence response in arbuscular mycorrhizal symbisos. J. Exp. Bot. 373, 1377–1386.

Gechev, T.S., Hille, J., 2005. Hydrogen peroxide as a signal controlling plant programmed cell death. J. Cell Biol. 168, 17–20.

Ghorbanli, M., Ebrahimzadeh, H., Sharifi, M., 2004. Effects of NaCl and mycorrhizal fungi on antioxidative enzymes in soybean. Biol. Plant. 48, 575–581.

Gill, S.S., Tuteja, N., 2010. Reactive oxygen species and antioxidant machinery in abiotic stress tolerance in crop plants. Plant Physiol. Biochem. 48, 909–930.

Gill, S.S., Anjum, N.A., Hasanuzzaman, M., Gill, R., Trivedi, D.K., Ahmad, I., et al., 2013. Glutathione and glutathione reductase: a boon in disguise for plant abiotic stress defense operations. Plant Physiol. Biochem. 70, 204–212.

Hajiboland, R., Aliasgharzadeh, N., Laiegh, S.F., Poschenrieder, C., 2010. Colonization with arbuscular mycorrhizal fungi improves salinity tolerance of tomato (*Solanum lycopersicum* L.) plants. Plant Soil. 331, 313–327.

Halliwell, B., 1981. Chloroplast Metabolism, the Structure and Function of Chloroplast in Green Leaf Cells. Charendon Press, Oxford.

Harinasut, P., Poonsopa, D., Roengmongkol, K., Charoensataporn, R., 2003. Salinity effects on antioxidant enzymes in mulberry cultivar. ScienceAsia 29, 109–113.

Havir, E.A., McHale, N.A., 1989. Enhanced-peroxidatic activity in specific catalase isozymes of tobacco, barley, and maize. Plant Physiol. 91, 812–815.

He, Z.Q., He, C.X., Zhang, Z.B., 2007. Changes of antioxidative enzymes and cell membrane osmosis in tomato colonized by arbuscular mycorrhizae under NaCl stress. Colloid. Surf. B. 59, 128–133.

Hodgson, R.A.J., Rasion, K.J., 1991. Superoxide production by thylakoids during chilling and its implication in the susceptibility of plant to chilling-induced photo-inhibition. Planta. 183, 222–228.

Huang, L.L., Yang, C., Zhao, Y., Xu, X., Xu, Q., Li, G.Z., et al., 2008. Antioxidant defenses of mycorrhizal fungus infection against SO_2-induced oxidative stress in *Avena nuda* seedlings. *Bull.* Environ. Contam. Toxicol. 81, 440–444.

Huang, Z., He, C.X., He, Z.Q., Zou, Z.R., Zhang, Z.B., 2010. The effects of arbuscular mycorrhizal fungi on ractive oxyradical scavenging system of tomato under salt tolerance. Agric. Sci. Chin. 9, 1150–1159.

Huseynova, I., Aliyev, J., 2012. Evaluation of free radicals and antioxidant properties of virus infected food crops in Azerbaijan. J. Life Sci. 6, 1307–1316.

Jiang, M., Zhang, J., 2001. Effect of abscisic acid on active oxygen species, antioxidative defence system and oxidative damage in leaves of maize seedlings. Plant Cell Physiol. 2, 1265–1273.

Khanna-Chopra, R., Selote, D.S., 2007. Acclimation to drought stress generates oxidative stress tolerance in drought-resistant than -susceptible wheat cultivar under field conditions. Environ. Exp. Bot. 60, 276–283.

Kocsy, G., Tari, I., Vankova, R., Zechmann, B., Gulyas, Z., Poor, P., et al., 2013. Redox control of plant growth and development. Plant Sci. 211, 77–91.

Koyro, H.W., Ahmad, P., Geissler, N., 2012. Abiotic stress responses in plants: an overview. In: Ahmad, P., Prasad, M.N.V. (Eds.), Environmental Adaptations and Stress Tolerance of Plants in the Era of Climate Change. Springer Science + business media, NY, USA, pp. 1–28.

Kreslavski, V.D., Los, D.A., Allakhverdiev, S.I., Kuznetsov, V.l.V., 2012. Signaling role of reactive oxygen species in plants under stress. Russ. J. Plant Physiol. 59, 141–154.

Lambais, M.R., Ríos-Ruiz, W.F., Andrade, R.M., 2003. Antioxidant responses in bean (*Phaseolus vulgaris*) roots colonized by arbuscular mycorrhizal fungi. New Phytol. 160, 421–428.

Lanfranco, L., Novero, M., Bonfante, P., 2005. The mycorrhizal fungus *Gigaspora margarita* possesses a CuZn superoxide dismutase that is up-regulated during symbiosis with legume hosts. Plant Physiol. 137, 1319–1330.

Li, J.Q., Zhang, B., Shi, W.W., Li, H.Y., 2008. Hydrogen peroxide in plants: a versatile molecule of the reactive oxygen species network. J. Integ. Plant Biol. 50, 2–18.

Li, T., Liu, R.J., He, X.H., Wang, B.S., 2012. Enhancement of superoxide dismutase and catalase activities and salt tolerance of euhalophyte *Suaeda salsa* L. by mycorrhizal fungus *Glomus mosseae*. Pedosphere. 22, 217–224.

Maher, P., 2005. The effects of stress and aging on glutathione metabolism. Ageing Res. Rev. 2, 288–314.

Mathur, N., Vyas, A., 1995. Changes in isozyme patterns of peroxidase and polyphenol oxidase by VAM fungi in roots of *Ziziphus* species. J. Plant Physiol. 145, 498–500.

May, M.J., Vernoux, T., Leaver, C.J., 1998. Glutathione homeostasis in plants: implication for environment sensing and plant development. J. Exp. Bot. 49, 649–667.

Mhamdi, A., Hager, J., Chaouch, S., Queval, G., Han, Y., Taconnat, L., et al., 2010. *Arabidopsis* glutathione reductase plays a crucial role in leaf responses to intracellular hydrogen peroxide and in ensuring appropriate gene expression through both salicylic acid and jasmonic acid signaling pathways. J. Plant Physiol. 153, 1144–1160.

Mittler, R., Vanderauwera, S., Gollery, M., Van Breusegem, F., 2004. The reactive oxygen gene network in plants. Trend. Plant Sci. 9, 490–498.

Moller, I.M., 2001. Plant mitochondria and oxidative stress: Electron transport, NADPH turnover, and metabolism of reactive oxygen species. Ann. Rev. Plant Physiol. Plant Mol. Biol 52, 561–591.

Movafegh, S., Hoseini, S.M., 2013. ROS in plant cell: a review. Int. J. Farm. All. Sci. 2, 23–26.

Muthukumar, T., Udaiyan, K., 2010. Growth response and nutrient utilization of *Casuarina equisetifolia* seedlings inoculated with bioinoculants under tropical nursery conditions. New For. 40, 101–118.

Muthukumar, T., Udaiyan, K., Rajeshkannan, V., 2001. Response of neem (*Azadirachta indica* A. Juss) to indigenous arbuscular mycorrhizal fungi, phosphate-solubilizing and symbiotic nitrogen fixing bacteria under tropical nursery conditions. Biol. Fert. Soil 34, 417–426.

Neill, S.J., Desikan, R., Hancock, J., 2002. Hydrogen peroxide signaling. Curr. Opin. Plant Biol. 5, 388–395.

Neill, S.J., Desikan, R., Clarke, A., Hurst, R.D., Hancock, J.T., 2002. Hydrogen peroxide and nitric oxide as signalling molecules in plants. J. Exp. Bot. 53, 1237–1247.

Noctor, G., Mhamdi, A., Chaouch, S., Han, Y., Neukermans, J., Marquez-Garcia, B., et al., 2011. Glutathione in plants: an integrated overview. Plant Cell Environ. 35, 454–484.

Palma, J.M., Longa, M.A., Arines, J., 1993. Superoxide dismutase in vesicular arbuscular-mycorrhizal red clover plants. Physiol. Plant. 87, 77–83.

Peterson, L., Massicotte, H.B., Melville, L.H., 2004. Mycorrhizas: Anatomy and Cell Biology. NRC Research Press, Ottawa.

Piterková, J., Luhova, L., Mieslerova, B., Lebeda, A., Petrivalsky, M., 2013. Nitric oxide and reactive oxygen species regulate the accumulation of heat shock proteins in tomato leaves in response to heat shock and pathogen infection. Plant Sci. 207, 57–65.

Porcel, R., Barea, J.M., Ruiz-Lozano, J.M., 2003. Antioxidant activities in mycorrhizal soybean plants under drought stress and their possible relationship to the process of nodule senescence. New Phytol. 157, 135–143.

Porcel, R., Aroca, R., Ruiz-Lozano, J.M., 2012. Salinity stress alleviation using arbuscular mycorrhizal fungi. A review. Agron. Sust. Dev. 32, 181–200.

Pospíšil, P., 2012. Molecular mechanisms of production and scavenging of reactive oxygen species by photosynthem II. Biochim. Biophys. Acta. 1817, 218–231.

Potters, G., Horemans, N., Bellone, S., 2004. Dehydroascorbate influences the plant cell cycle through a glutathione independent reduction mechanism. Plant Physiol. 134, 1479–1487.

Rasool, S., Ahmad, A., Siddiqi, T.O., Ahmad, P., 2013. Changes in growth, lipid peroxidation and some key antioxidant enzymes in chickpea genotypes under salt stress. Acta Physiol. Plant. 35, 1039–1050.

Rodríguez, M., Canales, E., Borras-Hidalgo, O., 2005. Molecular aspects of abiotic stress in plants. Biotechnol. Aplicada. 22, 1–10.

Roldán, A., Diaz-Wivancos, P., Harnández, J.A., Carrasco, L., Caravaca, F., 2008. Superoxide dismutase and total peroxidase activities in relation to drought recovery performance of mycorrhizal shrub seedlings grown in an amended semiarid soil. J. Plant Physiol. 165, 715–722.

Ruiz-Lozano, J.M., Azcón, R., Palma, J.M., 1996. Superoxide dismutase activity in arbuscular mycorrhizal *Lactuca sativa* L. plants subjected to drought stress. New Phytol. 134, 327–333.

Ruiz-Lozano, J.M., Collados, C., Barea, J.M., Azcón, R., 2001. Clonig of cDNAS encoding SODs from lettuce plants which show differential regulation by arbuscular mycorhizal symbiosis and by drought stress. J. Exp. Bot. 52, 2241–2242.

Ruiz–Lozano, J.M., Porcel, R., Azcón, C., Aroca, R., 2012. Regulation by arbuscular mycorrhizae of the integrated physiological response to salinity in plants: new challenges in physiological and molecular studies. J. Exp. Bot. 63, 4033–4044.

Sagi, M., Fluhr, R., 2006. Production of reactive oxygen species by plant NADPH oxidases. Plant Physiol. 141, 336–340.

Salzer, P., Corbiere, H., Boller, T., 1999. Hydrogen peroxide accumulation in *Medicago truncatula* roots colonized by the arbuscular mycorrhiza-forming fungus *Glomus intraradices*. Planta. 208, 319–325.

Sharma, P., Jha, A.B., Dubey, R.S., Pessarakli, M., 2012. Reactive oxygen species, oxidative damage, and antioxidative defense mechanism in plants under stressful conditions. J. Bot. ID. 217037, 1–26.

Sies, H., 1999. Glutathione and its role in cellular functions. Free Radical Biol. *Med.* 27, 916–921.

Signorelli, S., Arellano, J.B., Melo, T.B., Borsani, O., Monza, J., 2013. Proline does not quench singlet oxygen: Evidence to reconsider its protective role in plants. Plant Physiol. Biochem. 64, 80–83.

Smith, S.E., Read, D.J., 2008. Mycorrhizal Symbiosis. Academic Press, London.

Sun, J., Gu, J., Zeng, J., Han, S., Song, A.P., Chen, F.D., et al., 2013. Changes in leaf morphology, antioxidant activity and photosynthesis capacity in two different drought-tolerant cultivars of chrysanthemum during and after water stress. Sci. Hortic. 161, 249–258.

Tahat, M.M., Sijam, K., 2012. Mycorrhizal fungi and abiotic environmental conditions relationship. *Res.* J. Environ. Sci. 6, 125–133.

Tartoura, K.A.H., Youssef, S.A., 2011. Stimulation of ROS-scavenging systems in squash (*Cucurbita pepo* L.) plants by compost supplementation under normal and low temperature conditions. Sci. Hortic. 130, 862–868.

Tausz, M., Sircelj, H., Grill, D., 2004. The glutathione system as a marker in plant ecophysiology: is a stress-response concept valid. J. Exp. Bot. 55, 1955–1962.

Tian, G.Z., Li, H.F., Qiu, W.F., 2001. Advance on research of plant peroxidase. J. Wuhan Plant Res. 19, 332–344 (in Chinese with English abstract).

Torres, M.A., Dangl, J.L., Jones, J.D., 2002. *Arabidopsis* gp91^{phox} homologues *AtrbohD* and *AtrbohF* are required for accumulation of reactive oxygen intermediates in the plant defense response. Proc. Natl. Acad. Sci. 99, 517–522.

Torres, M.A., Jones, J.D., Dangl, J.L., 2005. Pathogen-induced, NADPH oxidase-derived reactive oxygen intermediates suppress spread of cell death in *Arabidopsis thaliana*. Nat. Genet. 37, 130–134.

Wang, B.S., Ulrich, L., Rafael, R., 2004. Specific regulation of SOD isoforms by NaCl and osmotic stress in leaves of the C3 halophyte *Suaeda salsa* L. Plant Physiol. 161, 285–293.

Wheeler, G.L., Jones, M.A., Smirnoff, N., 1998. The biosynthetic pathway of vitamin C in higher plants. Nature. 393, 365–369.

Wu, Q.S., 2011. Mycorrhizal efficacy of trifoliate orange seedlings on alleviating temperature stress. Plant Soil Environ. 57, 459–464.

Wu, Q.S., Zou, Y.N., 2009. Mycorrhiza has a direct effect on reactive oxygen metabolism of drought-stressed citrus. Plant Soil Environ. 55, 436–442.

Wu, Q.S., Zou, Y.N., 2011. Arbuscular mycorrhizal symbiosis alleviates oxidative stress in plants. In: Ahmad, P., Umar, S. (Eds.), Oxidative Stress: Role of Antioxidants in Plants. Studium Press Pvt. Ltd., New Delhi, India, pp. 269–282.

Wu, Q.S., Zou, Y.N., Xia, R.X., 2006a. Effects of water stress and arbuscular mycorrhizal fungi on reactive oxygen metabolism and antioxidant production by citrus (*Citrus tangerine*) roots. Eur. J. Soil Biol. 42, 166–172.

Wu, Q.S., Xia, R.X., Zou, Y.N., 2006b. Reactive oxygen metabolism in mycorrhizal and non-mycorrhizal citrus (*Poncirus trifoliata*) seedlings subjected to water stress. J. Plant Physiol. 163, 1101–1110.

Wu, Q.S., Xia, R.X., Zou, Y.N., 2006c. Effects of arbuscular mycorrhizal fungi on the growth and antioxidant enzymes of micropropagated citrus. Chin. J. Appl. Environ. Biol. 12, 635–639.

Wu, Q.S., Xia, R.X., Zou, Y.N., 2007a. Five *Glomus* species affect water relations of *Citrus tangerine* during drought stress. Bot. Stud. 48, 147–154.

Wu, Q.S., Zou, Y.N., Xia, R.X., 2007b. Effect of *Glomus versiforme* inoculation on reactive oxygen metabolism of *Citrus tangerine* leaves exposed to water stress. Front. Agric. Chin. 1, 438–443.

Wu, Q.S., Zou, Y.N., Liu, W., Ye, X.F., Zai, H.F., Zhao, L.J., 2010. Alleviation of salt stress in citrus seedlings inoculated with mycorrhiza: changes in leaf antioxidant defense systems. Plant Soil Environ. 56, 470–475.

Yan, J., Tsuichihara, N., Etoh, T., Iwai, S., 2007. Reactive oxygen species and nitric oxide are involved in ABA inhibition of stomatal opening. Plant Cell Environ. 30, 1320–1325.

Yang, S.S., Gao, J.F., 2001. Influence of active oxygen and free radicals on plant senescence. Acta Bot. Boreal-Occident Sin. 21, 215–220 (in Chinese with English abstract).

Zhao, B.L., 1999. Free Radical and Natural Antioxidant. Science Press, Beijing.

Chapter 16

Proline Protects Plants Against Abiotic Oxidative Stress: Biochemical and Molecular Mechanisms

Mohammad Anwar Hossain, Md. Anamul Hoque, David J. Burritt and Masayuki Fujita

16.1 INTRODUCTION

Plants experience multiple and inevitable abiotic and biotic stresses in their natural habitat and to cope, they have evolved intricate mechanisms that allow them to respond to adverse conditions. Abiotic stresses such as drought, salinity, temperature extremes (heat, chilling and frost), water logging, heavy metal toxicity, nutrient imbalances, ozone, and UV-B irradiation are the primary cause of crop loss worldwide (Pandy et al., 2011; Qin et al., 2011; Krasensky and Jonak, 2012; Naika et al., 2013). The frequency and duration of such abiotic stresses will increase in the near future due to global climate change (Lamb, 2012). Therefore, the development of stress tolerant crop varieties has become an urgent concern for many crop-breeding programs to ensure global food security. Modulation of adaptive mechanisms has been a cherished goal of plant breeders; however, these responses are under multigene regulatory control (Lopes et al., 2011). Thus the main obstacle to modern sustainable agricultural development is to develop advanced breeding and engineering tools that could help develop new varieties with desired agronomical traits (Le et al., 2007; Duque et al., 2013). Advancement in engineering and conventional breeding could be made through in-depth understanding of the molecular and biochemical mechanisms that are underlying various abiotic stress responses. Thorough knowledge could lead to advancement via execution of new tolerance strategies and pave new paths towards more sustainable agricultural production.

P. Ahmad (Ed): Oxidative Damage to Plants. DOI: http://dx.doi.org/10.1016/B978-0-12-799963-0.00016-2

Abiotic stresses perturb cellular redox homeostasis, alter metabolic interactions and impair key plant physiological processes (Krasensky and Jonak, 2012). One of the widespread phenomena and best documented attributes of plant responses to both abiotic and biotic stresses is the unrestrained accumulation of reactive oxygen species (ROS), such as singlet oxygen ($^1O_2$), superoxide anion ($O_2^{\bullet -}$), hydrogen peroxide (H_2O_2), and the hydroxyl radical ($^{\bullet}OH$), in different cellular organelles. This causes oxidative damage to many cellular components and structures, rendering metabolic disorder, and finally cell death unless the ROS are detoxified efficiently at their sites of production (Mittler et al., 2004; Ahmad et al., 2010; Gill and Tuteja, 2010; Hossain et al., 2011a; Hossain and Fujita, 2012). ROS accumulation under abiotic stress depends upon the rate of ROS production, levels of ROS detoxification, scavenging (Hossain and Fujita, 2010), and is also related to the developmental stage, severity of the stress and prevailing environments of the plant. Cellular ROS levels in plants are stringently controlled by sophisticated antioxidant defense systems that function as a signaling cascade to various abiotic stresses by regulating gene expression that is closely associated with plant abiotic stress tolerance under different plant physiological processes (Neill et al., 2002; Petrov and Van Breusegem, 2012).

A common physiological mechanism adopted by plants to counter the adverse effects of abiotic stresses is the *de-novo* synthesis of large quantities of low-molecular-weight organic compounds, exceptionally water soluble and nontoxic at millimolar concentration termed osmolytes; these include proline, glycine betaine, trehalose and others (Ashraf and Foolad, 2007; Szabados and Savouré, 2010; Verslues and Sharma, 2010; Hayat et al., 2012). Recent proteomic, genomic and metabolic studies have revealed that the function of proline is not as straightforward as initially believed. Research studies on plants, especially those on proline synthesis and catabolic genes, have demonstrated that the proline produced under stressful conditions can act as a compatible solute in osmotic adjustment, a free radical scavenger, a metal chelator, an activator of detoxification pathways, a cell redox balancer, a cytosolic pH buffer, a source of energy, nitrogen and carbon, a stabilizer for subcellular structures and membranes including photosystem II (PS II), or act as a signaling molecule (for reviews, see Hare and Cress, 1997; Matysik et al., 2002; Kavi Kishor et al., 2005; Sharma and Dietz, 2006; Trovato et al., 2008; Verbruggen and Hermans, 2008; Mattioli et al., 2009a; Szabados and Savouré, 2010; Hayat et al., 2012). Proline can also function as an electron sink mechanism under stressful conditions (Sharma and Dietz, 2006). Plants generally accumulate proline in response to abiotic stresses such as salinity and drought in the cytosol of the cellular organelles, where it contributes substantially to cytoplasmic osmotic adjustment (Ketchum et al., 1991). Another important function of proline is that it forms a hydration shell around delicate proteins and averts their deterioration under stressful conditions. Proline also reduces NaCl-induced potassium efflux and this is likely to increase K^+ homeostasis within plant cells, thus increasing the tolerance to abiotic stress (Cuin and Shabala, 2005, 2007). Sobahan et al., (2009) reported addition of proline restricted the apoplastic flow of Na^+ and induced stress tolerance in rice seedlings by maintaining higher K^+/Na^+ ratio in the leaves of rice seedlings growing under salt stress conditions in hydroponic medium. Proline accumulation might also play regulatory roles during plant growth and reproduction according to multiple research studies (Mattioli et al., 2008, 2009a).

Although the accumulation of proline and ROS is a common consequence of both abiotic and biotic stresses in plants, the exact molecular and biochemical mechanisms of proline mediated abiotic oxidative stress tolerance is still a matter of intensive research and recent plant molecular findings have shown the significance of this compatible solute in abiotic stress tolerance. In this chapter we will discuss and summarize the current understanding of ROS formation, proline biosynthesis and its accumulation in plants under various abiotic stresses, and the biochemical insights whereby proline enhances the capacity of plants to induce oxidative stress tolerance.

16.2 PATHWAYS OF PROLINE BIOSYNTHESIS AND DEGRADATION

Two biosynthetic pathways (Fig. 16.1) have been proposed for proline synthesis in higher plants, one the glutamate pathway and the other the ornithine pathway where glutamate (Glu) and ornithine (Orn) are used as a precursor, respectively (Verslues and Sharma, 2010; Hayat et al., 2012). Synthesis of proline from glutamate requires two enzymes (Δ^1-pyrroline-5-carboxylate synthetase (P5CS; EC 2.7.2.11) and Δ^1-pyrroline-5-carboxylate reductase (P5CR; EC 1.5.1.2) (Ashraf and Foolad, 2007; Burritt, 2012). First P5CS phosphorylates and reduces glutamate of glutamate-semialdehyde (GSA) which is spontaneously converted to Δ^1-pyrroline-5-carboxylate (P5C) and consequently the enzyme P5CR converted the P5C to proline (Delauney and Verma, 1993; Burritt, 2012). Recent plant metabolic studies revealed that most of the proline accumulated upon stress treatment is due to the higher production of glutamate of plants cells (Delauney and Verma, 1993; Hare and Cress, 1997; Saadia et al., 2012; Witt et al., 2012) and the glutamate pathway is the major pathway of proline under stress conditions (Lv et al., 2011). Additionally, proline biosynthesis from ornithine occurs through the enzymes ornithine-δ-aminotransferase (δ-OAT). The accumulation of proline through the ornithine pathway is very important at the time of seedling establishment and in a few plant species for stress-mediated proline augmentation (Armengaud et al., 2004; Xue et al., 2009). Proline biosynthesis steadily occurs in the cytosol, while it is augmented to the chloroplasts during stress conditions (Szabados and Savouré, 2010). As proline biosynthesis requires NADPH, the enhanced rate of its biosynthesis in chloroplasts maintains the

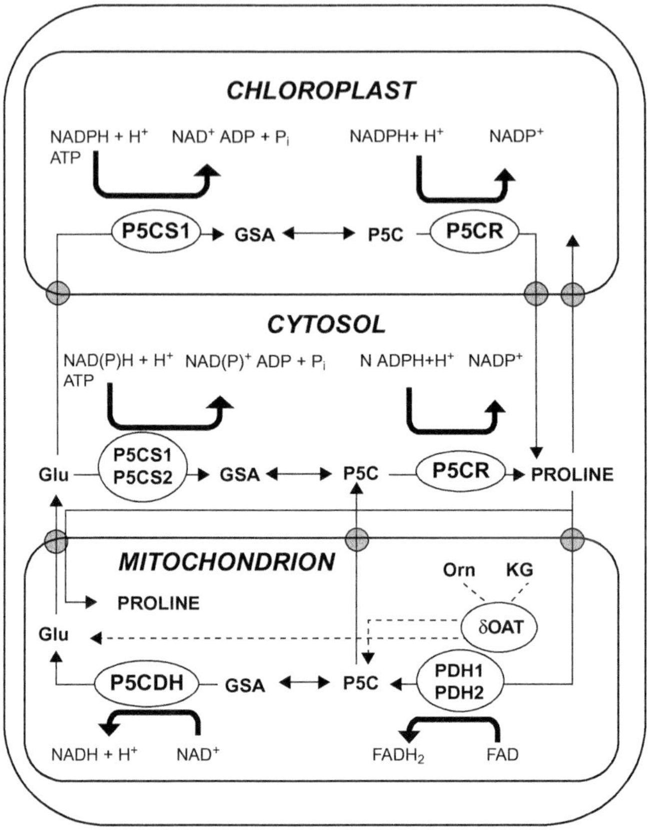

FIGURE 16.1 Biosynthetic pathways of proline metabolism in higher plants (adapted from Burritt, 2012). Abbreviations are defined in the text.

low NADPH: $NADP^+$ ratio, resulting in reduction of photoinhibition under high irradiance (Hare and Cress, 1997).

Proline accumulates in the cytosol or chloroplast whereas proline degradation (oxidation of proline to glutamate) in plants occurs in mitochondria (Krasensky and Jonak, 2012) through the consecutive action of two different enzymes. First the enzyme proline dehydrogenase (PDH; EC 1.5.99.8) converted the proline to pyrroline carboxylate (P5C) and the enzyme pyrroline-5-carboxylate dehydrogenase (P5CDH; EC 1.5.1.12) alters the P5C to glutamate. Accumulation of endogenous proline level under stressful conditions is the cumulative outcome of both, including a higher rate of synthesis and alteration of its degradation pathways (Delauney and Verma, 1993; Burritt, 2012); therefore, the intracellular proline content depends on its biosynthesis, degradation and transport from other plant parts (Kavi Kishor et al., 2005).

16.3 PROLINE ACCUMULATION AND ABIOTIC STRESS TOLERANCE

Increased endogenous proline levels are an important adaptive physiological attribute to counter deleterious effects of environmental stress in plants and was first observed in wilted ryegrass (Kemple and Macpherson, 1954). Generous research studies in plants have demonstrated that proline accumulation in plants under various abiotic stresses is positively correlated with oxidative stress tolerance (Anjum et al., 2012; Theocharis et al., 2012; Xu et al., 2012; Saeedipour, 2013). A general conserved mechanism of plants to various abiotic stresses causing cellular dehydration, such as salinity, drought, cold, heat, heavy metals, is the synthesis and accumulation of compatible solutes (Boscaiu et al., 2012; Hayat et al., 2012). Proline accumulation was found to occur when plants subjected to various abiotic stresses including drought (Choudhary et al., 2005; Türkan et al., 2005; Man et al., 2011; Anjum et al., 2012; Saiema et al., 2012; Witt et al., 2012; Saeedipour, 2013), high salinity (Yazici et al., 2007; Boscaiu et al., 2012; Sripinyowanich et al., 2013), chilling (Luo et al., 2011), waterlogging (Xu et al., 2012), high light and UV irradiation (Saradhi et al., 1995), heavy metals (Schat et al., 1997; Sharma and Dietz, 2006, 2009; Bao et al., 2011), gamma irradiation (Jan et al., 2012) and oxidative stress (Yang et al., 2009). Studies investigating resistivity and susceptibility of plants to abiotic stress have demonstrated the fundamental importance of proline in tolerance (Türkan et al., 2005; Man et al., 2011; Saiema et al., 2012; Saeedipour, 2013). Arunkumar et al. (2012) evidenced a varied response in heat stress with reference to proline accumulation, e.g. chickpea exhibited increased level of proline, rendering it more sensitive to heat stress probably by increased ROS production via the Pro/P5C cycle (Lv et al., 2011) but tobacco and *Arabidopsis* do not exhibit any increment in proline (Dobra et al., 2010; Hua et al., 2011; Lv et al., 2011). However, recently Li et al., (2013) showed that hydrogen sulfide induced heat tolerance in maize is associated with proline accumulation through higher P5CS activity and lower lipid degradation, during lower PDH activity. Importantly, exogenous proline application was also found to protect maize seedlings from high temperature stress by modulating their lipid peroxidation and growth activity. Comprehensive studies using transgenic plants or mutants have demonstrated that proline metabolism has a complex effect on development and stress responses (Mattioli et al., 2008; Szekely et al., 2008; de campos et al., 2011; You et al., 2012; Zarei et al., 2012). The possible involvement of endogenous proline in oxidative stress tolerance will be discussed later in more detail.

16.4 ROS FORMATION UNDER ABIOTIC STRESS

Reactive oxygen species are a group of O_2 derived reactive molecules produced in the chloroplasts, mitochondria, peroxisomes, plasma membrane

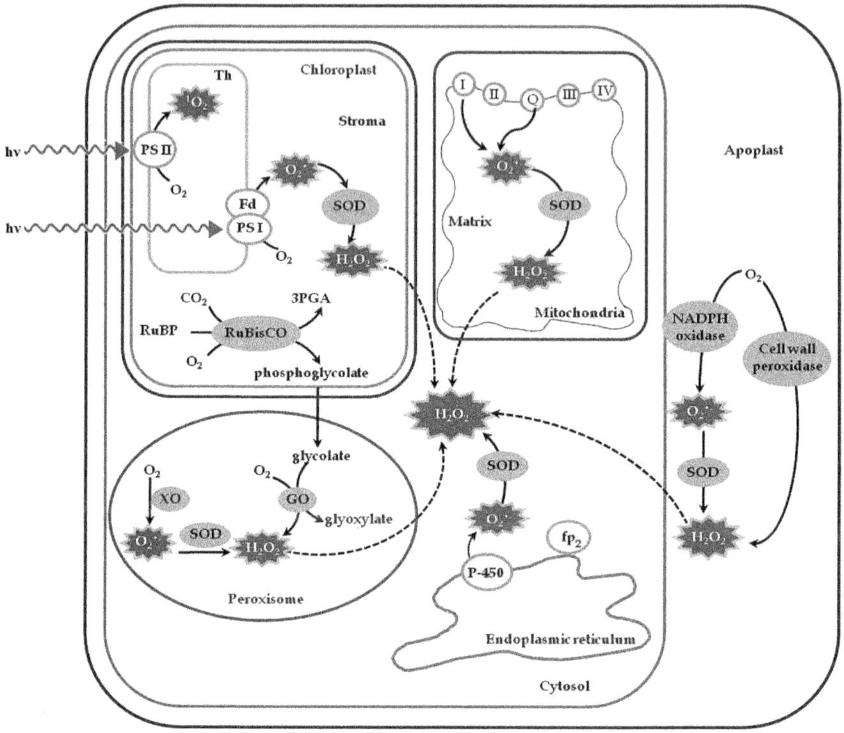

FIGURE 16.2 Major sites and sources of ROS production of plant cells (adapted from Hossain et al., 2011a).

and apoplast (Fig. 16.2; Møller, 2007; Ahmad et al., 2010; Sharma et al., 2012). Although normal plant physiological processes such as respiration and photosynthesis produce a limited amount of ROS, increased levels of ROS, such as $O_2^{\bullet-}$, H_2O_2, and $^{\bullet}OH$ accumulate under abiotic or biotic stress that results in oxidative stress (Mittler, 2002; Kotchoni et al., 2006). ROS are important signaling molecules involved in plant abiotic stress responses and tolerance (Foyer and Noctor, 2003; Vranova et al., 2002; Petrov and Van Verusegum, 2012). Elevated levels of ROS lead to the inactivation of proteins and inhibit the activity of multiple enzymes involved in metabolic pathways, and result in the oxidation of other macromolecules including lipids and DNA. The stress response of plant cells largely depends on the severity of the oxidative damage, and may range from stimulation of ROS detoxification systems to programmed cell death. Therefore ROS levels must be carefully controlled by maintaining adequate levels of antioxidants (Mittler, 2002).

Chloroplasts are the prime sources of ROS resulting in generation of $^1O_2$, $O_2^{\bullet-}$ and H_2O_2 generated from several locations, such as the electron transport chain (ETC), photosystem (PS I) and photosystem (PS II). Superoxide, which is produced mostly by electron leakage from Fe-S centers of PS I or reduced ferredoxin (Fd) to O_2 (Mehler reaction), is then converted to H_2O_2 by SOD (Gechev et al., 2006). Superoxide can also be produced by leaking of electrons to molecular oxygen from the ETCs of PS I and PS II (Sgherri et al., 1996). Singlet oxygen ($^1O_2$) is produced by excited-chlorophyll formation in the PS II reaction center and in the antennae system (Asada, 2006). In response to salinity stress plants decrease stomatal conductance, to reduce excess water loss, and this ultimately leads to a decrease in internal CO_2 concentrations, slows down the reduction of CO_2 by the Calvin cycle and induces photorespiration (Abogadallah, 2010). In response to drought stress, plants close their stomata to reduce excess water loss through transpiration. Blocking entry of CO_2 in the leaves due to stomatal closure is responsible for limiting photosynthetic carbon fixation (Cruz de Carvalho, 2008; Sanda et al., 2011). A negative consequence of drought stress is the acceleration of the photorespiratory pathway due to limited CO_2 fixation through RuBP oxygenation. More than 70% of total H_2O_2 accumulation occurs in photorespiration under drought stress (Noctor et al., 2002). An increase in temperature above the optimum level can also reduce the light activation of ribulose 1,5-bisphosphate carboxylase (RuBisCO), via heat-denaturation of RuBisCO activase (Eckardt and Portis, 1997; Feller et al., 1998). Additionally, increase in temperature suppresses the carboxylation reaction catalyzed by RuBisCO, by reducing the specificity of RuBisCO for CO_2 (Brooks and Farquhar, 1985; Kim and Portis, 2004; Kaushal et al., 2011). In response to low temperature stress the activities of Calvin cycle enzymes are slowed more than the energy-transducing reactions, causing $NADP^+$ depletion (Wise, 1995). The two essential metal ions Ca^{2+} and Mn^{2+} present in the PS II reaction center can be replaced, thereby limiting the photo system reaction and leading to the uncoupling of electron transport in the chloroplast via cadmium stress (Baszynski et al., 1980; Mohanty and Mohanty, 1988; Atal et al., 1991). Cadmium (Cd) was found to inhibit electron flow on the reducing side of PS I (Siedlecka and Baszyńaski, 1993). The negative consequence of heavy metal stress was also found in the carboxylating phase of photosynthesis. The two key enzymes required for CO_2 fixation, phosphoenol pyruvate carboxylase (PEPC) and RuBisCO, are the main targets of heavy metal damage. Substitution of Mg^{2+} ions, which are important co-factors of carboxylation reactions, by Cd^{2+} ions not only lowers the activity of RuBisCO and can damage its structure, but can also alter RuBisCO to favor the oxygenation reactions (Siedlecka and Baszyńaski, 1993; Siedlecka et al., 1998; Krantev et al., 2008; Hossain et al., 2011a). Limitation of CO_2 fixation and the induction of the oxygenase activity of RuBisCO enhances photorespiration, resulting in H_2O_2 production in peroxisomes (Jaspers and Kangasjarvi, 2010; Hossain et al., 2011a; Szarka et al., 2012).

Plant mitochondria can produce ROS at several sites along the ETC. The mitochondrial ETC is composed of numerous dehydrogenase complexes such as complex I, II, III and IV and the electron carrier ubiquinone (Q). Most of the ROS formation in mitochondria mainly occurs in the zone of complex I (NADH dehydrogenase) and ubiquininone (Blokhina and Fagerstedt, 2006). Several enzymes present in the mitochondrial matrix can also produce ROS, either directly or by feeding electrons to ETC (Sharma et al., 2012). ROS formation can occur during the normal respiration process, with the rate of ROS formation dramatically elevated under the condition of various abiotic and biotic stresses (Rhoads et al., 2006; Szarka et al., 2012). Upregulation of ROS production as a result of ETC perturbations has been reported in plants exposed to chilling (Prasad et al., 1994a, b; Purvis et al., 1995), salt stress (Hernandez et al., 1993; Mittova et al., 2003), high temperature (Schwarzlander et al., 2009), exposure to cadmium (Schwarzlander et al., 2009) and phosphate deficiency (Juszczuk et al., 2001; Parsons et al., 1999; Malusa et al., 2002). Different metal ions such as Fe, Cu and Zn are essential for proper functioning of the mitochondrial enzymes involved in the TCA cycle, synthesis of ATP, electron transport and antioxidant defense (Tan et al., 2010; Nouet et al., 2011). Transition metal ions are able to facilitate the propagation of oxidative reactions leading to oxidative stress (Keunen et al., 2011). Bi et al. (2009) showed a rapid induction of Cd induced ROS formation in *Arabidopsis thaliana* root cells due to the perturbations of mitochondrial ETC. Although mitochondrial ROS production is much lower when compared to chloroplast ROS production, mitochondrial ROS are important regulators of a number of cellular processes, including stress adaptations and programmed cell death (PCD) (Robson and Vanlerberghe, 2002). Mitochondria interact with chloroplasts and peroxisomes in the photorespiratory cycle; this cycle allows excess reducing equivalents produced during photosynthesis under conditions of restricted Calvin cycle to be eliminated, thus preventing an overreduction of the carriers of photosynthetic electron transport (Krömer, 1995). Furthermore, mitochondria may play a major role in interorganelle cross-talk under environmental/oxidative stress by signaling with chloroplasts (Millar et al., 2001) and by inducing altered nuclear gene expression through mitochondria-to-nucleus signaling (Rhoads et al., 2006).

Peroxisomes are major sites of H_2O_2 production due to their essentially oxidative metabolism. H_2O_2 is produced during photorespiration and also during β-oxidation of fatty acids, by the enzymatic reactions of flavin oxidases as well as by the disproportionation of $O_2^{\bullet-}$. Adverse environmental conditions that hamper the CO_2 fixation in chloroplasts cause glycolate to move to the peroxisomes, where it is oxidized by glycolate oxidase (GO) forming H_2O_2 (Gechev et al., 2006; Takashi and Murata, 2008). Xanthine oxidase (XO) coupled to SOD is also responsible for the production of H_2O_2 from O_2 in peroxisomes (Mhamdi et al., 2010). There are two sites of $O_2^{\bullet-}$ production in peroxisomes (del Rio et al., 2002), one in the organelle matrix and the other in the peroxisome membrane. Enzymes located in the apoplast, namely NADPH oxidases and cell-wall peroxidises, are the prime sources of $O_2^{\bullet -}$ and H_2O_2 and have been found to be stimulated during salt stress, contributing significantly to ROS generation (Sagi and Fluhr, 2006; Abogadallah, 2010). $O_2^{\bullet -}$ and H_2O_2 are also produced by other oxidases induced by abiotic and biotic stresses (Dat et al., 2000). Additionally, the extremely reactive $^{\bullet}OH$ is generated from H_2O_2 and $O_2^{\bullet -}$ in the presence of Fe and Cu ions via the Haber-Weiss or Fenton reactions (Halliwell, 2006). Direct enzymatic mechanisms of $^{\bullet}OH$ elimination are absent in plant cells and thus $^{\bullet}OH$ formation is restricted only through the action of nonenzymatic antioxidants or by enzymatic or nonenzymatic scavenging of H_2O_2 and $O_2^{\bullet -}$. While detoxification of ROS by antioxidants usually takes place at the site of production in most organelles, under conditions inducing stress the local ROS detoxification capacity is not able to manage the levels of ROS produced and H_2O_2 can spread to other cellular compartments, e.g. the cytosol (Figure 16.2).

16.5 ROS SCAVENGING AND DETOXIFICATION

To avoid the excessive accumulation of ROS and to protect from oxidative damage, plants are armed with sophisticated antioxidant defense systems: both enzymatic antioxidant (multiple superoxide dismutases (SOD), ascorbate peroxidase (APX), monodehydroascorbate reductase (MDHAR), dehydroascorbate reductase (DHAR), glutathione reductase (GR), catalase (CAT), glutathione peroxidase (GPX), glutathione *S*-transferase (GST), peroxidase (POX), guaiacol peroxidase (GPOX) and nonenzymatic antioxidant (ascorbate (AsA), glutathione (GSH), tocopherol, carotenoids, flavonoids, and proline) (Fig. 16.3; Ahmad et al., 2010; Gill and Tuteja, 2010; Hossain et al., 2011a; Sharma et al., 2012) in different subcellular organelles. The higher accumulation of ROS scavenger proteins in response to stress was found in different cellular locations such as the nucleus (Bae et al., 2003), mitochondria (Qiu et al., 2004), chloroplast (Giacomelli et al., 2006) plasma membrane (Komatsu et al., 2009) and cell wall (Nanjo et al., 2010). The presence of ROS detoxifying proteins in various cellular locations also supports the possibility that ROS could serve as signal molecules. The delicate balance between ROS production and scavenging that allows this duality in function to exist in plants is thought to be orchestrated by a large network of genes that tightly regulates ROS production and scavenging (Mittler et al., 2004; Miller et al., 2008; Petrov and Van Breusegem, 2012).

16.6 FUNCTION OF PROLINE IN STRESS RESISTANCE

Proline is an extensively studied molecule in the context of plant abiotic stress physiology. There is now strong evidence that proline has a plethora

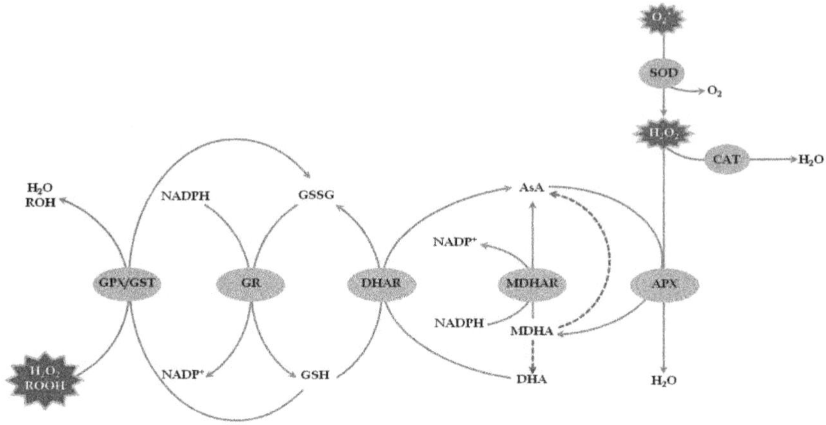

FIGURE 16.3 Reactive oxygen species detoxification systems in plants. Superoxide produced in different cell organelles rapidly converted to H_2O_2 by SOD. H_2O_2, in turn, is converted to H_2O by APX and CAT. The oxidation of AsA caused by ROS or by APX leads to the formation of monodehydroascorbate (MDHA) and dehydroascorbate (DHA). MDHA is reduced to AsA by MDHAR with the utilization of NADPH and DHAR converted DHA to AsA by the utilization of GSH. GR is responsible for recycling of GSSG to GSH by the expense of NADPH. GST and GPX catalyze the GSH-dependent reduction of H_2O_2 and organic peroxides, including lipid peroxides to H_2O or alcohols. Both AsA and GSH also serve as chemical scavengers of ROS in nonenzymatic reactions. Abbreviations are defined in the text.

of functions in both abiotic and biotic stress tolerance. Some important functions of proline related to plant abiotic stress tolerance are briefly described in the following subsections.

16.6.1 Osmotic Adjustment

The fundamental role of proline in osmotic adjustment and in increasing a plant's capability to resist cellular dehydration, caused by salinity, drought or extreme temperatures, has been well documented (Heuer, 1994; Chen et al., 2007). Accumulation of cytoplasmic osmolytes like proline is thought to aid in reducing the cellular water potential to a level below the external water potential; this enables water to move into the cell while at the same time minimizing potentially deleterious increases in ionic strength (Hare et al., 1998; Kavi Kishor et al., 1995; Burritt, 2012). In plants proline usually accumulates in the chloroplast stroma and in the cytoplasm. It has been reported that the proline content in the cytoplasm is proportionally much higher than that measured at the bulk tissue level, since the cytoplasm makes up only a small part of the total cellular volume. This suggests that small fluctuations in cytosolic proline levels could lead to profound effects on osmotic adjustment both directly and by augmenting the accumulation of potassium and other solutes in the cell vacuole (Verslues and Sharma, 2010).

16.6.2 Protection of Cellular Structure During Dehydration

Dehydration is the most crucial environmental constraint on plant growth and development. By modulating multiple protective mechanisms and by altering cellular metabolism, plants can adapt to and resist dehydration and osmotic stress. Proline has multiple functions during osmotic stress, including stabilization of subcellular structures and membranes, stabilization of proteins and maintenance of cellular functions by detoxifying ROS (Khedr et al., 2003; Okuma et al., 2004; Hoque et al., 2008; Banu et al., 2009). Proline accumulation improves stress tolerances without disrupting cellular structure. When the cellular water content decreases, proline and other compatible solutes can act as water substitutes to stabilize cellular structure through hydrophilic interactions and hydrogen bonding. Under osmotic or dehydration stress, membrane integrity must be maintained and proteins protected from denaturation. Proline can interact with enzymes to preserve protein structure and enzyme activities (Kavi Kishor et al., 2005). Proline also plays a critical role in protecting photosynthetic activity under stress and the mitochondrial electron transport complex II is protected by proline under NaCl stress (Hamilton and Heckathorn, 2001). Lehmann et al. (2010) suggests that high levels of proline can be found in pollen and seeds where it serves as a compatible solute, protecting cellular structures during dehydration.

16.6.3 Redox Buffering

During environmental stress redox cycling is a vital part of a plant's antioxidant defenses (Alia et al., 1991). The synthesis of proline generates $NADP^+$, whereas NADPH is produced during proline oxidation. Therefore, the proline biosynthesis and degradation cycle is crucial for buffering redox potential in different cellular organelles, such as the cytosol and plastids. Therefore proline is considered to play an important role in stress tolerance by ameliorating shifts in redox potential by replenishment of the $NADP^+$ supply (Hassine et al., 2008; Sharma et al., 2011). Proline accumulations to very high levels under drought stress is thought to act as an important redox buffer (Verslues and Sharma, 2010). Hare et al. (1999) and Szabados and Savouré (2010) reported that increased proline contents enhanced stress tolerances, as proline, stabilizing the redox status of cells. In tobacco cells and mung bean seedlings, proline improves salt and cadmium tolerances by maintaining the glutathione redox state (Hoque et al., 2008; Islam et al., 2009b; Hossain and Fujita, 2010; Hossain et al., 2010, 2011b). Proline may also act as an antioxidant, scavenging ROS and preserving the intracellular glutathione pool, a major redox buffer for cells (Krishnan et al., 2008); hence, proline metabolism may be more pivotal in maintaining redox homeostasis than previously thought, with proline exhibiting dual functions as a pro-oxidant via PDH and as a ROS scavenger. It has also been suggested that an increased rate of proline biosynthesis in chloroplasts during stress helps to maintain a low $NADPH/NADP^+$ ratio, stabilize redox balance, and reduce photoinhibition and damage of the photosynthetic apparatus (Hare and Cress, 1997).

16.6.4 Storage and Transfer of Reductants

In addition to its role as a compatible solute for osmotic adjustment, proline can play a pivotal role in the storage or transfer of energy and of reducing potential (Szabados and Savouré, 2010; Verslues and Sharma, 2010). This role depends on the spatial and temporal control of proline synthesis and catabolism to either take up or release reducing potential and energy at the correct place and time to meet the plant's needs. Verslues and Sharma (2010) suggested the significance of movement of proline within plant tissues and organs and separation of synthesis and catabolism between different tissues, with other amino acids also potentially participating in redox buffering as well as in energy transfer mechanisms during conditions of stress (Verslues and Sharma, 2010). Similarly, proline was considered as a storage compound supplying reductants, reduced nitrogen and carbon skeletons for post-stress growth recovery (Vartanian et al., 1992). However, a rapid decrease in proline contents after stress release is important for the recommencement of growth after stress and this is also a vital determinant of overall stress tolerance (Hayano-Kanashiro et al., 2009)

16.6.5 Proline a Potential Signaling Molecule

Although proline metabolism and its multiple functions under stressful conditions have been studied in plants for 45 years, information regarding the signaling pathways associated with proline metabolism is very limited. Recent plant proteomic and metabolomic studies have demonstrated that the metabolism of proline is largely regulated by stress signals. Although proline has long been considered a compatible osmolyte, recent studies have highlighted its multiple functions in stress adaptation, recovery and signaling. In response to a range of biotic and abiotic stresses, proline appears to function as a metabolic signal that regulates metabolite pools and redox homeostasis, regulating the expression of many stress responsive genes and affecting the development and growth of plants (Fig. 16.4; Sharma and Dietz, 2006; Banu et al., 2009; Szabados and Savouré, 2010). In addition, it has been suggested that under stressful conditions proline acts as an alternative energy source and a vital regulatory/signaling molecule able to accelerate the activation of the multiple responses that are part of the adaptation process (Maggio et al., 2002). Szabados and Savouré (2010) reported that proline can act as a signal molecule that modulates mitochondrial function,

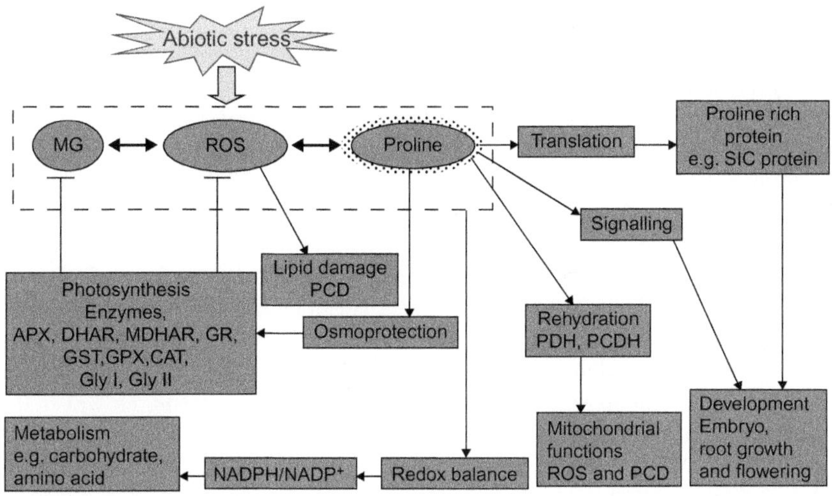

FIGURE 16.4 Function of proline in plant growth, development and stress tolerance. Proline is used for protein synthesis, has protective function as an osmolyte, contributes to the maintenance of redox balance by regulation of ROS and MG metabolism, enhances photosynthetic performance, can regulate development and is a component of metabolic signaling networks controlling mitochondrial functions, stress relief and development. Abbreviations: PCD programmed cell death; SIC, sickle protein (modified from Szabados and Savouré, 2010).

influences cell proliferation and cell death, and triggers expression of genes essential for recovery of plants from stress. In addition to its role in protein synthesis and plant environmental stress tolerance, proline also plays a role in flowering, and the growth and development of plants, functioning both as signal molecule and as a metabolite (Mattioli et al., 2009b). Hare and Cress (1997) proposed that proline might function as an osmoticum and regulatory signal at the same time. All of these findings clearly suggest that as a versatile signaling molecule proline has multiple functions in cells (Figure 16.4).

16.6.6 Reactive Oxygen Species Scavenging

Among the various compatible solutes, proline is the only one that has been shown to protect plant cells from oxidative damage by direct scavenging of ROS, such as singlet oxygen ($^1O_2$) and $^{\bullet}OH$. For a detailed discussion of the molecular mechanisms of ROS scavenging, please see Section 16.7.

16.6.7 Proline as Precursor for Other Antioxidant Molecules

Abiotic stress-induced accumulation of proline has also been found to influence the synthesis of the multifunctional antioxidant glutathione (GSH), probably because the two molecules have a common precursor, glutamate

(Kocsy et al., 2005). Genetic manipulation of proline levels has been found to change the GSH contents, plants with antisense transformants producing lower proline contents and higher GSH contents under drought stress, in relation to nontransformed plants. In contrast lower GSH contents were observed in sense transformants, producing more proline, probably due to a higher requirement of glutamate for biosynthesis of proline in the sense transgenic plants, and lesser employment of glutamate for proline biosynthesis in the antisense transgenic plants. Similarly, polyamine and proline biosynthesis were found to be linked, as they have a common precursor glutamate (Groppa and Benavides, 2008). Under heat stress conditions transgenic plants overproducing proline showed a mild increase in proline content whereas a meaningful enhancement of putrescine (both free and conjugated) was observed in the leaf tissues, probably due to *de-novo* synthesis. However the possibility that proline degradation may serve as a source of precursor of polyamine synthesis in the early phase of stress cannot be excluded (Cvikrová et al., 2012).

16.6.8 Proline as Metal Chelator

The higher constitutive level of proline and upregulation of proline biosynthesis under normal and stressful conditions in metal tolerant genotypes, as compared to nontolerant genotypes, suggests that proline could be involved in the chelation of metal ions (Farago and Mullen, 1979; Xu et al., 2009). It has been demonstrated that exogenous proline protects the enzymes nitrate reductase and glucose-6-phosphate dehydrogenase against Zn- and Cd-induced inactivation under *in-vitro* conditions (Sharma et al., 1998), with the protection possibly being due to the formation of proline metal complexes. The importance of the heavy metal chelating function of proline *in vivo* was also observed in Cu tolerant *Armaria maritima* (Farago and Mullen, 1979). However, proline may be of particular importance for binding metal ions that do not form complexes with phytochelatins (PC).

16.7 MOLECULAR MECHANISMS OF QUENCHING ROS BY PROLINE UNDER STRESS

As mentioned, abiotic stress often results in the production of highly reactive ROS, within plant cells, that can rapidly oxidize DNA, proteins and lipids, which in the latter case can result in the induction of autocatalytic lipid peroxidation. If not repaired, oxidative damage to these key macromolecules can result in membrane damage and dysfunction, loss of enzyme activities and DNA strand breaks. Two common ROS species found in plants under abiotic stress are singlet oxygen ($^1O_2$), dimolecular oxygen in its lowest energy excited state, which is generated in oxygenic photosynthetic organisms, and the hydroxyl radical ($^{\bullet}OH$), which is highly reactive and has a similar potential to cause oxidative damage to biological molecules as peroxides. Several published studies have concluded that proline can effectively quench both $^1O_2$ (Alia et al., 1999, 2001) and $^{\bullet}OH$ (Smirnoff and Cumbes, 1989). These results have led to the commonly held theory that proline can act as a nonenzymatic antioxidant in plant cells under abiotic stress.

Alia (Alia et al., 2001) demonstrated that the photochemical production of $^1O_2$, generated by irradiating a solution of toluidine blue and detected by the formation of the stable nitroxide radical, 2,2,6,6-tetramethylpiperidine-1-oxyl (TEMPO) by EPR spectrometry after reaction of the $^1O_2$ generated with 2,2,6,6-tetramethylpiperidine (TEMP), could be stopped by proline at concentrations of 20 mM or higher. They suggested that as amine compounds with low ionization potentials (IP) are very effective quenchers of $^1O_2$, forming a reversible charge-transfer complex (CT), proline as a cyclic secondary amine with a low IP could act as a powerful $^1O_2$ quencher. In a further publication, Matysik et al. (2002) suggested that proline could act as an effective scavenger of $^1O_2$ plants, but a rate constant has yet to be determined. Alia et al. (2001) suggested that proline could remove $^1O_2$ by physical quenching or by forming other ROS such as the superoxide radical ($O_2^{\bullet -}$) or the peroxide anion $[O\text{-}O]^{2-}$. However, if the latter mechanism was in fact correct the value of proline for the removal of $^1O_2$ in plant cells would be limited, as these ROS would in turn have to be removed. In a more recent study Signorelli et al. (2013) cast considerable doubt on the ability of proline to quench $^1O_2$. In this study the authors demonstrated that proline cannot quench $^1O_2$ in an aqueous buffer system and suggested that ability of proline to quench the $^1O_2$ produced in plant cells under stress should be reconsidered.

The role of proline as a scavenger of $^{\bullet}OH$ radicals appears less controversial at present. Smirnoff and Cumbes (1989) provided evidence that proline could scavenge $^{\bullet}OH$ radicals and a rate constant for this reaction ($4.8 \times 10^8\ dm^3\ mol^{-1}\ s^{-1}$) has been determined (Davies, 2005). Smirnoff and Cumbes (1989) stated that proline "could in principle, protect cells against internally generated hydroxyl radicals" and that "proline could perhaps provide extra protection" to plants that could not rapidly adapt to a stress that caused oxidative damage. Rustgi et al. (1977) suggested that proline could react with $^{\bullet}OH$, under hydrogen abstraction, to form a more stable radical. In addition, as most radicals generated in plant cells under stress are derived from the Fenton reaction, which involves the oxidation of metals, it is also possible that binding of proline to redox active metal ions could help protect against $^{\bullet}OH$ radicals (Matysik et al., 2002).

Proline could also play a less direct role in increasing the antioxidant capacity of plant cells under abiotic stress. For example, proline metabolism is an important regulator of the redox balance of cells (Matysik et al., 2002) and it has been suggested that stress-induced accumulation of amino acids like proline could provide cells with a large pool of precursors molecules required for the synthesis of other stress related molecules (Sanchez et al., 2008). For example, polyamines are a ubiquitous group of molecules that may have antioxidant capacity (Burritt, 2008). They can be synthesized from arginine or ornithine, and ornithine can be synthesized from glutamate, hence the pathways for proline and polyamine biosynthesis are interlinked (Groppa and Benavides, 2008). In addition, in a recent study on the nematode *Caenorhabditis elegans*, Zarse et al. (2012) demonstrated that mitochondrial proline catabolism induces a transient ROS signal important for the adaptive induction of stress defenses that have a positive influence on life span. Proline metabolism may also be important for abiotic stress signaling in plants, as it has already been shown to play a signaling role in plant reproduction (Mattioli et al., 2009a).

16.8 EXOGENOUS PROLINE ENHANCES OXIDATIVE STRESS TOLERANCE TO ABIOTIC STRESSES

Many studies on plants have confirmed that the application of exogenous proline increases abiotic oxidative stress tolerance. Exogenous proline application was found to increase the frost tolerance of *Solanum* species (Van Swaaij et al., 1985). Besides acting as a free radical scavenger and stabilizing membranes, exogenous proline can also act as a source of nitrogen and carbon, thereby improving seedling growth and the recovery of plants (mung bean) exposed to chilling stress (Posmyk and Janas, 2007). Exogenous application of proline to *Arabidopsis* roots resulted in reduced ROS levels and a ROS-induced K^+ efflux (Cuin and Shabala, 2007). Hoque et al. (2007a) investigated the protective role of exogenous proline against salinity induced growth reduction and on the activities of several ROS detoxifying enzymes in cultured tobacco cells. Exogenous application of proline alleviated the salinity induced growth reduction in tobacco (*Nicotiana tabacum* L.) BY-2 cells. Salt stress caused a profound decrease in the activities of SOD, CAT and POX whereas addition of proline in the culture medium favorably modulate the CAT and POX activities under salt stress, suggesting that exogenous proline mitigates, at least in part, the damaging effects of salinity through the modulation of antioxidant enzyme activities.

Hoque et al. (2007b) investigated the levels of antioxidants and the activities of enzymes involved in the AsA-GSH cycle in cultured tobacco cells treated with proline and exposed to salt stress. Salinity stress led to a significant reduction in the level of nonenzymatic antioxidant contents such as AsA and GSH. The activities of AsA-GSH cycle enzyme activities (APX, MDHAR, DHAR and GR) were also found to decrease in response to salt stress. Application of exogenous proline was found to upregulate the activities of APX, DHAR and GR. They suggested that proline offers protection against salt stress by increasing the activity of enzymes involved in the antioxidant metabolism.

Banu et al. (2010) also investigated the protective mechanisms of proline against NaCl stress. The levels of H_2O_2, nitric oxide (NO), and MG were found to be higher under both short- and long-term salinity stress, whereas $O_2^{\bullet -}$ level was higher in long-term stressed cells. The level of H_2O_2 was found to decrease by proline or betaine under either short- or long-term salt stress whereas MG levels decreased under long-term salt stress. Exogenous proline was also found to increase GPX transcription. They suggest that proline mitigates late responses to salt stress and the production of H_2O_2 and MG. The reduction in H_2O_2 and MG level by proline could be crucial to reduce cellular damage and to improve salinity tolerance. Proline also inhibits apoptosis-like cell death.

The beneficial role of exogenous proline under salt stress was also studied in Jinchum No. 2 (a salt-sensitive genotype of cucumber). Salt stress resulted in a decrease in plant dry weight, LRW, K^+ contents whereas the lipid peroxidation level in leaf was increased in addition with Na^+ and Cl^- levels. The activities of some antioxidant enzymes such as APX, CAT, SOD and POD were upregulated whereas proline synthesis was not affected. Exogenous foliar addition of proline reduces the growth reduction in connection with improved LRW water status and POD activity, higher endogenous proline level and lower lipid peroxidation level and SOD activity. Importantly, no meaningful differences in Na^+ and K^+ contents and the activities of APX and CAT were observed in proline treated and untreated seedlings subjected to salt stress. These findings showed that higher salinity tolerance in proline-supplemented plants was due to improved peroxidise activity and water status in the leaf (Huang et al., 2009).

In response to salt stress, olive plants (*Olea europaea* cv. Chemlali) activate their antioxidant defenses by upregulating APX, CAT and SOD activities. Proline supplementation mitigated the reduction of growth and photosynthetic activity of olive plants under salt stress. The exogenous application of proline increased the activities of the antioxidative enzymes SOD, CAT and APX in salt-stressed olive plants, as compared to control plants. Proline supplements seem to improve olive salt tolerance by increasing antioxidant enzyme activities and protecting the photosynthetic apparatus (Ben Ahmed et al., 2010).

The beneficial effects of exogenously applied proline (0.2 mM) were investigated in two melon (*Cucumis melo* L.) cultivars (cv. Yuhuang and cv. Xuemei) under salt stress where biomass, chlorophyll content, photosynthetic parameters, ROS and antioxidant enzymes activities were measured. Exogenous application of proline increased the fresh and dry weights of melon cultivars under NaCl stress, enhanced chlorophyll contents, net photosynthetic rate and actual efficiency of photosystem II as well as enhanced the activities of SOD, POD, CAT, APX, DHAR and GR in roots, and lowered superoxide anion radical, H_2O_2 and MDA levels. Proline treatment enhanced the salinity tolerance of both melon plants and alleviated the salt-mediated oxidative damage (Yan et al., 2011).

Reza et al. (2006) studied the effect of salinity and proline on antioxidant metabolism in two barley varieties (*Hordeum vulgare* L. var. Sahand and var. Makoui). Lipid peroxidation (as measured by MDA) levels of Makoui plants subjected to a range of salt concentrations were found to be the same, but MDA contents sharply enhanced in Sahand plants. Significant increases in antioxidant enzyme activities (CAT, APX, DHAR, and GR) in Makoui plants were observed with an increased level of salt stress. The lower level of MDA in Makouri plants was due to a concomitant increase in antioxidant activity in relation to salinity levels. The level of lipid peroxidation was found to be lower when plants were treated with proline, although the influence of proline on SOD and APX activities was not significant. Additionally, the decrease in DHAR activity and an increase in GR activity were observed in proline treated salt-stressed seedlings, indicating that proline is able to quench oxygen radicals both chemically and also by modulating ROS scavenging enzyme acidities. At 150 mM NaCl stress, PDH activity in Makoui variety decreased significantly whereas it was close to zero in plants subjected to 200 mM NaCl. Importantly, higher PDH activity was observed in the variety Sahand under varying levels of salt stress. In the variety Sahand, increased proline metabolism was observed under salt stress. Therefore, higher salt tolerance in Makouri variety could be due to proline synthesis and inhibition of its catabolism.

Islam et al. (2009a) demonstrated the beneficial role of exogenous proline on endogenous proline level, the level of lipid peroxidation and activities of ROS detoxifying enzymes in cultured tobacco BY-2 cells subjected to heavy metal stress (Cd). Upon imposition of heavy metal stress the growth of BY-2 cells was decreased. Surprisingly, addition of proline in the culture medium significantly reduced the growth inhibition. Cd stress leads to an accumulation of Cd and endogenous proline in cultured cells, and decreased the activity of SOD and CAT. Exogenous application of proline resulted in a decrease in lipid peroxidation and an increase in SOD and CAT activities without reducing Cd contents under Cd stress. Exogenous proline application increased and intracellular proline levels and showed enhanced tolerance to heavy metal stress by defending different cellular contents, and by modulating the activities of ROS detoxifying enzymes. Furthermore, Cd stress caused significant increases in Evans Blue-positive cells and impaired AsA-GSH cycle function (Islam et al., 2009b).

The protective effects of proline against Cd toxicity of callus and regenerated shoots of *Solanum nigrum* (a Cd-hyperaccumulator) were investigated by Xu et al., 2009. Prior to heavy metal stress treatment, exogenously applied proline was found to protect the callus membrane integrity and thereby enhanced metal tolerance of *Solanum*. Mass spectroscopy analysis depicted that addition of exogenous proline enhanced the Cd accumulation in callus and regenerated shoots of *S. nigrum*. Enzymatic and nonenzymatic analysis indicated that proline-induced heavy metal tolerance is associated with enhanced SOD and CAT activity and internal GSH pool.

The protective effects of proline were investigated by Nounjan and Theerakulpisut (2012) in two rice cultivars subjected to salinity (200 mM). Salt stress resulted in a reduction of growth, increase in Na^+ to K^+ ratio, high level of H_2O_2 content, and modulated the activities of several ROS detoxifying enzymes (SOD, POX, CAT and APX) in both cultivars of rice Pokkali (salt-tolerant) and Khao Dawk Mali 105 (salt-sensitive) under NaCl treatments. Exogenous proline reduced H_2O_2 in both cultivars of rice Pokkali (salt-tolerant) and Khao Dawk Mali 105 (salt-sensitive) during salt stress whereas proline enhanced the activity of APX in Pokkali, and POX, CAT and APX in Khao Dawk Mali 105 during both stress and recovery period (Nounjan and Theerakulpisut, 2012).

The oxidative damage caused by selenium (Se) was accompanied by elevation of lipid peroxidation and high levels of H_2O_2, and reduced levels of enzymatic (SOD, CAT, APX and GR) and nonenzymatic (AsA and GSH) antioxidants in Bean (Phaseolus vulgaris L.) seedlings (Aggarwal et al., 2011). Addition of proline reduced the toxic effects of Se and improved the growth of Bean seedlings. The stress-related damage was reduced significantly by proline treatment with simultaneous increases in enzymatic (SOD, CAT, APX and GR) and nonenzymatic (AsA and GSH) antioxidants.

The involvement of endogenous proline in copper (Cu) tolerance of four genotypes (RSG 888, CSG 144, CSG 104, RSG 44) of *Cicer arietinum* was investigated by Singh et al. (2010) by analyzing the antioxidant enzymes (SOD, APX and CAT), phytochelatins, Cu uptake, and proline accumulation. Increased SOD activity was observed with an increased level of Cu stress. APX activity was not increased at 25 and 50 μM Cu treatments whereas CAT activity increased at lower Cu concentrations but failed to stimulate at 50 μM Cu. Surprisingly, the order of proline accumulation in the chickpea genotypes was exactly similar to the order of Cu tolerance. The tolerant genotype (RSG 44) showed hyperaccumulation of proline and less lipid peroxidation. They concluded that higher proline accumulation of proline improves Cu tolerance in chickpea.

Although many studies have shown the beneficial effect of proline under stressful conditions, recently it was also reported that the application of exogenous proline under nonstress conditions also improved plant growth performance and metabolic activity in two *Brassica juncea* L. Cultivar namely Varma nad RH-30. Application of proline under natural conditions was found to increase plant growth, the rate of photosynthesis and the activities of ROS detoxifying enzymes when compared with the control seedlings (Wani et al., 2011).

The protective action of proline was investigated in two maize cultivar differences in their drought tolerance (Agaiti-02002, drought tolerant; EV-1098, drought sensitive) under irrigated and drought stress conditions. The concentrations of antioxidant compounds such as tocopherols, flavonoids, carotenoids, and tocopherol were found to enhance in maize seed on those plants that are treated with proline. Additionally, the level of aforesaid antioxidants and the ROS scavenging capacity was more pronounced in proline treated maize seedlings under drought stress (Ali et al., 2013).

Kaushal et al. (2011) showed that chickpea plants subjected to heat stress underwent oxidative damage having higher MDA and H_2O_2 contents, coupled with inhibition of antioxidant enzymatic (SOD, CAT, APX, GR) and levels of nonenzymatic (AsA, GSH, proline) antioxidants were observed. However, seedlings supplemented with 10 μM proline, accumulated proline up to 63 $\mu mol\ g^{-1}$ dry weight, and had reduced membrane damage, and higher chlorophyll and water contents under heat stress. Additionally, ROS mediated damage was considerably reduced which was associated with the modulation of antioxidant levels. They concluded that supplementation of proline induces heat tolerance to chickpeas by reducing cellular injury and protecting essential metabolic enzymes associated with carbon and oxidative metabolism, and that exogenous application of proline appears to mitigate damage due to elevated temperatures.

The effects of proline on antioxidant metabolism in grape (*Vitis vinifera* L.) plants exposed to oxidative stress induced by exogenous H_2O_2 was investigated by Ozden et al. (2009). The effects of exogenous proline application in alteration of endogenous proline, percentage EL, MDA concentrations, H_2O_2 and the activities of SOD, CAT, APX, and POD were measured under oxidative stress conditions. Inhibitory effects of H_2O_2 on antioxidant enzyme activities and increased levels of MDA and EL were found. In the presence of proline, SOD and CAT activities decreased, while POD and APX activities increased. A lower level of internal H_2O_2 content, MDA, and EL was also observed in proline pretreated plants, when the cellular concentration of proline was increased. The finding of the research depicts that both H_2O_2 and proline played significant roles in ROS mediated injury of grapevine leaves and synergistic effects of proline rendered the plants tolerant to oxidative stress through increased levels of antioxidants.

Sorkheh et al. (2012) studied the protective role of proline on antioxidant metabolism in wild almond (*Prunus* spp.) plants exposed to H_2O_2-induced oxidative stress. Oxidative stress led to a degradation of chlorophyll c, higher MDA levels and EL, but had no effect on carotenoid levels. H_2O_2 treated leaves showed significant decreases in the AsA and GSH pools and the activities of APX and POD, but total SOD and CAT activities were increased. However, the addition of proline + H_2O_2 to seedlings resulted in higher POD and APX activities compared to seedlings treated with H_2O_2 only. Additionally, proline + H_2O_2 treatments also caused a strong reduction in the cellular H_2O_2 and MDA contents and EL, indicating that proline protects against oxidative stress of wild almond plants by increasing antioxidant enzyme activities and by decreasing membrane damage.

Kumar and Yadav (2009) showed the modulation of ROS and MG detoxification systems by proline or betaine improves low temperature tolerance in tea plants. In response to cold stress (4°C) the level of MG and lipid peroxidation was found to increase in tea buds (youngest topmost leaf). Surprisingly, ted buds exposed to proline and cold stress showed less lipid peroxidation. Pretreatment of tea buds with betaine and proline increased the activities of GST and GR under the low temperature stress. Additionally the positive role of betaine and proline was also observed in the glyoxalase system enzymes. Finally, they stated that betaine or proline promoted better defense under conditions of low temperature stress by regulating MG levels through the stimulation or protection of some antioxidant and glyoxalase pathway enzymes.

Very recently, Soshinkova et al. (2013) reported that when proline was applied exogenously, it reduced ROS induced damage, as evidenced by reduced MDA levels in *Thellungiella salsuginea* plants and cultured cells under oxidative stress, induced by exogenous H_2O_2 (500 μM). They concluded that exogenous application of proline increased intracellular proline concentrations and changed the redox balance by modulating APX and SOD activities, under both normal and stress conditions.

The aforesaid examples clearly showed that, apart from the multiple functions of proline, the incorporation or addition of proline exogenously can make plants more tolerant to abiotic stress-induced oxidative damage through increased internal proline accumulation and also by modulating both enzymatic and nonenzymatic levels.

16.9 HIGHER ENDOGENOUS PROLINE ACCUMULATION AND ABIOTIC OXIDATIVE STRESS TOLERANCE

Many studies on plants involving stress sensitive and tolerant cultivated varieties or wild species have shown that endogenous increases in proline induces oxidative stress tolerance by modulating the activities of antioxidant enzymes (Türkan et al., 2005; Yazici et al., 2007; Luo et al., 2011; Man et al., 2011; Bao et al., 2011; Anjum et al., 2012; Xu et al., 2012; Saiema et al., 2012; Saeedipour, 2013). In plants, maintenance of PS II and PS I activities as well as electron flow through the photosynthetic electron transport chain is very important under stress conditions. Inhibition of the Calvin cycle and pentose phosphate pathways can channel NADPH, ATP and glutamate for proline synthesis in the chloroplasts. Thus, proline synthesis in the chloroplast may allow an efficient oxidation of photosynthetically produced NADPH providing the required $NADP^+$ for electron acceptors and avoiding the use of O_2 that leads to ROS generation (Hare and Cress, 1997; Szabados and Savouré, 2010). Türkan et al. (2005) studied the mechanisms of drought tolerance (induced by PEG-6000), and measured RWC, lipid peroxidation, stomatal conductance, proline and ROS metabolism in drought-sensitive common bean (*P. vulgaris*) and drought tolerant tepary bean (*P. acutifolius*). Thirty-five-day old seedlings were subjected to water stress for 14 d (osmotic potential of −0.40 MPa). Drought stress induced growth inhibition was more pronounced in *P. vulgaris* as compared to *P. acutifolius*. Superior stomatal conduction was observed in the tolerant genotype *acutifolius*. In *P. acutifolius* MDA levels were found to remain constant under the different stress situations. In addition higher inherent activities of SOD, CAT, APX and POX were found in P. *acutifolius* as compared to *P. vulgaris* and SOD, APX and GR activities were found to increase in *P. acitifolius* under drought stress. Endogenous proline synthesis was higher in P. acutifolius in response to drought as compared to *P. vulgaris*. They concluded that the better drought tolerance of *P. acutifolius* was due to higher constitutive antioxidative activities and increased proline levels.

Very recently, Anjum et al. (2012) studied the adaptation of two pepper (*Capsicum annuum* L.) cultivars (Shanshu-2001 and Nongchengjiao-2) under varying levels of progressive drought stress by investigating the leaf water status, proline accumulation, ROS and proline metabolism. Shanshu-2001 showed a higher RLWC, protein, and proline accumulation than Nongchengjiao-2 in all water regimes. Importantly, total soluble proteins and proline continued to increase with continued drought in cultivar Shanshu-2001, whereas in cultivar Nongchengjiao-2 only proline levels continued to increase, but at a lower rate than for the cultivar Shanshu-2001. Membrane lipid peroxidation and EL increased with prolonged drought, with higher rates in cultivar Nongchengjiao-2 than cultivar Shanshu-2001. The progression in drought enhanced the activities of CAT, POD and SOD initially, which were then decreased. Higher endogenous antioxidant enzyme activities (CAT, SOD and POD) were found in the cultivar Shanshu-2001 as compared to Nongchengjiao-2, with resulting better agronomic performance (growth and yield) in the cultivar Shanshu-2001. The cultivar Shanshu-2001 was considered more tolerant due to increased antioxidant metabolism, tissue water status and endogenous proline accumulation.

Man et al. (2011) studied the underlying mechanisms of drought tolerance in tall fescue [*Festuca arundinacea* (Schreb.)] cultivars. Two tall fescue cultivars (Van Gogh, drought tolerant and AST7002, drought sensitive) differing in drought tolerance were established and grown well watered, or under drought stress. Turfgrass quality, relative LWC, leaf indole-3-acetic acid (IAA) and cytokinin zeatin riboside (ZR) contents decreased in response to drought stress reduced, whereas proline and abscisic acid (ABA) contents increased. Under drought stress 'Van Gogh' had greater turfgrass quality rating, LWC, proline, ABA, and ZR content relative to 'AST7002'. At the end of drought stress, leaf proline, ZR, and ABA contents were 32%, 43%, and 50% higher in 'Van Gogh' relative to 'AST7002', respectively. No cultivar differences were observed under well-watered conditions. Their findings suggest that proline, ABA, and ZR contents are associated with drought tolerance. Selection and use of the cultivars with higher proline, ABA, and ZR contents under drought stress may be a practical approach to improve tall fescue drought tolerance.

The physiological and molecular relationships between salinity tolerance and antioxidant metabolism were investigated by Yazici et al. (2007), using salt and drought-tolerant purslane (*Portulaca oleracea* L.) plants. Plant growth significantly decreased at 140 mM NaCl stress, but although the RWC content increased after 18 d of salinity, a decrease in RWC content was observed after 30 d of salt stress. The levels of lipid peroxidation were unchanged under both levels of salt stress (70 and 140 mM) after 18 d, whereas higher lipid peroxidation was observed at 140 mM stress after 30 d. The activities of SOD, POX and CAT were found to decrease after 30 d of treatment at the higher level of salt stress (140 mM). The activities of GR showed profound increases after 18 d of salinity stress. The activities of APX decreased after 18 d of treatment, but sharply increased after 30 d. Surprisingly, a simultaneous increase in proline levels were found with increased salinity stress. This phenomenon suggests that the salt tolerance of purslane is due to higher biosynthetic capacity for proline, as well as higher antioxidant enzyme activities.

Recently, Saiema et al. (2012) studied the biochemical response of salt stress in two contrasting chickpea genotypes (SKU-06, salt tolerant and SKU-04, salt sensitive) by investigating total AsA, GSH, proline and protein contents. Plants (15-day-old) were treated with varying levels of salinity (0, 50 and 100 mM NaCl). In the salt tolerant genotype total AsA, GSH, proline and protein content increased with the increasing level of salt stress, as compared to control where a significant reduction of the above parameters were observed in the sensitive genotype, except for proline. They concluded that salt tolerance could be attributed to increased proline, protein, and AsA and GSH contents.

Luo et al. (2011) studied low temperature (0°C) induced ROS metabolism in two strawberry (*Fragaria ananassa* Duch.) cultivars (cv. Zoji and Toyonaka). Low temperature stress led to an enhancement of the activities of the antioxidant POD, SOD, CAT, APX, DHAR, GR as well as increases in $O_2^{\bullet -}$, H_2O_2, DHA and GSH, when compared with controls. However the levels of AsA and the chlorophyll content decreased. Among the two varieties cv. Toyonaka showed better performance, indicated by lower lipid peroxidation, as compared to cv. Zoji due to the enhanced activities of ROS detoxifying enzymes, and osmolyte and proline synthesis.

Bao et al. (2011) studied the impact of heavy metal (Cd) tolerance on ROS and proline metabolism using the Cd-hyperaccumulator (*Solanum nigrum*) and non-hyperaccumulator (*Solanum lycopersicum*). It was found that in the roots and leaves, SOD, POD and CAT activities of *S. nigrum* were significantly higher than those of S. *lycopersicum* in response to an increasing level of Cd stress (0.01, 0.1, 1 mg/l). Although MDA levels increased in both *S. nigrum* and *S. lycopersicum* under Cd stress a greater increase occurred in the tissues of *S. lycopersicum*. The concentrations of free proline in the leaves and roots of *S. nigrum* were higher than those in

S. lycopersicum across all the Cd treatments. These results suggest that the Cd-hyperaccumulator, *S. nigrum* had a greater capacity than *S. lycopersicum* to adapt to oxidative stress caused by Cd probably due to higher proline accumulation and ROS scavenging capacity.

Xu et al. (2012) studied the molecular mechanisms of waterlogging tolerance by using two waterlogging tolerant (WTG-2541, WTG-2413) genotypes, and one waterlogging sensitive (WSG-EZhi2) genotype of sesame, by assessing photosynthetic rates, proline contents, activities of antioxidant enzymes (POD, CAT and SOD) and lipid peroxidation levels (MDA). Their results showed the net photosynthetic rate (PN) of the three tested sesame seedlings decreased under water-logging stress. In response to flooding stress both sensitive and tolerant genotypes showed a profound increase in proline as compared to control but its level was significantly higher than the sensitive genotype. The MDA level in the tolerant genotype was similar to the control, but a significant increase was observed in the sensitive genotype. POD activities of the WTGs increased while activities decreased in the WSG. The opposite trend was observed for CAT, for which WTGs showed a smaller percentage decrease than that of the WSG. The SOD activities of the three tested genotypes increased in response to flooding, but greater increase was observed in the tolerant genotypes. They concluded that the WTGs of sesame showed substantial tolerance to flooding stress with increased proline contents and activities of antioxidant enzymes.

Recently, Saeedipour (2013) found that drought tolerant wheat (*Triticum aestivum* L.) exhibited higher grain yield under water deficit conditions as compared to sensitive genotype. Results showed that water stress significantly increased proline and ABA contents in the tolerant genotype whereas a slight increase was observed in the sensitive cultivar. Therefore, the more pronounced effect of both proline and ABA in Zagros cultivar renders it more productive under water stress conditions.

The above findings clearly demonstrate that endogenous accumulation of proline, induced by various abiotic stresses, plays a positive role in reducing cellular oxidative damage by modulating antioxidant enzymes activities and by increasing nonenzymatic antioxidant contents in addition to its potential role in osmotic adjustment.

16.10 MODULATION OF ROS AND METHYLGLYOXAL DETOXIFICATION SYSTEMS BY EXOGENOUS PROLINE INDUCES OXIDATIVE STRESS TOLERANCE

Plants exposed to abiotic or biotic stresses exhibit rapid accumulation of endogenous MG and its detoxification is one of the potential mechanisms for inducing tolerance to biotic and abiotic stresses (Yadav et al., 2005; Hossain et al., 2009, 2011a; Hossain and Fujita, 2009; Banu et al., 2010; Upadhyaya et al., 2011). MG is toxic to plant cells, causing inhibition of cell

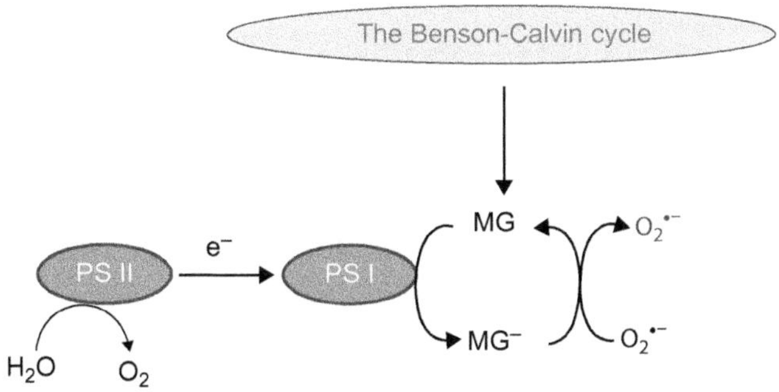

FIGURE 16.5 Superoxide production ($O_2^{\bullet-}$) by methylglyoxal in chloroplast (adapted from Saito et al., 2010).

proliferation, protein inactivation and inhibition of ROS detoxification systems and as a consequence disrupts cellular functions (Martins et al., 2001; Hoque et al., 2010, 2012a). However, signaling roles for MG in inducing abiotic stress tolerance have also been reported (Hoque et al., 2012b, c). Recently Saito et al. (2011) reported that MG accumulated in chloroplasts during the day from triose phosphates needs to be controlled by detoxification mechanisms, otherwise it will catalyze the photo reduction of O_2 to $O_2^{\bullet-}$ at PS I (Fig. 16.5). This increase in $O_2^{\bullet-}$ production could cause oxidative damage to plant cells. The glyoxalase system, the most important detoxification pathway of MG, is composed of two enzymes glyoxalase I (Gly I) and glyoxalase II (Gly II) that catalyze the detoxification of MG. Glyoxalase I converts MG to S-D-lactoylglutathione (SLG) by using GSH, while Gly II converts SLG to D-lactic acid, and via this reaction GSH is recycled in the system. Moreover, the glyoxalase system serves the prime function to detoxify highly reactive MG and in recycling of trapped GSH in plant antioxidant defense system to maintain a higher redox state (Creighton et al., 1988). Like MG, the SLG produced by Gly I was also found to be toxic at high cellular concentration (Thornalley, 1996). Plants showed tolerance reaction against abiotic or biotic stress by limiting over-accumulation of MG levels through the upregulation of Gly I and Gly II activities and also by modulating the GSH-based detoxification systems that ultimately lowered lipid peroxidation rate (Yadav et al., 2005; Singla-Pareek et al., 2006; Hossain et al., 2013a). The ubiquitous glyoxalase system played a pivotal role in plant abiotic stress tolerance by regulating MG levels and by regulating GSH-based ROS detoxification. Recent genetic and proteomic studies have shown the glyoxalase pathway has a profound effect on stress tolerance. The transcript abundance and activities of Gly I and Gly II are induced by various abiotic and biotic stresses (Espartero et al., 1995; Singla-Pareek et al., 2003, 2006; Mustafiz et al., 2006; Hossain et al., 2009; Lin et al., 2010) and also adverse conditions induced by exogenous MG and H_2O_2

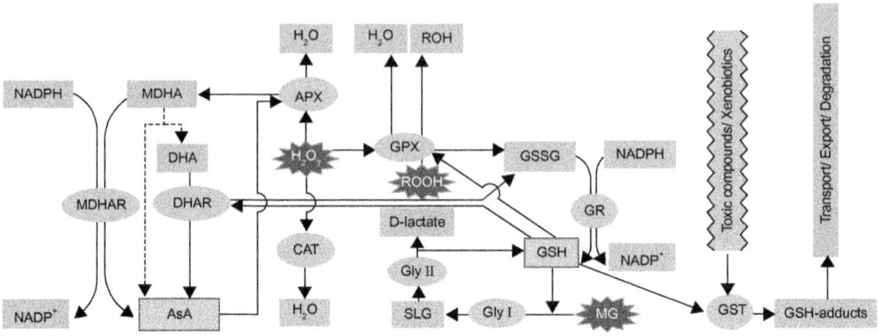

FIGURE 16.6 Metabolic interactions of AsA- and GSH-based antioxidative system and GSH-based glyoxalase system in plant cells (Hossain et al., 2011a). Dotted lines indicate nonenzymatic reactions.

(Wu et al., 2012). Wild-type stress tolerant studies have shown that the antioxidative and glyoxalase defense systems are closely linked and that the glyoxalase system has a direct influence on the ROS detoxification (Yadav et al., 2005; El-Shabrawi et al., 2010; Upadhyaya et al., 2011) and plants overexpressing either Gly I or Gly II gene enhances plant abiotic stress tolerance (Singla-Pareek et al., 2003, 2006, 2008; Lin et al., 2010; Wu et al., 2012; Viveros et al., 2013).

Several studies on plants have demonstrated that simultaneous induction of both glyoxalase and antioxidative systems renders plants more tolerant to oxidative stress (Fig. 16.6; Hoque et al., 2008; Kumar and Yadav 2009; Hossain and Fujita 2010; Hossain et al., 2010, 2011b, 2012a, b, 2013a, b). Seedlings primed with cold or heat-shock showed greater oxidative stress tolerance with modulations in both the ROS and MG detoxification systems (Hossain et al., 2013a, b) suggesting the involvement of both detoxification pathways in the cross-tolerance of plants under abiotic stresses. Recently, Upadhyaya et al. (2011) reported overexpression of *GalUR* gene in transgenic potato plants exhibited greater salinity tolerance and increased activities of the antioxidant enzymes APX, DHAR, GR, GST,GPX and the glyoxalase system enzymes (Gly I and Gly II) along with increment in GSH:GSSG ratios. Greater accumulation of AsA was found in the transgenic plants with a restricted increase in MG levels under salt stress. Additionally, a relatively higher GSH: GSSG ratio in these transgenic plants could also help to protect them from salinity-induced oxidative stress. Increased ROS and MG detoxification along with the favorable changes in the GSH and AsA redox state in these transgenic plants were thought to be the main reasons for enhanced salinity tolerance.

The protective role of exogenous proline or betaine in salt stress tolerance in relation to GSH utilizing and regenerating enzymes was studied by Hossain and Fujita (2010) in mung bean seedlings. The imposition of salt stress resulted in a marked increase in GSH pool within 24 h whereas a non-significant enhancement was observed after 48 h. Among the GSH metabolizing enzymes, upregulation of GPX, GST and Gly II activities and GR and Gly I was observed after 48 h stress and 24 h treatment, respectively. Addition of proline or betaine resulted in increment of the GSH pool, the GSH/GSSG ratio and the activities of GST, GPX, GR, Gly I and Gly II as compared to controls. Surprisingly, seedlings treated with salt alone resulted in elevated GSSG, H_2O_2 and MDA levels, indicating that either proline or betaine induces salt-induced oxidative stress tolerance by enhancing the ROS and MG detoxification systems.

We recently reported that supplementation with proline or betaine increases salinity and heavy metal-induced oxidative stress tolerance in *Vigna radiata* seedlings (Hossain et al., 2010, 2011b). Both salt and Cd stress showed higher lipid peroxidation and H_2O_2 levels as compared to control plants. Importantly, the inactivation or insufficient upregulation of ROS detoxification and glyoxalase pathway enzymes (APX, MDHAR, DHAR, CAT, GST, GPX, Gly and Gly II) and AsA and GSH contents were found in salinity- and Cd-stressed seedlings. Importantly, betaine or proline pretreatment or addition of proline/betaine in the plant nutrient solution favorably modulated the activities of these enzymes and the GSH/GSSG ratio, rendering plants more tolerant to NaCl- and Cd-induced oxidative stress.

Hoque et al. (2008) also reported the protective role of proline and betaine in modulating the ROS and MG detoxification pathways. Imposition of salt stress in tobacco BY-2 culture cells led to a marked increase in protein oxidation, the GSH/GSSG ratio and the activities of GST and Gly II. Addition of proline or betaine in salt-stressed tobacco cultured cells reduced the protein oxidation, increased the GSH content and the activities of GPX, GST and Gly I. The findings of this experiment suggest that proline acts to reduce salt-induced oxidative damage, as indicated by lower protein oxidation, through the activation of both the ROS and MG detoxification systems.

The aforesaid examples clearly demonstrated that proline under stressful conditions modulates multiple stress response pathways, making plants more tolerant to induced oxidative damage under abiotic stress.

16.11 PROLINE-ACCUMULATING TRANSGENIC PLANTS AND ABIOTIC OXIDATIVE STRESS TOLERANCE

The cloning of various genes from proline biosynthetic pathways in relation to its biosynthesis and catabolism has been well documented. The suppression of proline synthesis in transgenic plants containing the gene coding for P5CR in the antisense direction resulted in increased sensitivity to water deficit (de Ronde et al., 2000, 2001). Additionally *p5cs1* mutants showed strongly reduced proline accumulation in response to stress, concomitantly with reduced root growth, enhanced production of ROS in leaves, and a lower $NADP^+$ to NADPH ratio (Szekely et al., 2008; Sharma et al., 2011). Transgenic plants with higher proline synthesis display improved tolerance to various abiotic stresses including oxidative damage. To elucidate the possible involvement of proline in heavy metal stress tolerance, a comprehensive analysis of transgenic green microalga *Chlamydomonas reinhardtii* was reported by Siripornadulsil et al. (2002). Microalga overexpressing *P5CS* grow more rapidly in toxic Cd concentrations (100 μM), accumulate more than 2-fold proline and bind 4-fold more Cd than wild-type cells. Estimation of endogenous free proline, GSH/GSSG ratios and lipid peroxidation under cadmium stress showed Cd-tolerance of transgenic algae is positively correlated with free proline content and GSH redox state. Their finding denoted that the free proline likely acts as an antioxidant in Cd-stressed cells and the resulting enhanced level of GSH assist for higher rate of phytochelatin synthesis and sequestration of Cd, because GSH-heavy metal adducts are the substrates for phytochelatin synthase.

Parvanova et al. (2004) studied the reaction in the oxidative component of freezing in several tobacco lines transformed with the *P5CS* gene. In transgenic plants, the levels of MDA and H_2O_2 were significantly lower than the nontransformed plants after 24 h of freezing stress. Significant inhibition of CAT was observed in nontransformed plants, whereas transformed plants showed significantly less inhibition. It is speculated that the transfer of *P5CS* genes, which results in the accumulation of proline, is related to the reduction in freezing-induced oxidative damage, due to higher antioxidant enzyme activities.

Razavizadeh and Ehsanpour (2009) reported that tobacco (*Nicotiana tabacum* cv. Wisconsin) plant transformed with the gene *P5CS* showed an increase in APX, CAT activities and ultimately increases the tolerance to salinity. The findings of the experiment showed that *P5CS* is an inducible gene regulating the activities of CAT and APX and the accumulation of proline in plants subjected to salt stress.

De campos et al. (2011) studied the effects of the high endogenous proline levels on water relations, gas exchange and antioxidant enzymatic activity in leaves of transgenic 'Swingle' citrumelo rootstocks transformed with the *P5CSF129A* gene coding for proline biosynthesis. Under water stress conditions the leaf water content, xylem sap flow, osmotic pressure potentials, photosynthetic rates, stomatal conductance and lipid peroxidation levels as measured by MDA was evaluated in genetically modified plants with respect to control plants. Imposition of drought stress resulted in a reduction of SOD and APX activities in nontransformed plants, whereas the activity increases after re-watering. The activity of CAT was more active in nontransgenic plants as compared to transgenic plants under irrigated condition but the MDA level significantly increased. They concluded that transgenic plants were able to cope with drought stress better than controls since the high endogenous proline level acted not only by mediating osmotic adjustment, but also by contributing to gas exchange parameters and ameliorating the deleterious effects of drought-induced oxidative stress as indicated by lower level of MDA.

Transgenic soybean plants overexpressing the *P5CS* gene in both sense and antisense directions were evaluated in drought stress or combined heat and drought stress environments in relation to glutathione metabolism including other nonenzymatic and enzymatic antioxidant metabolisms (Kocsy et al., 2005). Antisense transformants showed severe ROS injury, as indicated by higher H_2O_2 and lipid hydrogen peroxide levels, while the least injury was observed in sense transformed plants due to lower H_2O_2 accumulation. Higher proline and AsA pool were observed in the plants that have sense direction of genes. Surprisingly, the higher GSH pool was found in antisense transformants, indicating that alteration of proline biosynthesis not only affects the GSH pool but also alters the antioxidant pool which has direct influences on ROS detoxification.

The potential role of proline accumulation in transgenic plants overexpressing the *AtP5CS1* gene was investigated under heat stress conditions by Lv et al., (2011). The activities of antioxidant enzymes SOD, POX and CAT, but not GR and APX, increased in all lines after heat treatment, but the increase was more significant in proline-overproducing seedlings. However, enhanced proline level in transgenic plants led to decrease in stress tolerance under high temperature due to higher ROS synthesis through the Pro/P5C cycle and interference in ethylene and ABA synthesis.

You et al. (2012) showed that genetically modified rice plants with OsOAT gene showed enhanced drought and osmotic stress tolerance. Transgenic plants showed higher δ-OAT activities and proline accumulation under normal growth conditions. In response to drought stress the transgenic plants showed higher GSH pool, and enhanced activities of GPX and POD. They concluded that higher ROS scavenging activity and proline accumulation are the main reasons for induced stress tolerance.

Molinari et al. (2007) studied transgenic sugarcane plants overexpressing *P5CS* gene under water-deficit conditions in relation to osmotic adjustment, chlorophyll content and oxidative protection. Proline content in transgenic plants was 2.5-fold higher as compared to control but no osmotic adjustment was observed. Importantly, the photosynthetic efficiency was 65% higher in the transformed plants. Lower levels of MDA and greater biomass were also found after 12 d of drought stress in transgenic plants. These results suggest that proline overaccumulation in transgenic plants not only acts in osmotic adjustment but also acts as a modulator of the ROS detoxification system. Kumar et al. (2010) also showed that a transgenic rice variety overexpressing a P5CSF129A gene showed higher salt stress tolerance and had reduced lipid peroxidation.

The above findings clearly demonstrate accumulation of endogenous proline regulates pivotal functions in inducing abiotic oxidative stress tolerance not only through osmoregulation but also by increasing ROS detoxifying enzyme activities and reducing lipid peroxidation.

16.12 PROLINE-ENHANCED TOLERANCE TO ABIOTIC OXIDATIVE STRESS

Numerous research studies in plants have demonstrated that the transformation of proline biosynthetic genes with higher endogenous proline showed higher tolerance to different abiotic stresses. Recent studies in plants including natural stress tolerant plant, mutants and transgenic plants overexpressing proline producing genes also showed that proline-mediated oxidative stress tolerance is also associated with the expression of ROS scavenging genes. The protective role of proline was investigated in the desert plant *Paneratium maritimum* under salt stress. Imposition of plants under higher salinity led to an inhibition of CAT and POX activity. However, exogenous proline application stabilizes the ROS enzymes and induces enhanced expression of dehydrin proteins. The induction of dehydrins by proline indicates that it might be a component of the regulatory process(s) that leads to the accumulation of dehydrins during salt stress (Khedr et al., 2003). A salt hypersensitive *Arabidopsis* mutant, *p5cs1*, showed severe oxidative stress as shown by increased level of lipid peroxidation, severe chlorophyll destruction, and elevated H_2O_2 content due to inhibition of the activities of the ascorbate-glutathione pathway enzymes (Szekely et al., 2008). Banu et al. (2009) also showed that the exogenous application of proline (20 mM) increased POX gene expression in NaCl-sensitive tobacco (*Nicotiana tabacum*) BY-2 cells subjected to salt stress (200 mM NaCl).

Radyukina et al. (2011) showed that *Salvia officinalis* plants exposed to UV-B radiation (10 min, 12.3 kJ/m^2) showed severe ROS-induced damage as depicted by higher MDA levels in leaf tissues. However, proline pretreatment (5 mM) inhibited ROS-induced damage as indicated by lower lipid peroxidation. The activities of three SOD isoforms – Mn-SOD, Fe-SOD, and Cu/Zn-SOD – were analyzed in leaf tissues and expression of the genes encoding these enzymes were also analyzed by RT-PCR. It was found that addition of proline, UV-B irradiation, or combination of both treatments regulated in a differential manner the activities of the SOD isoforms localized in various cell compartments. They found that proline involvement in the differential regulation of the various SOD isoforms largely depends on the severity of oxidative stress.

Recently, Nounjan et al., (2012) showed that exogenous application of proline (10 mM) to seedlings of Thai aromatic rice (cv. KDML105; salt-sensitive) during salt stress and subsequent recovery induces upregulation of genes encoding the antioxidant enzymes *Cu/ZnSOD, MnSOD, CytAPX, CatC*. Importantly, they also found an upregulation of proline synthesis genes (*P5CS* and *P5CR*) in response to exogenous proline application.

De Carvalho et al. (2013) showed that isoforms of four antioxidant enzyme genes (APX, CAT, SOD and GR) were differentially regulated in leaves of *Swingle citrumelo* transgenic plants with high endogenous proline accumulation submitted to water deficits and also under normal water supply condition. Proline caused a 2-fold change in the transcript levels of *APX1, APXcl, CAT2* and *Cu/ZnSOD2*, while during water deficit proline influenced mRNAs levels of *APXs* and *Cu/ZnSODs* isoforms, *MnSODmit* and *GRcl*. In addition to its known effects on diverse plant physiological and biochemical processes, high endogenous proline can also act as a regulatory/signaling molecule capable of altering the transcription levels of stress-related genes. Their findings added new information on the role of proline during drought conditions and, more importantly, without the potential confounding effects imposed by drought stress.

Therefore, proline overaccumulation in transgenic plants or the exogenous application is able to activate stress-specific genes. The functions of these genes contribute, at least in part, to a reduction of overaccumulation of ROS and stress protein expression under varied abiotic stress. Further studies to identify proline-inducible genes and the multiple functions of their products will advance our understanding of the proline-mediated abiotic stress tolerance of crop plants.

16.13 PROLINE CONTENT AS AN INDICATOR FOR BREEDING

Abiotic stress induced proline accumulation, and the quantitative variations in proline levels found in stress tolerant and susceptible plants, means that proline levels could possibly be used as a selection criterion for breeding programs (Ashraf and Harris, 2004; Man et al., 2011; Boscaiu et al., 2012). Overaccumulation of proline has been found to be activated under drought stress in various crop species such as rice, wheat, maize, rapeseed and pea, and in model legume *Lotus japonicus* (Rampino et al., 2006; Charlton et al., 2008; Norouzi et al., 2008; Betti et al., 2012; Sanchez et al., 2012; Witt et al., 2012). Similar increase in proline accumulation was also evidenced in strawberry cultivars under chilling stress (Luo et al., 2011) and in purslane under salt stress (Yazici et al., 2007). In Mediterranean environments proline levels were indeed recommended as a selection criterion for cereal cultivars (Richards, 1987). Recently, Man et al. (2011) suggested that proline level can be used as a metabolic measure to screen water stress tolerance in tall fescue. Wanichananan et al. (2003) showed that accumulation of proline was the highest in the salt-tolerant line (Leuang Tang Mo), among the 22 lines tested under salt stress conditions. A similar report of higher proline accumulation in response to salt stress was also observed in 12 rice genotypes (Pandy and Srivastava, 1990). The positive association between proline biosynthesis and the salinity tolerant index prompted said biochemical trait to be recommended as an important selection index for breeding salinity tolerance in rice. The extent of proline accumulation under salinity stress was also recommended for identifying salt tolerant genotypes (Refouli and Larher, 1989). While studying tolerant and susceptible genotypes, Yazici et al. (2007) also reported that endogenous proline overaccumulation is higher in the salt tolerant genotype as compared to its susceptible counterpart. Recently, Saiema et al. (2012) showed that proline accumulation under salt stress was higher in the salt tolerant chickpea genotype (SKU-06), whereas enhancement of endogenous proline levels in black gram cannot be considered as a tolerance indicator against salinity stress and therefore cannot be used for plant breeding programs (Ashraf, 1989). Finally, it can be summarized that overaccumulation of proline is species specific and generalization over diverse crop varieties is not always possible.

16.14 CONCLUSIONS

Plants can sense abiotic or biotic stresses and evolve remarkable mechanisms to uphold cellular homeostasis by triggering proper responses related to its growth, development and metabolism. Metabolic acclimation via the accumulation of proline is often regarded as a basic strategy for the protection and survival of plants under abiotic stress (Chen et al., 2007). Overaccumulation of ROS in response to stress is found to be the prime factor for the destruction of cellular function. Reactive oxygen species (especially H_2O_2) have dual function in plant cells: it acts as a signaling molecule that modulates the expression and activates multiple defensive gene responses when present at low concentration, whereas overaccumulation resulted in oxidative damage and even death of plants (Petrov and Van Breusegem, 2010). Therefore, plant cells must be able to maintain appropriate levels of ROS. Pragmatic evidence for the protective functions of proline under abiotic oxidative stress has been provided by studies using mutants and transgenic plants with proline deficiency and hyperaccumulation, and also through the exogenous application of proline. The potential biochemical mechanisms of proline-enhanced tolerance to oxidative stress include: (i) direct scavenging of ROS (e.g. singlet oxygen and hydroxyl radical), (ii) induction of ROS scavenging gene expression, (iii) favorable modulation of antioxidant enzymatic activities and nonenzymatic antioxidant contents, (iv) maintenance of PS II and PS I activities and low $NADPH/NADP^+$ ratio that condense photo inhibition and damage of photosynthetic apparatus, (v) enhancing stress protein expression. In addition to direct ROS scavenging, proline was also found to modulate the ROS and MG detoxification pathways, inducing abiotic oxidative stress tolerance. However, how proline accumulation influences particular regulatory pathways in response to stress is still not clear and requires further research. Despite the beneficial effects of endogenous and exogenous proline with respect to oxidative stress tolerance, overaccumulation or excessive application of proline was found to have negative consequences for plants (Roy et al., 1993; Lv et al., 2011). Further physiological and biochemical research integrating the protecting and growth inhibiting properties of proline is required. Therefore, it is worthwhile to identify the factors that influence the proline biosynthesis and degradation in plants subjected to various abiotic and biotic stresses. Furthermore, the identification of signaling components associated with proline synthesis, degradation and the coordination of gene expression events under stress, as well as during stress recovery, is of paramount importance. The chloroplast is the major source of ROS production in cells under stressful conditions; therefore genetic transformation technology could aim to upregulate the synthesis of proline in the chloroplast. Combining genetic analysis with metabolic profiling approaches could considerably increase our understanding pertaining to plant stress responses and the involvement of proline in plant stress adaptation, with special reference to abiotic oxidative stress. Modern and more powerful metabolic profiling tools might be helpful in understanding the regulation of proline mediated and proline-dependent signaling in plants. In the context of the ongoing climate changes, further study on proline metabolism under abiotic stress conditions will certainly supplement physiological knowledge about abiotic stress tolerance in crop plants. It is clear more significant effort is required to both complement and guide in breeding as well as in gene manipulation programs.

REFERENCES

Abogadallah, G.M., 2010. Antioxidative defense under salt stress. Plant Signal. Behav. 5, 369–374.

Aggarwal, M., Sharma, S., Kaur, N., Pathania, D., Bhandhari, K., Kaushal, N., et al., 2011. Exogenous proline application reduces phytotoxic effects of selenium by minimising oxidative stress and improves growth in bean (*Phaseolus vulgaris* L.) seedlings. Biol. Trace Elem. Res. 140, 354–367.

Ahmad, P., Jaleel, C.A., Salem, M.A., Nabi, G., Sharma, S., 2010. Roles of enzymatic and non-enzymatic antioxidants in plants during abiotic stress. Crit. Rev. Biotechnol. 30, 161–175.

Ali, Q., Anwar, F., Ashraf, M., Saari, N., Perveen, R., 2013. Ameliorating effects of exogenously applied proline on seed composition, seed oil quality and oil antioxidant activity of maize (*Zea mays* L.) under drought stress. Int. J. Mol. Sci. 14, 818–835.

Alia, A., Matysik, J., Mohanty, P., 1999. Singlet oxygen scavenging activity of proline: studies with spin trapping EPR spectroscopy. In: Vohora, S.B., Agarwal, V.P. (Eds.), Toxicology and Environmental Health. Society Biosciences, Jamia Hamdard University/Aptech Publications, New Delhi, India, pp. 121–124.

Alia, A., Saradhi, P.P., Mohanthy, P., 1991. Proline enhances primary photochemical activities in isolated thylakoid membranes of *Brassica juncea* by arresting photoinhibitory damage. Biochem. Biophys. Res. Commun. 181, 1238–1244.

Alia, A., Mohanty, P., Matysik, J., 2001. Effect of proline on the production of singlet oxygen. Amino Acids 21, 195–200.

Anjum, S.A., Farooq, M., Xie, X.Y., Lie, X.J., Ijaz, M.F., 2012. Antioxidant defense system and proline accumulation enables hot pepper to perform better under drought. Sci. Hort. 140, 66–73.

Armengaud, P., Thiery, L., Buhot, N., Grenier-De March, G., Savouré, A., 2004. Transcriptional regulation of proline biosynthesis in *Medicago truncatula* reveals developmental and environmental specific features. Physiol. Plant. 120, 442–450.

Arunkumar, R., Sairam, R.K., Deshmukh, P.S., Pal, M., Khetarpal, S., Pandya, S.K., et al., 2012. High temperature stress and accumulation of compatible solutes in Chickpea (*Cicer arietinum* L.). Indian J. Plant Physiol. 17, 145–150.

Asada, K., 2006. Production and scavenging of reactive oxygen species in chloroplast and their functions. Plant Physiol. 141, 391–396.

Ashraf, M., 1989. The effect of sodium choloride on water relations, chlorophyll and protein contents of two cultivars of balckgam (*Vinga mungo* L.). Plant Soil 119, 205–210.

Ashraf, M., Foolad, M.R., 2007. Roles of glycine betaine and proline in improving plant abiotic stress resistance. Environ. Exp. Bot. 59, 206–216.

Ashraf, M., Harris, P., 2004. Potential biochemical indicators of salinity tolerance in plants. Plant Sci. 166, 3–16.

Atal, N., Sardini, P.P., Mohanty, P., 1991. Inhibition of the chloroplast photochemical reactions by treatment of wheat seedlings with low concentrations of cadmium. Analysis of electron transport activities and changes in fluorescence yield. Plant Cell Physiol. 32, 943–951.

Bae, M.S., Cho, E.J., Choi, E.Y., Park, O.K., 2003. Analysis of the *Arabidopsis* nuclear proteome and its response to cold stress. Plant J. 36, 652–663.

Banu, M.N.A., Hoque, M.A., Watanabe-Sugimoto, M., Matsuoka, K., Nakamura, Y., Shimoishi, Y., et al., 2009. Proline and glycinebetaine induce antioxidant defense gene expression and suppress cell death in cultured tobacco cells under salt stress. J. Plant Physiol. 166, 146–156.

Banu, M.N.A., Hoque, M.A., Watamable-Sugimoto, M., Islam, M.A., Uraji, M., Matsuoka., M., et al., 2010. Proline and glycinebetaine ameliorated NaCl stress via scavenging of hydrogen peroxide and methylglyoxal but not superoxide or nitric oxide in tobacco cultured cells. Biosci. Biotechnol. Biochem. 74, 2043–2049.

Bao, T., Sun, T., Sun, L., 2011. Effect of cadmium hyperaccumulation on antioxidative defense and proline accumulation of *Solanum nigrum* L. Afr. J. Biotechnol. 10, 7198–7206.

Baszynski, T., Wajda, L., Król, M., Wolinska, D., Krupa, Z., Tukendorf, A., 1980. Photosynthetic activities of cadmium treated tomato plants. Plant Physiol. 98, 365–370.

Ben Ahmed, C., Ben Rouina, B., Sensoy, S., Boukhriss, M., Ben Abdullah, F., 2010. Exogenous proline effects on photosynthetic performance and antioxidant defense system of young olive tree. J. Agric. Food Chem. 58, 4216–4222.

Betti, M., Perez-Delgado, C., Garcia-Calderon, M., Diaz, P., Monza, J., Marquez, A.J., 2012. Cellular stress following water deprivation in the model legume *Lotus japonicas*. Cell 1, 1089–1106.

Bi, Y., Chen, W., Zhang, W., Zhou, Q., Yun, L., Xing, D., 2009. Production of reactive oxygen species, impairment of photosynthetic function and dynamic changes in mitochondria are early events in cadmium-induced cell death in *Arabidopsis thaliana*. Biol. Cell 101, 629–643.

Blokhina, O., Fagerstedt, K., 2006. Oxidative stress and antioxidant defenses in plants. In: Singh, K.K. (Ed.), Oxidative Stress, Disease and Cancer. Imperial College Press, UK, pp. 151–199.

Boscaiu, M., Lull, C., Llinares, J., Vicente, O., Boria, H., 2012. Proline as a biochemical marker in relation to the ecology of two halophytic *Juncus* species. J. Plant Ecol.. 10.1093/jpe/rts017.

Brooks, A., Farquhar, G.D., 1985. Effect of temperature on the CO_2/O_2 specificity of ribulose-1,5-bisphosphate carboxylase oxygenase and the rate of respiration in the light-estimates from gas-exchange measurements on spinach. Planta 165, 397–406.

Burritt, D.J., 2008. The polycyclic aromatic hydrocarbon phenanthrene causes oxidative stress and alters polyamine metabolism in the aquatic liverwort *Riccia fluitans* L. Plant Cell Environ. 31, 1416–1431.

Burritt, D.J., 2012. Proline and the cryopreservation of plant tissues: functions and practical applications. In: Katkov, I. (Ed.), Current Frontiers in Cryopreservation. INTECH-open access publisher, Croatia, pp. 415–430.

Charlton, A.J., Donarski, J.A., Harrison, M., Jones, S.A., Godward, J., Oehlschlager, S., et al., 2008. Responses of the pea (*Pisum sativum* L.) leaf metabolome to drought stress assessed by nuclear magnetic resonance spectroscopy. Metabolomics 4, 312–327.

Chen, Z., Cuin, T.A., Zhou, M., Twomey, A., Naidu, B.P., Shabala, S., 2007. Compatible solute accumulation and stress-mitigating effects in barley genotypes contrasting in their salt tolerance. J. Exp. Bot. 58, 4245–4255.

Choudhary, N.L., Sairam, R.K., Tyagi, A., 2005. Expression of delta1-pyrroline-5- carboxylate synthetase gene during drought in rice (*Oryza sativa* L.). Ind. J. Biochem. Biophys. 42, 366–370.

Creighton, D.J., Migiliorini, M., Pourmotabbed, T., Guha, M.K., 1988. Optimization of efficiency in the glyoxalase pathway. Biochemistry 27, 7376–7384.

Cruz de Carvalho, M.H., 2008. Drought stress and reactive oxygen species. Plant Signal. Behav. 3, 156–165.

Cuin, T.A., Shabala, S., 2005. Exogenously supplied compatible solutes rapidly ameliorate NaCl-induced potassium efflux from barley roots. Plant Cell Physiol. 46, 1924–1933.

Cuin, T.A., Shabala, S., 2007. Compatible solutes reduce ROS-induced potassium efflux in *Arabidopsis* roots. Plant Cell Environ. 30, 875–885.

Cvikrová, M., Gemperlová, L., Dobrá, J., Martincová, O., Prásil, I.T., Gubis, J., et al., 2012. Effect of heat stress on polyamine metabolism in proline-over-producing tobacco plants. Plant Biol. 182, 49–58.

Dat, J., Vandenabeele, S., Vranová, E., Van Montagu, M., Inze, D., Van Breusegem, F., 2000. Dual action of the active oxygen species during plant stress responses. Cell. Mol. Life Sci. 57, 779–795.

Davies, M.J., 2005. The oxidative environment and protein damage. Biochim. Biophys Acta 1703, 93–109.

de campos, M.K.S., de Carvalho, K., de Souza, F.S., Marur, C.S., Pereira, L.F.P., Filho, J.C.B., et al., 2011. Drought tolerance and antioxidant enzymatic activity in transgenic 'Swingle' citrumelo plants over-accumulating proline. Environ. Exp. Bot. 72, 242–250.

de Carvalho, K., de Campos, M.K., Domingues, D.S., Pereira, L.F., Vieira, L.G., 2013. The accumulation of endogenous proline induces changes in gene expression of several antioxidant enzymes in leaves of transgenic Swingle citrumelo. Mol. Biol. Rep. 40, 3269–3279.

de Ronde, J.A., Spreeth, M.H., Cress, W.A., 2000. Effect of antisense L-Δ1-pyrroline-5-carboxylate reductase transgenic soybean plants subjected to osmotic and drought stress. Plant Growth Regul. 32, 13–26.

de Ronde, J.A., Cress, W.A., Van Staden, J., 2001. Interaction of osmotic and temperature stress on transgenic soybean. S. Afr. J. Bot. 67, 655–660.

Delauney, A.J., Verma, D.P.S., 1993. Proline biosynthesis and osmoregulation in plants. Plant J. 4, 1037–1047.

del Rio, L.A., Corpas, F.J., Sandalio, L.M., Palma, J.M., Gómez, M., Barroso, J.B., 2002. Reactive oxygen species, antioxidant systems and nitric oxide in peroxisomes. J. Exp. Bot. 53, 1255–1272.

Dobra, J., Motyka, V., Dobrev, P., Malbeck, J., Prasil, I.T., Haisel, D., et al., 2010. Comparison of hormonal responses to heat, drought and combined stress in tobacco plants with elevated proline content. J. Plant Physiol. 167, 1360–1370.

Duque, A.S., de Almeida, A.M., da Silva, A.B., da Silva, J.M., Farinha, A.P., Santos, D., et al., 2013. Abiotic stress responses in plants: unraveling the complexity of genes and networks to survive. In: Vahdati, K., Leslie, C. (Eds.), Abiotic Stress-Plant Responses and Applications in Agriculture. INTECH-Open Access Publisher, Croatia, pp. 49–101.

Eckardt, N.A., Portis Jr., A.R., 1997. Heat denaturation profiles of ribulose-1,5-bisphosphate carboxylase/oxygenase (Rubisco) and rubisco activase and the inability of Rubisco activase to restore activity of heat-denatured Rubisco. Plant Physiol. 113, 243–248.

El-Shabrawi, H., Kumar, B., Kaul, T., Reddy, M.K., Singla-Pareek, S.L., Sopory, S.K., 2010. Redox homeostasis, antioxidant defense, and methylglyoxal detoxification as markers for salt tolerance in Pokkali rice. Protoplasma 24, 85–96.

Espartero, J., Aguayo, I.S., Pardo, J.M., 1995. Molecular characterization of glyoxalase-I from a higher plant; upregulation by stress. Plant Mol. Biol. 29, 1223–1233.

Farago, M.E., Mullen, W.A., 1979. Plant which accumulate metals. A possible copper-proline complex from the roots of *Armeria maritima*. Inorg. Chem. Acta 32, L93–L94.

Feller, U., Crafts-Brandner, S.J., Salvucci, M.E., 1998. Moderately high temperatures inhibit ribulose-1,5-bisphosphate carboxylase/oxygenase (Rubisco) activase-mediated activation of Rubisco. Plant Physiol. 116, 539–546.

Foyer, C.H., Noctor, G., 2003. Redox sensing and signaling associated with reactive oxygen in chloroplasts, peroxisomes and mitochondria. Physiol. Plant. 119, 355–364.

Gechev, T.S., Van Breusegem, F., Stone, J.M., Denev, L., Laloi, C., 2006. Reactive oxygen species as signals that modulate plant stress responses and programmed cell death. BioEssays 28, 1091–1101.

Giacomelli, L., Rudella, A., van Wijk, K.J., 2006. High light response of the thylakoid preoteome in *Arabidopsis* wild type and ascorbate deficient mutant vtc2-2. A comparative proteomics study. Plant Physiol. 141, 685–701.

Gill, S.S., Tuteja, N., 2010. Reactive oxygen species and antioxidant machinery in abiotic stress tolerance in crop plants. Plant Physiol. Biochem. 48, 909–930.

Groppa, M.D., Benavides, M.P., 2008. Polyamines and abiotic stress: recent advances. Amino Acids 34, 35–45.

Halliwell, B., 2006. Reactive species and antioxidants. Redox biology is a fundamental theme for aerobic life. Plant Physiol. 141, 312–322.

Hamilton, E.W., Heckathorn, S.A., 2001. Mitochondrial adaptations to NaCl. Complex I is protected by anti-oxidants and small heat shock proteins, whereas complex II is protected by proline and betaine. Plant Physiol. 126, 1266–1274.

Hare, P.D., Cress, W.A., 1997. Metabolic implications of stress induced proline accumulation in plants. Plant Growth Regul. 21, 79–102.

Hare, P.D., Cress, W.A., van Staden, J., 1998. Dissecting the roles of osmolyte accumulation during stress. Plant Cell Environ. 21, 535–553.

Hare, P.D., Cress, W.A., van Staden, J., 1999. Proline synthesis and degradation: a model system for elucidating stress-related signal transduction. J. Exp. Bot. 50, 413–434.

Hassine, A.B., Ghanem, M.E., Bouzid, S., Lutt, S., 2008. An inland and a coastal population of the Mediterranean xero-halophyte species *Atriplex halimus* L. differ in their ability to accumulate proline and glycinebetaine in response to salinity and water stress. J. Exp. Bot. 59, 1315–1326.

Hayano-Kanashiro, C., Calderon-Vazquez, C., Ibarra-Laclette, E., Herrera-Estrella, L., Simpson, J., 2009. Analysis of gene expression and physiological responses in three Mexican maize landraces under drought stress and recovery irrigation. PLoS ONE 4, e7531.

Hayat, S., Hayat, Q., Alyemeni, M.N., Wani, A.S., Pichtel, J., Ahmad, A., 2012. Role of proline under changing environments a review. Plant Signal. Behav. 7, 1–11.

Hernandez, J.A., Corpas, F.J., Gomez, M., Delrio, L.A., Sevilla, F., 1993. Salt-induced oxidative stress mediated by activated oxygen species in pea leaf mitochondria. Physiol. Plant. 89, 103–110.

Heuer, B., 1994. Osmoregulatory role of proline in water and salt-stressed proline accumulation in plants. In: Pessarakli, M. (Ed.), Handbook of Plant and Crop Stress. Marcel Dekker, Inc., New York, pp. 363–381.

Hoque, M.A., Okuma, E., Banu, M.N.A., Nakamura, Y., Shimoishi, Y., Murata, Y., 2007a. Exogenous proline mitigates the detrimental effects of salt stress more than exogenous betaine by increasing antioxidant enzyme activities. J. Plant Physiol. 164, 553–561.

Hoque, M.A., Banu, M.N.A., Okuma, E., Amako, K., Nakamura, Y., Shimoishi, Y., et al., 2007b. Exogenous proline and glycinebetaine ingresses NaCl-induced ascorbate glutathione cycle enzyme activities and proline improves salt tolerance more than glycinebetaine in tobacco Bright yellow-2 suspension-cultured cells. J. Plant Physiol. 164, 553–561.

Hoque, M.A., Banu, M.N.A., Nakamura, Y., Shimoishi, Y., Murata, Y., 2008. Proline and glycinebetaine enhance antioxidant defense and methylglyoxal detoxification systems and reduce NaCl-induced damage in cultured tobacco cells. J. Plant Physiol. 165, 813–824.

Hoque, M.A., Uraji, M., Banu, M.N.A., Mori, C., Namamura, Y., Murata, Y., 2010. The effects of methylglyoxal on glutathione S-tansferase from *Nicotiana tabacum*. Biosei. Biotechnol. Biochem. 74, 2124–2126.

Hoque, M.A., Uraji, M., Torii, A., Banu, M.N.A., Mori, I.C., Nakamura, Y., et al., 2012a. Methylglyoxal inhibition of cytosolic ascorbate peroxidase from *Nicotiana tabacum*. J. Biochem. Mol. Toxicol. 26, 315–321.

Hoque, T.S., Uraji, M., Ye, W., Hossain, M.A., Nakamura, Y., Murata, Y., 2012b. Methylglyoxal-induced stomatal closure accompanied by peroxidase-mediated ROS production in *Arabidopsis*. J. Plant Physiol. 169, 979–986.

Hoque, T.S., Uraji, M., Yuya, A., Nakamura, Y., Murata, Y., 2012c. Methylglyoxal inhibit seed germination and root elongation and up-regulates transcription of stress-responsive genes in ABA-dependent pathway in *Arabidopis*. Plant Biol.. 10.1111/j.1438–8677.2012.00607.x.

Hossain, M.A., Fujita, M., 2009. Purification of glyoxalase I from onion bulbs and molecular cloning of its cDNA. Biosci. Biotechnol. Biochem. 73, 2007–2013.

Hossain, M.A., Fujita, M., 2010. Evidence for a role of exogenous glycinebetaine and proline in antioxidant defense and methylglyoxal detoxification systems in mung bean seedlings under salt stress. Physiol. Mol. Biol. Plant 16, 19–29.

Hossain, M.A., Fujita, M., 2012. Regulatory role of components of ascorbate-glutathione (AsA-GSH) pathway in plant tolerance to oxidative stress. In: Anjum, N.A., Umar, S., Ahmed, A. (Eds.), Oxidative Stress in Plants: Causes, Consequences and Tolerance. IK International Publishing House Pvt. Ltd., INDIA, pp. 81–147.

Hossain, M.A., Hossain, M.Z., Fujita, M., 2009. Stress-induced changes of methylglyoxal level and glyoxalase I activity in pumpkin seedlings and cDNA cloning of glyoxalase I gene. Aust. J. Crop Sci. 3, 53–64.

Hossain, M.A., Hasanuzzaman, M., Fujita, M., 2010. Up-regulation of antioxidant and glyoxalase systems by exogenous glycinebetaine and proline in mung bean confer tolerance to cadmium stress. Physiol. Mol. Biol. Plant 16, 259–272.

Hossain, M.A., Teixeira da Silva, J.A., Fujita, M., 2011a. Glyoxalase system and reactive oxygen species detoxification system in plant abiotic stress response and tolerance: an intimate relationship. In: Shanker, A.K., Venkateswarlu, B. (Eds.), *Abiotic Stress*/Book 1. INTECH-Open Access Publisher, Rijeka, Croatia, pp. 235–266.

Hossain, M.A., Hasanuzzaman, M., Fujita, M., 2011b. Coordinate induction of antioxidant defense and glyoxalase system by exogenous proline and glycinebetaine is correlated with salt tolerance in mung bean. Front. Agric. China 5, 1–14.

Hossain, M.A., Piyatida, P., Teixeira da Silva, J.A., Fujita, M., 2012a. Molecular mechanism of heavy metal toxicity and tolerance in plants: central role of glutathione in detoxification of reactive oxygen species and methylglyoxal and in heavy metal chelation. J. Bot.. 10.1155/2012/872875Article ID 872875, 37 pages.

Hossain, M.A., Hossain, M.D., Rohman, M.M., Teixeira da Silva, J.A., Fujita, M., 2012b. Onion major compounds (flavonoids, organosulfurs) and highly expressed glutathione-related enzymes: possible physiological interaction, gene cloning and abiotic stress response. In: Aguirre, C.B., Jaramillo, L.M. (Eds.), Onion Consumption and Health. Nova Science Publishers Inc., NY, USA, pp. 49–90.

Hossain, M.A., Mostofa, M.G., Fujita, M., 2013a. Cross protection by cold-shock to salinity and drought stress-induced oxidative stress in mustard (*Brassica campestris* L.) seedlings. Mol. Plant Breed. 4, 50–70.

Hossain, M.A., Mostofa, M.G., Fujita, M., 2013b. Heat-shock positively modulates oxidative protection of salt and drought-stressed mustard (*Brassica campestris* L.) seedlings. J. Plant Sci. Mol. Breed. 2, 1–14.

Hua, X., Weitao, L.V., Bin, L., Zhang, M., 2011. Proline accumulation is inhibitory to *Arabidopsis* seedlings during heat stress. Plant Physiol.. 10.1104/p.11.

Huang, Y., Bie, Z., Liu, Z., Zhen, A., Wang, W., 2009. Protective role of proline against salt stress is partially related to the improvement of water status and peroxidase enzyme activity in cucumber. Soil Sci. Plant Nutr. 55, 698–704.

Islam, M.M., Hoque, M.A., Okuma, E., Banu, M.N.A., Shimoishi, Y., Nakamura, Y., et al., 2009a. Exogenous proline and glycinebetaine increase antioxidant enzyme activities and confer tolerance to cadmium stress in cultured tobacco cells. J. Plant Physiol. 166, 1587–1597.

Islam, M.M., Hoque, M.A., Okuma, E., Jannat, R., Banu, M.N.A., Jahan, M.S., et al., 2009b. Proline and glycinebetaine confer cadmium tolerance on tobacco Bright Yellow-2 cells by increasing ascorbate-glutathione cycle enzyme activities. Biosci. Biotech. Biochem. 73, 2320–2323.

Jan, S., Parween, T., Siddiqi, T.O., Mahmooduzzafar, 2012. Antioxidant modulation in response to gamma induced oxidative stress in developing seedlings of *Psoralea corylifolia* L. J. Environ. Radioact. 113, 142–149.

Jaspers, P., Kangasjarvi, J., 2010. Reactive oxygen species in abiotic stress signalling. Physiol. Plant. 138, 405–413.

Juszczuk, I.M., Wagner, A.M., Rychter, A.M., 2001. Regulation of alternative oxidase activity during phosphate deficiency in bean roots (*Phaseolus vulgaris*). Physiol. Plant. 113, 185–192.

Kaushal, N., Gupta, K., Bhandhari, K., Kumar, S., Thakur, P., Nayyar, H., 2011. Proline induces heat tolerance in chickpea (*Cicer arietinum* L.) plants by protecting vital enzymes of carbon and antioxidative metabolism. Physiol. Mol. Biol. Plant 17, 203–213.

Kavi Kishor, P.B., Hong, Z., Miao, G.H., Hu, C.A.A., Verma, D.P.S., 1995. Overexpression of D1-pyrroline-5-carboxylate synthetase increases proline production and confers osmotolerance in transgenic plants. Plant Physiol. 108, 1387–1394.

Kavi Kishor, P.B., Sangam, S., Amruth, R.N., Sri Laxmi, P., Naidu, K.R., Rao, K.R.S.S., et al., 2005. Regulation of proline biosynthesis, degradation, uptake and transport in higher plants: its implications in plant growth and abiotic stress tolerance. Curr. Sci. 88, 424–438.

Kemple, A.R., Macpherson, H.T., 1954. Liberation of amino acids in perennial rye grass during wilting. Biochem. J. 58, 46–49.

Ketchum, R.E.B., Warren, R.C., Klima, L.J., Lopez-Gutierrez, F., Nabors, M.W., 1991. The mechanism and regulation of proline accumulation in suspension cultures of the halophytic grass *Distichlis spicata* L. J. Plant Physiol. 137, 368–374.

Keunen, E., Remans, R., Bohler, S., Vangronsveld, J., Cypers, A., 2011. Metal-induced oxidative stress and plant mitochondria. Int. J. Mol. Sci. 12, 6894–6918.

Khedr, A.H.A., Abbas, M.A., Wahid, A.A.A., Quick, W.P., Abogadallah, G.M., 2003. Proline induces the expression of salt-stress-responsive proteins and may improve the adaptation of *Pancratium maritimum* L. to salt-stress. J. Exp. Bot. 54, 2553–2562.

Kim, K., Portis, J., 2004. Oxygen-dependent H_2O_2 production by Rubisco. FEBS Lett. 57, 124–128.

Kocsy, G., Laurie, R., Szalai, G., Szilagyi, V., Simon-Sarkadi, L., Galiba., G., et al., 2005. Genetic manipulation of proline levels affects antioxidants in soybean subjected to simultaneous drought and heat stresses. Physiol. Plant. 124, 227–235.

Komatsu, S., Wada, T., Abalea, Y., Nouri, M.Z., Nanjo, Y., Nakamura, N., et al., 2009. Analysis of plasma membrane proteome in soybean and application to flooding stress response. J. Proteome. Res. 8, 4487–4499.

Kotchoni, S.O., Kuhns, C., Ditzer, A., Kirch, H.H., Bartels, D., 2006. Over-expression of different aldehyde dehydrogenase genes in *Arabidopsis thaliana* confers tolerance to abiotic stress and protects plants against lipid peroxidation and oxidative stress. Plant Cell Environ. 29, 1033–1048.

Krantev, A., Yordanova, R., Janda, T., Szalai, G., Popova, L., 2008. Treatment with salicylic acid decreases the effect of cadmium on photosynthesis in maize plants. J. Plant Physiol. 165, 920–931.

Krasensky, J., Jonak, C., 2012. Drought, salt, and temperature stress-induced metabolic rearrangements and regulatory networks. J. Exp. Bot. 63, 1593–1608.

Krishnan, N., Dickman, M.B., Becker, D.F., 2008. Proline modulates the intracellular redox environment and protects mammalian cells against oxidative stress. Free. Radic. Biol. Med. 44, 671–681.

Krömer, S., 1995. Respiration during photosynthesis. Annu. Rev. Plant Physiol. Plant Mol. Biol. 46, 45–70.

Kumar, V., Yadav, S.K., 2009. Proline and betaine provide protection to antioxidant and methylglyoxal detoxification systems during cold stress and *Camellia sinensis* (L.) O. Kuntze. Acta Physiol. Plant 31, 261–269.

Kumar, V., Shriram, V., Kavi Kishor, P.B., Jawali, N., Shitole, M.G., 2010. Enhanced proline accumulation and salt stress tolerance of transgenic *indica* rice by over-expressing *P5CSF129A* gene. Plant Biotechnol. Rep. 4, 37–48.

Lamb, R.S., 2012. Abiotic stress responses in plants: a focus on the SRO family. In: Montanaro, G. (Ed.), Advances in Selected Plant Physiology Aspects. INTECH-Open Access Publisher, Rijeka, Croatia, pp. 1–21.

Le, B.H., Wagmaister, J.A., Kawashima, T., Bui, A.Q., Harada, J.J., Goldberg, R.B., 2007. Using genomics to study legume seed development. Plant Physiol. 144, 564–572.

Lehmann, S., Funck, D., Szabados, L., Rentsch, D., 2010. Proline metabolism and transport in plant development. Amino Acids 39, 949–962.

Li, Z.G., Ding, X.J., Du, P.F., 2013. Hydrogen sulfide donor sodium hydrosulfide-improved heat tolerance in maize and involvement of proline. J. Plant Physiol. 170, 741–747.

Lin, F., Xu, J., Shi, J., Li, H., Li, B., 2010. Molecular cloning and characterization of a novel glyoxalase I gene *TaGly I* in wheat (*Triticum aestivum* L.). Mol. Biol. Rep. 37, 729–735.

Lopes, M.S., Araus, J.L., van Heerden, P.D.R., Foyer, C.H., 2011. Enhancing drought tolerance in C_4 crops. J. Exp. Bot. 62, 3135–3153.

Luo, Y., Tang, H., Zhang, Y., 2011. Production of reactive oxygen species and antioxidant metabolism about strawberry leaves to low temperatures. J. Agric. Sci. 3, 89–95.

Lv, W.T., Lin, B., Zhang, M., Hua, X.J., 2011. Proline accumulation is inhibitory to Arabidopsis seedlings during heat stress. Plant Physiol. 15, 1921–1933.

Maggio, A., Miyazaki, S., Veronese, P., Fujita, T., Ibeas, J.I., 2002. Does proline accumulation play an active role in stress-induced growth reduction?. Plant J. 31, 699–712.

Malusa, E., Laurenti, E., Juszczuk, I., Ferrari, R.P., Rychter, A.M., 2002. Free radical production in roots of *Phaseolus vulgaris* subjected to phosphate deficiency stress. Plant Physiol. Biochem. 40, 963–967.

Man, D., Bao, Y.X., Han, L.B., Zhang, X., 2011. Drought tolerance associated with proline and hormone metabolism in two tall fescue cultivars. Hort. Sci. 46, 1027–1032.

Martins, T.M.B.T.S., Coedeiro, C.A.A., Freire, A.M.J.P., 2001. In situ analysis of methylglyoxal metabolism in *Saccromyces cerevisae*. FEBS Lett. 499, 41–44.

Mattioli, R., Marchese, D., D' Angeli, S., Altamura, M., Costantino, P., Trovato, M., 2008. Modulation of intracellular proline levels affects flowering time inflorescence architecture in *Arabidopsis*. Plant. Mol. Biol. 66, 277–288.

Mattioli, R., Costantino, P., Trovato, M., 2009a. Proline accumulation in plants: not only stress. Plant Signal. Behav. 4, 1016–1018.

Mattioli, R., Falasca, G., Sabatini, S., Altamura, M.M., Costantino, P., Trovato, M., 2009b. The proline biosynthetic genes *P5CS1* and *P5CS2* play overlapping roles in *Arabidopsis* flower transition but not in embryo development. Physiol. Plant. 137, 72–85.

Matysik, J., Alia, Bhalu, B., Mohanty, P., 2002. Molecular mechanisms of quenching of reactive oxygen species by proline under stress in plants. Curr. Sci. 82, 525–532.

Mhamdi, A., Queval, G., Chaouch, S., Vanderauwera, S., Van Breusegem, F., Noctor, G., 2010. Catalase function in plants: a focus on *Arabidopsis* mutants as stress-mimic models. J. Exp. Bot. 61, 4197–4220.

Millar, H., Considine, M.J., Day, D.A., Whelan, J., 2001. Unraveling the role of mitochondria during oxidative stress in plants. IUBMB Life 51, 201–205.

Miller, G., Shulaev, V., Mittler, R., 2008. Reactive oxygen signaling and abiotic stress. Physiol. Plant. 133, 481–489.

Mittler, R., 2002. Oxidative stress, antioxidants and stress tolerance. Trends Plant Sci. 7, 405–410.

Mittler, R., Vanderauwera, S., Gollery, M., Breusegem, F.V., 2004. Reactive oxygen gene network of plants. Trends Plant Sci. 9, 490–498.

Mittova, V., Tal, M., Volokita, M., Guy, M., 2003. Up-regulation of the leaf mitochondrial and peroxisomal antioxidative systems in response to salt-induced oxidative stress in the wild salt-tolerant tomato species *Lycopersicon pennellii*. Plant Cell Environ. 26, 845–856.

Mohanty, N., Mohanty, P., 1988. Cation effects on primary processes of photosynthesis. In: Singh, R., Sawheny, S.K. (Eds.), Advances in Frontier Areas of Plant Biochemistry. Prentice-Hall, Delhi, India, pp. 1–18.

Molinari, H.B.C., Marur, C.J., Daros, E., De Campos, M.K.F., De Carvalho, J.F.R.P., Filho, J.C. B., et al., 2007. Evaluation of the stress-inducible production of proline in transgenic sugarcane (*Saccharum* spp.): osmotic adjustment, chlorophyll fluorescence and oxidative stress. Physiol. Plant. 130, 218–229.

Mustafiz, A., Singh, A.K., Pareek, A., Sopory, S.K., Singla-Pareek, S.L., 2006. Genome-wide analysis of rice and *Arabidopsis identifies* two glyoxalase genes that are highly expressed in abiotic stresses. Funct. Integr. Genomics 11, 293–305.

Møller, I.M., Jensen, P.E., Hansson, A., 2007. Oxidative modifications to cellular components in plants. Annu. Rev. Plant Biol. 58, 459–481.

Naika, M., Shameer, K., Mathew, O.K., Gowda, R., Sowdhamini, R., 2013. STIFDB2: an updated version of plant Stress-responsive transcription factor database with additional stress signals, stress-responsive transcription factor binding sites and stress-responsive genes in *Arabidopsis* and rice. Plant Cell Physiol. 54, e8.

Nanjo, Y., Shultety, L., Ashraf, Y., Komatsu, S., 2010. Comparative proteomic analysis of early-stage soybean seedling response to flooding by using gel and gel-free technique. J. Proteome. Res. 9, 3989–4002.

Neill, S., Desikan, R., Hancock, J., 2002. Hydrogen peroxide signalling. Curr. Opin. Plant Biol. 5, 388–395.

Noctor, G., Veljovic-Jovanovic, S., Driscoll, S., Novitskaya, L., Foyer, C., 2002. Drought and oxidative load in the leaves of C_3 plants: a predominant role for photorespiration?. Ann. Bot. 89, 841–850.

Norouzi, M., Toorchi, M., Hosseini Salekdeh, G., Mohammadi, S.A., Neyshabouri, M.R., Aharizad, S., 2008. Effect of water deficit on growth, grain yield and osmotic adjustment in rapeseed. J. Food Agric. Environ. 6, 312–318.

Nouet, C., Motte, P., Hanikenne, M., 2011. Chloroplastic and mitochondrial metal homeostasis. Trends Plant Sci. 16, 395–404.

Nounjan, N., Theerakulpisut, P., 2012. Effects of exogenous proline and trehalose on physiological responses in rice seedlings during salt-stress and after recovery. Plant Soil Environ. 58, 309–315.

Nounjan, N., Nghia, P.T., Theerakulpisut, P., 2012. Exogenous proline and trehalose promote recovery of rice seedlings from salt-stress and differentially modulate antioxidant enzymes and expression of related genes. J. Plant Physiol. 169, 596–604.

Okuma, E., Murakami, Y., Shimoishi, Y., Tada, M., Murata, Y., 2004. Effects of exogenous application of proline and betaine on the growth of tobacco cultured cells under saline conditions. Soil Sci. Plant Nutr. 50, 1301–1305.

Ozden, M., Deirel, U., Kahraman, A., 2009. Effects of proline on antioxidant system in leaves of grapevine (*Vitis vinifera* L.) exposed to oxidative stress by H_2O_2. Sci. Hort. 119, 163–168.

Pandy, S.K., Nookaraju, A., Upadhyaya, C.P., Gururani, M.A., Venkatesh, J., Kim, D.H., et al., 2011. An update on biotechnological approaches for improving abiotic stress tolerance in tomato. Crop Sci. 51, 2303–2324.

Pandy, U.K., Srivastava, R.D.L., 1990. Salinity index in relation to nitrate reductase activity and proline accumulation in paddy genotypes. Indian J. Plant Physiol., 32175–32177.

Parsons, H.L., Yip, J.Y., Vanlerberghe, G.C., 1999. Increased respiratory restriction during phosphate-limited growth in transgenic tobacco cells lacking alternative oxidase. Plant Physiol. 121, 1309–1320.

Parvanova, D., Ivanov, S., Konstantinova, T., Karanov., E., Atanassov, A., Tsvetkov, T., et al., 2004. Transgenic tobacco plants accumulating osmolytes show reduced oxidative damage under freezing stress. Plant Physiol. Biochem. 42, 57–63.

Petrov, V.D., Van Breusegem, F., 2012. Hydrogen peroxide – a central hub for information flow in plant cells. AoB Plant. 10.1093/aobpla/pls014pls014.

Posmyk, M.M., Janas, K.M., 2007. Effects of seed hydropriming in presence of exogenous proline on chilling injury limitation in *Vigna radiata* L. seedlings. Acta Physiol. Plant 29, 509–517.

Prasad, T.K., Anderson, M.D., Martin, B.A., Stewart, C.R., 1994a. Evidence for chilling-induced oxidative stress in maize seedlings and a regulatory role for hydrogen-peroxide. Plant Cell 6, 65–74.

Prasad, T.K., Anderson, M.D., Stewart, C.R., 1994b. Acclimation, hydrogen-peroxide, and abscisic-acid protect mitochondria against irreversible chilling injury in maize seedlings. Plant Physiol. 105, 619–627.

Purvis, A.C., Shewfelt, R.L., Gegogeine, J.W., 1995. Superoxide production by mitochondria isolated from green bell pepper fruit. Physiol. Plant. 94, 743–749.

Qin, F., Shinozaki, K., Yamaguchi-Shinozaki, K., 2011. Achievements and challenges in understanding plant abiotic stress responses and tolerance. Plant Cell Physiol. 52, 1569–1582.

Qiu, Q.S., Guo, Y., Quintero, F.J., Prado, J.M., 2004. Regulation of vacuolar Na^+/H^+ exchange in *Arabidopsis thaliana* by the salt-overly-sensitive (SOS) pathway. J. Biol. Chem. 279, 207–215.

Radyukina, N.L., Shahukova, A.V., Makarova, S.S., Kuznetsoy, V.l.V., 2011. Exogenous proline modifies differential expression of superoxide dismutase genes in UV-B-irradiated *Salvia officinalis* plants. Russ. J. Plant Physiol. 58, 51–59.

Rampino, P., Pataleo, S., Gerardi, C., Mita, G., Perrotta, C., 2006. Drought stress response in wheat: physiological and molecular analysis of resistant and sensitive genotypes. Plant Cell Environ. 29, 2143–2152.

Razavizadeh, R., Ehsanpour, A.A., 2009. Effects of salt stress on proline content, expression of delta-1-pyrroline-5-carboxylate synthetase, and activities of catalase and ascorbate peroxidase in transgenic tobacco plants. Biotechnol. Lett. 46, 63–745.

Refouli, A., Larher, L., 1989. Compatible organic solutes and salt tolerance in seedlings of three annual Medicago species. C.R. Acad. Sci. Paris 308, 329–335.

Reza, S., Heidari, R., Zare, A., Norastehnia, A., 2006. Antioxidant response of two salt-stressed barley varieties in the presence or absence of exogenous proline. Gen. Appl. Plant Physiol. 32, 233–251.

Rhoads, D.M., Umbach, A.L., Subbaiah, C.C., Siedow, J.N., 2006. Mitochondrial reactive oxygen species. contribution to oxidative stress and interorganellar signaling. Plant Physiol. 141, 357–366.

Richards, R.A., 1987. Physiology and the breeding of winter-grown cereals for dry areas. In: Srivastava, J.P., Porceddu, E., Acevedo, E., Varma, S. (Eds.), Drought Tolerance in Winter Cereals. John Wiley and Sons Ltd, Chichester, pp. 133–150.

Robson, C.A., Vanlerberghe, G.C., 2002. Transgenic plant cells lacking mitochondrial alternative oxidase have increased susceptibility to mitochondria-dependent and independent pathway of programmed cell death. Plant Physiol. 129, 1908–1920.

Roy, D., Basu, N., Bhunia, A., Banerjee, S., 1993. Counteraction of exogenous L-proline with NaCl in salt-sensitive cultivar of rice. Biol. Plant. 35, 69–72.

Rustgi, S., Joshi, A., Moss, H., Riesz, P., 1977. ESR of spin-trapped radicals in aqueous-solutions of amino-acids – reactions of hydroxyl radical. Int. J. Radiat. Biol. 31, 415–440.

Saadia, M., Jamil, A., Akram, N.A., Ashraf, M., 2012. A study of proline metabolism in canola (*Brassica napus* L.) seedlings under salt stress. Molecules 17, 5803–5815.

Saeedipour, S., 2013. Relationship of grain yield, ABA and proline accumulation in tolerant and sensitive wheat cultivars as affected by water stress. PNAS India. 10.1007/s40011–012-0147-5.

Sagi, M., Fluhr, R., 2006. Production of reactive oxygen species by plant NADPH oxidases. Plant Physiol. 141, 336–340.

Saiema, R., Altaf, A., To, S., 2012. Differential response of chickpea genotypes under salt stress. J. Funct. Environ. Bot. 2, 59–64.

Saito, R., Yamamoto, H., Makino, A., Sugimoto, T., Miyake, C., 2011. Methylglyoxal functions as hill oxidant and stimulates the photoreduction of O_2 at photosystem I: a symptom of plant diabetes. Plant Cell Environ. 34, 1454–1464.

Sanchez, D.H., Siahpoosh, M.R., Roessner, U., Udvardi, M., Kopka, J., 2008. Plant metabolomics reveals conserved and divergent metabolic responses to salinity. Physiol. Plant. 132, 209–219.

Sanchez, D.H., Schwabe, F., Erban, A., Udvardi, M.K., Kopka, J., 2012. Comparative metabolomics of drought acclimation in model and forage legumes. Plant Cell Environ. 35, 136–149.

Sanda, S., Yoshida, K., Kuwano, M., Kawamura, T., Munekage, Y.N., Akashi, K., et al., 2011. Responses of the photosynthetic electron transport system to excess light energy caused by water deficit in wild watermelon. Physiol. Plant. 142, 247–264.

Saradhi, P.P., Alia, Arora, S., Prasad, K.V., 1995. Proline accumulates in plants exposed to UV radiation and protects them against UV induced peroxidation. Biochem. Biophys. Res. Commun. 20, 1–5.

Schat, H., Sharma, S.S., Vooijs, R., 1997. Heavy metal–induced accumulation of free proline in a metal-tolerant and a nontolerant ecotype of *Silene vulgaris*. Physiol. Plant 101, 477–482.

Schwarzlander, M., Fricker, M.D., Sweetlove, L.J., 2009. Monitoring the *in vivo* redox state of plant mitochondria: effect of respiratory inhibitors, abiotic stress and assessment of recovery from oxidative challenge. Biochim. Biophys. Acta 1787, 468–475.

Sgherri, C.L.M., Pinzino, C., Navari-Izzo, F., 1996. Sunflower seedlings subjected to increasing stress by water deficit: changes in $O_2^{\bullet-}$ production related to the composition of thylakoid membranes. Physiol. Plant 96, 446–452.

Sharma, P., Jha, A.B., Dubey, R.S., Pessarakli, M., 2012. Reactive oxygen species, oxidative damage, and antioxidative defense mechanisms in plants under stressful conditions. J. Bot. Article ID 217037, 26 pages.

Sharma, S., Villamor, J.G., Verslues, P.E., 2011. Essential role of tissue-specific proline synthesis and catabolism in growth and redox balance at low water potential. Plant Physiol. 157, 292–304.

Sharma, S.S., Dietz, K.J., 2006. The significance of amino acids and amino acid-derived molecules in plant responses and adaptation to heavy metal stress. J. Exp. Bot. 57, 711–726.

Sharma, S.S., Dietz, K.J., 2009. The relationship between metal toxicity and cellular redox imbalance. Trends Plant Sci. 14, 43–50.

Sharma, S.S., Schat, H., Vooijs, R., 1998. In vitro alleviation of heavy metal-induced enzyme inhibition by proline. Phytochemistry 49, 1531–1535.

Siedlecka, A., Baszyńaski, T., 1993. Inhibition of electron flow around photosystem I in chloroplasts of Cd-treated maize plants is due to Cd-induced iron deficiency. Physiol. Plant 87, 199–202.

Siedlecka, A., Samuelsson, G., Gardenstrom, P., Kleczkowski, L.A., Krupa, Z., 1998. The activatory model of plant response to moderate cadmium stress-relationship between carbonic anhydrase and Rubisco. In: Garab, G. (Ed.), Photosynthesis: Mechanisms and Effects. Kluwer Academic Publishers, Dordrecht, pp. 2677–2680.

Signorelli, S., Arellano, J.B., Melø, T.B., Borsani, O., Monza, J., 2013. Proline does not quench singlet oxygen: evidence to reconsider its protective role in plants. Plant Physiol. Biochem. 64, 80–83.

Singh, V., Bhatt, I., Aggarwal, A., Tripathi, B.N., Munjal, A.K., Sharma, V., 2010. Proline improves copper tolerance in chickpea (*Cicer arietinum*). Protoplasma 245, 173–181.

Singla-Pareek, S.L., Ray, M., Reddy, M.K., Sopory, S.K., 2003. Genetic engineering of the glyoxalase pathway in tobacco leads to enhanced salinity tolerance. PNAS USA 100, 14672–14677.

Singla-Pareek, S.L., Yadav, S.K., Pareek, A., Reddy, M.K., Sopory, S.K., 2006. Transgenic tobacco overexpressing glyoxalase pathway enzymes grow and set viable seeds in zinc-spiked soils. Plant Physiol. 140, 613–623.

Singla-Pareek, S.L., Yadav, S.K., Pareek, A., Reddy, M.K., Sopory, S.K., 2008. Enhancing salt tolerance in a crop plant by overexpression of glyoxalase II. Trans. Res. 17, 171–180.

Siripornadulsil, S., Traina, S., Verma, D.P.S., Sayre, R.T., 2002. Molecular mechanisms of proline-mediated tolerance to toxic heavy metals in transgenic microalgae. Plant Cell 14, 2837–2847.

Smirnoff, N., Cumbes, Q.J., 1989. Hydroxyl radical scavenging activity of compatible solutes. Phytochem 28, 1057–1060.

Sobahan, M.A., Arias, C.R., Okuma, E., Shimoishi, Y., Nakamura, Y., Hirai, Y., et al., 2009. Exogenous proline and glycinebetaine suppress apoplastic flow to reduce Na^+ uptake in rice seedlings. Biosci. Biotechnol. Biochem. 73, 2037–2042.

Sorkheh, K., Shiram, B., Khodambashi, M., Rouhi, V., Mosavei, S., Sofo, A., 2012. Exogenous proline alleviates the effects of H_2O_2-induced oxidative stress in wild almond species. Russ. J. Plant Physiol. 59, 788–798.

Soshinkova, T.N., Radyukina, N.L., Korolkova, D.V., Nosov, A.V., 2013. Proline and functioning of the antioxidant system in *Thellungiella salsuginea* plants and cultured cells subjected to oxidative stress. Russ. J. Plant Physiol. 60, 41–54.

Sripinyowanich, S., Klomsakul, P., Boonburapong, B., Bangyeekhun, T., Gu, H., Buaboocha, T., et al., 2013. The role of OsP5CS1 and OsP5CR gene expression during salt stress tolerance in indica rice (*Oryza sativa* L.). Environ. Exp. Bot. 86, 94–105.

Szabados, L., Savouré, A., 2010. Proline: a multifunctional amino acid. Trends Plant Sci. 15, 89–97.

Szarka, A., Tomasskovies, B., Banhegyi, G., 2012. The ascorbate-glutathione-α-tocopherol triad in abiotic stress response. Int. J. Mol. Sci. 13, 4458–4483.

Szekely, G., Abraham, E., Cseplo, A., Rigo, G., Zsigmond, L., Csiszar, J., et al., 2008. Duplicated P5CS genes of *Arabidopsis* play distinct roles in stress regulation and developmental control of proline biosynthesis. Plant J. 53, 11–28.

Takashi, S., Murata, N., 2008. How do environmental stresses accelerate photoinhibition? Trends Plant Sci. 13, 178–182.

Tan, Y.F., O'Toole, N., Taylor, N.L., Millar, A.H., 2010. Divalent metal ions in plant mitochondria and their role in interactions with proteins and oxidative stress-induced damage to respiratory function. Plant Physiol. 152, 747–761.

Theocharis, A., Clement, C., Barka, E.A., 2012. Physiological and molecular changes in plants grown at low temperature. Planta 235, 1091–1105.

Thornalley, P.J., 1996. Pharmacoly of methylglyoxal: formation, modification of proteins and nucleic acid, and enzymatic detoxification– a role in pathogenesis and antiproliferative chemotherapy. Gen. Pharmacol. 27, 565–573.

Trovato, M., Mattioli, R., Costantino, P., 2008. Multiple roles of proline in plant stress tolerance and development. Rendiconti Lincei 19, 325–346.

Türkan, I., Bor, M., Ozdemir, F., Koca, H., 2005. Differential responses of lipid peroxidation and antioxidants in the leaves of drought-tolerant *P. acutifolius* Gray and drought-sensitive *P. vulgaris* L. subjected to polyethylene glycol mediated water stress. Plant Biol. 168, 223–231.

Upadhyaya, C.P., Venkatesh, J., Gururani, M.A., Asnin, L., Sharma, K., Ajappala, H., et al., 2011. Transgenic potato overproducing L-ascorbic acid resisted an increase in methylglyoxal under salinity stress via maintaining higher reduced glutathione level and glyoxalase enzyme activity. Biotechnol. Lett. 33, 2297–2307.

Van Swaaij, A.C., Jacobsen, E., Feenstra, W.J., 1985. Effect of cold hardening, wilting and exogenously applied proline on leaf proline content and frost tolerance of several genotypes of *Solanum*. Physiol. Plant 64, 230–236.

Vartanian, N., Hervochon, P., Marcotte, L., Larher, F., 1992. Proline accumulation during drought rhizogenesis in *Brassica napus* var. *oleifera*. J. Plant Physiol. 140, 623–628.

Verbruggen, N., Hermans, C., 2008. Proline accumulation in plants: a review. Amino Acids 35, 753–759.

Verslues, P., Sharma, S., 2010. Proline metabolism and its implications for plant-environment interaction. *Arabidopsis* Book 8, e0140. 10.1199/tab.0140.

Viveros, M.F.A., Inostroza-Blancheteau, C., Timmermann, T., Gonzalez, M., Arce-Johnson, P., 2013. Over expression of Gly I and Gly II genes in transgenic tomato (*Solamum lycopersicum* Mill.) plants confer salt tolerance by decreasing oxidative stress. Mol. Biol. Rep. 40, 3281–3290.

Vranova, E., Inze, D., Van Brensegem, F., 2002. Signal transduction during oxidative stress. J. Exp. Bot. 53, 1227–1236.

Wani, A.S., Irfan, M., Hayat, S., Ahmad, A., 2011. Response of two mustard (*Brassica juncea* L.) cultivars differing in photosynthetic capacity subjected to proline. Protoplasma 249, 75–87.

Wanichananan, P., Kirdmanee, C., Vutiyano, C., 2003. Effect of salinity on biochemical and physiological characteristics in correlation to selection of salt tolerance in aromatic rice (*Oryza sativa* L.). Sci. Asia 29, 333–339.

Wise, R.R., 1995. Chilling-enhanced photooxidation – the production, action and study of reactive oxygen species produced during chilling in the light. Photosynth. Res. 45, 79–97.

Witt, S., Galicia, L., Lisec, J., Cairns, J., Tiessen, A., Araus, J.L., et al., 2012. Metabolic and phenotypic responses of greenhouse-grown maize hybrids to experimentally controlled drought stress. Mol. Plant 5, 401–417.

Wu, C., Ma, C., Pan, Y., Gong, S., Zhao, C., Chen, S., et al., 2012. Sugar beet M14 *glyoxalase I* gene can enhance plant tolerance to abiotic stresses. J. Plant Res. 126, 415–425.

Xu, F.Y., Wang, X.L., Wu, Q.X., Zhang, X.R., Wang, L.H., 2012. Physiological responses differences of different genotype sesames to flooding stress. Adv. J. Food Sci. Technol. 4, 352–356.

Xu, J., Yin, H.X., Li, X., 2009. Protective effects of proline against cadmium toxicity in micropropagated hyperaccumulator, *Solanum nigrum* L. Plant Cell Rep. 28, 325–353.

Xue, X., Liu, A., Hua, X., 2009. Proline accumulation a transcriptional regulation of proline biosynthesis and degradation in *Brassica napus*. BMB Rep. 42, 28–34.

Yadav, S.K., Singla-Pareek, S.L., Ray, M., Reddy, M.K., Sopory, S.K., 2005a. Methylglyoxal levels in plants under salinity stress are dependent on glyoxalase I and glutathione. Biochem. Biophys. Res. Commun. 337, 61–67.

Yan, Z., Guo, S., Shu, S., Sun, J., Tezuka, T., 2011. Effects of proline on photosynthesis, root reactive oxygen species (ROS) metabolism in two melon cultivars (*Cucumis melo* L.) under NaCl stress. Afr. J. Biotechnol. 10, 18381–18390.

Yang, S.L., Llan, S.S., Gong, M., 2009. Hydrogen peroxide-induced proline and metabolic pathway of its accumulation in maize seedlings. J. Plant Physiol. 166, 1694–1699.

Yazici, I., Türkan, I., Sekmen, A.H., Demiral, T., 2007. Salinity tolerance of purslane *(Portulaca oleracea* L.) is achieved by enhanced antioxidative system, lower level of lipid peroxidation and proline accumulation. Environ. Exp. Bot. 61, 49–57.

You, J., Hu, H., Xiong, L., 2012. An ornithine δ-aminotransferase gene OsOAT confers drought and oxidative stress tolerance in rice. Plant Sci. 197, 59–69.

Zarei, S., Ehsanpour, A.A., Abbaspour, J., 2012. The role of over expression of P5CS gene on proline, catalase, ascorbate peroxidase activity and lipid peroxidation of transgenic tobacco (*Nicotiana tabacum L.*) plant under *in vitro* drought stress. J. Cell Mol. Res. 4, 43–49.

Zarse, K., Schmeisser, S., Groth, G., Priebe, S., Beuster, G., Kuhlow, D., et al., 2012. Impaired insulin/IGF1 signaling extends life span by promoting mitochondrial L-proline catabolism to induce a transient ROS signal. Cell Metab. 15, 451–465.

Chapter 17

Trace Elements Tolerance Modulated by Antioxidant System in Plants

Marcelo Pedrosa Gomes, Marília Mércia Lima Carvalho Carneiro and Queila Souza Garcia

17.1 INTRODUCTION

Although some trace elements such as copper (Cu), zinc (Zn), and iron (Fe) are essential nutrients, plants are often exposed to phytotoxic levels of these elements (as well as to nonessential trace elements such as arsenic (As) and cadmium (Cd)) as a consequence of human activities (Li et al., 2005). Considerable losses of plant productivity and hazardous health effects in humans can be linked to trace element contamination of soils and water. Exposure to these toxic elements can intensify the production of reactive oxygen species (ROS). Some of those ROS are highly toxic and must be detoxified by cellular stress responses if the plant is to survive and grow (Gratão et al., 2005).

Different plant organs show different responses to abiotic stress such as trace element toxicity during their development. Seed coats provide some protection against trace element stress prior to germination, but they rupture or become more permeable during germination (Kranner and Colville, 2011). Metal(loid)s may affect seed germination through their toxicity or by inhibiting water uptake (Kranner and Colville, 2011). A number of studies have shown that trace element exposure resulted in concentration-dependent reductions in seed germination in many different plant species, e.g. lettuce (*Lactuca sativa*), spinach (*Spinacia oleracea*), Swiss chard (*Beta vulgaris*) (Bautista et al., 2013), *Dorycnium pentaphyllum* (Lefèvre et al., 2009) and *Eruca sativa* (Ozdener and Kutbay, 2009), which has been attributed to induced oxidative stress (Kranner and Colville, 2011; Gomes and Garcia, 2013). While these studies have highlighted both inter- and intraspecific variations in seed germination rates, the limited data currently available suggest

P. Ahmad (Ed): Oxidative Damage to Plants. DOI: http://dx.doi.org/10.1016/B978-0-12-799963-0.00017-4

that the seeds of metal(loid)-tolerant plants can generally germinate even when exposed to high trace element concentrations, and that essential micronutrients (e.g. Zn) can cause damage at relatively high concentrations (Maheshwari and Dubey, 2008).

Toxic concentrations of trace elements are known to produce growth reductions and nutritional disturbances and to affect the photosynthetic rates of seedlings by interfering with photosystem and pigment biosynthesis (Cuypers et al., 2001; Lefèvre et al., 2009; Dias et al., 2012; Peng et al., 2013). Trace element toxicity has also been attributed to their generation of ROS (Meharg and Hartley-Whitaker, 2002; Fang et al., 2013) that cause oxidative damage to biomolecules and eventual cell death (Gunes et al., 2007). Plants have evolved mechanisms to protect cells and subcellular systems from the effects of ROS (e.g. superoxide radicals, hydroxyl radicals, and hydrogen peroxide) through the synthesis of enzymatic and nonenzymatic antioxidants (Gunes et al., 2007), and the following pathway of hydrogen peroxide (H_2O_2) production and destruction has been proposed: superoxide dismutase (SOD) catalyze the conversion of highly reactive superoxide ($O_2^{\bullet -}$) to H_2O_2, and these molecules are in turn scavenged by catalase (CAT), ascorbate peroxidase (APX), and glutathione peroxidase (GPX); APX reduces H_2O_2 into water using ascorbate (AsA) as the electron donor, and the resulting dehydroascorbate (DHA) is cycled back to ascorbate using reduced glutathione (GSH) as the electron donor; the oxidized glutathione (GSSG) formed is converted back to GSH by NAD(P)H-dependent glutathione reductase (GR). Finally, GPX acts on H_2O_2 to form GSSH, which is then further reduced to GSH by GR (Gunes et al., 2009; Suzuki et al., 2012).

Our view of the role of ROS has greatly evolved in the light of recent studies. The derivatives of the reduction of oxygen as a superoxide, hydrogen peroxide, the hydroxyl radical ($^{\bullet}OH$), or singlet oxygen ($_1O^2$), were initially viewed as hazardous compounds, but they are now recognized as being important in the biological processes of plants. Reactive oxygen species appear to be central factors in plant adaptations to biotic and abiotic stress, exacerbating cell damage as well as signaling the activation of defense responses (Miller et al., 2008). Although high concentrations of ROS can lead to phytotoxicity, relatively low levels can function as signaling molecules under conditions of abiotic stress (Miller et al., 2008; Gill and Tuteja, 2010; Leymarie et al., 2012). In this context the activities of the systems modulating ROS content are quite important and have been intrinsically related to plant ability to cope with trace element toxicity (Gomes et al., 2013a, 2013b). As such, the present chapter evaluates the physiological responses of seeds and plants to trace elements in order to elucidate the underlying physiological mechanisms of their tolerance.

17.2 TRACE ELEMENTS AND PLANTS

Human activities such as agriculture and mining are known to eliminate the vegetation cover and release large quantities of trace elements into the environment. The widespread contamination of air, soil and water by trace elements has created a new stress scenario in which plants need to be able to grow and develop. Often, the exposition to high trace element amounts or to nonessential trace elements such as arsenic (As), cadmium (Cd), lead (Pb) and mercury (Hg) leads to plant intoxication and then death. However, some plants are able to cope with trace element phytotoxicity and their natural occurrence in contaminated areas has been verified (Gomes et al., 2013a, 2013b). This is a strong indicator that some plants show tolerance mechanisms which allow them to survive under the trace element stress. Moreover, these plant species are useful in programs of recuperation of degraded areas. Phytoremediation (the use of plants for trace element reclamation) has emerged as an environmentally friendly proposal and may be the most cost-effective treatment for metal-polluted soils, especially in cases of extensive contamination (Dary et al., 2010). As these programs are directed towards recovering vegetation structure, it is important to understand the trace element tolerance of potential phytoremediator species.

17.3 TRACE ELEMENTS AND SEED GERMINATION

Seeds are generally well-protected against most stress factors (Li et al., 2005), and many environmental variables affecting seed germination (light, temperature and moisture) have been widely studied (Kranner and Colville, 2011). However, the effects of soil contamination by trace elements on seed germination is not fully understood.

Although there is limited evidence for influxes of metals into intact mature seeds through their seed coats (Kranner and Colville, 2011), trace element influences on germination processes have been reported (Kranner and Colville, 2011; Duarte et al., 2012). Differences in seed coat permeability due to morphological variations (Moise et al., 2005) and environmental factors may lead into trace element sensitivity in seeds, and due to the different chemical interactions and toxicities of the different metal/loids, differential influences on biological processes such as seed germination are expected.

Trace elements such as As, Cd and Zn can negatively affect seed germination. Due to its chemical similarity to phosphorus (P), As (as arsenate, one of its inorganic forms – As(V)) replaces P on ATP, forming an unstable ADP-As complex that halts cell energy flow (Meharg and Hartley-Whitaker, 2002). As germination is intrinsically related to energy demands by seed embryos, As exposure will lead to germination inhibition (Duarte et al., 2012). Cadmium's deleterious effects on seed germination may be associated with the Cd influences on energy reserve mobilization (Mihoub et al., 2005) and seed metabolism (Bansal et al., 2000; 2002). Cadmium mimics other plant nutrients (Nelson, 1986; Clemens et al., 1998) interfering with their mobilization (Duarte et al., 2012). Moreover, Cd has been reported to cause alterations in the activities of α and β amylases as well as interfere with respiratory activity (Chugh and Sawhney, 1996) which will compromise reserve mobilization and embryo growth (Kabata-Pendias and Pendias, 2000). Essential trace elements, such as the micronutrient Zn, can also behave like a hazardous metal in the seed germination. Micronutrients have been reported to negatively impact germination only at relatively high concentrations (Maheshwari and Dubey, 2008). However, Duarte et al. (2012) noted total germination inhibition in *Cecropia hololeuca* and decreases in the germination rate of *Myracroduon urundeuva* seeds exposed to Zn doses from 50 mg L^{-1}, a typical Zn concentration in Brazilian soils, reinforcing the inter- and intraspecific nature of seed germination responses to metal stress (Kranner and Colville, 2011). Zinc phytotoxicity may be related to the generation of ROS and ROS-burst induced oxidative injures (Remans et al., 2012).

Some studies have also shown the stimulator effects of trace elements in seed germination (Li et al., 2007; Kranner and Colville, 2011; Duarte et al., 2012; Gomes et al., 2013a). This could be attributed to an overproduction of ROS and reactive nitrogen species (RNS) such as nitric oxide (NO) in plants challenged by trace elements, resulting in a slightly enhanced level of oxidative stress that stimulates germination – the oxidative window (Bailly et al., 2008; Lefèvre et al., 2009). In this context, a strictly regulated concentration of ROS is required for germination (Bailly et al., 2008) and so, the capacity of seeds to control ROS levels during germination could aid in neutralizing oxidative stress generated by the presence of trace elements (Gomes et al., 2013a).

17.3.1 Trace Elements Oxidative Damage/Signaling in Seeds

Many of the toxic effects of trace elements result from the oxidative damage arising from ROS generation (Gallego et al., 2012). Oxidative stress mediated by toxic trace elements concentration increase ROS production (Kranner and Colville, 2011; Gomes et al., 2013a), with losses of seed viability being attributed to increased lipid peroxidation (Hendry, 1993). Lipid peroxidation is the oxidative deterioration of lipids, which form the basis of many cellular organelles like mitochondria, plasma membranes, endoplasmatic reticulum, lysosomes, peroxisomes. Thus, the damage caused by lipid peroxidation is highly detrimental to the function of the cell and its survival (Raha and Robinson, 2000). Reactive oxygen species can also affect the genetic code in the embryo by reacting with nucleic acids, and can react with seed storage compounds such as proteins and lipids (Kranner and Colville, 2011).

In contrast, when actuating in the oxidative signaling pathway, ROS have important roles in germination as in the endosperm weakening, mobilization of seed reserves, protection against pathogens, and programmed cell death (PCD) (Bailly et al., 2008; Gomes and Garcia, 2013). The role of ROS in endosperm weakening has been proposed in analogy to its involvement in cell wall loosening in growing tissues (Liszkay et al., 2004). Reactive oxygen species can break down cell-wall–polysaccharides (Schweikert et al., 2002), regulate cellular calcium channels (Carol and Dolan, 2006), and mediate auxin functions in cell growth (Schopfer et al., 2002). In reserve mobilization processes, ROS can have a role in breaking down polysaccharides, DNA, RNA, and fatty acids (Buetler et al., 2004) and in protein carbonylation (which increases protein susceptibility to proteolytic cleavage), thus aiding the mobilization of those molecules during germination (Job et al., 2005). In protecting against pathogens, ROS may be used as weapons (as toxic compounds) or they may trigger hypersensitive reactions in the PCD of infected sites (Grant and Loake, 2000). Reactive oxygen species appears to induce plant cell death by initiating transductional pathways rather than through direct poisoning due to phytotoxicity (Van Breusegem and Dat, 2006). In this process, ROS trigger signal transduction cascades that lead to cell death by, for example, inhibiting *de novo* protein synthesis, blocking calcium fluxes, kinase/phosphatase activities, and oxidative bursts (Van Breusegem and Dat, 2006). Both $O_2^{\bullet-}$ and H_2O_2 are indisputably involved in genetically programmed cell death in plants (Wisniewski et al., 1999; Van Breusegem and Dat, 2006; Buetler et al., 2004).

The switching between the oxidative damage and oxidative signaling pathway of ROS is relative to the plant's ability to control internal amounts of ROS, and so, to the activities of the ROS scavenging systems, as shown in Fig. 17.1.

17.3.2 Trace Elements Tolerance in Seed Associated with Antioxidant Systems

Cellular antioxidant systems maintain intracellular redox homeostasis, preventing the accumulation of toxic ROS while allowing ROS-mediated signaling to occur (Foyer and Noctor, 2009; Ayer et al., 2012). The acquisition of tolerance has been seen to be closely related to the activation of antioxidant systems. For example, dried tolerant bean seeds showed high CAT and GR and low SOD and APX activities, while the opposite situation was seen in immature dried seeds (Bailly et al., 2001), suggesting that the acquisition of drying tolerance is related to reorientation of the enzymatic ROS-scavenging systems (Bailly et al., 2008). Similarly, ROS are also involved in the loss of seed viability (Bailly et al., 2008; Rajjou et al., 2008). Viability losses

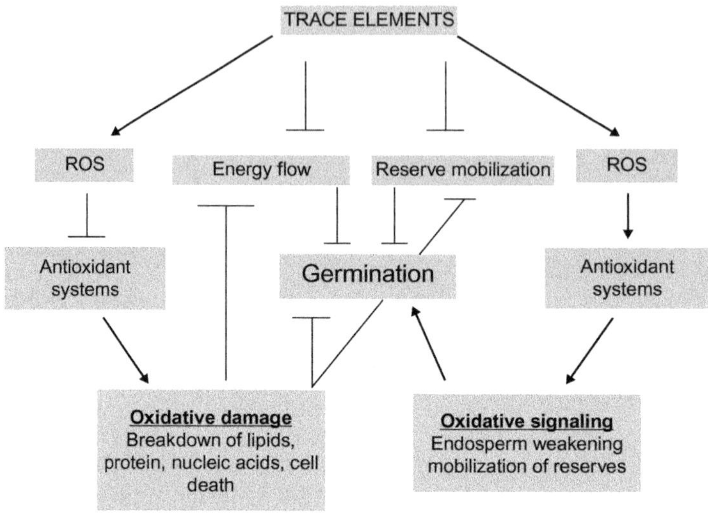

FIGURE 17.1 Trace elements modulating seed germination. The germination success under trace elements exposure is related to the seed's ability to cope with the ROS accumulation induced by trace elements. At these conditions, the switching between the oxidative damage and oxidative signaling pathway of ROS is relative on the activities of ROS scavenging systems. Inhibition of antioxidant systems leads to ROS accumulation and thus, to oxidative damage of cellular components, to blockage of energy flow and to interference in the process of reserve mobilization. In contrast, under effective activities of antioxidant systems, ROS act as signaling inducing seed germination by actuating, for example, in the endosperm weakening and by facilitating the mobilization of reserves.

during storage result from accumulations of ROS and its associated damage and a decrease in antioxidant mechanisms, which lead to oxidative damage during seed imbibition (Bailly et al., 2008). Losses of ROS-scavenging enzyme activities, the occurrence of oxidative reactions leading to membrane damage and lipid peroxidation have been reported with progressive seed aging (Bailly et al., 1996; Pukacka and Ratajczak, 2007; Munné-Bosch et al., 2011).

Gomes et al. (2013a) recently studied the systems modulating ROS content in the germinating seeds of two Brazilian savanna tree species exposed to trace elements. These authors observed increased H_2O_2 contents and lipid peroxidation activities in seeds treated with As and Zn, indicating that these trace elements had direct effects on seed germination through increased oxidative stress. Delays or inhibitions of seed germination were related to H_2O_2 accumulation and associated membrane damage (through lipid peroxidation). Interestingly, the germination percentages of *Anadenanthera peregrina* seeds treated with 50 mg Zn L^{-1} increased and they showed slightly higher H_2O_2 content than control seeds. Strictly regulated concentrations of ROS are currently viewed as being essential for germination (Bailly et al., 2008), and the

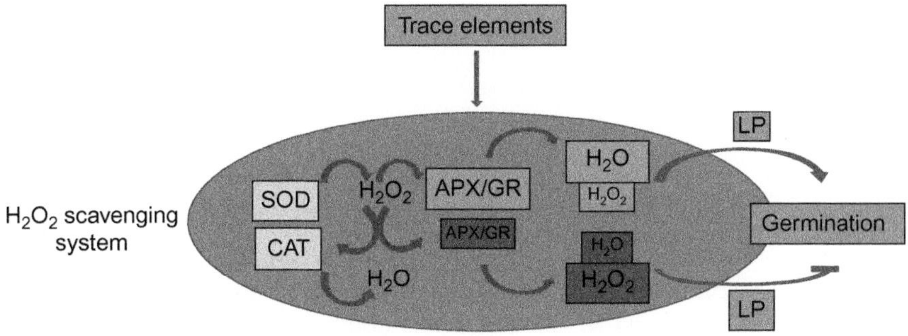

FIGURE 17.2 The system modulating H_2O_2 content in germinating seeds (in two Brazilian savanna species) exposed to trace elements (As and Zn), according to Gomes *et al.* (2013a). The activities of SOD and CAT did not show any type of As or Zn-induced inhibition; germinating seeds showed increased APX/GR activities and therefore decreased H_2O_2 contents and lipid peroxidation (LP) levels; decreased APX and GR activities (or their inhibition) contributed to increased H_2O_2 levels and to lipid peroxidation, resulting in decreased germination (or its total inhibition).

higher germination percentages of *A. peregrina* seeds in the aforementioned treatment was attributable to their enhanced oxidative status through ROS generation (Lefèvre et al., 2009). Cellular antioxidant systems thus have a central role in maintaining the fine-regulation of ROS contents that can act as signaling compounds in biological processes. Gomes et al. (2013a) noted the following strong relationships between germination percentages and antioxidant system levels (as also described in Fig. 17.2):

1. SOD activity was greater in both species studied (*Anadenanthera peregrina* and *Myracrodruon urundeuva*) in the presence of As and Zn. Higher SOD activity levels were positively related to the increases in H_2O_2 levels observed in the presence of the trace elements. CAT activity was similar to that of SOD, and neither enzyme showed any type of As- or Zn-induced inhibition;
2. APX was considered as a key germination enzyme as a positive relationship was observed between malondialdehyde (MDA) production (an indicator of lipid peroxidation) and APX activity. When seeds were able to germinate in the presence of As or Zn they showed increased APX activities and decreased H_2O_2 contents and lipid peroxidation levels. In contrast, when the trace elements studied induced decreases in germination (or total inhibition), the seeds showed decreased/inhibited APX activities and increased H_2O_2 contents and lipid peroxidation levels;
3. GR activity followed the same pattern as APX in being dramatically reduced by the highest As and Zn concentrations, with significant germination delays (or complete inhibition) being observed. GR activity also increased when seeds were exposed to As and Zn levels that allowed germination. The authors also suggested that as GR and APX are linked by

the gluthatione-ascorbate cycle (recycling oxidized ascorbate after catalysis by APX) (Gechev et al., 2006; Adi et al., 2012), decreased APX and GR activities (or their inhibition) will contribute to increasing H_2O_2 levels in As and Zn-treated seeds.

17.4 TRACE ELEMENTS AND SEEDLING ESTABLISHMENT

Seedlings are highly sensitive to environmental conditions after germination, and trace element contamination can constrain their establishment and/or development. Once fixed in the soil, roots come into contact with the soil solution and potential hazardous elements that can induce nutritional and metabolic disturbances in the growing seedlings. The ability to cope with these elements and to avoid their phytotoxicity determines whether the species will be able to survive in the presence of trace element contamination.

The presence of trace elements (in special nonessential elements) in the growth media may cause nutritional disturbances (Gomes et al., 2013d) directly competing with nutrient uptake and/or altering essential elements distribution and other metabolic processes (Sarwar et al., 2010). For example, Cd competition with macro- and micronutrients such as P, calcium (Ca), magnesium (Mg), iron (Fe), manganese (Mn), Cu, Zn, and nickel (Ni) for the same transmembrane carrier proteins has been reported (Rivetta et al., 1997). Because it is a chemical analogue, As may compete with P for uptake sites and exert its toxicity in plants by interfering with the many physiological functions performed by the nutrient (Meharg and Hartley-Whitaker, 2002; Puckett et al., 2012).

Trace elements are also known to interfere in plant nitrogen (N) and sulfur (S) nutrition which may contribute to the oxidative stress induced by trace elements (Gomes et al., 2012). Nitrogen and S are important components of antioxidant enzymes. Moreover, S is closely linked with phytochelatin metabolism (Hartley-Whitaker et al., 2001) which is involved in trace element tolerance (Jozefczak et al., 2012; Seth et al., 2012). Glutathione, a sulfur-containing tripeptide thiol with the formula γ-glutamate-cysteine-glycine and a precursor of phytochelatins (Rosen, 2002) is considered to be a very important antioxidant involved in the cellular defense against toxicants. Furthermore, redox-active metals (i.e. Cu) are also known to perform antioxidative protection (Cuypers et al., 2001). So, the maintenance of an adequate nutrition and distribution of nutrients associated to the avoidance of the uptake of high amounts of trace elements is an important strategy of tolerant seedlings.

Tolerant seedlings are provided with anatomical mechanisms which help to keep in control the amount of trace elements available to active cells and growing tissues, thus preventing the induced oxidative stress. Tolerant species often show an additional lignin deposit in the endodermis root cells. This contributes with a new layer of material that does not affect the radial transport of ions that usually move freely via the symplast (such as K^+ and PO_4^{3-}), but inhibits the movement of ions that flow mainly via the apoplastic route (Russel and Clarkson, 1975), which is the case for most trace elements (in particular, the metal ones). Additionally, as the root is a common deposition site for trace elements, expanded cell wall areas positively favor higher trace element retention (Gomes et al., 2011). Cell walls have been found to be the first barrier protecting the protoplast against the toxicity of trace elements such as Cu, Cd and Zn, and trace elements allocation in the cell walls is an important tolerance mechanism, especially in low concentrations and for short exposure periods (Wójcik et al., 2005; Zhao et al., 2012).

Significant increases in lipid degradation products (MDA) are often observed in the roots and leaves of plants exposed to trace elements (Martínez-Alcalá et al., 2013). The increased accumulation of lipid peroxides in seedlings is an indicative of enhanced production of toxic oxygen species, and lipid peroxidation may be a consequence of ROS generation. Both redox and non-redox active metal ions have been reported to increase lipid peroxidation via ROS generation in plants (Gallego et al., 2012). Cd-induced oxidative stress in sunflower leaves was observed to be mediated by ROS, as low levels of lipid peroxidation, the maintenance of GSH content, and increases in lipoxygenase activity were observed in the presence of free radical scavengers (Gallego et al., 1996).

Reduced photosynthetic activity is commonly observed in plants exposed to trace elements (Chugh and Sawhney, 1999; Dias et al., 2012). This has been attributed to the deleterious action of trace elements on chlorophyll synthesis (Stobart et al., 1985; Saygideger et al., 2013), on photosynthetic efficiency (Chugh and Sawhney, 1999; Austruy et al., 2013), on enzyme activity (Singh et al., 2006), and on water balance (Zhou and Qiu, 2005). Trace elements are capable of affecting chlorophyll contents, the activity of Calvin cycle enzymes, while PSII is extremely sensitive to these elements, in particular, the metal ions (Cagno et al., 1999; Cagno et al., 2001). It has long been recognized that H_2O_2 is a potent inhibitor of photosynthesis, because even at low concentrations (10 mM) it can inhibit CO_2 fixation by 50% because of the oxidation of the thiol modulated enzymes of the Calvin cycle (Foyer et al., 2011). Therefore, ROS accumulation induced by trace elements is an important factor disturbing photosynthesis. In association, whether directly or indirectly, these factors all contribute to reduced carbon assimilation rates.

17.4.1 Trace Elements Tolerance in Seedlings Associated with Antioxidant Systems

Plants are able to increase their antioxidant enzyme activities in order to reduce oxidative stress, thus increasing their trace element tolerance (Singh et al., 2006). Antioxidative defenses fall into two general classes: low molecular weight antioxidants, composed of lipidsoluble membrane-associated antioxidants (α-tocopherol and β-carotene) and water-soluble reductants (glutathione and ascorbate), and enzymatic antioxidants (SOD, CAT, GPX, and APX) (Cao et al., 2004). High activity levels of these antioxidant systems have commonly been reported in plants exposed to trace elements.

The responses of plant enzymatic antioxidant systems to exposure to trace elements can be highly variable, and will depend on the plant species, its age, the plant tissues under consideration, growth conditions, and the duration of trace element exposure. Studies of these responses in the leaves of tolerant Brazilian species have, however, revealed that SOD did not show any type of As-, Zn-, or Cd-inducing inhibition in the As-tolerant *Anadenanthera peregrina* (Gomes et al., 2013b), the Zn-tolerant *Myracrodruon urundeuva* (Gomes et al., 2013c), or the Cd-tolerant (and hyperaccumulator) *Pfaffia glomerata* (Gomes et al., 2013e) respectively; and SOD activity was always higher in treated seedlings. Increases in SOD activity have been linked to higher values of superoxide radical concentrations due to the synthesis of enzymes induced by superoxide-mediated signal transduction (Rossato et al., 2011). SOD represents the first line of antioxidant defense systems (Mittler, 2002) and has a marked role in ROS scavenging in seedlings subjected to trace element stress.

Once induced by SOD activity, H_2O_2 levels will be controlled by CAT and peroxidases. In both *M. urundeuva* and *P. glomerata* exposed to Zn and Cd, respectively, CAT showed a trend similar to that of SOD (Gomes et al., 2013c,d). CAT has a greater affinity (mM range) for H_2O_2 than do other peroxidases such as APX (μM range). This suggests that APX may be responsible for fine-tuning ROS for signaling, while CAT may simply be responsible for removing excess ROS during stress situations (Mittler, 2002). The high content of H_2O_2 produced by SOD during trace element exposure should be, therefore, degraded by CAT, thus justifying their coupled actions. APX activity in seedlings exposed to trace elements has also drawn considerable attention. *A. peregrina* and *M. urundeuva* seedlings exposed to As or Zn doses (respectively) causing growth reductions (indicative of phytotoxicity) demonstrated decreased APX activity and increased H_2O_2 contents and lipid peroxidation levels (Gomes et al., 2013b, 2013c); APX activity was not affected (or was higher than that seen in control plants) when exposed to nonphytotoxic doses of trace elements. Interestingly, the addition of phosphorus to the growth substrate increased As-tolerance in *A. peregrina* by increasing APX activity, which decreased H_2O_2 accumulation and helped prevent lipid peroxidation in its leaves (Gomes et al., 2013b). Similarly, *P. glomerata* seedlings exposed to Cd doses up to 90 mmol L^{-1} showed increased APX activity (Marques and Soares, 2011), or were not affected (Gomes et al., 2013e), and no growth reductions were observed under those conditions.

In regard to GR, when *M. urundeuva* and *P. glomerata* were exposed to Zn and Cd, respectively, GR demonstrated a trend similar to APX (Gomes et al., 2013c; Marques and Soares, 2011). These results would be expected as GR and APX are linked through the glutathione-ascorbate (AsA-GSH) cycle. APX uses two ascorbate molecules (AsA) in the AsA-GSH cycle to reduce H_2O_2 to water – with the generation of two molecules of monodehydroascorbate (MDHA) that can then be reduced to AsA by MDHA reductase (using NADPH as electron donors) (Chao et al., 2010). If it is not immediately reduced, the short-lifetime MDHA radical is oxidized to dehydroascorbate (DHA), and DHA is then reduced to AsA through the action of DHA reductase using GSH, which is generated from oxidized glutathione (GSSG) by NADPH-dependent GR activity (Chao et al., 2010).

Decreased APX and GR activities in plants exposed to toxic trace elements can be related to protein carbonylation induced by ROS accumulation, as reported by Romero-Puertas et al. (2002). This is an irreversible oxidative process in which the side-chains of Lys, Arg, Pro, and Thr are converted to aldehyde or keto groups, which inhibits or impairs multiple enzyme systems (Sohal et al., 2002). The same processes probably occur in *A. peregrina* and *M. urundeuva*, so that when H_2O_2 accumulation (provoked by the presence of toxic doses of trace elements) exceeds the tolerance limits of the seedlings, their APX and GR systems become inactivated.

The maintenance of the AsA-GSH metabolic cycle is another tolerance strategy in plants exposed to trace elements (López-Climent et al., 2013). In their study of the redox statuses (AsA and GSH) of *M. urundeuva* leaves, Gomes et al. (2013c) noted that their ascorbate and glutathione pools were higher when exposed to nonphytotoxic Zn doses than those of control plants. However, trace elements can also affect the AsA-GSH metabolic cycle, which is necessary for awarding H_2O_2 damage, mainly in the chloroplasts, by interfering with antioxidant enzyme activities (Chaoui et al., 1997) (Fig. 17.3). Gomes et al. (2013c), in fact, verified that toxic doses of Zn disturbed the glutathione balance and prevented cells from maintaining their ascorbate pools in a reduced state. The reduced GR activity led to the accumulation of GSSG; the reduced GSH content led, in turn, to less reducing power (GSH) that could be used in DHA reduction – resulting in the accumulation of the oxidized form of ascorbate (DHA) and higher DHA/AsA ratios. The starvation of AsA due to GR inhibition may therefore partly explain the observed decrease in APX activity.

The depletion of chloroplast AsA and the inactivation of chloroplast APXs are thought to limit photosynthetic efficiency (Ishikawa and Shigeoka, 2008) by inducing oxidative stress. Photosynthesis is a major source of ROS, which is generated through processes associated with energy transfer and electron transport operating in aerobic environments and indirectly through the initiation of the photorespiratory pathway (Foyer and Noctor, 2009). Ascorbate fulfills crucial roles in photosynthesis in chloroplasts by helping to prevent oxidation of stromal proteins and in protecting PSII by participating in the violaxanthin de-epoxidase reaction (Foyer and Noctor, 2009). AsA can also act as a temporary electron donor to PSII if the

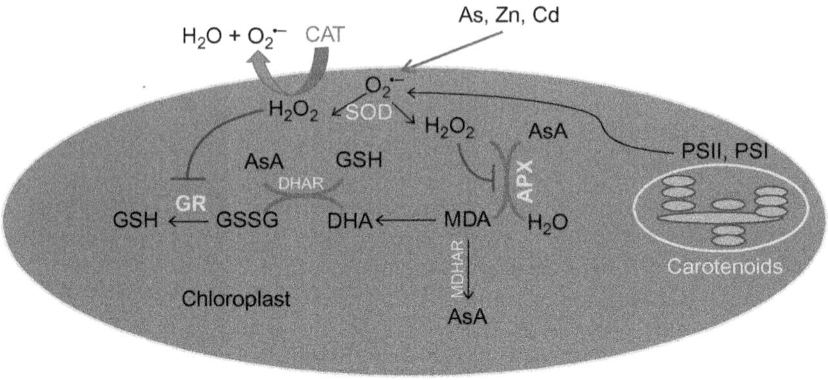

FIGURE 17.3 A schematic representation of the interactions between trace elements (As, Zn and Cd) and major sources of ROS generation and scavenging enzymes. Carotenoids presumably help protect photosynthetic systems against trace element-induced oxidative damage. Photosystems (PS) are sources of superoxide anions ($O_2^{\bullet-}$) through the Mehler reaction (primary source PSI). Trace elements can also induce $O_2^{\bullet-}$ production. Superoxide dismutase (SOD) does not show any type of As, Zn, or Cd-induced inhibition and is stimulated by these trace elements. H_2O_2 generated by SOD activity is degraded by catalase (CAT) (whose production is also stimulated by trace elements) external to the chloroplasts or by ascorbate peroxidase (APX). In addition to APX, glutathione reductase (GR) shows sensitivity to higher concentrations of H_2O_2, and reduced GR activities lead to the accumulation of oxidized glutathione (GSSG); the reduced glutathione (GSH) content leads, in turn, to less reducing power (GSH) that can be used in dehydroascorbate (DHA) – resulting in the accumulation of the oxidized form of ascorbate (DHA). Ascorbate (AsA) starvation due to GR inhibition may therefore explain the observed decrease in APX activity.

oxygen-evolving complex is inactivated (Tóth et al., 2009). Therefore, the larger AsA pools seen in *M. urundeuva* plants growing under nonphytotoxic doses may help in protecting the photosynthetic apparatus. Indeed, Gomes et al. (2013c) reported that *M. urundeuva* seedlings did not show disturbances in their photosynthetic levels when exposed to either toxic or nontoxic Zn doses. However, as mentioned above, the AsA pool decreased, and was maintained in an oxidized state, under toxic Zn doses. The question of why photosynthesis was not affected under these conditions was answered with the observation that the carotenoid contents of seedlings increased when exposed to toxic Zn doses. The increased carotenoid contents of plants exposed to high Zn doses presumably helps protect their photosynthetic systems against Zn-induced oxidative damage as Zn does not appear to disturb the photosynthetic apparatus of the seedlings. Lipophilic antioxidants such as carotenoids have essential roles in removing oxidants such as singlet oxygens that are generated within the thylakoid membranes (Foyer and Noctor, 2009). Singlet oxygen species generated by PSII are responsible for most (over 90%) of the observed nonenzymatic peroxidation of thylakoid membrane lipids (Triantaphylidès et al. 2008), and α-tocopherol and β-carotene operate synergistically to eliminate them from PSII (Trebst et al. 2002; Munné-Bosch, 2005).

The observation that ROS generators can induce carotenoid accumulation suggests that this carotenogenic response is mediated by ROS (Lu et al., 1998). ROS (or their products) may regulate carotenoid content by the direct activation of latent biosynthetic enzymes such as glutathione transferase (Aniya and Anders, 1992) and glutathione reductase (Miller and Claiborne, 1991). Alternatively, ROS may activate the expression of genes coding for carotenogenic enzymes (Bouvier et al., 1998). Carotenoids are usually involved in the detoxification of oxygen singlets but not other ROS species, although ROS can readily be enzymatically converted to oxygen singlets that can then react with carotenoids – thus protecting the plants from oxidative damage (Boussiba, 2000).

17.5 CONCLUSIONS

Exposure to these toxic elements can intensify the production of reactive oxygen species (ROS) that are continuously produced in both unstressed and stressed plant cells. Reactive oxygen species are important molecules that can play central roles in plant signaling as, for example, in seed germination. Plants are able to increase antioxidant enzyme activities to reduce oxidative stress, thus increasing their trace element tolerance. In this context, trace element tolerance in seeds and seedling is modulated by their antioxidant system activities, and APX was seen to be a central factor mediating that tolerance. There are also intrinsic interactions between the activities of APX and GR and the AsA-GSH metabolic cycle. Little is currently known, however, about ROS signaling pathways or their biological roles as harmful substances in seeds and seedlings exposed to trace elements. The interaction between ROS and plant hormones as signaling during environmental stresses should be better studied. Moreover, molecular studies, such as genomic silencing of APX, will certainly contribute to our better understanding the role of antioxidant systems in trace element tolerance and plant adaptive processes.

REFERENCES

Adi, M., Jens, P., Brotman, Y., Mikhail, K., Iris, S., Henryk, C., et al., 2012. Metabolomics: Open access stress responses to tomato yellow leaf curl virus (TYLCV) infection of resistant and susceptible tomato plants are different. Metabolomics S1, 1–13.

Aniya, Y., Anders, M.W., 1992. Activation of rat liver microsomal glutathione S-transferase by hydrogen peroxide: role for protein-dimer formation. Arch. Biochem. Biophys. 296, 611–616.

Austruy, A., Wanat, N., Moussard, C., Vernay, P., Joussein, E., Ledoigt, G., et al., 2013. Physiological impacts of soil pollution and arsenic uptake in three plant species: *Agrostis capillaris*, *Solanum nigrum* and *Vicia faba*. Ecotox. Environ. Safety 90, 28–34.

Ayer, A., Fellermeier, S., Fife, C., Li, S.S., Smits, G., Meyer, A.J., et al., 2012. A genome-wide screen in yeast identifies specific oxidative stress genes required for the maintenance of sub-cellular redox homeostasis. PloS one 7, e44278.

Bailly, C., Benamar, A., Corbineau, F., Côme, D., 1996. Changes in malondialdehyde content and in superoxide dismutase, catalase, and glutathione reductase activities in sunflower seeds as related to deterioration during accelerated ageing. Physiol. Plant 97, 104–110.

Bailly, C., Audigier, C., Ladonne, F., Wagner, M.H., Coste, F., Corbineau, F., et al., 2001. Changes in oligosaccharide content and antioxidant enzyme activities in developing bean seeds as related to acquisition of drying tolerance and seed quality. J. Exp. Bot. 357, 701–708.

Bailly, C., Maarouf-Bouteau, H., Corbineau, F., 2008. From intracellular signaling net-works to cell death: the dual role of reactive oxygen species in seed physiology. C. R. Biol. 331, 806–814.

Bansal, P., Sharma, P., Dhindsan, K., 2000. Individual and combined effects of Pb2+ and Cd2 + on seed germination and initial seedling growth of pea. Ann. Biol. 16, 11–16.

Bansal, P., Sharma, P., Goyal, V., 2002. Impact of lead and cadmium on enzyme of citric acid cycle in germinating pea seeds. Biol. Plant 45, 125–127.

Bautista, O.V., Fischer, G., Cárdenas, J.F., 2013. Cadmium and chromium effects on seed germination and root elongation in lettuce, spinach and Swiss chard. Agron. Colom. 2, 48–57.

Boussiba, S., 2000. Carotenogenesis in the green alga *Haematococcus pluvialis*: cellular physiology and stress response. Physiol. Plant 108, 111–117.

Bouvier, F., Keller, Y., d'Harlingue, A., Camara, B., 1998. Xanthophyll biosynthesis: molecular and functional characterization of carotenoid hydroxylases from pepper fruits (*Capsicum annuum* L.). Biochim. Biophys. Acta 1391, 320–328.

Buetler, T.M., Krauskopf, A., Ruegg, U.T., 2004. Role of superoxide as a signaling molecule. News Physiol. Sci. 19, 120–123.

Cagno, R., Guidi, L., Stefani, A., Soldatini, G.F., 1999. Effects of cadmium on growth of Helianthus annuus seedlings: physiological aspects. New Phytol. 144, 65–71.

Carol, R.J., Dolan, L., 2006. The role of reactive oxygen species in cell growth: Lessons from root hairs. J. Exp. Bot. 57, 1829–1834.

Cao, X., Ma, L.Q., Tu, C., 2004. Antioxidative responses to arsenic in the arsenic-hyperaccumulator Chinese brake fern (*Pteris vittata* L.). Environ. Pollut. 128, 317–325.

Cagno, R., Guidi, L., De Gara, L., Soldatini, G.F., 2001. Combined cadmium and ozone treatments affect photosynthesis and ascorbate defences in sunflower. New Phytol. 151, 627–636.

Chao, Y.Y., Hong, C.Y., Kao, C.H., 2010. The decline in ascorbic acid content is associated with cadmium toxicity of rice seedlings. Plant Physiol. Biochem. 48, 374–381.

Chaoui, A., Mazhoudi, S., Ghorbal, M.H., Ferjani, E.E., 1997. Cadmium and zinc induction of lipid peroxidase and effects on antioxidant enzyme activities in bean (*Phaseolus vulgaris* L.). Plant Sci. 127, 139–147.

Chugh, L.K., Sawhney, S.K., 1996. Effect of cadmium on germination, amylases and rate of respiration of germinating pea seeds. Environ. Pollut. 92, 1–5.

Chugh, L.K., Sawhney, S.K., 1999. Photosynthetic activities of *Pisum sativum* seedlings grown in presence of cadmium. Plant Physiol. Biochem. 37, 297–303.

Clemens, S., Antosiewez, D.M., Ward, J.M., Schatman, D.P., Schroeder, J.I., 1998. The plant cDNA LCT1mediates the uptake of calcium and cadmium in yeast. PNAS 95, 12043–12048.

Cuypers, A., Vangronsveld, J., Clijstes, H., 2001. The redox status of plant cells (AsA and GSH) is sensitive to zinc imposed oxidative stress in roots and primary leaves of *Phaseolus vulgaris*. Plant Physiol. Biochem. 39, 657–664.

Dary, M., Chamber-Pérez, M.A., Palomares, A.J., Pejuelo, E., 2010. "In situ" phytostabilisation of heavy metal polluted soils using *Lupinus luteus* inoculated with metal resistant plant-growth promoting rhizobacteria. J. Hazard Mater. 177, 323–330.

Dias, M.C., Monteiro, C., Moutinho-Pereira, J., Correia, C., Gonçalves, B., Santos, C., 2012. Cadmium toxicity affects photosynthesis and plant growth at different levels. Acta Physiol. Plant. 35, 1281–1289.

Duarte, D.M., Gomes, M.P., Barreto, L.C., Matheus, M.T., Garcia, Q.S., 2012. Toxic trace elements effects on seed germination of four Brazilian Savanna tree species. Seed Sci. Technol. 40, 425–432.

Fang, K.-M., Cheng, F.-C., Huang, Y.-L., Chung, S.-Y., Jian, Z.-Y., Lin, M.-C., 2013. Trace element, antioxidant activity, and lipid peroxidation levels in brain cortex of gerbils after cerebral ischemic injury. Biol. Trace El. Res. 152, 66–74.

Foyer, C.H., Noctor, G., 2009. Redox regulation in photosynthetic organisms: signaling, acclimation and practical implications. Antioxid. Redox Sig. 11, 861–905.

Foyer, C.H., Shigeoka, S., 2011. Understanding oxidative stress and antioxidant functions to enhance photosynthesis. Plant Physiol. 155, 93–100.

Gallego, S.M., Pena, L.B., Barcia, R.A., Azpilicueta, C.E., Iannone, M.F., Rosales, E.P., et al., 2012. Unravelling cadmium toxicity and tolerance in plants: Insight into regulatory mechanisms. Environ. Exp. Bot. 83, 33–46.

Gechev, T.S., Breusegem, F.V., Stone, J.M., Denev, I., Laloi, C., 2006. Reactive oxygen species as signals that modulate plant stress responses and programmed cell death. Bioessays 28, 1091–1101.

Gill, S.S., Tuteja, N., 2010. Reactive oxygen species and antioxidant machinery in abiotic stress tolerance in crop plants. Plant Physiol. Biochem. 48, 909–930.

Gomes, M.P., Garcia, Q.S., 2013. Reactive oxygen species and seed germination. Biologia. 68, 351–357.

Gomes, M.P., Carvalho, M., Carvalho, G.S., Garcia, Q.S., Guilherme, L.R.G., Soares, A.M., 2013b. Phosphorus improves arsenic phytoremediation by *Anadenanthera peregrina* by alleviating induced oxidative stress. Int. J. Phytorem. 15, 633–646.

Gomes, M.P., Duarte, D.M., Carneiro, M.M.L.C., Barreto, L.C., Carvalho, M., Soares, A.M., et al., 2013c. Zinc tolerance modulation in *Myracrodruon urundeuva* plants. Plant Physiol. Biochem. 67, 1–6.

Gomes, M.P., Duarte, D.M., Miranda, P.L.S., Barreto, L.C., Matheus, M.T., Garcia, Q.S., 2012. The effects of arsenic on the growth and nutritional status of *Anadenanthera peregrina,* a Brazilian savanna tree. J. Plant Nutr. Soil Sci. 175, 466–473.

Gomes, M.P., Carneiro, M.M.L.C., Nogueira, C.O.G., Soares, A.M., Garcia, Q.S., 2013a. The system modulating ROS content in germinating seeds of two Brazilian savanna tree species exposed to As and Zn. Acta Physiol. Plant. 35, 1011–1022.

Gomes, M.P., Marques, T.C.L.L.S.M., Oliveira, M.O.G.N., Castro, E.M., Soares, A.M., 2011. Ecophysiological and anatomical changes due to uptake and accumulation of heavy metal in *Brachiaria decumbens*. Sci. Agric. 68, 566–573.

Gomes, M.P., Marques, T.C.L.L.S.E.M., Soares, A.M., 2013d. Cadmium effects on mineral nutrition of the Cd-hyperaccumulator *Pfaffia glomerata*. Biologia. 68, 223–230.

Gomes, M.P., Marques, T.C.L.L.S.M., Martins, G.A., Carvalho, M.M.L., Soares, A.M., 2013e. Cd-tolerance markers of *Pfaffia glomerata* (Spreng.) Pedersen plants: anatomical and physiological features. Braz. J. Plant Physiol. 24, 293–304.

Grant, J.J., Loake, G.J., 2000. Role of reactive oxygen intermediates and cognate redox signaling in disease resistance. Plant Physiol. 124, 21–29.

Gratão, P.L., Polle, A., Lea, P.J., Azevedo, R.A., 2005. Making the life of heavy metal-stressed plants a little easier. Func. Plant Biol. 32, 481–494.

Gunes, A., Inal, A., Bagci, E.G., Pilbeam, D.J., 2007. Silicon-mediated changes of some physiological and enzymatic parameters symptomatic for oxidative stress in spinach and tomato grown in sodic-B toxic soil. Plant Soil 290, 103–114.

Gunes, A., Pilbeam, D., Inal, A., 2009. Effect of arsenic-phosphorus interaction on arsenic-induced oxidative stress in chickpea plants. Plant Soil 314, 211–220.

Hartley-Whitaker, J., Ainsworth, G., Vooijs, R., Bookum, W.M.T., Schat, H., Meharg, A.A., 2001. Phytochelatins are involved in differential arsenate tolerance in *Holcus lanatus*. Plant Physiol. 126, 299–306.

Hendry, G.A.F., 1993. Oxygen, free radical processes and seed longevity. Seed Sci. Res. 3, 141–153.

Hsu, F.H., Chou, C.H., 1992. Inhibitory effects of heavy metals on seed germination and seedling growth of *Miscanthus* species. Bot. Bull. Acad. Sin. 33, 335–342.

Ishikawa, T., Shigeoka, S., 2008. Recent advances in ascorbate biosynthesis and the physiological significance of ascorbate peroxidase in photo-synthesizing organisms. Biosci. Biotechnol. Biochem. 72, 1143–1154.

Job, C., Rajjou, L., Lovigny, Y., Belghazi, M., Job, D., 2005. Patterns of protein oxidation in *Arabidopsis* seeds and during germination. Plant Physiol. 138, 790–802.

Jozefczak, M., Remans, T., Vangronsveld, J., Cuypers, A., 2012. Glutathione is a key player in metal-induced oxidative stress defenses. Int. J. Mol. Sci. 13, 3145–3175.

Kabata-Pendias, A., Pendias, H., 2000. Trace Elements in Soils and Plants. CRC Press, London.

Kranner, I., Colville, L.E., 2011. Metals and seeds: biochemical and molecular implications and their significance for seed germination. Environ. Exp. Bot. 72, 93–105.

Lefèvre, I., Marchal, G., Corréal, E., Zanuzzi, A., Lutts, S., 2009. Variation in response to heavy metals during vegetative growth in *Dorycnium pentaphyllum* Scop. Plant Growth Regul. 59, 1–11.

Leymarie, J., Vitkauskaité, G., Hoang, H.H., Gendreau, E., Chazoule, V., Meimoun, P., et al., 2012. Role of reactive oxygen species in the regulation of *Arabidopsis* seed dormancy. Plant Cell Physiol. 53, 96–106.

Li, C.X., Feng, S.L., Shao, Y., Jiang, L.N., Lu, X.Y., Hou, X.L., 2007. Effects of arsenic on seed germination and physiological activities of wheat seedlings. J. Environ. Sci. 19, 725–732.

Li, W.Q., Khan, M.A., Yamaguchi, S., Kamiya, Y., 2005. Effects of heavy metals on seed germination and early seedling growth of *Arabidopsis thaliana*. Plant Growth Regul. 46, 45–50.

Liszkay, A., van der Zalm, E., Schopfer, P., 2004. Production of reactive oxygen intermediates, H2O2, and OH by maize roots and their role in wall loosening and elongation growth. Plant Physiol. 136, 3114–3123.

López-Climent, M.F., Arbona, V., Pérez-Clemente, R.M., Zandalinas, S.I., Gómez-Cadenas, A., 2013. Effect of cadmium and calcium treatments on phytochelatin and glutathione levels in citrus plants. Plant Biol. http://dx.doi.org/10.1111/plb.12006.

Lu, F., Vonshak, A., Zarka, A., Boussiba, S., 1998. Does astaxanthin protect *Haematococcus* against light damage? Z. Naturforsch. 53, 93–100.

Maheshwari, R., Dubey, R., 2008. Inhibition of ribonuclease and protease activities in germinating rice seeds exposed to nickel. Acta Physiol. Plant. 30, 863–872.

Marques, T.C.L.L.S.M., Soares, A.M., 2011. Antioxidant system of ginseng under stress by cadmium. Sci. Agric. 68, 482–488.

Martínez-Alcalá, I., Hernández, L.E., Esteban, E., Walker, D.J., Bernal, M.P., 2013. Responses of *Noccaea caerulescens* and *Lupinus albus* in trace elements-contaminated soils. Plant Physiol. Biochem. 66, 47–55.

Meharg, A.A., Hartley-Whitaker, J., 2002. Arsenic uptake and metabolism in arsenic resistant and non-resistant plant species. New Phytol. 154, 29–43.

Mihoub, A., Chaoui, A., El Ferjani, E., 2005. Changements biochimiques induits par le cadmium et le cuivreau cours de la germination des graines de petit pois (*Psidium sativum* L.) [Biochemical changes associated with cadmium and copper stress in germinating pea seeds]. C. R. Biol. 328, 33–41.

Miller, G., Shulaev, V., Mittler, R., 2008. Reactive oxygen signaling and abiotic stress. Physiol. Plant 133, 481–489.

Miller, H., Claiborne, A., 1991. Peroxide modification of monoalkylated glutathione reductase. Stabilization of an active-site cysteine-sulfenic acid. J. Biol. Chem. 266, 19342–19350.

Mittler, R., 2002. Oxidative stress, antioxidants and stress tolerance. Trends Plant Sci. 7, 405–410.

Moise, J.A., Han, S., Gudymaite-Savitch, L., Johnson, D.A., Miki, B.A., 2005. Seed coats: structure development composition and biotechnology. In Vitro Cell. Dev. Biol.–Plant 41, 620–644.

Munné-Bosch, S., 2005. The role of α-tocopherol in plant stress tolerance. J. Plant Physiol. 162, 743–748.

Munné-Bosch, S., Queval, G., Foyer, C.H., 2013. The impact of global change factors on redox signaling underpinning stress tolerance. Plant Physiol. 161, 5–19.

Nelson, M.T., 1986. Interactions of divalent cations with single calcium channels from rat brain synaptosomes. J. Gen. Physiol. 87, 201–222.

Ozdener, Y., Kutbay, H.G., 2009. Toxicity of copper, cadmium, nickel, lead and zinc on seed germination and seedling growth in *Eruca sativa*. Fresen. Environ. Bull. 18, 26–31.

Peng, H., Kroneck, P.M.H., Küpper, H., 2013. Toxicity and deficiency of copper in *Elsholtzia splendens* affect photosynthesis biophysics, pigments and metal accumulation. Environ. Sci. Technol. 47, 6120–6128.

Puckett, E.E., Serapiglia, M.J., DeLeon, A.M., Long, S., Minocha, R., Smart, L.B., 2012. Differential expression of genes encoding phosphate transporters contributes to arsenic tolerance and accumulation in shrub willow (*Salix* spp.). Environ. Exp. Bot. 75, 248–257.

Pukacka, S., Ratajczak, E., 2007. Age-related biochemical changes during storage of beech (*Fagus sylvatica* L.) seeds. Seed Sci. Res. 17, 45–53.

Raha, S., Robinson, B.H., 2000. Mitochondria, oxygen free radicals, disease and ageing. Trends Biochem. Sci. 25, 502–508.

Rajjou, L., Lovigny, Y., Groot, S.P.C., Belghaz, M., Job, C., Job, D., 2008. Proteome-wide characterization of seed aging in *Arabidopsis*: a comparison between artificial and natural aging protocols. Plant Physiol. 148 (620–641).

Remans, T., Opdenakker, K., Guisez, Y., Carleer, R., Schat, H., Vangronsveld, J., et al., 2012. Exposure of *Arabidopsis thaliana* to excess Zn reveals a Zn-specific oxidative stress signature. Environ. Exp. Bot. 84, 61–71.

Rivetta, A., Negrini, N., Cocucci, M., 1997. Involvement of Ca2+ -calmodulin in Cd2+ toxicity during the early phases of radish (*Raphanus sativus* L.) seed germination. Plant Cell Environ. 20, 600–608.

Romero-Puertas, M.C., Palma, J.M., Gómez, M., Del Río, L.A., Sandalio, L.M., 2002. Cadmium causes the oxidative modification of proteins in pea plants. Plant Cell Environ. 32, 677–686.

Rosen, B.P., 2002. Biochemistry of arsenic detoxification. FEBS Lett. 529, 86–92.

Rossato, L.V., Nicoloso, F.T., Farias, J.G., Cargnelluti, D., Tabaldi, L.A., Antes, F.G., et al., 2011. Effects of lead on the growth, lead accumulation and physiological responses of *Pluchea sagittalis*. Ecotoxicol. 7, 1–13.

Russel, R.S., Clarkson, D.T., 1975. Ion transport in root systems. In: Sunderland, N. (Ed.), Perspectives in Experimental Biology: Botany, vol 2. Pergamon Press, Oxford, UK, pp. 401–411.

Sarwar, N., Saifullah, Sukhdev, S.M., Munir, H.Z., Asif, N., Sadia, B., Ghulam, F., 2010. Role of mineral nutrition in minimizing cadmium accumulation by plants (Review). J. Sci. Food Agric. 90, 925–937.

Schopfer, P., Liszkay, A., Bechtold, M., Frahry, G., Wagner, A., 2002. Evidence that hydroxyl radicals mediate auxin-induced extension growth. Planta 214, 821–828.

Schweikert, C., Liszkay, A., Schopfer, P., 2002. Polysaccharide degradation by Fenton reaction- or peroxidase-generated hydroxyl radicals in isolated plant cell walls. Phytochem. 61, 31–35.

Scott, N., Hatlelid, K.M., MacKenzie, N.E., Carter, D.E., 1993. Reactions of arsenic(III) and arsenic(V) species with glutathione. Chem. Res. Toxicology 6, 102–106.

Seth, C.S., Remans, T., Keunen, E., Jozefczak, M., Gielen, H., Opdenakker, K., et al., 2012. Phytoextraction of toxic metals: a central role for glutathione. Plant Cell Environ. 35, 334–346.

Shetty, N., Jørgensen, H.L., Jensen, J., Collinge, D., Shetty, H.S., 2008. Roles of reactive oxygen species in interactions between plants and pathogens. In: Collinge, D., Munk, L., Cooke, B. M. (Eds.), Sustainable Disease Management in a European Context. Springer, Netherlands, pp. 267–280.

Singh, S., Eapen, S., D'Souza, S.F., 2006. Cadmium accumulation and its influence on lipid peroxidation and antioxidative system in an aquatic plant *Bacopa monnieri* L. Chemosphere 62, 233–241.

Sohal, R.S., Mockett, R.J., Orr, W.C., 2002. Mechanisms of aging: an appraisal of the oxidative stress hypothesis. Free Rad. Biol. Med. 33, 575–586.

Suzuki, N., Koussevitzky, S., Mittler, R., Miller, G., 2012. ROS and redox signalling in the response of plants to abiotic stress. Plant Cell Environ. 35, 259–270.

Tóth, S.Z., Puthur, J.T., Nagy, V., Garab, G., 2009. Experimental evidence for ascorbate-dependent electron transport in leaves with inactive oxygen-evolving complexes. Plant Physiol. 149, 1568–1578.

Trebst, A., Depka, B., Hollander-Czytko, H., 2002. A specific role for tocopherol and of chemical singlet oxygen quenchers in the maintenance of photosystem II structure and function in *Chlamydomonas reinhardtii*. FEBS Lett. 516, 156–160.

Triantaphylidès, C., Krischke, M., Hoeberichts, F.A., Ksas, B., Gresser, G., Havaux, M., et al., 2008. Singlet oxygen is the major reactive oxygen species involved in photooxidative damage to plants. Plant Physiol. 148, 960–968.

Van Breusegem, F., Dat, J.F., 2006. Reactive oxygen species in plant cell death. Plant Physiol. 141, 384–390.

Wisniewski, J.P., Cornille, P., Agnel, J.P., Montillet, J.L., 1999. The extensin multigene family responds differentially to superoxide or hydrogen peroxide in tomato cell cultures. FEBS Lett. 447, 264–268.

Wójcik, M., Vangronsveld, J., Haen, J.D., Tukiendorf, A., 2005. Cadmium tolerance in Thlaspi caerulescens II. Localization of cadmium in Thlaspi caerulescens. Environ. Exp. Bot. 53, 163–171.

Zhao, H., Wu, L., Chai, T., Zhang, Y., Tan, J., Ma, S., 2012. The effects of copper, manganese and zinc on plant growth and elemental accumulation in the manganese-hyperaccumulator Phytolacca americana. J. Plant Physiol. 169, 1243–1252.

Chapter 18

Plant Signaling under Environmental Stress

Mohammad Miransari

18.1 INTRODUCTION

Plants experience different environmental stresses and may be able to induce tolerance by utilizing morphological and physiological mechanisms. Various other mechanisms are also able to mitigate environmental stress in plants. For example, to increase their water efficiency under drought stress, plants decrease their leaf surface by rolling it or changing the angle, control the behavior of their stomata, and hence keep their water (Ford et al., 2011).

The physiological changes under stress include the production of different products, such as osmolytes and enzymes; activation of different signaling pathways; etc. Controlling plant growth under stress by the activation of signaling pathways, which is in fact a plant response to handle the stress, is of great importance. There are a set of signaling pathways activated under stress including mitogen-activated protein kinase signaling, reactive oxygen species and redox signaling, as well as hormonal signaling (Asai et al., 2002; Hirayama and Shinozaki, 2010; Munné-Bosch et al., 2013). The role of microRNAs under stress is also of significant importance, as they also enable the plant to respond to stress. Under stress, the methylation of DNA, remodeling of chromatin, histone methylation/acetylation, and processes related to small RNAs alone or combined may modify gene expression, rearrange the genome and hence influence plant tolerance to stress (Figs 18.1 and 18.2) (Jones, 2006; Guo and Lu, 2010; Lelandais-Briere et al., 2011; Grativol et al., 2012).

There are different techniques used to evaluate plant response under stress; one of them is proteomics. Using the proteomic technique it is possible to investigate which genes and hence proteins are activated and produced under stress (Evers et al., 2012; Swami et al., 2011). Accordingly, the production of tolerant plants under stress may be possible. Depending on the kind of stress, different genes and proteins are activated and produced.

P. Ahmad (Ed): Oxidative Damage to Plants. DOI: http://dx.doi.org/10.1016/B978-0-12-799963-0.00018-6

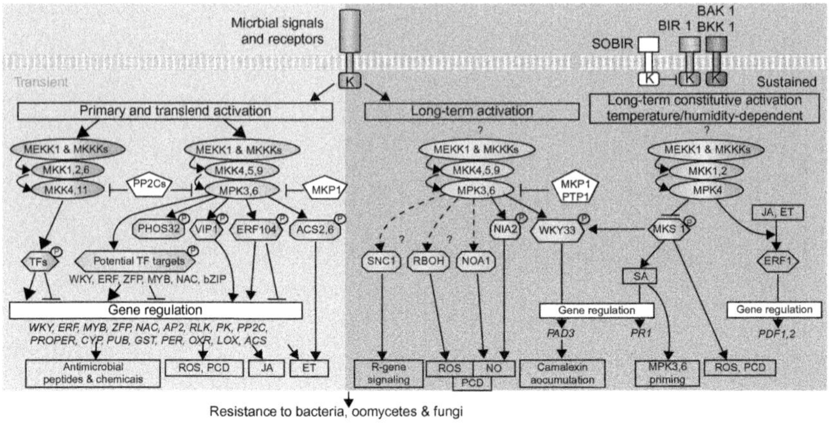

FIGURE 18.1 The network of MAPK signaling pathways in response to different biological and stress parameters, combined with other signaling pathways, including the reactive oxygen species and hormonal signaling. *(From Tena et al. (2011) with kind permission from Elsevier.)*

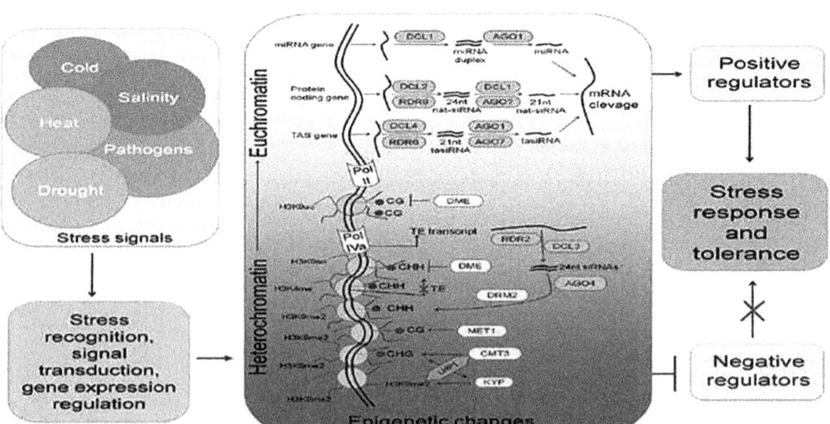

FIGURE 18.2 The regulation of response genes under stress by cross talk between epigenetic parameters including DNA methylation and histone modification resulting in plant tolerance under stress. *(From Grativol et al. (2012) with kind permission from Elsevier.)*

18.2 STRESS AND MITOGEN-ACTIVATED PROTEIN KINASE SIGNALING

Plant signaling pathways are among the most important plant responses to stress, enabling the plant to survive under stress. There are different signaling pathways activated under stress, including mitogen-activated protein kinase

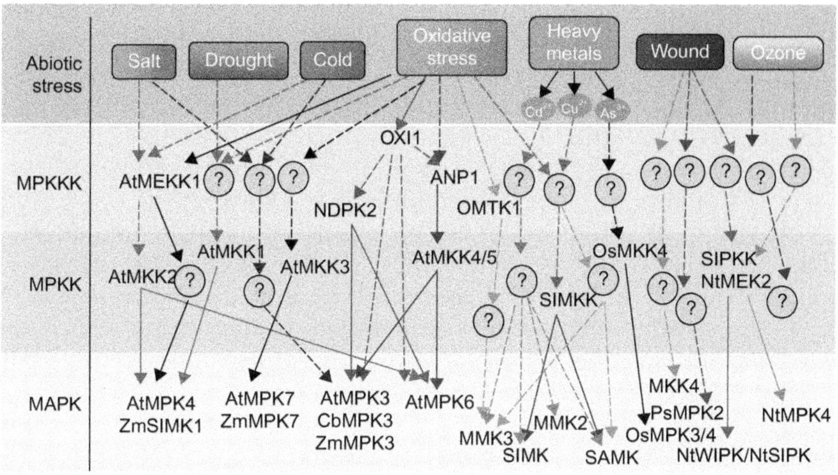

FIGURE 18.3 The cross talk of different MAPK signaling components, shown for different plants. Solid and dashed arrows indicate proven and postulated pathways, respectively. The question mark is the component, which has yet to be indicated. *(With kind permission from Sinha et al., 2011.)*

(MAPK) signaling. These are from the serine/threonine family, modifying plant response under stress. They are phosphorylated and activated by MAPK kinase in their cascade. MPAK kinase is also activated by MAPKK kinase. Different MAPK kinases have been indicated in plants including tobacco, alfalfa, *chorispora bungeana* and *Arabidopsis* (Fig. 18.3) (Zhang et al., 2006a, b, c; Sinha et al., 2011; Ahmad et al., 2011; Miransari et al., 2013).

Different activities in plants are regulated by MAPK signaling pathways, including: 1) modification of plant response to stress, 2) cellular cycles such as mitosis and cytokinesis, 3) development of stomata, 4) hormonal signaling, 5) plant immunity, and 6) plant tissue abscission (Wang et al., 2007b; Cho et al., 2008). Under stress the cellular surface receptors are activated, resulting in the production of intercellular signaling pathways, which are able to modify the cellular environment and hence handle the stress (Asai et al., 2002; Tor et al., 2009). Accordingly, the MAPK signaling pathway results in the transduction of signals to the nucleus and hence appropriate adjustment of cellular homeostasis (Sinha et al., 2011).

MAPK signaling is able to regulate the activity of MAP, affecting the dynamic of microtubules by phosphorylating the residues of serine and threonine at the time of microtubules binding. MAPK signaling is perceived by plant receptor protein kinases at the main plasma membrane (Beck et al., 2011). This kind of network can control the activities of transcription factors, hormones, enzymes, peptides, etc. in plants (Tena et al., 2011).

MAPK signaling is the plant response to different activities including the alleviation of stresses. Hence, genetic modification of MAPK signaling may be an effective method to increase plant resistance to stress. For example, research has indicated that under drought stress the activation of MAPK signaling results in the production of proline. The role of proline in plants under stress has been reported by many workers (Katare et al., 2012; Koyro et al., 2012; Rasool et al., 2013). The production of proline is controlled by different protein kinases under drought, salinity and cold stress (Shou et al., 2004a, b; Raghavendra et al., 2010; Krasensky and Jonak, 2012).

The role of MAPK signaling has been indicated in controlling the processes of mitosis and cytokinesis by affecting microtubule transition. Accordingly, Beck et al. (2011) used the mutant of *Arabidopsis taliana* to show such an effect. As a result, the processes of mitosis and cytokinesis were inhibited and bi- and multi-nucleate cells were formed, adversely affecting the vegetative cellular growth.

Pan et al. (2012) isolated *ZmMPK17*, a maize (*Zea mays* L.) MAPK gene, which is able to alleviate multiple stresses in plants. The gene transcription is responsive to plant hormones such as ethylene, jasmonic acid, abscisic acid, salicylic acid under suboptimal temperature, drought and salinity stress. The related transcription elements are regulated by Ca^{2+} and hydrogen peroxide under stress. When *ZmMPK17* was overexpressed, less reactive oxygen species were produced in plants under stress by affecting the process of antioxidant production. The transgenic plants were more tolerant to the stress, resulting in enhanced germination rate and production of proline and soluble sugars (Yang et al., 2013).

MAPK signaling pathways and reactive oxygen species are interactive; the production of reactive oxygen species in plants can result in the activation of MAPK signaling pathways and the pathways are able to control the production of such species. The exogenous use of hydrogen peroxide in *Arabidopsis thaliana* can activate the MAPK signaling pathways and the pathways are able to produce antioxidants (Pan et al., 2012).

Signaling by MAPK pathways may result in cellular activities including cell division, differentiation and cellular responses to stress (Mishra et al., 2006; Pan et al., 2012). Plants are able to form a network of MAPK signaling pathways, which eventually results in efficient cellular responses by the production of related stimuli (Mishra et al., 2006). Different molecules can act as the substrate for MAPK signaling pathways such as protein kinases, transcription factors and subsequent molecular cascade including MAPK2K and MAPK3K as well as related receptors (Whitmarsh and Davis, 1998; Cardinale et al., 2002).

There are 110 genes in *Arabidopsis thaliana* which are expressed during the production of the MAPK signaling pathways: 10 MAPK molecules, 20 MAP2K molecules and 80 MAP3K molecules (Pitzschke et al., 2009). The 10 MAPK molecules are able to phosphorylate 570 proteins, with a high number of transcription factors regulating plant growth and response to stress. Phosphorylation can influence the activity of transcription factors by affecting: the protein structure, their localization and activity and their interaction with other proteins (Fiil et al., 2009).

The *Arabidopsis* MKK1/MKK2-MPK4/MPK6 is able to regulate plant response to stresses such as salt and cold (Gao et al., 2008; Qiu et al., 2008). *Arabidopsis* MKK3 can affect plant immunity. MKK4/MKK5-MPK3/MPK6 are important signaling molecules in the regulation of plant development and response to stress (Asai et al., 2002; Wang et al. 2011). The regulation of cytokinesis and mitosis is by the activity of MKK6-MPK4/MPK11 (Beck et al., 2010, 2011). Plant systemic resistance is activated by MKK7 (Zhang et al., 2007). The MKK9-MPK3/MPK6 are important signaling molecules affecting ethylene signaling and leaf senescence (Xu et al., 2008).

18.3 STRESS AND REACTIVE OXYGEN SPECIES AND REDOX SIGNALING

Plants respond to stress by different morphological and physiological mechanisms. Among different physiological mechanisms is the production of different metabolites. For example, reactive oxygen species are produced under stress, as a result of metabolic by-products, which can adversely affect cellular structure and hence functioning. It is important for the plant cell to maintain cellular homeostasis for redox processes (Van Aken et al., 2009; Nick, 2013).

Under stress the production of reactive oxygen species can result in the degradation of cellular constituents such as carbohydrates, lipids, and proteins (Ahmad et al., 2010, 2011; Koyro et al., 2012). Oxidative stress can also have unfavorable effects on chlorophyll and carotenoid resistance, preventing the plant photosynthetic efficiency and respiratory processes (Yao et al., 2009). Plants respond by activating different signaling pathways and hence producing antioxidants molecules, which are able to degrade the products of cellular stress such as reactive oxygen species (Foyer and Shigeoka, 2011; Zhang et al., 2011). Enzymes such as superoxide dismutase, catalase and glutathione peroxidase are required for cellular activities and their enhanced levels under stress can catabolize the products of oxidative stress including the active oxygen species, resulting in the alleviation of stress (Sajedi et al., 2011; Ahmad et al., 2010, 2011; Koyro et al., 2012).

However, it has been indicated that low amounts of reactive oxygen species can act as signal molecules regulating plant response under stress by affecting Ca^{2+} signaling and hence ABA and stomata activity (Pei et al., 2000; Desikan et al., 2001; Zhang et al., 2001; Miller et al., 2008). ABA is able to affect plant response to stress by regulating the gene network of reactive oxygen species including catalase, ascorbate peroxidase and superoxide dismutase (Jiang and Zhang, 2002; Zhang et al., 2011; Fujita et al., 2013).

Different stresses including salinity, drought, cold, heat, high light, etc. may result in the production of reactive oxygen species, such as oxygen

(O_2), superoxide radical ($O_2^{\bullet -}$), hydroperoxyl radical, hydroxyl radical and hydrogen peroxide (H_2O_2), in plants and can regulate the activity of different transcription factors. However, at high amounts they adversely affect plant growth. Under stress, proxidases are produced, and hence the plant may respond by producing antioxidant enzymes as mentioned earlier (Mittler, 2002; Jones, 2006).

Airaki et al. (2012) investigated the effects of suboptimal temperature (8 °C) on the growth of pepper (*Capsicum annum* L.) and determined different reactive nitrogen and oxygen species, which were significantly affected by suboptimal temperature. They found that the production of products such as glutathione and ascorbate can help the plant to alleviate the stress by affecting cell redox potential.

18.4 STRESS AND HORMONAL SIGNALING

Plant hormones, including auxin, cytokinins, abscisic acid (ABA), gibberellins, ethylene, jasmonates, brassinosteroids and strigolactones, are able to regulate different functions in plant at cellular and molecular levels. There are different signaling pathways and interactions related to plant hormones, among which the role of hormonal signaling under stress can be of the greatest importance (Hirayama and Shinozaki, 2010; Miransari, 2012; Miransari et al., 2014). The response of plants under stress is regulated by plant hormones indicating that the presence of hormones can increase plant tolerance to stress. Production of hormones in plants may result in activation of different genes in plant and hence the regulation of different activities such as: 1) activation of different signaling pathways, 2) cell cycling, 3) plant water behavior, 4) plant response to stress, etc. (Wang et al., 2007a; Tuteja, 2007; Rahman, 2013).

The effects of auxin under stress can be through the induction of plant transcription factors related to genes such as *Aux/IAA*, *GH3*, and small auxin-up RNA (*SAUR*) genes. The auxin signaling pathways is mostly induced and regulated by transcription factors including auxin response factors (ARFs) and the Aux/IAA repressors (Han et al., 2009; Jain and Khurana, 2009).

The role of ABA under stress has also been indicated. Stresses such as salinity and drought result in the production of ABA. The activity of stomata under different conditions including stresses is regulated by ABA, which is its most important function in plants (Jia and Davies, 2007). Due to the various functions of ABA in plants it could be the most important signaling molecule among hormones. The expression of different genes by ABA and hence the subsequent plant response can result in the alleviation of stress in plants. For example, the expression of *nced* genes in plant is induced by ABA under stress (Wan and Li, 2006). The adverse effects on small RNA induce the production of ABA, indicating that there is a link between small RNA pathways and ABA signaling pathways in plant (Zhang et al., 2008).

The gene that produces cytokinin is *ipt* resulting in the production of isopentyl transferase and isopentenyladenosine-5′-monophosphate (McGraw, 1987). Among the important functions of cytokinin is the protection of photosynthesis under stress by interacting with the receptor proteins and activation of related signaling pathway. As a result the genes are expressed and miRNAs, electrons, carbon, photosynthesis related proteins, and the enzyme ribulose bisphosphate carboxylase/oxygenase are produced. By using the gene *ipt* it is possible to genetically modify plant response under drought stress as the process of leaf senescence is delayed (Rivero et al., 2007, 2009).

The gaseous plant hormone, ethylene, with the simplest structure as compared to other plant hormones, has some important functions in plants including the germination of seed, abscission and tissue senescence. Based on the related signaling pathways, ethylene is interactive with the ethylene receptors, which are two-component histidine protein kinases, located on the plasma membrane (Mount and Chang, 2002; Miransari and Smith, 2014).

The ethylene signaling pathway is among the best-known signaling pathways and has the important transcription factor ETHYLENE INSENSITIVE3. Under stress the production of the stress hormone ethylene increases, adversely affecting plant growth. Interestingly, it has been indicated that the use of plant growth promoting rhizobacteria (PGPR) may result in decreased production of ethylene by the production of the enzyme 1-aminocyclopropane-1-carboxylate (ACC) deaminase (Glick et al., 2007; Jalili et al., 2009).

The production of gibberellins in plants is catalyzed by the enzymes monooxygenases, dioxygenases and cyclases. The enhancing effect of gibberellins on plant growth is by degradation of DELLA proteins (Griffiths et al., 2006). DELLA proteins are able to modify plant response to stress by affecting the combined response of plant hormones to stress (Miransari, 2012).

Brassinosteroids are steroid products affecting different plant functions, including plant growth and development. So far about 70 brassinosteroids (Sasse, 2003; Yu et al., 2008) have been identified. During the production of brassinosteroids, molecular oxygen is required, indicating that this hormone can modify the effects of hypoxia on plant growth and development. The hormone is able to alleviate the unfavorable effects of different stresses in plants (Miransari, 2012).

The lipid hormones, jasmonates, are able to affect plant systemic resistance as well as plant growth and development (Schaller and Stintzi, 2009). Jasmonates are able to affect plant growth under stress by interacting with the other plant hormones, controlling the production of reactive oxygen species, calcium influx, and activating nitrogen protein kinase (Hu et al., 2009). The hormone has an important role in the process of nodulation in leguminous plants (Sun et al., 2006).

Among the most important effects of salicylic acid on plant growth is the regulation of plant systemic resistance, by the following mechanisms: 1) expression of different genes including the PAL and priming genes, 2) activation of phytoalexin related pathways, 3) deposition of callose and phenolic products, and 4) affecting the auxin signaling pathway (Chen et al., 2009).

Strigolactones are a new class of plant hormones affecting: 1) mycorrhizal fungi symbiosis with its plant host as hyphal branching factors, 2) shoot branching, and 3) germination of parasitic weed *Striga*. The important factor affecting the production of the hormone in plant is phosphorous starvation (Akiyama et al., 2005; Lopez-Raez et al., 2008; Miransari, 2011).

18.5 STRESS AND ROLE OF MIRNAS AND SIRNAS

Small RNAs, including the two major classes of microRNAs (miRNAs) and short-interfering RNAs (siRNAs), can regulate different plant functions including growth and development, phytohormone signaling, and flowering as well as plant adaptation to abiotic and biotic stresses (Chen, 2005; Yang et al., 2007; Shukla et al., 2008; Lelandais-Briere et al., 2011; Cuperus et al., 2011; Li et al., 2011). For the diverse performance of biological processes, they must be precisely regulated (Ji and Chen, 2012). The length of a mature miRNA ranges from 19 to 24 nucleotide and it can regulate the activity of post-transcriptional gene through paring with mRNA and the subsequent cleavage (Guo and Lu, 2010). Their structure can be modified and stabilized by the small RNA transferase and the methylate transferase (as its homologue) and siRNAs as well as the proteins bound to RNA (Chen et al., 2011; Ji and Chen, 2012).

Presently, miRNAs and their sequences have been indicated in 41 plant species (Griffiths-Jones, 2004; 2006). For the first time Subramanian et al. (2008) found 35 novel miRNA families in soybean [*Glycine max* (L.) Merrill] and researched their role in the symbiotic process between rhizobium and soybean. The presence of miRNAs has been reported by different workers, but very few have investigated their role under abiotic and biotic stresses (Wang et al., 2009; Chen et al., 2009).

Li et al. (2011) found new miRNAs, in *Populus euphratica*, which were responsive to drought stress. The sequencing of miRNA indicated the upregulation of 104 miRNAs and the downregulation of 27 miRNAs under drought stress. Such a finding can lead to the production of resistant plants under adverse environmental conditions. Under drought stress 22 miRNA were upregulated and 10 miRNA were downregulated. The substrates of miRNAs controlled different activities in plant including growth, protein functioning, nutrient conditions, etc. (Wang et al., 2011).

Using deep sequencing and data analysis Yu et al. (2011) reported five responsive miRNAs in *Brassica rapa* under heat stress. This indicates the important role of miRNAs in *Brassica rapa* under heat stress. The role of miRNAs under high levels of aluminum was investigated by Chen et al. (2012) in the model legume plant *Medicago truncatula*. The sequencing of miRNA showed that there were 23 responsive miRNAs under unfavorable levels of Al, most of which were downregulated by stress.

18.6 STRESS AND PLANT PROTEOMICS

It is important to indicate which genes and hence proteins are activated, expressed, and produced under stress so that the alleviation of stress and production of more tolerant plants become possible. Accordingly, different techniques have been developed and used including the use of "omics" such as "proteomics." Using proteomics, it may be possible to identify which proteins are produced under stress and hence how plants may respond under stress. The use of proteomic techniques under different conditions including stress has been investigated by many workers (Zhang et al., 2009; Zorb et al., 2009; Agrawal et al., 2012).

The use of the helix loop (OrbHLH2) improved the ability of *Arabidopsis thaliana* under salt and osmotic stress (Zhou et al., 2009). Swami et al. (2011) investigated the behavior of sorghum proteins in the leaf under salt stress (200 mM NaCl for 96 h). The expressed proteins under the stress were of signaling, transcriptional, metabolic and functioning significance. They recognized different proteins under the stress including RNA binding protein, putative inorganic pyrophosphatase, serine/threonine protein kinase, and indicated that under salt stress the plant has a special mechanism to alleviate the stress.

Using proteomics, Ford et al. (2011) investigated the expression of 159 wheat proteins under drought stress. With respect to the physiological properties and the level of tolerance in the tested cultivars, different numbers of proteins were changed during the stress. Yao et al. (2011) investigated the effects of phosphorous deficiency (concentration less than 5μM) on the protein collection of *Brassica napus* using proteomics. Proteins related to the transcription of genes, translation of proteins, metabolism of carbon, transfer of energy and plant growth were downregulated as a result of stress. However, root related proteins were upregulated.

The effects of cold and salt stress on the growth of potato (*Solanum tuberosum* L.) were investigated in a growth chamber experiment at both transcriptomic and proteomic levels. While a high number of genes were regulated by cold stress, salt stress resulted in significantly high number of proteins. Under both stresses the photosynthesis genes were downregulated, but cell rescue and transcriptional related genes were upregulated (Evers et al., 2012; Mansour, 2013).

Under stress plant response can be indicated by gene and protein expression; the production of reactive oxygen species and subsequent production of antioxidant enzymes is also a mechanism used by plants to alleviate the stress. Hence, relating plant behavior under stress at transcriptomic and proteomic level to the production of reactive oxygen species may be a good method to determine plant tolerance.

18.7 CONCLUSIONS

Plants are able to survive under stress as they modify their morphological and physiological mechanisms. Many signaling pathways such as mitogen-activated protein kinase signaling, reactive oxygen species and redox signaling, and hormonal signaling, as well as the small RNAs, are activated during stress. Use of appropriate techniques such as proteomics can also be important for the evaluation of plant response under stress. Indicating the activated signaling pathways and the related genes and proteins can be useful for the production of tolerant plants under stress.

The role of signaling under stress is of great importance and must be elucidated so that the production of tolerant plants may be likely at large scale. This chapter discussed some of the most important details regarding plant signaling pathways under stress. Accordingly, the related cellular components, genes and proteins, which are activated under stress have also been explained. However, for future prospects, biologists have to make the results of their research work more applicable: 1) The precision of new and suggested methods must be tested regularly so that the related signaling pathways are exactly elucidated; 2) cellular behavior, expressed genes and proteins must be investigated and evaluated precisely to make the use of alleviating strategies more applicable; 3) the cross talk and interactions between different signaling pathways can importantly indicate plant response under stress as well as the subsequent use of effective and required techniques; 4) the use of more sophisticated and precise instruments can be a useful tool to make the modification of plants under stress more possible; and 5) the literature being presented by researchers is also very effective, speeding up the rate of progress of scientific knowledge.

REFERENCES

Agrawal, G., Pedreschi, R., Barkla, B., Bindschedler, L.V., Cramer, R., Sarkar, A., et al., 2012. Translational plant proteomics: a perspective. J. Proteomics. 75, 4588–4601.

Ahmad, P., Jaleel, C.A., Salem, M.A., Nabi, G., Sharma, S., 2010. Roles of enzymatic and non-enzymatic antioxidants in plants during abiotic stress. Crit. Rev. Biotechnol. 30, 161–175.

Ahmad, P., Nabi, G., Jeleel, C.A., Umar, S., 2011. Free radical production, oxidative damage and antioxidant defense mechanisms in plants under abiotic stress. In: Ahmad, P., Umar, S. (Eds.), Oxidative Stress: Role of Antioxidants in Plants. Ed. Studium Press Pvt. Ltd., New Delhi, India, pp. 19–53.

Airaki, M., Leterrier, M., Mateos, R.M., Valderrama, R., Chaki, M., Barroso, J., et al., 2012. Metabolism of reactive oxygen species and reactive nitrogen species in pepper (*Capsicum annuum* L.) plants under low temperature stress. Plant Cell Environ. 35, 281–295.

Akiyama, K., Matsuzaki, K.I., Hayashi, H., 2005. Plant sesquiterpenes induce hyphal branching in arbuscular mycorrhizal fungi. Nature 435, 824–827.

Asai, T., Tena, G., Plotnikova, J., Willmann, M.R., Chiu, W.L., Gomez-Gomez, L., et al., 2002. MAP kinase signalling cascade in *Arabidopsis* innate immunity. Nature 415, 977–983.

Beck, M., Komis, G., Muller, J., Menzel, D., Samaj, J., 2010. *Arabidopsis* homologs of nucleus- and phragmoplast-localized kinase 2 and 3 and mitogen-activated protein kinase 4 are essential for microtubule organization. Plant Cell 22, 755–771.

Beck, M., Komis, G., Ziemann, A., Menzel, D., Samaj, J., 2011. Mitogen-activated protein kinase 4 is involved in the regulation of mitotic and cytokinetic microtubule transitions in *Arabidopsis thaliana*. New Phytol. 189, 1069–1083.

Cardinale, F., Meskiene, I., Ouaked, F., Hirt, H., 2002. Convergence and divergence of stress-induced mitogen-activated protein kinase signaling pathways at the level of two distinct mitogen-activated protein kinase kinases. Plant Cell 14, 703–711.

Chen, D., Meng, Y., Yuan, C., Bai, L., Huang, D., Lv, S., et al., 2011. Plant siRNAs from introns mediate DNA methylation of host genes. RNA 17, 1012–1024.

Chen, L., Wang, T., Zhao, M., Tian, Q., Zhang, W.H., 2012. Identification of aluminum-responsive microRNAs in *Medicago truncatula* by genome-wide high-throughput sequencing. Planta 235, 375–386.

Chen, R., Hu, Z., Zhang, H., 2009. Identification of microRNAs in wild soybean (*Glycine soja*). J. Integr. Plant. Biol. 51, 1071–1079.

Chen, X., 2005. MicroRNA biogenesis and function in plants. FEBS Lett. 579, 5923–5931.

Cho, S.K., Larue, C.T., Chevalier, D., Wang, H., Jinn, T.L., Zhang, S., et al., 2008. Regulation of floral organ abscission in *Arabidopsis thaliana*. Proc. Natl. Acad. Sci. U.S.A 105, 15629–15634.

Cuperus, J.T., Fahlgren, N., Carrington, J., 2011. Evolution and functional diversification of *MIRNA* genes. Plant Cell 23, 431–442.

Desikan, R., Mackerness, S.A.H., Hancock, J.T., Neill, S.J., 2001. Regulation of the *Arabidopsis* transcriptome by oxidative stress. Plant Physiol. 127, 159–172.

Evers, D., Legay, S., Lamoureux, D., Hausman, J.F., Hoffmann, L., Renaut, J., 2012. Towards a synthetic view of potato cold and salt stress response by transcriptomic and proteomic analyses. Plant Mol. Biol. 78, 503–514.

Fiil, B.K., Petersen, K., Petersen, M., Mundy, J., 2009. Gene regulation by MAP kinase cascades. Curr. Opin. Plant Biol. 12, 615–621.

Ford, K., Cassin, A., Bacic, A., 2011. Quantitative proteomic analysis of wheat cultivars with differing drought stress tolerance. Fron. Plant Sci. 2, 1–11.

Foyer, C.H., Shigeoka, S., 2011. Understanding oxidative stress and antioxidant functions to enhance photosynthesis. Plant Physiol. 155, 93–100.

Fujita, Y., Yoshida, T., Yamaguchi-Shinozaki, K., 2013. Pivotal role of the AREB/ABF-SnRK2 pathway in ABRE-mediated transcription in response to osmotic stress in plants. Plant Physiol. 147, 15–27.

Gao, M., Liu, J., Bi, D., Zhang, Z., Cheng, F., Chen, S., et al., 2008. MEKK1, MKK1/MKK2 and MPK4 function together in a mitogen-activated protein kinase cascade to regulate innate immunity in plants. Cell Res. 18, 1190–1198.

Glick, B.R., Todorovic, B., Czarny, J., Cheng, Z., Duan, J., 2007. Promotion of plant growth by bacterial ACC deaminase. Crit. Rev. Plant Sci. 26, 227–242.

Grativol, C., Hemerly, A., Cavalcanti, P., Ferreira, G., 2012. Genetic and epigenetic regulation of stress responses in natural plant populations. Biochim. Byophys. Acta. 1819, 176–185.

Griffiths, J., Murase, K., Rieu, I., Zentella, R., Zhang, Z.L., Powers, S.J., et al., 2006. Genetic characterization and functional analysis of the GID1 gibberellin receptors in *Arabidopsis*. Plant Cell 18, 3399–3414.

Griffiths-Jones, S., 2004. The microRNA registry. *Nucleic Acids Res.* 32 Database, D109–D111.

Griffiths-Jones, S., 2006. miRBase: the microRNA sequence database. Methods Mol. Biol. 342, 129–138.

Guo, L., Lu, Z., 2010. Global expression analysis of miRNA gene cluster and family based on isomiRs from deep sequencing data. Comput. Biol. Chem. 34, 165–171.

Han, H., Zhang, S., Sun, X., 2009. A review on the molecular mechanism of plants rooting modulated by auxin. Afr. J. Biotechnol. 8, 348–353.

Hirayama, T., Shinozaki, K., 2010. Research on plant abiotic stress responses in the post-genome era: past, present and future. Plant J. 61, 1041–1052.

Hu, X., Li, W., Chen, Q., Yang, Y., 2009. Early signal transduction linking the synthesis of jasmonic acid in plant. Plant Signal. Behav. 4, 696–697.

Jain, M., Khurana, J., 2009. Transcript profiling reveals diverse roles of auxin-responsive genes during reproductive development and abiotic stress in rice. FEBS J. 276, 3148–3162.

Jalili, F., Khavazi, K., Pazira, E., Nejati, A., Rahmani, H.A., Sadaghiani, H.R., et al., 2009. Isolation and characterization of ACC deaminase producing fluorescent pseudomonads, to alleviate salinity stress on canola (*Brassica napus* L.) growth. J. Plant Physiol. 166, 667–674.

Ji, L., Chen, X., 2012. Regulation of small RNA stability: methylation and beyond. Cell Res. 22, 624–636.

Jia, W., Davies, W.J., 2007. Modification of leaf apoplastic pH in relation to stomatal sensitivity to root-sourced abscisic acid signals. Plant Physiol. 143, 68–77.

Jiang, M.Y., Zhang, J.H., 2002. Water stress-induced abscisic acid accumulation triggers the increased generation of reactive oxygen species and up-regulates the activities of antioxidant enzymes in maize leaves. J. Exp. Bot. 53, 2401–2410.

Jones, D.P., 2006. Redefining oxidative stress. Antioxid Redox Signal 8, 1865–1879.

Katare, D.P., Nabi, G., Azooz, M.M., Aeri, V., Ahmad, P., 2012. Biochemical modifications and enhancement of psoralen content in salt-stressed seedlings of *Psoralea corylifolia* Linn. J. Funct. Environ. Bot. 2, 65–74.

Koyro, H.W., Ahmad, P., Geissler, N., 2012. Abiotic stress responses in plants: an overview. In: Ahmad, P., Prasad, M.N.V. (Eds.), Environmental Adaptations and Stress Tolerance of Plants in the Era of Climate Change. Springer, Science + business media, NY, USA, pp. 1–28.

Krasensky, J., Jonak, C., 2012. Drought, salt, and temperature stress-induced metabolic rearrangements and regulatory networks. J. Exp. Bot. 63, 1593–1608.

Lelandais-Briere, C., Sorin, C., Declerck, M., Benslimane, A., Crespi, M., Hartmann, C., 2011. Small RNA diversity in plants and its impact in development. Curr. Genomics. 11, 14–23.

Li, B., Qin, Y., Duan, H., Yin, W., Xia, X., 2011. Genome-wide characterization of new and drought stress responsive microRNAs in *Populus euphratica*. J. Exp. Bot. 62, 3765–3779.

Lopez-Raez, J.A., Charnikhova, T., Gomez-Roldon, V., Matusova, R., Kohlen, W., De Vos, R., et al., 2008. Tomato strigolactones are derived from carotenoids and their biosynthesis is promoted by phosphate starvation. New Phytol. 178, 863–874.

Mansour, M., 2013. Plasma membrane permeability as an indicator of salt tolerance in plants. Biol. Plant. 57, 1–10.

McGraw, B.A., 1987. Cytokinin biosynthesis and metabolism. In: Davies, P.J. (Ed.), Plant Hormones and their Role in Plant Growth and Development. Martinus Nijhoff, Dordrecht, The Netherlands, pp. 76–93.

Miller, G., Shulaev, V., Mittler, R., 2008. Reactive oxygen signaling and abiotic stress. Physiol. Plant. 133, 481–489.

Miransari, M., 2011. Interactions between arbuscular mycorrhizal fungi and soil bacteria. *Appl.* Microbiol. Biotechnol. 89, 917–930.

Miransari, M., 2012. Role of phytohormone signaling during stress. In: Ahmad, P., Prasad, M.N. V. (Eds.), Environmental Adaptations and Stress Tolerance of Plants in the Era of Climate Change. Springer, New York.

Miransari, M., Rangbar, B., Khajeh, K., et al., 2013. Salt stress and MAPK signaling in plants. In: Ahmad, P., et al., (Eds.), Salt Stress in Plants: Signalling, Omics and Adaptations. Springer Science + Business Media, New York, pp. 157–173.

Miransari, M., Smith, D.L., 2014. Plant hormones and seed germination. Environ. Exp. Bot. in press.

Miransari, M., Abrishamchi, A., Khoshbakht, K., Niknam, V., 2014. Plant hormones as signals in arbuscular mycorrhizal symbiosis. Critic. Rev. Biotechnol.

Mishra, N.S., Tuteja, R., Tuteja, N., 2006. Signaling through MAP kinase networks in plants. Arch. Biochem. Biophys. 452, 55–68.

Mittler, R., 2002. Oxidative stress, antioxidants and stress tolerance. Trends Plant Sci. 7, 405–410.

Mount, S.M., Chang, C., 2002. Evidence for a plastid origin of plant ethylene receptor genes. Plant Physiol. 130, 10–14.

Munné-Bosch, S., Queval, G., Foyer, C., 2013. The impact of global change factors on redox signaling. Plant Physiol. 161, 5–19.

Nick, P., 2013. Microtubules, signalling and abiotic stress. Plant J. 75, 309–323.

Pan, J., Zhang, M., Kong, X., Xing, X., Liu, Y., Zhou, Y., et al., 2012. ZmMPK17, a novel maize group D MAP kinase gene, is involved in multiple stress responses. Planta. 235, 661–676.

Pei, Z.-M., Murata, Y., Benning, G., Thomine, S., Klusener, B., Allen, G.J., et al., 2000. Calcium channels activated by hydrogen peroxide mediate abscisic acid signalling in guard cells. Nature. 406, 731–734.

Pitzschke, A., Schikora, A., Hirt, H., 2009. MAPK cascade signalling networks in plant defence. Curr. Opin. Plant Biol. 12, 421–426.

Qiu, J.-L., Fiil, B.K., Petersen, K., Nielsen, H.B., Botanga, C.J., Thorgrimsen, S., et al., 2008. *Arabidopsis* MAP kinase 4 regulates gene expression through transcription factor release in the nucleus. EMBO J. 27, 2214–2221.

Raghavendra, A.S., Gonugunta, V.K., Christmann, A., Grill, E., 2010. ABA perception and signalling. Trends Plant Sci. 15, 395–401.

Rahman, R., 2013. Auxin: a regulator of cold stress response. Physiol. Plant. 147, 28–35.

Rasool, S., Ahmad, A., Siddiqi, T.O., Ahmad, P., 2013. Changes in growth, lipid peroxidation and some key antioxidant enzymes in chickpea genotypes under salt stress. Acta Physiol. Plant. 35, 1039–1050.

Rivero, R.M., Kojima, M., Gepstein, A., Sakakibara, H., Mittler, R., Gepstein, S., et al., 2007. Delayed leaf senescence induces extreme drought tolerance in a flowering plant. Proc. Natl. Acad. Sci. U.S.A. 104, 19631–19636.

Rivero, R.M., Shulaev, V., Blumwald, E., 2009. Cytokinindependent photorespiration and the protection of photosynthesis during water deficit. Plant Physiol. 150, 1530–1540.

Sajedi, N., Ardakani, M., Madani, H., Naderi, A., Miransari, M., 2011. The effects of selenium and other micronutrients on the antioxidant activities and yield of corn (*Zea mays* L.) under drought stress. Physiol. Mol. Biol. Plants 17, 215–222.

Sasse, J.M., 2003. Physiological actions of brassinosteroids: an update. J. Plant Growth Regul. 22, 276–288.

Schaller, A., Stintzi, A., 2009. Enzymes in jasmonate biosynthesis – Structure, function, regulation. Phytochem. 70, 1532–1538.

Shou, H., Bordallo, P., Fan, J.B., Yeakley, J.M., Bibikova, M., Sheen, J., et al., 2004a. Expression of an active tobacco mitogen-activated protein kinase kinase kinase enhances freezing tolerance in transgenic maize. Proc. Nat. Acad. Sci. U.S.A. 101, 3298–3303.

Shou, H., Bordallo, P., Wang, K., 2004b. Expression of the *Nicotiana* protein kinase (NPK1) enhanced drought tolerance in transgenic maize. J. Exp. Bot. 55, 1013–1019.

Shukla, L.I., Chinnusamy, V., Sunkar, R., 2008. The role of microRNAs and other endogenous small RNAs in plant stress responses. Biochim. Biophys. Acta. 1779, 743–748.

Sinha, A.K., Jaggi, M., Raghuram, B., Tuteja, N., 2011. Mitogen-activated protein kinase signaling in plants under abiotic stress. Plant Signal. Behav. 6, 196–203.

Subramanian, S., Fu, Y., Sunkar, R., Barbazuk, W.B., Zhu, J.K., Yu, O., 2008. Novel and nodulation-regulated microRNAs in soybean roots. BMC Genomics. 9, 160.

Sun, J., Cardoza, V., Mitchell, D.M., Bright, L., Oldroyd, G., Harris, J.M., 2006. Cross talk between jasmonic acid, ethylene and Nod factor signaling allows integration of diverse inputs for regulation of nodulation. Plant J. 46, 961–970.

Swami, A.K., Alam, S.I., Sengupt, N., Sarin, R., 2011. Differential proteomic analysis of salt stress response in Sorghum bicolor leaves. Environ. Exp. Bot. 71, 321–328.

Tena, G., Boudsocq, M., Sheen, J., 2011. Protein kinase signaling networks in plant innate immunity. Curr. Opin. Plant Biol. 14, 519–529.

Tor, M., Lotze, M.T., Holton, N., 2009. Receptor-mediated signalling in plants: molecular patterns and programmes. J. Exp. Bot. 60, 3645–3654.

Tuteja, N., 2007. Abscisic acid and abiotic stress signalling. Plant Signal Behav. 2, 135–138.

Van Aken, O., Zhang, B., Carrie, C., Uggalla, V., Paynter, E., Giraud, E., et al., 2009. Defining the mitochondrial stress response in *Arabidopsis thaliana*. Molecul. Plant 2, 1310–1324.

Wan, X., Li, L., 2006. Regulation of ABA level and water stress tolerance of *Arabidopsis* by ectopic expression of a peanut 9-cisepoxycarotenoid dioxygenase gene. Biochem. Biophys. Res. Commun. 347, 1030–1038.

Wang, D., Pajerowska-Mukhtar, K., Culler, A.H., Dong, X., 2007a. Salicylic acid inhibits pathogen growth in plants through repression of the auxin signaling pathway. Curr. Biol. 17, 1784–1790.

Wang, H., Ngwenyama, N., Liu, Y., Walker, J.C., Zhang, S., 2007b. Stomatal development and patterning are regulated by environmentally responsive mitogen-activated protein kinases in *Arabidopsis*. Plant Cell 19, 63–73.

Wang, T., Chen, L., Zhao, M., Tian, Q., Zhang, W.-H., 2011. Identification of drought-responsive microRNAs in *Medicago truncatula* by genome-wide high throughput sequencing. Genomics. 12, 367.

Wang, Y., Li, P., Cao, X., Wang, X., Zhang, A., Li, X., 2009. Identification and expression analysis of miRNAs from nitrogen-fixing soybean nodules. Biochem. Biophys. Res. Commun. 378, 799–803.

Whitmarsh, A.J., Davis, R.J., 1998. Structural organization of MAP-kinase signaling modules by scaffold proteins in yeast and mammals. Trends Biochem. Sci. 23, 481–485.

Xu, J., Li, Y., Wang, Y., Liu, H., Lei, L., Yang, H., et al., 2008. Activation of MAPK kinase 9 induces ethylene and camalexin biosynthesis and enhances sensitivity to salt stress in *Arabidopsis*. J. Biol. Chem. 283, 26996–27006.

Yang, T., Xue, L., An, L., 2007. Functional diversity of miRNA in plants. Plant Sci. 172, 423–432.

Yang, Y., Yang, F., Li, X., Shi, R., Lu, J., 2013. Signal regulation of proline metabolism in callus of the halophyte *Nitraria tangutorum* Bobr. grown under salinity stress. Plant Cell Tissue organ Culture 112, 33–42.

Yao, X., Chu, J., Wang, G., 2009. Effects of drought stress and selenium supply on growth and physiological characteristics of wheat seedlings. Acta Physiol. Plant. 31, 1031–1036.

Yao, Y., Sun, H., Xu, F., Zhang, X., Liu, S., 2011. Comparative proteome analysis of metabolic changes by low phosphorus stress in two *Brassica napus* genotypes. Planta 233, 523–537.

Yu, X., Li, L., Li, L., Guo, M., Chory, J., Yin, Y., 2008. Modulation of brassinosteroid-regulated gene expression by jumonji domain-containing proteins ELF6 and REF6 in *Arabidopsis*. Proc. Natl. Acad. Sci. USA 105, 7618–7623.

Yu, X., Wang, H., Lu, Y., de Ruiter, M., Cariaso, M., Prins, M., et al., 2011. Identification of conserved and novel microRNAs that are responsive to heat stress in *Brassica rapa*. J. Exp. Bot. 63, 1025–1038.

Zhang, J.F., Yuan, L.J., Shao, Y., Du, W., Yan, D.W., Lu, Y.T., 2008. The disturbance of small RNA pathways enhanced abscisic acid response and multiple stress responses in Arabidopsis. Plant Cell Environ. 31, 562–574.

Zhang, L., Tian, L.H., Zhao, J., Song, Y., Zhang, C., Guo, Y., 2009. Identification of an apoplastic protein involved in the initial phase of salt stress response in rice root by two-dimensional electrophoresis. J. Exp. Bot. 57, 1537–1546.

Zhang, S., Seguin, A., Ellis, B.E., 2006a. Ancient signals: comparative genomics of plant MAPK and MAPKK gene families. Trends Plant Sci. 11, 192–198.

Zhang, T., Liu, Y., Yang, T., Zhang, L., Xu, S., Xue, L., et al., 2006b. Diverse signals converge at MAPK cascades in plant. Plant Physiol. Biochem. 44, 274–283.

Zhang, T., Liu, Y., Xue, L., Xu, S., Chen, T., Yang, T., et al., 2006c. Molecular cloning and characterization of a novel MAP kinase gene in *Chorispora bungeana*. Plant Physiol. Biochem. 44, 78–84.

Zhang, X., Zhang, L., Dong, F., Gao, J., Galbraith, D.W., Song, C.-P., 2001. Hydrogen peroxide is involved in abscisic acid-induced stomatal closure in *Vicia faba*. Plant Physiol. 126, 1438–1448.

Zhang, X., Dai, Y., Xiong, Y., DeFraia, C., Li, J., Dong, X., et al., 2007. Overexpression of *Arabidopsis* MAP kinase kinase 7 leads to activation of plant basal and systemic acquired resistance. Plant J. 52, 1066–1079.

Zhang, X., Wang, L., Meng, H., Wen, H., Fan, Y., Zhao, J., 2011. Maize ABP9 enhances tolerance to multiple stresses in transgenic *Arabidopsis* by modulating ABA signaling and cellular levels of reactive oxygen species. Plant Mol. Biol. 75, 365–378.

Zhou, J., Li, F., Wang, J., Ma, Y., Chong, K., Xu, Y., 2009. Basic helix-loop-helix transcription factor from wild rice (OrbHLH2) improves tolerance to salt- and osmotic stress in *Arabidopsis*. J. Plant Physiol. 166, 1296–1306.

Zorb, C., Herbst, R., Forreiter, C., Schubert, S., 2009. Short-term effects of salt exposure on the maize chloroplast protein pattern. Proteomics 9, 4209–4220.

Chapter 19

Hydrogen Peroxide (H_2O_2) Generation, Scavenging and Signaling in Plants

Ghader Habibi

19.1 INTRODUCTION

Stress factors trigger common reactions in plants and lead to cellular damages mediated by ROS (Mano, 2002). The ROS includes both free radical (superoxide radical ($O_2^{\bullet-}$), hydroxyl radical ($^{\bullet}OH$), hydroperoxyl radical ($HO_2^{\bullet}$), alkoxy radical ($RO^{\bullet}$) and peroxy radical ($ROO^{\bullet}$)) and nonradical (molecular) forms (singlet oxygen ($^1O_2$) and hydrogen peroxide (H_2O_2)), all of which are cytotoxic to plants (Dismukes et al., 2001). Among the different ROS, superoxide radicals are converted into H_2O_2 and O_2 by superoxide dismutase (SOD).

A class of cell membrane NADPH-dependent oxidases including respiratory burst oxidase homologues (RBOH) (Agrawal et al., 2003) and cell wall-associated peroxidases are the common source of H_2O_2 in plant cells. Photosystem I and II (in chloroplast), complex I, ubiquinone and complex III of mitochondrial electron transport chain (mETC) are the major sites for the generation of $O_2^{\bullet-}$, which are subsequently converted to H_2O_2. Oxidative stress refers to a serious imbalance between the production and removal of H_2O_2. Accumulation of H_2O_2 can damage critical organelles via lipid peroxidation and is capable of inducing damage to almost all cellular macromolecules including DNA, proteins and carbohydrates (Creissen and Mullineaux, 2002). It also inhibits the photochemical activities and decreases the activities of enzymes in the Calvin cycle (Monakhova and Chernyad'ev, 2002).

Both enzymatic (superoxide dismutase, catalase, ascorbate peroxidase, glutathione reductase, glutathione peroxidase, monodehydroascorbate reductase, dehydroascorbate reductase and glutathione-S-transferase) and nonenzymatic (ascorbic acid, glutathione, phenolic compounds, nonprotein amino acids and tocopherols) antioxidant defense systems protect plants against

P. Ahmad (Ed): Oxidative Damage to Plants. DOI: http://dx.doi.org/10.1016/B978-0-12-799963-0.00019-8

oxidative stress damages, acting in synergy to carry out H_2O_2 scavenging (Ahmad et al., 2011; Ahmad and Umar, 2011).

The balance between ROS production and scavenging has been regulated temporally and spatially and may determine whether ROS will act as damaging, protective or signaling factors (Gratao et al., 2005). H_2O_2 transfers across membranes passively or through water channels (Miller et al., 2010). H_2O_2 can also mediate the formation of membrane soluble lipid peroxides. Thus, H_2O_2 is a versatile signaling molecule with regard to its property and mobility within cells. In addition, H_2O_2 can interact with several different signaling pathways such as calcium and protein phosphorylation networks. It has been well documented that H_2O_2 is involved in the expression of a number of genes (Dalton et al., 1999). H_2O_2 plays an important role in plants, which may control many different physiological responses such as biotic and abiotic stress-response and systemic signaling. In this review, we provide an update of the new findings that support the biochemistry of H_2O_2 production, H_2O_2 in the scavenging antioxidant defense system, and H_2O_2 in sensing and signaling.

19.2 THE GENERATION OF H_2O_2

ROS are generated during normal metabolism in plants (Mittler, 2002; Scandalios, 2005; Halliwell, 2006), but under stress conditions, the balance between the production and scavenging of ROS is interrupted in cellular components of plants (Dismukes et al., 2001; Apel and Hirt, 2004; Karuppanapandian and Manoharan, 2008; Vellosillo et al., 2010; Karuppanapandian et al., 2011; Ahmad et al., 2011; Ahmad and Umar, 2011).

In plants, various environmental stresses such as high light intensity, UV radiation, temperature extremes, drought, salinity and heavy metals induce overaccumulation of H_2O_2 (Karuppanapandian and Manoharan, 2008; Krasensky and Jonak, 2012). H_2O_2 is a general signaling molecule, since H_2O_2 is diffusible through membrane. H_2O_2 plays a dual role in plants: at low concentrations, it acts as a signal molecule (Laloi et al., 2004; Mittler et al., 2004), but at high concentrations, it causes lipid peroxidation, membrane destruction, protein denaturation and DNA damage (Dat et al., 2000; Creissen and Mullineaux, 2002).

19.2.1 Sources of H_2O_2 in Plant Cells

During environmental stress, H_2O_2 production in organelles such as chloroplasts and mitochondria is elevated (Bhattacharjee, 2012). Photooxidation reactions, NADPH oxidase as well as xanthine oxidase (XOD) also produce H_2O_2 in plants. It is also generated during lignification and suberization. H_2O_2 production in each of these compartments is described below.

19.2.1.1 Chloroplasts

The reduction of NADP to NADPH by electron flow from the stimulated PSI occurs in the chloroplasts during normal conditions. $O_2^{\bullet-}$ is produced from O_2 generated in the chloroplasts during photosynthesis. Triplet chlorophyll, PSI and PSII are the major sites of ROS ($O_2^{\bullet-}$, $^1O_2$ and H_2O_2) production in chloroplasts (Fig. 19.1). The photosynthetic fixation of CO_2 can regulate the generation of ROS (Dinakara and Djilianov, 2012). Under strong light, the level of $NADP^+$ is reduced (Miller et al., 2010). The reduction of $NADP^+$ causes the transport of electrons from PSI to molecular oxygen, resulting in the production of H_2O_2 via $O_2^{\bullet-}$.

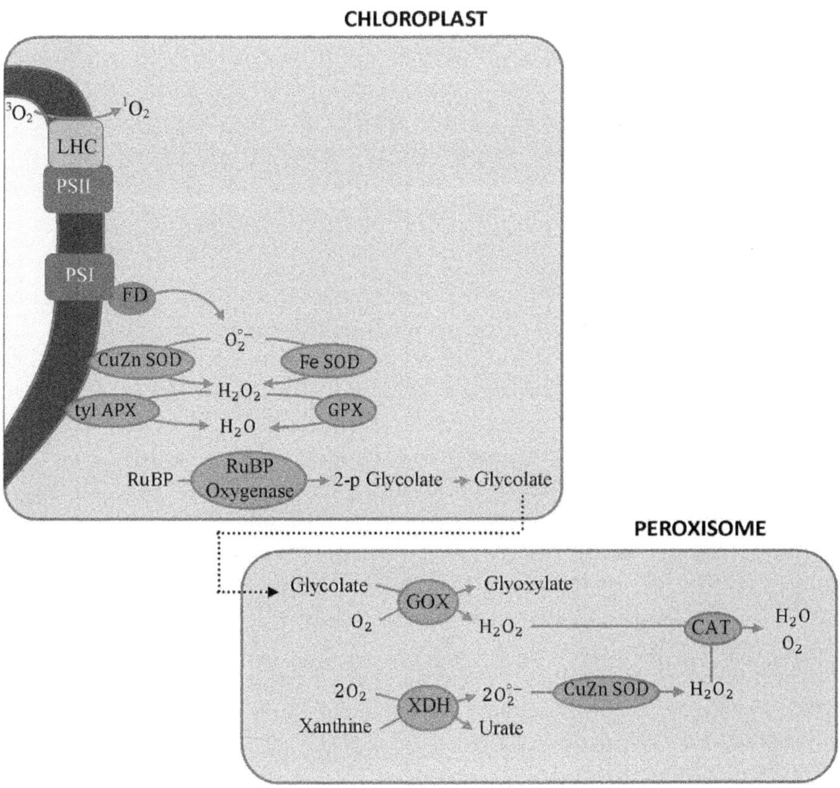

FIGURE 19.1 Localization of hydrogen peroxide (H_2O_2) generation and scavenging pathways in chloroplast and peroxisome.

Abbreviations: CAT: catalase, CuZn SOD: Copper-Zinc superoxide dismutase, GOX: glycolate oxidase, GPX: glutathione peroxidase, FD: ferridoxin, Fe SOD: iron superoxide dismutase, LHC: light harvesting complex, O_2^{-}: superoxide radical, $^1O_2$: singlet oxygen, $^3O_2$: triplet oxygen, PSII: photosystem II, PSI: photosystem I, 2-p glycolate: 2-phosphoglycolate, RuBP: ribulose-1,5-bisphosphate, tylAPX: thylakoid ascorbate peroxidase, XDH: xanthine dehydrogenase.

Under drought conditions, stomatal closure blocks the entry of CO_2 into leaves resulting in limited CO_2 fixation in daytime under high light conditions (Sanda et al., 2011). These limitations of CO_2 fixation suppress the carboxylation reaction of Rubisco and accelerate the photorespiratory pathway (Miller et al., 2010). During photorespiration, glycolate is produced in chloroplasts. The oxidation of glycolate by glycolate-oxidase occurs in the peroxisomes and leads to H_2O_2 production during photorespiration. Later studies have revealed that the acceptor side of ETC in PSII (QA and QB) transfers electron to O_2 producing $O_2^{\bullet -}$ (Yabuta et al., 2007). It is also found that excess photochemical energy that leads to photoinhibition of PSII causes significant enhancement in the generation of singlet oxygen (Hideg et al., 2002). $O_2^{\bullet -}$ is rapidly converted to H_2O_2 with the aid of superoxide dismutases (SODs) (Asada, 2000; Fryer et al., 2002). A membrane-attached copper/zinc superoxide (Cu/Zn SOD) in the proximity of PSI converts the superoxide radicals into H_2O_2. Afterwards a membrane-bound thylakoid ascorbate peroxidase (tyl APX) converts the H_2O_2 to water (Fig. 19.1). Treatment with exogenous H_2O_2 was shown to promote the oxidation of quinone A (QA), which decreased the generation of $^1O_2$ during the stress (Miller et al., 2010).

19.2.1.2 Mitochondria

Complex I and complex III in the mitochondrial electron transport chain (mtETC) are major sites of ROS production (Rhoads et al., 2006). Direct reduction of oxygen to $O_2^{\bullet -}$ occurs in the flavoprotein domain of NADH dehydrogenase segment (complex I) of mtETC (Arora et al., 2002). The reverse electron flow (from complex II to complex I) (Turrens, 2003) and ubiquinone-cytochrome region (complex III) of the ETC (Murphy, 2009) are other sites for ROS production in mitochondria. In plants, various stress factors lead to inhibition and modification of ETC and ATP syntheses components, leading to formation of H_2O_2 because of overreduction of electron carriers. Since $O_2^{\bullet -}$ is the primary ROS formed by monovalent reduction in the ETC, it is converted into H_2O_2 by the MnSOD (mitochondrial form of SOD) or APX. It has been reported that *Arabidopsis* plants with loss of function in mitochondrial alternative oxidase (AOX) exhibit altered expression of transcripts involved in antioxidant mechanisms in the mitochondria (Giraud et al., 2008).

19.2.1.3 Peroxisome

Peroxisomal glycolate oxidase (GO) produces glycolate in plant peroxisomes and uses O_2 as an electron acceptor to produce H_2O_2 (Fig. 19.1). In plant cells, the peroxisome is involved in β-oxidation. Acyl-CoA oxidase utilizes FAD as a cofactor to produce FADH, which then donates a hydride ion to molecular oxygen to generate H_2O_2. It was reported that xanthine oxidase uses O_2 as the electron acceptor and generates H_2O_2. In addition, urate oxidase uses O_2 as the electron acceptor and produces H_2O_2 (Mhamdi et al., 2012).

19.2.1.4 Endoplasmic Reticulum

In endoplasmic reticulum, NAD(P)H-dependent electron transport involving $CytP_{450}$ produces $O_2^{\bullet -}$ (Mittler, 2002). Cinnamate-4-hydroxylase is the best-characterized $CytP_{450}$ in plants, which operates in flavonoid and lignin biosynthesis, but also participates in gibberellin and sterol biosynthesis (Shi et al., 2010). Organic substrate, RH, reacts first with Cyt P_{450} and then is reduced by a flavoprotein to form a radical intermediate (Cyt P_{450} R^-). This radical intermediate can react with triplet oxygen ($^3O_2$) and produces oxygenated complex ($CytP_{450}$ ROO^-), which may be reduced by cytochrome b or sometimes the complexes may decompose releasing $O_2^{\bullet -}$. It has been suggested that the misfolded proteins accumulating in the ER may generate H_2O_2 via several possible mechanisms (Ceriotti, 2011).

19.2.1.5 Apoplast

Environmental stresses such as drought and salinity lead to production of H_2O_2 (Hernández et al., 2001; Zhu, 2001), which is mediated by enzymes, which includes NADPH oxidases (RBOHs), cell wall-associated oxidases, and pH-dependent cell wall POXs. Germin-like oxalate oxidases and polyamine oxidases have been proposed to generate H_2O_2 in the apoplast (Mittler, 2002). NADH oxidases have been suggested as playing an important role in the formation of H_2O_2 needed for lignifications (Hernández et al., 2001; Zhu, 2001; Rodríguez et al., 2004). Under several stress conditions, RBOHs regulate H_2O_2 signaling in response to heat, drought, cold, high light, salinity and wounding and also in biotic stresses and developmental processes (Suzuki et al., 2011).

19.2.1.6 Plasma Membranes and Cell Walls

The plasma membrane NADPH-dependent oxidase oxidizes NADPH at the cytosolic surface of the plasma membrane, and reduces O_2 to $O_2^{\bullet -}$ at the outer surface (Takeda et al., 2008; Torres and Dangl, 2005). H_2O_2 is produced by SOD-mediated dismutation of $O_2^{\bullet -}$. Later studies have revealed that NADPH oxidase is involved in plant defense reactions to biotic stresses (Sagi and Fluhr, 2001; Torres et al., 2002) and abiotic stresses (Kwak et al., 2003). Diphenylene iodinium (DPI) is an important inhibitor of NADPH oxidase. It has been well documented that DPI blocks or impairs H_2O_2 production during stress conditions in plants (Overmyer et al., 2003; Laloi et al., 2004).

Cell walls are sites of active metabolism and O_2 activation. Peroxidases using NADH as electron donor were also termed NADH oxidases and have been suggested to play an important role in the formation of H_2O_2 needed for lignifications. A cell wall malate dehydrogenase generates NADH, which gives rise to H_2O_2, possibly by the action of an NADH oxidase localized in the plasmalemma (Suzuki et al., 2011).

19.2.2 Role of Superoxide Dismutase (SOD) in H_2O_2 Production

Superoxide dismutases are the first line of defense against the oxidative stresses in virtually all oxygen-respiring organisms and catalyze reactive oxygen molecules to H_2O_2. H_2O_2 is consequently converted to water by catalase (Scandalios, 1997; Den et al., 2003). Four types of SODs have been identified. Copper-zinc superoxide dismutase is the most important enzyme of the oxygen-scavenging enzymes (Wang et al., 2005; Song et al., 2006; Ahmad et al., 2011; Ahmad and Umar, 2011). Some Cu/ZnSOD isozymes are found in cytoplasm and chloroplast (del Río et al., 2003). It has been well documented that Cu/Zn-SOD is closely related to resistance to stress in plants (Guo et al., 2005; Hu et al., 2008). Iron SOD has been found in prokaryotes, in algae and in some higher plant chloroplasts. The prokaryotic MnSODs are dimers, whereas MnSODs of mitochondria are tetramers. NiSOD (coupled Ni (II/III) at the active site) is found in the *Streptomyces genus* (Choudhury et al., 1999). Two forms of copper-zinc superoxide dismutase have been reported; one is cytosolic and the other is chloroplastic isoenzymes. Cytosolic copper-zinc superoxide dismutase is induced under adverse environmental conditions (Sheri and John, 1996). The SODs reduce one $O_2^{\bullet -}$ to H_2O_2 and oxidize another to molecular oxygen, through alternating reduction and oxidation of the active-site metal ion (Mattia et al., 2002).

Three FeSOD genes (*FSD1*, *FSD2* and *FSD3*), three Cu/ZnSOD genes (*CSD1*, *CSD2* and *CSD3*), and one MnSOD gene (*MSD1*) have been reported in *A. thaliana* genome (Kliebenstein et al., 1999). All forms of SOD are nuclear-encoded with an amino terminal targeting sequence that targeted to their respective subcellular compartments. Several forms of SOD have been cloned from a variety of plants (Scandalias, 1990). Biotic and abiotic stresses cause oxidative stress and implicate the upregulation of SODs. It has been documented that salt and drought stresses increase SOD activity in *Glycyrrhiza uralensis* Fisch, Cd treatment increases SOD activity in *Hordeum vulgare* (Guo et al., 2004) and *A. thaliana* (Skorzynska-Polit et al., 2004), and Cu stress in two cultivars of *Brassica compestris* causes significant increase in SOD activity (Li et al., 2009). A higher increase in activities of total leaf SOD and chloroplast SOD in *Trifolium repens* L. has also been detected following water stress. A general induction in SOD activity in *Anabaena doliolum* under NaCl and Cu^{2+} stress has also been reported (Srivastava et al., 2005). In transgenic *Arabidopsis* plants, overexpression of MnSOD increases salt tolerance (Wang et al., 2004). Later studies have revealed that the combined expression of Cu/Zn-SOD and APX in transgenic

Festuca arundinacea plants led to increased tolerance to H_2O_2, Cu, Cd and As (Lee et al., 2007). During early chloroplast development, heteromeric *FSD2* and *FSD3* work as H_2O_2 scavengers via protecting the chloroplast nucleoids from H_2O_2 (Myouga et al., 2008).

19.2.3 H_2O_2 and Oxidative Damage to Biomolecules

Enhanced level of H_2O_2 causes damage to biomolecules such as lipids, proteins and DNA and can alter intrinsic membrane properties like fluidity. Malondialdehyde (MDA) is one of the final products of peroxidation of unsaturated fatty acids in phospholipids and causes cell membrane damage (Halliwell and Gutteridge, 2007). Lipid peroxidation (LPO), membrane injury index and H_2O_2 production are elevated by water stress in leaves of *Phaleolus vulgaris* plants (Buettner, 1993). The incubation of pea leaf crude extracts with increasing H_2O_2 concentrations showed increase in carbonyl content (as a marker of protein oxidation) (Job et al., 2005; Moller et al., 2007). The oxidation of a number of protein amino acids, particularly Arg, His, Lys, Pro, Thr and Trp, produces free carbonyl groups. These carbonyl groups may inhibit or alter protein activities and increase susceptibility towards proteolytic attack (Shringarpure and Davies, 2002; Moller et al., 2007). ROS can cause base deletion, strand breakage and base alkylation and oxidation (Tuteja et al., 2001).

19.3 REMOVAL OF H_2O_2

At high concentration, H_2O_2 can inactivate enzymes by oxidizing their thiol groups, such as enzymes of the Calvin cycle (Dat et al., 2000). When hydrogen peroxide accumulates at levels of 10 μM, fructose-1,6-bisphosphatase, sedoheptulose-1,7-bisphosphatase and phosphoribulokinase, the enzymes in the Calvin cycle, lose 50% of their activity. At high concentration, H_2O_2 also oxidizes protein kinases, phosphatases and transcription factors containing thiolate residues (Dat et al., 2000).

The steady-state level of cellular H_2O_2, scavenging of other ROS and the redox status of the cell play a key role in elimination of H_2O_2 (Karpinski et al., 2003; Mateo et al., 2006). Different cellular compartments regulate and coordinate both H_2O_2 production and elimination in plant cells. The mechanisms of H_2O_2 scavenging are regulated by both nonenzymatic and enzymatic antioxidants.

19.3.1 H_2O_2 Scavenging Enzymes

19.3.1.1 Catalase (CAT)

H_2O_2 is generated by oxidases involved in β-oxidation of fatty acids, photorespiration and purine catabolism in peroxisomes and CATs dismutate H_2O_2

TABLE 19.1 Most Important Enzymatic Antioxidative Systems, Their Function and Localization

Antioxidant Enzymes	Function	Subcellular Localization
SOD	$O_2^{-} + O_2^{-} + 2H^+ \rightarrow H_2O_2 + O_2$	Cyt, Chl, Per and Mit
CAT	$2H_2O_2 \rightarrow O_2 + 2H_2O$	Per, Gly and Mit
APX	$H_2O_2 + ASC \rightarrow 2H_2O + DHA$	Cyt, Per, Chl and Mit
GR	$GSSG + NADPH \rightarrow 2GSH + NADP^+$	Cyt, Chl and Mit
MDHAR	$2MDHA + NADH \rightarrow 2ASC + NAD^+$	Chl, Mit and Cyt
DHAR	$DHA + 2GSH \rightarrow ASC + GSSG$	Chl, Mit and Cyt

Abbreviations: APX: ascorbate peroxidase, ASC: ascorbic acid, CAT: catalase, Cyt: cytosol, Chlo: chloroplast, DHA: dehydroascorbate, DHAR: dehydroascorbate reductase, Gly: glyoxisomes, GR: glutathione reductase, GSH: glutathione, GSSG: oxidized glutathione, H_2O: water, H_2O_2: hydrogen peroxide, MDHA: monodehydroascorbate, MDHAR: monodehydroascorbate reductase, Mit: mitochondria, O_2: oxygen, O_2^{-}: superoxide radical, Per: peroxisomes, SOD: superoxide dismutase.

into H_2O and O_2 (Table 19.1). Catalases can be divided into monofunctional (dismutating) and bifunctional (dismutating/peroxidatic) enzymes (Zamocky et al., 2008). CAT also reacts with some hydroperoxides such as methyl hydrogen peroxide (MeOOH).

The CAT isozymes in *Helianthus annuus* cotyledons and *Brassica* have been studied extensively (Polidoros and Scandalios, 1999; Azpilicueta et al., 2007). *CAT1* and *CAT2* are localized in peroxisomes and the cytosol, whereas *CAT3* is mitochondrial. CAT isozymes have been shown to be regulated temporally and spatially and may respond differentially to light (Scandalias, 1990; Willekens et al., 1994). *CAT1* expression is correlated with genes involved in β-oxidation of fatty acids and peroxisomal superoxide dismutase. *CAT3* shows high expression in senescent leaves and is explained as a senescence-associated gene (*SEN2*), while *CAT1* shows particularly high expression in male flower parts. It has been well documented that the H_2O_2 accumulation were dramatically reduced in the transgenic cassava plants that upregulate expression of CAT (Xu et al., 2013).

19.3.1.2 Ascorbate Peroxidase (APX)

APX is involved in scavenging of H_2O_2 and uses ASC as the electron donor (Table 19.1). The APX family consists of four different isoforms, viz: thylakoid (tAPX), glyoxisome membrane forms (gmAPX), chloroplast stromal soluble form (sAPX) and cytosolic form (cAPX) (Noctor and Foyer, 1998). Different stress factors enhance the expression of APX in plants. Cd stress elevates leaf APX activity in *Brassica juncea* (Mobin and Khan, 2007; Ahmad et al., 2011) and *T. aestivum* (Khan et al., 2007). Under mild drought conditions, plants have higher chloroplastic-APX activity than control grown plants but the activity declines following the imposition of severe water stress (Sharma and Dubey, 2005). Later studies have revealed that the cytosolic-APX plays a key role in protection of plants from a combination of drought and heat stress.

19.3.1.3 Glutathione Peroxidase (GPX)

Plants lack animal-type glutathione peroxidase (GPX) that was identified as a selenoenzyme, instead containing cysteine-dependent GPX (Eshdat et al., 1997; Rodríguez Milla et al., 2003) which are now thought to use thioredoxin (TRX) rather than GSH (Noctor et al., 2011). However, metabolism of H_2O_2 may occur through peroxidatic glutathione-S-transferases (GSTs) and glutaredoxin (GRX)-dependent PRX (Wagner et al., 2002; Jacquot et al., 2008) as well as through APX and GR activity. In the absence of selenium, H_2O_2 was eliminated primarily by the APX; however, the H_2O_2 was detoxified mainly by GPX following the application of this trace element (Ríos et al., 2009). In different plant cell organelles including cytosol, chloroplasts, mitochondria, peroxisome and apoplast, a large family of diverse GPX isozymes catalyze the reduction of H_2O_2 (Milla et al., 2003; Anjum et al., 2011). Higher plants contain only the monomeric and nonselenium type of GPX (Eshdat et al., 1997) compared to the four types of GPXs found in animal cells (Toppo et al., 2008; Lu and Holmgren, 2009). Among the *AtGPX1–AtGPX7* isoforms (Millar et al., 2003), *AtGPX1* and *AtGPX7* are localized in the chloroplasts, whereas the other GPXs isoforms are localized in the cytosol, mitochondria and the endoplasmatic reticulum (Milla et al., 2003). *AtGPX1* and *AtGPX7* regulate salicylic acid and H_2O_2 triggered plant immune responses (Chang et al., 2009).

19.3.1.4 Glutathione Reductase (GR)

GR plays an essential role in the defense system against ROS by regenerating the glutathione pool with NADPH as the electron donor in the ASC-GSH cycle. Thus, this enzyme is important for maintaining the reduced pool of GSH for H_2O_2 scavenging in the Halliwell-Asada pathway (Noctor and Foyer, 1998; Asada, 2000) (Table 19.1). It is localized predominantly in chloroplasts, which catalyze the NADPH-dependent formation of a disulphide bond in glutathione disulphide (GSSG). But a small amount of GR is also localized in the cytosol and mitochondria (Romero-Puertas et al., 2006; Gill and Tuteja, 2010). The accumulation of GSH causes an increase in GR activity in plants and results in stress tolerance. It has been well documented that abiotic stresses increase GR activity in pea (Hernández et al., 2001), cowpea (Contour-Ansel et al., 2006) and *Reaumuria soongorica*

(Bai et al., 2009). Abiotic stresses such as high light, mechanical wounding, high temperature, chilling, heavy metals and herbicides increase the expression of GR in plants (Apel and Hirt, 2004; Karuppanapandian et al., 2011; Ahmad et al., 2011; Ahmad and Umar, 2011). Transgenic plants that upregulate expression of GR have been found to be abiotic stress tolerant. It has been reported that both chloroplastic GR (*HvGR1*) and cytosolic GR (*HvGR2*) from barley showed *in vitro* GR activity, and the specific activity for *HvGR1* was three times higher than *HvGR2* (Bashir et al., 2007).

19.3.1.5 Monodehydroascorbate Reductase (MDHAR)

Monodehydroascorbate reductase (MDHAR) and dehydroascorbate reductase (DHAR) are involved in regeneration of ascorbate (ASC) in the ASC-GSH cycle (Sharma and Dubey, 2005). MDHAR uses photosynthetic NADPH directly to regenerate reduced ascorbate (ASC) from monodehydroascorbate (MDHA). MDHAR catalyzes the elimination of H_2O_2 in mitochondria and peroxisomes (del Río et al., 2002; Mittler, 2002). Sharma and Dubey (2005) documented that the activity of MDHAR was higher in drought-stressed rice seedlings. It has been well documented that tomato fruits with increased MDAR activity exhibit better tolerance to chilling (Stevens et al., 2008). Transgenic tobacco plants that exhibit overexpression of MDAR have been found to be salt and osmotic stress tolerant (Eltayeb et al., 2007).

19.3.1.6 Dehydroascorbate Reductase (DHAR)

DHAR plays an important role in defense system against H_2O_2 by regenerating the ascorbic acid in a reaction that requires GSH (Table 19.1). Thus, this enzyme is essential for maintaining the reduced pool of ASC for H_2O_2 scavenging in the ASC-GSH cycle (Eltayeb et al., 2007); however, it is only one of several routes for GSH oxidation (Chen et al., 2003). Transgenic plants that upregulate expression of DHAR have been found to be tolerant to Al stress by maintaining high ASC levels (Yin et al., 2010). In tobacco plants, the overexpression of DHAR causes an increase in the ASC pool and results in ozone toxicity tolerance (Chen and Gallie, 2005). In transgenic plants, overexpression of DHAR increases salt tolerance in *Arabidopsis* (Ushimaru et al., 2006) and drought and ozone stress tolerance in tobacco (Eltayeb et al., 2006).

19.3.2 Nonenzymatic Antioxidants

Plants have both enzymatic and nonenzymatic antioxidant defense systems against oxidative stress damages. The nonenzymatic antioxidant defense system is composed of antioxidants such as: ASC, α-tocopherol and GSH (Apel and Hirt, 2004; Ahmad et al., 2011; Ahmad and Umar, 2011). It was shown that ascorbate (vitamin C) can react with H_2O_2, $O_2^{\bullet -}$, $^{\bullet}OH$ and lipid peroxides (Smirnoff, 1996; Noctor and Foyer, 1998). Reduced glutathione

(GSH) is used to detoxify peroxides and to regenerate oxidized ascorbate and oxidized-SH groups of proteins (Noctor and Foyer, 1998). In addition, tocopherol (vitamin E) is considered as a scavenger of ROS such as $^1O_2$, and can react with lipid peroxides in the thylakoid membrane of chloroplasts (Hollander-Czytko et al., 2005). Thus, tocopherol is one of the crucial metabolites in plants which is considered to be the most important defense against ROS resulting in membrane stability. Moreover, it has been well documented that flavonoids and phenylpropanoids are used to scavenge ROS in the H_2O_2-scavenging system (Michalak, 2006).

19.3.2.1 Ascorbic Acid (Vitamin C)

In plants, ascorbate is a substrate of APX, and catalyzes the elimination of H_2O_2 (Smirnoff and Wheeler, 2000; Foyer and Noctor, 2011). DHA is very unstable and only ascorbate possesses antioxidant and free radical scavenger properties. DHA must be reduced back to ascorbate. Ascorbate is considered to be a ROS scavenger and donates electrons in a number of enzymatic reactions mediated by the enzymes of the ascorbate-glutathione cycle: APX, MDHAR, DHAR and GR.

19.3.2.2 Glutathione

Tripeptide glutathione (GSH) is one of the crucial metabolites against ROS because of its ability to regenerate ascorbate via the ASC-GSH cycle (Noctor and Foyer, 1998; Halliwell, 2006). It has been reported that GSH mostly remains available in reduced form. It is localized in different cell compartments: cytosol, endoplasmic reticulum, vacuole, mitochondria, chloroplasts, peroxisomes and apoplast (Jimenez et al., 1998). The high reductive potential of GSH, because of its nucleophilic cysteine residue, can provide protection against oxidative damage by directly scavenging the cytotoxic H_2O_2 (Anjum et al., 2012). Moreover, it can react nonenzymatically with other ROS, such as $^1O_2$ and $O_2^{\bullet -}$ (Noctor and Foyer, 1998; Wang et al., 2008).

As a result, the ascorbate-glutathione pathway, besides catalases, is involved in H_2O_2 metabolism and signaling (Mittler et al., 2004; Foyer and Noctor, 2011). In plants with genetically knocked down catalase activities, it has been reported that there is a good correlation between the intracellular H_2O_2 availability and the GSH-GSSG ratios (Queval et al., 2009). Later studies have revealed that the increased H_2O_2 availability decreases the status of GSH pools (Mhamdi et al., 2010).

19.4 H_2O_2 AS A SIGNALING MOLECULE

H_2O_2 is the only ROS that can diffuse through aquaporins in the membranes and over larger distances within the cell (Bienert et al., 2007) and is relatively stable compared to other ROS. It can act as a signal molecule involved in the controlling of many different physiological processes, both biotic (Neill et al., 2002) and abiotic (Suzuki et al., 2012) stress responses.

19.4.1 Perception of H_2O_2 Signals

The activation of heat shock transcription factor (HSF) is considered as one of the possible ROS-sensing mechanisms in plants and animals (Miller and Mittler, 2006). Two hypothetical models of HSFs action in plants have been reported: in the first, H_2O_2 acts as a transcriptional activator and changes the assembly of the HSFs homotrimer structure (Pucciariello et al., 2012), while in the second, H_2O_2 mediates the interaction of a particular HSF with a second HSF resulting in the gene expression (Miller and Mittler, 2006). However, whether H_2O_2 directly acts as activator in plants is still unclear (Liu et al., 2011). Gadjev et al. (2006) suggested that *HSFA2* is the most responsive to H_2O_2. Since HSFA4a shows a rapid response to H_2O_2, it is considered to be a direct ROS sensor.

In addition, a recent study suggested another alternative for H_2O_2 perception in plants: oxidation of methionine (Met) to methionine sulfoxide (MetSO) (Hardin et al., 2009) are mediated by H_2O_2 (Møller and Sweetlove, 2010; Rouhier, 2011) which may couple oxidative signals to changes in the protein phosphorylation state (Hardin et al., 2009).

It is well documented that following environmental stresses there is a good correlation between H_2O_2 signaling and the redox state of the plastoquinone pool in chloroplasts (Petrov and Van Breusegem, 2012). However, a sensor and/or signal transducer complex for H_2O_2 related proteins has not yet been recognized in plants.

19.4.2 H_2O_2 Transport

It is well documented that H_2O_2 needs to cross the hydrophobic membranes of the chloroplast and peroxisomes. There is a dramatic problem, due to its polar nature that limits its capacity to diffuse through hydrophobic membranes. It was shown that H_2O_2 can be transported through aquaporin channels (Bienert et al., 2007) to regulate protein activity and signaling in the cytoplasm (Paulsen and Carroll, 2013). It has been well documented that membrane aquaporin homologues of the TIP (tonoplast intrinsic protein) and PIP (plasma membrane intrinsic protein) families play an important role in the specific transport of H_2O_2. Moreover, recently, it has been reported that H_2O_2 transport might be mediated by PIP2 proteins in *Arabidopsis*, because the Ar/R (aromatic/arginine) regions within PIP2 proteins in *Arabidopsis* are involved in the selectivity transport of H_2O_2 (Dynowski et al., 2008). In addition, a recent study suggested another model for H_2O_2 transmitting in chloroplasts: transport in a light-intensity-dependent manner (Mubarakshina et al., 2010).

Since there is a good correlation between the increased intracellular H_2O_2 availability and the decreased GSH-GSSG ratios (Queval et al., 2009; Marty et al., 2009; Mhamdi et al., 2010), an important role for glutathione in transmitting H_2O_2 signals is probable. In addition, maintenance of low GSSG concentrations can provide high sensitivity in H_2O_2-triggered signal transduction under optimal conditions (Mhamdi et al., 2010).

19.4.3 Networking of H_2O_2 Signaling with Other Signaling Pathways

19.4.3.1 Mitogen-Activated Protein Kinases and H_2O_2

Later studies have revealed that the mitogen-activated protein kinase (MAPK) cascades are involved in the initiating of environmental stress responses during plant growth and development (Ichimura et al., 2000; Jonak et al., 2002; Mishra et al., 2006; Colcombet and Hirt, 2008). MPK6 is a well-characterized MAPK in *Arabidopsis* and is involved in an H_2O_2-dependent manner of pathogen signaling (Nakagami et al., 2005; Asai and Yoshioka, 2008; Qiu et al., 2008; Chang et al., 2012). The MAPK phosphatase 2 (MKP2) is a key regulator of MPK3 and MPK6 networks that are involved in controlling both abiotic and specific pathogen responses in plants (Lumbreras et al., 2010). Increase in the activation of guard cell anion channels during ABA signaling is triggered by MPK9 and MPK12 functioning downstream to H_2O_2 signals (Jammes et al., 2009; Mittler et al., 2011). In some cases, MPK3/6 responses to cadmium (Cd) treatment are mediated by the H_2O_2-signaling pathway, which is upregulated after an accumulation of H_2O_2. Hydrogen peroxide may also be involved in the MAP kinase 8 (MPK8) pathway, since expression of *RBOHD* is rapidly decreased by MPK8 resulting in negative regulation of H_2O_2 synthesis (Takahashi et al., 2011).

19.4.3.2 H_2O_2 and Ca^{2+}

Ca^{2+} is a central regulator in guard cell signaling in response to versatile stimuli like ABA, ROS and nitric oxide (NO) (Allen et al., 2000; Young et al., 2006; Marten et al., 2008). H_2O_2 and ABA can activate Ca^{2+} channels in the plasma membrane (Hamilton et al., 2000; Lemtiri-Chlieh et al., 2003). During pollen tube growth, it has been well documented that H_2O_2 manages pollen plasma membrane hyperpolarization activated Ca^{2+} channels (Wu et al., 2010). Calcium-dependent protein kinases (CDPKs) may be involved in biotic and abiotic stresses. H_2O_2 accumulation increases the expression of *CDPKs* in tomato and wheat leaves after treatment with H_2O_2 (Chico et al., 2002; Li et al., 2008).

Later studies have revealed that CAT can scavenge H_2O_2 through Ca^{2+}-dependent pathway in peroxisomes of *Arabidopsis* guard cells (Costa et al., 2010). Recent evidence proposes that RBOH-dependent H_2O_2 production

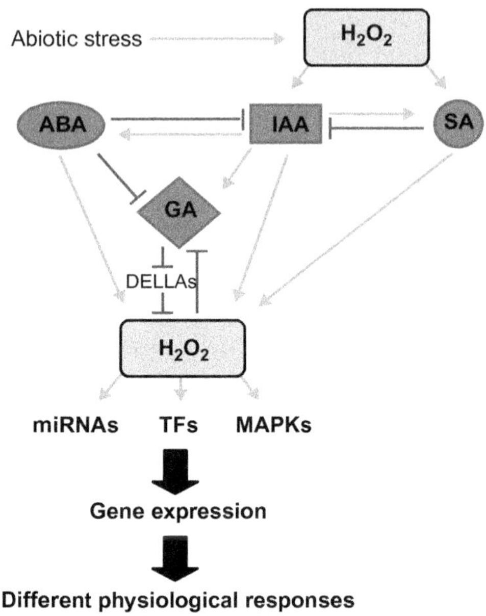

FIGURE 19.2 A simplified schematic model showing interactions between H_2O_2-transduction network and different hormonal signaling pathways in the plant cell. Pale gray arrows indicate activation, whereas dark lines indicate repression.
Abbreviations: ABA: abscisic acid, GA: gibberellin, IAA: auxin, MAPKs: mitogen-activated protein kinases, miRNAs: microRNAs, SA: salicylic acid, TFs: transcription factors.

might be mediated by Ca^{2+} homeostasis in *Arabidopsis* (Suzuki et al., 2011). In this case, cytoplasmic Ca^{2+} was shown to bind to the EF-hands of the N-terminal region of RBOH and thus promotes the activation of RBOH and producing H_2O_2 (Takeda et al., 2008) (Fig. 19.2).

19.4.3.3 H_2O_2 and Nitric Oxide

The properties of nitric oxide (NO) (free radical and highly diffusible across biological membranes) make it a very good factor to act as a signaling messenger in response to environmental stresses (Urszula and Rozalska, 2005). NO generated in response to abiotic stresses was found to mediate defense responses similar to those seen following H_2O_2 generation (Wang et al., 2013). Similarly, in *Arabidopsis*, H_2O_2 induces guard cells to synthesize NO, required if the induction of stomatal closure is to follow. Recently, He et al. (2013) reported that H_2O_2 and NO were involved in stimulation of stomatal closure in *Arabidopsis* leaves after ultraviolet B treatment. Removal of the H_2O_2 with antioxidant or inhibition of its synthesis by inhibiting NADPH oxidase activity prevents NO generation and stomatal closure. Clark et al.

(2010) reported that H_2O_2 and NO were involved in stimulation or inhibition of root hair growth in *Arabidopsis* after extracellular nucleotides application. Recent evidence proposes that H_2O_2-induced synthesis of NO might be mediated by MPK6 in *Arabidopsis* (Wang et al., 2010).

19.4.3.4 H_2O_2 and Hormonal Signaling

A highly complex mode of interaction between H_2O_2 signaling and hormonal signaling networks in plants is reported by several workers (Vlot et al., 2009; Ahmad et al., 2010). Similar to H_2O_2, salicylic acid (SA) is involved in defense and cell death responses and increased H_2O_2 levels can cause SA accumulation (Vlot et al., 2009; Petrov and Van Breusegem, 2012). Systemic acquired resistance (SAR) is a plant defense response induced after a local hypersensitive response to avirulent pathogens or by treatment with signal molecules such as SA (de León and Montesano, 2013). H_2O_2 signaling is integrated with SA signaling, which allows cytoplasmic H_2O_2 to inhibit the SA-induced transport of the SAR-associated transcription activator NPR1 (Nonexpressor of PR Genes) to the nucleus to regulate pathogenesis-related (PR) genes (Peleg-Grossman et al., 2010; Zhang et al., 2011).

ABA-induced H_2O_2 production in maize seedlings has been shown to influence expression of *CAT1, CAT2, CAT3, SOD4* and *SOD4a* genes (Scandalios, 2005). Recent evidence proposes that ABA-induced *CAT1* expression in *Arabidopsis* is mediated by a MAPK cascade-mediated ABA signaling in *Arabidopsis*, suggesting a tight link of H_2O_2 and ABA signaling because of *CAT1* may be part of the feedback regulation of H_2O_2 signaling (Xing et al., 2008; Michelet et al., 2013). It has been reported that ABA also induces calcium permeable channels via H_2O_2-signaling and stimulates stomatal closure (Murata et al., 2001; Mori et al., 2006; Desikan et al., 2008; Chan et al., 2013). AtGPX3 functions via the protein phosphatase ABA INSENSITIVE2 (ABI2) play an important role in regulation of H_2O_2 homeostasis (Miao et al., 2006). As a result, there is a tight link between H_2O_2 elevation patterns and Ca^{2+} oscillation during ABA signaling (Kwak et al., 2003; Young et al., 2006; Munemasa et al., 2007; Siegel et al., 2009; Dong et al., 2013).

Environmental factors cause the elevation of H_2O_2 and lead to integration of H_2O_2 with auxin signaling pathways resulting in morphological changes (Tognetti et al., 2010). These morphological changes might be mediated by H_2O_2 affecting auxin homeostasis during oxidative degradation, auxin catabolism, auxin transport (Krishnamurthy and Rathinasabapathi, 2013) and altered cellular location of PIN proteins. Later studies on leaves from *f19-01-9780127999630* mutant have revealed that thiol-disulphide status modulation may be one way through which increased H_2O_2 influences auxin-related responses (Mhamdi et al., 2010; Mittler et al., 2011). Recently, it has been reported that the ROS are central regulators of plant adaptation to submergence. H_2O_2 may also be involved in submergence adaptation, since it acts as a signaling intermediate in ethylene-controlled submergence adaptation (Steffens et al., 2013).

19.4.4 Other H_2O_2-Induced Factors

In plants, H_2O_2 signaling-related responses are mediated by different members of TFs, including NAC, ZAT, WRKY, DREB, bZIP and MYB. In *Arabidopsis*, 15 senescence-associated NAC TFs are upregulated by H_2O_2 (Balazadeh et al., 2010). Earlier studies have revealed that many ZIP, ERF, WRKY and MYB factors are involved in plant responses to environmental stresses (Eulgem and Somssich, 2007).

Recently, it has been reported that H_2O_2 application in rice seedlings changes the expression of seven miRNA families (Li et al., 2011). Five of the miRNA families are upregulated by H_2O_2 (miR169, miR397 (presumably involved in lignin biosynthesis), miR827, miR1425 and miR408-5p) and two are downregulated (miR528 and miR319a.2). It is also found that H_2O_2 stimulates the expression of miR169, miR397, miR827, miR1425 and miR408-5p but inhibits the expression of miR528 and miR319a.

19.4.5 H_2O_2 and Photosynthetic Mode of Carbon Assimilation

It was shown that the average concentration of H_2O_2 in CAM plants is about two-fold higher than in C_3 and C_4 plants (Li et al., 2001). In addition, later studies suggest that H_2O_2 signaling is integrated with CAM expression, thereby allowing plants to regulate photosynthetic plasticity, as well as adaptive responses to environmental stresses. In CAM plants, an increased O_2 level in Phase III accelerates the photorespiratory pathway and leads to formation of H_2O_2 (Lüttge, 2004). Recently, it has been reported that H_2O_2 can promote CAM expression when exogenously applied in *Mesembryanthemum crystallinum*, the intermediate C_3-CAM plant (Niewiadomska et al., 2004; Slesak et al., 2008).

In previous works we concluded that temporal and diurnal fluctuation patterns in the activity of antioxidant enzymes (SOD, CAT and APX) depended on duration of drought stress and was related to the mode of photosynthesis and degree of CAM induction (Habibi and Hajiboland, 2011; Habibi and Hajiboland, 2012). According to our results, *Sedum stoloniferum* developed a low degree of CAM activity, e.g. CAM-cycling metabolism under drought conditions, and the increase in the concentration of H_2O_2 that may act as a signal for C_3-CAM transition was higher in *Sedum album* (an intermediate C_3-CAM plant) compared with *S. stoloniferum* (Fig. 19.3).

It has been reported that H_2O_2 may lead to increased NO levels by stimulating the activity of NO-generating enzymes (Neill et al., 2002, 2008).

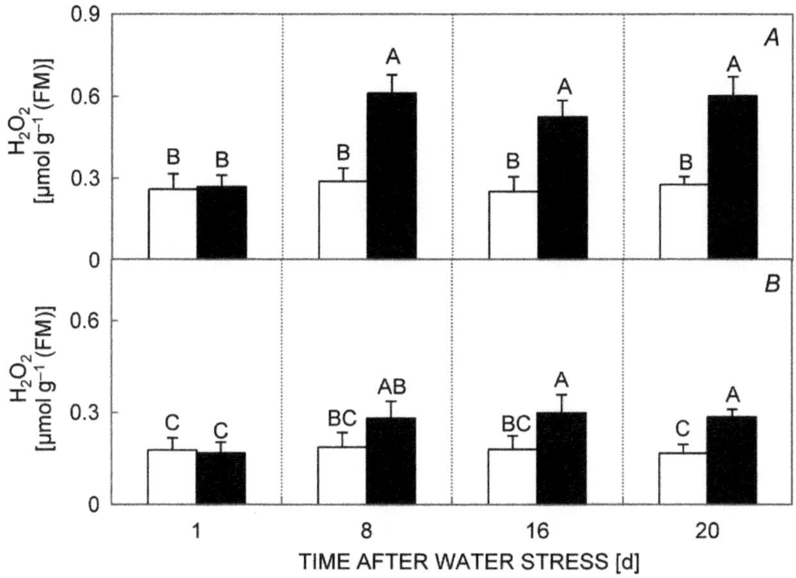

FIGURE 19.3 Leaf content of hydrogen peroxide (H_2O_2) in *Sedum album* (*A*) and *Sedum stoloniferum* (*B*) at different time intervals after drought treatment. Bars indicated with the same letter are not significantly different ($p < 0.05$). Values are means ± SD ($n = 4$) *(Adapted from Habibi and Hajiboland, 2012).*

Increased NO levels are triggered by H_2O_2 functioning downstream to oxidative stress, as well as CAM expression (Neill et al., 2008; Hao et al., 2008).

19.5 CONCLUSIONS

H_2O_2 can play a dual role in the cells. An enhanced level of H_2O_2 leads to damage to biomolecules such as lipids, proteins and DNA or even cell death, but at low concentrations H_2O_2 works as a signaling molecule. H_2O_2 plays an important role in plants, and may control many different physiological responses such as biotic and abiotic stress-response and systemic signaling. H_2O_2 signaling is highly integrated with other signaling pathways, thereby allowing plants to regulate developmental processes and responses to environmental stresses.

Many challenging questions are before plant biologists which are to be answered in the near future: e.g. (1) the production site of H_2O_2 under different stresses is still being debated; (2) a sensor and/or signal transducer complex for H_2O_2 related proteins has not yet been recognized in plants; (3) whether H_2O_2 directly acts as activator in plants is still unclear; (4) however, HSFs are sensors of H_2O_2, but this still awaits confirmation for plants. It is probable that solving the correlations between H_2O_2 oscillations and other signaling pathways, and specificity in H_2O_2 signaling will be two challenges for many years to come.

REFERENCES

Agrawal, G.K., Iwahashi, H., Rakwal, R., 2003. Small GTPase 'Rop': molecular switch for plant defense responses. FEBS Lett. 546, 173–180.

Ahmad, P., Umar, S., 2011. Oxidative Stress: Role of Antioxidants in Plants. Studium Press, New Delhi.

Ahmad, P., Jaleel, C.A., Salem, M.A., Nabi, G., Sharma, S., 2010. Roles of enzymatic and non-enzymatic antioxidants in plants during abiotic stress. Crit. Rev. Biotechnol. 30, 161–175.

Ahmad, P., Nabi, G., Ashraf, M., 2011. Cadmium-induced oxidative damage in mustard [Brassica juncea (L.) Czern. & Coss.] plants can be alleviated by salicylic acid. South Afr. J. Bot. 77, 36–44.

Allen, G.J., Chu, S.P., Schumacher, K., Shimazaki, C.T., Vafeados, D., Kemper, A., et al., 2000. Alteration of stimulus-specific guard cell calcium oscillations and stomatal closing in *Arabidopsis det3* mutant. Science 289, 2338–2342.

Anjum, N.A., Umar, S., Iqbal, M., Khan, N.A., 2011. Cadmium causes oxidative stress in moong-bean [*Vigna radiata* (L.) Wilczek] by affecting antioxidant enzyme systems and ascorbate-glutathione cycle metabolism. *Russ.* J. Plant Physiol. 58, 92–99.

Anjum, N.A., Ahmad, I., Mohmooda, I., Pachecob, M., Duartea, A.C., Pereiraa, E., et al., 2012. Modulation of glutathione and its related enzymes in plants' responses to toxic metals and metalloids-A review. Environ. Exp. Bot. 75, 307–324.

Apel, K., Hirt, H., 2004. Reactive oxygen species: metabolism, oxidative stress, and signal transduction. Annu. Rev. Plant Biol. 55, 373–399.

Arora, A., Sairam, R.K., Srivastava, G.C., 2002. Oxidative stress and antioxidative system in plants. Curr. Sci. 82, 1227–1238.

Asada, K., 2000. The water-water cycle as alternative photon and electron sinks. Phil. Trans. R. Soc. Lond. B. Biol. Sci 355, 1419–1431.

Asai, S., Yoshioka, H., 2008. The role of radical burst via MAPK signaling in plant immunity. Plant Signal Behav. 3, 920–922.

Azpilicueta, C.E., Benavides, M.P., Tomaro, M.L., Gallego, S.M., 2007. Mechanism of *CATA3* induction by cadmium in sunflower leaves. Plant Physiol. Biochem. 45, 589–595.

Bai, J., Gong, C.M., Chen, K., Kang, H.M., Wang, G., 2009. Examination of antioxidative system's responses in the different phases of drought stress and during recovery in desert plant *Reaumuria soongorica* (Pall.) Maxim. J. Plant Biol. 52, 417–425.

Balazadeh, S., Wu, A., Mueller-Roeber, B., 2010. Salt-triggered expression of the *ANAC092*-dependent senescence regultoin in *Arabidopsis thaliana*. Plant Signal. Behav. 5, 733–735.

Bashir, K., Nagasaka, S., Itai, R.N., Kobayashi, T., Takahashi, M., Nakanishi, H., et al., 2007. Expression and enzyme activity of glutathione reductase is upregulated by Fe-deficiency in graminaceous plants. Plant Mol. Biol. 65, 277–284.

Bhattachrjee, S., 2005. Reactive oxygen species and oxidative burst: roles in stress, senescence and signal transduction in plant. Curr. Sci. 89, 1113–1121.

Bhattacharjee, S., 2012. The language of reactive oxygen species signaling in plants. J. Bot. 10, 1–22.

Bienert, G.P., Møller, A.L.B., Kristiansen, K.A., Schulz, A., Møller, I.M., Schjoerring, J.K., et al., 2007. Specific aquaporins facilitate the diffusion of hydrogen peroxide across membranes. J. Biol. Chem. 282, 1183–1192.

Buettner, G.R., 1993. The pecking order of free radicals and antioxidants: lipid peroxidation, α-tocopherol, and ascorbate. Arch. Biochem. Biophys. 300, 535–543.

Ceriotti, A., 2011. Waste disposal in the endoplasmic reticulum, ROS production and plant salt stress response. Cell Res. 21, 555–557.

Chan, K.X., Wirtz, M., Phual, S.Y., Estavillo, G.M., Pogson, B.J., 2013. Balancing metabolites in drought: the sulfur assimilation conundrum. Trends in Plant Sci. 18 (1), 18–29.

Chang, C.C., Slesak, I., Jorda, L., Sotnikov, A., Melzer, M., Miszalski, Z., et al., 2009. *Arabidopsis* chloroplastic glutathione peroxidases play a role in cross talk between photooxidative stress and immune responses. Plant Physiol. 150, 670–683.

Chang, R., Jang, C.J.H., Branco-Price, C., Nghiem, P., Bailey-Serres, J., 2012. Transient MPK6 activation in response to oxygen deprivation and reoxygenation is mediated by mitochondria and aids seedling survival in *Arabidopsis*. Plant Mol. Biol. 78, 109–122.

Chen, Z., Gallie, D.R., 2005. Increasing tolerance to ozone by elevating folia ascorbic acid confers greater protection against ozone than increasing avoidance. Plant Physiol. 138, 1673–1689.

Chen, Z., Young, T.E., Ling, J., Chang, S.C., Gallie, D.R., 2003. Increasing vitamin C content of plants through enhanced ascorbate recycling. Proc. Natl. Acad. Sci. USA 100, 3525–3530.

Chico, J.M., Raíces, M., Téllez-Inón, M.T., Ulloa, R.M., 2002. A calcium dependent protein kinase is systemically induced upon wounding in tomato plants. Plant Physiol. 128, 256–270.

Choudhury, S.B., Lee, J.W., Davidson, G., Yim, Y.I., Bose, K., Sharma, M.L., et al., 1999. Examination of the nickel site structure and reaction mechanism in *Streptomyces seoulensis* superoxide dismutase. Biochemistry 38, 3744–3752.

Clark, G., Wu, M., Wat, N., Onyirimba, J., Pham, T., Herz, N., et al., 2010. Both the stimulation and inhibition of root hair growth induced by extracellular nucleotides in *Arabidopsis* are mediated by nitric oxide and reactive oxygen species. Plant Mol. Biol. 74, 423–435.

Colcombet, J., Hirt, H., 2008. *Arabidopsis* MAPKs: A complex signalling network involved in multiple biological processes. Biochem. J. 413, 217–226.

Contour-Ansel, D., Torres-Franklin, M.L., Cruz, H., de Carvalho, M., D'arcy-Lameta, A., Zuily-Fodil, Y., 2006. Glutathione reductase in leaves of cowpea: cloning of two cDNAs, expression and enzymatic activity under progressive drought stress, desiccation and abscisic acid treatment. Ann. Bot 98, 1279–1287.

Costa, A., Drago, I., Behera, S., Zottini, M., Pizzo, P., Schroeder, J.I., 2010. H_2O_2 in plant peroxisomes: an *in vivo* analysis uncovers a Ca^{2+}-dependent scavenging system. Plant J. 62, 760–772.

Creissen, G.P., Mullineaux, P.M., 2002. The molecular biology of the ascorbate-glutathione cycle in higher plants. In: Inzé, D., Van Montagu, M. (Eds.), Oxidative Stress in Plants. Taylor & Francis, UK, pp. 247–270.

Dalton, T.P., Shertzer, H.G., Puga, A., 1999. Regulation of gene expression by reactive oxygen. Annu. Rev. Pharmacol. Toxicol. 39, 67–101.

Dat, J., Vandenabeele, S., Vranova, E., Van Montagu, M., Inzé, D., van Breusegem, F., 2000. Dual action of the active oxygen species during plant stress responses. Cell Mol. Life Sci. 57, 779–795.

de León, I.P., Montesano, M., 2013. Activation of defense mechanisms against pathogens in mosses and flowering plants. Int. J. Mol. Sci. 14, 3178–3200

del Río, L.A., Corpas, F.J., Sandalio, L.M., Palma, J.M., Gomez, M., Barroso, J.B., 2002. Reactive oxygen species, antioxidant systems and nitric oxide in peroxisomes. J. Exp. Bot. 53, 1255–1272.

del Río, L.A., Sandalio, L.M., Altomare, D.A., Zilinskas, B.A., 2003. Mitochondrial and peroxisomal magnese superoxide dismutase: differential expression during leaf senescence. J. Exp. Bot. 54, 923–933.

Den, H.G.J., Haenen, G.R., Vegt, E., Vander, V.W.J., Bast, A., 2003. Superoxide dismutase: the balance between prevention and induction of oxidative damage. Chem. Biol. Interact. 145, 33–39.

Desikan, R., Horak, J., Chaban, C., Mira-Rodado, V., Witthoft, J., Elgass, K., et al., 2008. The histidine kinase AHK5 integrates endogenous and environmental signals in *Arabidopsis* guard cells. Plos. One 3, 2482–2491.

Dinakara, C., Djilianov, D., 2012. Photosynthesis in desiccation tolerant plants: Energy metabolism and antioxidative stress defense. Plant Sci. 182, 29–41.

Dismukes, G.C., Klimov, V.V., Baranov, S.V., Kozlov, Y.N., DasGupta, J., Tyryshkin, A., 2001. The origin of atmospheric oxygen on Earth: the innovation of oxygenic photosynthesis. Proc. Natl. Acad. Sci. U.S.A. 98, 2170–2175.

Dong, L., Wei, W., Chen, M., 2013. ATHK1 acts downstream of hydrogen peroxide to mediate ABA signaling through regulation of calcium channel activity in *Arabidopsis* guard cells. Chin. Sci. Bull 58, 336–343.

Dynowski, M., Schaaf, G., Loque, D., Moran, O., Ludewig, U., 2008. Plant plasma membrane water channels conduct the signalling molecule H_2O_2. Biochem. J. 414, 53–61.

Eltayeb, A.E., Kawano, N., Badawi, G.H., Kaminaka, H., Sanekata, T., Morishima, I., et al., 2006. Enhanced tolerance to ozone and drought stresses in transgenic tobacco overexpressing dehydroascorbate reductase in cytosol. Physiol. Plant. 127, 57–65.

Eltayeb, A.E., Kawano, N., Badawi, G.H., Kaminaka, H., Sanekata, T., Shibahara, T., et al., 2007. Overexpression of monodehydroascorbate reductase in transgenic tobacco confers enhanced tolerance to ozone, salt and polyethylene glycol stresses. Planta 225, 1255–1264.

Eshdat, Y., Holland, D., Faltin, Z., Ben-Hayyim, G., 1997. Plant glutathione peroxidases. Physiol. Plant 100, 234–240.

Eulgem, T., Somssich, I.E., 2007. Networks of WRKY transcription factors in defense signalling. Curr. Opin. Plant Biol. 10, 366–371.

Foyer, C.H., Noctor, G., 2011. Ascorbate and glutathione: the heart of the redox hub. Plant Physiol. 155, 2–18.

Fryer, M.J., Oxborough, K., Mullineaux, P.M., Baker, N.R., 2002. Imaging of photo-oxidative stress responses in leaves. J. Exp. Bot. 53, 1249–1254.

Gadjev, I., Vanderauwera, S., Gechev, T.S., Laloi, C., Minkov, I.N., Shulaev, V., et al., 2006. Transcriptomic footprints disclose specificity of reactive oxygen species signaling in Arabidopsis. Plant Physiol. 141, 436–445.

Gill, S.S., Tuteja, N., 2010. Reactive oxygen species and antioxidant machinery in abiotic stress tolerance in crop plants. Plant Physiol. Biochem. 48, 909–930.

Giraud, E., Ho, L.H., Clifton, R., Carroll, A., Estavillo, G., Tan, Y., et al., 2008. The absence of ALTERNATIVE OXIDASE1a in *Arabidopsis* results in acute sensitivity to combined light and drought stress. Plant Physiol. 147, 595–610.

Gratao, P.L., Polle, A., Lea, P.J., Azevedo, R.A., 2005. Making the life of heavy metal stressed plants a little easier. Funct. Plant Biol. 32, 481–494.

Guo, L.H., Wu, X.L., Gong, M., 2005. Roles of glutathione reductase and superoxide dismutase in heat-shock-induced cross adaptation in maize seedlings. Plant Physiol. Commun. 41, 429–432.

Guo, T., Zhang, G., Zhou, M., Wu, F., Chen, J., 2004. Effects of aluminum and cadmium toxicity on growth and antioxidant enzyme activities of two barley genotypes with different Al resistance. Plant Soil 258, 241–248.

Habibi, G., Hajiboland, R., 2011. Comparison of water stress and UV radiation effects on the induction of CAM and antioxidative defense in the succulent *Rosularia elymaitica* (Crassulaceae). Acta. Biol. Cracov. Bot. 53, 7–15.

Habibi, G., Hajiboland, R., 2012. Comparison of photosynthesis and antioxidative protection in *Sedum album* and *Sedum stoloniferum* (Crassulaceae) under water stress. Photosynthetica 50 (4), 508–518.

Halliwell, B., 2006. Oxidative stress and neurodegeneration: where are we now? J. Neurochem. 97, 1634–1658.

Halliwell, B., Gutteridge, J.M.C., 2007. Free Radicals in Biology and Medicine. Oxford University Press, Oxford, UK, pp. 1320–1325.

Hamilton, D.W.A., Hills, A., Kohler, B., Blatt, M.R., 2000. Ca^{2+} channels at the plasma membrane of stomatal guard cells are activated by hyperpolarization and abscisic acid. Proc. Natl. Acad. Sci. U.S.A. 97, 4967–4972.

Hao, G.P., Xing, Y., Zhang, J.H., 2008. Role of nitric oxide dependence on nitric oxide synthase-like activity in the water stress signaling of maize seedling. J. Integr. Plant Biol. 50, 435–442.

Hardin, S.C., Larue, C.T., Oh, M.H., Jain, V., Huber, S.C., 2009. Coupling oxidative signals to protein phosphorylation via methionine oxidation in *Arabidopsis*. Bioch. J. 422, 305–312.

He, J.M., Ma, X.G., Zhang, Y., Sun, T.F., Xu, F.F., Chen, Y.P., et al., 2013. Role and interrelationship of Gα protein, hydrogen peroxide, and nitric oxide in ultraviolet B-induced stomatal closure in *Arabidopsis* leaves. Plant Physiol. 161, 1570–1583.

Hernández, J.A., Ferrer, M.A., Jiménez, A., Barceló, A.R., Sevilla, F., 2001. Antioxidant systems and $O_2^{\bullet-}/H_2O_2$ production in the apoplast of pea leaves. Its relation with salt-induced necrotic lesions in minor veins. Plant Physiol. 127, 817–831.

Hideg, E., Barta, C., Kálai, T., Vass, I., Hideg, K., Asada, K., 2002. Detection of singlet oxygen and superoxide with fluorescent sensors in leaves under stress by photoinhibition or UV radiation. Plant Cell Physiol. 43, 1154–1164.

Hollander-Czytko, H., Grabowski, J., Sandorf, I., Weckermann, K., Weiler, E.W., 2005. Tocopherol content and activities of tyrosine aminotransferase and cystinelyase in *Arabidopsis* under stress conditions. J. Plant Physiol. 162, 767–770.

Hu, W.H., Song, X.S., Shi, K., Xia, X.J., Zhou, Y.H., Yu, J.Q., 2008. Changes in electron transport, superoxide dismutase and ascorbate peroxidase isoenzymes in chloroplasts and mitochondria of cucumber leaves as influenced by chilling. Photosynthetica 46, 581–588.

Ichimura, K., Mizoguchi, T., Yoshida, R., Yuasa, T., Shinozaki, K., 2000. Various abiotic stresses rapidly activate *Arabidopsis* MAP kinases ATMPK4 and ATMPK6. Plant J. 24, 655–665.

Jacquot, J.P., Lemaire, S., Rouhier, N., 2008. The role of glutathione in photosynthetic organisms: emerging functions for glutaredoxins and glutathionylation. Annu. Rev. Plant Biol. 59, 143–166.

Jammes, F., Song, C., Shin, D., Munemasa, S., Takeda, K., Gu, D., et al., 2009. MAP kinases MPK9 and MPK12 are preferentially expressed in guard cells and positively regulate ROS-mediated ABA signaling. Proc. Natl. Acad. Sci. USA 106, 20520–20525.

Jaspers, P., Kangasjarvi, J., 2010. Reactive oxygen species in abiotic stress signaling. Physiol. Plant 138, 405–413.

Jimenez, A., Hernandez, J.A., Pastori, G., del Río, L.A., Sevilla, F., 1998. Role of the ascorbate-glutathione cycle of mitochondria and peroxisomes in the senescence of pea leaves. Plant Physiol. 118, 1327–1335.

Job, C., Rajjou, L., Lovigny, Y., Belghazi, M., Job, D., 2005. Patterns of protein oxidation in *Arabidopsis* seeds and during germination. Plant Physiol. 138, 790–802.

Jonak, C., Okrísz, L., Bögre, L., Hirt, H., 2002. Complexity, cross talk and integration of plant MAP kinase signalling. Curr. Opin. Plant Biol. 5, 415–424.

Karpinski, S., Gabryś, H., Mateo, A., Karpinska, B., Mullineaux, P.M., 2003. Light perception in plant disease defence signalling. Curr. Opin. Plant Biol. 6, 390–396.

Karuppanapandian, T., Manoharan, K., 2008. Uptake and translocation of tri- and hexa-valent chromium and their effects on black gram (*Vigna mungo* L. Hepper cv. Co4) roots. J. Plant Biol 51, 192–201.

Karuppanapandian, T., Wang, H.W., Prabakaran, N., Jeyalakshmi, K., Kwon, M., Manoharan, K., et al., 2011. 2,4-dichlorophenoxyacetic acid-induced leaf senescence in mung bean (*Vigna radiata* L. Wilczek) and senescence inhibition by co-treatment with silver nanoparticles. Plant Physiol. Biochem. 49, 168–177.

Khan, N.A., Singh, S.S., Nazar, R., 2007. Activities of antioxidative enzymes, sulphur assimilation, photosynthetic activity and growth of wheat (*Triticum aestivum*) cultivars differing in yield potential under cadmium stress. *J. Agro.* Crop. Sci. 193, 435–444.

Kliebenstein, D.J., Dietrich, R.A., Martin, A.C., Last, R.L., Dangl, J.L., 1999. Regulates salicylic acid induction of copper zinc superoxide dismutase in *Arabidopsis thaliana*. Mol. Plant Microbe Interac. 12, 1022–1026.

Krasensky, J., Jonak, C., 2012. Drought, salt, and temperature stress-induced metabolic rearrangements and regulatory networks. J. Exp. Bot. 63, 1593–1608.

Krishnamurthy, A., Rathinasabapathi, B., 2013. Auxin and its transport play a role in plant tolerance to arsenite-induced oxidative stress in *Arabidopsis thaliana*. Plant Cell Environ. 36, 1–9.

Kwak, J.M., Mori, I.C., Pei, Z.M., Leonhardt, N., Torres, M.A., Dangl, J.L., et al., 2003. NADPH oxidase *AtrbohD* and *AtrbohF* genes function in ROS-dependent ABA signaling in *Arabidopsis*. EMBO J. 22, 2623–2633.

Laloi, C., Apel, K., Danon, A., 2004. Reactive oxygen signalling: the latest news. Curr. Opin. Plant Biol. 7, 323–328.

Lee, S.H., Ahsan, N., Lee, K.W., Kim, D.H., Lee, D.G., Kwak, S.S., et al., 2007. Simultaneous overexpression of both Cu/Zn superoxide dismutase and ascorbate peroxidase in transgenic tall fescue plants confers increased tolerance to a wide range of abiotic stresses. J. Plant Physiol. 164, 1626–1638.

Lemtiri-Chlieh, F., MacRobbie, E.A.C., Webb, A.A.R., Manison, N.F., Brownlee, C., Skepper, J.N., et al., 2003. Inositol hexakisphosphate mobilizes an endomembrane store of calcium in guard cells. Proc. Natl. Acad. Sci. U.S.A. 100, 10091–10095.

Li, A., Wang, X., Leseberg, C.H., Jia, J., Mao, L., 2008. Biotic and abiotic stress responses through calcium-dependent protein kinase (CDPK) signalling in wheat (*Triticum aestivum* L.). Plant Signal Behav. 3, 654–656.

Li, J., Zhao, X., Matsui, S., 2001. Hydrogen peroxide contents and activities of antioxidative enzymes among C_3, C_4 and CAM plants. J. Japan Soc. Hort. Sci. 70, 747–752.

Li, T., Li, H., Zhang, Y.X., Liu, J.Y., 2011. Identification and analysis of seven H_2O_2-responsive miRNAs and 32 new miRNAs in the seedlings of rice (*Oryza sativa* L. ssp. indica). Nucleic Acids Res. 39, 2821–2833.

Li, Z., Wakao, S., Fischer, B.B., Niyogi, K.K., 2009. Sensing and responding to excess light. Annu. Rev. Plant Biol. 60, 239–260.

Liu, H.C., Liao, H.T., Chang, Y.Y., 2011. The role of class A1 heat shock factors (HSFA1s) in response to heat and other stresses in *Arabidopsis*. Plant Cell Environ 34, 738–751.

Lu, J., Holmgren, A., 2009. Selenoproteins. J. Biol. Chem. 284, 723–727.

Lumbreras, V., Vilela, B., Irar, S., Sole, M., Capellades, M., Valls, M., et al., 2010. MAPK phosphatase MKP2 mediates disease responses in *Arabidopsis* and functionally interacts with MPK3 and MPK6. Plant J. 63, 1017–1030.

Lüttge, U., 2004. Ecophysiology of crassulacean acid metabolism (CAM). Ann. Bot. 93, 629–652.

Mano, J., 2002. Early events in environmental stresses in plants induction mechanisms of oxidative stress. In: Inzé, D., Van Montagu, M. (Eds.), Oxidative stress in plants. Taylor & Francis, UK, pp. 217–246.

Marten, H., Hyun, T., Gomi, K., Seo, S., Hedrich, R., Roelfsema, M.R.G., 2008. Silencing of *NtMPK4* impairs CO_2-induced stomatal closure, activation of anion channels and cytosolic Ca^{2+} signals in *Nicotiana tabacum* guard cells. Plant J. 55, 698–708.

Marty, L., Siala, W., Schwarzländer, M., Fricker, M.D., Wirtz, M., Sweetlove, L.J., et al., 2009. The NADPH-dependent thioredoxin system constitutes a functional backup for cytosolic glutathione reductase in *Arabidopsis*. Proc. Natl. Acad. Sci. U.S.A. 106, 9109–9114.

Mateo, A., Funck, D., Mühlenbock, P., Kular, B., Mullineaux, P.M., Karpinski, S., 2006. Controlled levels of salicylic acid are required for optimal photosynthesis and redox homeostasis. J. Exp. Bot. 57, 1795–1807.

Mattia, F., Peter, O.N., Maria, E.S., Alessandro, D., 2002. Superoxide dismutase kinetics. Methods Enzymol. 349, 38–49.

Mhamdi, A., Hager, J., Chaouch, S., Queval, G., Han, Y., Taconnat, L., et al., 2010. *Arabidopsis* GLUTATHIONE REDUCTASE1 plays a crucial role in leaf responses to intracellular hydrogen peroxide and in ensuring appropriate gene expression through both salicylic acid and jasmonic acid signaling pathways. Plant Physiol. 153, 1144–1160.

Mhamdi, A., Noctor, G., Baker, A., 2012. Plant catalases: peroxisomal redox guardians. Arch. Biochem. Biophys. 525, 181–194.

Miao, Y., Lv, D., Wang, P., Wang, X.C., Chen, J., Miao, C., 2006. An *Arabidopsis* glutathione peroxidase functions as both a redox transducer and a scavenger in abscisic acid and drought stress responses. Plant Cell 27, 49–66.

Michalak, A., 2006. Phenolic compounds and their antioxidant activity in plants growing under heavy metal stress. Pol. J. Environ. Stud. 15, 523–530.

Michelet, L., Roach, T., Fischer, B.B., Bedhomme, M., Lemaire, S.D., Krieger-Liszkay, A., 2013. Down-regulation of catalase activity allows transient accumulation of a hydrogen peroxide signal in *Chlamydomonas reinhardtii*. Plant Cell Environ. 36, 1204–1213.

Milla, M.A.R., Maurer, A., Rodriguez, H.A., Gustafson, J.P., 2003. Glutathione peroxidase genes in *Arabidopsis* are ubiquitous and regulated by abiotic stresses through diverse signaling pathways. Plant J. 36, 602–615.

Millar, A.H., Mittova, V., Kiddle, G., 2003. Control of ascorbate synthesis by respiration and its implications for stress responses. Plant Physiol. 133, 443–447.

Miller, G., Mittler, R., 2006. Could heat shock transcription factors function as hydrogen peroxide sensors in plants? Ann. Bot. 98, 279–288.

Miller, G., Suzuki, N., Ciftci-Yilmaz, S., Mittler, R., 2010. Reactive oxygen species homeostasis and signaling during drought and salinity stresses. Plant Cell Environ. 33, 453–467.

Mishra, N.S., Tuteja, R., Tuteja, N., 2006. Signaling through MAP kinase networks in plants. Arch. Biochem. Biophys. 452, 55–68.

Mittler, R., 2002. Oxidative stress, antioxidants and stress tolerance. Trends Plant Sci. 7, 405–410.

Mittler, R., Vanderauwera, S., Gollery, M., Van Breusegem, F., 2004. Reactive oxygen gene network of plants. Trends Plant Sci. 9, 490–498.

Mittler, R., Vanderauwera, S., Suzuki, N., Miller, G., Tognetti, B., Vandepoele, K., et al., 2011. ROS signaling: the new wave? Trends in Plant Sci. 16, 300–309.

Mobin, M., Khan, N.A., 2007. Photosynthetic activity, pigment composition and antioxidative response of two mustard (*Brassica juncea*) cultivars differing in photosynthetic capacity subjected to cadmium stress. J. Plant Physiol. 164, 601–610.

Møller, I.M., Sweetlove, L.J., 2010. ROS signaling-specificity is required. Trends Plant Sci. 15, 370–374.

Moller, I.M., Jensen, P.E., Hansson, A., 2007. Oxidative modifications to cellular components in plants. Annu. Rev. Plant Biol. 58, 459–481.

Monakhova, O.F., Chernyadév, I.I., 2002. Protective role of kartolin-4 in wheat plants exposed to soil drought. Appl. Biochem. Microbiol. 38, 373–380.

Mori, I.C., Murata, Y., Yang, Y., Munemasa, S., Wang, Y.F., Andreoli, S., 2006. CDPKs CPK6 and CPK3 function in ABA regulation of guard cell S-type anion- and Ca^{2+}-permeable channels and stomatal closure. PLoS. Biol. 4, 1749–1762.

Mubarakshina, M.M., Ivanov, B.N., Naydov, I.A., Hillier, W., Badger, M.R., Krieger-Liszkay, A., 2010. Production and diffusion of chloroplastic H_2O_2 and its implication to signalling. J. Exp. Bot. 61, 3577–3587.

Munemasa, S., Oda, K., Watanabe-Sugimoto, M., Nakamura, Y., Shimoishi, Y., Murata, Y., 2007. The coronatine-insentitive1 mutation reveals the hormonal signaling interaction between abscisic acid and methyl jasmonate in *Arabidopsis* guard cells specific impairment of ion channel activation and second messenger production. Plant Physiol. 143, 1398–1407.

Murata, Y., Pei, Z.M., Mori, I.C., Schroeder, J.I., 2001. Abscisic acid activation of plasma membrane Ca^{2+} channels in guard cells requires cytosolic NAD(P)H and is differentially disrupted upstream and downstream of reactive oxygen species production in *abi1-1* and *abi2-1* protein phosphatase 2C mutants. Plant Cell 13, 2513–2523.

Murphy, M.P., 2009. How mitochondria produce reactive oxygen species. Biochem. J. 417, 11–13.

Myouga, F., Hosoda, C., Umezawa, T., Iizumi, H., Kuromori, T., Motohashi, R., et al., 2008. Heterocomplex of iron superoxide dismutases defends chloroplast nucleoids against oxidative stress and is essential for chloroplast development in *Arabidopsis*. The Plant Cell 20, 3148–3162.

Nakagami, H., Pitzschke, A., Hirt, H., 2005. Emerging MAP kinase pathways in plant stress signalling. Trends Plant Sci. 10, 339–346.

Niewiadomska, E., Karpinska, B., Romanowska, E., Ślesak, I., Karpinski, S., 2004. A salinity-induced C_3-CAM transition increases energy conservation in the halophyte *Mesembryanthemum crystallinum* L. Plant Cell Physiol. 45, 789–794.

Neill, S., Barros, R., Bright, J., Desikan, R., Hancock, J., Harrison, J., et al., 2008. Nitric oxide, stomatal closure, and abiotic stress. J. Exp. Bot. 59, 165–176.

Neill, S.J., Desikan, R., Clarke, A., Hurst, R.D., Hancock, J.T., 2002. Hydrogen peroxide and nitric oxide as signalling molecules in plants. J. Exp. Bot. 53, 1237–1247.

Noctor, G., Foyer, C.H., 1998. Ascorbate and glutathione: keeping active oxygen under control. Annu. Rev. Plant Physiol. Plant Mol. Biol. 49, 249–279.

Noctor, G., Queval, G., Mhamdi, A., Chaouch, S., Foyer, C.H., 2011. Glutathione. The Arabidopsis Book. American Society of Plant Biologists, Rockville MD, pp. 119–142.

Overmyer, K., Brosche, M., Kangasjarvi, J., 2003. Reactive oxygen species and hormonal control of cell death. Trends Plant Sci. 8, 325–342.

Paulsen, C.E., Carroll, K.S., 2013. Cysteine-mediated redox signaling: chemistry, biology, and tools for discovery. Chem. Rev. 113, 4633–4679.

Peleg-Grossman, S., Melamed-Book, N., Cohen, G., Levine, A., 2010. Cytoplasmic H_2O_2 prevents translocation of NPR1 to the nucleus and inhibits the induction of PR genes in *Arabidopsis*. Plant signal. Behav 5, 1401–1406.

Petrov, V.D., Van Breusegem, F., 2012. Hydrogen peroxide: a central hub for information flow in plant cells. AoB Plants. 2012, pls014.

Polidoros, N.A., Scandalios, J.G., 1999. Role of hydrogen peroxide and different classes of antioxidants in the regulation of catalase and glutathione-S-transferase gene expression in maize (*Zea mays* L.). Physiol. Plant 106, 112–120.

Pucciariello, C., Banti, V., Perata, P., 2012. ROS signaling as common element in low oxygen and heat stresses. Plant Physiol. Biochem. 59, 3–10.

Qiu, J.L., Zhou, L., Yun, B.W., Yun, B.W., Nielsen, H.B., Fiil, B.K., et al., 2008. *Arabidopsis* mitogen-activated protein kinase kinases MKK1 and MKK2 have overlapping functions in defense signalling mediated by MEKK1, MPK4 and MKS1. Plant Physiol. 148, 212–222.

Queval, G., Thominet, D., Vanacker, H., Miginiac-Maslow, M., Gakiére, B., Noctor, G., 2009. H_2O_2-activated up-regulation of glutathione in *Arabidopsis* involves induction of genes encoding enzymes involved in cysteine synthesis in the chloroplast. Mol. Plant 2, 344–356.

Rhoads, D.M., Umbach, A.L., Subbaiah, C.C., Siedow, J.N., 2006. Mitochondrial reactive oxygen species. Contribution to oxidative stress and interorganellar signaling. Plant Physiol. 141, 357–366.

Ríos, J.J., Blasco, B., Cervilla, L.M., Rosales, M.A., Sanchez-Rodriguez, E., Romero, L., et al., 2009. Production and detoxification of H_2O_2 in lettuce plants exposed to selenium. Ann. Appl. Biol. 154, 107–116.

Rodríguez, A.A., Córdoba, A.R., Ortega, L., Taleisnik, E., 2004. Decreased reactive oxygen species concentration in the elongation zone contributes to the reduction in maize leaf growth under salinity. J. Exp. Bot. 55, 1383–1390.

Rodríguez Milla, M.A., Maurer, A., Huete, A.R., Gustafson, J.P., 2003. Glutathione peroxidase genes in *Arabidopsis* are ubiquitous and regulated by abiotic stresses through diverse signaling pathways. Plant J. 36, 602–615.

Romero-Puertas, M.C., Corpas, F.J., Sandalio, L.M., Leterrier, M., Rodriguez-Serrano, M., del Río, L.A., et al., 2006. Glutathione reductase from pea leaves: response to abiotic stress and characterization of the peroxisomal isozyme. New Phytol. 170, 43–52.

Rouhier, N., 2011. Plant glutaredoxins: pivotal players in redox biology and iron-sulphur centre assembly. New Phytol. 186, 365–372.

Sagi, M., Fluhr, R., 2001. Superoxide production by plant homologues of the $gp91^{phox}$ NADPH oxidase: modulation of activity by calcium and by tobacco mosaic virus infection. Plant Physiol. 126, 1281–1290.

Sanda, S., Yoshida, K., Kuwano, M., Kawamura, T., Munekage, Y.N., Akashi, K., et al., 2011. Responses of the photosynthetic electron transport system to excess light energy caused by water deficit in wild watermelon. Physiol. Plant 142, 247–264.

Scandalias, J.G., 1990. Response of plant antioxidant defense genes to environmental stress. Adv. Genet. 28, 1–41.

Scandalios, J.G., 1997. Oxidative stress and the molecular biology of antioxidant defenses. Cold Spring Harbor Laboratory Press, Cold Spring Harbor, NY, USA, pp. 2–11.

Scandalios, J.G., 2005. Oxidative stress: molecular perception and transduction of signals triggering antioxidant gene defenses. Braz. J. Med. Biol. Res. 38, 995–1014.

Sharma, P., Dubey, R.S., 2005. Modulation of nitrate reductase activity in rice seedlings under aluminium toxicity and water stress: role of osmolytes as enzyme protectant. J. Plant Physiol. 162, 854–864.

Sheri, P.K., John, G.S.A., 1996. Comparison of the structure and function of the highly homologous maize antioxidant Cu/Zn superoxide dismutase genes, *sod4* and *sod4A*. Genetics 144, 317–328.

Shi, R., Sun, Y.H., Li, Q., Heber, S., Sederoff, R., Chiang, V.L., 2010. Towards a systems approach for lignin biosynthesis in *Populus trichocarpa*: transcript abundance and specificity of the monolignol biosynthetic genes. Plant Cell Physiol. 51, 144–163.

Shringarpure, R., Davies, K.J., 2002. Protein turnover by the proteasome in aging and disease. Free Radic. Biol. Med. 32, 1084–1089.

Siegel, R.S., Xue, S., Murata, Y., Yang, Y., Nishimura, N., Wang, A., 2009. Calcium elevation dependent and attenuated resting calcium-dependent abscisic acid induction of stomatal closure and abscisic acid-induced enhancement of calcium sensitivities of S-type anion and inward-rectifying K^+ channels in *Arabidopsis* guard cells. Plant J. 59, 207–220.

Skorzynska-Polit, E., Drazkiewicz, M., Krupa, Z., 2004. The activity of the antioxidative system in cadmium-treated *Arabidopsis thaliana*. Biol. Plant 47, 71–78.

Slesak, I., Libik, M., Miszalski, Z., 2008. The foliar concentration of hydrogen peroxide during salt-induced C_3-CAM transition in *Mesembryanthemum crystallinum* L. Plant Sci. 174, 221–226.

Smirnoff, N., 1996. The function and metabolism of ascorbic acid in plants. Ann. Bot. 78, 661–669.

Smirnoff, N., Wheeler, G.L., 2000. Ascorbic acid in plants: biosynthesis and function. Crit. Rev. Biochem. Mol. Biol. 35, 291–314.

Song, F.,N., Yang, C.P., Liu, X.M., Li, G.B., 2006. Effect of salt stress on activity of superoxide dismutase (SOD) in *Ulmus pumila* L. J. For. Res. 17, 13–16.

Srivastava, A.K., Bhargava, P., Rai, L.C., 2005. Salinity and copper-induced oxidative damage and changes in antioxidative defense system of *Anabaena doliolum*. W. J. Microb. Biotechnol. 22, 1291–1298.

Steffens, B., Steffen-Heins, A., Sauter, M., 2013. Reactive oxygen species mediate growth and death in submerged plants. Front. Plant Sci. 4, 170–179.

Stevens, R., Page, D., Gouble, B., Garchery, C., Zamir, D., Causse, M., 2008. Tomato fruit ascorbic acid content is linked with monodehydroascorbate reductase activity and tolerance to chilling stress. Plant Cell Environ. 31, 1086–1096.

Suzuki, N., Miller, G., Morales, J., Shulaev, V., Torres, M.A., Mittler, R., 2011. Respiratory burst oxidases: the engines of ROS signaling. Curr. Opin. Plant Biol. 14, 1–9.

Suzuki, N., Koussevitzky, S., Mittler, R., Miller, G., 2012. ROS and redox signalling in the response of plants to abiotic stress. Plant Cell Environ. 35, 259–270.

Takahashi, F., Mizoguchi, T., Yoshida, R., Ichimura, K., Shinozaki, K., 2011. Calmodulin-dependent activation of MAP kinase for ROS homeostasis in *Arabidopsis*. Mol. Cell 41, 649–660.

Takeda, S., Gapper, C., Kaya, H., Bell, E., Kuchitsu, K., Dolan, L., 2008. Local positive feedback regulation determines cell shape in root hair cells. Science 319, 1241–1244.

Tognetti, B., Van Aken, O.V., Morreel, K., Vandenbroucke, K., van de Cotte, B., De Clercq Chiwocha, S., et al., 2010. Perturbation of indole-3-butyric acid homeostasis by the UDP-glucosyltransferase UGT74E2 modulates *Arabidopsis* architecture and water stress tolerance. Plant Cell 22, 2660–2679.

Toppo, S., Vanin, S., Bosello, V., Tosatto, S.C.E., 2008. Evolutionary and structural insights into the multifaceted glutathione peroxidase (GPX) superfamily. Antiox. Redox. Sig. 10, 1501–1513.

Torres, M.A., Dangl, J.L., 2005. Functions of the respiratory burst oxidase in biotic interactions, abiotic stress and development. Curr. Opin. Plant Biol. 8, 397–403.

Torres, M.A., Dangl, J.L., Jones, J.D.G., 2002. *Arabidopsis* gp91phox homologues *AtrbohD* and *AtrbohF* are required for accumulation of reactive oxygen intermediates in the plant defense response. Proc. Natl. Acad. Sci. USA 99, 517–522.

Turrens, J.F., 2003. Mitochondrial formation of reactive oxygen species. J. Physiol. 552 (2), 335–344.

Tuteja, N., Singh, M.B., Misra, M.K., Bhalla, P.L., Tuteja, R., 2001. Molecular mechanisms of DNA damage and repair: progress in plants. Crit. Rev. Biochem. Mol. Biol. 36, 337–397.

Ushimaru, T., Nakagawa, T., Fujioka, Y., Daicho, K., Naito, M., Yamauchi, Y., et al., 2006. Transgenic *Arabidopsis* plants expressing the rice dehydroascorbate reductase gene are resistant to salt stress. J. Plant Physiol. 163, 1179–1184.

Urszula, M., Rozalska, S., 2005. Nitric oxide and hydrogen peroxide in tomato resistance: Nitric oxide modulates hydrogen peroxide level in O-hydroxyethylorutin-induced resistance to *Botrytis cinerea* in tomato. Plant Physiol. Biochem. 43, 623–635.

Vellosillo, T., Vicente, J., Kulasekaran, S., Hamberg, M., Castresana, C., 2010. Emerging complexity in reactive oxygen species production and signaling during the response of plants to pathogens. Plant Physiol. 154, 444–448.

Vlot, A.C., Dempsey, D.A., Klessig, D.F., 2009. Salicylic acid, a multifaceted hormone to combat disease. Annu. Rev. Phytopathol. 47, 177–206.

Wagner, U., Edwards, R., Dixon, D.P., Mauch, F., 2002. Probing the diversity of the *Arabidopsis* glutathione-S-transferase gene family. Plant Mol. Biol. 49, 515–532.

Wang, J.P., Li, Y.L., Zhang, J.G., 2008. Effect of high-temperature and excessive-light stress on APX activity in apple peel. Acta. Agri. Boreali-Sin. 23, 144–147.

Wang, P., Du, Y., Li, Y., Ren, D., Song, C.P., 2010. Hydrogen peroxide–mediated activation of MAP Kinase 6 modulates nitric oxide biosynthesis and signal transduction in *Arabidopsis*. Plant Cell 22 (9), 2981–2998.

Wang, R.G., Chen, S.L., Liu, L.Y., Hao, Z.Y., Weng, H.J., Li, H., et al., 2005. Genotypic differences in antioxidative ability and salt tolerance of three poplars under salt stress. J. Beijing For Univ. 27, 46–52.

Wang, Y., Ying, J., Chen, X., Wang, C., 2004. Transgenic *Arabidopsis* overexpressing Mn-SOD enhanced salt-tolerance. Plant Sci. 167, 671–677.

Wang, Y., Lin, A., Loake, G.J., Chu, C., 2013. H_2O_2-induced leaf cell death and the crosstalk of reactive nitric/oxygen species. J. Integr. Plant Biol. 55, 202–208.

Willekens, H., Langebartels, C., Tire, C., Van Montagu, M., Inzé, D., Van Camp, W., 1994. Differential expression of catalase genes in *Nicotiana plumbaginifolia* (L.). Proc. Natl. Acad. Sci. U.S.A. 91, 10450–10454.

Xing, Y., Jia, W., Zhang, J., 2008. AtMKK1 mediates ABA-induced *CAT1* expression and H_2O_2 production via AtMPK6-coupled signaling in *Arabidopsis*. Plant J. 54, 440–451.

Xu, J., Duan, X., Yang, J., Beeching, J.R., Zhang, P., 2013. Enhanced reactive oxygen species scavenging by overproduction of superoxide dismutase and catalase delays postharvest physiological deterioration of cassava storage roots. Plant Physiol. 161, 1517–1528.

Yabuta, Y., Mieda, T., Rapolu, M., Nakamura, A., Motoki, T., Maruta, T., et al., 2007. Light regulation of ascorbate biosynthesis is dependent on the photosynthetic electron transport chain but independent of sugars in *Arabidopsis*. J. Exp. Bot. 58, 2661–2671.

Yin, L., Wang, S., Eltayeb, A.E., Uddin, M.I.Y., Tsuji, W., Takeuchi, Y., et al., 2010. Overexpression of dehydroascorbate reductase, but not monodehydroascorbate reductase, confers tolerance to aluminium stress in transgenic tobacco. Planta 231, 609–621.

Young, J.J., Mehta, S., Israelsson, M., Godoski, J., Grill, E., Schroeder, J.I., 2006. CO_2 signaling in guard cells: calcium sensitivity response modulation, a Ca^{2+}-independent phase and CO_2 insensitivity of the *gca2* mutant. Proc. Natl. Acad. Sci. U.S.A. 103, 7506–7511.

Zamocky, M., Furtmüller, P.G., Obinger, C., 2008. Evolution of catalases from bacteria to humans. Antioxid. Redox. Signal. 10, 1527–1548.

Zhang, F., Zhang, H., Xia, Y., Wang, G., Xu, L., Shen, Z., 2011. Exogenous application of salicylic acid alleviates cadmium toxicity and reduces hydrogen peroxide accumulation in root apoplasts of *Phaseolus aureus* and *Vicia sativa*. Plant Cell Reports 30, 1475–1483.

Zhu, J.K., 2001. Plant salt tolerance. Trends Plant Sci. 6, 66–71.

Chapter 20

Role of ROS as Signaling Molecules in Plants

Armin Saed-Moucheshi, Hassan Pakniyat, Hadi Pirasteh-Anosheh and MM Azooz

20.1 INTRODUCTION

Oxygen is one of the most important elements necessary for plant life. Plants require oxygen to be used in their cell organelles, such as the mitochondria and chloroplast, for producing sufficient amounts of energy to continue their life. Normally, plants introduce molecular oxygen (O_2) into the atmosphere by O_2-evolving photosynthetic organisms. In spite of the important roles of O_2, provoked and reactivated species of oxygen resulting from different plant cell metabolisms, and also from unfavorable conditions, have been the unwelcome companions of the aerobic life of plants (Abedi and Pakniyat, 2010; Zare and Pakniyat, 2012). During their evolution, plants have developed various mechanisms of acclimation, including various regulatory pathways for overcoming stress induced by activated oxygen.

Environmental stresses have many different effects on plants and the plants respond to these stresses with different physiological changes. One of the most important effects of environmental stresses on plants is overproduction and generating of *reactive oxygen species* (ROS), which are very reactive and interact with numerous metabolites such as DNA, proteins, RNA, lipids, pigments, and other vital cellular molecules, leading to serious damaging affects (Ashraf, 2009). Overproduction of ROS causes oxidative stress, where the capacity of cellular defenses to remove these toxic species is lower than their production. In other words, the balance between ROS generation and consumption shifts to higher ROS generation (Ahmad et al., 2008, 2010a,b, 2011a,b; Ahmad and Umar, 2011; Koyro et al., 2012). A plant's response to generating ROS under this situation is to produce ROS scavengers, consisting of enzymatic and nonenzymatic antioxidants.

In addition to damaging effects of ROS, new roles for ROS such as control and regulation of cell growth, cell cycle, programmed cell death,

P. Ahmad (Ed): Oxidative Damage to Plants. DOI: http://dx.doi.org/10.1016/B978-0-12-799963-0.00020-4

hormone signaling, plant responses to biotic and abiotic stress, and development of tissues have been reported (Foreman et al., 2003). The role of ROS as signaling molecules suggests that plants could control toxicity of ROS during evolution, and eventually use them as signaling molecules (Mittler et al., 2004). In this chapter, generation and control of ROS in plant cells are briefly discussed and the role as signaling molecules is highlighted.

20.2 ROS GENERATION AND DETOXIFYING IN PLANT CELLS

As mentioned above, the oxygen molecule (O_2) has the main function in aerobic reactions in plants. In plant chloroplasts, O_2 can be produced as a result of the oxidation of water by the photosynthetic electron transport chain (ETC). Reduction of O_2 results in reactive oxygen species (ROS) production, which can disturb a number of metabolic processes in plants (Asada, 1999). ROS are free radicals composed of oxygen which has at least one unpaired electron. Since unpaired electrons usually tend to be paired with other electrons in order to get a stable conformation, these forms of free radicals can quickly react with other molecules, generating additional free radicals (Foyer and Halliwell, 1976). When molecular reduction of O_2 takes place, it causes production of ROS, containing superoxide ($O_2^{\bullet-}$), hydrogen peroxide (H_2O_2), hydroxyl radical (($^{\bullet}OH$), and singlet oxygen ($^1O_2$) (Ahmad et al., 2008, 2010a,b, 2011a,b; Ahmad and Umar, 2011, Ahmad and Prasad, 2012a,b; Koyro et al., 2012; Ahmad et al., 2013). ROS is generated by a number of different mechanisms and are formed in different cell compartments, such as the apoplasm, mitochondria, peroxisomes, chloroplasts and endoplasmatic reticulum (Ashraf, 2009, 2011a,b; Ahmad and Umar, 2011; Ahmad and Prasad, 2012a,b; Koyro et al., 2012; Ahmad et al., 2013).

Plants have three main pathways to produce ROS, which act at different cells: 1) ETC in chloroplasts and mitochondria; 2) several peroxidases and oxidases processes (NADPH oxidase, NADH oxidase, xanthine oxidase, lipoxygenase, glycolate oxidase, amine oxidase, etc.); and 30 photosensitizer molecules containing chlorophyll (Blokhina et al., 2003). Photosynthetic processes producing light-driven products are the main ROS producers in chloroplasts. Reactions of Rubisco, which can catalyze carboxylase–oxygenase processes, causes both consumption and production of oxygen. Thus, the ETC in chloroplasts acts in a high concentration of oxygen, so that overloaded electrons leaking from ETC unavoidably lead to production of ROS. Furthermore, chlorophyll molecules cooperating with photosensitizers can facilitate the generating of ROS using the energy of sunlight (Mittler, 2002). Also, plant mitochondria as "energy factories" are believed to be a major site of ROS production, such as H_2O_2 and other ROS targets (Abedi and Pakniyat, 2010; Ahmad et al., 2011a,b; Ahmad and Umar, 2011; Ahmad and Prasad, 2012a,b; Koyro et al., 2012; Ahmad et al., 2013). Mitochondria of plants differ significantly from that of animals, as they have specific ETC apparatus and functions such as photorespiration. Moreover, because of having photosynthetic processes, creating O_2 and carbohydrate, the plant cellular mitochondria environment is also distinguishable from that in animals. In line with aerobic respiration, the free electrons resulting from mitochondrial ETC can reduce oxygen (O_2), resultiing in ROS generation in mitochondria.

The dynamic and rapid nature of ROS signaling in cells is a result of the different processes of ROS production and scavenging. Changing the balance between ROS production and scavenging results in generation of signals. In this section the most important ROS generation in cells and their scavenging will be discussed.

20.2.1 Singlet Oxygen ($^1O_2$)

Singlet oxygen ($^1O_2$) is a diamagnetic molecule formed of oxygen, which is less stable than its normal situation referred to as triplet oxygen (O_2). Depending on environmental circumstances, singlet oxygen, which has some unusual characteristics, can exist for about one hour at room temperature. Meanwhile, singlet oxygen differs from triplet oxygen related to its chemical properties and they have different electron shells. The destructive effects of sunlight on several organic materials such as polymers are regularly associated with singlet oxygen effects (Fig. 20.1).

The chlorophyll pigments attributed to the electron transport system are primary sources producing singlet oxygen. Singlet oxygen can be generated as a by-product resulting from activity of lipoxygenase as well. Like the hydroxyl radical and other ROS, $^1O_2$ is highly destructive and can react with most biological molecules. The excited singlet state of chlorophyll is used for the transfer of energy or electrons in ETC of photosynthesis. In addition to the production of energy in ETC, there are two other feasible pathways: one is fluorescence and other is production of $^1O_2$.

20.2.2 Superoxide ($O_2^{\bullet-}$) Production

Superoxide ($O_2^{\bullet-}$) is produced from one electron reduction of oxygen. Because of one unpaired electron in its electron shield, $O_2^{\bullet-}$ is known as a free radical with paramagnetic properties (Fig. 20.2).

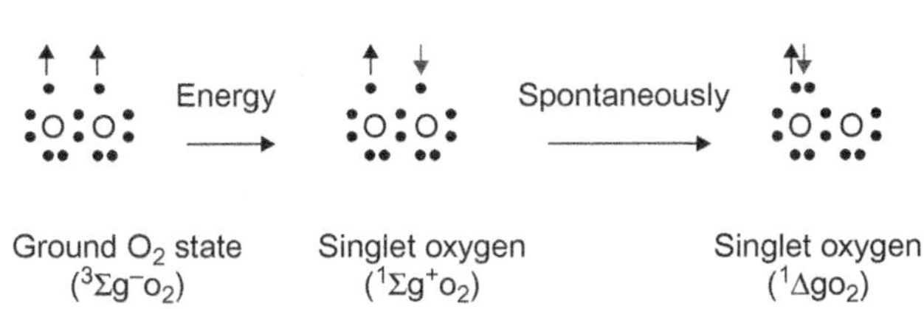

FIGURE 20.1 Formation of singlet oxygen (Saed-Moucheshi *et al.*, 2013)

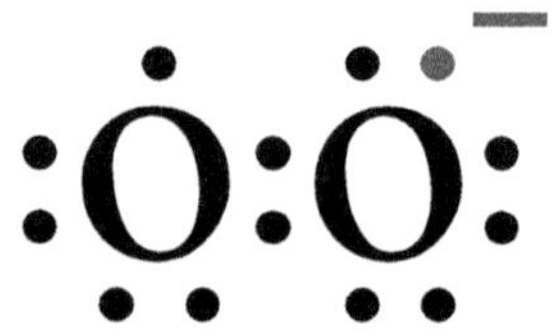

FIGURE 20.2 Superoxide.

$$2O_2 + 2Fd_{red} \rightarrow 2O_2^{\cdot -} + 2Fd_{ox,}$$

FIGURE 20.3 Formation of superoxide.

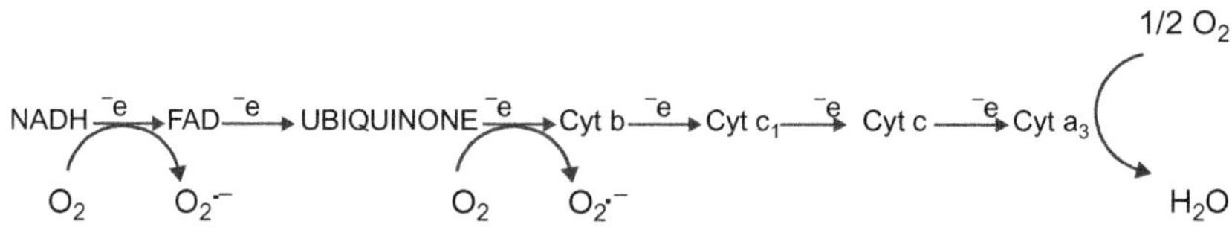

FIGURE 20.4 Superoxide radical and sites of its formation in mitochondrial electron transfer system. (Saed-Moucheshi *et al.*, 2013)

Photoreduction of O_2 in chloroplasts leads to production of acetaldehyde in presence of ethanol which generates hydrogen peroxide (H_2O_2). In most cases, electrons moving between PS-II and PS-I regulate the reduction phase of the PS-I acceptor side. This means that the redox state of PS-I acceptors does not significantly limit electron transport. The regulated activation of the Benson-Calvin cycle and the controlling rate of electron flow are important factors determining the redox phase of the ferredoxin pool. Ferredoxin and the electron carriers at the reducing side of PS-I have sufficiently negative electrochemical potential for providing electrons to O_2 (Asada et al., 1987) and forming $O_2^{\bullet -}$ radical (Fig. 20.3).

20.2.2.1 Generating Superoxide in Plant Mitochondria

Oxygen can be consumed in several pathways in plant mitochondria, which can be categorized in the three following groups:

1. Oxygen consumption through cytochrome oxidase, producing water. This process in mitochondria has responsibility for over 95% of oxygen consumption in normal conditions (Fig. 20.4).
2. Reducing oxygen and producing superoxide at the flavoprotein region of NADH dehydrogenase segment of the respiratory chain. The responsible component is probably flavoprotein (external or internal dehydrogenase) or maybe an iron–sulphur center (Fig. 20.4).
3. Reducing oxygen and producing superoxide in the ubiquinone–cytochrome region located in the respiratory chain. The process may be identified by its insensitivity to salicylhydroxamic acid and antimycin A,

FIGURE 20.5 Hydrogen peroxide.

$$2O_2^{\bullet -} + 2H^+ \rightarrow H_2O_2 + O_2,$$

$$2O_2^{\bullet -} + 2H^+ \xrightarrow{SOD} H_2O_2 + O_2$$

FIGURE 20.6 Dimutation of superoxide and formation of hydrogen peroxide.

its sensitivity to KCN and the sensitivity of the assayed rate to superoxide dismutase. So, fully reduced ubiquinone provides an electron to cytochrome C1 and leaves an unstable and highly reducing semiquinone species, which would normally reduce cytochrome b566. Since only a species at this site would have enough reducing potential for the reaction, it is likely that unstable semiquinone or an interacting species reduces the oxygen to produce superoxide (Fig. 20.4; Arora et al., 2002).

20.2.3 Hydrogen Peroxide (H_2O_2) Production

Hydrogen peroxide (H_2O_2) is a compound with an oxygen-oxygen single bond (Fig. 20.5). It is a clear liquid compound having more viscosity than water and is colorless in dilute solution as well.

H_2O_2 is produced by the superoxide radicals through dismutation in chloroplasts with a reaction commonly catalyzed by superoxide dismutase (Fig. 20.6; Asada et al., 1974).

20.2.4 Production of Hydroxyl Radical as a Highest Reactive Oxidation Radical in Plant Cells

The hydroxyl radical ($^{\bullet}OH$) is the neutral form of the hydroxide ion (OH^-). Hydroxyl radicals are highly reactive and short-lived in normal conditions. H_2O_2 and $O_2^{\bullet -}$ radicals are somewhat less destructive, but they can form species having highly destructive effects on cellular components including $^{\bullet}OH$, which can cause peroxidation of lipids and attack cellular molecules such as RNA, proteins, and DNA. Fenton in the past decades depicted the H_2O_2 oxidizing potential with ferrous salts. Then, Haber and Weiss recognized $^{\bullet}OH$ as the oxidizing species in these reactions (Fig. 20.7; Arora et al., 2002).

Accessibility of reduced ferrous in biological systems can limit the reaction, but ferric ion can be recycled to the reduced ferrous state by reducing agents such as superoxide. H_2O_2 and $O_2^{\bullet -}$ radicals form detrimental $^{\bullet}OH$,

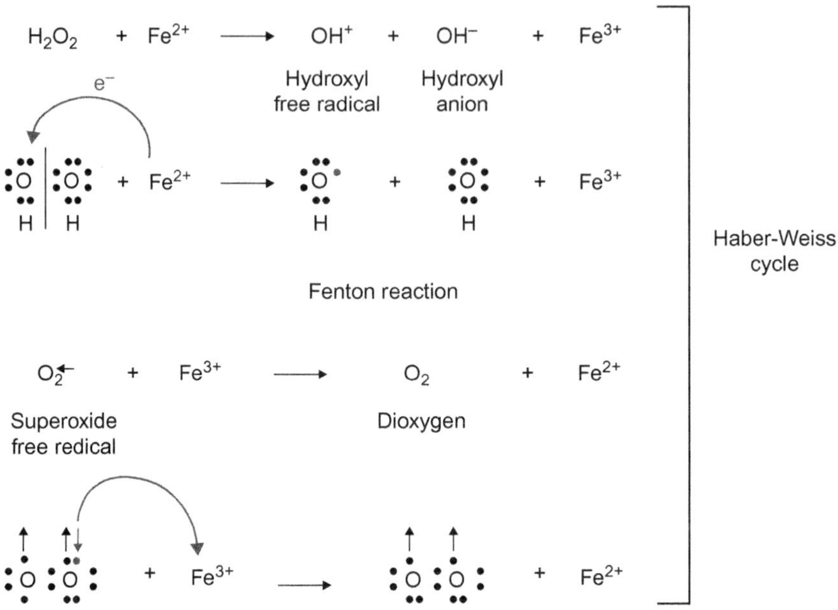

FIGURE 20.7 Fenton reaction and Haber-Weiss cycle in generation of hydroxyl radical (Saed-Moucheshi et al., 2013)

causing the oxidation of organic substances. Metal ions including Cu^{+}, Cu^{2+} can replace Fe^{2+}, Fe^{3+} in these reactions.

In short, oxygen molecules can get an electron and produce superoxide radicals. Superoxide radicals can react with two hydrogen molecules and get two electrons to generate hydrogen peroxide. Hydrogen peroxide could be detoxified and directly changed into water or be changed into hydroxyl radicals that finally will be scavenged to water (Fig. 20.8).

20.2.5 ROS Detoxification in Plants

There is universal agreement that ROS in plants are produced under normal conditions of growth while their content stays low (Polle, 2001). On the other hand, environmental stresses usually increase ROS production (Laloi et al., 2004; Ahmad and Umar, 2011; Ahmad and Prasad, 2012a,b; Zare and Pakniyat, 2012; Ahmad et al., 2013a,b; Rasool et al., 2013). Under physiological conditions, the concentration of H_2O_2 in the cell is usually less than 1 μM. Photo-induced ROS generation depends mainly on conditions of the ambient medium and on the physiological state of the photosynthetic apparatus (Foyer and Noctor, 2005, 2009). Even at high light, when the flow of electrons through the water-water cycle (oxygen reduction to water in the photosystem I (PSI) at expense of electrons generated in PSII due to photodegradation of water) increases, no substantial amounts of $^1O_2$ and H_2O_2 are accumulated, as

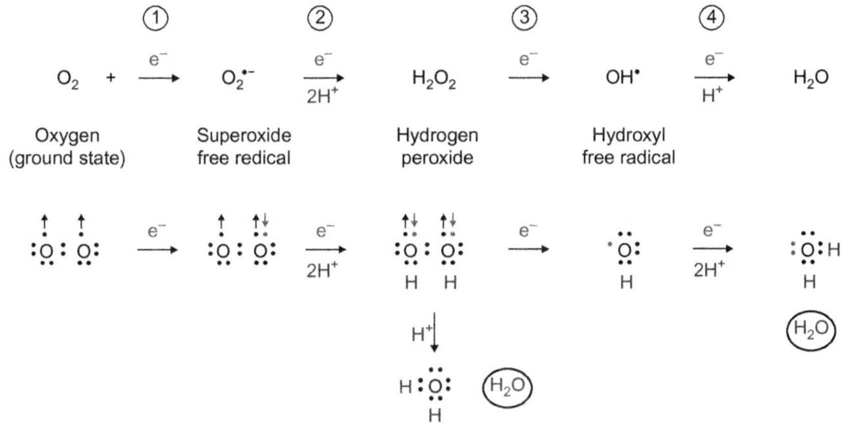

FIGURE 20.8 Summary of ROS production (Saed-Moucheshi *et al.*, 2013).

sufficient amounts of $NADP^+$ are present in the cell. At high light intensity and conditions of CO_2 assimilation saturation such as stomatal closure that limits CO_2 availability, increases occur in the rate of electron flow, and this leads to its redistribution. The rate of electron flow to $NADP^+$ reduces, and the rate of electron transfer to molecular oxygen (pseudocyclic electron transport) increases. This results in generation of superoxide radical in PSI, which is produced mainly on its acceptor side (Kreslavski et al., 2012).

During stress periods, ROS are generated in plants through pathways as mitochondrial respiration, photorespiration, and from the photosynthetic counterparts. Moreover, biotic or abiotic stresses can cause production of ROS by NADPH oxidases (Pei et al., 2000). Also, plants can scavenge ROS by producing antioxidants. Antioxidants are commonly grouped into two types, containing enzymatic and nonenzymatic antioxidants. Enzymatic antioxidants contain peroxidase (POD), catalase (CAT), superoxide dismutase (SOD), and some other enzymatic antioxidants that are in charge in the ascorbate-glutathione cycle, such as ascorbate peroxidase (APX), monodehydroascorbate reductase (MDHAR or MDAR), dehydroascorbate reductase (DHAR or DAR) and glutathione reductase (GR) (Ahmad et al., 2008, 2010a, b, 2011a,b). The commonly known nonenzymatic antioxidants are glutathione (GSH), ascorbate (AsA), carotenoids, tocopherols, flavones and anthocyanins (Gupta et al., 2005). Ascorbate and glutathione are the most important nonenzymatic antioxidant molecules, and are involved in the ascorbate-glutathione cycle as well (Ahmad et al., 2013; Rasool et al., 2013).

Enzymatic antioxidants containing SOD, CAT, APX, POD, GR and MDAR are recognized to decrease the levels of O_2^- and H_2O_2 in plants (Ali and Alqurainy, 2006; Ahmad et al., 2013; Rasool et al., 2013). SOD catalyzes the dismutation of O_2^- to molecular oxygen and H_2O_2 (Fig. 20.9.; Abedi and Pakniyat, 2010; Zare and Pakniyat, 2012). It is one of the most important

$$M^{(n+1)+}\text{-SOD} + O_2^{-} \rightarrow M^{n+}\text{-SOD} + O_2$$

$$M^{n+}\text{-SOD} + O_2^{-} + 2H^{+} \rightarrow M^{(n+1)+}\text{-SOD} + H_2O_2$$

Where M = Cu (n=1); Mn (n=2); Fe (n=2); Ni (n=2)

FIGURE 20.9 Functional mechanism of superoxide dismutase for detoxifying superoxide radical.

$$\text{Ascorbate} + \text{Hydrogen peroxide} \xrightarrow{\text{APX}} \text{Monodehydroascorbate} + \text{Water}$$

$$C_6H_8O_6 + H_2O_2 \xrightarrow{\text{APX}} C_6H_6O_6 + 2\,H_2O$$

FIGURE 20.10 Functional mechanism of ascorbate peroxidase for detoxifying hydrogen peroxide.

enzymatic antioxidants that plants are using against oxidative stresses, and it exists universally in every plant cell. SODs are multimeric metalloproteins based on the metal classes existing at their active sites. The best-known isoforms of SODs occurring in plants are copper–zinc (Cu/Zn-SOD), manganese (Mn-SOD), iron (Fe-SOD) and nickel (Ni-SOD) containing superoxide dismutase. Inducing of SOD in plant cells in response to altered stressful conditions shows its significant role in the plant's defense system. Stressful conditions commonly increase SOD activity to detoxifying superoxide radicals (Abedi and Pakniyat, 2010; Zare and Pakniyat, 2012).

Similar to SODs, catalases (CATs) have an important role in plant defense under oxidative stresses. CATs in plants are transcript by nuclear genes, and are tetrameric iron porphyrins. They are principally produced in peroxisomes and glyoxisomes in plants. CAT catalyzes a redox reaction in which dismuting of H_2O_2 changes its form to oxygen and water. However, CAT is specific to detoxification of H_2O_2. It can react with various organic hydroperoxides such as methyl hydrogen peroxide (MeOOH) also (Ali and Alqurainy, 2006).

Ascorbate peroxidase (APX) is also an important enzyme for detoxification of H_2O_2 in plants (Asada, 1999; Ahmad et al., 2013, Rasool et al., 2013). Hydrogen peroxide reacts with ascorbate to form monodehydroascorbate and water and the reaction is catalyzed by APX (Fig. 20.10). Four classes of APX are recognized in the plant cells: glyoxisome membrane (gmAPX), chloroplast thylakoid bound (tAPX), chloroplast stromal soluble (sAPX), and cytosolic (cAPX). Investigations showed increasing APX expression in plants in response to diverse abiotic stress conditions. APX overexpressing in chloroplasts of tobacco plants made them more tolerant to salinity stress and drought conditions (Abedi and Pakniyat, 2010; Zare and Pakniyat, 2012).

The enzymatic antioxidants working in the ascorbate-glutathione cycle also play a central role in stabilizing oxidative stresses in plants. Ascorbate

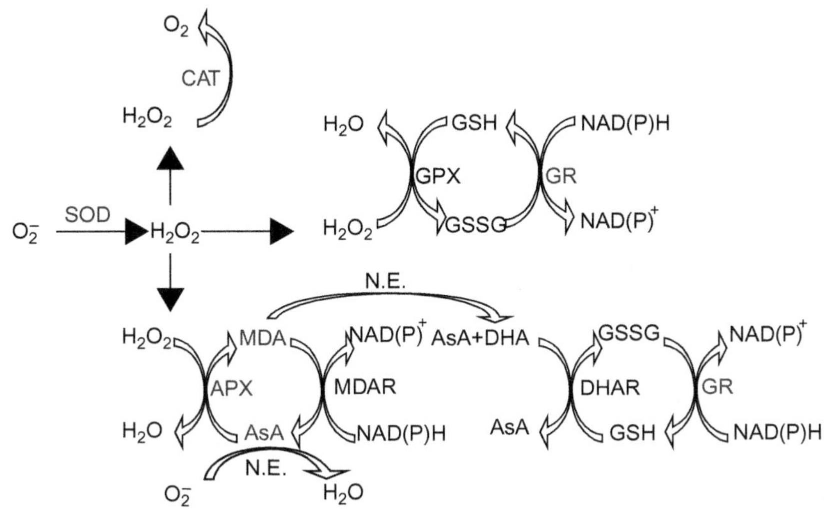

FIGURE 20.11 Ascorbate-glutathione cycle and some other metabolisms for scavenging ROS (Saed-Moucheshi *et al.*, 2013).

peroxidase (APX) catalyzes conversion of ascorbate to monodehydroascorbate (MDA). Revival of ascorbate from monodehydroascorbate for scavenging of H_2O_2 in chloroplasts is required. Monodehydroascorbate reduction in stroma is catalyzed by monodehydroascorbate reductase (MDAR). If reduced ferredoxin or MDAR could not transform MDA to ascorbate, dehydroascorbate (DHA) would be produced. Thiol enzyme DHAR causes ascorbate renewal from dehydroascorbate, although dehydroascorbate cannot produce as much ascorbate as MDAR (Asada and Takahashi, 1987; Fig. 20.11).

Glutathione reductase (GR) is another enzyme working in the ascorbate-glutathione system. It is localized primarily in chloroplasts, while a small amount can be found in cytosol and mitochondria (Creissen et al., 1994). GR caused glutathione reduction and antioxidative processes in plants.

Apart from enzymatic antioxidants, nonenzymatic antioxidants also have considerable roles in neutralizing oxidative stresses. The nonenzymatic antioxidants include tocopherols, glutathione, flavonoids, carotenoids, ascorbic acid, etc. (Schafer et al., 2002; Ahmad et al. 2008, 2010a,b, 2011a,b; Ahmad and Umar, 2011; Ahmad and Prasad, 2012a,b; Koyro et al., 2012; Ahmad et al., 2013; Rasool et al., 2013).

Glutathione (GSH) is a cysteine-containing tripeptide having important roles in plants. It appears in reduced form in plant tissues and is localized in almost all compartments of plant cells such as chloroplasts, apoplast, mitochondria, cytosol, vacuole, peroxisomes, and endoplasmic reticulum (Jimenez et al., 1998). Glutathione is vital for sustaining plant cells as it provides protection from all deleterious effects of oxidative stresses

(Wang et al., 2008). GSH plays a key role in the antioxidative defense system by regenerating another potential water soluble antioxidant, ascorbic acid, via the ascorbate-glutathione cycle (Foyer and Halliwell, 1976). Glutathione is the substrate of glutathione S-transferase (GST), which plays an important role in detoxifying of dehydroascorbate reductase (DHAR) and xenobiotics (Mendoza-Cozatl and Moreno-Sanchez, 2006). Glutathione conserves redox equilibrium in the cellular compartments by combining with its oxidized form (GSSG). This form of glutathione has a significant biological role for conserving the normal cellular redox system under stressful or normal situations (Wang et al., 2008).

Ascorbic acid (Vitamin C; AsA; AA) is one of the most important antioxidants among nonenzymatic antioxidants. Similar to glutathione, it is a water soluble metabolite and can be found in different organelles of the plant cell. Ascorbic acid is obtained frequently in its reduced form in plant leaves and chloroplasts under normal physiological conditions. Ascorbic acid is also available regularly in apoplast. Its concentration in plant cells can rise to millimolar range, and is generally greater than glutathione concentration (Smirnoff, 2005). Ascorbic acid is the most powerful ROS detoxification compound because of its capacity to provide electrons in many nonenzymatic or enzymatic reactions. Ascorbic acid can directly quench $O_2^{\bullet -}$, $^1O_2$ and also hydroxyl radicals. It can reduce H_2O_2 to water via the ascorbate peroxidase reaction (Foyer et al., 1997). Furthermore, ascorbic acid can revive tocopherols from tocopheroxyl radical and thus provide membrane protection. Therefore, raised endogenous ascorbic acid levels in plants are essential to balance damaging effects of oxidative stress (Smirnoff, 2005).

Carotenoids are pigments that have multiple functions in metabolism of the plants, including tolerance to oxidative stress. Carotenoids are lipophilic organic compounds that are available in plant chloroplasts. These kinds of plant pigments are generally mentioned as antenna molecules, which can capture photon light of the sun in the visible spectrum ranging between 450–570 nm and transport it to the plant chlorophylls (Taiz and Zeiger, 2006). Moreover, carotenoids have responsiblity for providing photoprotection to the photosynthetic structures and apparatus. Different forms of carotenoids are available in plant cells; but β-carotenes are the most important and principal carotenoids in higher plants. By means of quenching the triplet state, β-carotenes can efficiently avoid singlet oxygen being produced in chlorophyll molecules.

Tocopherols also can detoxify ROS and lipid radicals (Holländer-Czytko et al., 2005), and are available in all parts of the plant. Tocopherols are crucial metabolites in biological membranes, and they can have both roles as antioxidant and as nonantioxidant components. Four isomers (α-, β-, γ-, δ-) of tocopherols are recognized in plants (Ashraf, 2009). α-Tocopherol which is also referred to as vitamin E is a membrane-bound compound having the highest antioxidative activity among the above mentioned tocopherols, because of having three methyl groups in its molecular construction. It is well documented that chloroplast membranes belonging to the higher plants contain a considerable quantity of α-tocopherols, so that they are well secured against photo-oxidative destructions. Similar to carotenoids, tocopherols protect the thylakoid membranes and avoid the chain propagation step during lipid auto-oxidation (Serbinova and Packer, 1994).

Flavonoids are also plant pigments with antioxidant properties against a range of oxidizing compounds. They belong to the phenolics, a large class of organic compounds. Flavonoids are commonly found in plants, and are generally found in floral parts, pollens and plant leaves. These pigments are regularly accumulated in the vacuole of the plant as glycosides, but they also can be observed on the leaf surfaces, and other aerial parts of the plants as exudates. A number of flavonoids act as the potential inhibitor of the lipoxygenase enzyme, which converts polyunsaturated fatty acids to oxygen-containing derivatives (Nijveldt et al., 2001).

There are various other metabolites having antioxidizing activity such as alkaloids, phenolic acid, diterpenes, polyamines, proline and other amino acids and amines, but little is known about their antioxidative functions under stressful conditions (Saed-Moucheshi et al., 2013).

Antioxidative activities of enzymatic and nonenzymatic antioxidants are upregulated under stress conditions but their activities are diverse in different plant species (Smirnoff, 2005). Consequently, it can be observed that the effectiveness of plant antioxidant systems for detoxification of ROS and protecting plants against them depends on the plant species and genus, together with plant genetic background, stress intensity levels and also the growth stage of exposed plants. On the other hand, plant biologists have focused more on understanding the functions of different antioxidant in plant response to stresses. Because of significant differences in the protection process against ROS in plants, a general validity for effectiveness of antioxidants in plants' tolerance to stresses cannot be established (Ashraf and Harris, 2004), but it is well-known that they have a significant effect on plant metabolism pathways under normal or stress conditions.

20.3 ROS AS SIGNALING MOLECULES

ROS are extremely reactive and can react with most of the cellular organisms. It has been well documented that ROS initiate strong osmotic stress accompanied by injuries and death of the organism. The signaling molecules induce some molecular, biochemical, and physiological responses, which help to develop the adaptive mechanisms and improve tolerance in plants (Kreslavski et al., 2012). ROS can modify proteins, changing their structure and activities, in particular via action on thiol groups and iron-containing clusters in proteins (Spadaro et al., 2010). Because of such modifications, the detachment from the cell wall of weakly bound enzymes like isoforms of peroxidases (Kreslavski et al., 2012) and also changes in activities of some antioxidant enzymes like catalase and ascorbate peroxidases can occur (Shao et al., 2008). In addition, ROS could change the redox potential of redox-sensitive cell components (glutathione system, ascorbate system, plastoquinone pool, thioredoxin, etc.). Also, in response to changing normal environmental conditions, various compounds such as signaling intermediates (calcium ions, cAMP, nitrogen oxide), phytohormones (ethylene, ABA, salicylic acid), osmolytes (amino acids, sugar alcohols, tertiary amines), and some other metabolites in plant cells and tissues are accumulated (Kreslavski et al., 2012) and they interact with one another.

20.3.1 Redox Potential Related to ROS

It is obvious that the redox status of plant cells is critical for correct function of several enzymes, and it can be used to change activity of enzymes; therefore changing redox status of plant cells can be described as a signaling mechanism (Hancock et al., 2001). Along with regulatory effects of prooxidant compounds and the products of oxidative modification of macromolecules, in modern studies increasing attention is being paid to the *redox* control (Pfannschmidt et al., 2009). The models of this control are different and contain oxidation/reduction of flavins, iron-sulfur cores, thiol groups and hemes (Foyer and Noctor, 2005). This agrees with the fact that the cell redox homeostasis is mainly controlled by the presence of the large pools of thiol buffer-glutathione and NADPH/NADP+ , and also by the high concentration of ascorbic acid. Along with hydrophilic compounds, tocopherols can function as the lipophilic redox buffer; they protect lipids and other membrane components of chloroplasts by physical scavenging and chemical interaction with $^1O_2$ (Ivanov et al., 2007).

In the perception and transduction of the signal for osmotic stress inducing a transcriptional response, the main role belongs to various protein sensors. These redox-sensitive proteins are capable of reversible oxidation and reduction and serve as a molecular "switch" responding to redox changes in the cells (Kreslavski et al., 2012). ROS can oxidize redox-sensitive proteins directly or indirectly, via low-molecular redox-sensitive molecules, such as reduced glutathione or thioredoxin interacting with ferredoxin. Redox-sensitive enzymes are capable of directly modulating processes of the cell metabolism, whereas redox-sensitive signaling proteins (RSPs) function with the involvement of other components of the signaling pathways: MAPK, phosphatases, transcription factors, etc. (Pfannschmidt et al., 2009). A relevant case study for RSP association with redox signaling is the cytochrome-bf complex of photosynthetic electron transport chain localized in the chloroplast of higher plants. Since PSII and PSI have an in-tandem action with *z*-scheme in the electron transport chain system of photosynthesis, excess photochemical energy, abbreviated as EPE, is passed to two reaction centers and required to be regulated. If balance between light utilization and its dissipation is absent, toxic free radicals can be formed, resulting in oxidative injuries in plants (Bhattacharjee, 2005). When the amount of reactive oxygen species under excess photochemical energy conditions surpassed the capacity to which they can be detoxified, oxidative injuries to the photosynthetic system occurred and caused cell death. On the other hand, ROS could have a positive function in the plant response to excess photochemical energy by means of intensifying the degradation rate of D1 protein, which acts in PSII (Karuse, 1994). This leads to photoinhibition of photosynthetic electron transport, which may be a protective mechanism in such conditions. There is also a process of controlling dissociation of light-harvesting complex from PSII, which is controlled by phosphorylation. The kinase responsible for that phosphorylation is activated by a reduced plastoquinone pool, a signal that is transduced to kinase activation via structural change of Fe-S protein associated with cytochrome-bf complex51. Additionally, the redox state of plastoquinone controls the transcription of chloroplast genes encoding PSII and PSI reaction center apoprotein52 as well as mRNA stability and translation rate (Salvador and Klein, 1999).

However, the main source of ROS in photosynthesizing organisms is still believed to be chloroplast and primarily its photosynthetic ETC. Some researchers believe that one of the key ROS sources during stress is the redox regulators in the apoplast (Shao et al., 2007). At present, several possible sources of chloroplast signals, which can affect gene expression in the nucleus of the plant cell, are known. Among them are the biosynthesis of tetrapyrrol compounds, change in the redox potential of the photosynthetic ETC components, primarily plastoquinones, and also ROS generation. All these pathways could be interconnected; therefore, they are often considered as coupled factors. Unfortunately, the exact mechanism of transduction of signals about the accumulation of redox-sensitive compounds and ROS in the chloroplasts to the nucleus is poorly studied or unknown (Kreslavski et al., 2012).

One of the key redox sensors is a PQ, which is a factor of adaptation to light; it functions, in particular, via the regulation of plastid and nuclear genes, such as *LHCB*. The regulation of nuclear gene expression at the change in the redox potential of the PQ was demonstrated for genes *petE* (plastocyanin), *APX2* (ascorbate peroxidase 2), and *ELIP2* (early inducible protein), and also genes encoding PSI proteins (*psaD* and *psaF*). It should be noted that expression of *LHCB* is partially controlled by the redox state of PQ because the ATP synthesis and transmembrane potential also contribute to the redox state (Yang et al., 2001). It is supposed that PQ is the source of two signaling pathways, which are initiated under the influence of high and low light and activate expression of some plastid and nuclear genes. It might be that ROS arising on the acceptor side of PSII are some of regulators of these pathways. Changes in the redox potential on the PSI acceptor side also contribute to the regulation of the photosynthetic electron transport. This is primarily related to thioredoxin, which depends on the rate of electron transport from ferredoxin. The activities of numerous chloroplast enzymes depend on the redox state of these two compounds.

The analysis of changes in expression of so-called reporter genes plays a great role as these genes encode key proteins involved in the development of cell responses to changes in the redox state. In particular, the investigations of redox signaling between chloroplasts and nucleus were focused on the induction of genes of cytosolic ascorbate peroxidases *APX1* and *APX2*, genes *ZAT10* and *ZAT12* encoding zinc-finger transcription factors, and also *ELIP2* gene encoding a chlorophyll-binding protein. Both *ZAT10* and *ZAT12* genes favor the induction of gene clusters related to activation of the photosynthetic ETC under high light switching, including *APX1* and *APX2*. Treatment with hydrogen peroxide stimulated expression of *APX2, ZAT10*, and *ZAT12* genes. It may be assumed that ROS regulate the expression of genes presented in Fig. 20.12 by direct or indirect action on the chloroplast redox potential (Kreslavski et al., 2012).

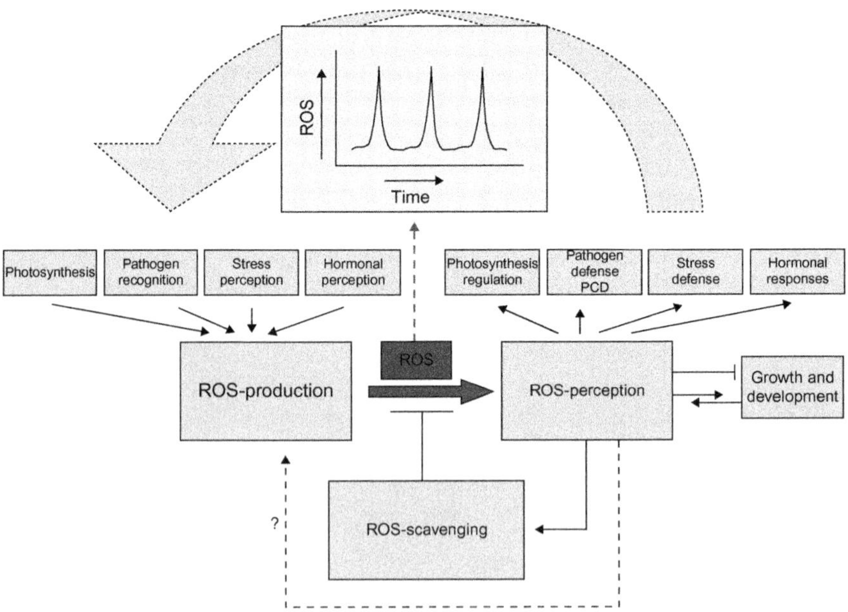

FIGURE 20.12 Modulating reactive oxygen species (ROS) signaling by the reactive oxygen-gene network of plants. Different cellular signals (e.g. pathogen recognition or stress perception) result in the enhanced production of ROS in cells by the ROS-producing pathways of the network. ROS are perceived by different ROS sensors and activate cellular responses (e.g. pathogen or stress defense). Modulation of ROS levels might also involve a positive feedback loop between ROS perception and ROS production (dashed line). In addition to activating or suppressing different cellular responses, ROS perception can affect growth and development (Mittler et al., 2004). Abbreviation: PCD, programmed cell death

Change in the redox potential triggers the mechanism of signal transduction from chloroplast into the cytosol and then into the nucleus. According to kinetic experiments, the redox signal arising in response to the change in light quality (spectral composition) is transmitted from the chloroplast to the nucleus for approximately 30 min, which corresponds to the time of signal transduction within plastids (Pfannschmidt et al., 2009). In particular, experiments in which high light effects on *Arabidopsis* plants were studied showed that nuclear genes *APX1* and *APX2*, encoding cytosolic peroxidases, were activated in 15–20 min. It was also found that activation of these genes was a part of the systemic response to excessive light energy. In experiments with isolated chloroplasts, it was shown that induction of chloroplast gene expression occurred in 15–20 min in response to changes in the redox potential in the organelles induced by changes in light quality (Kreslavski et al., 2012). It is possible to suggest that the same redox signals induced by changes in the light quality were transmitted to the nucleus for 30 min, which corresponds to the temporal range of the events in plastids. It is also possible that some components of the cascade signaling about changes in the ROS concentration in response to stress are present in the cells under optimum conditions, but not synthesized in response to stress. This explains the fact that the time of signal transduction about ROS appearance corresponds to the time of transduction of other intracellular stimuli (Kreslavski et al., 2012).

Accumulating redox-active compounds, such as free radicals, inside the plant chloroplast is associated with the photosynthetic electron transport rate. Redox-active systems, like thioredoxin and PQ, could function via definite pathways of signal transduction or known second messengers appeared in the cytosol in response to abiotic stressors or interact with them. Signals arising during ETC functioning could also change expression of nuclear genes by generation of systemic signals in the process known as systemic acquired resistance. PK-protein kinases; Ph-phosphatases; DAG-diacylglycerol; PhLC-phospholipase C; IP3-inosytol 1,4,5-triphosphate; *ZAT10* and *ZAT12*-genes encoding transcription factors; *LHCB*-gene encoding a protein of LHG in PSII; *ELIP2*-gene encoding early light-inducible protein; *CHS*-gene encoding chalcone synthase; *AOX1*-gene encoding mitochondrial alternative oxidase are known secondary messengers. The question mark indicates a hypothetic sensor of the cell redox state. The scheme is built on the basis of data from Kreslavski et al. (2012).

20.3.2 ROS and Plant Development

ROS have been shown to play an important role in plant development. ROS that are produced by NADPH oxidases generate the superoxide radical ($O_2^{\bullet -}$). The NADPH act as an electron donor and are the most responsible in this plant regulatory pathway. As it is mentioned before, NADPH oxidase proteins in plants are like the mammalian enzymes responsible for the respiratory burst, which are available in activated mammalian neutrophils (Segal and Abo, 1993). This group of genes in *Arabidopsis* is mentioned as *Arabidopsis* respiratory burst oxidase homologs (Atrboh; Keller et al., 1998; Torres et al., 1998). The activity of three members of this family has been shown to be involved in different aspects of root growth (Foreman et al., 2003; Kwak et al., 2003). The root hair defective2 (RHD2)/AtrbohC protein is required for root elongation. In the *rhd2* mutant for *Arabidopsis* plants losing function of RHD2, levels of ROS in their roots have decreased and they were 20% shorter than the wild type, indicated that cell expansion is defective in these mutant plants (Foreman et al., 2003; Renew et al., 2005). This deficiency results in altered uptake of Ca^{2+} and distraction for expansion of root cells. Furthermore, quenching ROS directly caused inhibition of root elongation (Laloi, et al., 2004). Applying some inhibitors such as diphenylene iodonium (DPI) showed that NADPH oxidases-derived ROS regulates expansion of cells in maize roots (Liszkay et al., 2004), indicating that ROS contributing and mediating in growth is not a phenomenon only related to *Arabidopsis*. Since DPI is a common flavin enzymes inhibitor, being treated with DPI might affect the other protein activities in addition to NADPH oxidases (Moulton et al., 2000). Consequently, without direct genetic proof for contribution of NADPH oxidases in this pathway, the role of NADPH oxidases requires confirmation in maize.

The roots of plants lacking both AtrbohD and AtrbohF (*atrbohd*, *atrbohf* double mutants) are indistinguishable from the wild type, indicating that they are not involved in growth under standard conditions, but the double-mutant roots are less sensitive to the inhibitory effects of ABA on root elongation (Kwak et al., 2003). The fact that AtrbohD and AtrbohF are also required for the stomatal response to ABA, it suggests that ROS derived from the D and F proteins are involved in the ABA-signaling mechanism that controls plant growth responses in drought conditions. It could be shown that there are at least two distinct ROS-requiring mechanisms that occur during root growth and elongation in *Arabidopsis*, containing a requirement of RHD2/AtrbohC for elongation and an ABA-related growth inhibition process that requires AtrbohD and AtrbohF.

Whereas there is clearly a role for ROS in root elongation, there is also evidence that NADPH oxidases-derived ROS are required during the growth of other organs. During leaf expansion, a wave of ROS-dependent cell growth sweeps through the leaf (Rodriguez et al., 2002). This local expansion zone is the site for accumulation of ROS, and inhibition of ROS formation by treating with DPI inhibits leaf growth. This indicates that not only ROS are involved in growth, but also a flavin-containing oxidase such as NADPH oxidases is required for their production. Furthermore, the accelerated elongation that occurs upon auxin treatment is accompanied by the formation of higher levels of ROS in comparison to coleoptiles grown without auxin treatment (Schopfer et al., 2002) suggesting that the rate of cell growth may be proportional to the amount of ROS produced in growing organs.

Although this evidence suggests that NADPH oxidases control cell growth, there is some evidence to suggest that these enzymes control other aspects of development, such as apical dominance and leaf shape (Sagi et al., 2004). Transgenic plants harboring NADPH oxidases-antisense constructs not only had reduced NADPH oxidases activity and ROS levels, but exhibited some morphological defects, including reduced apical dominance, leading to an increase in branching; reduced leaf lobing; and curled leaflets. These phenotypes suggest that ROS control more processes than cell expansion (Gapper and Dolan, 2006). Collectively, these results indicate that ROS production by catalytic subunits of several plasma-membrane NADPH oxidases may control plant development by activating Ca^{2+} channels. *Arabidopsis* possesses ten different Atrboh genes, which are differentially expressed, so it is conceivable that different Rboh, either alone or in various combinations, may contribute to the release of signals in a variety of different cell types and in response to numerous endogenous and/or exogenous cues (Kwak et al., 2003).

The cell wall plays an important role in cell expansion; loosening the wall allows cells to expand, whereas wall cross-linking can inhibit expansion. There is evidence that ROS are involved in both processes. They have been implicated in the loosening of cell walls in growing tissues and also in making cell walls hard and difficult to bend as growth ceases and cells differentiate (Gapper and Dolan, 2006).

At least part of the stimulatory effect of ROS on expansion is through the ability of the hydroxyl ($^{\bullet}OH$) radical to promote cell growth. $^{\bullet}OH$ radicals are present in the walls of cells in elongating organs *in vivo* (Renew et al., 2005). One way in which $^{\bullet}OH$ radicals regulate growth is through the schisming xyloglucan polymers allowing walls to be stretched. If, as is likely to be the case, the $^{\bullet}OH$ radical has the same effect *in vivo*, then it is likely that it contributes to growth by decreasing the resistance of the wall to the pressure from the expanding protoplast through xyloglucan schisming.

20.3.3 Hydrogen Peroxide (H_2O_2) as Signaling Molecule

The most long-living ROS species is H_2O_2, which is also very reactive. Even at the concentration of 10 μM, hydrogen peroxide could inhibit CO_2 fixation by 50%, mainly due to the oxidation of SH groups of the Calvin-cycle enzymes (Foyer and Noctor, 2005, 2009). $^1O_2$ can be produced due to charge recombination in the PSII reaction centers (Krieger-Liszkay, 2005), whereas H_2O_2 is produced mainly in PSI at formation of $O_2^{\bullet -}$. H_2O_2, being uncharged and a rather stable molecule, can pass great distances and penetrate across membranes but it is less toxic than other ROS. In this relation, H_2O_2 is a most suitable candidate for the role of intra- and interorgan messenger

(Hung et al., 2005). Plant treatment with H_2O_2 induces the expression of chloroplast and nuclear genes (Fig. 20.12) related to plant protective responses to stress (Foyer and Noctor, 2009). In particular, H_2O_2 was found to activate genes encoding many antioxidants and signaling proteins such as ascorbate peroxidase, glutathione reductase, catalase, MAPK, and phosphatases.

Some data were published concerning the regulatory role of H_2O_2 in both functioning of the photosynthetic apparatus and the development of its stress protective systems (Kuznetsov and Shevyakova., 1999). It is assumed that H_2O_2 of chloroplast origin can serve a redox signal triggering the expression of the gene encoding cytoplasmic ascorbate peroxidase. H_2O_2 produced on the cytoplasmic membrane or in the apoplast can also play the role of the redox signal, possibly functioning together with ABA (Bechtold et al., 2008). H_2O_2 appearance in the cytoplasm can be a signal for chloroplasts. H_2O_2 molecules produced in the chloroplasts can diffuse outside the organelles, inducing signaling processes in the cytoplasm (Mubarakshina et al., 2010), switching activation of the MAP-kinase cascade, which activates nuclear genes in the cell, in particular, the gene encoding cytoplasmic APX (Yabuta et al., 2004; Fig. 20.12).

H_2O_2 is also involved in the induction of expression of some light-responsive genes. Thus, plant treatment with H_2O_2 stimulated expression of the *APX2* gene encoding ascorbate peroxidase and *ZAT10* and *ZAT12* genes encoding transcription factors, whereas catalase addition resulted in the reduced expression of *APX2* and *ZAT10* (Davletova et al., 2005).

The interaction between signaling pathways of H_2O_2 and $^1O_2$ is possible via modulation of the redox state of PQ. H_2O_2 induces oxidation of the primary electron acceptor QA, which enhances electron transport and reduces a probability of singlet oxygen formation under high light (Laloi et al., 2007).

Although sensors of H_2O_2 in plant cells are unknown, experiments performed on the cyanobacterium *Synechocystis* (oxidative stress was created by the addition of H_2O_2 (0.25 mM) to the cells for 20 min) showed that in cyanobacteria histidine kinases could serve as H_2O_2 sensors. Mutations after the genes of histidine kinases Hik34, Hik16, Hik41, and Hik33 prevented expression induction of genes stimulated by hydrogen peroxide (Kanesaki et al., 2007). These histidine kinases controlled 26 of 77 genes induced by hydrogen peroxide. Histidine kinase Hik34, which has been earlier characterized as the regulator of gene expression in heat, salt, and hyperosmotic stresses, regulated expression of *htpG* gene at oxidative stress as well. In addition, Hik34 was subjected to autoregulation in the presence of H_2O_2. The pair of histidine kinases Hik16–Hik41 regulated the genes *sll0967* and *sll0939* with unknown functions. These histidine kinases controlled the same two genes under salinity and hyperosmotic stress. Hik33 controls 22 genes; among them are *ndhD2* encoding NADH dehydrogenase, three *hli* (high-light-inducible) genes, *pgr5* encoding ferredoxin-plastoquinone reductase, genes *nblA1* and *nblA2* involved in phycobilisome degradation, and others. It should be noted that osmotic stress induces also HSP genes, *hspA, dnaJ, dnaK2, clpB1, ctpA*, and *sigB*. It is true that, in this case, these genes were activated without the involvement of histidine kinases but by some other mechanism, in spite of the fact that the repressor of HSP genes, Hik34, is one of the regulators of osmotic stress genes (Kreslavski et al., 2012).

In incompatible plant-pathogen interactions H_2O_2 has been implicated in the elicitation of a variety of defense responses. Increase in H_2O_2 production (as a result of oxidative burst) has been noticed by action of plasma-membrane associated NADPH-dependent superoxide synthase together with apoplastic SOD during hypersensitive response (Varnova et al., 2002). Since H_2O_2 is an endogenous oxidant that accumulates under stress situations and senescence, a central role of this metabolite as a diffusible signal molecule for selective induction of defense genes has been envisioned. So, an important contribution to the redox state of plant cells by H_2O_2 has been implicated, which ultimately plays a crucial role in stress tolerance (Bhattacharjee, 2005).

There are several lines of evidence suggesting strongly that H_2O_2 initiates a signal transduction process for acquisition of tolerance to abiotic and biotic stresses. A transient increase in H_2O_2 was suggested to signal activation of the protective mechanism for acclimation to chilling. Exogenous application of H_2O_2 can induce tolerance to chilling, high temperature and biotic stress, all of which cause elevated endogenous H_2O_2 production (Matusda et al., 1994). Treatment of winter wheat with low concentrations of H_2O_2 and catalase inhibitor induced the synthesis of polypeptides similar to those found when plants were exposed to low temperature (Knight et al., 1996). Prasad et al. (1994) also demonstrated that maize seedlings became more chilling-resistant following treatment with low concentrations of H_2O_2. A transient increase in H_2O_2 was suggested to signal activation of protective mechanisms for acclimation to chilling. In *Arabidopsis*, treatment with H_2O_2 produced altered cytosolic Ca^{2+} concentrations similar to those observed during chilling acclimation. In *Arabidopsis* suspension culture, application of harpin (a proteinaceous elicitor) not only induces oxidative burst that requires protein phosphorylation and Ca^{2+} influx, but also causes expression of defense-related genes (Bhattacharjee, 2005).

H_2O_2 production is correlated with the stiffening of cell walls as growth ceases in a number of cell types. For example, H_2O_2 is produced at the same time as lignification in the walls of *Zinnia elegans* xylem tracheary elements and required for the formation of the complex cross bridges found in the walls of this cell type (Ros-Barcelo et al., 2002). Similar H_2O_2-induced stiffening of the cell wall by increasing cross-linking of polymers occurs as growth ceases at the onset of differentiation in many cell types (Brisson et al., 1994). A variation on this theme is the oxidative H_2O_2-mediated cross-linking of cell wall proteins, which reinforces the cell wall, thereby blocking the pathogen spread that occurs upon infection (Brisson et al., 1994).

H_2O_2 has also been shown to be involved in differentiation of the cellulose-rich cell wall in cotton fibers. An increase in H_2O_2 production coincides with deposition of the secondary walls of cotton (*Gossypium hirsutum*) fibers (Potikha et al., 1999). Furthermore, incubation of developing cotton fibers with H_2O_2 scavengers blocks differentiation, whereas application of H_2O_2 stimulates secondary wall formation. Its precise role in the wall differentiation process is unknown, but it may act by stimulating the activity of cellulose synthases that are active during the growth of the cotton fiber (Gapper and Dolan, 2006).

20.3.4 Singlet Oxygen ($^1O_2$) as Signaling Molecule

Under aerobic oxidation, plant capacity to resist oxidative injury is important for physiological acclimation of plants and their photosynthetic apparatus to varying environmental conditions. One key factor inducing photooxidative damage is the singlet oxygen radical. It is produced mainly in PSII under high light when absorption of light energy exceeds a possibility for its usage in photosynthesis (Kreslavski et al., 2012).

Singlet oxygen is formed due to an energy transfer from the excited chlorophyll molecule in triplet state to the basic triplet state of molecular oxygen (Krieger-Liszkay, 2005). $^1O_2$ is very reactive molecule, damaging proteins of PSII. In many cases, damaging effects are induced not only by $^1O_2$ but also by products of polyunsaturated fatty acid (PUFA) oxidation, which can serve as second messengers in $^1O_2$ signaling (Galvez-Valdivieso and Mullineaux, 2010). Singlet oxygen is characterized by short life time; therefore, its activity demands the involvement of other signaling components; among them, proteins Executer 1 and 2 (EX1, EX2). Lipid peroxides induced by $^1O_2$ and their derivatives can also serve as signal molecules releasing chloroplasts. In *Chlamydomonas reinhardtii* cells, it is assumed that the $^1O_2$ molecules can exit from chloroplasts directly into the cytosol and attain the nucleus, inducing expression of the nuclear *GPXH* gene encoding glutathione peroxidase. However, the proportion of "mobile" $^1O_2$ is evidently very small. In fact, such effect was manifested only under high light and so far observed only in the cells of this microalga (Fischer et al., 2007). It is also not excluded that the oxidation products of PUFA produced during $^1O_2$ generation in the lipid fraction of thylakoid membranes exit into the cytosol. Oxidation of lipoperoxides can result in generation of singlet oxygen in the cytosol.

The signaling pathway of $^1O_2$ leading to cell apoptosis interacts tightly with signaling pathways of hormones and other ROS. This is possible due to the cell capacity to scavenge singlet oxygen by the increase in the amount of lipid-soluble antioxidants and also the acceleration of reduction of photo-damaged D1 protein in PSII RC. This pathway counteracts cell apoptosis along the EXE1 and EXE2 pathways. During $^1O_2$ generation, the signaling pathways controlled by salicylic and jasmonic acids are also activated;

they result in changes of the expression of numerous genes related to stress-protective systems (Kreslavski et al., 2012).

Singlet oxygen not only induces transcription of genes of general response to stress but also activates expression of genes related to specialized mechanisms of acclimation (Apel and Hirt, 2004). When plants generate $^1O_2$ they showed the occurrence of specific responses on the transcription level. However, information about transcription factors and functions of genes protecting cells against $^1O_2$ is still scarce (Kreslavski et al., 2012).

20.3.5 Superoxide Anion Radical as Signaling Molecule

Superoxide ($O_2^{\bullet -}$) anion radical has a short life time, which does not exceed several milliseconds, and a small radius of action at neutral pH values. Therefore, it is rather difficult to distinguish between $O_2^{\bullet -}$ signaling and that of $O_2^{\bullet -}/H_2O_2$. Using transgenic plants in combination with various compounds favoring $O_2^{\bullet -}$ generation, methyl viologen in particular, could especially help researchers in studying this signaling. However, knowledge of this type of signaling is not complete and our understanding is rather limited (Kreslavski et al., 2012).

Some proofs of the $O_2^{\bullet -}$ signaling function were obtained in the analysis of gene expression using DNA microarrays and also of data about $O_2^{\bullet -}$ accumulation in plants deficient in Cu/Zn-SOD, and the expression is usually induced by $O_2^{\bullet -}$ generation in chloroplasts (Kreslavski et al., 2012). Since many genes studied with the usage of DNA microarrays are induced only after the appearance of $O_2^{\bullet -}$ but not other ROS species, the conclusion is that chloroplast systems of gene expression are specifically sensitive to $O_2^{\bullet -}$ accumulation in thylakoids.

$O_2^{\bullet -}$ reacts rapidly with nitric oxide, producing peroxinitrite. Peronitrite is likely synthesized in chloroplasts, where it can also fulfill signaling functions (Foyer and Shigeoka, 2011).

The $O_2^{\bullet -}$ radical has a negative charge in animal cells under normal conditions (its $p\kappa = 4.8$, whereas blood $p\kappa = -7.4$). However, at low pH, sometimes existing in plant tissues, this radical is protonated and in the form of hydroperoxyl (HO_2) can pass across membranes. In this case radicals produced in the cell wall due to the activation of plant NADPH-oxidase can penetrate into the cytoplasm and be scavenged by cytoplasmic SOD (Kreslavski et al., 2012).

20.3.6 ROS Signaling Associated with Nitric Oxide Signaling

Similar to ROS, nitric oxide (NO) is another reactive oxygen derivative produced. In recent years, NO has been demonstrated to be an intra- and intercellular signaling molecule involved in the regulation of diverse biochemical and physiological processes in plants (Corpas et al., 2006). NO interplays with ROS in plant and animal tissues in different ways associated with distinct physiological processes, and also in response to stress conditions. In the plant cell, ROS can directly cause strengthening of host cell walls via cross-linking of glycoproteins (Lamb and Dixon, 1997), or lipid peroxidation and membrane damage (Montillet et al., 2005). One of the most studied interplays in plants is NO–ROS cooperation during the hypersensitive reaction (HR), which is characterized by a programmed cell death that contributes to plant resistance to biotic stresses (Zaninotto et al., 2006). Less information is available on ROS–NO cooperation under abiotic stress conditions. An increase of both ROS and NO has been reported to take place under high temperature, osmotic stress, salinity, and mechanical stress (Garcês et al., 2001). However, a reduction of NO accumulation was observed after long periods of treatment with heavy metals. NO seems to work in conjunction with ROS in the potentiating of the pathogen-induced cell death. Cytological studies show that ROS and NO are associated with cell death adjacent to infected cells and both signals modulate each other's accumulation (Zeier et al., 2004). Interestingly, both ROS and NO collaborate to mediate abscisic acid (ABA)-induced stomata closure. NO synthesis and stomata closure in response to ABA are severely reduced in the NADPH oxidase double mutant of *Arabidopsis*, atrbohD atrbohF, suggesting that endogenous H_2O_2 production elicited by ABA is required for NO synthesis. Collectively, these data suggest that the interplay between these molecules mediates a variety of physiological responses (Torres et al., 2006). Despite growing knowledge on the role of NO and ROS in signaling and cellular response to biotic and abiotic stress, less attention has been paid to the topological distribution of ROS and NO production in plant tissues. Information on the contribution of different types of cells to the accumulation of ROS and NO would increase our knowledge on the cellular response in plants to adverse conditions. Analysis of cross sections of principal roots of pea plants treated with Cd by confocal laser microscopy shows a Cd-dependent increase of $O_2^{\bullet -}$ accumulation in the cell walls from xylem vessels and pericycle and, to a lesser extent, in the epidermis and cortex. Under physiological conditions, ROS production has been associated with lignification of the cell wall in xylem vessels, but ROS generation in cortex cells can be associated with oxidative damages imposed by the heavy metal. The analysis of NO accumulation in principal roots with DAF-2 DA showed opposite results to those observed for ROS production, with a strong decrease of NO-dependent fluorescence by Cd treatment. As a positive control, Cd-treated plants are incubated with an NO donor (10 mM SNP), which produces an increase in fluorescence similar to that observed in control plants. In cross sections of principal roots from control plants, the production of NO takes place in the cortex, xylem, and, to a lesser extent, in phloem. Under Cd treatment, the fluorescence attributed to cell walls of cortex cells is reduced considerably, but the fluorescence of xylem does not change. The overlay of both peroxides/ $O_2^{\bullet -}$ and NO accumulation images under Cd stress allows one to observe an increase of ROS production in those places where NO is reduced, mainly in the cortex. NO and ROS were reported to participate in the osmotic tolerance of wheat seedlings by stimulating ABA biosynthesis and NO evolution in stressed reed plants impart osmotic tolerance through its protection against oxidative damage (Misra et al., 2010).

20.3.7 ROS and Hormonal Signaling in Plant

A new notion about ROS action as signaling intermediates appeared first during studying hormonal signaling and regulation of gene expression involved in plant protection against pathogen attack. The interaction of ROS with salicylic acid and nitrogen oxide plays the important role in alleviating stress (Kreslavski et al., 2012). Considering the short lifetime of ROS, logically mechanisms including secondary messengers and signaling proteins with the longer lifetime can initiate or mediate ROS signaling pathways in plant. In particular, it is supposed that proteins of EX1 and EX2 are involved in the signal transduction (Lee et al., 2007). This pathway may function with signals induced by ethylene, salicylic acid, and jasmonic acid.

Accumulating evidence supports the concept that cellular redox signaling and hormone signaling pathways form an integrated redox-hormone network that regulates plant growth and defense pathways (Fig. 20.13). The efficient operation of this network requires extensive metabolic crosstalk and multiple points of reciprocal control. For example, stomatal closure is controlled by a number of hormones including ABA, brassinosteroids and ethylene and it requires redox regulation through NO and ROS production (Foyer et al., 2008) but the abundance of ascorbate in the guard cells modulates hormone action (Chen and Gallie, 2004). Root architecture is also tightly controlled by the integrated action of redox and hormone-related signals (Guo et al., 2009). ABA is a positive regulator of leaf senescence that accumulates in response to stresses such as drought, salt or temperature extremes, leading to a reprogramming of gene expression and adaptive responses such as stomatal closure and accumulation of osmo-compatible solutes (Chandrasekar et al., 2000). ABA triggers NADPH oxidase dependent ROS production that is important in mediating the closure of stomata and the regulation of MAPKinase signaling cascades (Zhang et al., 2001). ABA-induced H_2O_2 production by the plasma membrane NADPH oxidases, RbohD and RbohF, leads to the activation of calcium-permeable channels, the increase in cytosolic Ca^{2+} causing stomata closure (Kwak et al., 2003). This activation is impaired in the ABA-insensitive gca2 mutants (Pei et al., 2000). Moreover, high ascorbic acid concentrations in guard cells made the stomata less responsive to addition of H_2O_2 or ABA (Chen and Gallie, 2004).

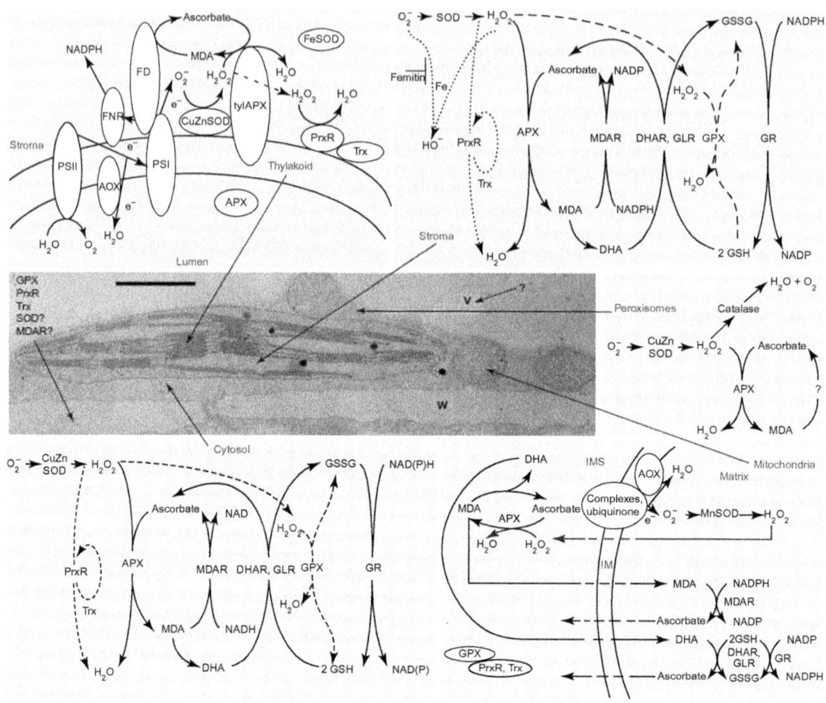

FIGURE 20.13 A transmission electron micrograph of a portion of a plant cell is used to demonstrate the relative volumes of the different cellular compartments and their physical separation (middle left). The enzymatic pathways responsible for ROS detoxification are shown. The water-water cycle detoxifies O_2^- and H_2O_2, and alternative oxidase (AOX; Immutans) reduces the production rate of O_2^- in thylakoids (top left; in some plants iron superoxide dismutase (FeSOD) might replace CuZnSOD in the chloroplast). ROS that escape this cycle and/or are produced in the stroma undergo detoxification by SOD and the stromal ascorbate-glutathione cycle. Peroxiredoxin (PrxR) and glutathione peroxidase (GPX) are also involved in H_2O_2 removal in the stroma (top right). ROS produced in peroxisomes during photorespiration, fatty acid oxidation or other reactions are decomposed by SOD, catalase (CAT) and ascorbate peroxidase (APX) (middle right). SOD and other components of the ascorbate-glutathione cycle are also present in mitochondria. In addition, AOX prevents oxidative damage in mitochondria (bottom right). In principle, the cytosol contains the same set of enzymes found in the stroma (bottom left). However, these are encoded by a different set of genes and the major iron-chelating activity in the cytosol responsible for preventing the formation of HO^- radicals is unknown. The enzymatic components responsible for ROS detoxification in the apoplast and cell wall (W) are only partially known, and the ROS-scavenging pathways at the vacuole (V) are unknown. Membrane-bound enzymes are depicted in white, GPX pathways are indicated by dashed lines and PrxR pathways are indicated by dotted lines in the stroma and cytosol. Although the pathways in the different compartments are mostly separated from each other, H_2O_2 can easily diffuse through membranes and antioxidants such as glutathione and ascorbic acid (reduced or oxidized) can be transported between the different compartments (Mittler et al., 2004).
Abbreviations: DHA, dehydroascrobate; DHAR, DHA reductase; FD, ferredoxin; FNR, ferredoxin NADPH reductase; GLR, glutaredoxin; GR, glutathione reductase; GSH, reduced glutathione; GSSG, oxidized glutathione; IM, inner membrane; IMS, IM space; MDA, monodehydroascorbate; MDAR, MDA reductase; PSI, photosystem I; PSII, photosystem II; Trx, thioredoxin; tyl, thylakoid

ABA-induced H_2O_2 production by RbohD and RbohF is also important in the induction of plant defense responses (Torres and Dangl, 2005).

Programmed cell death is an important plant defense response that prevents the proliferation of pathogens. Salicylic acid (SA), jasmonic acid (JA) and ethylene (ET) fulfill essential roles in mediating pathogen responses and they interact in the control of ROS generation and antioxidant enzyme activities to control stress responses and cell suicide pathways (Ashraf et al., 2010). The threshold for such signaling events is strongly influenced by cellular redox buffering capacity, as illustrated in the low ascorbate mutants, which show enhanced resistance to biotrophic pathogens (Mukherjee et al., 2010). Conversely, plants overexpressing ascorbate oxidase have more oxidized ascorbate redox state in the apoplast which showed decreased pathogen resistance.

Salicylic acid (SA) is a plant-signaling molecule involved in defense responses, both local and systemic defenses, to pathogen attack (Durrant and Dong, 2004). SA levels increase dramatically in cells surrounding infection sites (Enyedi et al., 1992). ROS was proposed to act synergistically in a signal amplification loop with SA to drive the hypersensitive response (HR) and establishing systemic defenses (Draper, 1997). This model was based on experiments using submaximal doses of both exogenous H_2O_2 and pathogen to drive SA accumulation, and the results showed subsequent increases in SA enhanced ROS production (Torres et al., 2006). SA accumulation can also downregulate those ROS-scavenging systems that, in turn, can contribute to increase in overall ROS levels following pathogen recognition (Klessig et al., 2000). However, ROS and SA antagonize action of each other in the regulation of cell death (Torres et al., 2006). Similarly, SA and jasmonic acid also either synergize or antagonize in the signaling functions at different concentrations. Synergy, in this case, drives ROS production and cell death (Mur et al., 2006).

The development of the osmotic stress under the influence of SA is shown in several studies. There is evidence that exogenous SA enhances plant cold tolerance. This effect is attributed to the inhibition of catalase and, as a consequence, with the osmotic stress determined by the accumulation of H_2O_2. On the example of two maize genotypes, it was shown that the cold-resistant line had a molecular form of catalase, which was inhibited by salicylic acid in the sensitive line (Horváth et al., 2002).

SA can be synthesized either from phenylalanine via cinnamic acid or from chorismate by isochorismate synthase. SA can act as a central regulator of cell fate by the reprogramming of gene expression, a process that involves the activation of plasma membrane-bound NADPH oxidases. Together with cell wall peroxidases, NADPH oxidases are responsible for the oxidative burst and accompanying cytosolic Ca^{2+} release that occurs in the apoplast in response to the perception of biotic and abotic stresses. The apoplastic oxidative burst and resultant ROS accumulation in the extracellular space is characteristic of plant cells exposed to abiotic stresses including physical and chemical shocks, insects and herbivores, symbiotic microorganisms and pathogens. Class III peroxidases may participate in SA-induced ROS metabolism. It has been suggested that SA could act as an e^- (electron) donor for Prx, a process that would generate SA radicals (Kawano et al., 2004). Some of the SA-dependent ROS production in plant cells might therefore depend on the interaction between SA and the activation of NADPH oxidases (Almagro et al., 2009). SA is also considered to be an inhibitor of the respiratory alternative oxidase (Hayat et al., 2007).

At present, the notion that ROS can function as secondary messengers in the transduction of hormonal signals is actively developed. There is evidence that auxin can affect gene expression using ROS as secondary messengers, which simultaneously regulate activity and expression of glutathione transferase, fulfilling the antioxidant function. There are some data that phospholipase D can stimulate H_2O_2 production in *Arabidopsis* leaves via generation of phosphatidic acid, fulfilling the role of lipid messenger (Sang et al., 2001).

The balance between ROS and antioxidant signals regulates auxin biology at multiple levels from biosynthesis, conjugation/oxidation, and transport to signal transduction (Tognetti et al., 2010). For example, ROS function as downstream components in auxin-mediated signal transduction to control gravitropism responses in roots (Joo et al., 2001). Genes encoding antioxidant enzymes are among primary auxin-response genes, suggesting a role for auxin in plant stress and defense responses (George et al., 2010). Any exogenous or endogenous stimulus that perturbs cellular redox balance can activate auxin homeostasis because NADPH oxidase dependent ROS production influences polar auxin transport (Joo et al., 2001).

20.3.8 Interaction between ROS and Other Signaling Systems

In spite of the previously mentioned signals that are related to ROS signaling, there is increasing evidence of a close connection between ROS and other various signaling systems in the plant cell, such as Ca^{2+} dependent signaling pathways (Glyan'ko, 2011). In particular, H_2O_2 stimulates a rapid increase in the intracellular Ca^{2+} concentration. The development of osmotic stress also controls the activity of several isoforms of calmodulin (a calcium-binding protein present in all nucleated cells – it can mediate a variety of cellular responses to calcium) and, conversely, an increase in the Ca^{2+} ion concentration results in the activation of ROS generation in mitochondria (Kim et al., 2009). Thus, oxidative and calcium dependent signaling pathways are closely interconnected. Increases in cytosolic Ca^{2+} is also one of the fastest responses upon pathogen infection, and the use of specific inhibitors showed that Ca^{2+} influx is required for ROS production after elicitation. Ca^{2+} can activate an Rboh protein *in vitro*, and all plant Rboh proteins contain two

EF-hands in their N-terminal region that may account for this Ca^{2+} regulation (Torres et al., 2006). On the other hand, ROS appears to be required to prime Ca^{2+} influx after elicitation. Therefore, Ca^{2+} fluxes appear to function both upstream and downstream of ROS production, indicating a complex spatiotemporal Ca^{2+} regulation of these signaling networks. Phosphorylation events have also been proposed to occur both upstream and downstream of ROS production in response to pathogens (Apel and Hirt, 2004). ROS generated via the NADPH oxidase and subsequent Ca^{2+} channel activation may represent a common signaling link in many plant responses. For example, ROS functions as an intermediate in ABA signaling during stomata closure through the activation of Ca^{2+} channels in guard cells (Pei et al., 2000). Thus, activation of Ca^{2+} channels represents a common signaling cassette in response to at least ABA and pathogen response. ROS may be the crucial signal in each system, since fungal elicitors induce both elevation of free cytosolic Ca^{2+} and stomata closure in guard cells.

Ethylene can induce programmed cell death and senescence. With the development of superoxide or ozone induced programmed cell death, ethylene dependent metabolic pathways can act as a signaling molecule. In addition, H_2O_2 induces the phosphoinositide cycle that leads to switching signaling pathways associated with the secondary messengers, IP3 and DAG (Munnik et al., 1998). Both ROS and ethylene have been implicated in signaling in response to viral infection. Interestingly, the ethylene receptor ETR1 can function as an ROS sensor, mediating stomatal closure in response to H_2O_2. Thus, this protein may constitute a node mediating crosstalk between ethylene and H_2O_2. Thus, ROS signaling interacts with many other regulatory events in a complex network of signals that govern the response to pathogens and other factors of the environment as well as developmental cues. This crosstalk may account for the multiplicity of responses mediated by ROS and explain why ROS produced by the same mechanism exert variable effects in different contexts.

Strigolactones (SLs) are signaling molecules that are synthesized from carotenoids in plastids mainly in the lower parts of the stem and in the roots in response to metabolic and environmental triggers (Domagalska and Leyser, 2011). They are important in the control of interactions with other organisms in the environment such as mycorrhizal fungi (Xie et al., 2010) and parasitic plants (*Striga* sp. and *Oro-branche* sp). SLs are transported by members of the ATP-binding cassette (ABC) transporter family (Kretzschmar et al., 2012) and they function downstream of auxin in the control shoot and root branching (Gomez-Roldan et al., 2008). They are considered a secondary messenger in auxin signaling pathways that interact with auxin in a dynamic feedback loop for the control of organ development. By restricting auxin transport in a systemic and local manner, they cause auxin accumulation to levels that inhibit growth, for example, in buds to control the axillary shoot branching. They also influence senescence and photomorphogenesis (Gomez-Roldan et al., 2008). It is possible that the SL signaling pathway, like that of auxin and other hormones, produces ROS as secondary messengers. Several studies have indicated that SLs interact directly with the redox signaling network (Woo et al., 2004) but the precise nature of this interaction is not yet understood. For example, the delayed senescence mutant, ore9, is more tolerant to oxidative stress than the wild type. ORE9 is a homologue of MAX2, which is a component of the SL signaling pathway (Woo et al., 2001).

20.4 CONCLUSIONS

An unavoidable consequence of adapting life to an oxygen-containing environment was the continuous production of reactive oxygen species (ROS) as the by-products of metabolism. ROS are generated by electron transport activities of chloroplast, mitochondria, and plasma membrane or as a by-product of various metabolic pathways localized in different cellular compartments. Under normal growth conditions, ROS production in various cell compartments is low. However, various environmental stresses such as drought, salinity, chilling, metal toxicity, and UV can disturb the cellular homeostasis and enhance the production of ROS. An enhanced level of ROS causes oxidative damage to lipid, protein, and DNA leading to altered events such as disrupting ion transport, loss of enzyme activity, protein crosslinking, inhibition of protein synthesis, DNA damage, ultimately resulting in cell death. In order to avoid the oxidative damage, higher plants possess a complex antioxidative defense system comprising nonenzymatic and enzymatic components. It is possible that the evolution of mechanisms for ROS detoxification progressed gradually, with those signaling of ROS at various situates where ROS were produced. Later cells had sophisticated mechanisms to produce and use ROS as secondary messengers for various tasks ranging from coping with environmental challenges to making developmental decisions. ROS signaling from chloroplasts might be one of the key regulators for signal transductions in the cell because it regulates both retrograde signaling and systemic signaling under stress conditions. By uncovering novel biological roles for ROS it becomes clear that ROS signaling is more than superficial and that a number of questions await clarification for full understanding to take place. It may be that different signaling cascades of ROS originate for different tissues and internal cellular sites. Probably different types of intracellular ROS induce distinct biological outputs in the same cellular system. Future progress will help in clarifying biochemical networks of ROS signaling involved in cellular responses to oxidative stress. Improved understanding of these will be helpful in producing plants with built-in capacity of enhanced levels of tolerance to ROS using plant breeding approaches.

REFERENCES

Abedi, T., Pakniyat, H., 2010. Antioxidant enzyme changes in response to drought stress in ten cultivars of oilseed rape (*Brassica napus* L.). Czech J. Genet. Plant Breed 46, 27–34.

Aguirre, J., Rios-Momberg, M., Hewitt, D., Hansberg., W., 2005. Reactive oxygen species and development in microbial eukaryotes. Trends Microbiol. 13, 111–118.

Ahmad, P., Prasad, M.N.V., 2012a. Environmental Adaptations and Stress Tolerance in Plants in the Era of Climate Change. Springer Science + Business Media, LLC, New York.

Ahmad, P., Prasad, M.N.V., 2012b. Abiotic Stress Responses in Plants: Metabolism, Productivity and Sustainability. Springer Science+Business Media, LLC, New York.

Ahmad, P., Umar, S., 2011. Antioxidants: Oxidative Stress Management in Plants. Studium Press Pvt. Ltd, New Delhi, India.

Ahmad, P., Sarwat, M., Sharma, S., 2008. Reactive oxygen species, antioxidants and signaling in plants. J. Plant Biol. 51 (3), 167–173.

Ahmad, P., Jaleel, C.A., Salem, M.A., Nabi, G., Sharma, S., 2010a. Roles of enzymatic and non-enzymatic antioxidants in plants during abiotic stress Crit. Rev. Biotechnol 30, 161–175.

Ahmad, P., Umar, S., Sharma, S., 2010b. Mechanism of free radical scavenging and role of phytohormones during abiotic stress in plants. In: Ashraf, M., Ozturk, M., Ahmad, M.S.A. (Eds.), Plant Adaptation and Phytoremediation. Springer, Dordrecht Heidelberg London, New York, pp. 99–108.

Ahmad, P., Nabi, G., Jaleel, C.A., Umar, S., 2011a. Free radical production, oxidative damage and antioxidant defense mechanisms in plants under abiotic stress. In: Ahmad, P., Umar, S. (Eds.), Oxidative Stress: Role of Antioxidants in Plants. Studium Press Pvt. Ltd, New Delhi, India, pp. 19–53.

Ahmad, P., Nabi, G., Ashraf, M., 2011b. Cadmium-induced oxidative damage in mustard [*Brassica juncea* (L.) Czern. & Coss.] plants can be alleviated by salicylic acid. S. Afr. J. Bot. 77, 36–44.

Ahmad, P., Ashraf, M., Azooz, M.M., Rasool, S., Akram N.A. (2013). Potassium starvation-induced oxidative stress and antioxidant defense responses in Brassica juncea. J. Plant Interactions, DOI:10.1080/17429145.2012.747629.

Ali, A.A., Alqurainy, F., 2006. Activities of antioxidants in plants under environmental stress. In: Motohashi, N. (Ed.), The Lutein-Prevention and Treatment for Diseases. Transworld Research Network, India, pp. 187–256.

Almagro, L., Gómez Ros, L.V., Belchi-Navarro, S., Bru, R., Ros Barceló, A., Pedrêno, M.A., 2009. Class III peroxidases in plant defence reactions. J. Exp. Bot. 60, 377–390.

Apel, K., Hirt, H., 2004. Reactive oxygen species: metabolism, oxidative stress, and signal transduction. Annu. Rev. Plant Biol. 55, 373–399.

Arora, A., Sairam, R.K., Srivastava, G.C., 2002. Oxidative stress and antioxidative system in plants. Current Sci. 82, 1227–1238.

Asada, K., 1999. The water–water cycle in chloroplasts: scavenging of active oxygens and dissipation of excess photons. Annu. Rev. Plant Physiol. Plant Mol. Biol. 50, 601–639.

Asada, K., Takahashi, M., 1987. Production and scavenging of active oxygen in photosynthesis. In: Kyle, D.J., Osmond, C.B., Arntzen, C.J. (Eds.), Topics in Photosynthesis: Photoinhibition, 9. Elsevier, Amsterdam, pp. 227–288.

Asada, K., Kiso, K., Yoshikawa, K., 1974. Univalent reduction of molecular oxygen by spinach chloroplasts on illumination. J. Biol. Chem. 249, 2175–2181.

Ashraf, M., 2009. Biotechnological approach of improving plant salt tolerance using antioxidants as markers. Bio. Adv. 27, 84–93.

Ashraf, M., Harris, P.J.C., 2004. Potential biochemical indicators of salinity tolerance in plants. Plant Sci. 166, 3–16.

Ashraf, M., Akram, N.A., Arteca, R.N., Foolad, M.R., 2010. The physiological, biochemical and molecular roles of brassinosteroids and salicylic acid in plant processes and salt tolerance. Critic. Rev. Plant Sci. 29, 162–190.

Bechtold, U., Richard, O., Zamboni, A., Gapper, C., Geisler, M., Pogson, B., et al., 2008. Impact of chloroplastic- and extracellular-sourced ROS on high light-responsive gene expression in *Arabidopsis*. J. Exp. Bot. 59, 121–133.

Bhattacharjee, S., 2005. Reactive oxygen species and oxidative burst: roles in stress, senescence and signal transduction in plants. Current Sci. 89, 1113–1121.

Blokhina, O., Virolainen, E., Fagerstedt, K.V., 2003. Antioxidants, oxidative damage and oxygen deprivation stress: a review. Ann. Bot. 91, 179–194.

Brisson, L.F., Tenhaken, R., Lamb, C., 1994. Function of oxidative cross-linking of cell wall structural proteins in plant disease resistance. Plant Cell 6, 1703–1712.

Chandrasekar, V., Sairam, R.K., Srivastava, G.C., 2000. Physiological and biochemical responses of hexaploid and tetraploid wheat to drought stress. J. Agro. Crop. Sci. 185, 219–227.

Chen, Z., Gallie, D.R., 2004. The ascorbic acid redox state controls guard cell signaling and stomatal movement. Plant Cell 16, 1143–1162.

Corpas, F.J., Barroso, J.B., Carreras, A., Valderrama, R., Palma, J.M., León, A.M., et al., 2006. Constitutive arginine-dependent nitric oxide synthase activity in different organs of pea seedlings during plant development. Planta 224, 246–254.

Creissen, G.P., Broadbent, P., Kular, B., Reynolds, H., Wellburn, A.R., Mullineaux, P.M., 1994. Manipulation of glutathione reductase in transgenic plants: implications for plant responses to environmental stress. Proc. R. Soc. Edin. 102, 167–175.

Davletova, S., Schlauch, K., Coutu, J., Mittler, R., 2005. The zinc-finger protein ZAT12 plays a central role in reactive oxygen and abiotic stress signaling in *Arabidopsis*. Plant Physiol. 139, 847–856.

Domagalska, M.A., Leyser, O., 2011. Signal integration in the control of shoot branching. Nat. Rev. Molec. Cell Biol. 12, 211–221.

Draper, J., 1997. Salicylate, superoxide synthesis and cell suicide in plant defense. Trends Plant Sci. 2, 162–165.

Durrant, W.E., Dong, X., 2004. Systemic acquired resistance. Annu. Rev. Phytopathol. 42, 185–209.

Enyedi, A.J., Yalpani, N., Silverman, P., Raskin, I., 1992. Localization, conjugation and function of salicylic acid in tobacco during the hypersensitive reaction to tobacco mosaic virus. Proc. Nat. Acad. Sci. U.S.A. 89, 2480–2484.

Fischer, B.B., Krieger-Liszkay, A., Hideg, E., Snyrychová, I., Wiesendanger, M., Eggen, R.I., 2007. Role of singlet oxygen in chloroplast to nucleus retrograde signaling in *Chlamydomonas reinhardtii*. Febs. Lett. 581, 5555–5560.

Foreman, J., Demidchik, V., Bothwell, J.H.F., Mylona, P., Miedema, H., Angel, M., et al., 2003. Reactive oxygen species produced by NADPH oxidase regulate plant cell growth. Nature 422, 442–446.

Foyer, C.H., Halliwell, B., 1976. Presence of glutathione and glutathione reductase in chloroplasts: a proposed role in ascorbic acid metabolism. Planta 133, 21–25.

Foyer, C.H., Noctor, G., 2005. Redox homeostasis and antioxidant signaling: a metabolic interface between stress perception and physiological responses. Plant Cell 17, 1866–1875.

Foyer, C.H., Noctor, G.D., 2009. Redox regulation in photosynthetic organisms: signalling, acclimation, and practical implications. Antiox. Redox. Signal 11, 861–905.

Foyer, C.H., Shigeoka, S., 2011. Understanding oxidative stress and antioxidant functions to enhance photosynthesis. Plant Physiol. 155, 93–100.

Foyer, C.H., Lopez-Delgardo, H., Dat, J.F., Scott, I.M., 1997. Hydrogen peroxide and glutathione-associated mechanisms of acclimatory stress tolerance and signaling. Physiol. Plant. 100, 241–254.

Foyer, C.H., Trebst, H., Noctor, G., 2008. Signaling and integration of defense functions of tocopherol, ascorbate and glutathione. In: Demmig-Adams, B., Adams III, W.W., Mattoo, A.K. (Eds.), Photoprotection, Photoinhibition, Gene Regulation, and Environment. Springer, pp. 241–268.

Galvez-Valdivieso, G., Mullineaux, P.M., 2010. The role of reactive oxygen species in signalling from chloroplasts to the nucleus. Physiol. Plant 138, 430–439.

Gapper, C., Dolan, L., 2006. Control of plant development by reactive oxygen species. Plant Physiol. 141, 341–345.

Garcês, H., Durzan, D., Pedrosa, M.C., 2001. Mechanical stress elicits nitric oxide formation and DNA fragmentation in *Arabidopsis thaliana*. Ann. Bot. 87, 567–574.

George, S., Venkataraman, G., Parida, A., 2010. A chloroplast-localized and auxin-induced glutathione S-transferase from phreatophyte *Prosopis juliflora* confer drought tolerance on tobacco. J. Plant Physiol. 167, 311–318.

Gidrol, X., Sabelli, P.A., Fern, Y.S., Kush, A.K., 1996. Annexin-like protein from *Arabidopsis thaliana* rescues Δ*oxyR* mutant of *Escherichia coli* from H_2O_2 stress. Proc. Natl. Acad. Sci. U.S.A. 93, 11268–11273.

Glyan'ko, A.K., 2011. Review of: Kolupaev, Yu.Ye., Karpets, Yu.V. Formation of Adaptive Reactions of Plants on the Action of Abiotic Stress Factors. Kiev: Osnova, 2010 [monograph]. Appl. Biochem. Microbiol. 47 (3), 328–331.

Gomez-Roldan, V., Fermas, S., Brewer, P.B., Puech-Pages, V., Dun, E.A., Pillot, J.P., et al., 2008. Strigolactone inhibition of shoot branching. Nature 455, 189–194.

Guo, K., Kong, W.W., Yang, Z.M., 2009. Carbon monoxide promotes root hair development in tomato. Plant Cell Environ. 32, 1033–1045.

Gupta, K.J., Stoimenova, M., Kaiser, W.M., 2005. In higher plants, only root mitochondria, but not leaf mitochondria reduce nitrite to NO, in vitro and in situ. J. Exp. Bot. 56, 2601–2609.

Hancock, J.T., Desikan, R., Neill, S.J., 2001. Role of reactive oxygen species in cell signaling pathways. Biochem. Soc. Trans. 9 (2), 345–350.

Hayat, S., Ali, B., Ahmad, A., 2007. Salicylic acid: biosynthesis, metabolism and physiological role in plants. In: Hayat, S., Ahmad, A. (Eds.), Salicylic Acid: A Plant Hormone. Springer, New York, pp. 1–14.

Holländer-Czytko, H., Grabowski, J., Sandorf, I., Weckermann, K., Weiler, E.W., 2005. Tocopherol content and activities of tyrosine aminotransferase and cystine lyase in *Arabidopsis* under stress conditions. J. Plant Physiol. 62, 767–770.

Horváth, E., Janda, T., Szalai, G., Páldi, E., 2002. *In vitro* salicylic acid inhibition of catalase activity in maize: differences between the isozymes and a possible role in the induction of chilling tolerance. Plant Sci. 163, 1129–1135.

Hung, S.H., Yu, C.W., Lin, C.H., 2005. Hydrogen peroxide functions as a stress signal in plants. Bot. Bull. Acad. Sinica. 46, 1–10.

Ivanov, B.N., Mubarakshina, M.M., Khorobrykh, S.A., 2007. Kinetics of the plastoquinone pool oxidation following illumination. Oxygen incorporation into photosynthetic electron transport chain. Febs Lett. 581, 1342–1346.

Jimenez, A., Hernandez, J.A., Pastori, G., del Río, L.A., Sevilla, F., 1998. Role of the ascorbate–glutathione cycle of mitochondria and peroxisomes in the senescence of pea leaves. Plant Physiol. 118, 1327–1335.

Joo, J.H., Bae, Y.S., Lee, J.S., 2001. Role of auxin-induced reactive oxygen species in root gravitropism. Plant Physiol. 126, 1055–1060.

Kanesaki, Y., Yamamoto, H., Paithoonrangsarid, K., Shoumskaya, M., Suzuki, I., Hayashi, H., et al., 2007. Histidine kinases play important roles in the perception and signal transduction of hydrogen peroxide in the cyanobacterium, *Synechocystis* sp. PCC 6803. Plant J. 49, 313–324.

Karuse, G.H., 1994. The role of oxygen in photoinhibition and photosynthesis. In: Foyer, C.H., Mullineauxe, P.M. (Eds.), Causes of Photooxidative Stress and Amelioration of Defense Systems in Plants. CRC Press, Boca Raton FL USA, pp. 43–76.

Kawano, T., Furuichi, T., Muto, S., 2004. Controlled salicylic acid levels and corresponding signaling mechanisms in plants. Plant Biotech. 21, 319–335.

Keller, T., Damude, H.G., Werner, D., Doerner, P., Dixon, R.A., Lamb, C., 1998. A plant homolog of the neutrophil NADPH oxidase $gp^{91(phox)}$ subunit gene encodes a plasma membrane protein with Ca21 binding motifs. Plant Cell 10, 255–266.

Kim, M.C., Chung, W.S., Yun, D., Cho, M.J., 2009. Calcium calmodulin-mediated regulation of gene expression in plants. Mol. Plant 2, 13–21.

Klessig, D.F., Durner, J., Noad, R., Navarre, D.A., Wendehenne, D., Kumar, D., et al., 2000. Nitric oxide and salicylic acid signaling in plant defense. Proc. Natl. Acad. Sci. U.S.A. 97, 8849–8855.

Knight, H., Trewavan, A.J., Knight, M.R., 1996. Cold calcium signaling in *Arabidopsis* involves two cellular pools and a change in calcium signature after acclimation. Plant Cell 8, 489–503.

Kojima, K., Oshita, M., Nanjo, Y., Kasai, K., Tozawa, Y., Hayashi, H., et al., 2007. Oxidation of elongation factor G inhibits the synthesis of the D1 protein of photosystem II. Mol. Microbiol. 65, 936–947.

Kojima, K., Motohashi, K., Morota, T., Oshita, M., Hisabori, T., Hayashi, H., et al., 2009. Regulation of translation by the redox state of elongation factor G in the cyanobacterium *Synechocystis* sp. PCC 6803. J. Biol. Chem. 284, 18685–18691.

Koyro, H.W., Ahmad, P., Geissler, N., 2012. Abiotic stress responses in plants: an overview. In: Ahmad, P., Prasad, M.N.V. (Eds.), Environmental Adaptations and Stress Tolerance of Plants in the Era of Climate Change. Springer, Science+business media, NY, USA, pp. 1–28.

Kreslavski, V.D., Los, D.A., Allakhverdiev, S.I., Kuznetsov, V., 2012. Signaling role of reactive oxygen species in plants under stress. Russ. J. Plant Physiol. 59, 141–154.

Kretzschmar, T., Kohlen, W., Sasse, J., Borghi, L., Schlegel, M., Bachelier, J.B., et al., 2012. A petunia ABC protein controls strigolactone-dependent symbiotic signalling and branching. Nature 483, 341–345.

Krieger-Liszkay, A., 2005. Singlet oxygen production in photosynthesis. J. Exp. Bot. 56, 337–346.

Kuznetsov, V.V., Shevyakova, N.I., 1999. Proline under stress: biological role, metabolism, and regulation. Russ. J. Plant Physiol. 46, 274–288.

Kwak, J.M., Mori, I.C., Pei, Z.M., Leonhardt, N., Torres, M.A., Dangl, J.L., et al., 2003. NADPH oxidase AtrbohD and AtrbohF genes function in ROS-dependent ABA signaling in *Arabidopsis*. EMBO 22, 2623–2633.

Laloi, C., Apel, K., Danon, A., 2004. Reactive oxygen signaling: the latest news. Curr. Opin. Plant Biol. 7, 323–328.

Laloi, C., Stachowiak, M., Pers-Kamczyc, E., Warzych, E., Murgia, I., Apel, K., 2007. Cross-Talk between singlet oxygen- and hydrogen peroxide-dependent signaling of stress responses in *Arabidopsis thaliana*. Proc. Natl. Acad. Sci. U.S.A. 104, 672–677.

Lamb, C., Dixon, R.A., 1997. The oxidative burst in plant disease resistance. Annu. Rev. Plant Physiol. Plant Mol. Biol. 48, 251–275.

Lee, K.P., Kim, C., Langraf, F., Apel, K., 2007. EXECUTER1- and EXECUTER2-dependent transfer of stress-related signals from the plastid to the nucleus of *Arabidopsis thaliana*. Proc. Natl. Acad. Sci. U.S.A. 104 (10), 270–275.

Lin-Mark, Y.F., Aarts, G.M., 2012. The molecular mechanism of zinc and cadmium stress response in plants. Cell Mol. Life. Sci. 69, 3187–3206.

Liszkay, A., van der Zalm, E., Schopfer, P., 2004. Production of reactive oxygen intermediates $O_2^{\bullet -}$, H_2O_2, and OH by maize roots and their role in wall loosening and elongation growth. Plant Physiol. 136, 3114–3123.

Liu, H., Colavitti, R., Rovira, I.I., Finkel, T., 2005. Redox-dependent transcriptional regulation. Circul. Res. 97, 967–974.

Matusda, Y., Yokuda, T., Sagisaka, S., 1994. Regulations of protein synthesis by hydrogen peroxide. Biosci. Biotechnol. Biochem. 58, 906–909.

Mendoza-Cozatl, D.G., Moreno-Sanchez, R., 2006. Control of glutathione and phytochelatin synthesis under cadmium stress. Pathway modeling for plants. J. Theor. Biol. 238, 919–936.

Misra, A.N., Misra, M., Singh, R., 2010. Nitric oxide: a ubiquitous signaling molecule with diverse role in plants. Afric. J. Plant Sci. 5, 57–74.

Mittler, R., Vanderauwera, S., Gollery, M., Breusegem, F.V., 2004. Reactive oxygen gene network of plants. Trends Plant Sci. 9, 490–498.

Mittler., R., 2002. Oxidative stress, antioxidants and stress tolerance. Trends Plant Sci. 7, 405–410.

Montillet, J.L., Chamnongpol, S., Rusterucci, C., Dat, J., van de Cotte, B., Agnel, J.P., et al., 2005. Fatty acid hydroperoxides and H2O2 in the execution of hypersensitive cell death in tobacco leaves. Plant Physiol. 138, 1516–1526.

Moulton, P., Martin, H., Ainger, A., Cross, A., Hoare, C., Doel, J., et al., 2000. The inhibition of flavoproteins by phenoxaiodonium, a new iodonium analogue. Eur. J. Pharmacol. 401, 115–120.

Mubarakshina, M.M., Ivanov, B.N., Naidov, I.A., Hillier, W., Badger, M.R., Krieger-Liszkay, A., 2010. Production and diffusion of chloroplastic H_2O_2 and its implication to signalling. J. Exp. Bot. 61, 3577–3587.

Mukherjee, M., Larrimore, K.E., Ahmed, N.J., Bedick, T.S., Barghouthi, N.T., Traw, M.B., et al., 2010. Ascorbic acid deficiency in *Arabidopsis* induces constitutive priming that is dependent on hydrogen peroxide, salicylic acid and the NPR1 gene. Molec. Plant Mic. Interac. 23, 340–351.

Munnik, T., Irvine, R.F., Musgrave, A., 1998. Phospholipid signaling in plants. Biochim. Biophys. Acta. 1389, 222–272.

Mur, L.A., Kenton, P., Atzorn, R., Miersch, O., Wasternack, C., 2006. The outcomes of concentration-specific interactions between salicylate and jasmonate signaling include synergy, antagonism, and oxidative stress leading to cell death. Plant Physiol. 140, 249–262.

Neill, S.J., Desikan, R., Clarke, A., Hurst, R.D., Hancock, J.T., 2002. Hydrogen peroxide and nitric oxide as signalling molecules in plants. J. Exp. Bot. 53, 1237–1247.

Nijveldt, R.J., Van Nood, E., Van Hoorn, D.E., Boelens, P.G., Van Norren, K., Van Leeuwen, P.A., 2001. Flavonoids: a review of probable mechanisms of action and potential applications. Am. J. Clin. Nutr. 74, 418–425.

Nishiyama, Y., Allakhverdiev, S.I., Murata, N., 2011. Protein synthesis is the primary target of reactive oxygen species in the photoinhibition of photosystem II. Physiol. Plant. 142, 35–46.

Pei, Z.M., Murata, Y., Benning, G., 2000. Calcium channels activated by hydrogen peroxide mediate abscisic acid signaling in guard cells. Nature 406, 731–734.

Pfannschmidt, T., Bräutigam, K., Wagner, R., Dietzel, L., Schröter, Y., Steiner, S., et al., 2009. Potential regulation of gene expression in photosynthetic cells by redox and energy state: approaches towards better understanding. Ann. Bot. 103, 599–607.

Pnueli, L., Liang, H., Rozenberg, M., Mittler, R., 2003. Growth suppression, altered stomatal responses, and augmented induction of heat shock proteins in cytosolic ascorbate peroxidase (Apx1)-deficient *Arabidopsis* plants. Plant J. 34, 187–203.

Polle, A., 2001. Dissecting the superoxide dismutase–ascorbate–glutathione pathway by metabolic modeling: computer analysis as a step towards flux analysis. Plant Physiol. 126, 445–462.

Potikha, T.S., Collins, C.C., Johnson, D.I., Delmer, D.P., Levine, A., 1999. The involvement of hydrogen peroxide in the differentiation of secondary walls in cotton fibers. Plant Physiol. 119, 849–858.

Prasad, T.K., Anderson, M.D., Martin, B.A., Stewart, C.R., 1994. Evidence for chilling induced oxidative stress in maize seedlings and a regulatory role of H_2O_2. Plant Cell. 6, 65–74.

Rasool, S., Ahmad, A., Siddiqi, T.O., Ahmad, P., 2013. Changes in growth, lipid peroxidation and some key antioxidant enzymes in chickpea genotypes under salt stress. Acta. Physiol. Plant. 35 (4), 1039–1050.

Remans, T., Opdenakker, K., Smeets, K., Mathijsen, D., Vangronsveld, J., Cuypers, A., 2010. Metal-specific and NADPH oxidase dependent changes in lipoxygenase and NADPH oxidase gene expression in *Arabidopsis thaliana* exposed to cadmium or excess copper. Funct. Plant Biol. 37 (6), 532–544.

Renew, S., Heyno, E., Schopfer, P., Liszkay, A., 2005. Sensitive detection and localization of hydroxyl radical production in cucumber roots and *Arabidopsis* seedlings by spin trapping electron paramagnetic resonance spectroscopy. Plant J. 44, 342–347.

Rizhsky, L., Liang, H., Mittler, R., 2003. The water-water cycle is essential for chloroplast protection in the absence of stress. J. Biol. Chem. 278, 38921–38925.

Rizhsky, L., Davletova, S., Liang, H., Mittler, R., 2004. The zinc finger protein Zat12 is required for cytosolic ascorbate peroxidase 1 expression during oxidative stress in *Arabidopsis*. J. Biol. Chem. 279, 11736–11743.

Rodriguez., A.A., Grunberg, K.A., Taleisnik, E.L., 2002. Reactive oxygen species in the elongation zone of maize leaves are necessary for leaf extension. Plant Physiol. 129, 1627–1632.

Ros-Barcelo, A., Pomar, F., Lopez-Serrano, M., Martinez., P., Pedreno, M.A., 2002. Developmental regulation of the H_2O_2-producing system and of a basic peroxidase isoenzyme in the *Zinnia elegans* lignifying xylem. Plant. Physiol. Biochem. 40, 325–332.

Saed-Moucheshi, A., Shekoofa, A., Pessarakli, M., 2013. Reactive oxygen species (ROS) generation and detoxifying in plants. J. Plant. Nut. In Press.

Sagi, M., Davydov, O., Orazova, S., Yesbergenova, Z., Ophir, R., Stratmann, J.W., et al., 2004. Plant respiratory burst oxidase homologs impinge on wound responsiveness and development in *Lycopersicon esculentum*. Plant Cell. 16, 616–628.

Salvador, M.L., Klein, U., 1999. Redox state regulates RNA degradation in chloroplast of *Chlamydomonas reinhardtii*. Plant Physiol. 121, 1367–1374.

Sang, Y., Cui, D., Wang, X., 2001. Phospholipase D and phosphatidic acid-mediated generation of superoxide in *Arabidopsis*. Plant Physiol. 126, 1449–1458.

Schafer, R.Q., Wang, H.P., Kelley, E.E., Cueno, K.L., Martin, S.M., Buettner, G.R., 2002. Comparing carotene, vitamin E and nitric oxide as membrane antioxidants. Biol. Chem. 383, 671–681.

Schopfer, P., Liszkay, A., Bechtold, M., Frahry, G., Wagner, A., 2002. Evidence that hydroxyl radicals mediate auxin-induced extension growth. Planta. 214, 821–828.

Segal, A.W., Abo, A., 1993. The biochemical basis of the NADPH oxidase of phagocytes. Trends Biochem. Sci. 18, 43–47.

Serbinova, E.A., Packer, L., 1994. Antioxidant properties of α-tocopherol and α-tocotrienol. Meth. Enzymol. 234, 354–366.

Shao, N., Krieger-Liszkay, A., Schroda, M., Beck, C.F., 2007. A reporter system for the individual detection of hydrogen peroxide and singlet oxygen: Its use for the assay of reactive oxygen species produced *in vivo*. Plant J. 50, 475–487.

Shao, N., Beck, C.F., Lemaire, S.D., Krieger-Liszkay, A., 2008. Photosynthetic electron flow affects H2O2 signaling by inactivation of catalase in *Chlamydomonas reinhardtii*. Planta 228, 1055–1066.

Smirnoff, N., 2005. Antioxidants and Reactive Oxygen Species in Plants. Blackwell Publishing.

Spadaro, D., Yun, B.W., Spoel, S.H., Chu, C., Wang, Y.Q., Loake, G.J., 2010. The redox switch: dynamic regulation of protein function by cysteine modifications. Physiol. Plant. 138, 360–371.

Taiz, L., Zeiger, E., 2006. Plant Physiology, fourth ed. Sinauer Associates, Sunderland, Massachusetts.

Tausz, M., Ircelj, H., Grill, D., 2004. The glutathione system as a stress marker in plant ecophysiology: is a stress-response concept valid?. J. Exp. Bot. 55, 1955–1962.

Tognetti, V.B., Van Aken, O., Morreel, K., Vandenbroucke, K., van de Cotte, B., De Clercq, I., et al., 2010. Perturbation of indole-3- butyric acid homeostasis by the UDP-glucosyltransferase UGT74E2 modulates *Arabidopsis* architecture and water stress tolerance. Plant Cell. 22, 2660–2679.

Torres, M.A., Dangl, J.L., 2005. Functions of the respiratory burst oxidase in biotic interactions, abiotic stress and development. Curr. Op. Plant. Biol. 8, 1–7.

Torres, M.A., Onouchi, H., Hamada, S., Machida, C., Hammond-Kosack, K.E., Jones, J.D., 1998. Six *Arabidopsis thaliana* homologues of the human respiratory burst oxidase (gp^{91phox}). Plant J. 14, 365–370.

Torres, M.A., Dangl, J.L., Jones, J.D.G., 2002. *Arabidopsis* $gp^{91(phox)}$ homologues *AtrbohD* and *AtrbohF* are required for accumulation of reactive oxygen intermediates in the plant defence response. Proc. Natl. Acad. Sci. U.S.A. 99, 517–522.

Torres, M.A., Jones, J.D.G., Dangl, J.L., 2006. Reactive oxygen species signaling in response to pathogens. Plant Physiol. 141, 373–378.

Varnova, E., Inzé, D., Van Breusegem, F., 2002. Signal transduction during oxidative stress. J. Exp. Bot. 53, 1227–1236.

Wang, S., Assmann, S.M., Fedoroff, N.V., 2008. Characterization of the *Arabidopsis* heterotrimeric G. protein. J. Biol. Chem. 283, 13913–13922.

Woo, H.R., Chung, K.M., Park, J.H., Oh, S.A., Ahn, T., Hong, S.H., et al., 2001. ORE9, an F-box protein that regulates leaf senescence in *Arabidopsis*. Plant Cell. 13, 1779–1790.

Woo, H.R., Kim, J.H., Nam, H.G., Lim, P.O., 2004. The delayed leaf senescence mutants of *Arabidopsis*, ore1, ore3 and ore9 are tolerant to oxidative stress. Plant Cell Physiol. 45, 923–932.

Xie, X., Yoneyama, K., Yoneyama, K., 2010. The strigolactone story. Annu. Rev. Phyt. 48, 93–117.

Yabuta, Y., Maruta, T., Yoshimura, K., Ishikawa, T., Shigeoka, S., 2004. Two distinct redox signaling pathways for cytosolic APX induction under photooxidative stress. Plant Cell Physiol. 45, 1586–1594.

Yang, D.H., Andersson, B., Aro, E.M., Ohad, I., 2001. The redox state of the plastoquinone pool controls the level of the light-harvesting chlorophyll *a/b* binding protein complex II (LHC II) during photoacclimation. Photosynth. Res. 68, 163–174.

Zaninotto, F., La Camera, S., Polverari, A., Delledonne, M., 2006. Cross-talk between reactive nitrogen and oxygen species during the hypersensitive disease resistance response. Plant Physiol. 141, 379–383.

Zare, S., Pakniyat, H., 2012. Changes in activities of antioxidant enzymes in oilseed rape in response to salinity stress. Int. J. Agri. Crop. Sci. 4, 398–403.

Zeier, J., Delledonne, M., Mishina, T., Severi, E., Sonoda, M., Lamb, C.J., 2004. Genetic elucidation of nitric oxide signaling in incompatible plant-pathogen interactions. Plant Physiol. 136, 2875–2886.

Zhang, L., Paakkarinen, V., van Wijk, K.J., Aro, E.M., 2000. Biogenesis of the chloroplast-encoded D1 protein: regulation of translation elongation, insertion and assembly into Photosystem II. Plant Cell. 12, 1769–1782.

Zhang, X., Zhang, L., Dong, F., Gao, J., Galbraith, D.W., Song, C.P., 2001. Hydrogen peroxide is involved in abscisic acid-induced stomatal closure in *Vicia faba*. Plant Physiol. 126, 1438–1448.

Zhou, Z.S., Huang, S.Q., Yang, Z.M., 2008. Bioinformatic identification and expression analysis of new microRNAs from *Medicago truncatula*. Biochem. Biophys. Res. Commun. 374, 538–542.

Zhu, C., Ding, Y., Liu, H., 2011. MiR398 and plant stress responses. Physiol Planta. 143, 1–9.

Index

Note: Page numbers followed by "*f*", and "*t*" refers to figures and tables respectively.

A

D

E

F

G

H

I

J

K

L

M

N

O

P

Q

R